17th International Conference on Heat Transfer and Renewable Sources of Energy (HTRSE-2018)

E3S Web of Conferences Volume 70 (2018)

Miedzyzdroje, Poland
2 - 5 September 2018

Editors:

Tomasz Kujawa
Aleksander A. Stachel
Zbigniew Zapalowicz

ISBN: 978-1-5108-7592-0

Printed from e-media with permission by:

Curran Associates, Inc.
57 Morehouse Lane
Red Hook, NY 12571

Some format issues inherent in the e-media version may also appear in this print version.

For additional information, please contact EDP Sciences – Web of Conferences
at the address below.

EDP Sciences – Web of Conferences
17, Avenue du Hoggar
Parc d'Activité de Courtabœuf
BP 112
F-91944 Les Ulis Cedex A
France

Phone: +33 (0) 1 69 18 75 75
Fax: +33 (0) 1 69 28 84 91

contact-edps@webofconferences.org

Additional copies of this publication are available from:

Curran Associates, Inc.
57 Morehouse Lane
Red Hook, NY 12571 USA
Phone: 845-758-0400
Fax: 845-758-2633
Email: curran@proceedings.com
Web: www.proceedings.com

TABLE OF CONTENTS

XVII Międzynarodowe Sympozjum
WYMIANA CIEPŁA I ODNAWIALNE
ŹRÓDŁA ENERGII
17th International Conference
EAT TRANSFER AND RENEWABLE
SOURCES OF ENERGY
HTRSE - 2018
Ciepło i prąd

XVII Międzynarodowe Sympozjum
WYMIANA CIEPŁA I ODNAWIALNE
ŹRÓDŁA ENERGII
17th International Conference
HEAT TRANSFER AND RENEWABL
CES OF ENERGY
RSE - 2018
Heat Transfer
and Renewable
Sources of Energy
2018

XVII Międzynarodowe Sympozjum
WYMIANA CIEPŁA I ODNAWIALNE
ŹRÓDŁA ENERGII
17th International Conference
Ciepło i prąd

P-18

edp sciences

Web of Conferences

Statement of Peer review

In submitting conference proceedings to *Web of Conferences*, the editors of the proceedings certify to the Publisher that

1. They adhere to its **Policy on Publishing Integrity** in order to safeguard good scientific practice in publishing.
2. All articles have been subjected to peer review administered by the proceedings editors.
3. Reviews have been conducted by expert referees, who have been requested to provide unbiased and constructive comments aimed, whenever possible, at improving the work.
4. Proceedings editors have taken all reasonable steps to ensure the quality of the materials they publish and their decision to accept or reject a paper for publication has been based only on the merits of the work and the relevance to the journal.

Title, date and place of the conference

Heat Transfer and Renewable Sources of Energy, HTRSE-2018
2. - 5. September 2018, Międzyzdroje, Poland

Proceedings editor(s):

Tomasz KUJAWA, Aleksander A. STACHEL, Zbigniew ZAPAŁOWICZ

Date and editor's signature 19. November 2018

17, avenue du Hoggar - PA de Courtabœuf – BP 112 - 91944 Les Ulis Cedex A (France)
Tél. : 33 (0)1 69 18 75 75 - Fax : 33(0)1 69 07 45 17 – www.edpsciences.org

17TH International Conference HEAT TRANSFER AND RENEWABLE SOURCES OF ENERGY HTRSE-2018

02. – 05. September 2018, Międzyzdroje, Poland

Organizers:
Department of Heat Engineering, West-Pomeranian University of Technology, Szczecin, Poland.

Scientific patronage:
The Committee of Thermodynamics and Combustion, Polish Academy of Science.
The Committee of Energy Problems, Polish Academy of Science.

Preface

The HTRSE conferences organized every two years have an established position and are an important place to exchange information and experiences between universities and research centers, bringing together a large group of participants. For many years, they have been serving as an important forum for propagating new ideas and sharing information on the latest achievements and intentions in research and implementation. They are also a place for establishing cooperation between various scientific centers. Their aim is to exchange views between representatives of science, including practice, on issues related to broadly understood energy in the aspect of existing technical, economic, ecological and social needs.

The topic of the Conference is traditionally associated with renewable and unconventional energy sources, which is of particular importance in the era of increasing energy and environmental problems. In addition, it covers a wide range of issues related to heat transfer, energy acquisition and processing, energy management, environmentally friendly energy technologies and environmental protection.

The thematic scope of the conference includes:

- new trends and methods of research and calculation regarding heat exchange issues;
- the issues of sourcing, processing and use of energy from renewable and conventional sources in power engineering, heating and other technical fields;
- pro-ecological energy systems using renewable energy sources;
- cooperation of conventional energy systems with installations using renewable energy sources in the aspect of energy management efficiency;
- low-temperature ORC power plants;
- cogeneration, centralized heating systems, refrigeration, air conditioning;

Honorary Committee of Conference

- Prof. Gerard Besler	Wroclaw University of Science and Technology
- Prof. Marian Czapp	Koszalin University of Technology
- Prof. Achim Dittmann	Technische Universität Dresden
- Prof. Jan Krysiński	Lodz University of Technology
- Prof. Jochen Lehmann	Fachhochschule Stralsund
- Prof. Jarosław Mikielewicz	Institute of Fluid-Flow Machinery Gdańsk, Member of PAS
- Prof. Władysław Nowak	West Pomeranian University of Technology, Szczecin
- Prof. W. Roetzel	Universität der Bundeswehr Hamburg
- Prof. Janusz Terpiłowski	Military Technical Academy, Warszawa
- Prof. Marian Trela	Institute of Fluid-Flow Machinery PAS, Gdańsk
- Prof. Leonid L. Vasiliev	Luikov Heat & Mass Transfer Institute, Minsk
- Prof. Andrzej Ziębik	Silesian University of Technology

https://doi.org/10.1051/e3sconf/20187000001

Scientific Committee of Conference

- Prof. Ryszard Białecki	Silesian University of Technology
- Prof. Tadeusz Bohdal	Koszalin University of Technology
- Prof. Dariusz Butrymowicz	Bialystok University of Technology
- Prof. Tadeusz Chmielniak	Silesian University of Technology, Member of PAS
- Prof. Dorota Chwieduk	Warsaw University of Technology
- Prof. Janusz Cieśliński	Gdansk University of Technology
- Prof. Roman Domański	Warsaw University of Technology
- Prof. Ingwer Ebinger	HS für Angewandte Fahrzeugtechnik, Hamburg
- Prof. Jacek Eliasz	West Pomeranian University of Technology, Szczecin
- Prof. Abdelmajid Ghannouchi	Ecole Nationale d'Inqenieurs de Tunis
- Prof. Zbigniew Gnutek	Wroclaw University of Science and Technology
- Prof. Beata Kępińska	Mineral and Energy Economy Research Institut, Kraków
- Prof. Henryk Kaproń	Warsaw University of Technology
- Prof. Władysław Kryłłowicz	Lodz University of Technology
- Prof. Thomas Luschtinetz	Fachhochschule Stralsund
- Prof. Jan Łach	Bialystok University of Technology
- Prof. Dariusz Mikielewicz	Gdansk University of Technology
- Prof. Jens Morgenstern	Hochschule für Technik und Wirtschaft, Dresden
- Prof. Andrzej Nowak	Silesian University of Technology
- Prof. Mieczysław Poniewski	Warsaw University of Technology
- Prof. Ewa Radziemska-Klugman	Gdansk University of Technology
- Prof. Kazimierz Rup	Cracow University of Technology
- Prof. Jan Składzień	Silesian University of Technology
- Prof. Aleksander Stachel	West Pomeranian University of Technology, Szczecin
- Prof. Władysław Szaflik	West Pomeranian University of Technology, Szczecin
- Prof. Janusz Szymczyk	Fachhochschule Stralsund
- Prof. Jan Taler	Cracow University of Technology
- Prof. Wiesław Zima	Cracow University of Technology
- Prof. Zbigniew Zapałowicz	West Pomeranian University of Technology, Szczecin

Organizing Committee of Conference

- Prof. Aleksander Stachel	Chairman
- Prof. Zbigniew Zapałowicz	Scientific Secretary
- Dr Tomasz Kujawa	Organizing Secretary

Address of Organization Committee

West Pomeranian University of Technology, Szczecin
Department of Heat Engineering
al. Piastów 17, PL 70310 Szczecin, Poland
tel. +48 91 4494272; email: htrse@zut.edu.pl

Detailed information about the HTRSE Conferences is available on the website: www.htrse.zut.edu.pl

We invite you to participate in our next conferences.

Organizers

E3S Web of Conferences 70, 01001 (2018) https://doi.org/10.1051/e3sconf/20187001001
HTRSE-2018

The possibility of achievement of the nZEB state of a refurbished detached house powered solely by renewable energy sources in Southern Poland

Jacek Biskupski[1,*]

[1]AGH University of Science and Technology, Dep. of Energy and Fuels, Al. Adama Mickiewicza 30, PL- 30-059 Cracow, Poland

Abstract. This paper gives a thorough description of the two different scenarios of retrofit of an existing detached country house (with high primary energy demand) to a net zero energy building (nZEB) or near zero energy (nearZEB) by using energy form on-site RES. Using a designed piece of modelling software author pointed out two possible solutions. First one, based on a bio boiler and small on-site PV generator (on-grid) and the other based on large PV generator and three heat pumps. A 24 months test was performed in order to find out the output of both scenarios. In first period, the bio boiler delivered energy for space heating and DHW, while energy from PV was used to cover all electricity needs of the household during the 12 months testing period. In the later, the energy received from 10 kWp PV of was partly used to cover the current needs of the entire household (switchable on/off-grid system), and surplus was stored in the national electricity grid and regained later in the winter for the space heating (by a GSHP) and ventilation (ASHP) and DHW (dedicated ASHP). In both cases the system proofed the possibility to achieve the nZEB (nearZEB in first scenario) state of the household, as all (in the first near all) energy needs were covered by renewable energy produced on-site.

1 Preface

The EU Directives and guidelines, (especially EPBD), stats clear that all new buildings in Europe shall be nearly or net Zero Energy Buildings (nZEB) from 2020 onwards. The exact definition of nZEB and nearZEB are to be developed by members states of EU, but general definition describing them as "buildings with very low energy demand, covered mostly by energy from RES harnessed locally" which could be used in both cases. In Poland the netZEB is understand as state of zero nett energy building (in polish "nZEB"), where the amount of entire energy delivered to the boundary is equal the entire energy used by entire household. The definition of nearZEB (in polish "niemalZEB") is a case of a building with yearly energy demand approaching the state where all energy needs [2] are reduced beyond the state minimum demand (95 kWh/m2 year as per year 2017).

In Poland there are ~ 4 million existing houses and at least 150.000 new ones are raised every year [1]. The vast majority (>85%) of existing detached buildings use coal and natural gas to deliver energy for space heating and DHW. Some of new building go for bio-boilers (mainly wooden pellets) and heat pumps. The PV systems are at the initial stage but getting wind [4]. From the year 2016 polish prosumerism started, as new bill on net metering for PV weas introduced [3]. According to this regulation, every household in Poland is entitled to install a on-grid PV generator (up to 40 kWp), and grid company is forced to intake the surplus of produced in-site energy without any constrains. As it is net-metering model, the grid company has to give it back to the household in a period of a year (with a 20 % deduction volume charge) so it is net metering rated 1:0,8 instead of 1:1. It means once one household transfers to the grid 10 MWh in summer, is entitled to get back 8 MWh into the end of the 365 days period.. Not collected (in 365 days) energy surplus is not transferable, neither returned back in any form. In practise 10 MWh transfered to the grid and redeemed by household only 6 MWh means 2 MWh given out to the grid company for free (as the 20 % werer gained by the grid as their "fee"). Despite this, the amount of individual PV installations in Poland doubles every year [4]. Poland has quite a good solar radiation, similar to one in Germany, even better to the GB's one. PV systems are relatively cheap, partly manufactured locally and most of suppliers give for them longer warranty than for example ST panels. Poland is gaining momentum with PV solar Energy.

The roof top PV generator seems to be the cheapest and most reliable solution in a way to receive sustainable energy to create a nZEB household in Poland. The problem is, that generally polish detached houses have space heating and DHW systems based on coal/natural gas boilers with relatively high demand for the final energy (only 11% of houses have less than 150 kWh/m2year[1]) at high heating medium temperatures. In order to meet such a high demand, the PV roof top

* Corresponding author: jacek.biskupski181@gmail.com

https://doi.org/10.1051/e3sconf/20187001001

systems are not able to deliver sufficient amount of energy (simply the roofs are too small and efficiency too low to harness desired amount of electric energy). On the other hand, there is lot of bio fuels available in Poland. The wood chips as well as pellets are widely manufactured here (Poland is one of important exporters of bio fuels in EU). There are some researchers opting for oats as potential bio fuel [2] as we had long history of planting oats. Historians claim that before World War II, there were over one million hectares planted oats, reduced to some 200 thousand nowadays. There are other RES potentials like heat pumps (also manufactured locally), which already gained wide acceptance. Author seeks for sets of reliable solutions for retrofitting polish typical detached households to a nZEB houses.

2 Potential solutions considered

Author has considered many heating systems solutions as a potential retrofit of an existing detached house into a nZEB where HVAC system is currently powered by a coal/natural gas boiler. In the research we have focused on detached and semi-detached houses, claiming that MFH creates completely different challenges. First observation lead to a differentiation related to actual positioning of the building and its urbanistic relation to the environment. Table 1 shows this consideration.

Table 1. Potential retrofit solutions toward nZEB of existing detached house depending on location.

Main source of energy	City	Suburb	Village	Env. Impact
Natural gas	+	+*	+*	Modereate
Bio	-	+**	+	Low
Heat Pump	-*	+*	+	Modereate /Low
PV (as DC heat source)	+	+	+	Very low
LPG	-	+	+	Modereate

Source: Author's PHD research [2]

* - where available/possible, not every location has the potential to install it.

** - solid state powered boilers are currently banned in many polish cities and some city suburbs has also this restriction.

Data in Table 1 [2] shows some limitations in wide use of solid fuels boilers and boilers nationwide. Due to high level of pollutions in Poland, there are many places where a total ban on solid fuels was introduced including bio pellets (eg. city of Cracow). Worth to quote here, that in a list of 50 the most polluted cities in EU, as many as 30 are located in Poland. Natural gas is not accessible everywhere (due to lack of the piping) and the LNG containers are so far too expensive for detached households, so far. The oil and LPG are not considered as potential eco-retrofit, mainly due to high prices of installation and questionable environmental impact. Environmentally speaking, in comparison of the natural gas and coal, the coal burning (in-site) always brings carbon dioxide and some pollutions (sometimes causing smog), while natural gas brings "only" carbon dioxide and water vapor. This become vital when cost of carbon dioxide emissions, are taken in account. Many bio boilers may bring significant emissions with practically very low carbon dioxide emission (the fuel is considered to be planted locally). On the other hand, the local bans do not allow to use them (bio boilers/stoves) widely for retrofit.

Heat pumps (ASHP and GSHP) are very promising, but they still show demand for lot of primary energy (Poland's in European cold climate area). Especially in Poland any electricity based heating system cause a problem in achieving limits of primary energy per house, as primary energy coefficient electricity for electricity is very high (CC=3). Only HP's with high performance of SCOPnet => 3,5 are suitable here. PV as main source of energy – in Poland – has the main disadvantage of sesonal harvesting, eg 80% of energy coming in summer period, when most of the energy needs comes in winter heating season. Thus direct PV DC heating (be electric DC heaters) seems to be not suitable as the main heating system of the house. Very interesting solution lays potentially in PV generation (as renewable primary energy source) and use it as main source of energy for heat pumps. From the environmental impact point of view such a solution could be assigned as "Moderate" when the grid energy is used to run a HP, or "Low" when the on-site PV energy is used to power a compressor and shunt pumps in a HP. Comparing to grid electricity, Author come to conclusion, that PV energizing HP seems to be the most suitable solution for nZEB. Such a solution use minimum of primary energy and generates minimum pollution in LC. Also bio fuels planted locally, like wood chops and oats could potentially demand lowest amount of primary energy, especially the derived from fossil fuels. The biggest disadvantage of the such a natural bio harvesting seemed to be the amount of human work demanded from the residents or the neighbours. As far as PV is concern, the PV generators working together with the grid as storage media, seemed to be very attractive – even with the cost of 20-30% (depending on nominal ower of PV generator). The first result from this research was, that in Poland the bio and PV with HP's are potentially the best solutions toward sustainable retrofit of and existing building toward a nZEB one (with some restrictions to the building's location). It's obvious that some improvements in house insulation as well as ventilation systems should be made first. For the insulation the certain level of heat loses of the building envelope must be achieved first, which is more important for HP's solution than for the bio boilers. Also the efficiency and accuracy of ventilation system seemed to be very important factor from the very beginning of planning. First modelling results showed clearly, that natural (gavity driven) ventilation system generates too much energy loses to be suitable for nZEB or even is too difficult for nearZEB. Also in both Variants (1 and 2) the retrofit house ventilation to a mechanical system with heat high efficiency heat recovery were designed.

3 Methodolgy and Results

3.1. Applied Solutions

Fig. 1. The test building surrounded by oats fields and PV/ST panels used as proposed in the first retrofit scenario.

3.1.1 The bio boiler as main source of energy for HVAC system

The country house, located south of Poland (near Cracow), inhabited with a eco orientated family of 4 people was the laboratory. The idea of applying a bio boiler was the first to come for house like this, showing very high oats planting potential around. At the first stage Author decided [2] to retrofit in this house (a natural gas combustion boiler (24 kW) to a new, 30 kW bio boiler (class 5) powered by pellet or oats as a fuel (switchable). At first stage of experiment (from 03.05.2016 till 02.05.2017) the local manufactured boiler was the main source of space heating, using this same house distribution heat system as previously with the gas boiler (heat exchanger and radiators). Also DHW system stayed unchanged, still powered from the main space heating boiler (also heated in summer from PV-ST). The auxiliary energy demand as well as home appliances were retrofit to be powered solely from 4,5 PV generator connected on/off-grid (switchable by BMS). BMS system was redesigned to handle the new HVAC management.

3.1.2 Heat Pumps as main source of heat and cool and energy for DHW

In the second stage of experiment (from 03.05.2017 till 02.05.2018) the bio boiler was turned off and three heat pumps were deployed to deliver space heating (GSHP 14kW), ventilation and cooling (ASHP 3,5kW) and DHW preparation (ASHP 3kW). This solution for nZEB building was already recommend in European Standard [4]. Worth to mention, that the HP's ground exchanger was placed just below the oats filed (the oats was still planted but not used as a fuel).

3.1.3 Photovoltaic generator in the house

The all house appliances were turned to electricity and taken in account in energy demand calculation (counted as auxiliary energy incl. shunt pumps, lights, ITC units). In order to power it all, the first variant roof top PV generator of 4,5 kWp was redesigned and repowered with new PV modules. In the stage the PV generator was enlarged using some extra space (like roof of the garage and a piece of the surrounding ground) at the south side of the house (Fig 2). Finally, the on-site PV generator was enlarged to the potential of 10 kWp delivering some 10 MWh yearly. The energy was partly used by the house appliances directly (DC and AC via inverters) and partly exported (mainly in the summer) to the grid and consequently redeemed in the winter.

4 The theoretic base

The theoretical base for the yearly energy calculation for such a solution is drawn in a formula [1] proposed by the Author.

$$U * \int_A (PSupp(t) - PDem(t))dt + \int_{R\setminus A}(PDem(t) - PSupp(t))dt = 0 \quad (1)$$

where: *PSupp* – is stream of energy delivered to the household from RES (single carrier), *PDem-* is the total energy demand of household, *U* =1/Es, where Es is efficiency coefficient of the energy storage, in Poland it is currently equal to 0,8 or 0,7 (for PV installations up to 10 kWp and above 10 kWp respectively). The idea was first introduced in [2], similar to work in [6].

At the first stage of experiment, the space heating and DHW was calculated separately as it was covered by the energy from bio boiler and partially from existing ST. The electricity demand was covered partly by on-site PV. The PV energy was (in every hour of the year) instantly used by the household devices to maximum possible level and surplus "exported" to the grid (managed by the BMS). The two like (1) equitation's were used (one for heat and the other for electricity). In the second scenario, (heat pumps) yearly energy demand of the entire household was converted to electricity and calculated according to equitation (1) as received from RES.

The idea of net zero energy demand of the entire household is based on comparison of input stream of energy from RES and output energy (demand and supply) during the course of a year, taking on account the storage of energy surplus in the grid Es (1). The initial calculation of electricity demand (in the first scenario) showed that 4.5 kWp PV generator should be able to deliver desired electricity in this house, excluding the space heating (bio boiler fueled by oats) and cooling (not existed in the first scenario). For the other scenario (Variant 2) the modeling showed, that using designed HP's the energy harnessed In the other scenario, modelling software based on (1) showed that the RES system should be enlarged to 10 kWp.

Fig. 2. An extra PV panels deployed in the second scenario at the roof top of a garage.

4.1 Building calculation modelling

In order to establish the energy demand two steps modelling process was used

1) The software calculation of energy demand of the entire household (in silico)

2) The real energy consumption measurement performed in real time by the Unihome® BMS System (in situ) was compared to the calculated values.

3) The differences were reported and analysed on-line and discussed every month with the inhabitants.

In the first scenario (bio boiler + 4,5kWp PV) the total energy demand was calculated for 5 Mg of bio fuel (sun dried oats at 18MJ/kg) and around 4500 kWh combined (self use and some "accumulated" in the grid) electricity. In the second scenario around 8-10 MWh was calculated as enough to meet the yearly demand (10 kWp delivering 10 MWh of electricity, and some energy given out to the grid system). The cooling part of PV (PVT) was used to preheat the DHW in most time of the "warmer seasons".

The test performed in 24 months aimed to compare the operational rating with the modelling calculations in two possible retrofit scenarios. In both cases Author expected energy balance effects closed to modelling measured by the real consuming of delivered energy, as the operational rating of the household was well known prior to the experiment (separately for space heating and DHW and for el. appliances – from individual meters). The energy performance as well as space comfort was measured every hour and discussed with residents periodically. The unknown part was in total emissions of GHG triggered by the retrofitted system, and the inhabitant's reaction to new challenges related to the new HVAC systems introduced. More about it in the chapter **Troubleshooting.**

5 Discussion

5.1 The sun proved to be potentially the main source of energy for entire HVAC system

The first performed scenario (bio+PV) proved to be closed to the modelling. Following the theoretical calculation, author expected some 5 Mg of oats to be burned in the boiler and the energy from 4,5 kWp PV plus existing ST (PVT) would cover the total demand of energy (heat&electricity) in a year. In real, due to problems with the existing leaks ventilation the boiler burned 5,8 Mg of oats in heating season. The demand for electricity was closed to the calculated modelling value of 4400-4600 kWh (+/- 100 kWh). Nevertheless, the building showed demand for more energy than calculated This gap was bridged mainly by thermal part of PVT panels and grid electricity, and went for ventilation and DHW. In order to limit the final demand, in the middle step, some modifications were made between phase one and two. First (before performing the second scenario), the air tightness of the building was tested and improved, fixing air leakage in attic area. Some windows which showed leaks, were also fixed. The domestic hot water system was completely redesigned getting rid of circullation. Some minor changes in management software (BMS) were made (eg better information system regarding the energy balance for every day of the year). The biggest change was in improving the HRS (energy recovery in ventilation system), and to existing HRS system an extra ground exchanger (gravel) was added. Finally, in the second part of the performed test (Variant 2), the building was more closed to the calculated data at modelling stage. The final generation of 9650 kWh of electricity (including some heat from specially designed for the purpose PVT panels) was enough to meet the year demand of the entire household of 4 people and ensure the internal climate comfort during the whole year. The very important comfort up-grade in the second stage was new feature of climate control (low air temperature inflow), which brought a new level of comfort to this country house.

5.2 The financial investment and LCC consideration

In Table 2, there is a matrix of financial and environmental aspects of considered retrofits.

The financial investments (CAPEX) was widely discussed with the home owners, but not in the entry stage. In order to leave them not biased by the real investment costs, the test were performed using leased HVAC systems instead of purchased (fixed small rate per month). The final costs were shown to the residents after the 24 months period. At the stage of testing, the initial analyse was presented to them (described in Table 2). Due to high environmental responsibility of the entire "family to be tested" they preferred the second scenario from the very beginning, but also liked idea of planting fuels by themselves (oats). Finally, after 24 months test, they were deeply surprised by the real CAPEX of solutions (the second appeared to be much more expensive) but presented LCC was for them encouraging. The perspective of 15 years without any expenditure for HVAC and activity from their side was finally convincing. Additional surplus of potential planted oats (extra income from selling out the crop) become a main argument for the second solution. Despite tree fold higher cost of the hardware for the second scenario they went for the heat pumps solution. The bio boiler was dismounted after the test.

Table 2. Financial and environmental analyse.

Space heat +DHW	Env.impact	LCC Cost	Capex	Primary energy Consumption
Bio boiler	**Low**	Low	Moderate	Acceptable
Gas boiler	Moderate	High	Low	Too high
Pv+HP	Low	Low	High	Acceptable
PV + gas	Moderate	Moderate	Moderate	Acceptable

Fig 3. The house with PV mounted at the roof at the first stage of experiment.

5.3 Troubleshooting

Fig 4. The ST and PV arrays in front the house in winter.

5.3.1 The human factor

As they had seemed to be excited at the beginning they were a bit confused at the end of first 12 months test.
The residents found that the main disadvantage of the first scenario was – unexpected from the beginning, significant demand of activity in heating season from their side (eg. fuel uploading, ash downloading, chimney cleaning, etc). Burning almost 6 tons of oats demanded a lot of effort, power and they seemed not be ready for it. Finally they liked much more the other scenario where heat pumps were deployed and no manpower was demanded. The other problem occurred with the ventilation system. They have already used the HRS, and when it proved not to be efficient they were not happy to build another (ground) heat exchanger. Fortunately, it fitted in a basement of just raising garage. The garage was also used as a base for extra PV solar panels essential in the second scenario (Fig.2). Unfortunately, the two roofs (of building and garage) were together still too small to hold all necessary PV arrays. Some extra ground had to be deployed for that purpose what was a surprise for residents. The problem was that initially PV modules of 360 Wp were designed for the project, but due to overall costs they were reduced to 280-320Wp power each, so the amount of them increased significantly (only 2 pcs of 360 Wp were deployed on a tracker). Finally some 36 PV modules instead of 28 initially designed were installed.

5.3.2 The snow and cooling problem of PV modules

Also formerly widely used ST (solar thermal panels) in this house seemed not to deliver as much energy as they were expected in wintertime (due to the snow, see Fig3), and the problem concerned also PV modules. To overcome the problem the new version of PVT was designed and tested in this project (patent pending). The cooling part of PV modules was used in winter to warm them up and get rid of snow layers (first right at Fig 4).

Fig 5. The PV arrays standing on the ground and the top of garage.

5.3.3 The residents fromer habits problem

One of the most interesting results of the second stage of the experiment was the diagnosed problem of necessary changes in residents habits in order to meet the nZEB state. As the energy demand was constnlty measured and displayed (by Unihome®BMS), it appears that "regular" habbits of some residents (like heavy use of hot water) have to be changed. This was difficult, but widely discussed was corrected in the second Variant by the residents themselves. As the family liked very much the idea of fully energy independence and zero emmisions they have adopted to the new conditions of hot water usage. Surprisingly enough the disc washing habits have to be adjusted. Finally a compromise of exchanging the disc washer to new model with much less energy and water demand solved the problem. Also some changes in heating habits and cooling habits have to be introduced. Resistance from window opening as well as better temperature adjustment (individual for every room) solved the problem at stage two.

6 Summary

To sum up, the performed 24 months test showed potential in both bio and PV technologies as possible retrofitting scenarios of existing houses into nZEB's in Poland. The test found important limitations and preferences of such refurbishing, mainly related to human beings living in such a building. Some economic calculations performed show that not the investment cost but LCC should be considered as real cost in such a retrofit. Finally the solution based on improvement of insulation of the building envelope, ventilation retrofiting system and introduction of HP's as main source of heat and cool energized from locally harnessed solar energy (PV and PVT) proved to be the optimal solution. Nevertheless the achievement of nZEB state was possible only thanks to possibility of relatively cheap energy storage in the national grid.

References

1. *Finansowanie poprawy efektywności energetycznej budynków w Polsce* Report Buildings Performance Institute Europe (BPIE) www.bpie.eu 2016
2. J. Biskupski, *Techniczne, ekonomiczne i środowiskowe aspekty zastosowania OZE w budynkach mieszkanych w Polsce* (PhD Thesis Politechnika Krakowska 2018)
3. Ustawa o OZE – tekst jednolity z lipca 2018 https://narodowyprogram.pl/oze/ustawa-oze-2018
4. Raprot PV in Poland : posted on a web site : http://polskapv.pl/polski-rynek-fotowoltaiczny-w-liczbach-stan-na-31-grudnia-2017r/
5. EN 15603:2013, "European Standard: draft Overarching standard EPBD," CEN (European Committee for Standardization), Avenue Marnix 17, B-1000 Brussels, 2013.
6. I. Sartori, *Methodology for optimal energy system design of Zero Energy Buildings using mixed-integer linear programming,* Elsvier 2016.

Analysis of thermal and electrical efficiency of photovoltaic/thermal – PV/T modules operating in moderate climate at microscale

Dorota Chwieduk[1,*], and *Jarosław* Bigorajski[1]

[1]Institute of Heat Engineering, Faculty of Power and Aeronautical Engineering, Warsaw University of Technology, Nowowiejska 21/25, 00-665 Warsaw, Poland

Abstract. The paper presents an application concept for PV/T - Photovoltaic Thermal Technology in moderate climates (such as the Polish climate), at a micro scale, i.e. for a single family house. The paper analyses the operation of a PV/T system applied to Domestic Hot Water – DHW heating and electricity production. A mathematical model of the system operation has been developed. The paper focuses on modeling thermal and electrical efficiency of photovoltaic/thermal - PV/T modules. It also briefly presents the governing equations for the thermal energy balance of a storage tank, where thermal stratification effects take place. Some selected results of the numerical simulation of the PV/T system operation are described. Daily distribution of hourly averaged thermal and electrical efficiency of the PV/T modules without cover and with one glazing are presented. The PV/T systems do not give significant thermal energy output in winter. PV/T modules without glazing do not supply heat at all for three winter months, their highest thermal efficiency is in summer and it can be nearly 15%. In the same period glazed modules have efficiency equal to nearly 24%. However, the unglazed modules can give much more electrical energy in summer than those with glazing, and the electrical efficiency can reach the levels of 11.4% and 9.4%, respectively. In winter the difference is smaller, i.e. for unglazed the efficiency is 12.2%, and for glazed 11.2%.

1 Introduction

Utilization of solar energy in buildings has become of great interest to construction companies and developers dealing with innovative technologies, especially nowadays when European buildings should be nearly zero–energy according to the EU Directive on the energy performance of buildings [1]. Building-related solar energy technologies are connected with the architectural and civil engineering concepts, which include the application of solar energy systems. The solar energy systems are usually installed on buildings' roofs, but they can be also incorporated into the envelope of a building, constituting its integral elements. In such a case BIST - Building Integrated Solar Thermal - and BIPV - Building Integrated Photovoltaics systems are used in modern low-energy buildings [2]. A building becomes a complex structure of integrated architectural and energy elements, which include different types of solar passive and active systems.

The active solar systems convert solar radiation into a useful form of energy: thermal or electric using different devices in an active way. The conversion of solar energy takes place on the spot, for self-use by building's occupiers. However, the surface of a building envelope is limited and usually there is not enough area of south façade and south roof to incorporate solar thermal and photovoltaic systems to supply most of heat and electricity needed by the building's users. One of possibilities to solve the problem is to use the PV/T Photovoltaic Thermal systems, which combine both functions in one device, i.e. in a PV/T module.

It can be underlined that PV/T Photovoltaic Thermal technology can be used for new buildings, as well as for existing buildings, which can be suitably renovated, following energy sustainability measures. In order to reach nearly zero energy consumption, or nearly zero CO_2 emissions targets of new buildings or buildings under thermal refurbishment and modernization, or going even further to reach the goal of PEDs – Positive Energy Districts [3], it is necessary to develop mechanisms to enhance development and implementation of solar systems characterized by the high fraction of solar energy supplied to thermal and electrical systems in buildings.

*Corresponding author: Dorota.Chwieduk@itc.pw.edu.pl

PV/T Photovoltaic Thermal systems are believed to be such systems.

2 PV/T modules and systems

PV/T modules can work with different electrical and thermal efficiencies, and give different power and heat output. It depends on the ambient conditions in which they gain energy, mainly on solar irradiation and air temperature, and how the energy gained is used, from point of view of modes of operation of the system and distribution of energy needs in time.

With a decrease of solar irradiance the power of the PV cells and modules is reduced, caused by the decrease of the current (the voltage reduces only slightly). The power and efficiency of PV cells and modules decrease with the increase of their temperature. The temperature increase is caused by the absorbed solar radiation that is not converted into electricity, but is converted into heat (which is a consequence of the increase of internal energy). With the increase of the temperature of the PV cells the voltage is evidently decreasing (current is only slightly increasing) causing reduction of power and efficiency. This undesirable effect can be partially avoided by the heat extraction from the PV cells and modules through a fluid circulation along their bottom surfaces. Such solution is technically realized by the hybrid photovoltaic/thermal - PV/T module (collector) [3].

A PV/T module is the main element of the solar PV/T energy system, which provides electrical and thermal energy, thus achieving a higher energy conversion rate of the absorbed solar radiation, than both of such systems operating separately. The PV modules are coupled to heat extraction devices, in which air or water (or other liquid) of lower temperature than that of the PV modules is flowing through. As a result the cooling fluid is heated whilst the PV module temperature is reduced. In the case of air, the contact with the PV panels can be direct (flow over the surface). While for liquids, the contact is through a heat exchanger in a form of piping or channels in a plate. The thermal part of the PV/T modules, i.e. heat exchangers of the modules are usually connected to a water storage tank. When water is used as a cooling medium of the PV/T modules it can circulate in an open loop between the PV/T modules and the storage tank. When antifreeze mixture circulates between PV/T modules and the tank the loop is closed. Heat gained by the PV/T modules is stored in a tank and used when is needed. If temperature of the water stored in a tank is lower than required then an auxiliary heater is used.

Electric energy produced by PV/T modules can be used through inverters directly by the electrical energy appliances installed at the building or transmitted to the grid. In a case of the on-grid system the grid is treated as a store. When there is an excess of energy gained by the PV/T panels then this energy can be sent to the grid. An electrical system can also be equipped with batteries of different size depending on their function, e.g. they can be used as emergency store (for few hours breaks of electrical energy supplied to the building from the grid) or a few days store in a case of an off-grid system.

Combined PV/T systems with appropriate size of modules and energy storage should assure provision of the total energy needs: heating (DHW) and electricity of buildings in moderate climates in summer. These systems should operate in summer in monovalent mode using only one energy source –solar energy. However, there are some obstacles to reaching the goal of 100% solar energy fraction supplying thermal energy to a building in summer. Of course, knowing the level of solar energy availability in winter, it is obvious that solar systems need coupling to other energy systems to fulfill energy requirements. In PV/T systems the production of electricity is the priority, therefore it is necessary to operate the PV modules at low temperature in order to keep the PV cell and the PV modules electrical efficiency at a sufficient level. This requirement limits the effective operation of the PV/T thermal unit to low temperatures. Thus, the extracted heat can be not enough to supply all DHW heating needs in summer.

3 PV/T systems for DHW heating and electricity production at the micro scale

3.1 Description of the PV/T system configuration and operation

The paper presents an idea of application the PV/T - Photovoltaic Thermal technology in moderate climates, (such as the Polish climate), at the micro scale, i.e. for a single family house. The paper analyses operation of the PV/T system applied to Domestic Hot Water – DHW heating and electricity production. The considered system is presented in Fig. 1.

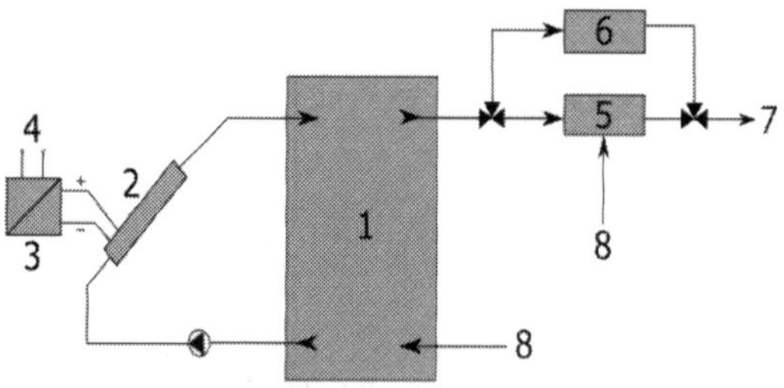

Fig. 1. PV/T system applied for Domestic Hot Water – DHW heating and electricity production
1 - storage tank, 2 - photovoltaic/thermal module, 3 - inverter
4 - electric energy consumption, 5 – mixer, 6 - auxiliary heater
7 - DHW consumption, 8 - filling with cold water from mains.

The thermal part of the PV/T system is a typical active solar water heating system used for DHW heating and working with an conventional auxiliary heater. This system is used to meet the demand of the owner. The collector loop is open. In a climate characterized by low ambient temperature, low solar irradiance and frequent overcast, as in Poland, it is recommended to use a close collector loop [4]. In such climatic conditions for all-year applications an antifreeze mixture (glycol–water) is applied and the collector loop and heat storage tank are separated by a heat exchanger. However, the preliminary analysis of the PV/T system operation showed that the thermal part of the system would not work in winter, due to small solar irradiation level and negligibly small amount of energy converted into thermal energy. As was mentioned, the priority is for electricity production in considered system. If the system is not used in winter for DHW then the water can circulate in a solar collector loop at the rest of the year, being a heat transfer and a heat storage medium at the same time.

Heat is stored in water, and storage is accomplished by using the thermal capacity of water. It is a short-term thermal storage system. Thermal stratification phenomena take place in the storage tank. It means, that the coldest water is located at the bottom and the temperature rises along with tank height, reaching maximum at its very top. The hot water outlet is located at the top part of the tank. The cold water inlet and outlet are at the bottom. The cold water outlet supplies cold water to PV/T modules to cool them and the heat gained is transferred from the outlet of the modules to upper part of the water storage tank. Heat stratification is beneficial, because it allows user to choose an outlet nozzle position to meet the specific temperature heating needs. Full effectiveness of DHW heating is achieved if the temperature at the tank bottom does not fall below certain value and temperature difference between top and bottom layers is also maintained at the required level.

The thermal efficiency of a solar system depends on a number of individual factors related to the design and construction of PV/T modules and the entire solar system, local insolation conditions, mode of operation of the solar system, and its method of its cooperation with storage and auxiliary or peak heating device/system. When solar radiation is available and the heat gained by the PV/T modules is above the minimum value, the heat is stored in a storage tank. The following are the basic modes of supplying heat to the user: If there is a heat demand and the amount of heat and temperature of water stored in the tank is high enough then thermal energy is supplied directly from the tank to the user. If the amount of heat and the temperature of water stored in a tank are too low then the water from the tank flows through the auxiliary heater to be warm up to the required level. If the temperature of water in a storage tank is at the minimum level (usually it is 10°C, which is equal to the water temperature of the supply network), then water from the supply network can flow directly to the auxiliary heater.

The PV/T system considered in the paper is an on-grid system without batteries. When PV/T modules operate the electrical energy produced is used directly or transmitted to the grid. The grid is used as a store of energy, where according to Polish regulations [5] the accumulation takes place with 80% efficiency. The considered PV/T system is an on-grid system and because of that electricity can be taken from the grid all the time.

3.2 Mathematical model of the system operation

To analyze the solar energy availability for the surfaces of PV/T modules of southern orientation and 30° of inclination the solar irradiance has been calculated for given time using the isotropic solar radiation model of Liu-Jordan [6]. Thus, the solar irradiance G_s on an inclined surface of the PV/T is given by:

$$G_s(t) = G_b(t)R_b(t) + G_d(t)R_d + \\ +[G_b(t) + G_d(t)]R_r\rho \qquad (1)$$

where:
G_b - direct solar irradiance on a horizontal surface in a given time, (W/m^2);
G_d - diffuse solar irradiance on a horizontal surface in a given time, (W/m^2);
R_b - correction factor for direct solar irradiance on an inclined surface in a given time;
R_d - correction factor for diffuse solar irradiance on an inclined surface;
R_r - correction factor for reflected solar irradiance on an inclined surface;
ρ - reflectivity of the ground.

Direct and diffuse solar irradiance on a horizontal surface at a given time can be taken from a meteorological data base (e.g. [7]). Correction factors for diffuse and reflected solar irradiances are constant for given inclination and orientation of the surface under consideration. Reflectivity of the ground can be assumed as a constant. However, the correction factor for the direct solar irradiance on an inclined surface must be calculated for every moment of the considered time (usually for every hour) [6].

The useful heat rate Q_u gained by the PV/T module can be described by the Hottel–Whiller Bliss correlation [6]:

$$\dot{Q}_u(t) = A_{PVT}F'(t)\begin{bmatrix}(\tau\alpha)G_s(t) + \\ -U_L(t)(\bar{T}_f(t) - T_a(t))\end{bmatrix} \qquad (2)$$

where:
A_{PVT} - area of PV/T (m^2)
$(\tau\alpha)$ – transmittance - absorptance product of the thermal part (just heat exchanger) of the PV/T module;
F' - PV/T module efficiency factor;
G_s - solar irradiance on inclined module plane determined by equation (1), (W/m^2);

U_L - thermal loss coefficient (W/m^{2}K);
T_f - arithmetic average of fluid inlet and outlet temperature (K);
T_a – ambient air temperature (K).

The thermal efficiency of the PV/T module is expressed as ratio of the useful heat rate (given by equation (2)) to the incident solar irradiance (determined by equation (1)) for a given surface area of the PV/T modules.

As mentioned, the heat rate Q_u gained by the PV/T module is stored in a storage tank. The tank has been modelled as a tank with thermal stratification and has been divided into n layers. Numbering of layers goes from the top to the bottom. The energy balance for the i layer can be written as:

$$(\rho c_w V_i)\frac{dT_{s,i}}{dt} = \dot{Q}_u(t) - \dot{Q}_{HW}(t) - \dot{Q}_L(t) + \\ +\dot{Q}_{S,i-1}(t) - \dot{Q}_{S,i+1}(t) \quad (3)$$

where:
ρ - density of water of layer i, [kg/m^3],
c_w - specific heat of water of layer i, [J/(kg K)],
V_i - volume of water of layer i, [m^3],
$T_{s,i}$ - temperature of water of layer i, [K],
t - time, [s],
Q_u - useful heat rate gained by the PV/T module and supplied to the layer i, [W],
Q_{HW} - heat rate supplied to the user flowing out of the layer i, [W],
Q_L - heat rate losses of the layer i, [W],
$Q_{S,i-1}$ - heat rate transfer between layer i and layer $i-1$ above, [W],
$Q_{S,i+1}$ - heat rate transfer between layer i and layer $i+1$ *below*, [W].

To solve the problem the model of the end use pattern of DHW has to be assumed. The time step for the end use of water is usually equal to one hour. Extracting heat out of the store at the top to supply heat for the end use keeps the stratification effect. This is beneficial for supplying the coldest water from the bottom of the tank to the PV/T module to extract heat during its operation. When the energy balance equations are formulated for every layer of the tank then control functions are defined: a PV/T module control function to determine which layer receives warm water from the module, and a load return control function for the liquid returning from the load [6]. To solve the governing equations a numerical simulation was applied on a base of a purpose-written numerical code formulated in Visual Basic [8]. The daily distribution of hot water consumption hour by hour was assumed [9]. The distribution assumed is not the best one for solar energy heating systems, because there is no use of hot water during the day time for 6 hours. Solar irradiation was calculated using equation (1) and as an input averaged hourly solar radiation data on horizontal surface for averaged days of averaged months of the year [6] for Warsaw.

Electric power output of the PV/T module can be described by a standard formula:

$$P(t) = V_{el}(t)I_{el}(t) \quad (4)$$

In case of the PV/T modules, as in regular PV modules, the efficiency is changing, even very rapidly in time. According to the standard definition of efficiency the electric efficiency of the PV/T module is expressed as ratio of the electric power output (given by equation (2)) to the energy rate input, i.e. incident solar irradiance for a given surface area of the PV/T modules. As already mentioned the electrical efficiency of the PV and as a result of the PV/T strongly depends on solar irradiance, ambient air temperature and the module temperature. Therefore, very often the electrical efficiency of a PV module is determined as a function of all of them. The same situation applied to determination of the electrical efficiency of the PV/T module. There are different formulae in the literature to determine the PV module temperature under different operation conditions [10] - [14]. However, up to now there is no universal formula recommended for analysis of the efficiency of the PV system operating in different conditions. What is more there is no formula for the temperature of a PV module *(T_{PV})* being a part of a PV/T module. One study [8] performed, among others, was aimed at giving some input for determination of those formulas.
One of the formula to determine the electrical efficiency of the PV module, which can be used for the PV/T module is following [10]:

$$\eta_{el}(t) = (\tau\alpha)(t)\eta_{ref}\left(1 - \beta_{ref}\left(T_{PV}(t) - T_{ref}\right)\right) \quad (5)$$

where:
($\tau\alpha$) – transmittance - absorptance product;
η_{ref} – efficiency of the PV module in the STC (Standard Test Conditions);
β_{ref} – temperature efficiency factor, [1/K];
T_{ref} – temperature of the PV module in the STC (Standard Test Conditions) equal to 25°C.

In a specific PV/T module the PV module must have the transmittance - absorptance product, the efficiency of the PV module, the temperature efficiency factor and the reference temperature of the PV determined for the STC - Standard Test Conditions. These parameters should be given by the PV producers. The temperature of the PV module (T_{PV}) should be determined. It is changing in time and depends on parameters of the system operation. This temperature is different in a standard PV module and in a PV/T module.

The temperature of the PV module (T_{PV}) is usually described by different approximation formulas, which take into account the solar irradiance and ambient air temperature. Some of them refer to NOCT – Nominal Operating Cell Temperature Conditions. It is difficult to find a correlation, which determines the temperature of the PV in the PV/T module. As has been described, in such a situation the temperature of the PV module depends not only on solar radiation and ambient air conditions, but also on heat transfer conditions between the PV module and fluid circulating in a heat exchanger in the PV/T module.

Tripanagnostopoulos [10] proposed to call the temperature of the PV in the PV/T module as $(T_{PV})_{eff}$. To determine this temperature he suggested to use Lasnier and Ang formula [15] for the regular PV module temperature and correct it through introduction of the temperature difference between the T_f arithmetic average of the fluid inlet and outlet temperature at the PV/T module and the ambient air T_a temperature. Thus, to calculate the electrical efficiency of the PV/T module it is necessary to use in equation (5) instead the value of the T_{PV} temperature the value of $(T_{PV})_{eff}$, which can be expressed as follows:

$$(T_{PV})_{eff}(t) = 30 + 0.0175(G_s(t) - 300) + \\ +1.14(T_a(t) - 25) + \left(T_f(t) - T_a(t)\right) \qquad (6)$$

The daily distribution of electricity consumption hour by hour was assumed [16].The distribution assumed is similar to that of DHW consumption, assuming that for 6 hours during the day time electricity consumption is very small (only stand-by mode of electrical appliances is used). Such a situation is not good for solar energy utilization, because the solar energy availability and electrical energy use are not coherent in time. It means that during the day time, energy produced by the PV/T modules will be send to the grid and electricity from the grid will be used when there is no solar radiation or its level is low.

3.3 Results of simulation studies

Some selected results of numerical simulation of the PV/T system operation at a micro scale in Polish climatic conditions are presented in this section. Simulation studies have been performed for different sizes of the PV/T system used to supply heating energy for the DHW system and electric energy for lighting and other electrical appliances. Daily consumption of heating energy for the DHW system is about 8.15 kWh and electricity about 8.22 kWh. As mentioned, daily distribution of energy consumption changes during the day time, but states the same for every day of the year. The maximum energy consumption is in the evening from 5 p.m. to 11 p.m. and in the morning from 7 a.m. to 9 a.m., but at that time is much smaller.

The PV/T system selected for presentation in the paper seems to be the most suitable for electrical energy consumption of the building. There are PV/T modules of A_{PV} area equal to 22.2 m^2 with installed capacity of 3 kW$_p$. They are located on a south roof inclined at 30°. The volume of the water storage tank is 1.11 m^3 (height 2.34 m, diameter 0.78 m). Two types of the PV/T modules have been analyzed: with a glazed cover and without.

Figures 2 and 3 present the daily distribution of hourly averaged thermal efficiency of the PV/T modules without cover and with one cover, respectively. As could be expected, the glazed (one cover) PV/T modules are characterized by much higher thermal efficiency than the unglazed. The highest thermal efficiency is in summer and can reach 15% in the case of unglazed modules and nearly 24% in the case of glazed modules. In winter for three months: December – February the unglazed modules do not work and in November for very short time, about 2 hours, with low efficiency (maximum 5%). The glazed PV/T modules do not operate in December and in January only for about 1 hour (maximum efficiency 5%). However, results present in Fig. 4 and 5 show the opposite situation in the case of the daily distribution of electrical efficiency of the considered PV/T modules. Electrical efficiency is the highest in winter and the lowest in summer. In a case of PV/T modules without covers the highest electrical efficiency is in winter in the morning and reaches 12.2%, in a case of modules with a cover it reaches 11.2%. In summer minimal electric efficiency is about 2 p.m. and for the PV/T modules without cover it can drop to 10.1% and for modules with a cover even much more to 8%. It is evident, that the cover is a good solution for improving thermal efficiency of the PV/T module but it cannot be recommended for the efficient operation of the PV part of the module.

4 Conclusions

Simulation studies performed show that PV/T systems with modules without a cover can operate with relatively high electrical efficiency during the whole year, and especially in winter, in micro systems supplying heating energy for a DHW system and electricity in moderate climates. Systems with glazed modules have lower electrical efficiencies, especially in summer. As has been already mentioned the opposite situation can be noticed from the point of view of thermal efficiency, which is larger for the glazed modules. As simulation studies showed, changes of the share of solar energy in supplying energy needs annually have the same tendency (these results are not presented in the paper, but can be found in literature [8]). For the considered size of the PV/T systems and energy needs the modules with glazing can fully cover the DHW demand during 5 months, from May to the end-of September, but they cannot cover the electricity demand, which can be at level of 67% in September to 93% in June. At the same time the modules without glazing can cover the heating demand only from 62% in September to 90% in July, but they can cover all the electricity demand from May till the end of August (using the grid as storage). What is more, during all these four months there is surplus of energy sent to the grid, which (according to existing regulations including net metering) can be taken out of the grid later e.g. in winter months.

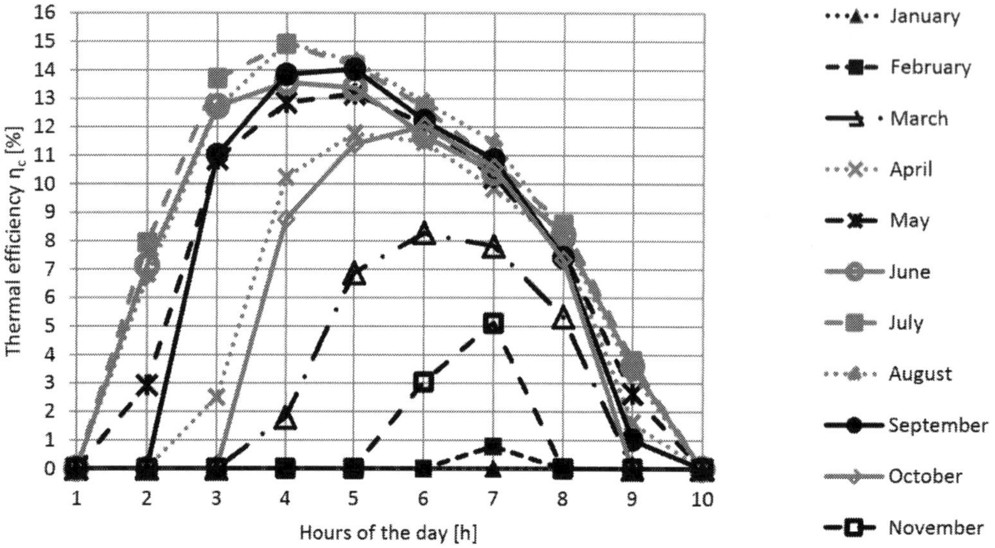

Fig. 2. Daily distribution of hourly averaged thermal efficiency of the PV/T modules without cover.

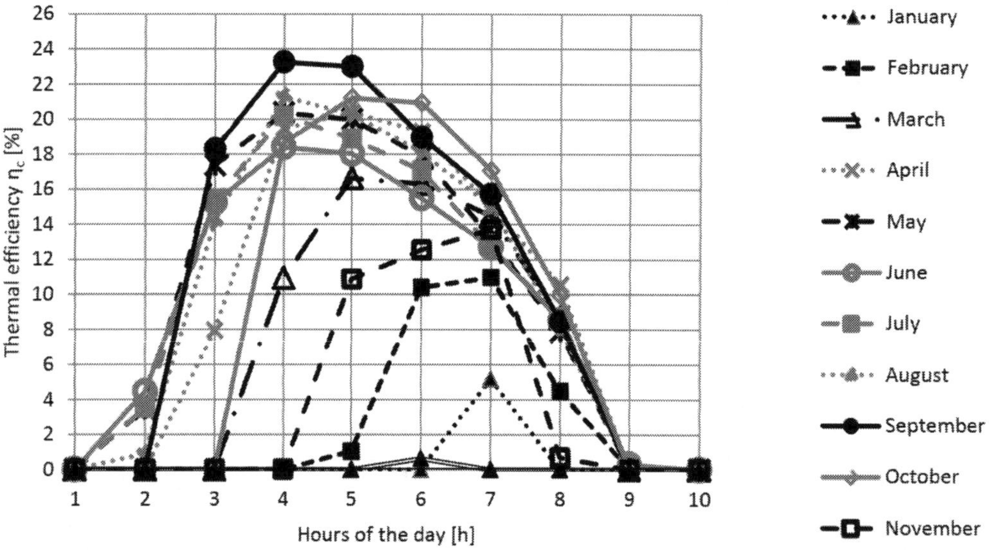

Fig. 3. Daily distribution of hourly averaged thermal efficiency of the PV/T modules with a cover.

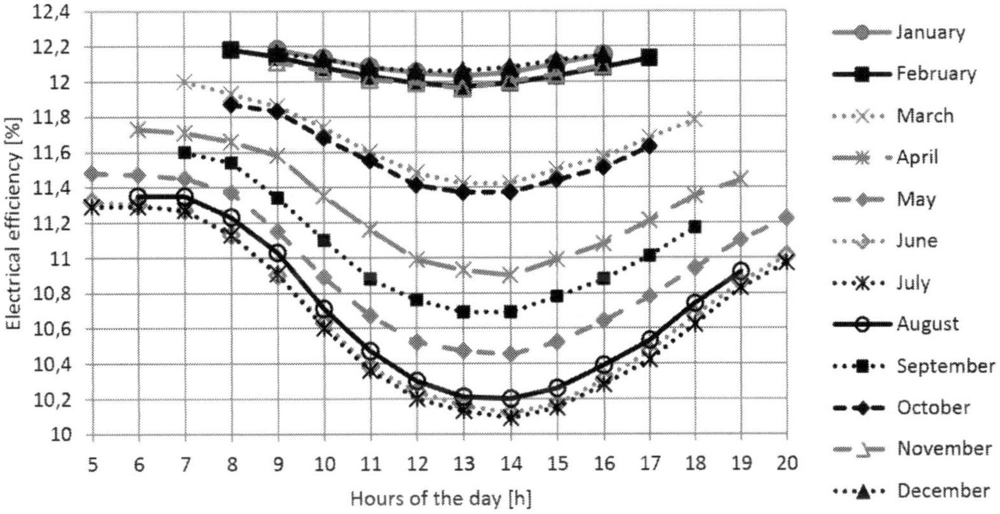

Fig. 4. Daily distribution of hourly averaged electrical efficiency of the PV/T modules without a cover.

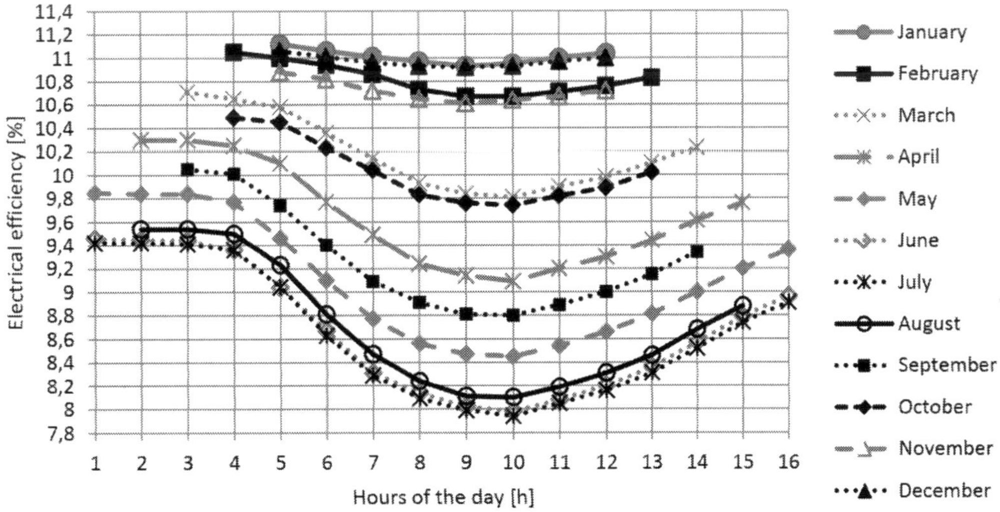

Fig. 5. Daily distribution of hourly averaged electrical efficiency of the PV/T modules with a cover.

All this shows that micro PV/T systems can be recommended for moderate climates. Because of their electrical efficiency perhaps it is better to use the PV/T modules without covers. In such a case they do not operate in winter as thermal modules, so without danger of freezing the water can be used as working fluid for the rest of the year. However, if more heat could have been extracted in summer from the modules, because of much larger heating needs (e.g. swimming pools, underground thermal energy storage), then also the electrical efficiency would not drop so much. It can be stated that the PV/T technology can assure more effective operation of the PV modules if they are cooled during their operation, which in a moderate climate is mainly needed in summer and spring. There are many questions to be answered. The PV/T technology is a new one and it is very difficult to formulate the final recommendation. Much more theoretical and experimental research is needed, including optimization studies.

Building-integrated photovoltaic/thermal systems (BIPV/T) should become popular in buildings in moderate climates. They can be integrated with building envelope, primarily its main façade and roof on the southern side. They can be treated as a kind of a building envelope ("construction") element, electricity generator (device which simultaneously generate electricity) and heating device. Depending on energy consumption of the building and taking into account the pattern of the heating energy use, part of the envelope can be constructed as the BIPV/T and a part just as the BIST – Building Integrated Solar Thermal System.

References

1. *Directive 2010/31/EU of the European Parliament and of the Council of 19 May 2010 on the energy performance of buildings*.
2. E. Biyik, M. Araz, A. Hepbasli, et al., *A key review of building integrated photovoltaic (BIPV) systems*, Engineering Science and Technology, an International Journal, Vol. 20, 833-858, (2017).
3. A. Monti & Pesch, Dirk & Ellis, Keith & Mancarella, Pierluigi: *Energy Positive Neighborhoods and Smart Energy Districts: Methods, Tools, and Experiences from the Field*, (2016).
4. D. Chwieduk, *Energetyka słoneczna budynku*, Wydawnictwo Arkady, Warszawa, (2011).
5. *Act on Renewable Energy Sources* ("RES Act", Dz.U. 2015 poz. 478)
6. J.A. Duffie, W.A. Beckman, *Solar Engineering of Thermal Processes*, Wiley, (2013).
7. Official website of the Ministry of Investment and Development, *Typical meteorological and statistical climate data*, https://www.miir.gov.pl (access: 13/07/2018).
8. J. Bigorajski, D. Chwieduk (in press), *Analysis of a micro photovoltaic/thermal – PV/T system operation in moderate climate*, Renewable Energy, DOI: 10.1016/j.renene.2018.01.116, (2018).
9. E. Fuentes, L. Arce, J. Salom, *A review of domestic hot water consumption profiles for application in systems and buildings energy performance analysis, Renewable and Sustainable Energy Reviews, Vol.* 81, 1530-1547, (2018).
10. T.M. Pavlović, Y. Tripanagnostopoulos, et al., *Solar Energy in Serbia, Greece and the Republic of Srpska, Academy of Sciences and Arts of the Republic of Srpska*, Department of Natural-Mathematical and Technical Sciences Monograph, Vol. XXVI, Banja Luka, (2015).
11. E. Skoplaki, A.G. Boudouvis, J.A. Palyvos, A simple correlation for the operating temperature of photovoltaic modules of arbitrary mounting, Solar Energy Materials & Solar Cells, Vol. 92, 1393-1402, (2008).
12. E. Skoplaki, J.A. Palyvos, *On the temperature dependence of photovoltaic module electrical performance: A review of efficiency/power correlations*, Solar Energy, Vol. 83, 614-624, (2009).
13. E. Skoplaki, J.A. Palyvos, *Operating temperature of photovoltaic modules: A survey of pertinent correlations*, Renewable Energy, Vol. 34, 23-29, (2009)
14. Y. Tripanagnostopoulos, *Photovoltaic/Thermal Solar Collectors [in:] Comprehensive Renewable Energy*, Newnes, t. 3, 255-300, (2012).
15. F. Lasnier, T.G. Ang, *Photovoltaic Engineering Handbook*, Princeton, NJ: Adam Higler, (1990).
16. B. Anderson, S. Lin, A. Newing, et al.: *Electricity consumption and household characteristics: Implications for census-taking in a smart metered future*, Computers, Environment and Urban Systems, Vol. 63, 58-67, (2017).

LabVIEW based measurements and control system of a micro-ORC installation with Scroll Expander

Albrecht Eicke[1,*], and *Sławomir* Smolen[1]

[1]City University of Applied Sciences Bremen, JRM Institute of Energy Engineering, Neustadtswall 30, Bremen, 28199, Germany

Abstract. An innovative ORC-installation with a scroll engine as expansion device has been developed and designed at the City University of Applied Sciences in Bremen. This new transportable installation has the specific capability to be connected to an alternative power source, such as solar power. Pentafluoropropane (R245fa) has been selected as the working fluid because it is suitable for the required temperature ranges and exhibits some additional advantages related to this specific application. The details of the ORC-Cycle have been presented in another paper and are not the main topic of the presented work. The core matter of this paper is the measurement and control system of the installation, which enables the thermodynamic evaluation of the obtained experimental data, some back coupling, and control tasks. The data acquisition and matching procedure, connected with the operational measurement system, identifies the aggregate and operational states of working fluid in characteristic points of the thermodynamic cycle. The algorithm for searching for the points from the table of parameters take so much time, that the control loop cannot work correctly in time. This problem was solved with the producer-consumer architecture in LabVIEW. Data acquisition and quick response in the control loops are done in the producer-while loop. To reduce the number of values, data are written only every five seconds in a queue. From this queue the consumer-while loop reads values, if the computer has enough capacity. This technique ensures that no data are lost. The matching procedure searches in the parameter file for the thermal properties and displays the cycle in a T-s diagram. Some simple calculation procedures make it possible to evaluate the thermal efficiency and other cycle parameters while operating. At the end of this sequence, the data are saved in a file. In the next step, the producer-loop is executed directly in a Field Programmable Gate Array, so that the control can be executed stand-alone and the computer is only used for visualization.

1 Introduction

One of the most effective and efficient modes of energy extraction from previously unsuitable sources is the Organic Rankine Cycle (ORC). Through the ORC, energy from a medium to low temperature (enthalpy) source can be transformed into high-grade energy used for several purposes. Of importance is that energy from a renewable source like wind and solar energy can be converted into high grade energy, which is used for several applications. This system has low maintenance requirements and is highly flexible. Additionally, the system is environmentally friendly because it does not consume any more fuel [1].

The major focus of this paper is the LabVIEW program that was created to analyse, and process the data [2]. The measurement and control system of the installation is the core mater of this paper; it enables thermodynamic evaluation of the data received, back coupling, and control tasks. In short, the LabVIEW program consists of five sections:

- Data acquisition and measurement system
- Control loops for the heating system and the generator
- Matching procedure for the thermal properties of the working fluid
- Thermodynamic evaluation
- Saving results

2 Test installation

This ORC-installation is comprised of a scroll expander, a preheater, an evaporator, a condenser, and a feed pump. In order to improve the practical operation, a refrigerant subcooler may be incorporated into the system. This is added after the storage tank and promotes operational quality by ensuring there are no leftover bubbles before the cycle begins again. In Figure 1 below, the ORC-installation with its additional component is shown. To present the full cycle in a T-s-diagram and a h-s-diagram, the temperature and the pressure are measured before (1) and after the expander (2), after the condenser (3), before the subcooler (4), before the feed pump (5), before the preheater (6), and before the evaporator (7). This LabVIEW code was firstly developed within a master project and afterwards improved and optimized continuously in the Laboratory

* Corresponding author: Albrecht.Eicke@hs-bremen.de

for Energetics at Bremen City University of Applied Sciences and is tailored to run tests and change the parameters while running the micro-ORC installation.

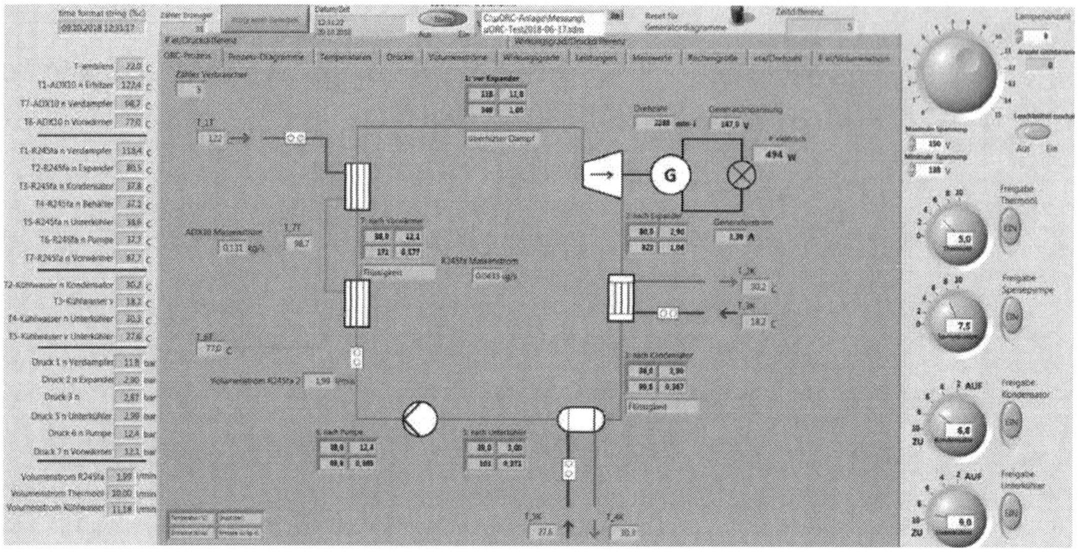

Fig. 1. Micro-ORC LabVIEW Process Diagram Screenshot.

3 Producer-Consumer-Problem

In this program, two different tasks are processed; on the one hand a fast data acquisition and quick response in the control loop and on the other hand a lot of calculations which need much calculation time, that the first loop cannot answer in time. To solve this problem, the Producer-Consumer-Architecture in LabVIEW is used shown in Figure 2. The two tasks are separated in two independent loops. The producer-loop can answer in a certain time and writes only every five seconds the values in queue to reduce the number of values to be shown in the diagram. The consumer-loop reads the values from the queue, as long as the computer has enough capacity to do so. This ensures no data are lost.

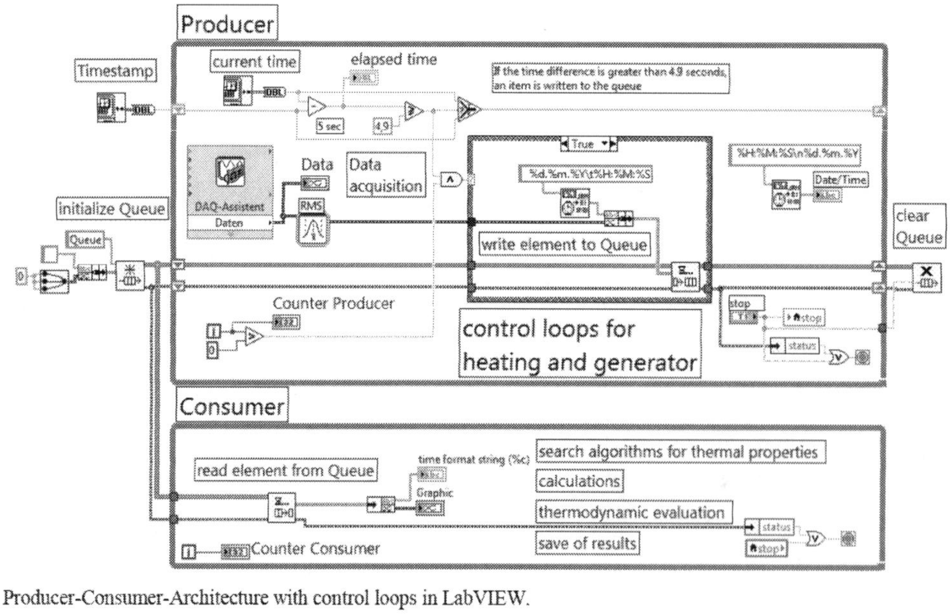

Fig. 2. Producer-Consumer-Architecture with control loops in LabVIEW.

4 Producer While-Loop with measurements and control

4.1 Data Acquisition of process values

In the producer loop thermal data of temperatures, pressures and volume flow rates of the three cycles are measured. The cycles are thermo-oil heating loop, the refrigerant cycle and the water cooling cycle. Although physical and electrical characteristics like revolution per minute, voltage and current flow are acquired. 500 values are averaged in each second so that the reaction time for the control of the generator and the heating system is pretty short.

Feed pump and circulation pump are adjusted over a frequency conformer by the user. The condenser performance is adjusted over an electric valve but until now not controlled automatically.

4.2 Control of the heating system with Pulse Width Modulation

This process had to be adopted to control heating power. The temperature t1 at the input of the expander should be held constant. If the fluid is a few degrees superheated it makes it easier to control the whole cycle and to reach and keep a steady state. The LabVIEW program is able to accomplish this through Pulse Width Modulation. A Mechanical Switch Relay is employed, and the time duration after each step is chosen to 50 seconds, which results in a temperature difference of 3 K. This time was chosen so long, so as not to endanger the mechanical switch by to high switching cycles. The original built in two-point switch causes a temperature difference of 30 K. In the future, a shorter time between each step would reduce the difference of temperatures in the refrigerant. However, too short a step-time would melt the switch. To realize this reduced step-time in future, a solid-state relay could be employed. With this change, the time period between each switch could be, for example, one second instead of the current 50 seconds.

4.3 Control of the load of the generator

Generator control refers to the ability to set the voltages to control the generator. In the case that less energy than required comes from the expander, the load from the generator would have to be lowered to maintain the speed of the generator and expander. The voltage and speed of the scroll engine are set by the user, for example, within a range of 138 Vmin to 150 Vmax. The control loop would set one lightbulb to either turn on or off to maintain the speed of both the generator and scroll engine.

5 Consumer While-Loop with calculations

In this loop, the calculations are done to prepare the visualisation and evaluation of the cycle. To calculate the latent heat of vaporisation it is necessary to determine the enthalpy and the entropy of the characteristic points of the process. To make the main program clearer, the matching procedure and some calculations are done in a sub-program. Finally, the results are saved every ten seconds. In LabVIEW these virtual instruments are called sub-VI.

5.1 Matching procedure for the thermal properties of the working fluid

One sub-VI searches in a table of thermal properties for enthalpy and entropy with the input of temperature and pressure. At first, it is checked if the measured values are in the array from 0.12 MPa to 1.4 MPa and from 20 °C to 140 °C as shown in Table 1. The table was created with the program Genetron Properties V1.3 from the company Honeywell based on the data from the National Institute of Standards and Technology (NIST) Refprop 9.1 ASHRAE [2]. The values are grouped by a temperature from 20 °C in increment of 1 °C to 140 °C and constant pressure. The pressure is then increased by 0.01 MPa and the next group is attached. The maximum pressure is set to 1.4 MPa.

Table 1. Thermal properties for R245fa from 0.12 MPa to 1.4 MPa and from 20 °C to 140 °C.

No.	temperature	pressure	density	enthalpy	entropy	specific heat
	t	p	ρ	h	s	c_p
	°C	MPa	kg/m³	kJ/kg	kJ/(kg*K)	kJ/(kg*K)
0	20.0	0.12	6.98	268.99	0.9488	0.9338
1	21.0	0.12	6.95	268.92	0.9520	0.9335
..	..	..	..	..	..	..
15123	139.0	1.40	65.81	370.76	1.1021	1.2222
15124	140.0	1.40	65.49	371.98	1.1050	1.2206

This array is separated into two tables: one for temperature and one for pressure. The row numbers which correspond with the searched temperature and the searched pressure are each written in an array. In the second step, the row numbers are checked that they are equal. The values for density, enthalpy, entropy, and specific heat of this found out row are given back to the main program. This is done for all the points described above until the point 4 in the storage tank where no phase change occurs.

Especially the points of evaporation and condensation, the saturated states, are very interesting for the evaluation of the cycle. A second sub-VI searches in this case in table 2 with the properties of the saturation line also with the input of temperature and pressure for enthalpy and entropy. But in this case the table includes both, saturated liquid point, marked with h' or s' , and the saturated vapour point, marked with h" or s".

Table 2. Saturation curve for R245fa from 0 °C (0.0 MPa) to 146.8 °C (3.2 MPa).

No.	temperature	pressure	density'	density"	enthalpy'	enthalpy"	entropy'	entropy"
	t	p	ρ	ρ	h'	h'	s	s
	°C	MPa	kg/m³	kg/m³	kJ/kg	kJ/kg	kJ/(kg*K)	kJ/(kg*K)
0	0.0	0.053	1403.97	3.23	49.76	254.22	1.1967	0.9453
1	0.1	0.053	1403.72	3.24	49.88	254.30	1.1972	0.9453
..	..	..	..	..	..	..	..	..
1467	146.7	3.198	795.26	264.52	274.48	335.98	0.8361	0.9826
1468	146.8	3.203	793.96	265.62	274.73	335.91	0.8366	0.9823

This sub VI selects the pressure of saturation for the measured temperature from this table and compares it with the measured pressure. If the values are equal within the measurement accuracy, the condition point is in the wet steam area.

If the pressure value from table 2 for the measured temperature is higher than the pressure, then the steam was superheated. To find the correct point on the evaporation line, the pressure for a lower temperature in the row above is compared with the measured pressure iteratively until this difference is zero - see the right side on figure 3.

In the third case, if the pressure value from table 2 for the measured temperature is lower than the pressure, the fluid was in a liquid area. Now the pressure corresponding with a higher temperature in the row underneath is compared with the measured pressure iteratively until this difference is zero. In all three cases both values for enthalpy and entropy are returned to the main program and can be used for the visualisation and further calculations.

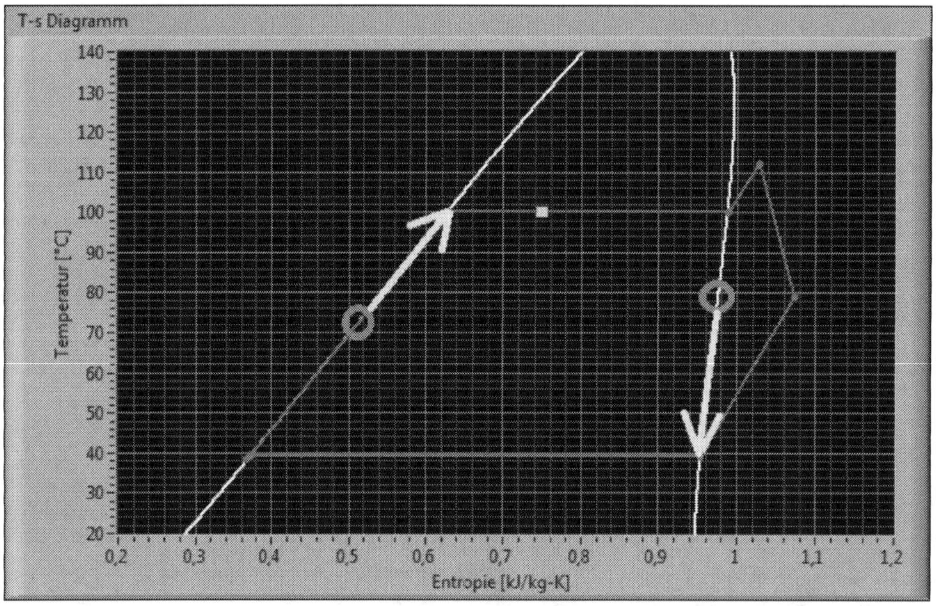

Fig. 3. T-s-diagram with the Organic-Rankine-Cycle finding the saturation points.

5.2 Thermodynamic calculations of energy balance

In the heat exchanger, it is possible to calculate the balance of the heat flow between the thermo-oil heating cycle and the cycle of the working fluid to determine the energy input of the ORC-installation, shown in Figure 1. In the thermo-oil section of the plate heat exchanger the heat flow is calculated by the multiplication of the mass flow rate, the specific heat capacity and the temperature difference between input and output. In the working fluid, a phase change is expected to occur, so that the calculation here must be done with enthalpies. In the refrigerant section of the heat exchanger, the heat flow is calculated by the multiplication of mass flow rate and enthalpy difference between inlet and outlet. With this knowledge it is possible to calculate the vapour mass fraction.

5.3 Calculation of the vapour mass fraction

The refrigerant heating takes place in two steps. A preheater increases the temperature from 40°C to 104°C. The second heat exchanger evaporates the fluid and superheats it to 115°C at a pressure of 1.1 MPa. An additional challenge is the determination of the aggregate state between the preheater and the evaporator. In order to visualise this state a sight glass was installed at this point. For quantitative determination of the relevant state the vapour mass fraction should be computed.

In a T-s-diagram shown in Figure 4 the area under the curve of a cycle can be expressed as a heat flow. Now the heat flow calculated in the heat exchanger by readings is compared with the area in the diagram. In an additional sub-VI the area of the heat from 40°C to 104°C (approximately a triangle) and the evaporation (a rectangle) are summed up in small steps and multiplied with the mass flow iteratively until this calculated heat power is equal to the power transferred in the preheater. The same calculation is done for the evaporator. But in this case, the calculation starts at the state point t1 with the triangular area of superheating and adds the rectangle area of evaporation from right to left until the heat flow is balanced. So the entropy for this state point is computed on to different ways. Now the mass fraction can be calculated by dividing this value by the entropy difference for the latent heat.

The increment of the entropy in the range of evaporation is 0.005 kJ/(kg*K). In the integral calculus for preheating, values from the saturation line (table 2) are used. The integral for superheating is based on the thermal properties from table 1 so that the step width is given from these tables. These triangle areas are approximated with many small rectangle areas which are always lower than the correct value, but because of the accurate tables, the differences are small.

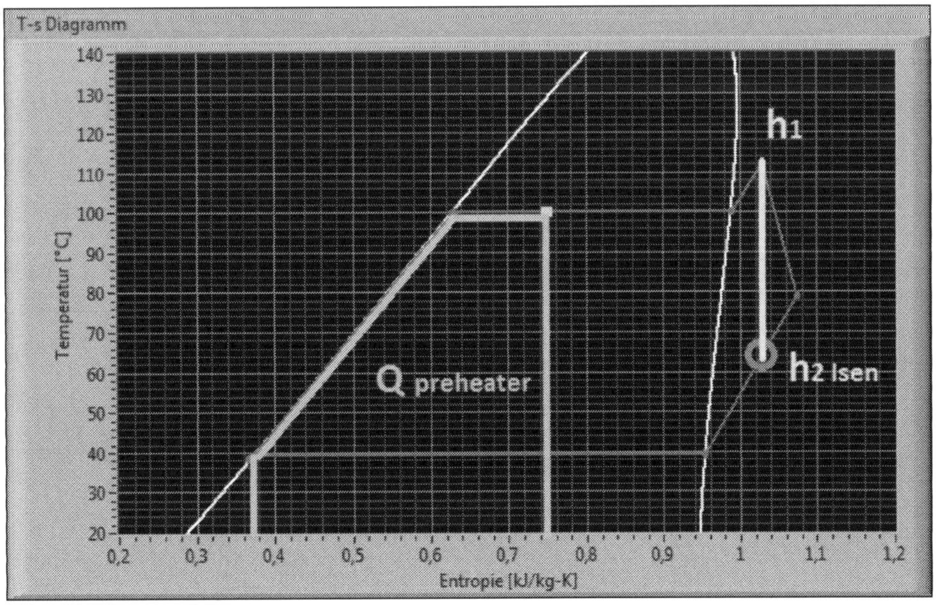

Fig. 4. T-s-diagram determines vapour mass fraction and thermal efficiency.

5.4 Performance and efficiency

The internal thermal efficiency of the expander is calculated by the real difference of the enthalpy divided by the difference of the enthalpy in isentropic expansion (see Figure 4). This requires the knowledge of the enthalpy h2isentropic at the output if an isentropic expansion occurs.

Here, nearly the same sub-VI, described above, is used for searching the state point. The database is again the thermal properties from table 1. However, in this case, the inputs are the pressure p2 at the end of the expansion and instead of the temperature the entropy s1 at the beginning of the expansion. The uncertainty for the searching procedure for the entropy is +/-0.0023 kJ/(kg*K). The row numbers which matched the searched entropy and the searched pressure are compared. If the row numbers are equal, it is returned to the main program. Now the internal efficiency of the expansion machine can be calculated.

In addition, it is also possible to calculate and visualise the general energy balance, the energy losses, the overall efficiency between the heat input of the thermo-oil into the system and the electric power of the generator.

6 Conclusions and prospects

A producer-consumer-architecture with queuing data makes it possible to run the control-loop very quickly but also gives the possibility to search for every relevant point of the refrigerant cycle in a huge parameter file of 15125 values.

The sensitivity analysis shows that the increment of 0.01 MPa is too accurate. The control of heating section can be improved through a solid-state relay, which would reduce the time between each step of the Pulse Width Modulation. Improvement in the producer-loop section can be achieved through its execution in a Field Programmable Gate Array, so that the control can be executed stand-alone and the computer will be used only for visualization.

References

1. P. Haberkorn, L. Herrmann, M. Sax, B. Speckmann, D. Zoche, KAI-ORC, Technical Project Documentation, University of Applied Sciences Bremen (Hochschule Bremen, Advisor: Smolen, S., Eicke, A.), Februar 2015, not published
2. S. Smolen, A. Eicke, Theroretical and practical investigation of a μORC installation with Scroll Expander, 23rd International Symposium Research – Education – Technology, 2017, ISBN: 978-3-9817740-2-3
3. Genetron Properties V1.3, Honeywell, database National Institute of Standards and Technology (NIST) Refprop 9.1

https://doi.org/10.1051/e3sconf/20187001004

New means of hydrogen storage – the potentials of methanol as energy storage for excessive windpower in North Germany

Johannes Gulden[1,*], *Andreas* Sklarow[1], and *Thomas* Luschtinetz[1]

[1]Stralsund University of Applied Sciences, Institute of Renewable EnergySystems, 18435 Stralsund, Germany

Abstract. The aim of the presented project is the technological development of hydrogen storage in methanol. This technology enables the carbon dioxide-based chemical storage of renewable energies as well as a decentralized supply of energy and hydrogen. Additional advantages are the very good compatibility with the existing infrastructure for liquid energy storage as well as the temporal decoupling of energy production and consumption. The latter can be managed independently, thus taking into account the fluctuating nature of wind and solar energy. The centrepiece is the use of new catalysts and processes that enable the chemical reactions in the methanol cycle under mild conditions.

1 Introduction

Within the frame of a project funded by the German BMWI the Friedrich-Alexander-University Erlangen-Nürnberg (FAU), the University of Applied Science Stralsund, the Center for Fuelcell Technology in Duisburg (ZBT), the ATI Küste GmbH, and the EnviTec Biogas AG under the coordination of the Leibniz-Institute for Catalysis in Rostock (LIKAT) want to enable the decentralized indirect storage of energy in methanol.

The methanol produced is converted to demand-oriented electrical energy in the sense of a temporal and / or interregional energy exchange via the stage of a low-temperature hydrogen release in a fuel cell. The feasibility of the storage technology according to Figure 1, which also shows the responsibilities in the project consortium, is shown through pilot plants. For the first time, already established technologies such as wind power utilization, electrolysis, CO_2-based methanol synthesis, which is the focus in this article, will be combined with the new processes low-temperature reforming of aqueous methanol and direct use of the resulting hydrogen-carbon dioxide gas mixture in a fuel cell.

First a wind turbine with an electrolyzer and the direct dynamic conversion of the hydrogen to methanol is coupled in a conventional synthesis reactor in Stralsund. After storage and transport in a complementary pilot plant at the Erlangen site, the methanol obtained here is converted back into hydrogen and carbon dioxide in a low-temperature process and used directly with a connected PEM fuel cell, including heat extraction, to generate electricity. This symbolizes the possible temporal and spatial separation of energy storage and utilization and minimizes the costs in view of otherwise necessary transport of the reforming plant. Various CO_2 sources are being evaluated for their suitability for methanol synthesis, and biogas-based carbon dioxide is used as an example.

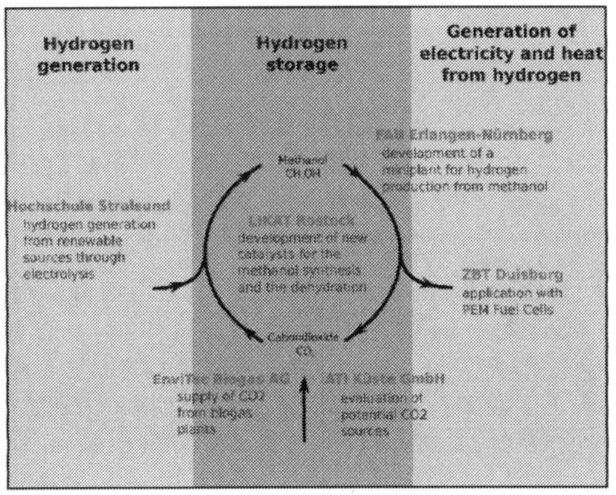

Fig. 1. Schematic drawing of the concept and the partners in the metha cycle project.

The aim of the project partners from Stralsund is a modular simulation of the entire system including the methanol production as well as the reconversion into electrical energy. The new results from within the project consortium will be included to analyse their influence on the system. In the final step a methanol reactor will be built up in Stralsund to demonstrate an efficiency optimized methanol cycle process.

The production of methanol from hydrogen and CO_2, is already possible today in processes with heterogeneous catalysts such as copper-zinc oxide-alumina at relatively high temperatures of 170-265 °C and a pressure of 10 to 80 bar. Examples include the "Silicon Fire Methanol Process" (80 bar and 265 °C) and the "George Olah CO_2 to Renewable Methanol Plant" of

* Corresponding author: johannes.gulden@hochschule-stralsund.de

https://doi.org/10.1051/e3sconf/20187001004

Carbon Recycling International (ORMP) [1,10]. The Swiss Silicon Fire AG operates since 2010 a mobile demonstration plant for the production of 50 liters of methanol per day with a purity of 99.99% [2]. Since the end of 2011, ORMP has been producing methanol in Iceland with a current capacity of approx. 4 million liters per year [3]. Further activities exist in Japan at the company Mitsui Chemicals (pilot analysis with 100 annual tons) and in Germany [4]. The dynamic operation with a direct coupling to a wind turbine has never been demonstrated before.

2 Analysis of the production potential

The first step is to analyse the production potential for methanol produced from renewable energies. It makes sense to use wind power, which cannot be included in the transmission grid at certain times. Due to the enormous power generation possibilities of wind power in the region, the focus is placed on the direct coupling of wind energy and methanol synthesis. Since there are currently no economic alternatives to the conventional synthesis (see above), the dynamic operation of such a reactor should be adapted and simulated to the wind potentials, which can be adapted in a second step with newly developed catalysts.

Analysing the potential for a future hydrogen infrastructure has to start with the renewable energy potential. According to the Einsman reports [5] 317.57 GWh of electricity were not fed into the grid in 2016 in Mecklenburg Western Pomerania alone. In 2015, there were almost 4725 GWh of renewable electricity in Germany cut due to limitations in the grid. Even the planned grid upgrades will not be sufficient, for a planned corridor of more than 2.8 GW [6] installed wind power per year, to transport the power generated to the consumer. Different storage technologies are therefore in discussion, with power-to-gas being the key technology [7,8]. The subsequent storage in liquid methanol is, due to the high energy density and the compatibility with conventional fuels, a possible way to operate a wind-hydrogen infrastructure economically.

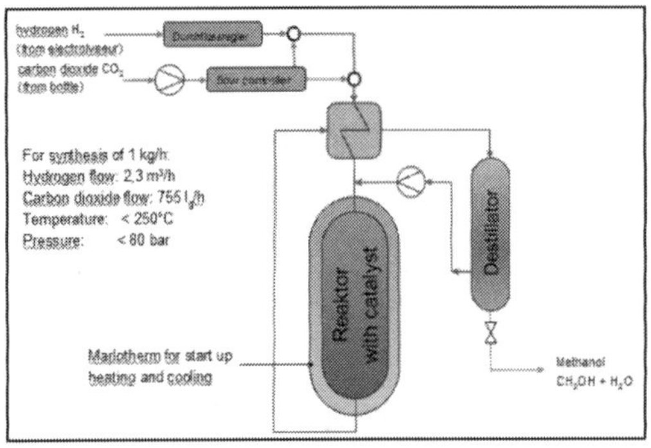

Fig. 2. Schematic drawing of the concept for the methanol synthesis.

With one GWh of electricity, with an efficiency of 75% in the electrolysis process, 750 MWh of hydrogen will be produced. This corresponds to a volume of 1,250 m3 at 200 bar. If this hydrogen is now synthesized in an ideal reactor with 83,472 Nm³ CO2 to methanol, about 138,800 liters of synthetic fuel are produced. The reaction proceeds exothermically at 80 bar and 265 ° C [2].

Figure 2 shows the concept of a planned synthesis plant at the University of Applied Sciences Stralsund. The system should be designed so that different operating points are possible. A special focus here is the use of catalysts, which are developed by the project partners.

Following the assumption that the entire non-fed regenerative wind energy will be converted into methanol, future production potentials of up to 34.8 kilotons of methanol will result annually, in Mecklenburg Western Pomerania alone. It is well known that the economics of technologies scale with the number of full load hours, so it can be assumed that the potentials are even greater, because the methanol synthesis plants, in addition to peak shaving, may also contribute to regional value creation and the downstream hydrogen economy.

3 Simulation of the methanol storage cycle

The project started to model the methanol storage cycle in a model. The goal is to simulate the entire methanol storage cycle including the regenerative energy generator side and the hydrogenation process as well as the dehydrogenation process and the reconversion. On the basis of the interface and system parameters derived from the simulation and the results of the project partners, the overall system is to be scaled in advance.

In a MATLAB Simulink model, each individual component is simulated as a block, which communicate with one another via suitable interfaces. This makes it possible to customize each component individually and refine the model iteratively. The input for the simulation will be real data from wind turbines and other regenerative power generators.

The results of the project partners are to be included in the simulation as well as the results of the methanol reactor to be built, which will provide for the first time measurement results from the direct coupling with the wind hydrogen chain. The obtained simulation parameters will help to design optimized operating strategies of an economical methanol storage cycle.

4 Summary and outlook

The increasingly successful implementation of the energy transition and the associated increase in the share of renewable energies in the coverage of primary energy demand force the development of additional centralized and decentralized long-term power storage technologies in the order of up to 18 GW / 7.5 TWh, due to the fluctuating occurrence of wind and solar energy and the resulting difficulty to calculate a prediction of these forms of energy. Already at about 40% share of renewable energies in electricity consumption, which were planned in the Federal Government's Energy Concept 2050 for the year 2015, existing and new transmission and distribution grids alone are unable to compensate for the temporal and regional imbalances in production and consumption [9]. In addition, individual regions in Germany, such as Mecklenburg Western Pomerania, will achieve up to 100% coverage of renewable energy in the electricity sector already much earlier.

For the long-term storage of large amounts of energy chemical storage systems are predestined. The generation of hydrogen in electrolyzers using renewable energy sources, such as wind energy and photovoltaic systems is a first step in this context. However, this creates the downstream problem of storing the regeneratively generated hydrogen. Previous technologies such as cryogenic, pressure or storage in metal hydrides have low gravimetric energy densities and the energy required for liquefaction and compression leads to additional energy losses of up to 30%.

A promising alternative is liquid high-energy chemical storage systems. Following the example of nature, where solar energy is largely bound in the form of carbohydrates, suitable substances can take over this task in an alternative material and energy cycle based on renewable hydrogen. Particularly suitable for this purpose is methanol accessible by hydrogenation of carbon dioxide.

References

1. G. Olah, *Angewandte Chemie*, **125**, 112 (2013)
2. http://www.silicon-fire.com/projekte-text.php, last visited on 13/07/2018
3. G. K. S. Prakash, Workshop *Sustainable Fuels from Renewable Energies*, IASS Potsdam, Germany (2013)
4. F. Pontzen, W. Liebner, V. Gronemann, M. Rothaemel, B. Ahlers, *Catal. Today*, 171, 242 *(2011)*
5. https://www.bundesnetzagentur.de Quartalsbericht Gesamt 2016
6. EEG 2017
7. J. Lehmann, T. Luschtinetz, *Wasserstoff und Brennstoffzellen – Unterwegs mit dem sauberen Kraftstoff*, ISBN 978-3-642-34668-2, Springer (2014)
8. NOW-Elektrolyse-Studie, short version, 36 ff (2011)
9. Verband der Elektrotechnik-Elektronik-Informationstechnik (VDE), *Energiespeicher für die Energiewende* (2012)
10. M. He, Y. Sun, B. Han, *Angewandte Chemie*, **125**, 9798 (2013)

Multi-objective analysis of an influence of a brine mineralization on an optimal evaporation temperature in ORC power plant

Marcin Jankowski[1,*], *Sławomir* Wiśniewski[1], and *Aleksandra* Borsukiewicz[1]

[1]West Pomeranian University of Technology, Chair of Thermal Engineering, Al. Piastów 19, 70-310, Szczecin, Poland

Abstract. The fact that Organic Rankine cycle system is very promising technology in terms of electricity production using low grade heat sources, necessitates constant research in order to determine the best cycle configuration or choose the most suitable working fluid for certain application. In this paper, multi-objective optimization (MOO) approach has been applied in order to conduct an analysis that is to resolve if there is an influence of a mineralization of a geothermal water on an optimal evaporation temperature in ORC power plant with R1234yf as the working fluid.

1 Introduction

Organic Rankine Cycle system (ORC) is constantly developing technology in past several decades. The main advantage of such systems is the availability of effective conversion low grade heat into power. Moreover, they are characterised by simple construction, low negative environmental influence and relatively high effectiveness. In view of the above advantages, a lot of research is still being conducted in order to ensure the most efficient work of designed ORC system. These studies are taking into account different technical and ecomomic performance indicators, called objective functions, which are to be minimized or maximized. Discussed criteria include: energy efficiency η_I (first law efficiency), exergy efficiency η_{II} (second law efficiency), net power output W_{net}, heat transfer area A, overall heat transfer coefficient k, overall cost of the system C, total exergy destruction rate δB_{tot} [1] or size parameter SP, which allows to initially evaluate the size of a turbine [2]. Some authors also used different combinations of these quantities. Shengjun et. al [3] concucted thermo-economic optimization of subcritical ORC and transcritical power cycle system for low temperature geothermal power generation using several objective functions: ratio of heat transfer area and net power output, also known as *APR* parameter, or levelized energy cost *LEC*. Roy et. al [4] carried out performance analysis of an ORC with superheating under different heat source temperature conditions using so called availability ratio Φ, which is defined as the ratio of the available energy (difference between heat obtained from the thermal source and total exergy destruction in the system) to the total energy obtained from the thermal source. Guo et. al [5] conducted thermodynamic analysis of waste heat power generation system. One of the applied objective function was so called sustainability index *SI*, which is also used in this paper.

Most of the conducted studies converge single-objective analysis, which are focused on optimizing one of the above performance indicators. Much less attention has been paid to multi-objective optimization (MOO). This approach allows to carry out more comprehensive analysis, using more than one objective function, which means that ORC system can be optimized from both thermodynamic and ecomonic perspectives. Moreover, the selection process of the working fluid can be also optimised including several criteria [6].

Wang et. al [7] carried out multi-objective optimization of an ORC for low grade waste heat recovery using evolutionary algorithm. The authors used exergy efficiency and capital cost as performance indicators of the system. Xiao et. al [8] conducted multi-objective optimization of evaporation and condensation temperatures for subcritical ORC, using sustainability index *SI* and the ratio of the cost of the system *C* to net power output as objective functions. The authors solved optimization problem using linear weighted evaluation function. Toffolo et. al [9] made the optimal selection of working fluid and design parameters in ORC system, using multi-criteria approach. The objective functions that the authors applied, was levelized cost of electricity *LCOE*, which considers the annual energy production of the system or specific investment cost *SIC* defined as the ratio between total investment cost and net power output.

In several cases, it is very important to not only seeking to improve all thermodynamic and economic indexes, but also include some inconveniences of the system. Therefore, in case of geothermal water, which is the energy source in the system analysed in this paper, the level of mineralization M of a brine should be considered. In order to solve this problem, multi-objective optimization has been carried out using weighted sum method [10] with global function G, which is to be minimized.

* Corresponding author: marcin.jankowski@zut.edu.pl

 https://doi.org/10.1051/e3sconf/20187001005

2 Calculation model of a system

2.1 System description

The energy source of the system has a significant impact on the overall performance and effectiveness of ORC power plant and it is important to carry out complex analysis in order to determine the optimal working parameters of the ORC. In case of considered system, geothermal water (brine) has been used to supply ORC power plant. One of the crucial property of such fluid is mineralization *M*. It is worth to emphasise that this parameter can quite considerably vary depending on local hydrogeothermal conditions. The level of salinity of the brine affects its values of basic thermophysical properties, such as: specific heat c_{pl} and density ρ_l, so in calculation procedure, these parameters should be considered as function of either temperature and mineralization.

The ORC system, as shown in figure 1 and 2, works as classical power plant. Slightly superheated (point 1) vapour flows into the expander and its enthalpy is converted to useful work of the turbine. The low pressure vapour (point 2) is cooled (to the state corresponding to point 2") and phase change occurs in condenser (cold water is a coolant). Then the saturated liquid (3) is pumped to higher pressure and goes to the inlet of vapour generator, where it is heated up, evaporated and, optionally, superheated.

The choice of the working fluid affects substantially the overall performance of ORCs and a lot of analysis have been done in order to choose the best one for certain configuration of the system [11], [12], [13]. In this paper, R1234yf has been chosen as one of the most promising working fluid, because of its beneficial impact on an effectiveness of ORC, as well for a favourable environmental indicators, such as: *ODP=0* and *GWP=4* [14].

As more attention in calculation procedure has been devoted to the evaporation process of the working fluid, in figure 3 only vapour generator (VG) and its *T-ΔH* diagram are presented. The VG is considered as countercurrent plate heat exchanger (PHE). It is worth mentioning that the evaporator section of the VG is divided into finite small elements in order to assume constant thermophysical properties of evaporating fluid for each element. The values of the parameters shown in figure 3 are provided in table 1.

Table 1. Parameters of ORC system.

Energy source			Organic Rankine cycle			Heat sink		
brine			organic fluid			cold water		
$\dot{V}_1$	[m³h⁻¹]	10	$\dot{m}_2$	[kgs⁻¹]	*	$\dot{m}_3$	[kgs⁻¹]	*
t_{1w}	[°C]	105	t_{ev}	[°C]	53-68	t_{1c}	[°C]	10
t_{2w}	[°C]	60	t_{con}	[°C]	30	t_{2c}	[°C]	20
M	[kg/kg]	0.00-0.16	ΔT_p	[K]	*	-		
-			ΔT_{sup}	[K]	5			
Star sign '*' means that the quantity is a result of calculation. Pinch point temperature ΔT_p varies in the interval: 7,5-16,5 K.								

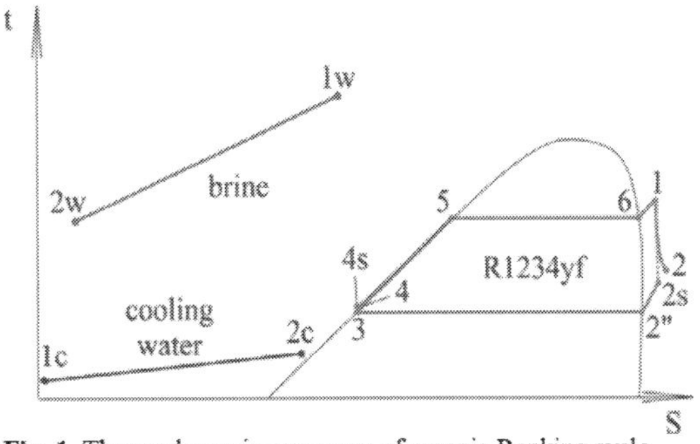

Fig. 1. Thermodynamic processes of organic Rankine cycle.

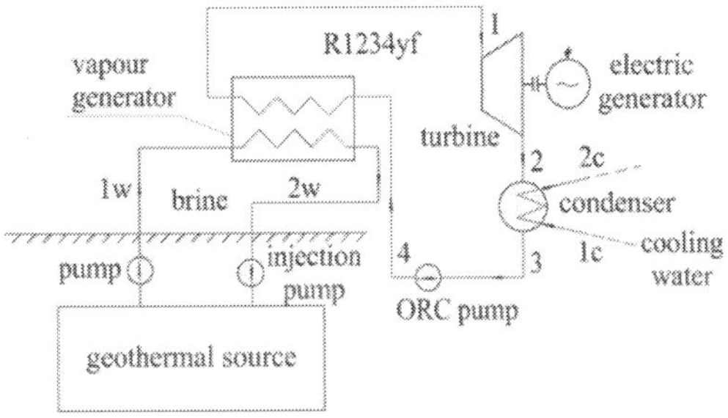

Fig. 2. ORC power plant.

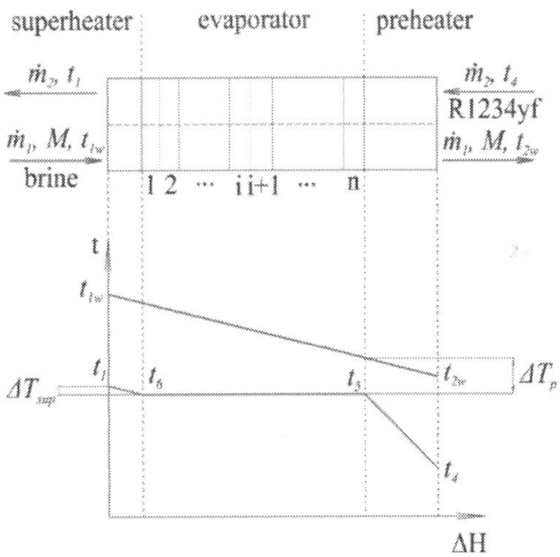

Fig. 3. Temperature field in a vapour generator.

2.2 Calculation model

The proposed model is created in order to reflect if there exists significant influence of the salination of the geothermal water on the optimum evaporation temperature of the working fluid in ORC power plant. However, it is difficult to make a comprehensive optimization using only single objective functions, since, during designing process of ORCs, it is impornant to include several factors, such as: thermodynamic, economic or environmental. In that model, multi-objective optimization (MOO) approach has been proposed with multi-objective function *G(X)*, which consists of the quantities that are most substantially affected by decision variables, i.e.: the evaporation temperature t_{ev} and the mineralization M of the brine. Component quantities are the following: heat exchanger area of the vapour generator A_{VG}, net power output of ORC system N_{out}, overall exergy destruction rate δB_{tot} in the system and exergy drop ΔB_w of the geothermal water at the inlet and outlet .

Calculations have been made using *MATLAB 2017b* software [15] with *REFPROP 9.1* [16] as database of the thermophysical properties of R1234yf fluid.

2.2.1 Optimization model

The optimization model has been solved using weighted sum method with aforementioned *G(X)* as objective function, which, in this case, is the composition of two single objective functions:

$$G(X) = w_1 f_1^{\,n} + w_2 f_2^{\,n} \tag{1}$$

Single objective functions, which are to be minimised, can be written as follows:

$$f_1 = \frac{A_{VG}}{N_{out}} \tag{2}$$

$$f_2 = \frac{\delta B_{tot}}{\Delta B_w} \tag{3}$$

Function f_1 is an economic indicator, while function f_2 is called sustainability index *SI*, which determines the influence on the environment. However, in equation (1), normilized forms of the functions (2) and (3) have to be considered, as units and ranges of their values are different. In general, function f can be transformed to normalized form using the following formula:

$$f^n = \frac{f(x) - f_{\min}}{f_{\max} - f_{\min}} \tag{4}$$

where: f_{min} and f_{max} are minimum and maximum values of the objective function f in considered range of decision variable x. The intervals of decision variables, i.e. evaporation temperature t_{ev} and mineralization M of the brine are provided in table 1.

The assigned values of weights are the same: $w_1=w_2=0.5$, as both of the component objective functions f_1 and f_2 are considered to have the same level of significance.

2.2.2 Thermodynamic and heat transfer model

Heat transfer area of vapour generator A_{VG}

A_{VG} in this model is considered to be the sum of preheater, evaporator and superheater, which, using heat transfer formula, can be written in the following form:

$$A_{VG} = \frac{Q_{pre}}{k_{pre}\Delta T_{\log pre}} + \sum_{i=1}^{n} \frac{Q_i}{k_i \Delta T_{\log i}} + \frac{Q_{\sup}}{k_{\sup}\Delta T_{\log \sup}} \tag{5}$$

It is worth to remind that the evaporator section (middle part of the (5) equation) is divided into finite small elements (see figure 3), so each quantity is determined with respect to i-th fininte element.

Heat fluxes for preheater, evaporator and superheater sections can be obtained from energy balance equation. On the assumption that ORC power plant is in steady state as well as there is no heat loss, corresponding heat fluxes can be calculated from the following formulas:

$$Q_{pre} = m_2 (h_5 - h_4) \tag{6}$$

$$Q_i = m_2 (h_{i+1} - h_i) \tag{7}$$

$$Q_{\sup} = m_2 (h_1 - h_6) \tag{8}$$

where enthalpies corresponding to states of working fluid (see figure 1 or 2) are determined using *Refprop* database. Enthalpy h_4 is obtained using equation (23). Mass flow rate of the R1234yf can be calculated from the following equation:

$$m_2 = \frac{Q_1}{q_{4-1}} = \frac{\rho_1 V_1 c_{p1}(t_{1w} - t_{2w})}{h_1 - h_4} \tag{9}$$

As has been mentioned before, density ρ_1 and specific heat c_{p1} of the geothermal water should be considered as function of temperature as well as mineralization of the brine. Equations of these properties provided in [17] are as follows:

$$\begin{gathered} \rho_1(t, M) = a_1 + a_2 t + a_3 t^2 + a_4 t^3 + a_5 t^4 + \\ + b_1 M + b_2 M t + b_3 M t^2 + b_4 M t^3 + b_5 M^2 t \end{gathered} \tag{10}$$

where: a_1=9.999·10^2, a_2=2.034·10^{-2}, a_3=-6.162·10^{-3}, a_4=2.261·10^{-5}, a_5=-4.657·10^{-8}, b_1=8.020·10^2, b_2=-2.001, b_3=1.677·10^{-2}, b_4=-3.060·10^{-5}, b_5=-1.613·10^{-5} for the range of applicability: $0<t<180$ °C, $0<M<0.16$ kg·kg^{-1}

$$c_{p1}(T, M) = A + BT + CT^2 + DT^3 \tag{11}$$

where:

$$A = 5.328 - 9.760 \cdot 10^{-2} \cdot M + 4.040 \cdot 10^{-4} \cdot M^2$$

$$B = -6.913 \cdot 10^{-3} + 7.351 \cdot 10^{-4} \cdot M - 3.150 \cdot 10^{-6} \cdot M^2$$

$$C = 9.600 \cdot 10^{-6} - 1.927 \cdot 10^{-6} \cdot M + 8.230 \cdot 10^{-9} \cdot M^2$$

$$D = 2.500 \cdot 10^{-9} + 1.666 \cdot 10^{-9} \cdot M - 7.125 \cdot 10^{-12} \cdot M^2$$

for the following range of applicability: 273.15<T<453.15K, 0<M<180 g·kg^{-1}. It is mandatory to use units of temperature and mineralization such as in given ranges. Resulting density and specific heat are in kg·m^{-3} and kJ·kg^{-1}K^{-1} respectively.

The logarithmic mean temperature difference ΔT_{log} can be calculated using equation :

$$\Delta T_{\log} = \frac{t_1^{'} - t_2^{"} - \left(t_1^{"} - t_2^{'}\right)}{\ln \frac{t_1^{'} - t_2^{"}}{t_1^{"} - t_2^{'}}} \quad (12)$$

where superscripts: ' and " indicates inlet and outlet temperatures respectively.

Overall heat transfer coefficient k can be obtained from the general formula:

$$k = \left(\frac{1}{h_1} + \frac{\delta}{\lambda} + \frac{1}{h_2}\right)^{-1} \quad (13)$$

δ=0.6mm and λ=18$Wm^{-1}K^{-1}$ are the assumed values of thickness and thermal conductivity of the heat exchanger plate respectively. Depending on which section of the vapour generator is being analysed, different correlations of Nusselt number Nu_2 for organic working fluid side are used. Definition of Nusselt number allows to calculate heat transfer coefficient h:

$$h = \frac{Nu \lambda_l}{d_h} \quad (14)$$

where d_h is calculated as double distance h_0 between plates (d_h=2·1.5=3mm).

In case of geothermal water side, the correlation for Nu_1 is the same for all sections and has the following form [18]:

$$N_1 = C\left(\mathrm{Re}_1^{\left(0.728+0.0543\sin\left(\frac{2\pi\beta}{90}\right)+3.7\right)}\right)\mathrm{Pr}_1^{1/3}\left(\frac{\mu_1}{\mu_{wa}}\right)^{0.14} \quad (15)$$

where:

$$C = 0.2668 - 0.006967\beta + 7.244 \cdot 10^{-5}\beta^2 \quad (16)$$

The so called chevron angle is assumed to be equal to: β=60°. The Reynolds number Re is defined as follows:

$$\mathrm{Re} = \frac{v d_h \rho}{\mu} \quad (17)$$

Superheater section

The following Nusselt number correlation [19] for organic working fluid side has been used:

$$Nu_{\sup} = 0.2946 \cdot \mathrm{Re}_{\sup}^{0.7} \mathrm{Pr}_l^{1/3} \quad (18)$$

Evaporator section

Yan and Lin [20] proposed the following correlation for two-phase flow boiling process:

$$Nu_{ev} = 1.926 \mathrm{Re}_{eq} Bo_{eq}^{0.3} \mathrm{Re}_l^{-0.5} \mathrm{Pr}_l^{1/3} \quad (19)$$

where boiling number is defined as:

$$Bo = \frac{q}{Gr} \quad (20)$$

Detailed description of the equation is provided in aforementioned article.

Preheater section

In case of preheating process of the R1234yf, the same correlation (14) for Nusselt number was used as for water side.

Net power output N_{out}

The general formula for N_{out} is the following:

$$N_{out} = \dot{m}_2\left[h_1 - h_2 - \left(h_4 - h_3\right)\right] \quad (21)$$

where the enthalpies h_2 and h_4 are calculated using definitions of isentropic efficiencies (equations (22) and (23)), which assumed values for a turbine and a pump are equal to: η_{iT}=80% and η_{iP}=75%.

$$h_2 = h_1 - \eta_{iT}\left(h_1 - h_{2s}\right) \quad (22)$$

$$h_4 = h_3 + \frac{h_{4s} - h_3}{\eta_{iP}} \quad (23)$$

Overall exergy destruction rate δB_{tot}

The total exergy destruction in analysed ORC system can be presented as the sum:

$$\delta \dot{B}_{tot} = \delta \dot{B}_{VG} + \delta \dot{B}_{con} + \delta \dot{B}_T + \delta \dot{B}_P \quad (24)$$

The components of equation (24) correspond to all elements of ORC system are shown in figure 2 and are explained below.

Vapour generator

The exergy destruction in every component can be calculated using Gouy-Stodola equation, which in case of VG has the following form:

$$\delta \dot{B}_{VG} = T_r \sum \Delta \dot{S}_{VG} \quad (25)$$

where subscript r means reference state, which level is characterized by temperature T_r=298.15K and pressure p_r=101325Pa. The sum of the entropy fluxes can be found from the entropy balance equation:

$$\sum \Delta \dot{S}_{VG} = \dot{m}_1\left(s_{2w} - s_{1w}\right) + \dot{m}_2\left(s_1 - s_4\right) \quad (26)$$

Condenser

Similarly to the vapour generator, the equations of the exergy destruction and the entropy balance for condenser are the following:

$$\delta \dot{B}_{con} = T_r \sum \Delta \dot{S}_{con} \quad (27)$$

$$\sum \Delta \dot{S}_{con} = \dot{m}_3\left(s_{2c} - s_{1c}\right) + \dot{m}_2\left(s_3 - s_2\right) \quad (28)$$

Turbine

For the turbine, Gouy-Stodola equation has the following form:

https://doi.org/10.1051/e3sconf/20187001005

$$\delta B_T = T_r \sum \Delta S_T \qquad (29)$$

where:

$$\sum \Delta S_T = m_2(s_2 - s_1) \qquad (30)$$

Pump

Analogically, for the pump:

$$\delta B_P = T_r \sum \Delta S_P \qquad (31)$$

$$\sum \Delta S_P = m_2(s_4 - s_3) \qquad (32)$$

Exergy drop ΔB_w of the geothermal water

Exergy drop of the brine between inlet and the outlet of vapour generator can be calculated from:

$$\Delta B_w = m_1(b_{1w} - b_{2w}) \qquad (33)$$

Using the definition of exergy, b_{1w} and b_{2w} are calculated as:

$$b_{1w} = h_{1w} - h_r - T_r(s_{1w} - s_{r1}) \qquad (34)$$

$$b_{2w} = h_{2w} - h_r - T_r(s_{2w} - s_{r1}) \qquad (35)$$

3 Results and discussion

In figures 4-10 overall results of thermodynamic and optimization analysis are presented. As can be seen in figure 4, heat transfer area A_{VG} of the vapour generator increases with the increase of evaporation temperature, which is, mostly, due to the fact that with the increase of t_{ev}, heat flux $\dot{Q}_i$ (see equation (7)) of the working fluid R1234yf also increases, which clearly involves A_{VG} (see equation (5)). However, with the increase of salination of a brine, there can be seen the opposite dependence. This fact is mainly influenced by the variability of the geothermal water heat flux $\dot{Q}_1$, as it is the function of mineralization M. Decreasing values of the heat flux $\dot{Q}_1$ with increasing mineralization M stems from an influence of the M on the thermophysical properties of the brine, i.e.: density ρ_1 and specific heat c_{p1}. It turns out that the density ρ_1 is increasing while the specific heat c_{p1} is decreasing with the increase of mineralization M, however – the overall effect, which comes from the product of ρ_1 and c_{p1} (see equation (9)), leads to the decrease of $\dot{Q}_1$. Moreover, the changes in heat flux $\dot{Q}_1$ affect heat fluxes transfered in preaheater, evaporator and superheater (see equations (6),(7),(8)), since the enegy balance equation must be conserved, thus decreasing values of $\dot{Q}_1$ lead to the less values of A_{VG}.

Net power output N_{out}, as can be seen in figure 5, is increasing with the evaporation temperature, since the mass flow rate $\dot{m}_2$, which affects N_{out}, depends on previously discussed $\dot{Q}_1$ (see equations (9) and (21)). For an analogical reason, N_{out} is decreasing with the mineralization M, as higher values of M correspond to the lower values of $\dot{m}_2$.

Figure 6 presents the results for the first objective function f_1, i.e. ratio of heat transfer area of the vapour generator A_{VG} and net power output N_{out}.

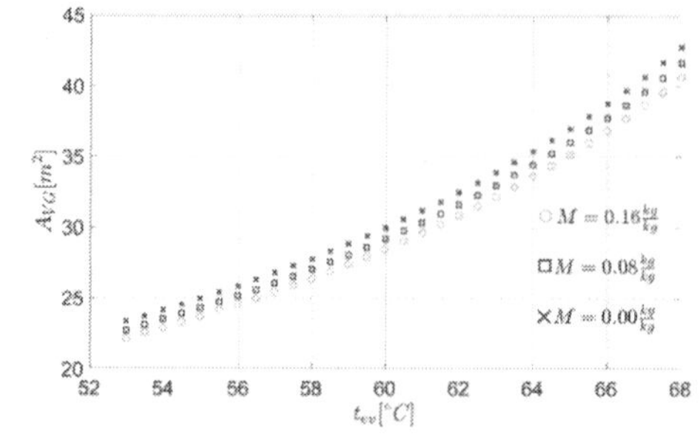

Fig. 4. Vapour generator area A_{VG} as the function of evaporation temperature t_{ev} for different levels ofmineralization M.

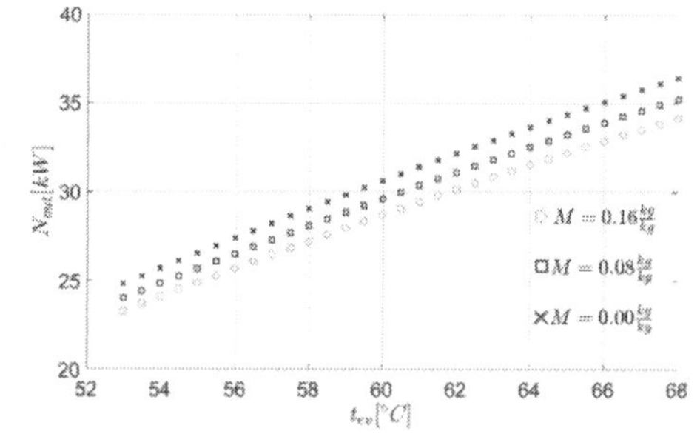

Fig. 5. Net power output N_{out} as the function of evaporation temperature t_{ev} for different levels of mineralization M.

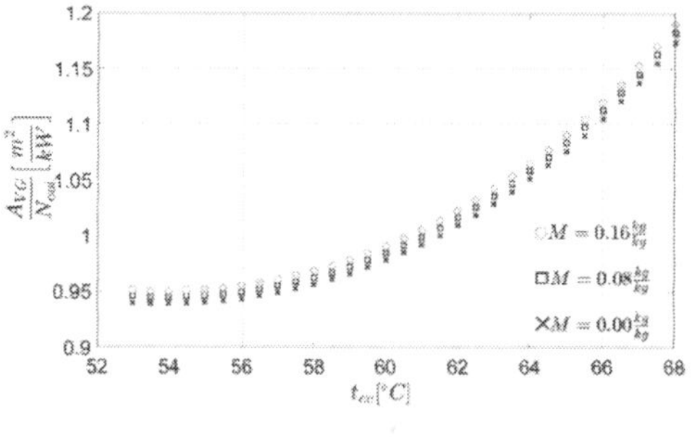

Fig. 6. The ratio of vapour generator area to the net power output as the function of evaporation temperature t_{ev} for different levels of mineralization M.

It is worth noticing that there exists optimum evaporation temperature of the function f_1 and it is located approximetaly at the temperature of 54°C. The influence of mineralization is relatively small, but negative – increasing values of M lead to increasing values of f_1 which is to be minimised.

Overall exergy destruction rate δB_{tot} is decreasing with the increase of evaporation temperature, as can be seen in figure 7. It is worth mentioning that the greatest exergy destruction occurred in vapour generator. Moreover, these losses are changing most rapidly with respect to evaporation temperature t_{ev} and mineralization *M*. Therefore, it should come as no surprise that exergy destruction in the system is decreasing , as higher values of t_{ev} lead to smaller temperature differences between geothermal water and working fluid and therefore – to less entropy generation (see for instance equations (25) and (26)). The exergy destruction also decreases with respect to mineralization, which can be explained using equation (26). As has been mentioned before, mass flow rate of the working fluid $\dot{m}_2$ is decreasing with the increase of mineralization. Simultaneously, mass flow rate of the geothermal water $\dot{m}_1$ is increasing, because of the increase of the density ρ_1. The entropy drops of the water and working fluid are negative and positive respectively, thus the overall effect leads to the decrease of exergy destruction.

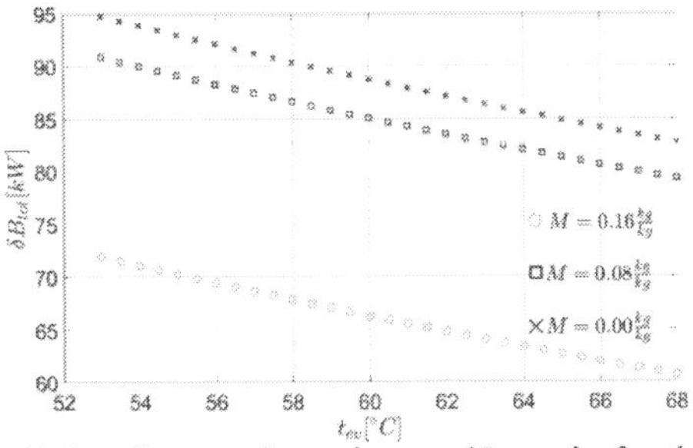

Fig. 7. Overall exergy destruction rate δB_{tot} as the function of evaporation temperature t_{ev} for different levels of mineralization *M*.

In figure 8, exergy drop of the geothermal water is changing with respect to only mineralization *M*, as quantities in equation (33) are independent on the evaporation temperature t_{ev} of the working fluid. The less values of ΔB_w for higher levels of mineralization can result from decreasing values of specific heat c_{p1}, which in turn affect the enthalpies that are the components in equations (34) and (35).

Figure 9 presents the results for the second objective function f_2, which is the ratio of total exergy destruction δB_{tot} to exergy drop of the geothermal water ΔB_w. This function is also known as sustainability index *SI*. Lower values of that indicator correspond to lower negative influence on an environment. The drop in of the values of the objective function with respect to evaporation temperature t_{ev} is obvious, since the exergy destruction from figure 7 has the same tendency and exergy drop of water from figure 8 is not changing. The mineralization *M* of the brine affect the objective function f_2 analogically as on its components. Although, that function does not indicate optimum evaporation temperature t_{ev} of working fluid, however, there can be seen that higer saturation temperatures would be more beneficial in this case.

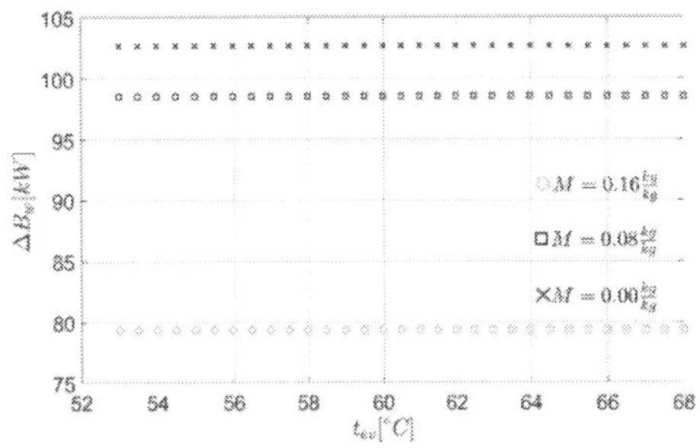

Fig. 8. Exergy drop ΔB_w of the geothermal water as the function of evaporation temperature t_{ev} for different levels of mineralization *M*.

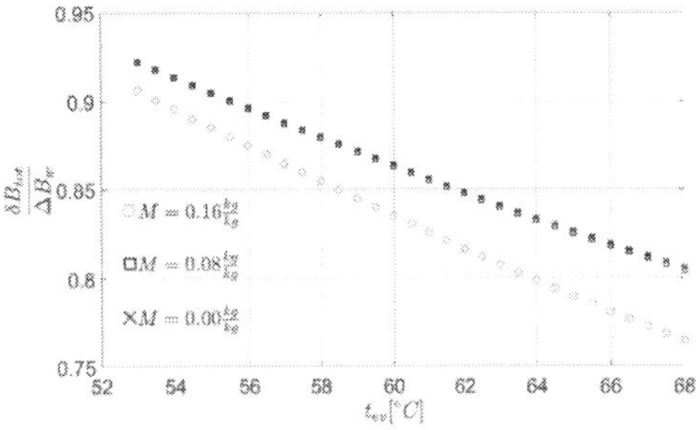

Fig. 9. Sustainability index *SI* as the function of evaporation temperature t_{ev} for different levels of mineralization *M*.

Finally, figure 10 presents the results for multi-objective function *G*. On the assumptions that have been made, i.e. equivalent importance (w_1=w_2=0.5) of both objective functions f_1 and f_2 and using evaporation temperature t_{ev}, and mineralization *M* as decision variables, it turns out that optimum evaporation temperature equals to approximately 61°C, but more importantly, it is the same for all analysed levels of mineralization of the brine. This means that mineralization *M* has no influence on the choice of

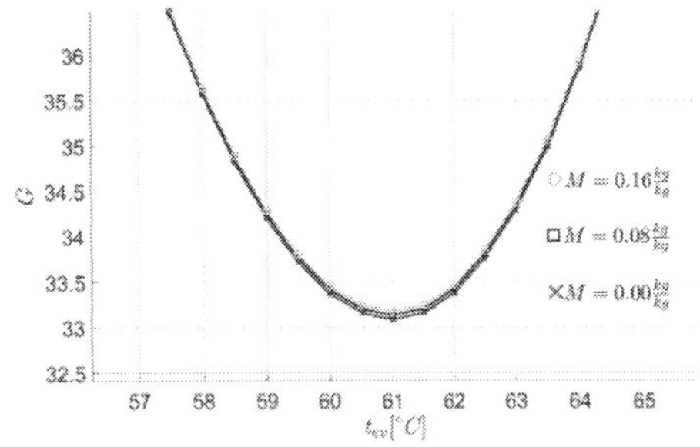

Fig. 10. Multi-objective function *G* as the function of evaporation temperature t_{ev} for different levels of mineralization *M*.

optimum evaporation temperature. Moreover, it can be seen that minimum values of the function *G* have been obtained for mineralization *M=0.00*. However, each of three curves almost coincide with each other, which means that the overall influence of the mineralization M is almost negligible, which can be explained by the fact that single-objective functions f_1 and f_1 have presented the opposite relationship with respect to salination – f_1 was increasing and f_2 was decreasing for certain evaporation temperatures.

4 Conclusions

In the article, multi-objective function G has been created using performance indicators of ORC power plant, i.e. ratio of heat transfer area of vapour generator A_{VG} to net power output N_{out} (objective function f_1) and SI index (objective function f_2), which are to be minimised. The proposed calculation model allowed to conduct an analysis of the results from both: single- and multi-objective optimization perspective. The main aim of the paper was to resolve, if there is a significant influence of the mineralization on individual (f_1 and f_2) and global (G) ORC performance indicators as well as on the optimal evaporation temperature of the working fluid R1234yf. The presented results are important and can be used by scientists as well as by engineers, since they reflect, if there is a high risk of drop in economic (function f_1), environmental (function f_2) or overall (function G) index for different levels of mineralization M. The multi-objective problem has been solved using weighted sum method. The following conclusions were drawn:

- there is a low impact of the mineralization M of the brine on the single-objective functions f_1 and f_2, however in case of the first one the influence is negative (function f_1 takes higher values for higher levels of mineralization M), in the second one: positive (the values of f_2 are lower with respect to increasing levels of mineralization M),
- the multi-objective function G takes almost the same values for each of the analysed levels of mineralization M in considered interval of the evaporation temperature t_{ev}, which can result from the aforementioned opposite impacts of single-objective functions,
- the evaporation temperature t_{ev} corresponding to minimum value of G is the same (t_{ev}=61°C) for each of the mineralization M levels, i.e. for the adopted criteria, there is no influence of such parameter on the optimum evaporation temperature of R1234yf in considered ORC system.

Nomenclature

A – heat exchanger area [m^2]
Bo – boiling number
$\dot{B}$ – exergy flux [kW]
b_0 – width of plate [m]
c_p – specific heat capacity [J kg^{-1} K^{-1}]
d – diameter [m]
f – single-objective function
h – heat transfer coefficient [W m^{-2} K^{-1}] or specific enthalpy [Jkg^{-1}]
h_0 – pitch between plates [m]
G – density mass flux [kg s^{-1} m^{-2}] or multi-objective function
k – overall heat transfer coefficient [W m^{-2} K^{-1}]
L – length [m]
M – mineralization [kg kg^{-1}]
N_{out} – net power output [kW]
Nu – Nusselt number
$\dot{m}$ – mass flow rate [kg s^{-1}]
p – pressure [Pa]
q – density heat flux [W m^{-2}]
Re – Reynolds number
r – latent heat of vaporization [J kg^{-1}]
Pr – Prandtl number
S – entropy [kJ K^{-1}]
s – specific entropy [kJ kg^{-1} K^{-1}]
t – temperature [°C]
V – volume flowrate [m^{3}s^{-1}]
v – velocity [m s^{-1}]
w – weight

Greek symbols

β – chevron angle [°]
$\Delta\dot{B}_w$ – exergy drop of geothermal water [kW]
ΔH – enthalpy difference [kW]
ΔT – temperature difference [K]
δ – plate thickness [m]
$\delta\dot{B}$ – exergy destruction rate [kW]
η – efficiency
λ – thermal conductivity [W m^{-1} K^{-1}]
μ – dynamic viscosity [Pa s]
ρ – density [kg m^{-3}]

Sub- or superscripts

‘ – correspond to inlet
“ – correspond to outlet
1 – correspond to brine
2 – correspond to working fluid
3 – correspond to cooling water
c or con – condenser
ev – evaporation
h – hydraulic
i – correspond to section or isentropic
l – liquid
log – logarithmic
n – normalized
P – pump
p – pinch point
pre – preheater
r – reference state
VG – vapour generator
sup – superheater
T – turbine
w – water
wa – wall

References

1. I.H. Aljundi, *Effect of dry hydrocarbons and critical point temperature on the efficiencies of organic Rankine cycle*, Renew. En. **36** (2011)

2. A.A. Lakew, O. Bolland, *Working fluids for low-temperature heat source*, App. Ther. Eng. **30** (2010)
3. Z. Shengjun, W. Huaixin, G. Tao, *Performance comparison and parametric optimization of subcritical Organic Rankine Cycle (ORC) and transcritical power cycle system for low-temperature geothermal power generation*, App. En. **88** (2011)
4. J.P. Roy, M.K. Mishra, A. Misra, *Performance analysis of an Organic Rankine Cycle with superheating under different heat source temperature conditions*, App. En. **88** (2011)
5. J. Guo, M. Xu, L. Cheng, *Thermodynamic analysis of waste heat power generation system*, En. **35** (2010)
6. Bendig M., Favrat D., Marechal F., *Methodology for Identification of Suitable ORC-Cycle and Working-Fluid using Integration with Multiple Objectives*, Chem. Eng. Trans. **39** (2014)
7. J. Wang, Z. Yan, M. Wang, M. Li, Y. Dai, *Multi-objective optimization of an organic Rankine cycle (ORC) for low grade waste heat recovery using evolutionary algorithm*, En. Conv. and Man. **71** (2013)
8. L. Xiao, S.-Y. Wu, T.-T. Yi, C. Liu, Y.-R. Li, *Multi-objective optimization of evaporation and condensation temperatures for subcritical organic Rankine cycle*, En. **83** (2015)
9. A. Toffolo, A. Lazzaretto, G. Manente, M. Paci, *A multi-criteria approach for the optimal selection of working fluid and design parameters in Organic Rankine Cycle systems*, App. En. **121** (2014)
10. T.R. Marler, J.S. Arora, *The weighted sum method for multi-objective optimization: new insights*, Struct. Multidisc. Optim. **41** (2010)
11. H. Chen, Y.D. Goswami, E.K. Stefanakos, *A review of thermodynamic cycles and working fluids for the conversion of low-grade heat*, Renew. and Sustain. En. Rev. **14** (2010)
12. J. Bao, L. Zhao, *A review of working fluid and expander selections for organic Rankine cycle*, Renew. and Sustain. En. Rev. **24** (2013)
13. A.I. Papadopoulos, M. Stijepovic, P. Linke, *On the systematic design and selection of optimal working fluids for Organic Rankine Cycles*, App. Ther. Eng. **30** (2010)
14. M.T. Nasir, K.C. Kim, *Working fluid selection and parametric optimization of an Organic Rankine Cycle coupled Vapor Compression Cycle (ORC-VCC) for air condidioning using low grade heat*, En. and Build., **129** (2016)
15. The MathWorks Inc, MATLAB version R2017b. 2017.
16. E.W. Lemmon, M.L. Huber, M.O. McLinden, *Reference thermodynamic and transport properties – Refprop NIST standard reference base 23,* Version 9.1
17. M. Miecznik, *Error in the estimation of the potential for electricity generation in a binary ORC systems associated with variation of thermodynamic parameters of geothermal water*, Geotermia, Zrównoważony Rozwój nr 2/2013
18. J.T. Cieśliński, A. Fiuk, K. Typiński, B. Siemieńczuk, *Heat transfer in plate heat exchanger channels: Experimental validation of selected correlation equations*, Archives of Thermod. **37** (2016)
19. Z. Janković, J. Sieres, F. Cerdeira, B. Pavković, *Analysis of the impact of different operating conditions on the performance of a reversible heat pump with domestic hot water production*, Inter. Journ. of Refri. **86** (2018)
20. Y.Y. Yan, T.F. Lin, *Evaporation heat transfer and pressure drop of refrigerant R134a in a plate heat exchanger*, ASME J. Heat Transfer. **121** (1999)

 https://doi.org/10.1051/e3sconf/20187001006

Analysis of the possibility of utilization of waste heat from the marine engine in the ORC power plant

Radomir Kaczmarek[1], and *Aleksander* Stachel[1,*]

[1]West-Pomeranian University of Technology, Szczecin, Department of Heat Enginnering, al. Piastów 17, PL-70-310 Szczecin, Poland

Abstract. Modern marine diesel engines are characterized by ~50% conversion of chemical energy of fuel into useful work. The remaining amount of heat is lost to the environment as the so-called waste heat, whereby its part can be used in various marine heating installations. In the ship's engine the biggest source of waste heat, potentially useful for further use, both due to the amount of available energy and its quality, are exhaust emissions from the main engine. Taking into account the need for rational energy management, the paper presents a preliminary analysis and assessment of the possibility of using the heat of hot exhaust gases to generate electricity in the so-called low-temperature ORC installation. Based on the available data, the operation of ORC installation has been evaluated, taking into account the influence of engine operating parameters on the obtained electrical power and the cycle efficiency. The aim of the analysis was to demonstrate that there is a real possibility to use the waste heat from marine engine exhausts to drive the ORC and generate electricity.

1 Introduction

The naval power plant is consisting of a set of machines and devices generating various forms of energy necessary to fully meet the ship's energy demand, which in sea conditions must be a self-sufficient unit in this respect. On the ship, the useful energy is obtained as a result of conversion of chemical energy of fuels. As a result mechanical, electric and thermal energy is generated .

Due to its duties, the naval power plant is divided into:

- principal power station, the main part of which is the major engine, and its task is to generate mechanical energy for the propulsion of the ship (e.g. propulsion of the propeller shaft);
- auxiliary power plant, the purpose of which is to generate energy necessary for the operation of devices and mechanisms servicing the main propulsion system, as well as all other devices and receivers installed on the ship.

Due to the used solutions, the power plants are divided into combustion, steam and combined plants, for example, combustion-steam ones.

One of the most commonly used types of main propulsion are reciprocating diesel engines, also used in generating sets and other ship equipment. Main propulsion engines are built as high-pressure, two-stroke, low-speed or as high-pressure, four-stroke, medium-speed ones. Both groups are adapted to burning diesel fuels and the so-called heavy fuels. In addition to piston engines, ship turbines also use steam turbines and gas turbines (turbogas power plants).

An important element of the main engine operation, including the entire engine room, is the rationalization of fuel consumption, which translates into, among others, reduction of costs and reduction of emissions. The cutback on fuel consumption can be achieved by improving the energy efficiency of the power station and other ship equipment, as well as the utilization of the so-called waste energy (heat). Due to the amount of power generated, this applies in particular to the main propulsion engine (Fig. 1). The heat contained in the exhaust gases leaving the engine, as well as in other heat carriers, such as cooling water and charge air, can be used for various purposes, e.g. heating.

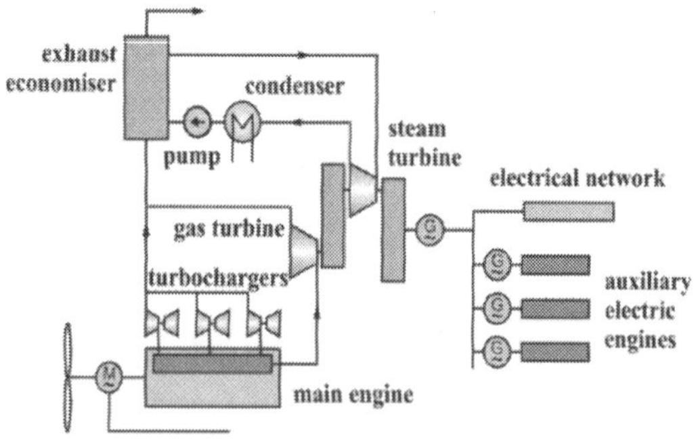

Fig. 1. An exemplary scheme of a ship's power plant with the utilization system of exhaust waste heat [1].

* Corresponding author: andrzej.stachel@zut.edu.pl

https://doi.org/10.1051/e3sconf/20187001006

2 Waste heat sources in the naval power plant

Internal combustion engines used as the main propulsion of ships, are characterized by about 50% efficiency in the use of chemical energy in the fuel. Apart from the produced useful mechanical energy, significant rates of the so-called waste heat contained in various heat carriers are released to the surroundings.

The basis for determining the waste energy resources is the thermal balance of the engine being a combination of energy flows supplied in the fuel (q_d) and produced work (L_u) as well as the sum of individual losses (Σq_s):

$$q_d = L_u + \sum q_s \tag{1}$$

where:

$$q_s = q_c + q_w + q_t + q_{od} + q_{pd} + q_d + q_{os} + q_{sw} + q_r$$

Heat losses are associated with cooling of cylinder heads and cylinder sleeves (q_c), injectors (q_w), pistons (q_t), charge air (q_{pd}), turbochargers (q_d) and engine lubricating oil (q_{os}) and finally the turbochargers (q_{od}). They also include exit losses (q_{sw}) and the so-called the residual losses (q_r), which include the losses due to radiation.

The extent of individual balance terms are variable and depend, among others, on the engine load (developed power), rotational speed, fuel type and ambient conditions as well as on the design and used cooling fluids. The real balances for specific engine types, provided by the manufacturers, are the basis for calculating the motor and ship related equipment and installations [1].

In the Figures 2 and 3 are presented, developed on the basis of catalog data [2], simplified heat balances of low-speed piston engines manufactured by MAN D&T, as well as the values of basic temperatures of waste energy carriers, with the specification of minimum and maximum values. The graphs show that the main source of waste energy on the ship power plants are exhaust emissions from the main engine, both due to the amount of energy contained in them as well as their quality.

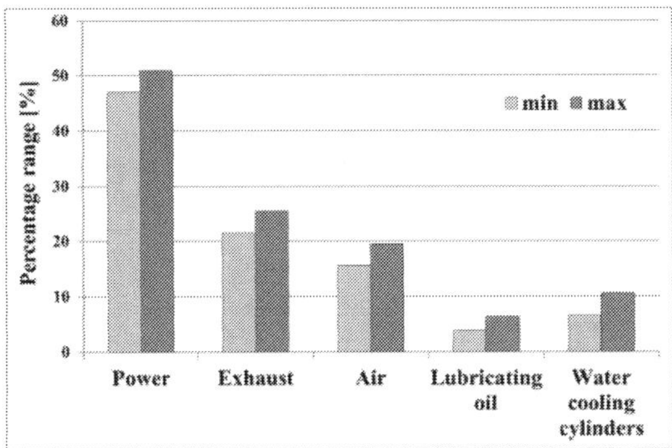

Fig. 2. Average percentage shares of power and rates of waste heat in low-speed engines (minimum – maximum values) [2].

It should be noted, however, that the introduced constructional and operational changes of the engines (e.g. larger piston stroke, increase of the boost pressure, lowering the rotational speed, etc.) affect the reduction of some waste energy streams (e.g. exhaust gases), while increasing others (e.g. air recharging).

In practice, not all available waste energy is valuable for further use. Its rate should be adequately large, and the temperature, pressure and mass flow of the heat carrier (e.g. exhaust gases) are of decisive importance. Knowledge of these values allows to determine the amount of energy and exergy of exhaust gases in individual places of the disposal system. Thus, it allows to make a rational decision regarding the selection of one of many possible variants of the waste energy utilization system.

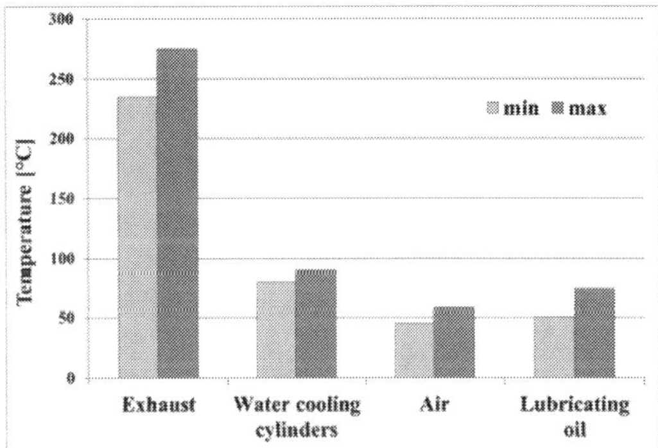

Fig. 3. Average ranges of temperatures of waste heat energy carriers in low-speed main engines (minimum – maximum values) [2].

In the case of internal combustion engines, a significant impact on the thermal parameters of the exhaust gas has the engine load related to the size of the generated power and the speed of rotation. It is also influenced by a number of other factors, such as the type of speed controller, which is related to, among others, the type of propeller (the propeller with a fixed or adjustable pitch). Examples of temperature changes (ΔT_s) and rate of exhaust gas ($\Delta \dot{m}_s$) resulting from the change of engine load in the engine type L70MC-C are shown in Figure 4 and Figure 5.

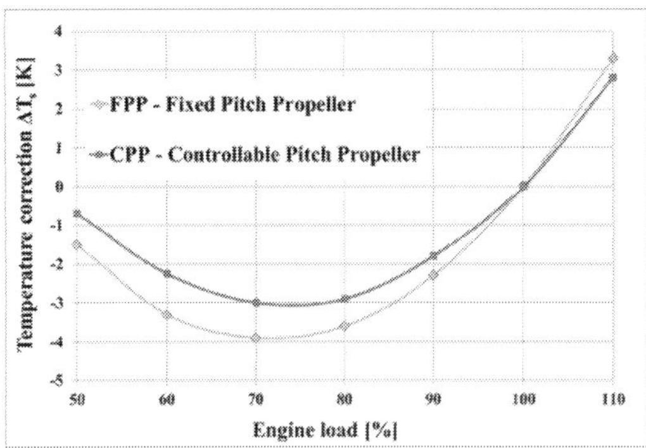

Fig. 4. Changes in exhaust gases temperature in relation to the engine load [3].

The operating loads of main engines (engine power) correspond usually to 85 – 90 % of the so-called contractual power, and the corresponding rotational speed is 95 - 97% of the contractual speed, respectively. It can be seen from the graphs that the correct design of utilization systems requires consideration of various engine operating states, including those corresponding to partial loads in a wide range, thus taking into account the relevant parameters of exhaust gases.

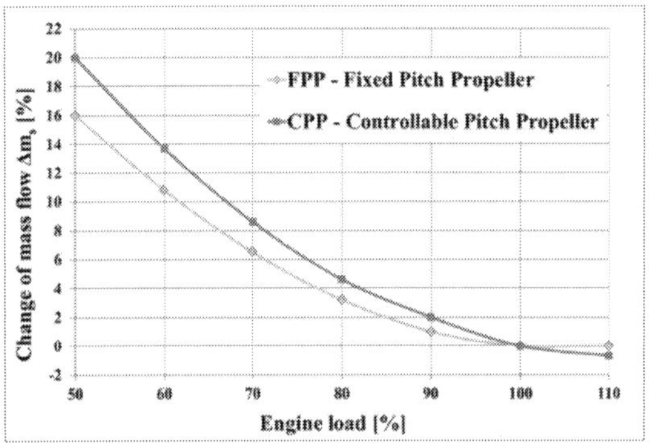

Fig. 5. Changes in exhaust gases mass flow rates in relation to the engine load [3].

Fig. 6 shows exemplary flue gas temperature changes, respectively for behind the engine cylinders and in the collectors before and after the turbochargers, depending on the load of the ship piston diesel engine.

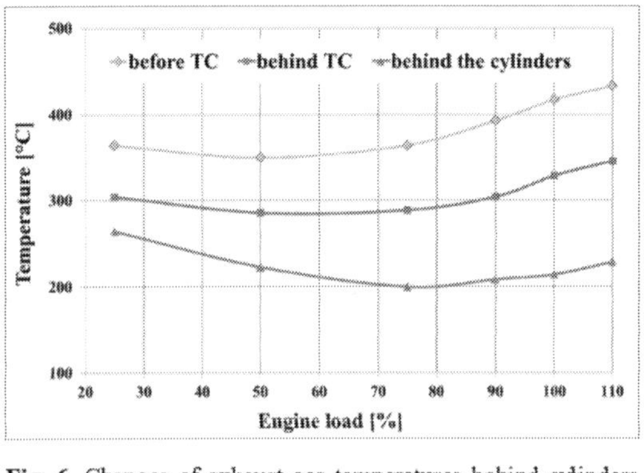

Fig. 6. Changes of exhaust gas temperatures behind cylinders and in collectors before turbochargers and after turbochargers, as a function of load of the engine 6L70MC [2].

3 Ways of utilizing waste heat on ships

The demand by the modern ships for various forms of energy (electric, heat and other), determined in relation to the main propulsion capacity, reaches values up to 40% for reefers, up to 25% for container vessels, up to 15% for bulk carriers, and depends on the period from their commissioning as well as the technical condition [4]. Some of these needs may be covered from waste energy sources.

Analyzing the solutions of modern marine combustion power plants, one can distinguish the main waste energy utilization systems used for:

- meeting the ship's heating demand,
- meeting electrical power demand,
- supporting the main propulsion of the ship,
- production of fresh water from sea water,
- meeting the cooling demand and others.

In the vast majority of power plants, the effect of waste energy utilization is useful heat, electricity and freshwater produced in evaporators. The easiest way - is to use that energy in the so called utilization boilers of steam / hot water for heating purposes. The waste energy used for this purpose is most often the heat from the exhaust gases from the motors.

Due to the development in recent years of so-called low-temperature power plant technology, one of the concepts of utilizing the waste energy from the main engine of the ship assumes the generation of electricity in the so-called ORC power plant working with the organic working fluid [5,6]. The operating principle of the ORC power plant is based on the classic comparative cycle of the steam water power plant (C-R cycle). The essential difference is the thermodynamic fluid used in circulation, which instead of water is a low-boiling organic substance.

4 The concept of using waste heat from engine in ORC installation

Analysis of the possibilities of using the ORC installation was related to the MAN D&T L70MC-C8-TII marine engine [3]. It is a low-speed, two-stroke, 6-cylinder, in-line combustion engine, with mechanical control of the camshaft. The cylinder diameter is 700 mm, whereas the piston stroke is 2360 mm.

Depending on the size of the ratio of piston stroke to cylinder diameter, it is built in two variants marked first with the symbol L-MC (mean stroke, S/D = 3.2) and secondly as S-MC-C (long stroke S/D = 4.0). The engine is used as a main drive on ships such as container ships, tankers, bulk carriers and gas carriers. It can be powered with light fuel oil and heavy fuel oil (HFO). Selected basic engine operating parameters, determined at the so-called standard operating conditions (air temperature 25°C, pressure 100 kPa, humidity 30%), are given in the Table 1.

Table 1. Standard operation conditions of the combustion engine 6L70MC-C8 [3].

Parameter	Unit	Value
Engine power	kW	19.620
Revolutions of the engine	min^{-1}	108
Piston velocity	m/s	8,3-8,5
Mean useful pressure	bar	19.0
Temperature of cooling water at the engine inlet	°C	32-36
Temperature of cooling water at the engine outlet	°C	80-90
Flow rate of cooling water	m^3/h	155
Temperature of exhaust gases including the turbocompressor	°C	265
Mass flowrate of exhaust gases	kg/s	44,5

4.1 ORC power plant supplied from the exhaust gases

A simplified diagram of the ship engine system - ORC power plant is shown in Fig. 7. The installation consists of a heat exchanger, turbine with a generator, condenser and a circulating pump. The source of heat is the rate of exhaust gases leaving the engine available at temperature T_{s1}, supplied to the heat exchanger (evaporator - heater), where the thermal energy contained in them is transferred to the working fluid in the ORC system, causing its heating to saturation conditions and subsequent evaporation. This results in the use of heat contained in the flue gases, and thus a decrease in their temperature (enthalpy), and as a result in a reduction of the exit losses. The exhaust gases, after giving away a part of the thermal energy contained in them, are discharged into the environment. Due to their relatively high thermodynamic potential, they still can be used for other purposes.

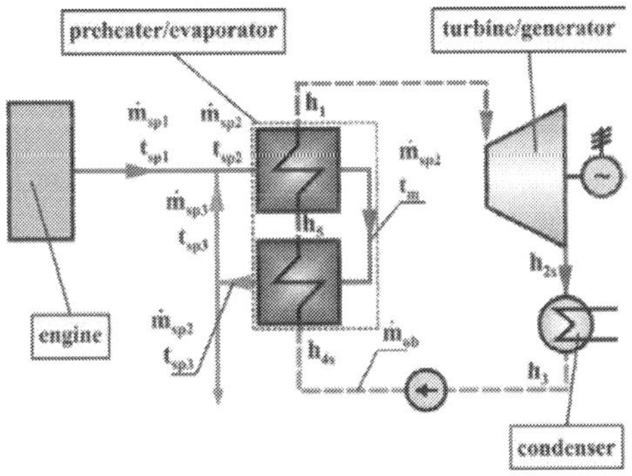

Fig. 7. Schematic of ORC system with an evaporator – preheater heat exchanger, using a reversal of the part of the exhaust gases rate of heat.

In the considered variant of the power plant, implemented is the reversal of the part of the exhaust gases from behind the exchanger (heater) to the inlet to the exchanger (evaporator). This allows to regulate the temperature of the exhaust gas supplied to the evaporator, so as to obtain the assumed temperature difference between the heating fluid (exhaust gas) and the operating fluid in the ORC system. This is of great importance in the case when the temperature of exhaust gases is changing, resulting from changes in the load of the engine.

The saturated or superheated vapour produced in the evaporator (depending on the operating conditions and the working fluid used) - it is directed to the turbine, where it expands and generates work used to drive the electric generator. The vapour after expansion in the turbine flows to the condenser, where it condenses. The condensed liquid is pumped back into the heater - evaporator assembly of heat exchangers.

4.2 Methodology of calculations

In order to determine the possibility of conversion of exhaust energy into the useful one (electric) in the ORC plant, the calculations of basic quantities characterizing its operation (power, efficiency) were made using the methodology given below.

The power plant capacity is defined as the difference in the work of the expansion of the working fluid in the turbine L_t and the work of pumping in the pump L_p:

$$N_{C-R} = \dot{L}_t - \dot{L}_p \tag{2}$$

where the work of expansion and work of pump can be expressed as products of the mass flow rate of working fluid in circulation ($\dot{m}_{ORC}$) and corresponding enthalpy changes:

$$L_t = \dot{m}_n (h_1 - h_{2s}) \tag{3}$$

$$L_p = \dot{m}_n (h_4 - h_3) \tag{4}$$

In the case of the ORC power plant, depending on the working fluid and pressures resulting from the operational temperature of the cycle, the pumping work may constitute up to 15% of the work generated in the turbine.

The thermal efficiency of the power plant (C-R cycle) is defined by the relation:

$$\eta_{C-R} = \frac{\dot{L}_t - \dot{L}_p}{\dot{Q}} \tag{5}$$

where $\dot{Q}_d$ – heat supplied to the cycle:

$$\dot{Q}_d = \dot{m}_n (h_1 - h_4) \tag{6}$$

Heat removed from the cycle in the condenser:

$$\dot{Q}_w = \dot{m}_n (h_{2s} - h_3) \tag{7}$$

Electric power generated in the ORC power plant is defined by the formulae:

$$N_{el} = N_{C-R} \cdot \eta_m \cdot \eta_g \cdot \eta_i \tag{8}$$

In order to determine the above quantities, the following values were determined:

- from the energy balance for the heat exchanger – mass flow rate of working fluid circulating in the power plant cycle:

$$\dot{m}_n = \frac{\dot{m}_{sp2} \cdot c_{sp} \cdot (t_{sp2} - t_{sp3})}{h_1 - h_{4s}} \tag{9}$$

- from the energy balance of exhaust gases – mass flow rate of exhaust gases flowing through the exchanger (evaporator - preheater):

$$\dot{m}_{sp2} = \frac{\dot{m}_{sp1} \cdot (t_{sp1} - t_{sp3})}{t_{sp2} - t_{sp3}} \tag{10}$$

Utilizing the energy balance equations defined for the preheater (11) and the preheater - evaporator system (12):

$$\dot{m}_{sp2} \cdot c_{sp} \cdot t_m + \dot{m}_n \cdot h_{4s} = \dot{m}_{sp2} \cdot c_{sp} \cdot t_{sp3} + \dot{m}_n \cdot h_5 \quad (11)$$

$$\dot{m}_{sp2} \cdot c_{sp} \cdot t_{sp2} + \dot{m}_n \cdot h_{4s} = \dot{m}_{sp2} \cdot c_{sp} \cdot t_{sp3} + \dot{m}_n \cdot h_1 \quad (12)$$

Determined was the temperature of exhaust gases leaving the heat exchangers (also that returning to the evaporator):

$$t_{sp3} = \frac{\Delta h \cdot t_m - t_{sp2}}{\Delta h - 1}, \text{ where: } \Delta h = \frac{h_1 - h_{4s}}{h_5 - h_{4s}} \quad (13)$$

In order to determine the values of the flow rate of exhaust gases from behind the preheater into the inlet of evaporator ($\dot{m}_{sp3}$), which enables obtaining of the assumed temperature difference between the exhaust gases and the circulation fluid in ORC system there has been used energy balance equation for the node „A":

$$\dot{m}_{sp1} \cdot c_{sp} \cdot t_{sp1} + \dot{m}_{sp3} \cdot c_{sp} \cdot t_{sp3} = \dot{m}_{sp2} \cdot c_{sp} \cdot t_{sp2} \quad (14)$$

and the mass flow rate balance:

$$\dot{m}_{sp2} = \dot{m}_{sp1} + \dot{m}_{sp3} \quad (15)$$

In the calculations, it was assumed that the temperature of the flue gases at the evaporator inlet is 15 K higher, and on the evaporator outlet by 5 K from the circulating fluid evaporating temperature in the ORC power plant. An exemplary temperature distribution for the evaporator - heater unit is shown in the Fig. 8.

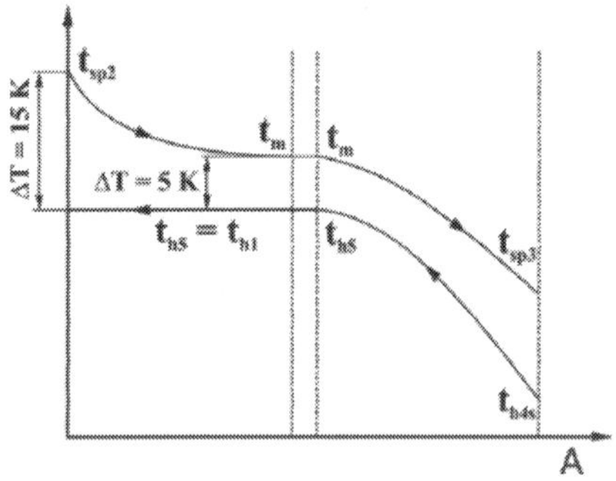

Fig. 8. Temperature distributions in the heat exchanger of the evaporator/preheater type in the ORC installation in case of dry fluids.

4.3 Assumptions for calculations

It was assumed that the considered ORC power plant operates in a subcritical cycle, and selected working fluids belong to the group of so-called dry fluids, which means that their isentropic expansion in the turbine, starting on the dry saturated vapour line, ends in the area of superheated vapour.

The selection of the working fluid, which can be used in the ORC system, depends on the temperature of the energy carrier (exhaust) and requires careful consideration, both due to the efficiency of the circulation work and the operational possibilities. The working fluid should be characterized by:

- appropriate thermal and physical properties,
- chemical stability at high temperatures,
- lack of harmful effects on the environment and elements of the installation.

In addition, in the case of ORC marine installations, the working fluid should meet the operational safety requirements, in particular its non-flammability. This condition quite significantly limits the number of substances that can be used in the marine ORC power plant.

Detailed recommendations for selection of working fluids and the requirements set for them are discussed in numerous works on the ORC power plants [5,6,7].

Taking the above into consideration, five working fluids from the group of dry substances were assumed for the analysis, namely: R245fa (1,1,1,3,3-pentafluoro-propane, $CF_3CH_2CHF_2$); perfluoropentane (dodeca-fluoro-pentane, C_5F_{12}); R123 (2,2-dichloro-1,1,1-trifluoro-ethane, $CHCl_2CF_3$) (Table 2).

Table 2. Selected organic working fluids and their fundamental parameters [8].

Working fluid	Molar mass	Range of applicability		Critical point	
	[kg/kmol]	T_{min} [°C]	T_{max} [°C]	T_{kr} [°C]	P_{kr} [MPa]
R245fa	134,05	-102,1	166,85	154,01	3,651
Perfluoro-pentane	288,03	-124,8	226,85	147,41	2,045
R123	152,93	-107,1	326,85	183,68	3,662

An important element of the analysis of the operation of ORC installation under consideration is the determination of evaporating temperature of the working fluid depending on the temperature of the exhaust gas supplied to the evaporator. The temperature of the exhaust depends on the main engine operating parameters, that is on its load and exhibits serious fluctuations within relatively large limits. In order to ensure stable operation of the ORC plant, it was assumed that the power plant is supplied with a constant temperature flue gases at the inlet to the evaporator, which was obtained by returning parts of the cooled exhaust gases leaving the exchanger and mixing them with fresh exhaust gases supplied from the engine.

Data describing mass flow rates and flue gas temperatures as a function of 6L70MC-C main engine load, determined on the basis of relations given in the manufacturer's catalog [3], are presented in the Table 3.

Table 3. Exhaust gas parameters assumed for calculations which result from the load of the main engine.

Engine load	Engine power	$\Delta m_{\%}$	ΔT_M	M_{sp}	T_{sp}
		[%]	°C	kg/s	°C
0,60	11772	5.11	-30.65	30.80	214.35
0,65	12753	4.31	-25.85	33.11	219.15
0,70	13734	3.57	-21.40	35.40	223.60
0,75	14715	2.88	-17.26	37.68	227.74

Table 3: continuation.

0,80	15696	2.23	-13.39	39.94	231.61
0,85	16677	1.63	-9.75	42.18	235.25
0,90	17658	1.05	-6.32	44.41	238.68
0,95	18639	0.51	-3.08	46.63	241.92
1,00	19620	0.00	0.00	48.83	245.00

5 Results of calculations

Calculations of the ORC power plant have been made for different values of mass flow rates and different exhaust temperatures, resulting from the main ship engine (its load) generated power. The extent of the mass flow rates and temperature were determined on the basis of the relations given in the manufacturer's catalog materials the MAN D&T [3].

Values of enthalpy of the working fluid, necessary for calculating the individual states of the ORC cycle, specified for the assumed temperature and evaporating pressure, were determined on the basis of the Refprop 9.1 database (Table 4) [8].

Table 4. Enthalpy values for selected working fluids in characteristic points of the ORC cycle (for $T_{par} = 140°C$).

Working fluid		R245fa	Perfluoro-pentane	R123
P_{par}	MPa	2,8287	1,7690	1,7563
h_1	kJ/kg	488,59	168,27	457,94
h_{2s}		439,41	141,99	410,70
h_3		239,10	0,27	230,26
h_{4s}		241,11	1,31	231,39
h_5		409,80	136,75	353,92

The sample results of the ORC plant calculations related to the nominal engine operation parameters (100% load) and selected working fluids are presented in Table 5. The calculations were made taking into account the nominal power N_{nom} = 1962 kW, rotational velocity n_{nom} = 108 rpm and mass flow rate flue gases m_{nom} = 48.83 kg/s.

The results of calculations made for the whole analyzed range of loads of the main engine and considered working fluids are presented graphically in the following figures. In the Fig. 9 presented is the attainable extent of power, whereas in Fig. 10 the values of thermodynamic efficiency of the Clausius-Rankine cycle. It stems from these graphs that for the assumptions set out in the work, in case when the engine was operated with the nominal load, the highest power was obtained when using R245fa as the working fluid, and the lowest one in case of considering the fluid named as perfluoropentane. At partial loads of the motor, the ORC power is correspondingly lower and shows an increasing trend as the load increases. The lowest values of thermodynamic efficiency of the Clausius-Rankine cycle were observed for perflupropentane, while the highest values for R123, where in the system under consideration for each of these working fluids, these values are constant, regardless of the engine load.

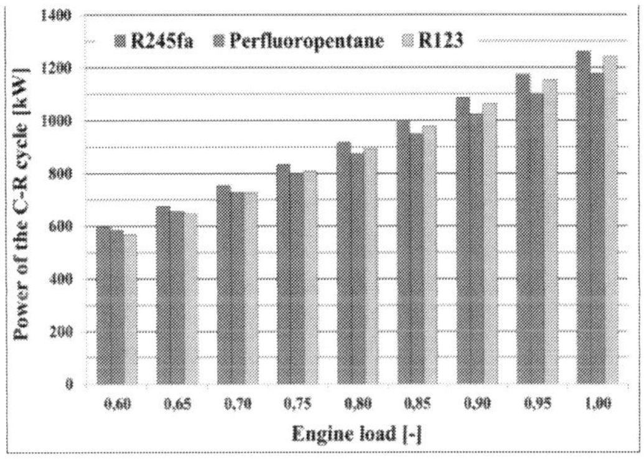

Fig. 9. The power of the C-R cycle in function of the engine load, for selected working fluid.

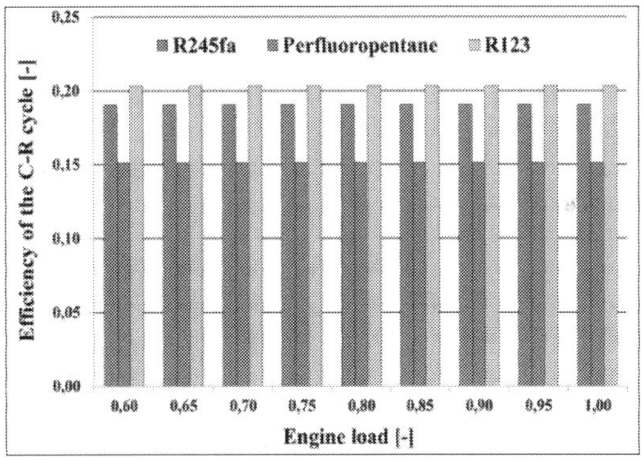

Fig. 10. Efficiency of the C-R cycle in function of main engine load, for selected working fluids.

Table 5. Results of calculation of the ORC system supplied with heat of flue gases from the 6L70MC-C engine (operation with nominal power).

Working fluid *)	t_{sp1}	t_{sp2}	t_{sp3}	t_m	m_{sp1}	m_{sp2}	m_{sp3}	m_{ob}	N_{OR}	η_{C-R}	N_{el}	η_{el}
	°C	°C	°C	°C	kg/s	kg/s	kg/s	kg/s	kW	%	kW	%
R245fa			123,59			188,76	139,92	26,74	1261,13	19,1	880,78	13,3
Perfluoropentane	245,00	155	102,03	145	48,83	131,81	82,97	46,67	1177,82	15,1	822,59	10,6
R123			133,22			250,63	201,80	26,89	1239,86	20,4	865,92	14,2

*) R245fa: 1,1,1,3,3-pentafluoropropane; Perfluoropentane: dodecafluoropentane; R123: 2,2-dichloro-1,1,1-trifluoroethane

Fig. 11 contain the results of calculations of the electric power that can be generated in the ORC plant are shown, for the case of various fluids considered in the analysis and for different loads of the main engine. In the calculations, the following efficiencies were assumed: turbine η_i = 0.75, mechanical η_m = 0.97 and electric generator η_g = 0.96

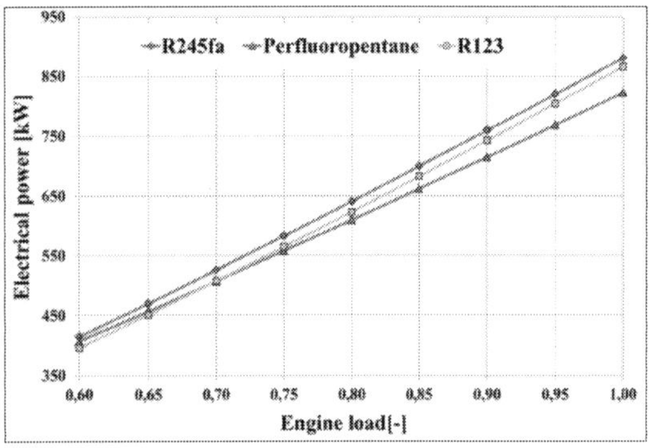

Fig. 11. Generated electric power in the ORC power plant powered by the exhaust gases from a marine engine (as a function of engine load, for selected working fluids).

6 Conclusions

The presented analysis is a part of the work carried out by the authors, regarding the use of the ORC power plant to generate electricity using low-temperature heat sources, including waste energy sources.

The obtained results of calculations show that there is a real possibility of using waste heat from the marine engine to power the low-temperature ORC power plant. They also show that the type of selected working fluid and the thermal parameters of the exhaust gases, resulting from the operation of the ship engine, have a significant impact on the energy effect of the plant operation.

The power generated by the ORC plant, defined as the power of the C-R cycle, at nominal load of the main engine, reaches the value of 1177.82 kW (for the working fluid perfluoropentane) to 1261,13 kW (for R245fa), which corresponds to 6,00 up to 6,43% of the nominal engine power. On the other hand, the plant efficiency defined as the efficiency of the C-R cycle varies from 15.1% (perfluoropentane) to 20.4% (R123).

The theoretical electric power of the ORC plant obtained in the calculations achieves, depending on the working fluid, the value from 822.59 kW to 880.78 kW, which constitutes from 4.19 to 4.49% of the nominal power of the main engine. The powers indicated above relate to the operation of the engine with nominal load. When operating with a partial load, the attainable powers are correspondingly smaller.

Summarizing, it can be concluded that the use of the ORC installation supplied with heat from the exhaust gases from the main marine engine allows for generation of additional electricity and thus improves the energy efficiency of the ship's engine.

References

1. A. Balcerski, *Ship power plants: thermodynamics bases, engines and main drives, auxiliary devices, installations* (Gdańsk University of Technology Publishing House, Gdańsk, 1986, in Polish)
2. R. Michalski, W. Zeńczak, S. Żmudzki, *Selected problems of energy conversion technology in marine energy systems* (Kaprint, Lublin, 2012, in Polish).
3. MAN B&W L70MC-C8-TII Project Guide Camshaft Controlled Two-Stroke Engines (MAN Diesel & Turbo, Copenhagen, Denmark, 2010).
4. C. Behrend, Problems of determining and using waste heat resources in marine energy systems. Scientific Journals of the Polish Naval Academy, 10 **(82)** pp. 31-40 (2006, in Polish).
5. P. Weisser, P. Skotnicki, *Perspectives of using ORC systems for low temperature waste energy.* Publisher: Instytut Śląski (2010)
6. R. DiPippo, *Geothermal power plants. Principles, applications, case studies and environmental impact* (Elsevier, Amsterdam, 2008)
7. W. Nowak, A. Borsukiewicz-Gozdur, P. Klonowicz, A. Stachel, P. Hanausek, W. Klonowicz, Preliminary results of testing the prototype system of mini-water plants with ORC fed with water at 100°C. Geological Rev. **58**, 7 (2010).
8. Refprop 9.1 Standard Reference Database 23, Version 9.1, Reference Fluid Thermodynamic and Transport Properties. National Institute of Standarts and Technology (Gaithersburg, USA, 2013).

Nomenclature

c_w	specific heat capacity	kJ/(kgK)
h	specific enthalpy	kJ/kg
$\dot{m}$	mass flow rate	kg/s
N	power	kW
$\dot{Q}$	heat flow	kW
t	temperature	°C
η	efficiency	%

Subscripts:

CR	refers to the Clausius-Rankine cycle
el	refers to the power plant
n	refers to the working medium
p	refers to the pump
s	refers to geothermal water (heating water)
sp	refers to the exhaust gases
t	refers to a turbine
1,2*s*, ..	refers to the characteristic points of the CR cycle

Preliminary assessment of the efficiency of ORC plant operation supplied with hot water from the HDR system

Radomir Kaczmarek[1], and *Aleksander* Stachel[1,*]

[1]West-Pomeranian University of Technology, Szczecin, Department of Heat Enginnering, al. Piastów 17, PL-70-310 Szczecin, Poland

Abstract. The paper presents the results of a theoretical analysis of the low-temperature ORC power plant operation fed with hot water obtained in a hot dry rock (HDR) extraction system. Issues related to the energy content of the Earth interior and its resources are discussed. The principles of acquiring heat in HDR installation are described and examples of installations commissioned in Europe as well as around the world are presented. The appropriate locations for building HDR installations in Poland were also characterized. Presented also is the concept of the ORC power plant supplied with water heated in the HDR rock exchanger. The analysis and evaluation of the ORC/HDR power plant operation supplied with water with parameters possible to obtain in selected locations throughout the country was accomplished. It has been shown that it is possible to build an ORC power plant powered from the HDR resources.

1 Introduction

Among the so-called renewable energy sources an important one is the geothermal energy defined as the heat resource contained in the interior of the Earth, while the operational significance has the heat accumulated in the rocks forming the Earth's crust. These rocks accumulate heat coming from the Earth's core, as well as the heat from radioactive elements decay processes. As a result, the Earth's crust is a rock structure with a significant energy potential, characterized by high temperature reaching several hundred degrees Celsius [1].

Due to the way in which this energy is accumulated, we divide the resources into:

- hydrothermal - where heat is accumulated in the unrestricted underground reservoir fluids,
- petrothermal - where heat is stored in the hot rocks (also salt dredges).

The heat resources contained in hot liquids (water, steam) can be used for heating purposes, and in the case of high thermal parameters - for generating electricity. Petrothermal resources are considered as prospective source of heat, and of significant importance are energy resources accumulated in the so-called hot dry rocks.

2 Technology of hot dry rocks

Hot dry rocks are defined as solid, not very strongly fractured rock structures devoid of reservoir properties and constituting a natural heat reservoir with a temperature above 150°C. In these layers there are no natural occurrence of water and / or steam resources. The energy potential accumulated in them is possible to be utilised using the appropriate technology, known as the hot dry rock technology, HDR.

In practice, this means acquiring (and using) the heat accumulated in solid crystalline rocks, mainly granite rocks, located at significant depths (up to 5 km and more), covered with insulating rock layers which poorly conduct the heat.

According to the definition of the Energy Research and Development Administration (US ERDA), HDR heat is defined as "accumulated in rocks at a depth of up to 10 km from the Earth's surface, from which it can be economically extracted using the natural hot water or steam" [2] .

Due to the efficiency of the process of obtaining heat, the areas useful for creating the HDR systems are:

- young volcanic areas;
- intense heat flow from the inside of the globe;
- significant concentration of radioactive elements constituting spontaneous radioactive sources of heat.

The concept of obtaining energy from hot rock structures was developed in the 1970s and uses natural or artificially created fracture systems developing in the rock mass in a kind of the "non-permeable" heat exchanger connected to the surface by access channels (injection and production wells) [1,3].

Acquisition of heat is attained by injecting a high-pressure water into rock structures, which, flowing through an extensive network of cracks picks up heat from the hot rocks. Hot water flows out of the extraction well to the surface, where it transmits the heat it contains to the recipients (Fig. 1). This heat can be used for heating or industrial purposes (e.g. electricity generation). After the heat is removed, the water is again injected back into the bed.

* Corresponding author: aleksander.stachel@zut.edu.pl

The HDR extraction system encompasses a bed with a sufficiently high porosity, with an extended heat transfer surface, ensuring good permeability of the injected liquid and its intensive heating. In the case where the rock structure is solid, or if the natural cracks and pores occurring in it do not ensure appropriate flow of the injected water, the permeability of the bed can be improved in the process of so-called hydraulic fracturing.

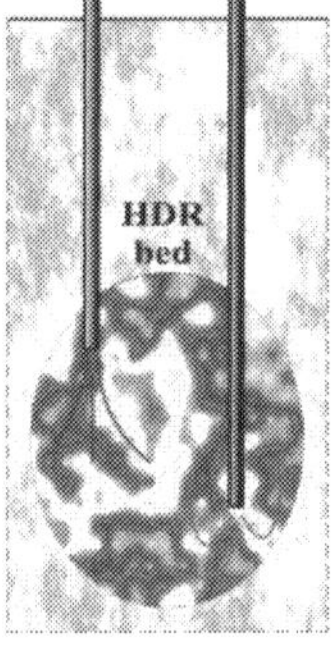

Fig. 1. A concept of the heat exchanger in HDR technology [1].

It should be remembered that the rocks forming the "non-permeable" heat exchanger transmit energy in the transient heat transfer process, while the rate of heat decreases over time as the temperature of the rocks decreases. An important problem is also the mineralization of the outgoing hot water, caused by the dissolution of chemical compounds (salts) contained in the rock massif.

3 HDR installations worldwide

The first installation acquiring the heat from dry hot rocks was commissioned in 1973 in Fenton Hill, New Mexico (USA) and was an attempt to develop a full-scale HDR heat exchanger. The installation was based on two boreholes, injecting and extracting ones, reaching a hot, sealed, crystalline granite formation featuring 185°C, located at a depth of ~2600 m. The rock heat exchanger made it possible to heat the stream of injected water (23-57 m^3/h) to a temperature of 135-140°C (Fig. 2) [4].

In the following years, a number of other HDR projects were implemented, including the Hijiori project in Japan [4]. The system was based on two horizontally fractured rock deposits at two different depths, i.e. 1800 m and 2200 m respectively, separated by a layer of solid impermeable rock with a thickness of 400 m. Heat reservoirs with a temperature of 250°C (upper) and 270°C (lower), have been connected with the surface of the Earth by means of injection and extraction ducts. The conducted tests, apart from the hydraulic testing of the HDR system, also included the assessment of the possibility of converting the energy of obtained hot water into electric energy with the aid of ORC installation.

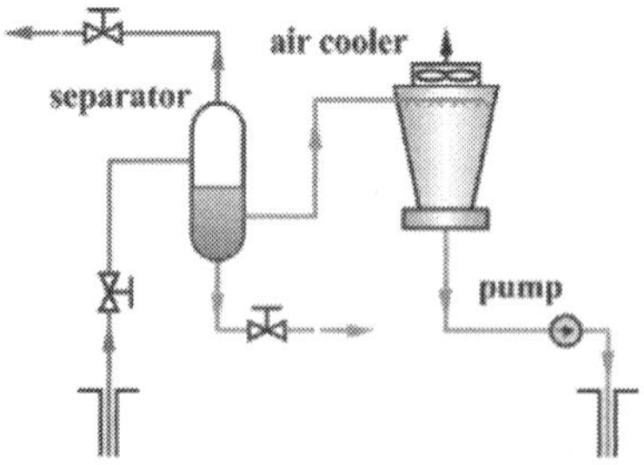

Fig. 2. Simplified schematic of the Fenton Hill HDR testing installation [4,5].

In Europe, the largest research project in the area of utilisation of hot dry rocks is the installation in Soultz-sous-Fortes (Alsace, France), the construction of which was initiated in 1987. Its location resulted from the favorable geothermal conditions of the geological and tectonic formation known as the Burgundy Trough, including the forecasted high temperature of rocks (~200 °C at a depth of ~5 km). The source of heat are the deposits of crystalline rocks covered with a sedimentary rock layer up to 1000 m thick. The average geothermal gradient of sedimentary rocks is 0.08-0.1 K/m and decreases to 0.028-0.05 K/m below the granite rocks [4].

Several years of work and research led to the development of the installation whose extraction part consists of one injection well (GPK3) and two extraction wells (GPK2, GPK4), reaching a depth of ~5000 m, where the rock temperature was found to be ~200°C. The heads of the extraction holes are located near to the head of the injection well, while their lower ends are deviated in opposite directions from it by about 600 m. The heat transfer area of the HDR exchanger created by hydraulic fracturing is estimated at about 3 km^2, while its volume - at 2-3 km^3.

The installation in Soultz was launched in 2006. The water introduced with the GPK3 injection well is heated from the hot rocks, flowing through the system of cracks and gaps, and then flows to the surface through the GPK2 and GPK4 wells. The heat obtained from the rocks supplies the power plant operating as the organic Rankine cycle, in which isobutane is used as a working fluid (Fig. 3). The ORC power plant was modernized in 2015 and produces the electric power in the amount of 1.7 MW at the efficiency of 11.4%. It is supplied with the flow rate of water (33.3 dm^3/s) at a temperature of 160°C. Water is introduced back to the bed at a temperature of 70°C.

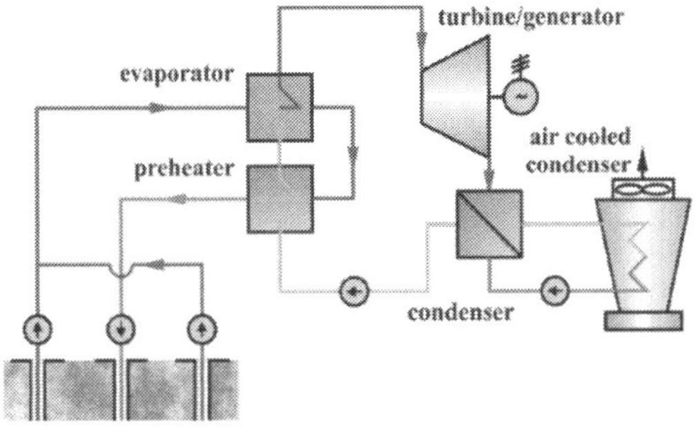

Fig. 3. Simplified schematic of the Soultz-sous-Forets power plant [4,5]: GPK3 – injection well, GPK2, GPK4 – production wells.

4 Geological conditions of HDR installations in Poland

Also in Poland, activities are underway to search for the possibility of building an installation for acquiring the heat from hot dry rocks [6,7]. In recent years accomplished has been amongst the others the research grant with the purpose to indicate regions with favorable conditions for the construction of a geothermal installation based on the HDR technology [8]. The project indicates the useful for this purpose locations of hot dry rocks, which should be characterized by a maximum temperature and be placed at the smallest possible depth. In addition, they should feature adequate porosity and permeability, high specific heat and high thermal conductivity, and be susceptible to fracturing processes. These properties have a significant impact on the installation location, as well as its subsequent operation.

Rock structures that are useful for this purpose are related to the occurrence of, among others igneous rocks. Rocks of this type with relatively large heat transfer surfaces and volumetric capacities exist mainly in the Sudety and the Tatra Mountains. Due to environmental reasons, their use is more realistic within the Sudety area, and therefore this area is perceived as the perspective region for the HDR technology. This applies in particular to the Karkonosze granite massifs in which significant heat availability from the inside of the Earth (Cieplice) was determined to be significant, as well as the locally increased emission of radiogenic heat. This applies in particular to the area of Szklarska Poręba, where the heat flux released in the process of radioactive decay (the presence of uranium and thorium isotopes) is extremely high and reaches 4.5 - 5.0 $\mu W/m^3$, and according to some sources even 7.0 $\mu W/m^3$. The temperature distribution model developed for this area, taking into account relatively high values of the geothermal gradient, estimates the rock temperatures from 145°C (at a depth of 3.5 km), to 225°C in the north-east zone (at a depth of 5.5 km). This renders that the location of HDR installations within the Karkonosze area seem to be exceptionally promising [7,8,9].

It is believed that the second constructive area for obtaining heat from hot dry rocks may be the area to the west of Gorzów Wielkopolski, in particular the vicinity of Lubuskie Dębno, characterized by a heat flux of geothermal heat equal to 100 mW/m^2 and a high value of geothermal gradient (about 40 K/km). This allows to conclude that at the depth of 3-5 km there can be expected a relatively high temperature.

However, it should be borne in mind that the indicated rock structures may be subject to a significant error, both due to geological conditions and predicted temperatures, associated with the variable value of geothermal gradient and measurement uncertainties. Therefore, final location decisions should be supported by in situ research, enabling the verification of geological, thermal, physical and chemical conditions.

5 A concept of ORC / HDR power plant for the Polish conditions

Apart from geological analyzes, important due to the specific requirements facing the rock structures and useful for the construction of HDR systems, works are also underway to assess the feasibility and rationale of construction of the ORC power plant powered by the hot dry rock energy.

Due to the estimated temperatures of the water supplied from the HDR system, it is assumed that the hypothetical power plant should operate according to the Clausius-Rankine (C-R) thermodynamic cycle with the low boiling point organic fluid. The choice of technology is dictated by the relatively low temperature of the feed water. The conversion of heat obtained in a traditional power plant, in which the water / steam is the working fluid, would be ineffective from the energy conversion point of view.

Schematic diagram of the ORC power plant is shown in Fig.4. The power plant consists of two heat exchangers, namely the preheater (PH) and evaporator (EV), a turbogenerator (TG), a condenser (CO) and a circulating pump (P). Hot water from the HDR system is fed to the evaporator (EV), where the heat contained in it is transferred to the working fluid of the cycle, causing its evaporation. Part of the water flow rate (m_{s1}) leaving the evaporator (EV) is directed to the preheater (PH). Both streams (m_{s1} + m_{s2}), after recombination, are injected into the HDR structure.

The working fluid, heated in the preheater and vaporised in the evaporator, is directed to the turbine, where it expands and generates work. After expansion and condensation in the condenser, the fluid is again pumped into the heater - evaporator system.

In ORC power plants, various organic substances with a low boiling point and appropriate thermo-physical, operational and environmental properties are used as working fluids. The recommendations for their

 https://doi.org/10.1051/e3sconf/20187001007

selection are given in numerous literature on the subject, for example in [4,5,10].

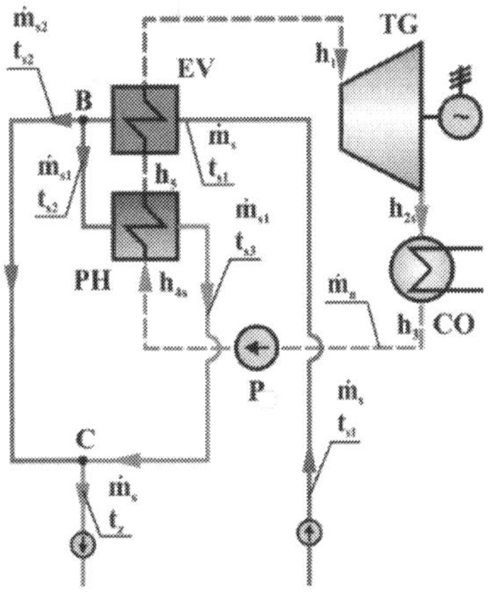

Fig. 4. A schematic diagram of the ORC power plant supplied with heat from the HDR deposits.

6 Analysis of ORC power plant

Initial evaluation of the operation of a hypothetical ORC power plant fed with hot water from the HDR deposit, was referred to the installation shown schematically in Fig. 5 A number of assumptions have been made in the calculations, among others that:

- power plant works according to the subcritical cycle;
- temperature of water supplied to the evaporator is t_{s1} = 130 - 180°C, which results from the predicted thermal parameters of HDR systems, possible to achieve in Poland;
- condensation of the working fluid vapor in the condenser proceeds at a pressure corresponding to the condensing temperature, t_{h3} = 30°C;
- temperature differences between heating and working fluids in the heater and evaporator are defined in Fig. 5;
- calculations of C-R cycle power were related to the unit water flow rate supplied from the HDR system, $\dot{m}_s = 1$ kg/s;
- geothermal water mineralization, S = 120 g/kg;
- five working fluids are included in the calculations.

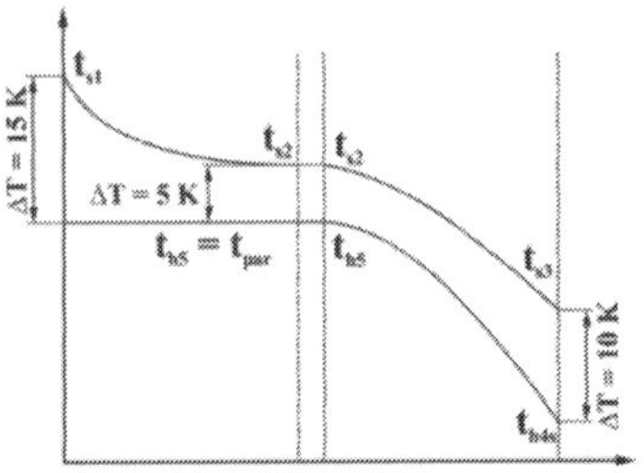

Fig. 5. Temperature distribution in evaporator and preheater.

On the basis of the assumptions made, the assessment of the power plant operation was carried out, specifying, among others, the power and efficiency of the C-R cycle, including:

- type of working fluid,
- temperature changes of geothermal water,
- extent of the flow rate and the degree of mineralization of water supplied to the system $\dot{m}_s$.

Thermodynamic parameters of the considered working fluids at individual nodes of the C-R cycle were determined using the Refprop software [11].

7 Methodology of calculations

The purpose of calculations of the hypothetical ORC power plant was to determine the achievable power and efficiency of the Clausius-Rankine cycle, at the assumed feeding and operation parameters. The calculations were based on the following relations.

The mass flow rate of the low-boiling point fluid in the boiler leg was determined from the evaporator energy balance equation:

$$\dot{m}_n = \frac{\dot{m}_s c_w (t_{s1} - t_{s2})}{h_1 - h_5} \tag{1}$$

The mass flow rate of water $\dot{m}_{s1}$ from the evaporator energy balance for the water leg yields:

$$\dot{m}_{s1} = \frac{\dot{m}_n (h_5 - h_{4s})}{c_w (t_{s2} - t_{s3})} \tag{2}$$

whereas the mass flow rate of unused water directed to the HDR system was determined from the mass balance for the node „B":

$$\dot{m}_{s2} = \dot{m}_s - \dot{m}_{s1} \tag{3}$$

Temperature of injected water was determined on the basis of the energy balance equation for the node „C":

$$t_z = \frac{\dot{m}_{s1} t_{s3} + \dot{m}_{s2} t_{s2}}{\dot{m}_s} \tag{4}$$

Subsequent quantities characterizing the ORC power plant have been referenced to the unit water flow rate acquired from the HDR system (1 kg/s), by means of:

- heat rate supplied to the ORC system in evaporator and heater:

$$\dot{Q}_d = \dot{m}_n (h_1 - h_{4s}) \tag{5}$$

- heat rate removed from the system in the condenser:

$$\dot{Q}_w = \dot{m}_n (h_{2s} - h_3) \tag{6}$$

- theoretical power of turbine:

$$N_t = \dot{m}_n (h_1 - h_{2s}) \tag{7}$$

- theoretical power of the circulation pump:

$$N_p = \dot{m}_n (h_{4s} - h_3) \tag{8}$$

On that basis calculated was:

- unit power of ORC cycle referred to the assumed flow rate of water acquired in the HDR system HDR (1 kg/s):

$$N_{CR} = N_t - N_p \tag{9}$$

- theoretical efficiency of C-R cycle:

$$\eta_{CR} = \frac{N_{CR}}{\dot{Q}_d} 100\% \tag{10}$$

In addition, in the calculations considered was the influence of specific heat of water supplied from the HDR system on the resulting power of the C-R cycle. Specific heat depending on the degree of its mineralization, that is, the amount of dissolved compounds (salts).

8 Results of calculations

Based on the assumed output data, the power and efficiency calculations of a hypothetical ORC power plant cooperating with the HDR system were made. The calculations were made for five selected working fluids to be used in the power cycle (isohexane, isopentane, pentane, R245ca, R365mfc), as well as for different values of the supply water temperature delivered from the HDR exchanger. The aim was to demonstrate the influence of various working fluids as well as various power parameters on the efficiency of installation operation. The calculations were carried out with the assumption of a unit flow rate of water feeding the power plant cycle.

Sample results of calculations related to one of the considered working fluids in the ORC (iso-hexane) power plant are presented in Table 1.

Table 1. Sample results of calculations of ORC power plant (isohexane, T_{skr} = 30°C, T_{kr} = 224,55°C).

t_{ev}	c_w	h_1	h_{2s}	h_3	h_{4s}	h_5
°C	kJ/ (kg·K)	kJ/kg	kJ/kg	kJ/kg	kJ/kg	kJ/kg
115	3,663	417,85	334,04	-70,58	-69,95	140,6
125	3,665	435,41	343,04	-70,58	-69,79	168,21
135	3,668	452,90	352,24	-70,58	-69,59	196,50
145	3,671	470,23	361,53	-70,58	-69,36	225,51
155	3,675	487,28	370,81	-70,58	-69,10	255,31
165	3,679	503,90	379,93	-70,58	-68,79	285,99

Table 1: continuation.

t_{ev}	t_{s1}	t_{s2}	t_{s3}	t_z	$\dot{m}_{s1}$	$\dot{m}_{s2}$	$\dot{m}_n$
°C	°C	°C	°C	°C	kg_s/s	kg_s/s	kg_n/kg_s
115	130	120	40	112,41	0,09	0,91	0,13
125	140	130	40	121,09	0,10	0,90	0,14
135	150	140	40	129,62	0,10	0.90	0,14
145	160	150	40	137,95	0,11	0,89	0,15
155	170	160	40	146,01	0,12	0,88	0.16
165	180	170	40	153,72	0,13	0,87	0.17

Table 1: continuation.

t_{ev}	$\dot{Q}_d$	$\dot{Q}_w$	N_t	N_p	N_{CR}	η_{CR}
°C	kW/ (kg_s/s)	kW/ (kg_s/s)	kW/ (kg_s/s)	kW/ (kg_s/s)	kW/ (kg_s/s)	%
115	64,44	53,45	11,07	0,08	10,99	17,05
125	69,30	56,74	12,67	0,11	12,56	18,13
135	74,74	60,49	14,40	0,14	14,26	19,08
145	80,94	64,82	16,31	0,18	16,12	19,92
155	88,14	69,92	18,45	0,24	18,21	20,67
165	96,68	76,05	20,93	0,30	20,63	21,33

The influence of the type of used working fluid and the feed water temperature on the extent of C-R cycle power output is shown in Fig. 6. For all considered feedwater temperatures, the highest cycle power was obtained for R245ca whereas the smallest one for isohexane.

The C-R circuit power increases with the increase of the feeding temperature (increase of the evaporating temperature of the fluid). The highest values of power were obtained at 180°C (evaporating temperature 155°C), whereas the lowest ones at 130°C (evaporating temperature 115°C). It should be noted that in the case of the working fluid R245ca, as the temperature rises, a much larger increase in power is observed compared to other working fluids. It can be therefore concluded that both the parameters of water supplied from the HDR system (temperature) and the type of working fluid used have a significant impact on the operation of the power plant and the power obtained in it.

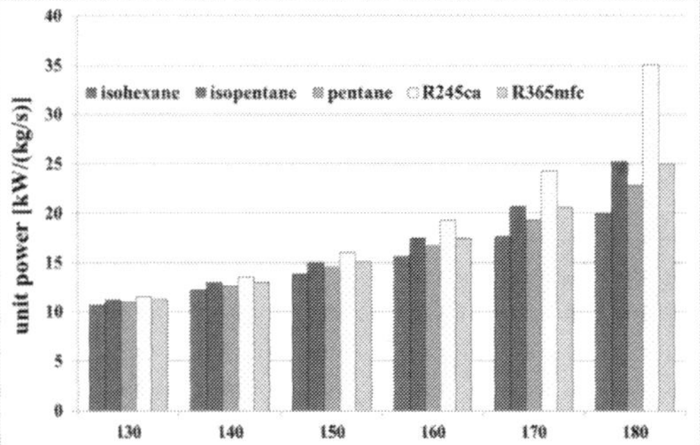

Fig. 6. Attained ORC power in function of the type of working fluid and temperature of water supplied from the HDR system (with reference to 1 kg/s of water).

In Fig. 7 presented is the distribution of ORC's efficiency on the type of working fluid used and the temperature of the feed water.

The highest theoretical efficiency of the C-R cycle was obtained for pentane, while the lowest for the R365mfc working fluid. In turn, the fluid R245ca, which achieves the highest power in the entire range of considered temperatures, shows at the same time the lowest efficiencies. It should be also noted that with the same conditions of supply (temperature), the efficiencies of the power plant operating with various fluids are close to each other. At the same time, they increase with temperature of the supplied water, and thus the evaporation temperature.

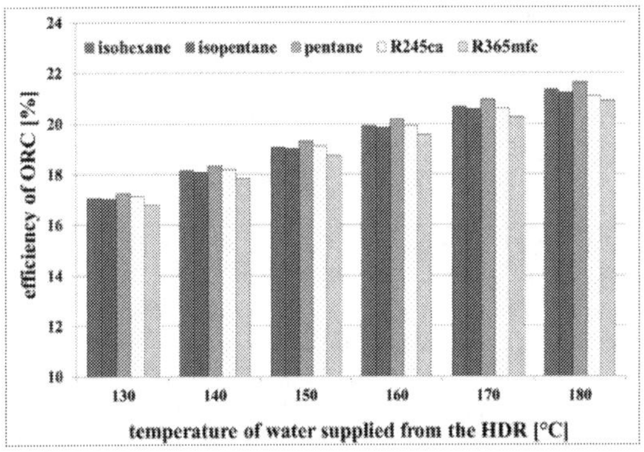

Fig. 7. Efficiency of ORC in function of working fluid and temperature of supply water from the HDR system (with reference to 1 kg/s of water).

The next graph (Fig. 8) shows the dependence of the C-R cycle power on the degree of mineralization of water supplied to the power plant from the HDR exchanger. The specific heat of water depends on mineralization (salinity), the lower the salinity, the higher its value. Mineralization of water results from the process of dissolving chemical compounds contained in rocks in water injected into the HDR bed, during its flow through the cracks and slots of the rock structure. Increasing mineralization causes a decrease in the flow of the working fluid circulating in the turbine cycle, which in turn results in the reduction in the power of the cycle.

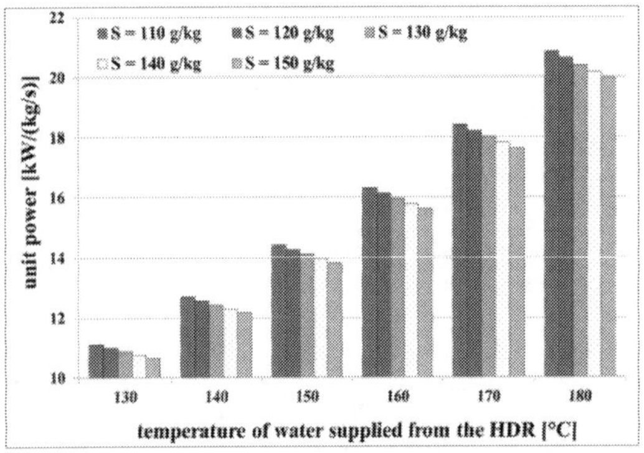

Fig. 8. Influence of mineralization of the supply water from the HDR exchanger on the power of ORC (for isohexane).

An important element of the ORC power plant operation is also the condensation temperature of the vapour leaving the turbine. Heat removal in the condenser is most often carried out with the aid of cooling water supplied from the natural watercourses (river, lake). In the absence of water, so-called cooling towers are used. Cooling with water allows for lower condensing temperatures than in the case of cooling towers. Lower condensing temperature allows to attain higher power installation as well as higher cycle efficiency (Fig. 9).

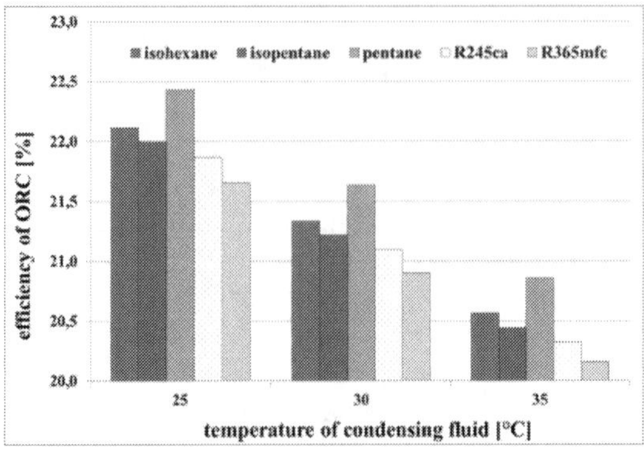

Fig. 9. Influence of condensing temperature on the efficiency of C-R cycle in relation to the type of working fluid (evaporation temperature of the working medium: 165 °C).

The last important element of the ORC power plant operation evaluation is the influence of the extent of the water flow rate supplied from the HDR deposit. Increasing the water flow rate allows to get more power, and therefore more electricity. The decisive factor here are here the geological and hydraulic properties of the HDR rock exchanger.

9 Summary and final conclusions

The work discusses the principles of obtaining heat in HDR systems. Examples of such installations have been characterized. The required geological conditions were discussed and potential locations for construction of such types of installations in Poland were presented. On the basis of the assumed water supply parameters, the power and efficiency calculations of a hypothetical ORC power plant supplied from the HDR system were made, taking into account various working fluids as well as different temperatures of water supplied from the geothermal exchanger.

The conducted analyzes show that in Poland there exist favorable conditions for acquisition of heat from the so-called hot dry rocks [11]. In turn, the calculations and the results obtained prove that it is possible to build a geothermal power plant based on the low-temperature ORC cycle, supplied with heat from the HDR system. The theoretical power of such a power plant, supplied with water with the flow rate of 50 kg/s and temperature 180° C, is estimated at N_{CR} = 1500 kW.

However, it should be taken into account that the construction of the ORC power plant cooperating with the HDR system is a complex process, especially in the part concerning the construction of the HDR exchanger. Therefore, the investment decision should be preceded by thorough in-situ testing, enabling the verification of geological, thermal, physical and chemical conditions, as well as detailed analyzes of the fluids affecting the effectiveness and profitability of the planned project.

https://doi.org/10.1051/e3sconf/20187001007

References

1. W. Nowak, R. Sobański, M. Kabat, T. Kujawa, *Systems for acquiring and utilizing geothermal energy* (Publishing House of Szczecin University of Technology, 2000, in Polish)
2. H. Gupta, S. Roy, *Geothermal energy: an alternative resource for the 21st century* (Elsevier, Amsterdam, 2007)
3. D.W. Brown, D.V. Duchane, G. Heiken, V. Hriscu, *Mining the earth's heat: hot dry rock geothermal energy* (Springer Verlag, Berlin, 2012)
4. R. DiPippo, *Geothermal power plants. Principles, applications, case studies and environmental impact* (Elsevier, Amsterdam, 2008)
5. A. A. Stachel, *The use of the energy of the interior of the Earth* (Publishing House of West Pomeranian Univ. of Technology, Szczecin, 2013, in Polish)
6. M. Plewa, S. Plewa, Analysis of the possibilities of using the geothermal energy of dry hot rocks in Poland, [in:] *Possibilities of using the geothermal water in Poland, with particular emphasis on the Mogilno-Łódź synklinorium.* (AGH, *Ślesin-Kra*ków, 1990, pp. 171–181, in Polish)
7. R. Skrzypczak: Proposals for research of location for the needs of geothermal hot dry rock technology in the Sudety region. Technika PGGZR, **50**, 1–2, pp. 93–108 (2011, in Polish)
8. *Assessment of potential, thermal balance and prospective geological structures for the needs of closed geothermal systems (Hot Dry Rocks) in Poland. Final report.* The work commissioned by the Ministry of Environment of the Republic of Poland (Warszawa / Kraków, 2013 (in Polish)
9. J. Szewczyk, D. Gientka, Terrestrial heat flow density in Poland - a new approach. Geological Quarterly, **53**, 1 (2010).
10. W. Nowak, A. Stachel, A. Borsukiewicz-Gozdur: *Applications of renewable energy sources* (Publishing House of Szczecin Univ. of Technology, 2008, in Polish).
11. Refprop 9.1, Standard Reference Database 23, v. 9.1, Reference Fluid Thermodynamic and Transport Properties. National Institute of Standards and Technology, Gaithersburg, MD, USA 2013.

Nomenclature

c_w	specific heat capacity	kJ/(kgK)
h	specific enthalpy	kJ/kg
$\dot{m}$	mass flow	kg/s
N	power	kW
$\dot{Q}$	heat flow	kW
S	mineralization	g/kg
T, t	temperature	K, °C
η	efficiency	%

Subscripts:

CR	refers to the Clausius-Rankine cycle
d	refers to the heat supplied
ev	refers to the evaporation process
n	refers to the working medium
p	refers to the pump
s	refers to geothermal water (heating water)
t	refers to a turbine
w	refers to the heat removed
1,2*s*, ..	refers to the characteristic points of the CR cycle

The use of the T-history method to estimate thermal capacity and latent heat for RT15 and RT18 materials

Paulina Kapica[1,*], *Jarosław* Karwacki[1], and *Sebastian* Bykuć[1]

[1]The Szewalski Institute Of Fluid Machinery, Polish Academy of Sciences, Fiszera 14, 80-231 Gdańsk, Poland

Abstract. The paper presents theoretical assumptions and implementation of the T-history method, which was used to estimate the thermal capacity and latent heat for phase change materials (PCMs). The research equipment, measurement system and obtained results were discussed. The T-history method was used to investigate the heating and cooling process of two PCMs, sold under the trade names of RT15 and RT18. Obtained results were compared with the manufacturer's data. The analysis of the research results has led to the conclusion that the presented measurement method made it possible to perform a preliminary verification of the enthalpy–temperature dependency of the PCMs. The method can be also used for studying different commercial materials and estimating this dependency for materials of yet unknown properties.

1 Introduction

The knowledge of thermophysical properties of heat storage materials is essential when designing heat and cold storage systems. In the case of solids in which only sensible heat is used, it is sufficient to know their specific heat and thermal conductivity. When liquids are used for heat storage purposes, the process of convective heat transfer should also be taken into account in the majority of design solutions. A much more difficult task is to design a heat store in which a PCM (phase change material) is employed. In such a case, it is necessary to know the specific heat for both the liquid and solid states ($c_{p,s}$ and $c_{p,l}$ respectively) as well as the specific heat of fusion, which is related to the solid-liquid phase change. In most cases, these parameters can be determined using the well-known differential scanning calorimetry (DSC) technique. However, this method requires expensive laboratory equipment. Another (much simpler) solution is to use the T-history method, which was presented for the first time in publication [1]. This method has been developed and improved over many years [2]. Due to the possibility of using samples of PCMs with masses greater than 50 mg in the tests, the obtained results may be closer to reality.

In this paper, the T-history method was used to investigate specific heat and specific heat of fusion of two PCMs, sold under the trade names of RT15 and RT18. Fig. 1 and Fig. 2 presents the partial enthalpy distribution for those materials, that is the amount of heat /cold from unit of mass corresponding to a given temperature.The presented research had the following objectives:

- developing the test rig and measurement procedures for the implementation of the T-history method,
- validating the measurement method based on data given for materials sold on the market,
- determining further actions that can be taken in order to improve the reliability of the obtained results.

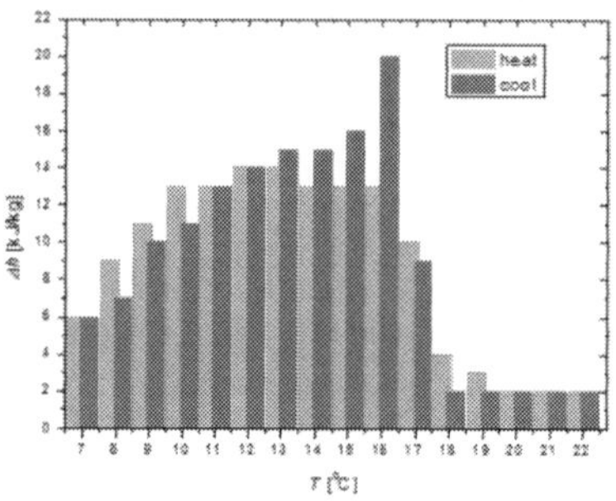

Fig. 1. Partial enthalpy distribution of phase change material - RT15.

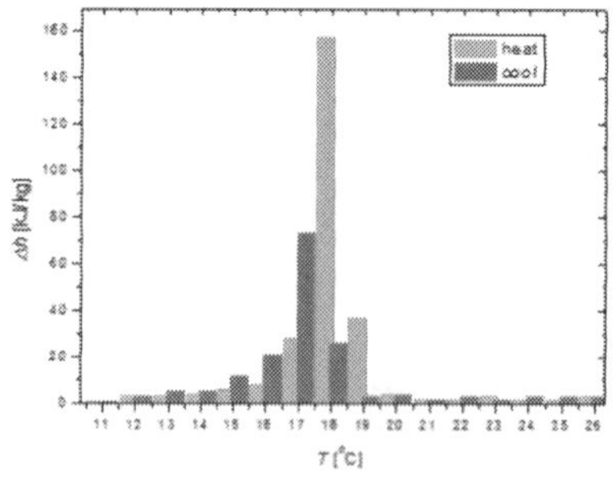

Fig. 2. Partial enthalpy distribution of phase change material - RT18.

* Corresponding author: paulina.kapica@imp.gda.pl

2 Calculation method

The T-history method is a thermal analysis method that is based on registering temperatures of the investigated materials within a certain time period. Material samples of the proper diameter and length should be used. It is necessary that at least one of the materials to be tested has well-known thermal properties. Such a material is called "reference material". In the temperature range of 0 °C – 100 °C, water is usually used as a reference material. A comparison with the reference material allows to determine the phase transition temperature and then the thermal capacity as well as the latent heat of PCMs.

When using the above-mentioned method, the test tube with a material to be tested should be immersed in a liquid whose temperature significantly differs from the phase transition temperature. After the thermal equilibrium is reached, the tested materials are usually immediately exposed to ambient temperature. Temperature is measured using sensors inserted directly into material samples. In the T-history method, material samples are heated after their placement in the so-called "water bath", as shown in Fig. 3.

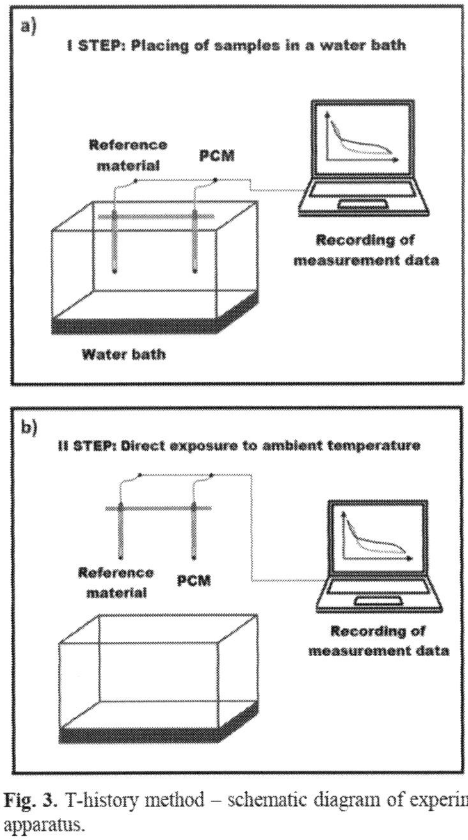

Fig. 3. T-history method – schematic diagram of experimental apparatus.

A prerequisite for the use of the T-history method is to make ensure that the material to be tested will have even temperature distribution along its entire cross-section. This prerequisite is met if the Biot number is lesser than 0.1. This is done by choosing appropriate dimensions of test tubes and paying a close attention to the adequate conduct of the experiment.

The first step of the T-history method is to place test tubes inside the hot water tank (Fig. 3a). Afterwards, they are heated until the temperature reaches T_0, namely the temperature that is greater than the melting point of the PCM (T_1). The test tubes are removed from the tank as soon as a steady state is reached. After the removal, they are kept at the ambient temperature (T_a), which usually is the temperature of the ambient air. In the configuration described herein, the tested materials were taken out from the water bath at the appropriate moment (Fig. 3b). It is important that the ambient temperature (T_a) should be significantly lower than the freezing point of the PCM (T_s). As soon as the heated test tubes (with the PCM and the reference material) are exposed to ambient temperature the cooling process commences. The temperature of both materials was measured, which made it possible to plot the graphs presenting their temperature as functions of time. The measurement started at a temperature of T_0 and ended after a temperature of T_e was reached. The temperature denoted by T_e has to be lower than the temperature denoted by T_s (solidification point of the PCM). As a result, temperature profiles of the PCM and the reference material (T_m and T_r respectively) are obtained.

On the graph that shows the cooling process curves one can differentiate the three following temperature ranges:

1 – from T_0 (start of the measurement) to T_1 that is the melting point of the PCM (in this temperature range the PCM is in the liquid state),

2 – from T_1 to T_s that is the solidification point of the PCM (the range in which the phase change takes place),

3 – from Ts to Te which is the temperature registered at the end of the experiment (in this temperature range the PCM is in the solid state).

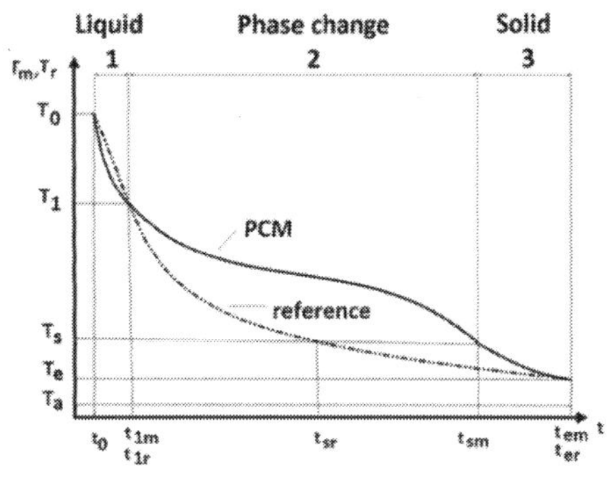

Fig.4. Cooling process curves for PCM and reference material with division into PCM phase zones.

In the first zone marked in Fig. 4, the heat transfer in both test tubes is defined by the following equations [3]:

$$(m_{tm}c_{ptm} + m_m c_{pml})(T_0 - T_1) = h_1 S_{tm} A_{1m} \quad (1)$$

$$(m_{tr}c_{ptr} + m_r c_{pr})(T_0 - T_1) = h_1 S_{tr} A_{1r} \quad (2)$$

where:
m_m, m_r – mass of the PCM / reference material
m_{tm}, m_{tr} – mass of the test tube in which the PCM / reference material was inserted
c_{pml}, c_{pr} - specific heat of the PCM in the liquid state / reference material
c_{pt} – specific heat of the test tube
h –convective heat-transfer coefficient
S_t – convective heat transfer area

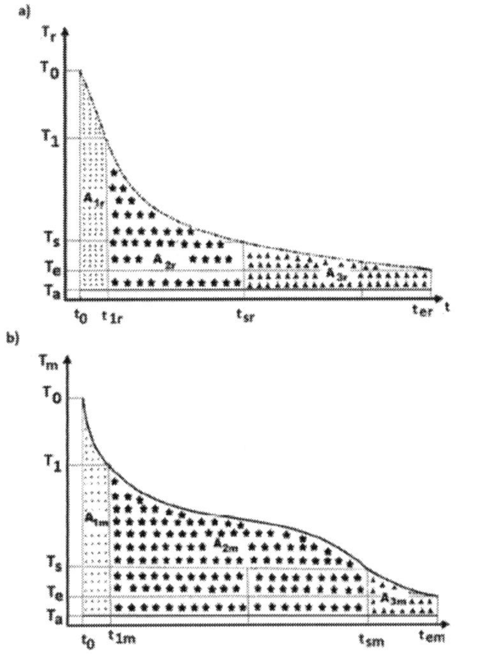

Fig.5. Thermal capacity of the reference material (a) and PCM (b) at temperature intervals.

The next zone is the PCM phase change zone and the heat transfer can be expressed by equations (3) and (4).

$$m_{tm}c_{ptm}(T_1 - T_s) + m_m H_m = h_2 S_{tm} A_{2m} \quad (3)$$

$$(m_{tr}c_{ptr} + m_r c_{pr})(T_1 - T_s) = h_2 S_{tr} A_{2r} \quad (4)$$

The last zone corresponds to the temperature range in which the PCM occurs in the solid state. The heat transfer balance between both materials (PCM and reference material) and their surroundings is given by equations (5) and (6).

$$(m_{tm}c_{ptm} + m_m c_{pms})(T_s - T_e) = h_3 S_{tm} A_{3m} \quad (5)$$

$$(m_{tr}c_{ptr} + m_r c_{pr})(T_s - T_e) = h_3 S_{tr} A_{3r} \quad (6)$$

$$A_{1m} = \int_{t_0}^{t_{1m}} (T_m(t) - T_a)dt \quad (7)$$

For the purposes of the above equations, the magnitude of the areas shown in Fig. 5 can be calculated using the following integrals:

$$A_{1r} = \int_{t_0}^{t_{1r}} (T_r(t) - T_a)dt \quad (8)$$

$$A_{2m} = \int_{t_{1m}}^{t_{2m}} (T_m(t) - T_a)dt \quad (9)$$

$$A_{2r} = \int_{t_0}^{t_{sr}} (T_r(t) - T_a)dt \quad (10)$$

$$A_{3m} = \int_{t_{sm}}^{t_{em}} (T_m(t) - T_a)dt \quad (11)$$

$$A_{3r} = \int_{t_{sr}}^{t_{er}} (T_r(t) - T_a)dt \quad (12)$$

Phase transition temperature values of a tested PCM are determined on the basis of the derivative of the cooling process curve. On the graph presenting this derivative, one should be able to notice two temperature peaks that correspond to the start and end moments of the phase transition. The first peak (T_1), which occurs in the temperature range from t_0 to t_{1m}, corresponds to the beginning of phase change. The second peak occurs in the range from t_{1m} to t_{em} and denotes the temperature (T_s) at which the phase change is already over. The temperature denoted by T_e is the temperature at which the experiment ended. At this temperature, the value which was close to the ambient temperature, the thermal equilibrium of the tested materials was reached.

Taking into account the calculated magnitude of the areas (formulas 7-12), dimensions of the test tubes, particular masses as well as the specific heat of the reference material and of the material from which test tubes are made, one can calculate the values of the following parameters:

- specific heat of the PCM in the liquid state:

$$c_{p,ml} = M\frac{A_{1m}}{A_{1r}} - N \quad (13)$$

- specific heat of the PCM in the solid state:

$$c_{p,ms} = M\frac{A_{3m}}{A_{3r}} - N \quad (14)$$

- latent heat:

$$H_m = \left(\frac{MA_{2m}}{MA_{2r}} - N\right)(T_1 - T_s) \quad (15)$$

where:

$$M = (m_{tr}c_{p,tr} + m_r c_{p,r})\frac{S_{tm}}{m_m S_{tr}} \quad (16)$$

$$N = m_{tm}\frac{c_{p,tm}}{m_m} \quad (17)$$

The above-described procedure concerns the classical implementation of the T-history method that was proposed by Z. Yinping in 1999 [1]. As far as the research described in the following part of this article is concerned, material samples were not preheated to a suitably high temperature. On the contrary, they were cooled in the container with ice water so that the PCMs were completely solidified. After removing the samples from the water, their temperatures were registered until the materials reached the ambient temperature. This procedure resulted from the fact that the temperature at which the phase transition was complete is lower than the room temperature. It should be noted that formulas (14) to (17) remain valid also in such a case.

3 Description of the experimental set-up

The test rig, on which the RT15 and RT18 materials were tested, consists of the following elements:

- six test tubes made of glass
- thermally insulated container with ice water
- five thermocouples (CZAKI TP232) with a diameter of 1 mm
- measurement system.

Initially, during the research were used the following test tube types: three cylindrical glass test tubes with an internal diameter of 16 mm and a length of 190 mm and three test tubes with an external diameter of 15 mm and a length of 160 mm. The first two of the smaller test tubes were filled with materials that were to be tested (i.e. RT15 and RT18) and the last one of them with water that served as a reference material. The internal and external diameters of the test tubes as well as masses of the used materials are given in Table 1.

Table 1. Internal and external diameters of the test tubes and masses of materials inserted into tchem.

d_{in}	d_{ext}	m_t	m_m	m_r	m_{tm}	m_{tr}
mm	mm	kg	kg	kg	kg	kg
13	15	0.023	0.0092	0.0119	0.032	0.035

A thermocouple was inserted into each test tube. All three test tubes were fixed to a triangular plate made of a polycarbonate, equipped with a metal handle. Since the test tubes were equidistant from each other, a uniform temperature distribution was obtained during measurements. The larger test tubes were attached to the lid of the thermally insulated container. That way, smaller test tubes can be easily inserted into those larger ones (Fig.6).The thermally insulated container was filled with ice water of a temperature of approx. 0 °C. Polystyrene foam was used as an insulation material in the container.

The temperature measurement was conducted using three thermocouples inserted in test tubes with tested materials and one thermocouple placed inside the container with ice water. During the research, the ambient temperature was measured as well. The NI SCXI data acquisition and signal conditioning system, as well as the software created in the LABView environment, were employed for recording and archiving the measured temperatures.

After analysing the research results, it turned out that it was necessary to modify the test rig (Fig. 7a). Placing tested samples in the container with air and then inserting this container into the thermally isolated container with ice water allowed to cool the samples as a result of heat transfer. That way, the heat was freely exchanged between the warm samples and the cooled air surrounding them. It was decided to attach the container filled with air to the lid of the rectangular container, so that test samples could be placed therein (Fig. 7b). A special moisture absorbing material was put at the bottom of the smaller container. Other components of the test rig were left unchanged.

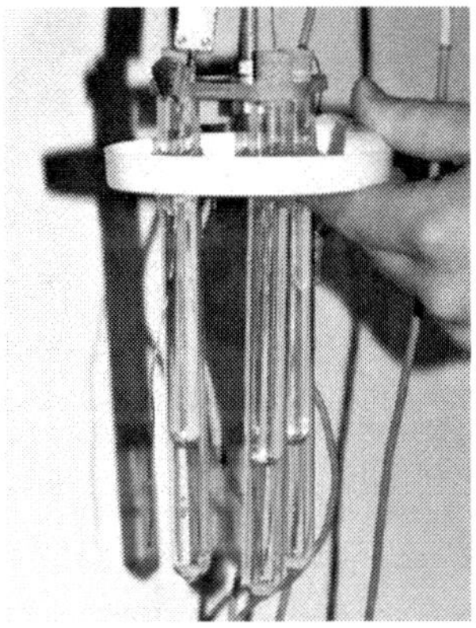

Fig.6. Placing smaller test tubes inside bigger ones that are attached to the lid of the container.

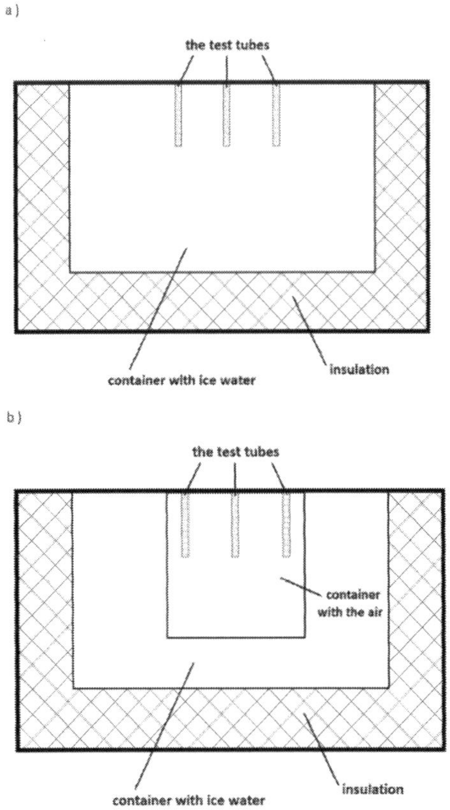

Fig.7. Cross-section of the test rig before (a) and after (b) its modification.

4 Methodology of tests

The tests were started by filling smaller test tubes with water as well as RT15 and RT18 materials to the same volume (12 ml). This was done at the ambient temperature conditions (at a temperature of T_a).

As mentioned, the way in which the research on RT15 and RT18 materials was carried out differs from the original T-history method [1]. In the original T-history method (and in its other implementations as well [2,3,4,6]), containers with materials are heated to a temperature much higher than the ambient temperature, and afterwards, they are abruptly exposed to the ambient temperature, causing them to cool down. To carry out a phase transition for some PCMs, a very wide temperature range is required. In practice, this involves using specialized equipment to achieve high temperatures.

In the experiment discussed herein, material samples are cooled by submerging them in ice water, in contrast to the usually performed heating up. At the beginning of the experiment, the tested materials had a temperature equal to the ambient temperature. The samples were placed in test tubes that were put in a thermally insulated container with ice water. The ice water had a lower temperature than the solidification point of the tested PCMs. About one hour later PCMs were completely solidified.

Then, after the temperature of the tested substances was stabilised, the samples were taken out of the insulated container and exposed to the ambient temperature. Heating started. During heating, the fusion process took place. The heat transfer between the samples and the surrounding air was not as intensive as during cooling because it lasted about four hours. During the heating process, a phase transition occurred since the ambient temperature was higher than the fusion temperatures of the PCMs.

As a result of the test, the full course of the phase transition was registered and temperature profiles over time were archived. The registration of measurement data was carried out in two stages. In the first stage, the temperature versus time was recorded during cooling of the samples in the ice water bath until a steady state was achieved. The second stage of recording started from the moment the samples were taken out from the ice water and ended after they reached stable temperatures equal to the ambient temperature. The results obtained from the measurement allowed to plot heating curves for RT15, RT18 and water.

In Fig. 9a, the heating curves concerning all materials (including water) abruptly changed their courses at a temperature of approx. 18 °C. This is probably related to the formation of water droplets (as a result of condensation) on the walls of the test tubes just after removing them from the ice water bath and then their evaporation. That is the reason why an analysis of the properties of PCMs based on the heating process was abandoned.

The Biot number was only determined for the cooling process and its value was calculated according to the following formula:

$$Bi = \frac{\alpha \cdot \frac{V}{A}}{\lambda_{PCM}} \quad (18)$$

Characteristic dimension in Biot number is equal to V/A (volume by surface area). The thermal conductivity coefficient (λ_{PCM}) was 0.2 W/m·K (based on the manufacturer's data [8]). The convective heat-transfer coefficient (α) was calculated using the following formula [5]:

$$\alpha = \frac{Nu \cdot \lambda}{l} \quad (19)$$

where:

$$Nu = 0.54 \cdot (\mathrm{Gr} \cdot \mathrm{Pr})^{\frac{1}{4}} \quad (20)$$

$$Gr \cdot Pr = \frac{g\beta(T_a - T_z)l^3}{v^2} \cdot Pr \quad (21)$$

$$\beta = \frac{1}{0.5(T_z + T_a)} \quad (22)$$

The following parameters (at given temperatures) were read from thermodynamic tables contained in the REFPROP program: Prandtl numer (*Pr*), thermal conductivity coefficient (λ) and coefficient of kinematic viscosity (v).The Grasfoh number was calculated by knowing the ambient temperature T_a, the temperature inside the sample T_z, gravity *g* and kinematic viscosity *v*. The mean value of the diameter of the samples (using in calculation) was 14 mm. The Biot number was calculated for two values of the characteristic length (*l*), namely 150 mm (corresponding to the length of the test tube) and 95 mm (corresponding to the height of the tested substance). Due to the lack of precision in the method of presenting the characteristic length (*l*) in publications [1,4], it was decided to present the Biot number for both values of *l*. The obtained results differ, which means that the value of the characteristic length has an impact on the value of the Biot number. Therefore, it is crucial to correctly interpret its value. Nevertheless, setting the value of the characteristic length (l) to the value equal to the height of the tested substance seems to be the most logical choice.

The equations presented above show that the Biot number is dependent (although not directly) on the obtained temperatures. In publications [1,3], there is no information about the exact range of the Biot number during the carried out measurements. Their authors only ensured that the Biot number criterion was fulfilled (Bi<0.1).

In the conducted experiment, the calculated value of the Biot number for the initial test rig was higher by several orders of magnitude than the value given in the criterion and that was the case throughout the whole duration of the cooling process. Obtaining such a great value of the Biot number (in comparison with the value given in the criterion) was caused by the occurrence of the convective heat-transfer coefficient (α) in the range of 100-350 W/m^2·K (depending on the temperatures obtained) for almost the entire duration of the test. This is the reason why the cooling process of samples was so intense. The obtained value of the Biot number during the cooling process for the initial test rig precluded the use of the T-history method and a modification of the test rig became necessary.

After upgrading the test rig, the material samples were tested once again. This time the samples were cooled down using cooled air. The ice water served for cooling the air surrounding the tested materials. After the cooling finished, the samples were heated in the same way as it had been done before the modification of the test rig.

The Biot number during the cooling process that took place on the modified test rig was calculated using formula (18) and its values for several selected temperatures are shown in Table 2.

Table 2. Biot number values for chosen temperatures after the modification of the test rig.

T_a	T_z	Nu_1	Nu_2	α_1	α_2	Bi_1	Bi_2
°C	°C	[-]	[-]	W/m2K	W/m2K	[-]	[-]
0.84	22.2	30.46	21.62	4.89	5.48	3.66	2.60
10	18	22.96	16.30	3.79	4.25	2.84	2.02
0.84	1	8.92	6.34	1.43	1.61	1.07	0.76

1 - value calculated for the characteristic length l = 150 mm
2 - value calculated for the characteristic length l = 95 mm

Even though the modification of the test rig brought an intended result, namely the Biot number had a significantly lower value, it still exceeded the value specified in the relevant criterion (Bi<0.1).

5 Discussion of results

The carried out research on RT15 and RT18 materials allowed to plot heating and cooling curves. The curves presented for the cooling process show the measured temperature versus time. In Figs. 8a and 8b, the phase transition of the RT18 material, which occurred in the temperature range of 13–20°C, is clearly visible. As a matter of fact, the phase transition of the RT18 material occurred after a significantly shorter time during the experiment performed on the test rig before its modification. This testifies that the intensive cooling of RT18 before the modification contributed to shortening the duration of the phase transition. After the modification of the test rig, RT18 required more time to change its state. The duration of the phase transition is important when determining the thermophysical parameters of the materials tested. Therefore, both the cooling conditions of samples and the heat exchange process play a significant role. As far as the RT15 material is concerned, both graphs clearly indicate that the phase transition started at a temperature of approx. 20 °C. However, it is hard to say when the change of state was complete. Perhaps RT15 should be cooled by using a temperature lower than 0 °C, which is the cause of the lack of a clear indication of the end of the phase transition. The cooling of RT15 lasted longer than the cooling of RT18 and water and both PCMs finally solidified completely.

During the heating process (Fig. 9), water was the first to reach the ambient temperature. Although the RT15 material needed the longest charging (cooling) time in the ice water, it reached the ambient temperature much faster than the RT18 material. It is also worth noting that at the beginning of the heating process, the temperature of the RT18 material increased faster than the temperature of the water. Once the RT18 reached a temperature of 13.5 °C, the phase transition process began.

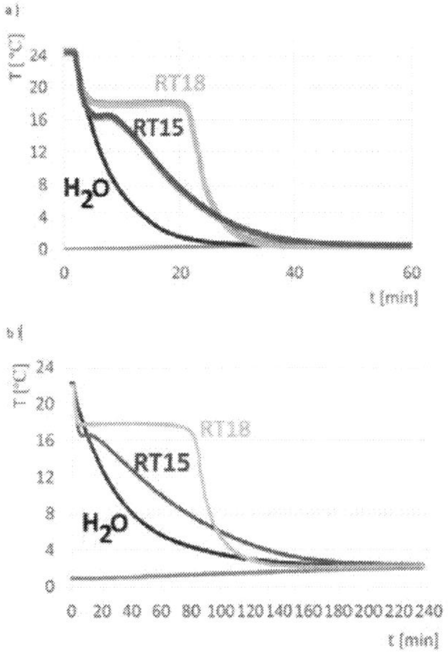

Fig. 8. Cooling curves of the test samples before (a) and after (b) the modification of the test rig.

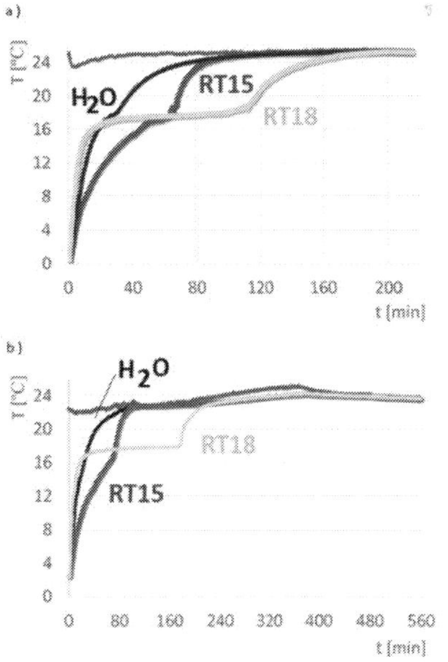

Fig. 9. Heating curves of the test samples before (a) and after (b) the modification of the test rig.

The value of the Biot number exceeded the value specified in the criterion (Bi<0.1) both before and after the modification of the test rig. In this case, the assumed criterion was not met. Thus, the T-history method should not be used. The Biot number criterion (Bi<0.1) assumes that there is an even temperature distribution in a sample. Exceeding a value of 0.1 causes a disruption of this distribution. For this reason, the results concerning the estimation of the latent heat during the cooling process are not reliable and thus the T-history method should not be used for their. However, for the purposes of this article, it was decided to estimate the specific heat of selected PCMs in the ranges in which they occur in the liquid and solid states as well as the generated latent heat on the basis of the T-history method. Calculation of derivatives of the temperature curves of the tested materials yielded two peaks for the RT18 material (at temperatures of 20 °C and 13.5 °C), as shown in Fig. 10, and one peak for the RT15 material (at a temperature of 20.3 °C), as shown in Fig. 11.

According to the assumptions, the obtained temperature peaks were assumed as moments of the start and end of the phase transformation. Thus, the phase transition of the RT18 material occurred (according to the experimental results) in the range of 13.5 °C – 20 °C. For the RT15 material, it can be only determined that the temperature at the beginning of the phase transition was 20.3 °C. However, it is not possible to ascertain when this transition ended, which makes it difficult to determine the latent heat. The phase change range for RT18 obtained as a result of the experiment differs from the range given by the manufacturer.

Table 3. Comparison of the research results obtained for RT15 and RT18 materials during the cooling process with the manufacturer's data – Rubitherm GmbH.

		RT15		RT18	
		1	2	1	2
$c_{p,ml}$	kJ/kgK	5.68	2	5.68	2
$c_{p,ms}$	kJ/kgK	3.69	2	2.54	2
H_m	kJ/kg	93.2	155	264.32	260
T_l	°C	20.3	17	20	19
T_s	°C	13.7	10	13.5	17
T_0	°C	22.2	22	22.2	26
T_e	°C	0.9	7	0.9	11
T_a	°C	0.9	-	0.9	-
Temperature range of phase change	°C	13.7-20.3	10-17	13.5-20	17-19
Range of measurement error	kJ/kg	-	±11.6	-	±19.5

1 – the result of the experiment
2 – the result provided by the manufacturer

The temperature range in which the test was carried out and the fact that the criterion of the Biot number (Bi<0.1) was not met both have an impact on the obtained results. The manufacturer tested materials in the temperature range of 11 °C – 26 °C and the temperature range in the experiment described herein was 0.8 °C – 22.3 °C (which is wider by 6.5 °C). For the temperature range of 11 °C – 26 °C, the latent heat of RT18 material given by its manufacturer is 268 kJ/kg. On the other hand, the latent heat of the same material obtained in the experiment for the temperature range of 0.8 °C – 22.2 °C is 264.3 kJ/kg. Considering that the temperature range in which the research was carried out in the experiment is wider than the temperature range of the manufacturer's research, the latent heat calculated on the basis of the experiment would hypothetically have a higher value than if it was calculated for the temperature range that is given by the manufacturer.

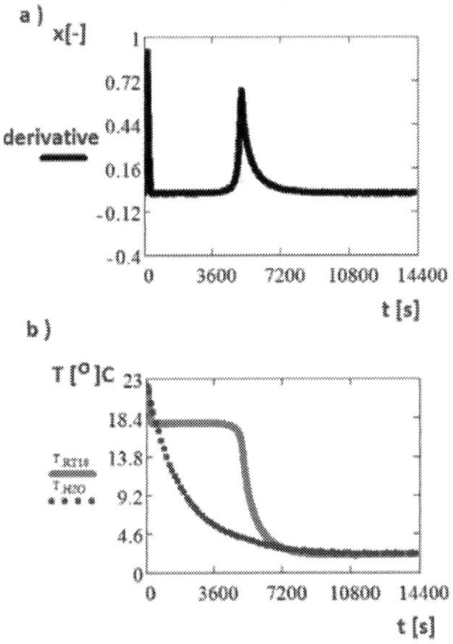

Fig. 10. Derivative of the RT18 temperature vs. time (a) and cooling curves of RT18 and water (b).

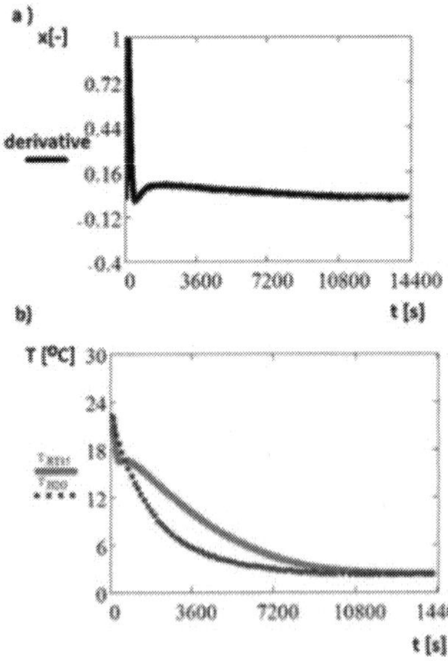

Fig. 11. Derivative of the RT15 temperature vs. time (a) and cooling curves of RT15 and water (b).

When analysing the results obtained, it should be borne in mind that they are vitiated by an error resulting from exceeding the value of the criterion (Bi <0.1). Is the error in the estimation of the amount of the latent heat (resulting from the fact that the criterion Bi<0.1 was not met) falls within the range of the manufacturer's measurement error? At this stage, it is very difficult to determine.

6 Conclusions

The described T-history method is used to estimate the latent heat and specific heat of a PCM in the liquid and solid state. The use of its modified version for the purposes of the experiment allowed to investigate two PCMs that are available on the market (i.e. RT15 and RT18). These materials were tested as samples with a capacity of 12 ml each (which corresponds to a mass of about 9.2 g). The research carried out on large samples of PCMs allows determining the latent heat and specific heat of those materials in a given state of aggregation, and thus the parameters necessary for designing heat and cold storage systems.

The experimental research was performed on RT15 and RT18 materials. The temperatures at which the phase transformation began and ended were clearly identified for the RT18 material. However, the research failed to determine the moment at which the phase transformation was over for the RT15 material. Therefore, it should be investigated whether this moment can be observed if the temperature before commencing the heating process would be lower than 0 °C.

Moreover, it was noticed that the characteristic of the RT15 material given by the manufacturer does not cover temperatures lower than 7 °C. It can, therefore, be assumed that the phase change process covers a wider temperature range than that presented by the manufacturer. As a matter of fact, the conducted research can help in acquiring valuable information regarding the determination of the thermal capacity of heat and cold storage systems in which the RT15 material is used.

A comparison of the manufacturer's data regarding the latent heat with the results obtained from the experiment is possible only for the same temperature range. For this reason, it is planned to determine the enthalpy-temperature relation that will enable to perform such a comparison.

The prerequisite for the use of the T-history method was discussed (i.e. the Biot number criterion has to be fulfilled, Bi<0.1), which is of the utmost importance when calculations are carried out. Therefore, before calculating the latent heat, one has to make sure that the criterion is met. Otherwise, the latent heat will be vitiated by a large error.

The preliminary research carried out allowed to determine the scope of further actions aimed at improving the accuracy and reliability of the measurement methodused. At a further stage of the research, the following activities are planned:

- conducting tests using smaller test tubes that will allow for the fulfilment of the Biot number criterion (Bi<0.1) in the whole temperature range,
- determining the enthalpy-temperature relation for the examined materials,
- applying the method proposed in publication [7], which will allow obtaining more precise results of the latent heat in cases where the Biot number criterion will not be met (in other words, in cases where the Biot number will be greater than 0.1).

References

1. Z. Yinping, J. Yi, A simple method, the T-history method, of determining the heat of Ffusion, specific heat and thermal conductivity of phase- change materials, Measurement Science and Technology, 10, 201-205 (1999)
2. H. Hong, S.K. Kim, *Accuracy improvement of T-history method for measuring heat of fusion of various materials,* International Journal of Refrigeration, **27**, 360-366 (2004)
3. E.D. Kravvaritis, K.A. Antonopoulos, C. Tzivanidis, *Improvements to the measurement of the thermal properties of phase change materials,* Measurement Science and Technology, **21**, 9 (2010)
4. H. Mehling, H.P. Ebert, P. Schossig, *Development of standards for materials testing and quality control of PCM,* 7th IIR Conference on Phase Change Materials and Slurries for Refrigeration and Air Conditioning, (2006)
5. W. Pudlik, *Wymiana i wymienniki ciepła*, Wydanie V – cyfrowe, Wydawnictwo Politechniki Gdańskiej, Gdańsk (2012), published in "Pomorska Biblioteka Cyfrowa"(20.05.2018):http://pbc.gda.pl/Content/4404/wymiana-i-wymienniki-final.pdf
6. A. Solé, L. Miró, C. Barreneche, I. Martorell, L.F. Cabeza, *Review of the T-history method to determine thermophysical properties of phase change materials (PCM)*, Renewable and Sustainable Energy Reviews, **26**, 425-436 (2013)
7. B. Delcroix, M. Kummert, A. Daoud, J. Bouchard, *Influence of experimental conditions on measured thermal properties used to model phase change materials,* Building Simulation, **8**, 637-650 (2015)
8. https://www.rubitherm.eu/en/index.php/productcategory/organische-pcm-rt

Presented work was supported by SuPREME project. This project has received funding from the European Union's Horizon 2020 research and innovation programme under grant agreement No. 692197.

The range of application of ORC power plant in power stations fed with low and medium thermal sources

Tomasz Kujawa[1,*], and *Władysław* Nowak[1]

[1]West-Pomeranian University of Technology, Szczecin, Department of Heat Engineering, al. Piastów 17, PL-70-310 Szczecin, Poland

Abstract. The work presents the diagrams of the applied power plant solutions with a description of their operation and calculation algorithm. In the latter case, the cycle of thermodynamic transformations of the Rankine cycle presented in the *T-s* coordinate system was used, which reflects the subsequent transformations occurring in individual devices of the power plant, including the method of determining unit quantities characterizing their operation. The energy balance equations of the evaporator and the heater are presented, which among other things will be the basis for determining the coefficient that plays an important role in determining the range of applications of the single-circuit sub-critical power plant in the power plant/CHP plant. The coefficient ϕ was defined as the ratio of the mass stream of water flowing through the counter-current heat exchanger in which the liquid of the working medium is heated within the limits of condensation and evaporation temperatures, to the mass stream of water leaving the evaporator according to the dependence $\phi = \dot{m}_{s1} / \dot{m}_s$. It has been assumed that a given mass stream of water $\dot{m}_s$ at a given temperature T_{s1} is heated in a heat exchanger which is the upper heat source. The calculations of the power plant for selected working fluids (R227ea, RC318, R1234ze and R1234yf) were made for different evaporation temperatures taking into account near-subcritical temperatures for selected supply water temperatures. On this basis, the figures $\phi = f(T_{EVAP})$ illustrating the application ranges of a single-circuit power plant in a power plant/CHP plant were drawn up. The analysis of the above figures became the basis for formulating final conclusions.

1 Introduction

The starting point of the considerations presented in the work is an organic power plant, in which a subcritical single-circuit power plant with a dry working medium was used. The power plant is supplied with geothermal water (Fig. 1) [1-6].

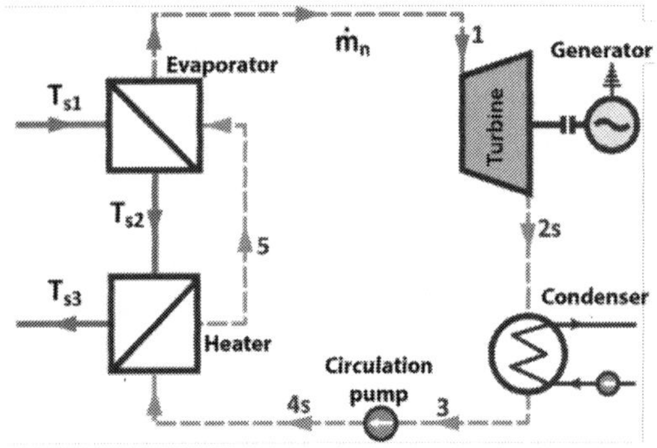

Fig. 1. Diagram of the plant installation working with dry medium.

The principle of operation of the ORC power plant is the same as for a steam power plant. Figure 2 presents cycles of thermodynamic transformations of the power plant, also the temperatures of the energy carrier in the relevant characteristic points are marked.

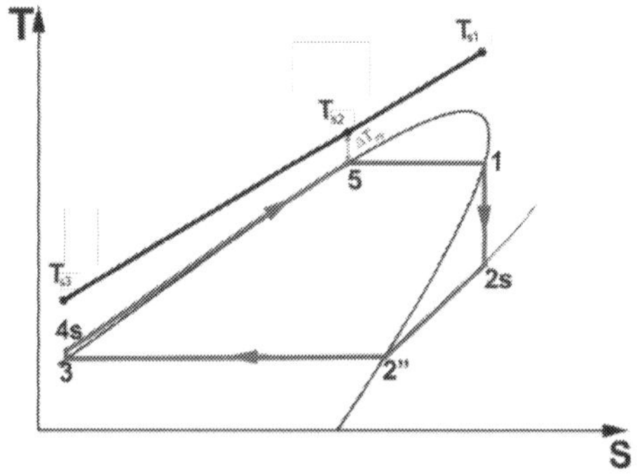

Fig. 2. Cycle of thermodynamic transformations of subcritical ORC cycle with dry factor [2].

ORC power plant with dry medium as a working medium consists of an evaporator, a heater, a turbine, a condenser and a circulation pump of an organic medium.

* Corresponding author: tomasz.kujawa@zut.edu.pl

The curves shown in Fig. 2 reflect subsequent changes taking place in the individual devices of the system:

- transformation 1-2s - isentropic vapour expansion of the organic agent in the turbine,
- transformation 2s-3 –cooling and condensation of the medium vapour in the condenser,
- transformation 3-4s – isentropic pumping of an organic agent in a circulating pump,
- transformation 4s-5 – heating the organic medium in a heater-type exchanger from the condensation temperature to the evaporation temperature,
- transformation 5-1 – evaporation of the organic agent in the evaporator.

It has been assumed that a given mass stream of water $\dot{m}_s$ at a given temperature T_{s1} is heated in a heat exchanger which is the upper heat source.

2 Variants of the power plant operation and calculation algorithm

For calculations the energy balance equations of the evaporator and the heater were used.

The equation for the energy balance of the evaporator can be written as follows:

$$\dot{Q}_{EVAP} = \dot{m}_s \cdot \left[h'_{s1}(T_{s1}) - h'_{s2}(T_{s2})\right] = \dot{m}_n \cdot \Delta h_{EVAP}, \quad (1)$$

while the equation of the energy balance of the heater in the form:

$$\dot{Q}_{HEAT} = \dot{m}_s \cdot \left[h'_{s2}(T_{s2}) - h'_{s3}(T_{s3})\right] = \dot{m}_n \cdot \Delta h_{HEAT}. \quad (2)$$

The coefficient ϕ was defined as the ratio of the mass stream of water $\dot{m}_{s1}$ flowing through the counter-current heat exchanger in which the liquid of the working medium is heated within the condensing and evaporating temperatures to the mass stream of water $\dot{m}_s$ leaving the evaporator according to the dependence:

$$\phi = \dot{m}_{s1} / \dot{m}_s . \quad (3)$$

The tests show that for the above-mentioned power plant the coefficient ϕ generally fulfills the condition $\phi \le 1$.

If $\phi = 1$, the above single-circuit power plant is applicable only to the power plant.

In this case, the water jet leaving the evaporator $\dot{m}_s$ is equal to the mass flow of water $\dot{m}_{s1}$ which feeds the heater, that means that $\dot{m}_s = \dot{m}_{s1}$.

The size of the coefficient ϕ (at the same stream of feed water $\dot{m}_s$ and at the same temperature T_{s1}) is influenced by: the type of dry working medium of the power plant and the type of operation of the evaporator in the power plant.

The latter includes the following temperature differences:

- drop in feed water temperature

$$\Delta T_s = T_{s1} - T_{s2} \quad (4)$$

equal to the difference in water temperature at the inlet T_{s1} and at the outlet T_{s2} of the evaporator,

- temperature difference

$$\Delta T_{s1} = T_{s1} - T_{EVAP} \quad (5)$$

as the temperature difference between the water supplying the evaporator T_{s1} and the evaporating temperature of the working medium T_{EVAP},

- temperature difference

$$\Delta T_{s2} = T_{s2} - T_{EVAP} \quad (6)$$

as the temperature difference between the water leaving the evaporator T_{s2} and the evaporating temperature of the working medium T_{EVAP}.

It is easy to show that between the given temperature differences ΔT_s, ΔT_{s1}, ΔT_{s2} there is a dependence:

$$\Delta T_s = \Delta T_{s1} - \Delta T_{s2} \quad (7)$$

If it is not possible to utilize heat for technological and heating purposes, the power plant solution for $\phi < 1$ was also used in the power plant (Figure 3). It allowed to take into account the continuity of changes ϕ in the characteristics of the power plant operation and to develop a full range of the power plant operation characteristics in the power plant for $\phi \le 1$ which is the basic area for this work characteristic. If $\phi = 1$ then $\dot{m}_s = \dot{m}_{s1}$, which means that $\dot{m}_{s2} = 0$.

In this case, the cable connecting the node A to the node B in Fig. 3 is not needed.

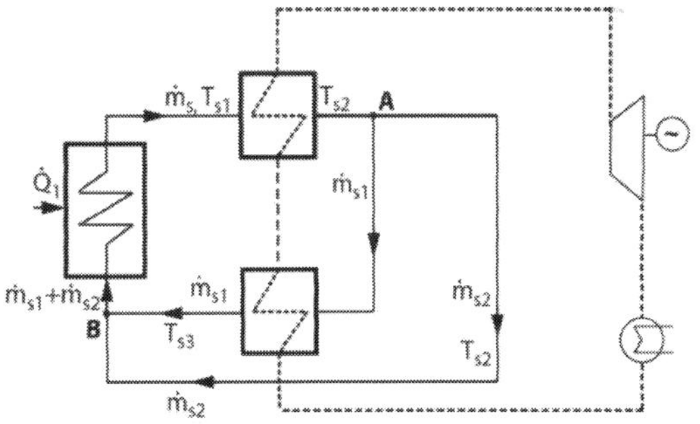

Fig. 3. Diagram of a single-circuit ORC plant used in a power plant for $\phi \le 1$.

The work carried out at the Department of Heat Engineering [1, 2] showed that the coefficient ϕ may be equal to and less than one ($\phi \le 1$), and also greater than one ($\phi > 1$) and then covers the second area of the operating range of the power plant in a hybrid power plant. In this second area of the power plant performance at the power plant, just like in the case of a power plant for $\phi = 1$, the mass flow of water $\dot{m}_s$ at the temperature

T_{s2} leaving the evaporator is directed to the fluid heater of the power plant working medium (Fig. 4). This stream of water $\dot{m}_s$ after lowering its temperature to T_{s3} is directed to the first heat exchanger, which is the upper heat source of the installation. This stream, after reaching the temperature T_{s1}, is directed to the power plant evaporator again.

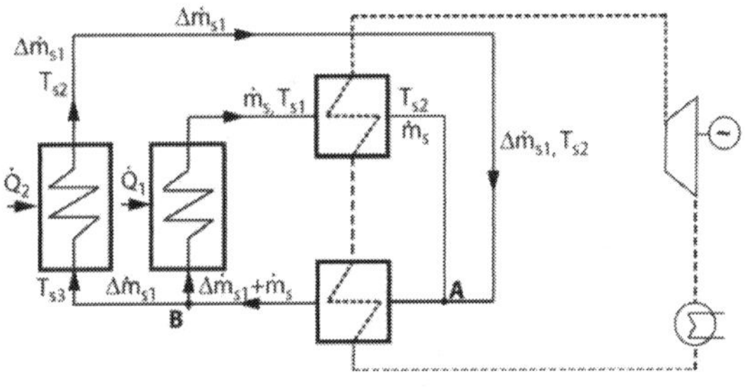

Fig. 4. Diagram of a single-circuit ORC plant used in a power plant for $\phi > 1$.

The required stream of water $\dot{m}_{s1}$, which should be fed to the fluid heater of the working medium, is determined from the definition of coefficient ϕ (3). The stream is:

$$\dot{m}_{s1} = \dot{m}_s \cdot \phi . \tag{8}$$

Considering that the mass flow of water $\dot{m}_s$ from the evaporator was directly supplied to the fluid heater of the working medium of the plant, after taking into account dependence (8), an additional mass stream of water $\Delta\dot{m}_{s1}$ should be fed to the liquid heater of the working medium, whose size can be determined from the dependence below:

$$\Delta\dot{m}_{s1} = \dot{m}_{s1} - \dot{m}_s = \dot{m}_s(\phi - 1), \tag{9}$$

for $\phi > 1$.

This stream ($\Delta\dot{m}_{s1}$) after increasing the temperature to the value T_{s2} is supplied to the fluid heater of the working medium of the power plant, in which after lowering the temperature to the value T_{s3} it is directed again to the second heat source.

After taking into account the balance of the evaporator and the condenser, the value of the coefficient ϕ can be determined from the dependence:

$$\phi = \frac{(\Delta T_{s1} - \Delta T_{s2})}{(T_5 - T_{4s})} \cdot \frac{\Delta h_{HEAT}}{\Delta h_{EVAP}} = \frac{\Delta T_s}{T_{PAR} - T_{4s}} \cdot \frac{\Delta h_{HEAT}}{\Delta h_{EVAP}} . \tag{10}$$

3 Calculation results and conclusion

Calculations of the power plant were made for various operating factors for selected feed water temperatures for the condensing temperature $T_{COND} = 29°C$. Evaporation temperatures T_{EVAP} were changed from 60°C to the subcritical temperature value, the value of which depends on the type of working medium.

Based on this, after the calculation figures $\phi = (T_{EVAP})$ illustrating the application ranges of the single-circuit power plant within the power plant were developed (Figs. 5-8).

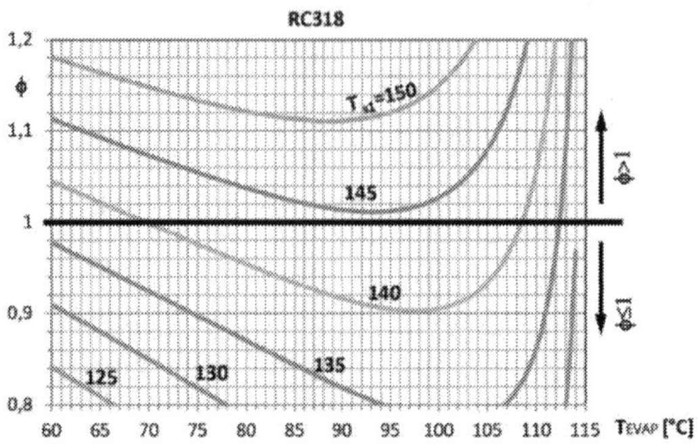

Fig. 5. Variability of coefficient ϕ as a function of evaporating temperature for RC318 refrigerant and water temperature feeding the evaporator T_{s1} (range: $\phi = 0.8 \div 1.2$).

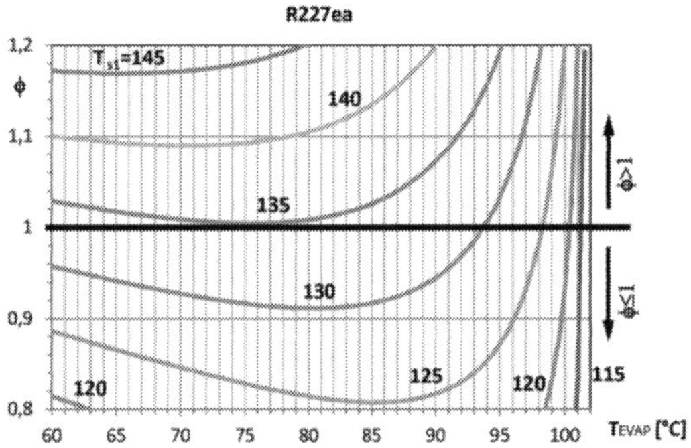

Fig. 6. Variability of coefficient ϕ as a function of evaporating temperature for R227ea refrigerant and water temperature feeding the evaporator T_{s1} (range: $\phi = 0.8 \div 1.2$).

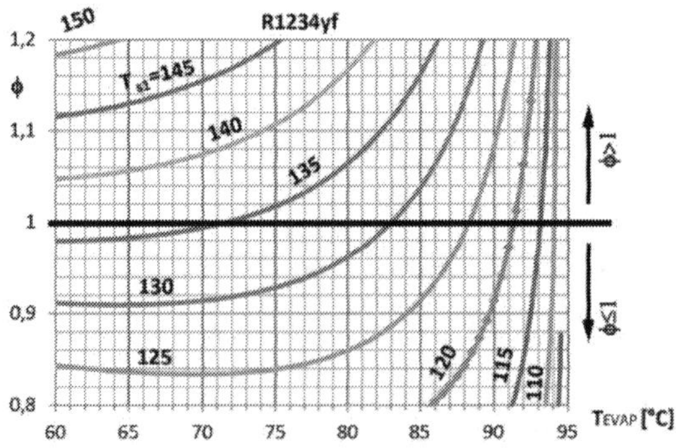

Fig. 7. Variability of coefficient ϕ as a function of evaporating temperature for R1234yf refrigerant and water temperature feeding the evaporator T_{s1} (range: $\phi = 0.8 \div 1.2$).

 https://doi.org/10.1051/e3sconf/20187001009

The figures clearly show that the scope of work of the power plant (case $\phi \le 1$ and $\phi > 1$) depends on the type of fluid in the ORC cycle of the power plant, the temperature feeding the evaporator T_{s1}, heater T_{s2} and evaporation temperature T_{EVAP}. The variant of the hybrid

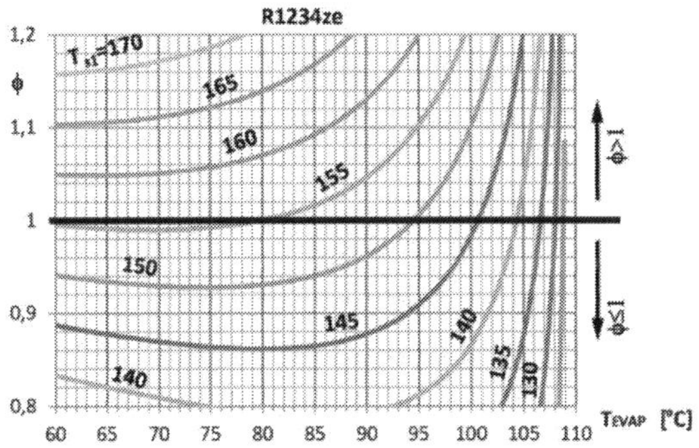

Fig. 8. Variability of coefficient ϕ as a function of evaporating temperature for R1234ze refrigerant and water temperature feeding the evaporator T_{s1} (range: $\phi = 0.8 \div 1.2$).

power plant operation (case $\phi > 1$) in the entire evaporation temperature range can be obtained only and exclusively for certain temperatures of the water feeding the evaporator, the value of which depends on the type of ORC fluid in the power plant circuit. For example, for R227ea fluid, the minimum value of the water temperature in the evaporator, enabling this variant of the power plant operation is 135°C, for RC318 – 145°C, for R1234yf – about 137°C, and for R1234ze – about 156°C. Operation of the power plant in this range is also possible for lower evaporator temperatures T_{s1}, but may occur only and exclusively for evaporation temperatures close to the subcritical value, e.g. for R227ea fluid in the range T_{EVAP}= 90 ÷ 103°C the operating range $\phi > 1$ for T_{s1} < 130°C will be achieved, while for R1234ze fluid in the range T_{EVAP}=85 ÷ 108°C the operating range $\phi > 1$ for T_{s1} < 155°C will be achieved. The calculations and analysis may be helpful to select the appropriate type of the ORC medium in the power plant, depending on what variant of power plant operation will be used (case $\phi \le 1$ and $\phi > 1$).

In order to bring closer the possibilities of using the range of applications of the developed ORC plant operating characteristics in the power plant for $\phi > 1$, presented in Figures 5-8, additional calculation variants have been made. The results of these calculations, presented in Table 1, relate to the variant $\phi = 1$ using the given working fluids (RC318, R227ea, R1234yf, R1234ze) for appropriately selected water temperatures T_{s1} supplying the evaporator and for properly selected evaporation temperatures T_{EVAP} that will allow to determine the temperature of the water feeding the heater $T_{s2} = T_{EVAP} + \Delta T_{s2}$ of the additional mass stream $\Delta \dot{m}_{s1}$ directed to the liquid heater of the working medium. It allows to determine the values characterizing the efficiency of the hybrid power plant, i.e. the size of power and efficiency, if $\phi > 1$.

The use of R1234ze is characterized by the highest efficiency and power of the ORC plant. The smallest values are achieved using R227ea (Table 1).

Thermodynamic parameters of the considered working fluids at individual nodes of the C-R cycle were determined using the Refprop software [7].

Table 1. Obtained power and efficiency of the ORC power plant using selected fluids (variant $\phi = 1$, amount of ORC fluid $\dot{m}_n = 1$ kg/s).

ORC circulating fluid	$N_{C\text{-}R}$	$\eta_{C\text{-}R}$
	kW	%
RC318	51,26	11,00
R227ea	43,42	10,00
R1234yf	44,70	10,26
R1234ze	66,54	13,08

Nomenclature

h	specific enthalpy	kJ/kg
$\dot{m}$	mass flow rate	kg/s
$\Delta \dot{m}_{s1}$	additional mass flow rate	kg/s
N	power output ORC	kW
$\dot{Q}$	heat flow	kW
T	temperature	°C
ϕ	coefficient	-
η	efficiency	%
Δ	refers to the working medium	

Subscripts:

CR	refers to the Clausius-Rankine cycle
n	refers to the working medium
s	refers to geothermal water (heating water)
1,2*s*, ..	refers to the characteristic points of the CR cycle

Literature

1. R. Mazurek, W. Nowak, A. Borsukiewicz-Gozdur, *Ocena wpływu charakterystyki parowacza na efektywność pracy elektrowni z jednoobiegową klasyczną siłownią ORC zasilaną wodą geotermalną*, Cieplne Maszyny Przepływowe, No. 143, pp.137-146, 2013.
2. W. Nowak, i inni, *Analiza i ocena wpływu blisko-krytycznych warunków wrzenia czynników obiegowych na poprawę efektywnoi sci pracy elektrownii geotermalnej*, sprawozdanie z projektu badawczego NN513393436, Szczecin 2012.

3. W. Nowak, R. Sobański, M. Kabat, T. Kujawa, *Systems for acquiring and utilizing geothermal energy* (Publishing House of Szczecin University of Technology, 2000, in Polish)
4. A.A. Stachel, S. Wiśniewski, Influence of the type of working fluid in the lower cycle and superheated steam parameters in the upper cycle on effectiveness of operation of binary power plant, Archives of Thermodynamics, Vol. 36, no. 1, pp 111-123, (2015)
5. A. Borsukiewicz-Gozdur, Dual-fluid-hybrid power plant co-powered by low-temperature geothermal water. Geothermics 39, 170-176, (2010)
6. S. Wiśniewski, W. Nowak, ORC power plant effectiveness as affected by the cycle working fluid evaporation at narrow subcritical conditions, 3rd International Conference on Contemporary Problems of Thermal Engineering, CD, CPOTE 2012, 18-20 September 2012, Gliwice, Poland
7. Refprop 9.1 Standard Reference Database 23, Version 9.1, Reference Fluid Thermodynamic and Transport Properties. National Institute of Standarts and Technology (Gaithersburg, USA, 2013).

Designs of PCM based heat exchangers constructions for thermal energy storage tanks – examples and case study for selected design

Marta Kuta[1,*], *Dominika* Matuszewska[1], *Tadeusz Michał* Wójcik[1]

[1]AGH University of Science and Technology, Department of Energy and Fuels, al. Mickiewicza 30, 30-059 Krakow, Poland

Abstract. Increasing energy consumption in residential and public buildings requires development of new technologies for thermal energy production and storage. One of possibilities for the second listed need is the use of phase change materials (PCMs). This work is focused on solutions in this area and consists of two parts. First one is focused on different designs of thermal energy storage (TES) tanks based on the phase change materials. The second part is the analysis of tests results for TES tank containing shelf and tube heat exchanger and filled with phase change material. Thermal energy storage tank is analyzed in order to use it in domestic heating and hot utility water installations. The aim of this research was to check the applicability of phase change material for mentioned purpose. Results show that using phase change materials for thermal energy storage can increase amount of stored heat. The use of properly selected PCM and heat exchanger enables the process of thermal energy storing and releasing to become more efficient.

1 Introduction

It is noticeable that continuous economic and social growth forces development of energy around the world, including Poland. Attention is focused not only on the electricity, but also on the production and storage of thermal energy.

It is particularly important to store heat in situation when amount of heat produced at a given moment by heating sources does not coincide with the current demand. Excess thermal energy can be stored and used later, when the needs are bigger. It enables the reduction of waste heat, limits additional heat production, it also has a positive impact on the environment and allows financial savings.

Many methods of thermal energy storage have been used over the years. One of them is based on the use of phase change materials (PCMs) and its latent heat. Phase change materials are substances characterized by the ability to store large amount of heat within small temperature range. This feature can be applied to store thermal energy or to regulate the ambient temperature. This paper will focus on the first option.

There are some requirements which have to be met for PCM based thermal energy storage systems. It is important to consider all problems related to phase change materials, such us: low thermal conductivity, changes in properties after many work cycles, changes in volume during the phase transition, possibility of corrosion during the contact with some materials. It is also important to select PCM suitable for the application, considering phase transition temperature range and enthalpy. Additionally it is possible to improve work of the system by design right heat exchanger's construction what is discussed in this paper.

2 Thermal energy storage systems based on phase change materials - requirements, advantages and disadvantages of this type of solution

Application of phase change materials has a great contribution to development of new thermal energy storage technologies and at the same point - reduction of waste heat and increasing the efficiency of using renewable energy sources. All requirements, phase change material characteristic, advantages and disadvantages should be taken into consideration at the design stage.

2.1 Phase change materials

2.1.1 Phase change materials for low temperature TES systems

Thermal energy storage with the use of phase change materials is based on the latent heat. It means that large amount of heat can be stored and released during the phase transition process at almost constant temperature.

Characteristics of PCM differs for material from organic (paraffins, polymers, alcohols, fatty acids), inorganic (i.e. salt, salt hydrates) and eutectics group. Phase change materials can be classified based on the type of the transition: solid-liquid, solid-solid, liquid-gas.

* Corresponding author: marta.kuta@agh.edu.pl

There are some proprieties required for PCMs used in TES systems[1, 2]:

- melting point according to the needs of the application – in this case suitable for central heating or hot water,
- high heat of fusion,
- high thermal conductivity in both phases,
- non-toxicity, non-flammability and lack of explosive properties during the normal use,
- compatibility with container material,
- small volume change during the transition,
- stability after many work cycles,
- small temperature difference between the melting and solidification points,
- reasonable price.

It is not possible to find material having all of the listed features, but it is important to choose one that meets the most important assumptions for the application.

2.1.2 Phase change materials properties improvements

Enhancement of thermal conductivity is one of the biggest challenge when it comes to PCM based applications. This parameter has to be analyzed and modified to improve process of charging and discharging thermal energy storage tanks filled with PCM.
There are some solutions investigated and applied for thermal conductivity improvement. Selected solutions used for PCM tanks are listed below:

- addition of high conductivity particles: silver, copper, aluminum, graphite, carbon nanotubes,
- incorporation of metal structures,
- macroencapsulation,
- heat exchanger designed taking into account all requirements.

In the next section selected designs of heat exchangers currently used in PCM based tanks and its advantages and disadvantages will be analyzed.

3 Overview of different PCM based storage tank configurations

3.1. Examples of different heat exchangers designs

Tay et al. [3] present results of experiment conducted on tube-in-tank filled with PCM for cold storage application. They compared tanks equipped with: one-, two- and four- tubes systems. Latent heat energy storage capacity slightly decreased with the increase of tubes amount, but at the same – time of melting process decreased as well (PCM melted faster). Both phenomena were related to the fact that tank parameters did not change, Authors concluded that the tube-in-tank design can deliver a high energy storage density.

Gil et al. [4] present experimental study of heat transfer enhancement of tube based storage tank. They compared two identical tanks equipped with tubes through which flows heat transfer fluid. One of them consists of 196 transversal squared fins and another one - does not. Authors concluded that addition of fins can increase effectiveness of thermal conductivity between 4,11% and 25,83% for experiments with highest and lowest thermal power supplied to the PCM.

Ahmadi et al. [5] investigated spiral coil tube heat exchanger. They also performed numerical simulation to predict effect of changes of parameters like: tube parameters, spiral coil diameter, heat transfer fluid parameters. Authors concluded that spiral coil diameter plays an important role in melting process and heat transfer. It is possible to reduce total melting time up to 71.4% by increasing coil diameter from 50 mm to 70 mm. They presented that tube diameter has little effect on melting process. They indicated that for larger spiral coil diameter, increasing tube diameter has opposite effect for reduction of melting time.

Kabbara et al. [6] studied experimentally latent heat energy storage system with coil – in –tank heat exchanger. They observed that natural convection played a significant role during the charging stage, but during the discharging stage the most important turned out to be conduction. They also found out that the inlet temperature of heat transfer fluid has bigger impact on system performance than heat transfer flow rates. They concluded that general thermal behavior of the PCM in tested tank is similar as for another published thermal storage geometries. However authors emphasize that heat transfer results for this specific geometry are valuable for future designs of PCM based heat exchangers and for establishing design rules for this kind of heat exchangers.

Prieto et al. [7] present results for thermal energy storage system based on PCM plate heat exchangers in comparison with a water tank. They are testing two different PCMs (Palmitic Acid and RT60 paraffin) characterized by similar transition temperature. This storage system is a part of micro – cogeneration system applied to the office space. The three TES systems were compared in terms of charges and discharges stages duration and in terms of the appropriate heat transfer rate. Authors concluded that the best system among considered in the research described in the paper is the one using Palmitic Acid. This system showed better performance during discharging, greater storage capability and lower cost for mCHP heating system. Authors did not indicated significant problems with this kind of heat exchanger design.

Guo et al. [8] present results of tests for direct contact thermal energy storage container. Authors commented that this kind of solution shows a good charging and discharging performance except the blocking by the deposited phase change material (PCM) at the initial stage of the charging process. They propose solution in the form of forming the quick channels with electric heaters and deal with the blocking issue. They concluded that results present that it is a good method to solve mentioned problem for the direct contact mobile thermal energy storage system, especially with a low rate of heat release.

https://doi.org/10.1051/e3sconf/20187001010

Kabbara and Abdallah [9] investigated phase change material based thermal energy storage system with PCM encapsulated in cylinders. They analyzed the system in terms of its storage capacity and the heat transfer rate to the PCM. Authors highlighted that inlet temperature was the most important parameter in this experiment. They concluded that there is evident that not all material inside tubes are melting and for further experiment there should be made some improvement in delivering heat to PCM from heat transfer fluid.

Solutions described above are only examples of designs used for thermal energy storage systems based on phase change materials. Fig. 1 presents graphically the appearance of exemplary heat exchangers designs described above and few additional.

4 Experimental setup

4.1 Phase change materials

The main assumption of the tests was the use of a PCM - thermal energy storage system for hot water and low-temperature heating. For this test purpose, commercially available organic PCM: CrodaTherm 53 has been used. Material used for tests is non-toxic, easily biodegradable and has low flammability. Figure 2 presents CrodaTherm53 before first phase transition, figure 3 presents material in liquid phase and figure 4 presents the appearance of the material after melting and after re-solidification. PCM has initially the form of white gleaming flakes, after melting - dense, almost transparent liquid, after re-solidification it becomes white, solid block of material. Tests in smaller (200 ml) and larger volume (60 l) did not show any problems with melting and solidification process (i.e. air voids).

Phase change transition temperature and heat storage capacity of this PCM specified by the manufacturer is respectively: 53°C (melting area), 51°C (congealing area) and 226 kJ/kg. More properties of this PCM has been presented in table 1.

Table 1 Properties of phase change material: CrodaTherm53.

Properties	**CrodaTherm53**
Melting range, °C	53
Congealing range, °C	51
Heat storage capacity, kJ/kg (kWh/kg)	226 (0.062)
Heat storage capacity, kJ/m^3 (kWh/m^3)	187354 (52)
Density liquid, kg/m^3	829
Density solid kg/m^3	-
Heat conductivity (liquid), W/(m·K)	0.16
Heat conductivity (solid), W/(m·K)	0.28

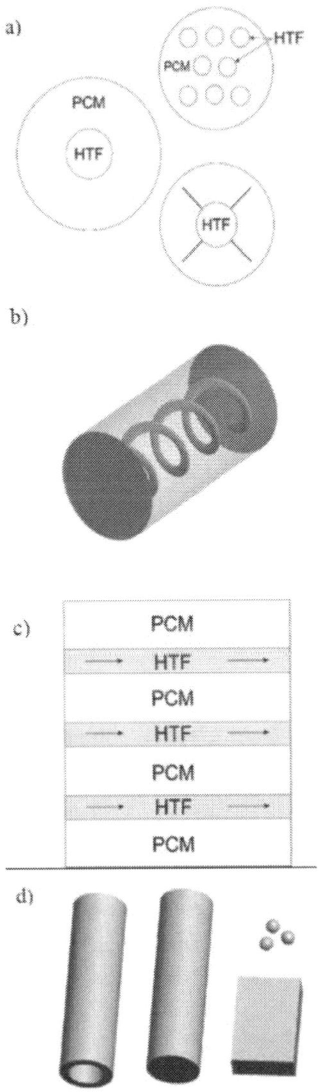

Fig. 1. Examples of PCM based heat exchangers : a) tube based storage tank (shell and tube,tube with fins heat exchangers): b) spiral coil tube heat exchanger, c) PCM plate heat exchanger, d) capsules filled with PCM ;HTF –heat transfer fluid.

Fig. 2. CrodaTherm53 before first phase transition.

Fig. 3. CrodaTherm53 after melting – liquid phase.

Fig. 4. CrodaTherm53 after re-solidification – solid phase.

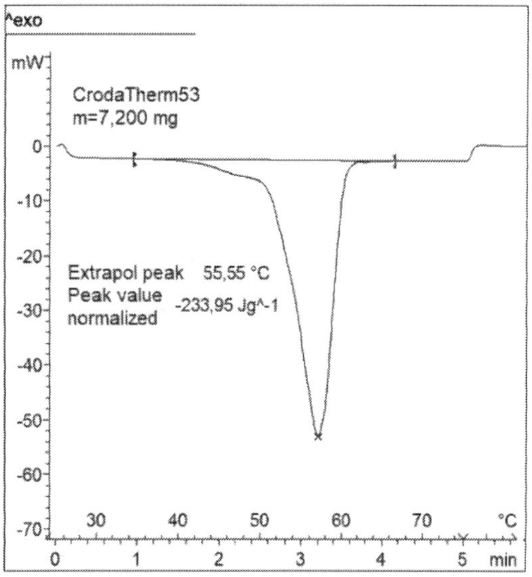

Fig. 5. DSC curve for CrodaTherm 53 – melting.

Figure 5 and 6 present results of the measurements made with the use of differential scanning calorimeter (DSC) for CrodaTherm 53: Fig. 5 – melting stage and Fig. 6: solidification stage. The DSC analysis shows that the transition temperature is around 55°C for melting and for congealing is a few degrees lower. (Data sheet: $T_{melting}$ = 53°C, $T_{Congealing}$= 51°C. The phase change heat approx. 230 kJ/kg (data sheets: 225 kJ/kg). Obtained results are similar to data indicated in the product sheet.

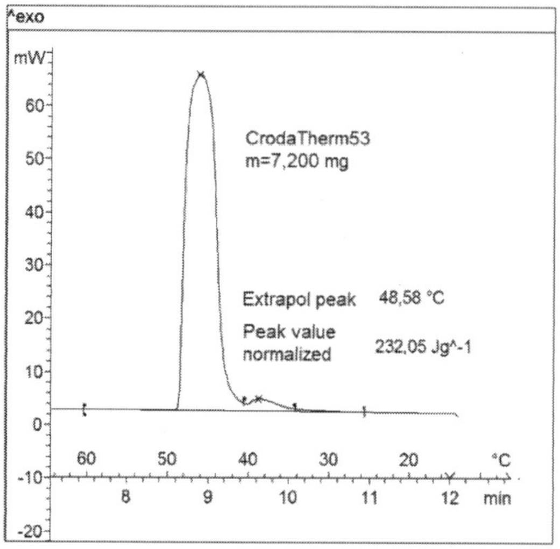

Fig. 6. DSC curve for CrodaTherm 53 – solidification.

4.1 Test rig description

Test rig consists of thermal energy storage tank filled with water with nominal capacity 200 l. Its general view has been presented in Fig. 7. The tank is made of steel with a thickness of 5 mm, protected against corrosion. It is possible to open the tank to place an insert filled with phase-changing material inside it. The tank is equipped with measuring connectors (PT 100, B class) which allow temperature measurement inside both - the magazine and the insert. Measuring connectors were distributed along the length and breadth of the tank and insert to allow observation of the temperature distribution in both the central and the external parts of the tank and insert. Figure 8 presents the insert - shell and tube heat exchanger. It is made of steel and has a system of 26 steel pipes with a diameter of 25 mm. For testing purposes 80% of its volume has been filled with PCM (60 l).

The test rig is powered by a heating system based on a 10 kW flow-through electric heater. The flow of heat transfer fluid (in this case – water) is forced by a circulation pump. Heating system is located in an electrical box attached to the tank. Measurement data is automatically downloaded and saved at 2-minute intervals using a data acquisition system.

Fig. 7. Thermal energy storage tank.

Fig. 8. PCM based insert for thermal energy storage tank.

5 Results and discussion

Figure 9 and 10 present comparison of temperature curve for a TES tank containing empty PCM cartridge and exactly the same tank, but containing 60 l of PCM: CrodaTherm53. Figure 9 presents the heating stage. It can be observed that heating process for tank filled with PCM runs much slower (about 6 times in the temperature range: 30-65°C). It is related to PCM phase transition which can be observed in the temperature range: 51-55°C.

Figure 10 presents cooling stage. Also in this case, the flattening of the temperature curve in the area of 51 - 55°C can be observed. The heat dissipation from the tank runs smoothly.

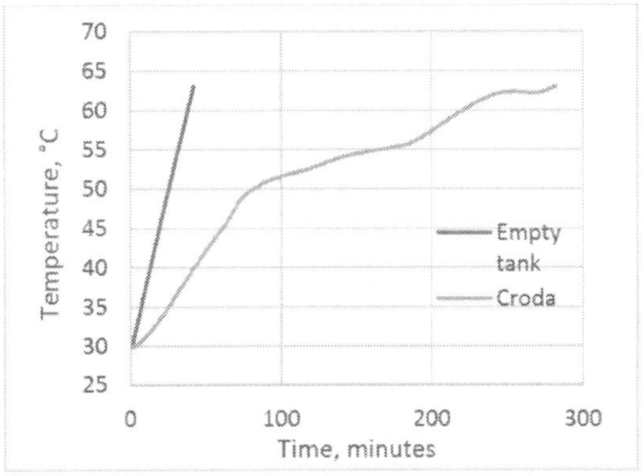

Fig. 9. Heating process for empty tank and tank with PCM – insert; Measuring connectors located in the middle of the tank and in the middle of the insert.

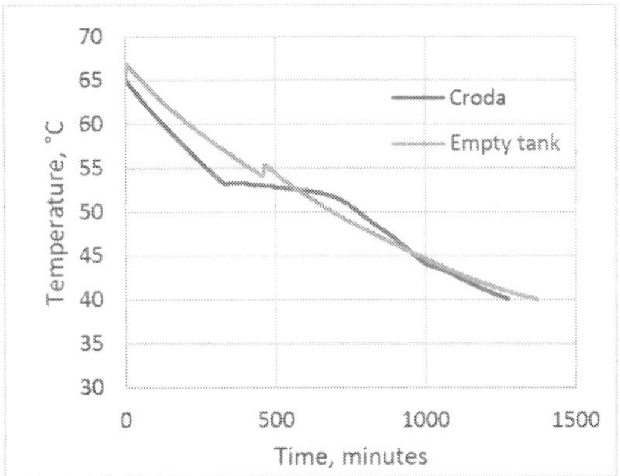

Fig. 10. Cooling process for empty tank and tank with PCM – insert; Measuring connectors located in the middle of the tank and in the middle of the insert.

Figure 11 and 12 present comparison of results for heating stage (Fig.11) and cooling stage (Fig.12) for the tank consist of PCM-insert. Temperatures in three parts of the insert are presented on the diagram: temperature next to the left wall, next to the right wall and in the middle of the insert. There is also presented the temperature of the water inside the tank, flowing around the PCM insert. It can be observed in the Fig. 11 that temperature of the water started decreasing after turning the heater off. When temperature was close to 50°C, the heater has been turned on. This process has been repeat one more time to enable observation of the efficiency of heat transfer from water to phase change material. It can be observed that transition occurs in a similar way in the entire insert. Curves for all analyzed points – the center and both walls run similarly and they follow the water curve. When the temperature of 55°C is exceeded, the PCM temperature starts to rise quickly and evenly with the water temperature – this is the information that phase transition occurred and PCM in the form of liquid is storing sensible heat.

Figure 12 presents cooling process for the points discussed above – left wall, right wall and the middle of the insert. It can be observed that PCM release heat efficiently. Water temperature curve runs close to PCM temperatures.

Results shows that heat is stored evenly throughout the PCM – insert.

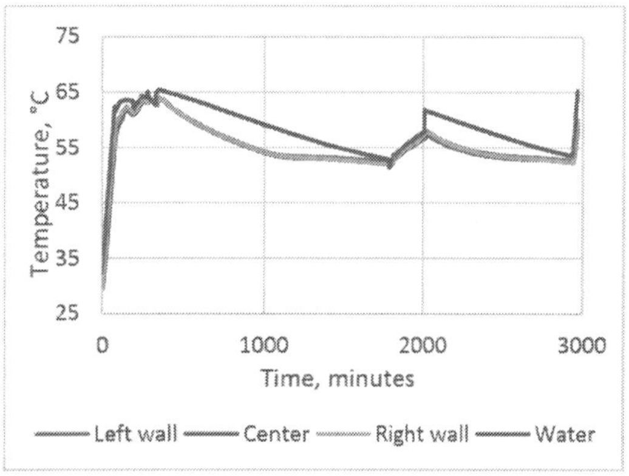

Fig. 11. Temperature distribution inside the tank – heating.

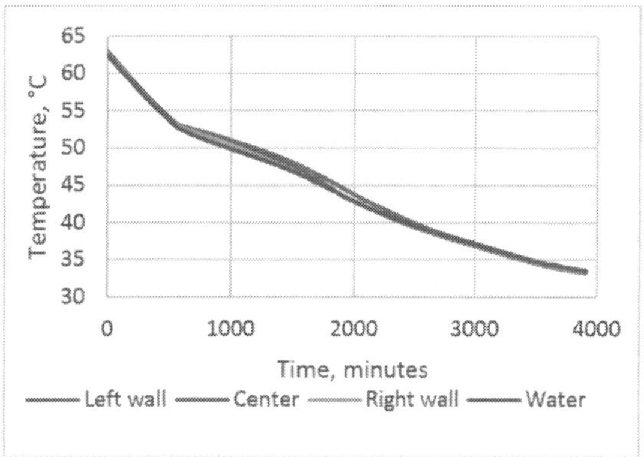

Fig. 12. Temperature distribution inside the tank – cooling.

6 Conclusions

This paper presents examples of PCM based heat exchangers applied for thermal energy storage tanks with its advantages and disadvantages. Analysis shows that heat exchanger has to be designed based on its application and used phase change materials. Many aspects need to be considered, for example: required temperature, used PCM (organic/inorganic), heat release time, volume of PCM. This paper presents also results from experimental study of shell and tube heat exchanger used for 200 - litres thermal energy storage tank. Results shows that:

- It is possible to use phase change materials to modify TES tank and achieve better results – store more heat in the same volume.
- Using shell and tube heat exchanger enables efficient supply of heat to the entire volume of PCM.
- Phase change material CrodaTherm 53 responds to the requirements of quick heat transfer when used with an appropriate heat exchanger.
- Test results for PCM do not differ significantly from the information provided by phase change materials manufacturer.
- In order to store large amounts of heat, the tested inset should be replaced with a larger one - fully filling the tank or heat exchanger tubes should be installed directly in the tank and tank should be filled with PCM.

The aim of the research has been successfully accomplished, but it was also observed that there is still a large area to be developed in the area of PCM based heat exchangers (related to both: construction of heat exchanger and properties of PCM).

Acknowledgements

Authors would like to express the gratitude to innogy Polska S.A. for the opportunity to perform research at the AGH INNOGY STORAGE LAB located at the AGH UST Center of Energy.

This work was supported by AGH – University of Science and Technology (Project 11.11.210.376).

References

1. A. Sharma, V.V Tyagi, C.R. Chen, D. Buddhi, Renewable and Sustainable Energy Reviews **13**, 318-345 (2009)
2. D. Fernandes, F. Pitié, G. Cáceres, J. Baeyens, Energy **39**, 246-257 (2012)
3. N.H.S. Tay, M. Belusko, F.Bruno, Applied Energy **90**, 288-297 (2012)
4. A. Gil, G. Peiró, E. Oró, L.F. Cabeza, Applied Thermal Engineering **142,** 736-744 (2018)
5. R. Ahmadi, M.J. Hosseini, A.A. Ranjbar, R. Bahrampoury, Sustainable Cities and Society **38**, 145-157 (2018)
6. M. Kabbara, D. Groulx, A. Joseph, International Journal of Thermal Sciences **130**, 395-405 (2018)
7. M.M. Prieto, B. Gonzalez, E. Granado, Energy and Buildings **122**, 89-97 (2016)
8. S. Guo, J. Zhao, W. Wang, G. Jin, X. Q.An. Wang, W. Gao, Applied Thermal Engineering 78, 556-564 (2015)
9. M.J. Kabbara, N.B. Abdallah, Procedia Computer Science **19,** 694-701 (2013)

E3S Web of Conferences **70**, 01011 (2018) https://doi.org/10.1051/e3sconf/20187001011
HTRSE-2018

Power to X – green hydrogen for electrical energy and fuel, for production and products

Jochen Lehmann[1,*], *Thomas* Luschtinetz[1], and *Johannes* Gulden[1]

[1]Stralsund University of Applied Science, Institute of Renewable Energy Systems, 18435 Stralsund, Germany

Abstract. Basing on the figure "Power to Hydrogen / Power to Gas", shown by the authors at the last HTRSE conference, this time it will be illustrated, that green hydrogen - produced with renewable energy - has the potential to become a basic material in the economy at general instead of fossil one. Synergies are available. But the low price of hydrogen produced via steam reformation of natural gas prevents to reach business cases for environment friendly products as long as the European laws and regulations do not support production and use of green hydrogen for instance by a tax for CO_2 emission.

1 Introduction

It seems that not only Germany will miss the goals of reduction of CO_2-emissions by 2025. Up to now the main ways to try it have been: The reduction of special production ways, saving heat, using more and more natural gas, renewable energy sources, biomass and biogas as well.

Much more possibilities will become opened by the use of green hydrogen according to Fig.1. Large effects could be activated besides of the energy sector. (All the green hydrogen issues one can find easily in the compendium edited by J. Töpler [1]).

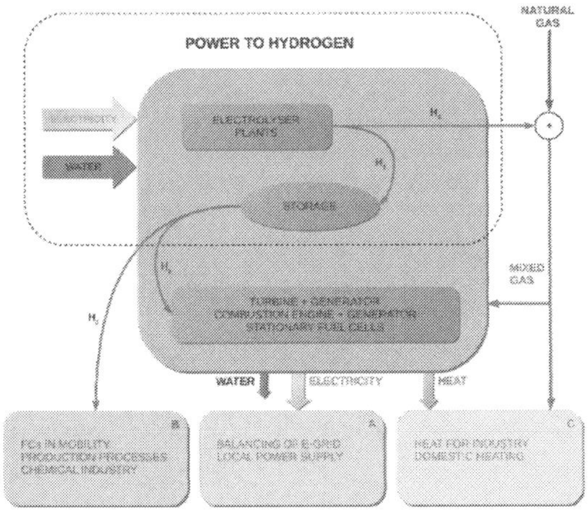

Fig. 1. Power to X on the base of electrolysis.

2 The Field of Electricity (A)

Mostly the harvest of renewable energies happens as electricity: Photovoltaic and wind (in growing parts), water power (not with a volatile characteristic, the enlargement of its part is complicated), high temperature thermo solar in hot countries, wave energy, and some others. In case of feed in of changing current into the grid, we know, more than 30% of it will make the voltage and the frequency inconstant. We have to equalize the changing, and we have to sorrow for backup power in case of deficits (night, calm, etc.). For that purpose, back-up power stations are used on the base of fossils up to now.

To reduce the CO_2, we should do it in a clean way in the future. And in very large scales (may be seasonal equalisation). That will be possible via power to hydrogen only [2]. On the other hand, that method is an ideal one for building stand-alone systems and for a clean emergency supply. To improve the efficiency, the process heat should be used.

All the components are available or become developed such as low and high temperature fuel cells, hydrogen burning intern combustion engines and gas turbines.

3 Traffic and Production (B)

The need of seasonal hydrogen storage means, that there will become constructed hydrogen factories (Fig.2).

These have to electrolyse and to store hydrogen, and to act like back-up power stations delivering electricity into the grid in case of a lack between demand and supply.

For their own stability the factories could and should sell a second product - hydrogen as a material.
Besides of the Stralsund installation as a model there are two "H_2-factories" in the north-east of Germany: Prenzlau (500 kW electrolysis) since 2011 and Grapzow (1 MW electrolysis) opened 2013. Three years later started the plant near Mainz with three 1.5-MW-PEM-electrolysers new developed by Siemens [4]. All they are not able to operate rentable, according to the current law,

* Corresponding author: jochen.lehmann@hochschule-stralsund.de

https://doi.org/10.1051/e3sconf/20187001011

that both the electricity from the windmills and the delivering of reconverted current as backup power have to be taxed.

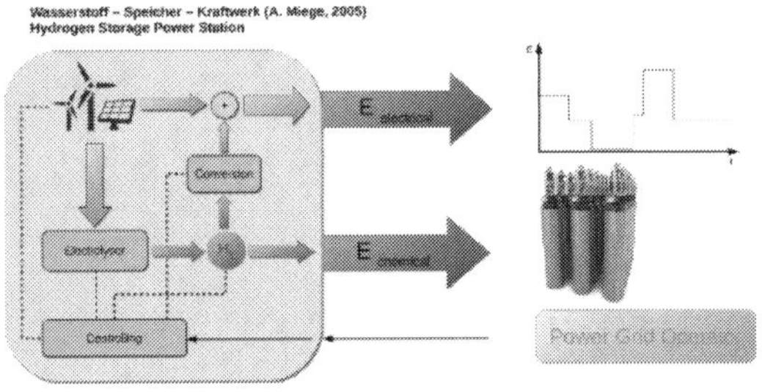

Fig. 2. Hydrogen Storage Power Station – "H_2-Factory" [3].

A similar problem the green hydrogen finds in case of selling it as a fuel. But some consumers are asking for the hydrogen fuel already:

In the year 2013 Hyundai started the series production of its fuel-cell-VAN, Toyota and Honda followed. FC-busses one can buy in China. The competing enterprises hesitate with a larger number of such vehicles.

The first FC-train started in the German public railway service this year. Some more will follow definitely. And the use of fuel cells for powering trucks and heavy street engines seems to come. Established on the market are FC powered fork lifters produced by Linde and Toyota.

One of the general problems of hydrogen as a fuel in the mobility is the small density of hydrogen connected with a small volumetric energy density. Therefore, the laboratories developed liquids contending much hydrogen as partner in the compounds under the topic "Power to Liquid".

On board of the consumer one need a reformer, setting free hydrogen for fuelling. The discharged material has to be transported back to the producer for forming new the LOHC-material. The advantage of the LOHC's is the best hydrogen density next of the metal hydrides. [5]

All together: That is a good development, but a quite slow one.

Therefore, the working group "performing energy" (a part of the German Hydrogen and Fuel Cell Association [6]) is consequently engaged to find more possibilities for the use of green hydrogen. For instance, the production of Diesel with green hydrogen instead of fossil one would be a case for support by the Brussels rules. On the other hand: Such a large demand of green hydrogen would help to develop a powerful electrolyser industry as the base for the future hydrogen economy.

A second idea is directed to the same goal: Within the steel production the ore has to become reduced by burning coke or natural gas in a newer procedure. Also this step could be realised by hydrogen [7].

Hydrogen is a partner in lots of reactions and compounds. Making all these "greener" means that all the prices will raise. The consumers should be open to pay it, finally a CO_2-tax they will have to pay too.

4 Heat for Industry and Houses (C)

Hydrogen coming out of the electrolyser fed in directly into a natural gas pipe - that is the shortest way to the heat market. According to actual regulations less than 3 % are allowed. That should not bring any problems for the heating systems; the burning value is a little bit lower. For intern combustion engines about 30 % hydrogen in the mixture should be acceptable, the power of the engines becomes reduced. A problem could be to hold a constant relation from natural gas and hydrogen in a large pipe system.

Powering heat turbines with mixed gas is thinkable for a period, but in the end turbines should be proof for 100% hydrogen.

Of course: It is possible to add a methanisation step and put the product into the natural gas pipe. The disadvantage would be the reduction of the total efficiency.

Producing green heat one should use the thermo solar technique, the burning of biomass, and the geothermal energy, of course. Under special conditions catalytic hydrogen burners for cooking and heating are good options [8]. In that case the fuel is pure hydrogen.

At general it is to notice, that the greenest heat is saving heat. Better constructing and more isolating will bring the best effects for heating and cooling. But that is a very large task.

By the way: One of the basic ideas of the world first wind-hydrogen-project by Olof Tegström (1985/6) has been to produce hot water for heating the single house, the second one the electrolysis and hydrogen storing for cooking and powering the car (reported in [9]). But the general proposal to use "surplus" wind power for heating - "Power to Heat" - seems not to be the best, storing it as hydrogen opens better possibilities.

5 Summary

"Power to X" is a very complex connection of procedures basing all together on the conversion of renewable energies into green hydrogen by electrolysis. This procedure is good for the smallest up to the largest scale.

The gas hydrogen is storable in four different ways: Pressured or liquefied, insides of metal hydride alloys or as a partner of chemical compounds. These characteristics and the fact, that hydrogen and water reactions are building a circuit as the main feature of the sustainability, are making the gas to a material of universal interest for the future economy.

It should be able to better the world, making lower CO2-emissions, creating jobs and national wealth, reducing the demand of natural oil and gas, and de-escalate the political situation.

References

1. Johannes Töpler et. al., Hydrogen and Fuel Cell, ISBN 978-3-662-44971-42007, Springer, 2016

https://doi.org/10.1051/e3sconf/20187001011

2. F. Crotogino, R. Hamelmann, „Wasserstoff-speicherung zur Glättung des Windangebots", XIV. Energie-Symposium, Stralsund, Proc. 11-17 (Quoted and discussed in: HTRSE-2014, Proc. 309), 2007
3. J. Lehmann, O. Luschtinetz, A. Miege, H. Sandlaß, C. Sponholz, Vergleichsmäßigungsstudie im Auftrag Energieministerium MV, Hochschule Stralsund, 2005
4. M. Kopp u.a., "Energiepark Mainz - Betriebserfahrungen", XXIII. Energie-Symposium, Stralsund, Proc. 80-84, 2016
5. J. Gulden, et. al., "New means of hydrogen storage", to be held at the HTRSE 2018
6. Performing energy, www.dwv-info.de, 2018
7. M. Hölling et. al., "Stahlerzeugung mit grünem Wasserstoff", HZwei 03/2018, 34/35, 2018
8. Ulf-Peter Großmann u.a., "A non-stationary hydrogen cooker with MH storage", Int.J.of Hydrogen Energy 25-1, 87-90, 2000
9. L. Voss, et. al., "Erinnerung", XXIV. Energie-Symposium, Stralsund, Proc. 229-234, 2017

All the proceedings of the Stralsund conferences you can find:
https://www.hochschule-stralsund.de/forschung-und-transfer/institute/institut-fuer-regenerative-energiesysteme/

Multi-objective optimization of ORC geothermal conversion system integrated with life cycle assessment

Dominika Matuszewska[1,*], *Marta* Kuta[1], and *Jan* Górski[1]

[1]AGH University of Science and Technology, Faculty of Energy and Fuels, Department of Thermal and Fluid Flow Machines, al. A. Mickiewicza 30, 30-059 Krakow, Poland

Abstract. This paper details the development of a systematic methodology to integrated life cycle assessment (LCA) with thermo-economic models and to thereby identify the optimal exploitation schemes of geothermal resources. Overall geothermal systems consist of a superstructure of geothermal exploitable resources, a superstructure of conversion technology and multiple demand profiles for Swiss city. In this paper, an enhanced geothermal system has been chosen as exploitable resources. The energy conversion technology used in modelling is an organic Rankine cycle, which can be used to supply heat and electricity. In the Swiss case four demand profiles periods are considered: summer, interseason, winter and extreme winter, the city Nyon serving for the example case study. The multi-objective optimization system, that uses an evolutionary algorithm, is employed to determine the optimal scheme for some of the prepared models, with exergy efficiency and environmental impact as objectives.

1 Introduction

The use of geothermal resources for supplying simultaneously different energy services such as electricity, district heating and district cooling has recently gained interest. The optimal design of such systems requires knowledge of the geothermal resources that will be used, the selection of appropriate energy conversion technologies, and a specification of component operating conditions. This can be achieved by using process integration techniques and multi-objective optimization in a multi-period time perspective, with due regard to economic, thermodynamic and environmental criteria. While economic and thermodynamic analysis methods have been widely applied in this framework, the dynamic evaluation of environmental impact in a process design context is still relatively new. The environmental evaluation of geothermal systems should be performed from a life cycle perspective since the impact from drilling and construction is likely to be significant. Therefore, a life cycle assessment is the appropriate method to quantitatively evaluate and compare the different system configurations, including resources, conversion technologies and the services to be supplied. However, to serve as an effective design and evaluation tool, for the optimal configuration of geothermal systems, a life cycle inventory has to be fully integrated in the process design framework. It also has to be operated in a multi-period time perspective, to account for the variation in energy supply levels throughout the year.

1.1 Literature review

Energy conversion systems have been widely described in literature, starting from the description of individual technologies for the utilization of geothermal heat [1,2], through working fluid selection in ORC systems [3], up to detail problems, which occur during exploitation [4,5].

In case of life cycle assessment (LCA), Keoleian paper [6] in general has presented the use of LCA in product design context. Unfortunately, that work does not include the methodology to conduct LCA in a computer-aided optimization. The trade-offs between environmental and economic performance indicators in multi-objective optimization have been presented in work of Kniel, Delmarco, and Petrie [7], Azapagic and Clift [8], as well as Alexander, Barton, Petrie, and Romagnoli [9]. The biggest problem with these works is that they do not take under consideration multiperiod approach in the energy system design. Papandreou and Shang [10] present work about conducting LCA with multi-objective optimization in the field of energy system. The work is focused only on the energy based on fossil fuel sources and thus takes into account only gaseous emissions. The use of LCA in a multi-objective optimization with consideration of energy integration techniques is presented by Li, Maréchal, Bure, and Favrat [11], Bernier, Maréchal, and Samson [12].The work presents the relationship between the levelized cost of electricity and the life cycle global warming potential as performance indicators. The trade-off between the minimizing the environmental impact and increasing the

* Corresponding author: dommat@agh.edu.pl

efficiency has been described there as well as the usage of LCA in CO_2 tax evaluation.

A systematic methodology for identification the optimal exploitation scheme of geothermal conversion systems has been presented in studies [13]. Thermo-economic modeling was used to connect the exploitable geothermal resources with conversion technologies to provide multiple energy services. However, the LCA methodology was not considered in this evaluation. Report [14] provide the data for life cycle inventory for models of the geothermal exploitable resources and conversions technologies using Ecoinvent database. Although there were several studies on the LCA use in a process design context the first time the methodology for integration the LCA into thermo-economic models have been presented on the basis of biofuels [15].

2 Research methodology

Research methodology consisted of several steps. The first step included the integration of LCI data - specified in the Ruiz Dern study [14] - into the available geothermal resources, technologies and demand profile models. The integration was based on specific LCA methodology for process design integration, as has been presented below. The next step involved defining input data for various scenarios: geological data (geological profile with associated potential geothermal resources) for one or more given locations (for example in Switzerland), and yearly demand profiles for energy services. The next step involved defining the functional unit and systems limits for the life cycle assessment evaluation. The life cycle inventories have been integrated with realistic data to define a base case scenario. The base case scenario has been used to analyze the impact of the different life cycle stages and periods for the different geothermal system configurations, and identify the main impact contributions. In the next step a multi-objective optimization have been carried out to identify the trade-offs and so arrived at the optimal configuration for the geothermal system. This would then be followed by an analysis of the influence of the life cycle assessment data on the general optimal configuration. Finally, the configuration has been compared to the base case scenario.

2.1 Specific LCA methodology

The thermo-economic design approach was based on a computational platform, which connected different models that were representative for the energy system design. To model interactions between each operational unit, involved the use of energy flow models of the process unit operations as well as integration techniques. The thermo-economic model for geothermal energy conversion systems, as well as the economic and environmental layers of the LCA model have also need to be added. The model with the LCA is presented in the figure below (fig. 1).

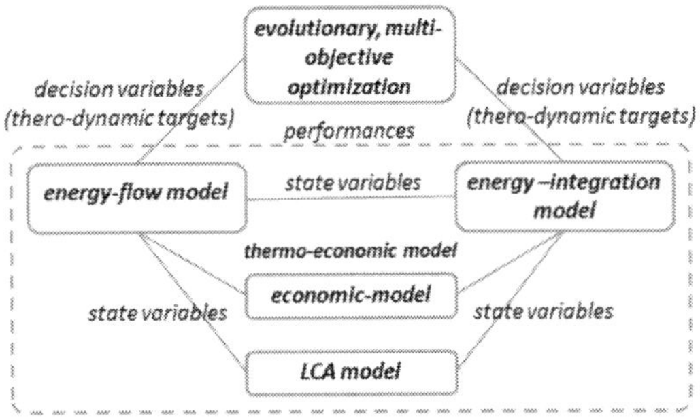

Fig. 1. Architecture of the process model [15].

The simulation results, such as mass and energy flows, were taken to create an energy integration model. The post-calculation phase was based on thermodynamic states and flow rates of the energy flow and energy integration models. A post-calculation phase has been used to size the equipment, estimate the cost and evaluate the performances of the process configuration. The Life Cycle Inventory (LCI) has been performed based on reference data in the EcoSpold format from the ecoinvent® life cycle inventories database and the simulation results [16,17]

Figure 2 presents previously developed methodology for the integration of LCA within a thermo-economic model.

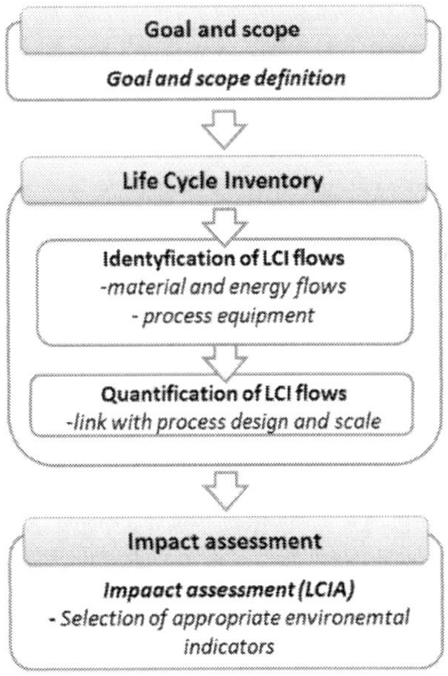

Fig. 2. Developed general methodology for LCA model conception [15].

Note that the LCA connection with the process design has been singled out in the figure. The crucial matter of listing LCI flows was to identify to which process units the flows were linked and their function. This identification was essential because flows were mathematically expressed as a function of the decision variable of thermo-economic model. The scaling of

https://doi.org/10.1051/e3sconf/20187001012

impacts due to changes in operating conditions and sizes of the process equipment also need to be considered. The mathematical expressions for LCI flows and impacts based on process equipment are included in the LCA, which makes it possible to calculate the whole LCI for a given process configuration. The Ecoinvent® life cycle inventories database has been used in the life cycle inventories process and impact assessment.

This paper put emphasis on preparing the proper methodology for integration the LCA with thermo-economic model. However, from the LCA side the main goal of this work is to identify the most environmental friendly process configuration. The environmental impact is treated as one of the performance indicators, which is minimized during optimization.

The functional unit of the system is 1kWh available from the exploitable geothermal resources. The 'cradle-to-grave' approach determined the boundaries of the system. During the evaluation, three phases (construction, operation and dismantling) of lifetime of geothermal power plant was considered.

Ecological Scarcity method has been used in evaluation, due to the fact of using 'distance to target' principle in evaluation of life cycle inventories. It used eco-factors, which are expressed as eco-point per unit of pollutant emissions or resources extraction. The eco-factors has been determined by both current emission situation and the political target of the Swiss environmental policy [18] In this method the varies tapes of eco-factor are aggregated t a present pollution level, and compared with level defined by Swiss environmental policy. This is the main reason why the ecological scarcity method should be used only for Swiss context problem [19].

3 Physical model development

The possible combinations of three sub-systems that were modelled separately are considered. The overall geothermal conversion systems consist of the superstructure of potential exploitable geothermal resources, the superstructure of conversion technologies, the superstructure of conversion technologies and the multi-period demand profiles that define the demand for services.
The possible combination of the exploitable geothermal resources, conversion technologies, demand profiles and life cycle phases has been presented in fig. 3.

3.1.1 Geothermal resource model

The superstructure of geothermal resources is divided into three types: Enhanced Geothermal System (EGS), deep aquifer and shallow aquifer. During the modelling of geothermal resources the emphasis was put on the information needed to evaluate these kinds of resources, such as geological data (thermal gradient, water mass flow rate, depth, etc.), geotechnical data (exploitation mode, drilling techniques, number of wells). Thermal gradient changes for each resource, were assumed to be linear. For residential areas of Nyon in Switzerland, it was set at 3.8°C for each 100m of drilling depth.

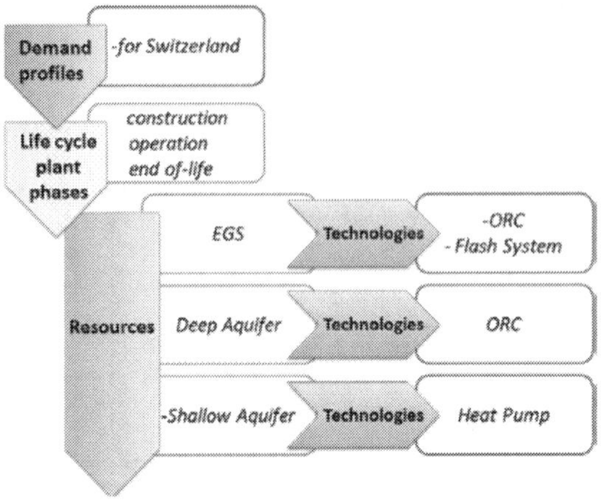

Fig. 3. The scheme of geothermal system combinations.

3.1.1 Geothermal resource model

The superstructure of geothermal resources is divided into three types: Enhanced Geothermal System (EGS), deep aquifer and shallow aquifer. During the modelling of geothermal resources the emphasis was put on the information needed to evaluate these kinds of resources, such as geological data (thermal gradient, water mass flow rate, depth, etc.), geotechnical data (exploitation mode, drilling techniques, number of wells). Thermal gradient changes for each resource, were assumed to be linear. For residential areas of Nyon in Switzerland, it was set at 3.8 °C for each 100m of drilling depth.

3.1.2 Conversion technologies

There are a number of ways of converting geothermal heat into useful energy. The flash systems, organic Rankine cycles and heat pumps are taken under consideration. The ways these technologies have been used depends mainly on geothermal conditions as well as the demand for heating and, or cooling in a particular region. A conversion technology superstructure is presented in fig. 4 below.

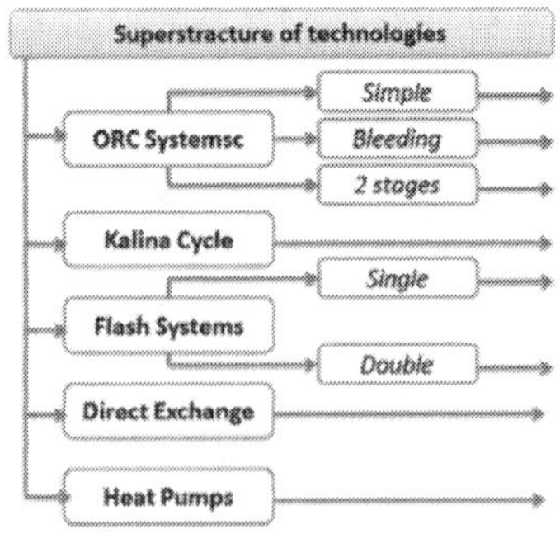

Fig. 4. Geothermal Conversion Technologies.

3.1.3 Energy services demand

This study focuses on energy services demand under Swiss conditions (in the city of Nyon). In the Swiss case, four periods are covered. It is assumed that district cooling is not included in the evaluations. Electricity is also not included in the demands profiles – it is assumed that any electricity generated will be sold to the grid.

The Swiss demand profiles were based on data for the city of Nyon, Switzerland (Methodology for the Urban Energy Conversion Concept – An Example at Nyon, n.d.), and divided into four periods: summer, interseason, winter and extreme winter. The supply and return temperatures and load for each period are show in table 1 below.

Table. 1. The Swiss demand profiles.

Demand profile	Load [kW]	Supply temperature [°C]	Return temperature [°C]	Operating time [h]
Summer	200	60	25	525
	300	25	21	
Interseason	150	60	32	
	500	32	26	3942
	300	28	20	
Winter	100	60	50	
	1200	50	40	
	100	40	34	4205
	700	34	28	
	100	28	10	
Extreme winter	100	75	65	
	1750	65	50	
	150	50	40	88
	1000	40	32	
	100	32	10	

4 System resolution

The one-run system resolutions for default values for each modelled system were launched to check for any technical problems with the models and model conversation. Further, this paper will focus only on date from the multi-objective optimization which was run for enhanced geothermal system with organic Rankine cycle for seven different fluids: benzene (BZ), toluene, iso-butane (IC4), n-butane (N4), cycle-butane (CYC4), iso-pentane (IC5), n-pentane (N5). The environmental impact was calculated from single points chosen from sets of optimized system solutions.

4.1 Objective function

The models can be optimized by taking into account thermodynamic, economic and environmental performance indicators. In this study, only thermodynamic and environmental criteria are considered. The exergy efficiency and environmental impact are selected as objective functions.

The exergy efficiency was chosen because it considers the environment during evaluation. Exergy efficiency is defined as the ratio of the output exergy to the input exergy:

$$\eta = \frac{\sum_{p=1}^{n_p} \dot{E}_p^- t_p + \sum_{p=1}^{n_p} \dot{E}x_p^- t_p}{\sum_{p=1}^{n_p} \dot{E}x^+ t_p} \tag{1}$$

where: $\dot{E}^-$ - the electricity produced by the overall system during period p, t_p – the duration of the period p, $\dot{E}x^+$ - the exergy transfer to the district heating during period p.

The specific impact of district heating per kWh generated was chosen as the environmental indicators. It is calculated as the total life cycle impact for the overall system divided by the amount of district heating supplied during the lifetime of the system [20]:

$$I_Q = \frac{\sum_{p=1}^{n_p} \sum_{i=1}^{n_{eo}} I_{O_{i,p}} + \sum_{i=1}^{n_{ec}} \max(I_{C_i})_p + \sum_{i=1}^{n_{ee}} max(I_{E_i})_p}{\sum_{p=1}^{n_p} \dot{Q}_p^- t_p \, t_{life}}$$

where: I_O - is the impact due to the operation phase for period p of the LCI element i, n_{eo} – the total number of LCI elements from the operation phases, I_C – is the impact of the construction phase of the element I for period p, n_{ec} –is the total number of LCI elements from the construction phases, I_E – is the impact due to the end-of-life phase of the element I for period p, n_{ee} – is the total number of LCI elements from the end-of-life phases and t_{life} – is the overall lifetime of the geothermal system (in this study it was assumed to be 40 years).

5 Validation and optimization

The validated models were optimized using a genetic algorithm. The multi-objective optimizations were run to show the trade-off between the thermodynamic and environmental criteria. The decision variables chosen for organic Rankine cycles were the evaporation temperature in the 80 - 120°C range, the superheating temperature in the 120-230°C range, the depth in the 3000-6000m range and the reinjection temperature in the 70-110°C range. The multi-objective optimization of enhanced geothermal system with organic Rankine simple cycle has been launched for seven different working fluids and evaluated with use of Ecological Scarcity impact assessment method.

To determine the optimal configuration of geothermal conversion system, two objective function were considered: exergy efficiency and impact on environment. The optimal solution for analyzed working fluids has been presented as Pareto curve in fig. 5.

The optimization gave two sets of solution for substance like iso-butane, n-butane and cyclo-butane. The first set has been composed by solution with low exergy efficiency, while the second one by the higher one. It can be seen, that the solution with lower exergy efficiency

 https://doi.org/10.1051/e3sconf/20187001012

have lower impact on environmental than those with higher exergy efficiency. However, both sets have harmful influence on environment (all obtained results for those substances are above x-axis). The impact on environment has been less harmful in case of low exergy efficiency set of solutions, due to the fact that in those case there is no need for deep drilling (blue points in figure 5). The increase of drilling depth results in growth

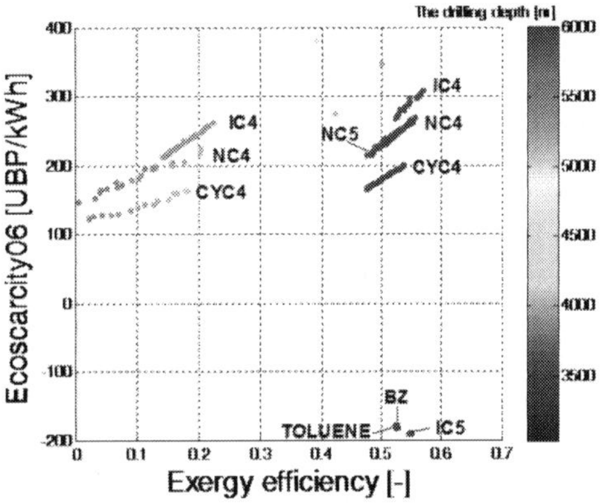

Fig. 5. Pareto curve foe Ecoscarcity06 and exergy efficiency with varying the drilling depth for EGS ORC simple system with different working fluids.

of emissions and extraction (among others from steel for casing, diesel, starch, bentonite used for drilling mud, etc.). In addition, the drilling depth has got an effect on costs of investment – they grow as the depth of a drilling well increases. Considering that the drilling costs in this type of facilities are one of the most significant, further work should take this into account.

The sets of solution depend strongly not only on drilling depth but also with another decision variable – which strongly correlates with depth – a superheating temperature. Deeper drilling is aimed at obtaining higher temperature of geothermal brine at the outlet from a production well, and thus later – superheating temperature (and ultimately exergy efficiency). The growth of superheating temperature in ORC system is correlated with growth of exergy efficiency. However, low heat exchange coefficient for used working fluids can lead to expensive and large heat exchanger, which should be included in future development. Only three working fluids from seven analyzed showed beneficial impact on environment and they are: benzene, toluene and iso-pentane. In addition, these working fluid obtained exergy efficiency from top of range and can be recommended in analyzed case.

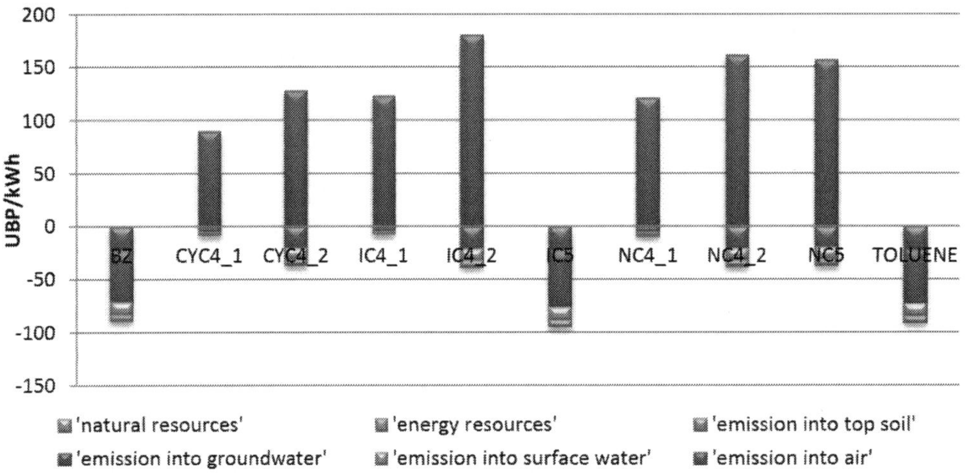

Fig. 6. Environmental impact based on Ecoscarcity06 calculated by LCIA categories.

The multi-objective optimization gave a set of points with optimal system configurations for different working fluids. The single points for each LCIA method have been selected from this set. The parameters of these points were chosen as constant to launch one-run system resolutions. In based of that the analysis of particular systems configuration on environment were provided. The results are presented to show that the given problem can be presented in many ways. The one-rum system resolutions for selected optimal results has been presented in fig. 6.

From optimal results for ecological scarcity method, different set of points was chosen to show the impact of each category. The method shows that the most significant contribution into impact have emissions into air, deposited waste and emissions into surface water. It is connected with the substance used for reservoir enhancement as well as working fluid losses. For butane substances, performances for two points were chosen to one-run system resolution what is presented in fig. 6. The low efficiency point is marked with index 1, while the high efficiency - with index 2.

6 Conclusion

In this work, the methodology for integration the life cycle assessment (LCA) in thermo-economic models has been presented. The identification of the optimal exploitation scheme geothermal resources is presented. The emphasis of the work is put on environmental criteria. The overall geothermal conversion system was modeled as three separate sub-systems: the exploitable geothermal resources, conversion technologies and demand profiles of useful services. The methodology of integration LCA into thermo-economic model was applied to the real case for Swiss city Nyon. For this purpose, the superstructure of organic Rankine cycle and enhanced geothermal system was used. The main aim of optimization was determination the most suitable working fluid in the light of objective functions. Seven different working fluids were considered: benzene, cyclo-butane, iso-butane, iso-pentane, n-butane, n-pentane and toluene.

The environmental impact of the system was evaluated. based on LCA Ecological Scarcity methodology. The influence on results had a couple of factors e.g. approach used in each method, the way of normalized and weighting the impact categories, the choice of categories, etc.

The results obtained by ecological scarcity have harmful influence on environment. Iso-butane, n-butane and cycl-butane belongs to this group of working fluid, which negatively effect on environment. The beneficial impacts have toluene, benzene and iso-pentane. The ecological scarcity method was designed in based on Swiss environmental policy. In fact this method, is the most suitable one to analysis the problems in Swiss context.

Analyzing the obtained results the quantification of the effect on environment of a particular technology can be evaluated using different LCA methods. To create the process model, it is possible to use process integration techniques (e.g. pinch point analysis, costing or life-cycle analysis). This action helps to define the optimal configuration of the system. The choice of optimal performances depends on decision-maker and a given problem. The optimization returns a set of optimal solutions, which can be used optionally.

This work focuses on optimization with thermodynamic and environmental performances indicators. However, it is important to remember that nowadays costs are crucial matter in making decision, so the optimization with economic performance indicator should be done. It would be desirable to show if there is any trade-off between beneficial and environmental friendly projects.

This work was supported by AGH – University of Science and Technology (Project 11.11.210.376).

References

1. R. DiPippo, Geothermal Power Plants, 4th Edition (Butterworth-Heinemann, **2015**)
2. W. Nowak, A. Borsukiewicz-Gozdur, Transactions of the Institute of Fluid-Flow Machinery **114**, 57-65 (2003)
3. D. Mikielewicz, J. Mikielewicz, Applied Thermal Engineering **30(16)**, 2357-2362 (2010)
4. M. Lukawski, J. W. Tester, R. DiPippo, Sustainable Energy & Fuels **5** (2017)
5. D. Matuszewska, K. Sztekler, J. Gorski, MATEC Web of Conferences **18**, 03005 (2014)
6. G. Keoleian, Journal of Cleaner Production **1**, 143-149 (2003)
7. G. Kniel, K. Delmarco, J. Petrie, Environmental Progress **15**, 221-228 (1996)
8. A. Azapagic, R. Clift, Computer & Chemical Engineering **23**, 1509-1526 (1999)
9. B. Alexander, G. Barton, J. Petrie, J. Romagnoli, Computer & Chemical Engineering **24**, 1195 – 1200 (2000)
10. V. Papandreou, Z. Shang, Computer & Chemical Engineering **32**, 1589-1602 (2008)
11. H. Li, F. Maréchal, M. Burer, D. Favrat, D, Energy **31**, 3117-3134 (2006)
12. G. Bernier, F. Maréchal., R. Samson, Energy **35**, 1121-1128 (2010)
13. M. Hoban, L. Gerber, F. Maréchal, *Integrate Thermo-economic Modeling of Geothermal Resources for Optimal Exploitation Scheme Identification*. http:/www.ecos2010.ch (2010)
14. A. M. Ruiz Dern, *Life Cycle Assessment of Geothermal Energy Conversion Systems*. (Master's thesis. Industrial Energy System Laboratory of the Swiss Technology Institute of Lausanne. Lausanne, Switzerland, 2010)
15. L. Gerber, M. Gassner, F. Maréchal, *Integration of LCA in a thermo-economic model for multi-objective process optimization of SNG production from woody biomass*. 19th European Symposium on Computer Aided Process Engineering – ESCAPE19 (2009)
16. R. Frischknecht, N. Jungbluth, *Overview and Methodology, ecoinvent report No 1. Tech. rep. ecoinvent center* (2007)
17. R. Frischknecht, N. Jungbluth, H-J. Althaus, G. Doka, R. Dones, T. Heck, S. Hellweg, R. Hischier, T. Nemecek, G. Rebitzer, M. Spielmann, International Journal of Life Cycle Assessment **10**, 3-9 (2005)
18. R. Frischknecht, R. Steiner, N. Jungbluth, *The Ecological Scarcity Method - Eco-Factors 2006*, (Federal Office for the Environment FOEN. Switzerland, 2009)
19. R. Bolliger, *EnergyTechnologies Dacumentation – Version 2.1 Revision 2113* (Industrial Energy Systems Laboratory, Ecole Polytechnique Fédérale de Lausanne. Switzerland, 2010)
20. G. Leda, F. Maréchal, Applied Thermal Engineering **43**, 29-41 (2012)

https://doi.org/10.1051/e3sconf/20187001013

Improving the energy efficiency of lighting systems by the use of solar radiation

Sławomir Sowa[1,*]

[1]Institute of Heat Engineering, Faculty of Power and Aeronautical Warsaw University of Technology, Poland

Abstract. The paper presents two ways of using the solar radiation to limit consumption of the electric power for lighting, i.e., increasing the energy efficiency of lighting systems. The solar radiation energy is a generally available type of renewable energy that to a large extent may influence a reduction in consumption of the electric power for lighting. Facilities designers and architects aim at ensuring the highest possible availability of natural lighting in rooms. Usually, lighting systems are not provided with an appropriate control system using the solar radiation. A lack of means for a correct control of the artificial lighting intensity is a factor hindering an improvement of the energy efficiency of the lighting systems. Implementation of the correct control system is possible when distribution of the lighting intensity in a room is known. The studies conducted in this area are a valuable source of information which can be used to develop control algorithms or determine optimum locations and parameters for installation of lighting fixtures. The second way for improving the energy efficiency of the lighting systems is a use of the solar radiation energy converted to the electric power to supply lighting systems.

1 Solar radiation potential

The solar radiation reaching the Earth surface is composed mainly from the visible light and near-infrared wavelengths. In residential and public utility buildings light plays an important role, significantly influencing a comfort of room use, occupants mood, and quality and safety of performed activities. The magnitude of light intensity reaching a room is also of importance for a type and power of installed lighting fixtures. When an appropriate intensity of natural light is ensured in a given room, then, using the solar radiation we can reduce the electric power consumption by switching off or dimming artificial lighting. The amount of the solar radiation reaching a room depends on many factors. A basic obvious factor is a geographical location of a facility, its orientation, and direct environment. Another aspect resulting from laws of physics is a season and a time of a day. Weather conditions also have a significant influence on intensity of light reaching a room. Other conditions influencing the amount of the solar radiation reaching the room are associated with building architecture.

Relevant factors for delivery of the solar radiation are specified in the standard PrEN 15193 [1]. This standard specifies factors which can be used as a basis for classification of rooms with small, medium and large availability of light. The solar radiation is not just the electromagnetic radiation within the visible spectrum, which we use to light rooms. The solar radiation is also power that can be converted in photovoltaic cells into the electric power. In that case, possibilities for using the solar radiation energy are greater because photovoltaic systems are installed in optimal locations enabling maximum use of the available solar energy. This effective use of the solar radiation for lighting rooms and producing the electric power can significantly reduce consumption of the electric power used for lighting. The use of the solar radiation potential to the highest extent possible will depend on an appropriate selection of a system for control of the lighting system and installation of an efficient photovoltaic system. Ways for using the solar radiation on a basis of results of tests and measurements proving that effectively used solar radiation resources can significantly influence an improvement in the energy efficiency of lighting systems are presented in further sections of this paper. It should also be remembered that the solar radiation exceeding lighting needs of rooms is a negative phenomenon. In these cases we must limit the solar radiation access, preventing negative phenomena, i.e., for example, glare or excessive heating of a room. A reduction in an excessive solar radiation reaching a room can be controlled, for example, by using shutters or blinds.

2 Tests and measurements location

Tests of the lighting intensity distribution and measurements of the generated electric power were conducted at an educational facility, the School Complex No. 1 in Swarzędz near Poznań. In a period of 18 months, measurements of the lighting intensity distribution were conducted and the energy generated in the photovoltaic system was recorded. The

* Corresponding author: Slawomir@sowa.poznan.pl

https://doi.org/10.1051/e3sconf/20187001013

measurements were conducted in lecturing rooms facing different, opposite directions.

2.1 Studied facilities

2.1.1 Building location

The school building is a detached facility located near Poznań, of the following coordinates: latitude of 52.3975 N and longitude of 17.0641 E. No external obstacles in form of buildings or landscape features that could limit the flow of the solar radiation to the studied rooms were present at the locations where lighting intensity was measured. This advantageous location ensures reliable measurements independent of external conditions, e.g., trees.

2.1.2 Facilities architecture

The school facilities where the studies were conducted consist of three two-storey wings forming a letter U. The two twin wings contain lecture rooms and technical laboratories. The rooms have access to the solar radiation from S and N. The central halls on the first and the second floor do not have access to the daylight. The third wing, joining the twin wings of similar size and layout of rooms, contains a traffic route, an auditory and technical rooms.

All lecture rooms have a layout typical for education facilities, with the external wall consisting of window joinery in ca. 80%. A view of an architecture of the whole School Complex No. 1 in Swarzędz and its orientation in terms of the Sun movement are presented in Figure 1. Currently, a construction of a new educational part that will form an integral part of the whole facilities is being completed. It is outlined in Figure 1 with a dashed line.

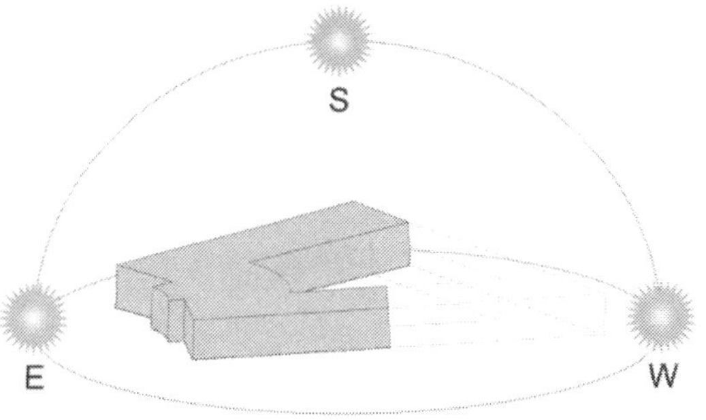

Fig. 1. An illustrative architecture of the ZS1 object in Swarzędz with the apparent solar movement shown.

2.2 Photovoltaic system

The photovoltaic system was installed on a roof of the middle part of the school facilities. It consists of 4 polycrystalline panels, of 250 W each. A construction, on which all PV panels were installed, was prepared for installation of a tracking system. Solar radiation intensity sensors send a message to a control device. Following a data analysis, the controller send information to a system controlling motors of the horizontal and the vertical axes to direct the set of panels towards the largest intensity of the solar radiation. Another device included in the photovoltaic system is the inverter. At this stage, a charging controller and a battery pack were not installed. A construction of power storage capacities is foreseen in the next stage of the solar system expansion. The system operates in the off-grid system, meaning, the generated electric power is not supplied to an external grid. The energy demand of the whole facility is so large that even at a peak of the power generation, during which the panels can produce the power of 1 kW, the whole electric power is used by devices operating at the facilities. Figure 2 shows an example of a diagram of the installed solar system. The dashed lines show components that will be connected and started in the future, that is, the tracker controller, the charging regulator, and batteries.

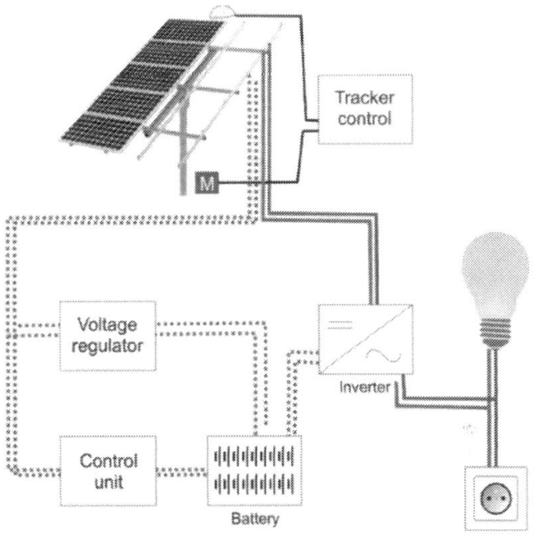

Fig. 2. PV system diagram.

3 Measurements of light intensity in rooms

Measurements were conducted in rooms located on S and N sides. The rooms are located on a second floor and are the same in terms of window and floor surface areas. The light intensity data was recorded in different weather conditions, for 18 months. This way an extensive range of data was obtained which was used to determine characteristics of the light intensity distribution in rooms. Simultaneous measurements conducted in rooms located on opposite sides allowed to compare intensity distributions. The obtained results are presented and discussed in section 3.2 of this paper.

3.1. Measurement method and location

Light intensity data was recorded at 16 points in two rows, at 1 m intervals, separately for each room. This measurement method was selected to eliminate possible measurement errors. The measurements were conducted manually, using a professional light meter. A disadvantage of this method is low measurement precision resulting from a need to precisely operate the light sensor, and a long time of measurement. Currently, a measurement matrix was developed and works on starting the whole measurement system are conducted. The measurement matrix will consist of light intensity sensors installed on 16 stable measurement supports. The light intensity values will be recorded simultaneously for all 16 sensors. This way errors occurring mainly during measurements of light intensity on very cloudy days will be eliminated.

3.2 Measurement results

The conducted measurements of light intensity distributions show similar courses of characteristics for rooms located on N and on S. This means that the light intensity distribution in a room does not depend on the building orientation. The obtained results can be used to draw conclusions concerning the intensity of external light ensuring the required lighting intensity inside, in a given place in the room. This data is useful in development and use of appropriate algorithms for control of lighting for ex. in smart buildings [2]. Figure 3 shows examples of the light intensity distribution in the room on the S side. The characteristics show light intensity distributions in various weather conditions., The measurements were carried out in March 2017. Analogous data for the room on the N side is shown in Figure 4. On the charts showing the illumination in the rooms from the south and north sides, the values for illumination demanded by law were marked. For such rooms it is 500 lux. By comparing a course of characteristics in both charts it can be said that the room orientation does not significantly influence the light intensity distribution in a room. The charts show that in worse weather conditions the course of the light intensity distribution is similar as in good weather conditions. An obvious difference concerns higher variations in the intensity at the initial stage, for the large intensity. Whereas, for example, in a case of the cloudy sky, when the solar radiation intensity is lower, the course of the characteristic course is smoother at the initial phase. The obtained characteristics can also be used to conclude that the change in the light intensity deeper in the room is significantly lower that at the initial distance near the window. Problems with measurements in a case of significant variability in the solar radiation intensity, for example, on cloudy days, should also be mentioned. Differences in the light intensity can reach an amplitude even of several thousand luxes in just a few seconds.

Recordings of the electric power generated in the installed photovoltaic panels showed that the photovoltaic system in the school building generated 937 kWh of the electric power during 12 months. Of course, starting of the tracker control system will possibly increase the amount of the generated energy.

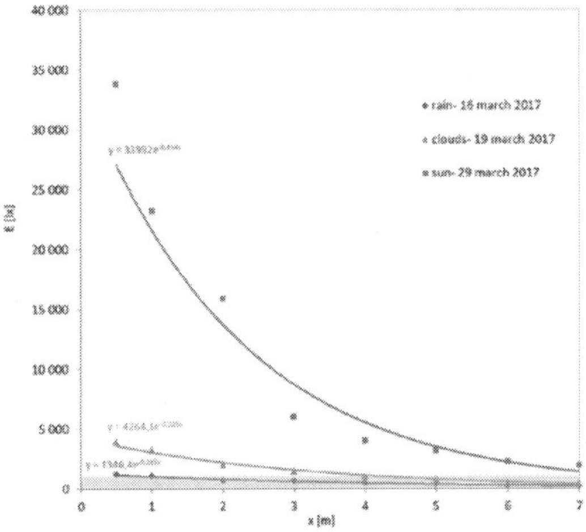

Fig. 3. Distribution of daylight in the room from the South side.

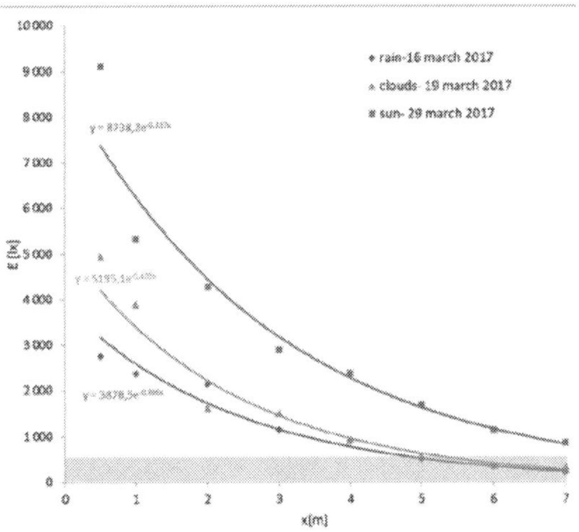

Fig. 4. Distribution of daylight in the room from the North side.

4 Savings resulting from the use of the solar radiation

Energy consumption for lighting represents from 10 to 20 % of total energy costs at facilities [3]. Therefore, actions aiming at savings in energy consumption by lighting systems are a very important issue. The conducted studies on the light intensity distribution in rooms allow to estimate savings that can be achieved by using the solar radiation. The obtained results were used to determine the value of the solar radiation intensity at which we can switch artificial lighting off or reduce its power. Limit values for the solar radiation intensity at which the required lighting intensity is ensured at a relevant depth of a room are shown in Table 1. It includes a classification into the light intensity outside, E_{out}, and inside, E_{in}. The parameter x expressed as [m] means a room depth measured from the window. The obtained results were used to conclude that at the light intensity below 7000 lx outside or of 1200 lx inside near a window, the required light intensity is not ensured at the whole depth of the room, that is, it is necessary to switch all installed sources of artificial lighting on. Whereas, when the solar radiation intensity exceeds 12000 lx outside and 3200 lx inside near a window, the required light intensity is provided in the whole room, and all installed sources of artificial lighting can be switched off. Other intensity values for which the required lighting intensity is ensured at a relevant depth of a room are provided in Table 1. The lighting control system, which will switch the artificial lighting system on/off and control its power, should be selected according to these criteria.

The obtained measurement results allowed to estimate savings that can be obtained by implementation of an appropriate control system. The conducted calculations show that the use of solar radiation for lighting of rooms can reduce the electric power consumption for lighting by 45–48%, depending on a control system used [3]. Use of appropriate lighting control algorithms is important, as they will ensure an optimal control of selected sources of artificial lighting in a room [4]. The existing control systems in the venues are also used for effective energy consumption in smart buildings [5].

Table 1. Values of the solar radiation intensity ensuring a required light intensity at a specified distance from a window.

X [m]	E_{out} [lx]	< 7000	7000 - 9000	9000 - 1200	> 12000
	E_{in} [lx]	< 1200	1200 - 2300	2300 – 3200	>3200
2		ON	OFF	OFF	OFF
4		ON	ON	OFF	OFF
6		ON	ON	ON	OFF

5 Conclusions

The solar radiation has a high energy potential that can be used to improve the energy effectiveness of lighting systems by reducing the electric power consumption for lighting purposes. The first method presented in the paper concerns the use of the daylight to light rooms, with simultaneous reduction in artificial lighting [6]. The required light intensity in a room can be ensured by switching relevant selected sources of artificial lighting on. The solar radiation intensity distributions in a room can be useful for development of the lighting control algorithms. The obtained study results allow us to ensure appropriate power control and regulation for individual lighting fixtures. In consequence, we can achieve savings reaching nearly 50% of the total power consumption for lighting. However, it should be remembered that when no lighting control system is used, the availability of the solar radiation will not influence savings in the electric power consumption. A problem of lighting system control in conditions of significant variability of the radiation intensity still remains to be solved. In such cases, lack of an appropriate control algorithm can result in flickering of artificial lighting, and this is an undesirable phenomenon having adverse effects both on operation of the lighting system, and on people.

The second method for use of the solar radiation energy is its conversion into the electric power in a photovoltaic system. The solar radiation cannot be stored directly. However, we can store the solar energy transformed into the electric power. When a battery pack is used, it can supply the electric power at times when the PV panels do not generate power, e.g., at night. The system developed this way can lead to significant reductions in the electric power consumption for lighting. According to the current Directives [7], this approach is a recommended action for improvement in the energy efficiency of the facilities, and, in consequence, reduces consumption of the primary energy, and this directly translates to a greater protection of the natural environment.

References

1. Pr EN:15193 Energy performance of buildings Energy requirements for lighting (2017)
2. M. Kalinowski, T. Zarębski, 2012, Lighting control in smart buildings, Wiadomości elektrotechniczne, 39-42, **10** (2012)
3. S. Sowa, A. Kamińska, 2018. Predicted reduction of energy consumption in the school building by lighting control with KNX system. Przegląd Elektrotechniczny, 193 – 197, **2** (2018)
4. S. Sowa, J. Gielniak, Lighting control algorithms in a school venue using KNX system. Electrical Engineering, 153-162, **96** (2018)
5. M. Horyński, 2013, The application of dispersed processing networks in order to optimize the energy consumption in contemporary buildings, Przegląd elektrotechniczny, 293-295, **4** (2013)
6. E. Plikunas, M. Ziajkowski, Use of solar radiation in shaping the energy efficiency in public buildings, Electrical Engineering, **87** (2016)
7. Directive 2012/27/EU of the European Parliament and of the Council of 25 October 2012 on energy efficiency 2012/27/UE, Brussels (2012)

The study on the efficiency of a photovoltaic cell with a system of supercapacitors and batteries

Dariusz Strąk[1,*], *Kinga* Strąk[1], and *Magdalena* Piasecka[1]

[1]Kielce University of Technology, Faculty of Mechatronics and Mechanical Engineering, al. 1000-lecia P.P. 7, 25-314 Kielce, Poland

Abstract. The main aim of this work is to propose a new system of a hybrid photovoltaic system working with batteries and supercapacitors and to analyze its efficiency. The preliminary results of the study on the operation of the system are discussed. The results obtained while testing the operation of a hybrid system and a PV system working separately with batteries and supercapacitors are compiled and compared. The tests covered the systems efficiency for the following electrical loads: heater warming up water in a tank, and lighting - a LED light. The work of batteries and supercapacitors during discharge has been analysed. The use of a hybrid system made it possible to increase energy storage efficiency and the system operation flexibility, compared to solar systems offered by the sector.

1 Introduction

One of the key trends of today's world economy is the shift from the marginal to mainstream use of renewable energy as an alternative to fossil-fuel based energy production. The last several years have seen extensive research related to the renewable energy technology, with solar photovoltaic (PV) regarded as one of the most promising. The main role of PV system deployment is to support residential electrical, central heating and hot water systems as a supplementary energy source, or to serve as a standalone power source at sites where grid-connection is impossible or impractical, such as street lighting or luminous road signs. An example of PV deployment is energy generation from micro-grids using three monocrystalline solar panels with a rated power 585 W in a residential building [1], where the PV system pays for itself. Another example is the PV system (26.6 KWp) installed to provide electricity for a bus repair shop [2]. The system has twofold benefits; one is cost-related - reduced amount of energy purchased from the current supplier and the other one is ecological - reduced emission of carbon dioxide into the atmosphere.

A need for energy storage is created by increased demand for charging batteries (Bs) and supercapacitors (SCs) through photovoltaic systems. Solar energy acquisition in regions with inadequate solar irradiation is problematic. In conventional systems, PV energy is stored only by batteries that, at low power density, have high energy density and use high currents or high frequency currents for charging. The life of the battery (Bat) in PV systems is quite short. In comparison with Bs, SCs allow a hundred times higher power density, with ten times lower energy density. Undergoing frequent charge/discharge cycles at high current and frequency does not shorten SCs life. The use of hybrid energy storage enhances the reliability of the system at various variable current conditions and increases the efficiency of the system under variable solar irradiation [3].

Researchers in [4] present an embedded energy share method between batteries (the energy storage system) and supercapacitors (the auxiliary energy storage system) in which supercapacitors improve the life of the batteries and reduce their stress by providing/absorbing peaks currents as demanded by the load. Simulation results illustrate the performance of the proposed control strategy.

The use of supercapacitors (SCs) to improve battery life and reduce the life cycle cost of a standalone photovoltaic PV/Bat system for a remote agricultural farm power application is discussed in [5]. The authors analyze the performance of a PV/Bat and a PV/Bat/SC systems in a typical remote agricultural feedmill.

Another scheme to control the power sharing between batteries and supercapacitors is reported in [6]. The researchers introduce an innovative method for sizing the energy storage system together with the hybrid distribution based on the photovoltaic power curves to reduces the stress levels on the battery, extend its life, and provide constant power injection to the grid over a defined time interval.

In [7] a system able to integrating a photovoltaic generation system, including the energy storage to moderate fluctuations in the energy generated, by using a set of supercapacitors are presented. The system consists of a voltage source inverter and a bidirectional direct current converter regulating the charge of the supercapacitor.

A simple novel control strategy for a hybrid energy storage system is elaborated and analyzed in [8]. In the method proposed by the authors, batteries are used to

* Corresponding author: dstrak@tu.kielce.pl

counterbalance the slow changing power surges. A supercapacitors (SCs) are applied to equiponderate the fast changing power surges. The slow response of battery system is overcome by diverting the power surges to the SCs system and reduce the current stresses on battery.

In [9] an nonlinear flatness control applied to supercapacitors contribution in hybrid power systems by using photovoltaic source and batteries is shown. Two systems are desribed to manage energy flows: the first with a photovoltaic (PV) source considered and lead-acid batteries used as storage unit and the second hybrid system with the PV source, lead-acid batteries, and supercapacitors. In the system control, the nonlinear flatness control is used to manage energy. Under conditions of high discharges, supercapacitors allow batteries to operate under reasonable conditions limiting strong solicitations and keeping them in a higher state of charge. The use of supercapacitors in a hybrid system increases sources lifetimes battery.

A simple scheme of the supercapacitor based on (RC) circuit by using experimental methods is modeled and characterized in [10]. It is an example of hybrid photovoltaic/supercapacitor stand-alone system. There is applied the maximum power point tracking (MPPT) control for photovoltaic and the supercapacitor state of charge (SOC) control. A simulation model is developed for the considered hybrid energy system by using MATLAB/Simulink. The results present the main role of the supercapacitor when the load changes rapidly.

In [11] the role of the supercapacitor in a PV energy control unit (ECU) by using Matlab/Simulink models is discussed. According to authors, the hybrid storage system can achieve higher specific power than the battery storage system. An alternative way of supplying large bursts of current is to combine VRLA batteries and supercapacitors to form a hybrid storage system where the battery can supply continuous energy and the supercapacitor can supply the instant power to the load.

The storage of photovoltaic energy by supercapacitors is studied in [12]. The experimental setup consists of a photovoltaic array, a module of supercapacitors in series, electrical relays, and a resistive load. Analog to digital converters are used, and an interface has been developed with Labview software. The second approach consists in simulating photovoltaic energy storage by supercapacitors with a model composed of solar irradiance evaluation, equivalent electrical circuit for photovoltaic conversion, and a multibranch circuit for supercapacitor. Both the experimental and calculated results are confronted, and an error of 1% on the stored energy is found with a correction largely within % of the transmission line capacitance according to temperature.

In [13] is proposed a hybrid generator with a photovoltaic energy conversion system supercapacitors and lead-acid batteries. The task of this system is to provide active power to the electricity grid. By designing a control system it is possible to ensure a high battery state of charge and overcharge security. A continuous dynamic model is used and a control design of the power system is studied.

Use of supercapacitors as energy storage systems is evaluated in [14]. Supercapacitors are compared with other technologies such as: compressed air, pumped hydro, superconductors and flywheels. The supercapacitors are studied in detail, presenting device structures, how they can be modeled, the balancing, their useful life and principal applications.

In [15] intelligent systems data collection and processing in the different applications where supercapacitors are used, such as: automatic highway traffic management, area surveillance, and geological activity monitoring are discussed. Key and basic field systems that are away from the power grid infrastructure energy are self-sustainability. Instead of the conventional battery-based energy storage, the buffer solar energy in supercapacitors has the advantages of precise energy lifetime awareness, low maintenance, and operational robustness. The load can also consume energy directly from PV Solar Energy Harvesting Circuit built around connected via SEPIC DC-DC converter.

A system that supercapacitors are an emerging choice for energy buffering in field systems is studied in [16]. Supercapacitors offer advantages compared to rechargeable batteries for energy buffering due to their energy charge and discharge efficiency as well as environmental friendliness.

A special class of converters as the energy transfer mechanism DC-DC are described in [17]. The most commonly used DC-DC converter designs are Buck and Boost or Buck-Boost.

According to [18], SEPIC is a reasonably efficient solution for a an energy transfer circuit that is providing current into a supercapacitor. It can be used with systems built with supercapacitors.

In [19] presents of the high frequency DC/DC converter built on the basis of silicon carbide elements (SiC). The converter has the ability to bi-directionally lower and increase the voltage, which allows the energy flow in both directions regardless of the input voltage to the output voltage. SiC elements used are characterized by higher efficiency about devices from silicon elements.

The concepts of the battery-supercapacitor PV system applied in a system of energy generation and storage in public and long-distance transport vehicles and for unmanned aerial vehicle are proposed in [20, 21], respectively.

The main aim of this work is to propose a new system study of a photovoltaic cell with a system of supercapacitors and batteries and to analyses its efficiency. The preliminary results of the study on the operation of a hybrid photovoltaic system working with batteries and supercapacitors were discussed.

2 Experimental stand and methodology

Figure 1 shows a diagram of a storage system for energy from photovoltaic panels, which use batteries and supercapacitors to store energy.

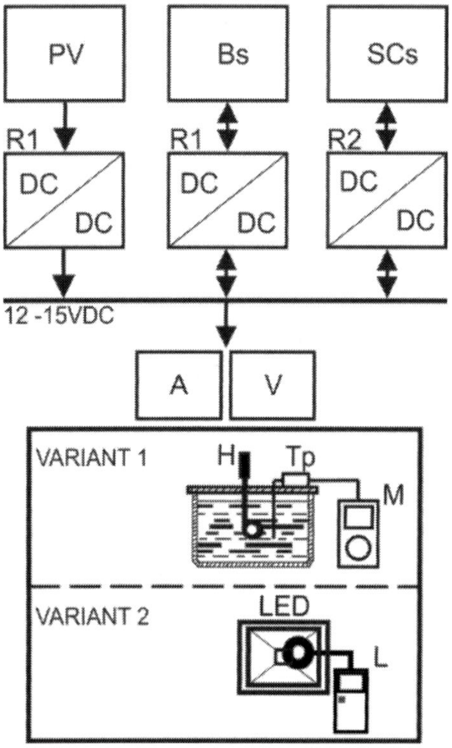

Fig. 1. Photovoltaic system working with a 12V load, using SCs and Bs: PV- photovoltaic panel, Bs - batteries, SCs-supercapacitors, R1-PWM charge controller, R2-pulse type charge controller, A-ammeter, V-voltmeter, L-luxmeter, H-heater, LED-a LED light, M-multimeter, T-thermocouple.

Direct current generated in PV cells is delivered to solar voltage controller PWM (R1), and to the energy storage unit. The SCs are provided with a progressive pulse type voltage charge controller (R2) connected to the Bs, which is to charge the SCs and to protect against overcharging and polarity reversal in a system of connected SCs. As soon as the SCs are charged, the energy is transferred using *Schottky* diode with nominal operating parameters: 300 A and 30 V, which secures the system against SCs polarity reversal to the output to electrical loads. At the output from the diode system current is measured using an ammeter 400 A (A) and voltmeter (V). Rated parameters of the PV panels, SCs, Bs and voltage charge controllers are specified in Table 1. Current collection system is provided with heater (H) with rated operating parameters: voltage - 12 V and power - 200 W, or a LED light - power 50 W and voltage 12 V. A luxmeter, *Voltcraft*, type PL-110SM (L) was used to measure light intensity of the lighting system and the PV. Ambient temperature and temperature of water in the tank was measured with K -type thermocouples.

The experimental stand allows parallel charging of both SCs and Bs. In case if Bs are fully charged, the energy is transferred to the SCs charging system. The system contains conductors made of soft copper. The SCs system is connected in series, the aggregate value of resistance for the entire system is ca. 4.2 mΩ (not including resistance of load connected to the system), and power is 5 kW. Voltage at the output of fully charged SCs system is 15 V, and current intensity is 700 A.

Table 1. Rated parameters of a solar system, the complex of Bs and SCs.

Charge controllers		
Parameters	Charge controller R1	Pulse type charge controller R2
Rated voltage, V	12/24	12-15
Min. input voltage, V	10.5	6
Max. charge current, A	60	30
Max. load current, A	60	30
Charging technology	PWM	Pulse
Charging time	1 h to 24 h	from 100 s
Charging to output current, A	-	400
Rated parameters of a system containing 4 PV panels		
Parameters	PV	
Model	SGM-170 W	
Type	monocrystals	
Power tolerance, %	± 3	
System voltage, V	12	
Rated voltage, V	17.8	
Charging current, A	4x9	
Power of the PV system, W	4x169.9	
Cell efficiency, %	18.43	
Operating parameters of the Bs and SCs system		
Parameters	Bs	SCs
Qty. pcs.	2	6
Model	MWL 120-12 h	BCAP - 3000
Capacity	240 Ah	90 000 F
Voltage, V	12	15
E, Wh	2 000	18.225
Energy density, Wh/kg	84.80	60.75

For the purposes of analysing the results of the system operation and evaluating its efficiency, loading of the system consisting of the Bs, SCs and PV panels is used at the experimental stand, applied in two variants: analysis of water heating time in an insulated tank (capacity 1 dm^3) with a 200 W heater up to the temperature of 45 °C (variant I), and measurement of obtained power and lighting intensity of a 50 W LED light (variant II).

The tests were performed in laboratory conditions in Kielce on May 15, 2018, at atmospheric pressure of 992.62 hPa and ambient temperature 27.7 °C.

3 Results

3.1 Energy balance for the employed hybrid system

Energy generated by a complex of photovoltaic panels, stored in Bs, SCs and used by electrical loads, has been defined as:

$$E_{OB} = E_{PV} - E_B - E_{SC} \quad (1)$$

$$E_{PV} = \int_0^{24} f(x)dx \quad (2)$$

$$E_{SC} = \int_{x_1}^{x_2} \left(f(x) - g(x) + \frac{C}{10}\right) \quad (3)$$

$$E_B = \int_{x_3}^{x_4} (f(x) - g(x) - E_{SC}) \quad (4)$$

where: E_{OB} - energy delivered to the loads, E_{PV} - energy generated by the complex of photovoltaic panels, E_{SC} - energy stored in SCs, E_B - energy stored in Bs, C - capacity of Bs.

Figure 2 presents electrical output power plot for a complex of 680 W photovoltaic panels, similar as in [4].

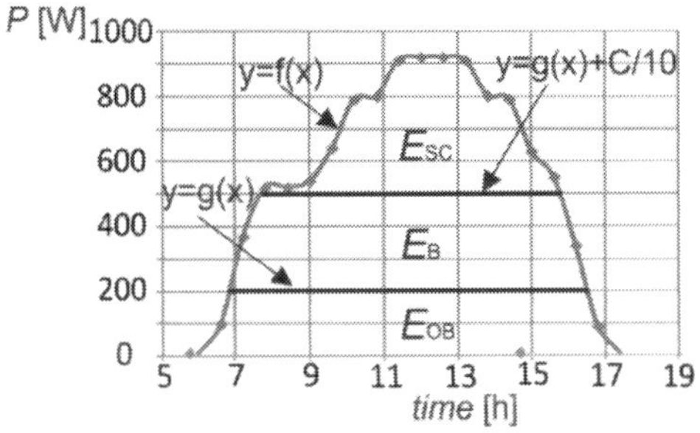

Fig. 2. Output power characteristics for a complex of photovoltaic panels; E_{SC} - energy stored in SCs, E_B - energy stored in Bs, E_{OB} - energy consumed by the load.

3.2 Testing of the supercapacitors and batteries charging characteristics

Figure 3 shows voltage in function of time, and Fig. 4 shows current intensity in function of time, during the SCs and Bs charging.

The SCs system is characterised by very fast charging process, from 5 to 8 minutes, low internal resistance (even below 0.9 mΩ), minor internal energy losses, high efficiency (exceeding 95%), and short recharging times. For the Bs system charging time is few hours only. Analysis of the results shown in Fig. 3 shows that the time of SCs charging to 15 V is 7 minutes, (Fig. 3a), while for the Bs complex it takes ca. 4.5 hours to reach 13.4 V (Fig. 3b). There is a big difference between the Bs and SCs charging time. While analysing diagrams of the SCs charging with the intensity of current taken from the Bs in relation to time, Fig. 4a, it is visible that at the initial stage of charging, current intensity was 0.5 A, and then it increased to 16 A. After few seconds of this work, the third stage began, during which the R2 controller started to increase current intensity gradually to 27 A. Then, current intensity remained at constant level in time.

Figure 4b shows that for Bs charging, current intensity remained at constant level (10 A) between 0 and 3 hours. Then, current intensity dropped with increasing battery charge time to 3.8 A.

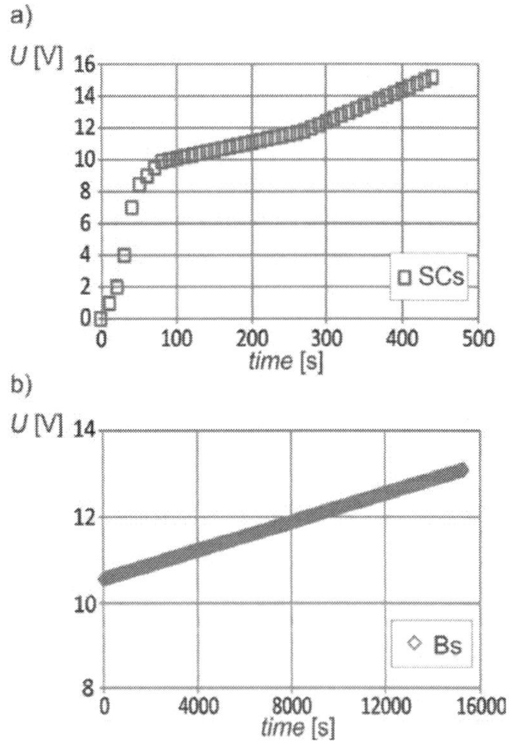

Fig. 3. The voltage versus time, while charging SCs (a) and Bs (b).

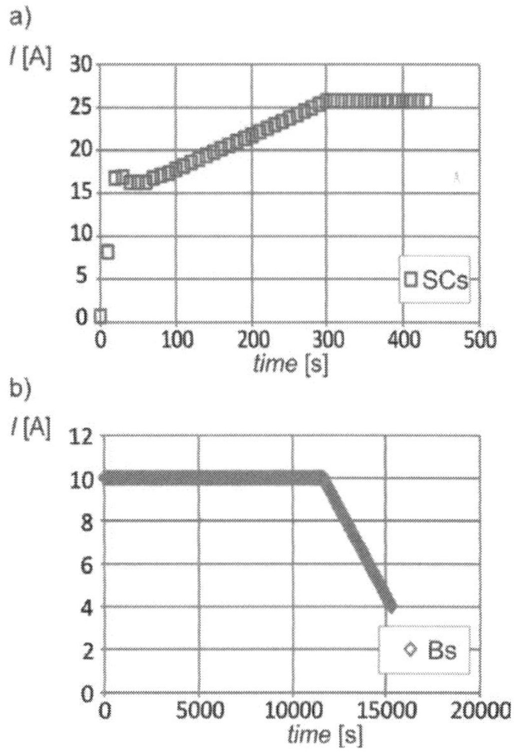

Fig. 4. The current intensity versus time, while charging SCs (a) and Bs (b).

3.3 Results of tests with electrical load, variant 1

Figure 5 presents operation characteristics for the system with heater (variant I), in form of the following relationships: heater power consumption in function of time (Fig. 5a), and water heating rate in time (Fig. 5b) for the PV system working only with the SCs, or only with the Bs. The batteries were being discharged for 25 minutes (Fig. 5a), without recharging by the PV panels at high current intensity reaching 10 A (heater connected). In this time, heater power consumption indicated drop in battery charge power from 84% to 60%. The temperature of water in the tank increased from 23.4 °C to 45 °C within approximately 17 minutes (Fig. 5b). The results are much the same both for Bs and SCs.

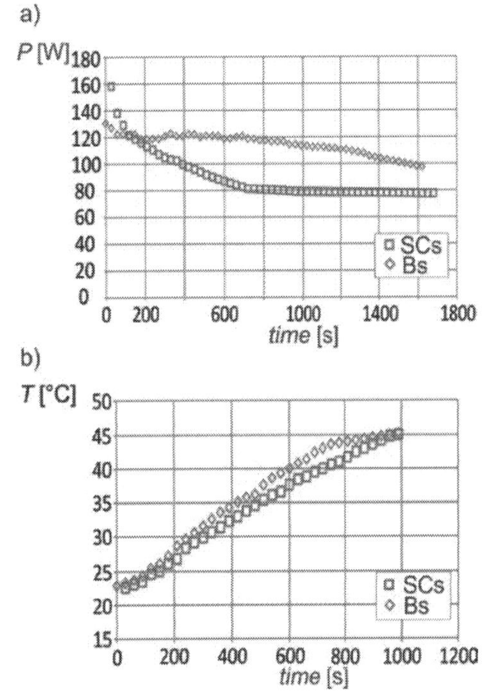

Fig. 5. a) Heater power consumption in function of time, b) the rate of water heating in time in a 1 dm^3 tank; data shown for SCs and Bs, separately.

Figure 6 shows drop of voltage and current intensity for the heater in function of time for the hybrid system, where the SCs were recharged from the Bs. SCs voltage remained close to 10.5 V after 23 minutes, while for Bs - after 30 minutes, Fig. 6a. Initial current intensity for the SCs and Bs was approximately 10.5 A, Fig. 6b. For the Bs, current intensity remained stable (9.8 A) after 3 minutes. For the SCs, gradual drop in current intensity to 7.8 A was observed within 14 minutes. Later, amperage remained stable.

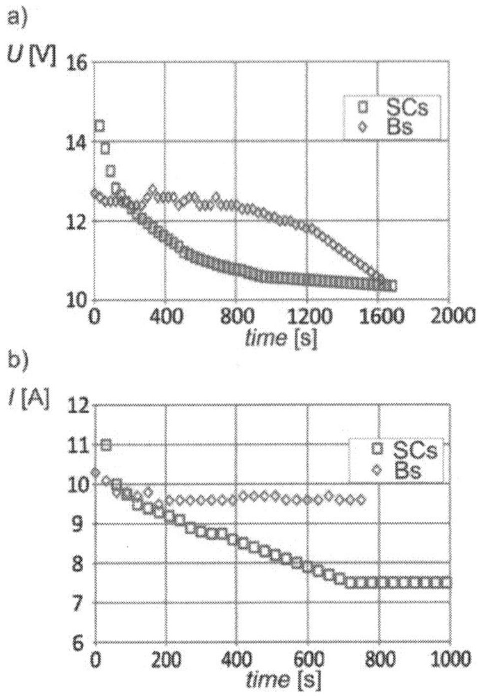

Fig. 6. Relationship of heater loading in relation to time for: a) voltage, b) current intensity; achieved for the hybrid system (PV with SCs and Bs).

3.4 Results of tests with electrical load, variant II

Figure 7 presents the relationships for the hybrid system in function of time, while the SCs were charged by R2 with the LED light connected as a load for:

- Bs voltage during charging and discharging, Fig. 7a,
- percent discharge of Bs during the SCs charging, Fig. 7b.

Time of the SCs recharging at regular intervals was ca. 6 minutes. Initial Bs charge was 78%, and the SCs charge controller activated charging process every 5 minutes, then power in the Bs was cyclically dropping to 70%, Fig. 7a. In the hybrid system with the SCs, voltage was slightly dropping during the SCs charging. The LED light intensity was approximately 150 lux and remained stable in time.

Figure 8a shows current intensity in function of time for the Bs and the SCs during discharge, for a hybrid system. Initial power consumption for the Bs was 13.4 A, and for the SCs - 2.5 A. In 28 minutes amperage dropped from 4 A to 0.5 A. For the SCs load the LED light stopped working after 50 minutes (Fig. 8b). Whereas, for the Bs load the lamp went off at 10.5 V. Therefore, even at as low voltage as 4 V the SCs reach current intensity allowing operation of the LED light.

https://doi.org/10.1051/e3sconf/20187001014

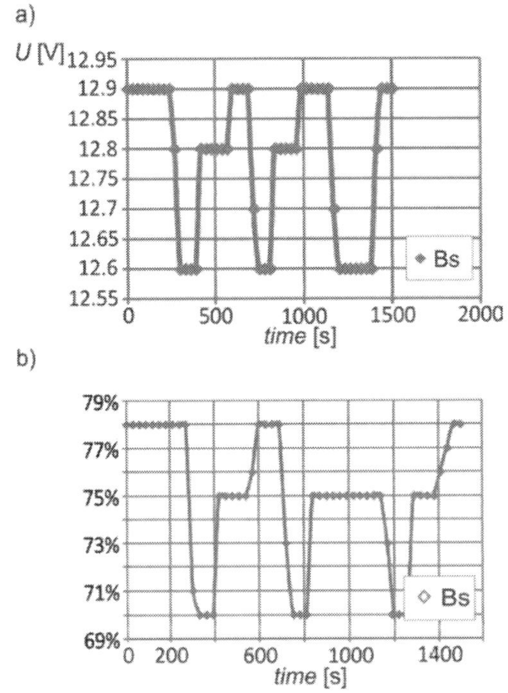

Fig. 7. Relationships for the hybrid system in function of time, while the SCs were charged by the R2 with the LED light connected as the load: a) voltages for the Bs during charge and discharge, b) percent discharge of the Bs while charging the SCs.

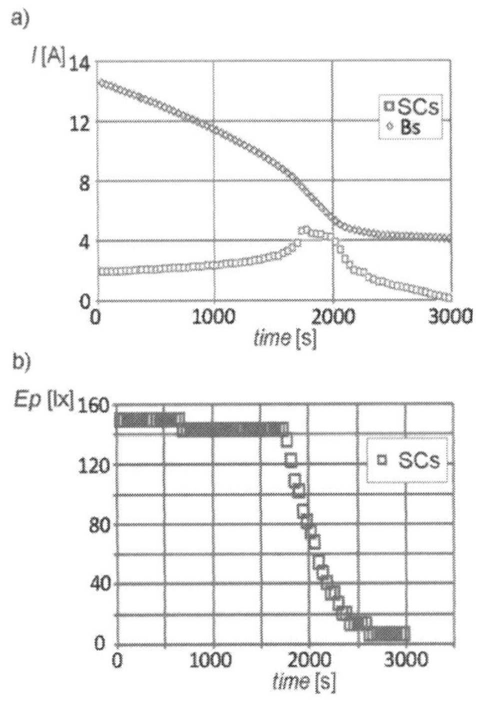

Fig. 8. Relationships for the hybrid system in function of time: a) current intensity drawn by the LED light, b) intensity of light emitted by the LED light during the SCs loading.

3.5 Comparison between operation of batteries and supercapacitors

The results of tests and measurement characteristics of the system allow stating that the SCs system is the most prospective energy storage system. Its advantage is high power density, with short charging times (recharging) of only few minutes, compared to the Bs. Batteries are sensitive to excessive discharge, which reduces their capacity and service life. The SCs have lower self-discharge current values compared to the Bs.

The hybrid system proved to have better operating effects than the Bs system with PV. For variant I, with heater as an electrical load, the hybrid system was faster to warm up water to 45 °C, while for variant II with the LED light it considerably improved smooth exchange of accumulated energy and stabilised light intensity at 150 lux. The demonstrated hybrid energy accumulation system allows reducing operation time of batteries used during the tests, while original system efficiency is maintained. Improved parameters of the system buffer have been reached after introducing the SCs connected to the Bs. In this setup the SCs are the final energy storage units. In power consumption system the SCs receive energy from the Bs. In this system, when there is surplus solar energy available, the Bs and SCs are charged in parallel while the system is loaded, and the energy is delivered by photovoltaic panels. When hybrid system in not loaded, the energy is fully accumulated in batteries and supercapacitors. Using the SCs in this system made it possible to obtain energy accumulation reached 95%. In a similar situation, accumulation energy was estimated to be 78% for the Bs. In the examined hybrid system the loads receive energy from the SCs first, and they consume only insignificant portion of energy from the Bs.

4 Conclusions

The paper presents results of a preliminary study on a photovoltaic system working with the system consisting of SCs and Bs (a hybrid system). Moreover, the research involved comparison of the work of Bs and SCs during discharge for two electrical loads: a heater warming up water, and lighting - a LED light. The Bs charge power (while charging the SCs) was dropping from approx. 80% to 70%, which would considerably improve smooth exchange of accumulated energy, compared to direct connection between the load and the Bs. Batteries are sensitive to excessive discharge, their capacities decrease in time, and their service life is shorter. The system consisting of SCs is characterised by very fast charging process, high efficiency, short recharging times, and lower self-discharge current values. When heater was an energy consuming load, the SCs heated up water to preset temperature faster than the Bs.

Generally, the use of a hybrid system made it possible to increase energy storage efficiency and the system operation flexibility, compared to solar systems offered by the sector.

References

1. J. Dąbrowski, E. Hutnik, Energy Market **3**, 102-108 (2015) (in Polish)
2. A. Guźda, N. Szmolke, BUSES - Technol. Oper. Transp. Syst. **6**, 188-192 (2017) (in Polish)
3. M Harasimczuk, Electr. Eng. Pozn. Univ. Technol. Acad. Journals **87**, 109-118 (2016)
4. Z. Cabrane, M. Ouassaid, M. Maaroufi, *Int. Renew. Sustain. Energy Conf.* (IRSEC), 185–190 (2014)
5. T. R. Ayodele, A. S. O. Ogunjuyigbe, B. E. Olateju J. Renew. Sustain. Energy **10**, 13503 (2018)
6. V. M. Miñambres-Marcos, M. Á. Guerrero-Martínez, F. Barrero-González, M. I. Milanés-Montero, Sensors **17**(8), 1856 (2017)
7. M. Á. Guerrero-Martínez, E. Romero-Cadaval, V. Minambres-Marcos, M. I. Milanés-Montero, J. Microelectron. Electron. Components Mater. **44**(1), 40-52 (2014)
8. K S. K.ollimalla, M. K. Mishra, N. L. Narasamma, 2014, IEEE Trans. Sustain. Energy **5** (4), 1137-1144 (2014)
9. M. Benaouadj, A. Aboubou, M. Y. Ayad, M. Becherif, Energy Procedia **50**, 333-341(2014)
10. N. Benyahia, H. Denoun, M. Zaouia, S. Tamalouzt, M. Bouheraoua, N. Benamrouche, T. Rekioua, S. Haddad, Energy Procedia **42**, 539-548 (2013)
11. M. E. E. Glavin, P. K. W Chan S. Armstrong, W. G. G. Hurley, *13th Int. Power Electron. Motion Control Conf. EPE-PEMC,* 10288042 (2008)
12. P.-O. Logerais, O. Riou, M. A. Camara, J.-F. Durastanti, J. Sol. Energy, 659014, 9 pages (2013)
13. H. Fakham, D. Lu, B. Francois, IEEE Trans. Ind. Electron. **58**(1) 85-94 (2011)
14. M.,Guerrero, E. Romero, F. Barrero, M. Milanés, E. González, Electr. Rev.**85**(10), 188-195 (2009)
15. A. Fahad, T. Soyata, T. Wang , G. Sharma, W. Heinzelman, K. Shen, *Conference: SOC Conference (SOCC), 2012 IEEE International,* 13229941 (2012)
16. M. Hassanalieragh, T. Soyata, A. Nadeau, G. Sharma, *27th IEEE International System on Chip Conference (SOCC),* 14719240 (2014)
17. A. Pressman, K. Billings, T. Morey, Switching Power Supply Design, Mc Graw-HilI, (2009)
18. http://www.ti.com/lit/an/slyt309/slyt309.pdf
19. W. Matelski, L. Wolski, S. Abramik, IAPGOS **3**, 64-69 (2016)
20. D. Strąk, D. Michalski, M. Piasecka, BUSES - Technol. Oper. Transp. Syst. **12**, 428-433 (2017) (in Polish)
21. D. Strąk, M. Miesikowska, M. Piasecka, BUSES - Technol. Oper. Transp. Syst. **12**, 434-437 (2017) (in Polish)

https://doi.org/10.1051/e3sconf/20187001015

Design of thermal energy storage unit for Compressed Air Energy Storage system

Maciej Szybiak[1], and *Maciej* Jaworski[1,*]

[1]Warsaw University of Technology, Faculty of Power and Aeronautical Engineering, Nowowiejska 24, 00-665 Warsaw, Poland

Abstract. The aim of this paper is to present a new concept of a high-temperature thermal energy storage (TES) for the application in the compressed air energy storage (CAES) systems. The proposed storage unit combines the advantages of pressurized containers with packed beds, e.g. of rocks, with the strengths of non-pressurized systems such as those encountered in CSP plants. Designed TES unit consists of the heat exchanger located inside a high-temperature thermocline-type vessel with molten *HITEC®* salt used as a heat storing material. In terms of the geometry of the designed heat exchanger, a tube-in-tube helical coil type was chosen due to its higher convective heat transfer coefficients in comparison with straight tubes. To find the most suitable case, four helical coils with different dimensions (diameter, pitch) were considered. Heat transfer and pressure drop analysis for each configuration were conducted. In particular, convective and overall heat transfer coefficients as well as friction factors were computed based on the empirical correlations. To verify the obtained results, the analysis based on numerical approach has been carried out with the use of ANSYS Fluent software for the most suitable case.

1 Introduction

Nowadays we are facing the necessity of making the transition from energy systems based on fossil fuels to ones operating mainly on renewable energy sources. The most serious challenge in this matter is alleviating the consequences of their intermittency. Energy storage seems to be the only technologically simple solution to these problems. Among many techniques of energy storage Advanced Adiabatic Compressed Air Energy Storage (AA-CAES) is a promising and feasible solution on the system level as it provides as large storage capacities as the most often used Pumped Hydro Storage (PHS). Also, power available during discharge of these storage plants is very high, reaching hundreds of MW.

AA-CAES comprises of 4 main components: compressor stage(s), thermal energy storage (TES), compressed air storage (CAS) and turbine stage(s). As compressors and turbines are mostly off-the-shelf products and the choice for CAS systems is rather limited to underground cavities, the main research area is the development of a high efficient TES.

The basic principle how a CAES system works is fairly simple. In times of electricity surplus, it is used to compress ambient air by an electrically driven compressor. Compressed air is cooled down and stored in a pressurized container or underground cavity, e.g. in an underground cavern. If the electrical energy shall be recovered, the air is released, heated up and expanded in a turbine. The turbine in turn drives an electric generator which supplies electrical energy back to the grid. A simplified diagram of the AA-CAES plant is shown in Fig. 1 [1, 2].

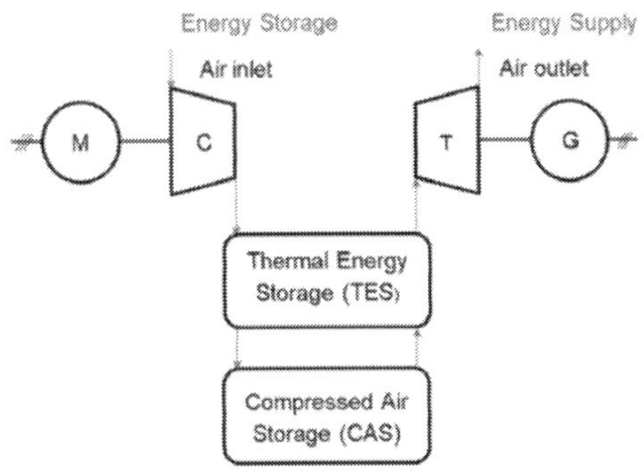

Fig. 1. Simplified diagram of the AA-CAES plant; M – electric motor driving the compressor, C – air compressor, T – turbine, G – electric generator.

So far, there have been three major CAES concepts: diabatic (D-CAES), (advanced) adiabatic ((A)A-CAES) and isothermal (I-CAES) technologies. All of them differ from one another only in the way how the heat generated during the air-compressing phase is managed. In the case of diabatic CAES, heat is irreversibly lost in the environment. Therefore, an additional source of energy is required to heat the air up before it expands in a turbine. On the other hand, AA-CAES preserves the heat and uses it to raise the temperature of the air during the discharge phase. In the isothermal CAES concept, the air is enabled to constantly exchange heat with the

* Corresponding author: mjawo@itc.pw.edu.pl

environment and so to ensure quasi-isothermal conditions.

The diabatic technology is the simplest one to put into effect but, on the other hand, it requires a significant amount of energy from an additional source. In the vast majority of cases fossil fuels play this role. Dwindling reserves and their contribution to the emission of CO_2 and other harmful substances have enhanced the interest for AA-CAES technology, with the TES being the most dynamic field of research. A TES unit designated for an AA-CAES systems must fulfil numerous requirements with the main ones being:

- operating in the temperature range of 50-600°C,
- stabilizing the outlet air temperature at a given level,
- minimizing heat losses to the environment,
- minimizing air pressure drops.

A number of different high-temperature TES concepts have been proposed and experimentally investigated for the use with air, such as packed beds of rocks or ceramic pieces, large concrete blocks with embedded heat exchangers [3-7]. They remain the most common type of high-temperature TES units. This is because rocks are cheap, easily accessible and they are characterized by a relatively large heat transfer area. However, a pressure vessel must withstand enormous temperature and pressure gradients, and thus, its walls have to be of a significant or even prohibitive thickness. Additionally, loosely stuffed crushed rocks contribute to high pressure drops and periodic thermal stresses, which may lead to attrition of the TES material and, therefore, the necessity of installing air filters.

Another concept of the TES unit, commonly employed in commercial Concentrated Solar Power (CSP) plants, consists of two tanks: one with cold and one with hot oil or molten salt as the heat-transfer fluid (HTF) and thermal energy storage medium [3]. According to [8], especially nitrate salts and their eutectic mixtures are suitable to use as the HTF because of their appropriate thermal properties and no corrosive effects. In addition, they have relatively low melting point, even around 150°C, and stay stable to over 550°C. In this temperature range they neither develop toxic gases nor are caustic [8]. A prime example here is the *HITEC®* salt – an eutectic mixture of potassium nitrate (53%), sodium nitrate (7%) and sodium nitrite (40%) [9]. The main advantages of the two-tank concept are: the pressure part of the heat exchanger is limited only to the air domain, air temperature is kept steady at the outlet, air pressure drops are low. However, such a two-tank system is relatively less compact, requires a greater use of materials and so greater investment costs.

An alternative to the two-tank concept is to store TES medium (hot/cold salt) inside a single tank volume with a separation of the region of different temperatures due to the difference in density of the salt. This concept is called thermocline [10]. In order to increase the thermal capacity of such storage tank inclusion of rocks or macrocapsules containing phase change materials (PCMs) are also considered [11, 12]. TES unit based on high temperature PCM (eutectic alloy of aluminium, cupper and silicon) for CAES application is described in [7].

The aim of this paper is to present a concept for a high-temperature TES combining the advantages of pressurized packed beds like thermal stratification, with the strengths of non-pressurized systems, in particular those regarding safety issues. The concept of TES unit is not new, similar heat storage unit with steam generator embedded inside molten salt container for the application in CSP is analysed in [13]. In the study reported in this paper the main goal was to check the possibility of developing compact enough heat exchanger that fulfils requirements for cooling/heating of compressed air in a relatively high range of temperature, and that could be enclosed inside heat storage unit. The final effect of the research analysis includes a design of a lab-scale prototype of a thermal energy storage unit for AA-CAES with molten *HITEC* salt used as a storing material.

2 Prototype design of the TES unit

A scheme of the lab-scale prototype of TES unit integrated with heat exchanger is shown in Fig. 2.

A counter-flow, helical coil heat exchanger is used to transfer heat from the air to the molten salt (*HITEC*). The heat exchanger is located inside a high-temperature thermocline-type thermal energy storage unit with molten salt used as a storing material.

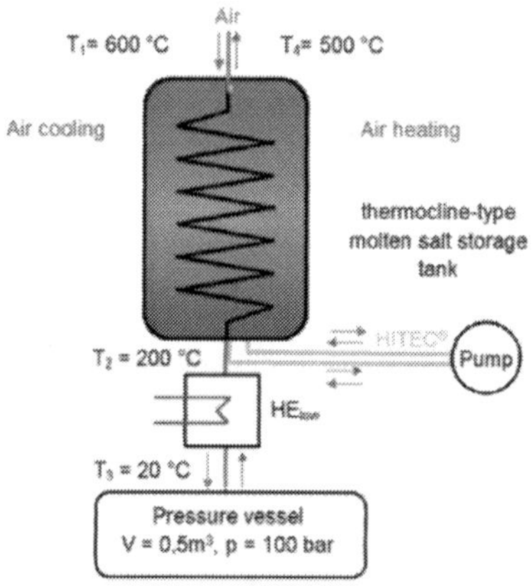

Fig. 2. The lab-scale prototype scheme; HE_{low} – low temperature heat exchanger; T_1, T_2 – air temperatures at the inlet and outlet of the TES unit/heat exchanger, T_3 – final air temperature during charging, T_4 – final air temperature during discharging.

Compressed hot air flows through a tube (grey line) to the tank filled with molten salt. In the upper part of the tank begins the counter-flow, helical coil tube-in-tube heat exchanger (black polyline) with the air flowing in the inner tube and molten salt in the outer annulus. At this point, the molten salt may freely flow in and out of the heat exchanger (short yellow line) – the direction depends on whether heat is being absorbed or released. After cooling down to around 200°C, air leaves the tank and flows through the low temperature heat exchanger

 https://doi.org/10.1051/e3sconf/20187001015

(e.g. pressurised water – air heat exchanger) to reach the pressure vessel, where it is stored at approximately 20°C and pressure 100 bar. The counter-flow of the molten salt shall be provided by an outer pump. The pump shall be connected to the stub of the helical coil heat exchanger as well as directly to the tank's inside (yellow lines).

As for the geometry of the designed heat exchanger, a tube-in-tube helical coil type was chosen due to its higher convective heat transfer coefficients in comparison with straight tubes. To find the most suitable case, four helical coils with different dimensions (diameter, pitch) were considered – heat exchangers available on the market were taken into account. Their dimensions and a scheme are shown in Table 1 and in Fig. 3.

Table 1. Dimensions of the tubes in the tube-in-tube heat exchanger (in mm).

Configuration #		1	2	3	4
Inner tube (air)	$d_{i,o}$	6.4	9.5	12.7	19.1
	$d_{i,i}$	4.6	7.1	9.3	15.7
Outer tube (molten salt)	$d_{o,o}$	12.7	19.1	25.4	31.8
	$d_{o,i}$	10.9	16.7	22	28.4
Coil diameter	D	101,6	152,8	203,2	254,4
Pitch	H	15	24	30	37

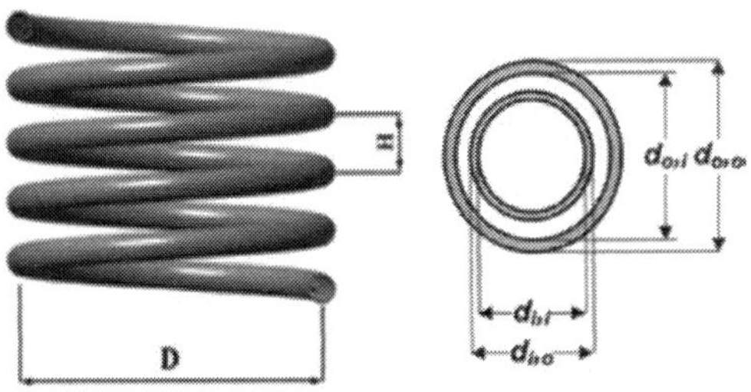

Fig. 3. A scheme of a tube-in-tube helical coil [14].

3 Heat transfer calculations

The specific aim of the work was to design a heat exchanger (heat storage unit) of a given internal structure that would ensure the cooling of air by 400 degrees (from 600°C to 200°C), while warming up the *HITEC* salt from 160°C to 530°C. With the assumed mass flow rate of air (0.01 kg/s) and, thus, resulting from a simple heat balance calculations mass flow rate of the storage medium (0.0074 kg/s), the size of heat exchanger is the main result of the analysis.

The first step in the heat transfer analysis included identification of the thermo-physical properties of air, *HITEC* salt and stainless steel 304H in the range 150-600°C. Although these were easily accessible for *HITEC* salt [15] and stainless steel 304H [16], that was not the case for air because the vast majority of sources provide information for pressure $p_a = 1$ bar. As it was reported by [17], a relative error caused by applying an ideal gas model and its properties for air instead of exact values is less than 5%. Thus, all of the air properties at pressure p = 100 bar were calculated on the basis of the ideal gas model.

All the calculations were carried out using empirical correlations. These in turn require parameters' values at the given temperatures. In such a wide temperature range it is unacceptable to employ single values at e.g. 350°C (arithmetic average) as it would lead to some serious errors. Thus, in order to enhance the accuracy of calculations, the whole heat exchanger was divided into 20 sections – with the assumption that air temperature drops by 20°C in each section. Although inlet and outlet temperatures at the air side are fixed, these still have to be determined for the *HITEC* salt side. Here, a simple law of energy conservation for an isolated system was employed:

$$T_{H,in,i} = T_{H,out,i} - \frac{\dot{m}_a c_{a,i}(T_{a,in,i} - T_{a,out,i})}{\dot{m}_H c_{H,i}} \tag{1}$$

In addition, log mean temperature difference was computed:

$$\Delta T_{ln,i} = \frac{(T_{a,in,i} - T_{H,out,i}) - (T_{a,out,i} - T_{H,in,i})}{\ln\left(\frac{T_{a,in,i} - T_{H,out,i}}{T_{a,out,i} - T_{H,in,i}}\right)} \tag{2}$$

where: $T_{out,i}$ – outlet temperature at the i-th section, °C; $T_{in,i}$ – inlet temperature at the i-th section, °C; $\dot{m}$ – mass flow rate, $\dot{m}_a = 0{,}01$ kg/s; $\dot{m}_H = 0{,}0074$ kg/s; C_i – specific heat for an average temperature of the fluid in the i-th section, J/(kg·K). It was assumed that $T_{a,in,1}$ = 600°C and $T_{H,out,1}$ = 530°C as the maximum suggested operating temperature for air and for the *HITEC* salt [15]. Indices 'a' and 'H' refer to the air and *HITEC* salt, respectively.

For each section a computational analysis was performed which included sequential calculation of:

- characteristic dimensionless numbers,
- convective heat transfer coefficients,
- heat transfer coefficients related to the unit length,
- required lengths of the sections,
- overall heat transfer coefficients.

Based on the value of Reynolds number, Nusselt number for every section was calculated using one of the two following empirical correlations [18, 19]:

- turbulent flow, for $10^4 < \mathrm{Re} < 5 \cdot 10^6$ and $0.6 < \mathrm{Pr} < 2500$:

$$Nu = 0{,}021 Re_f^{0,8} Pr_f^{0,43} \left(\frac{Pr_f}{Pr_w}\right)^{0,25} \varepsilon_L \tag{3}$$

- laminar flow, for Re < 2000:

$$Nu = 0{,}15 Re_f^{0,33} Pr_f^{0,43} Gr_f^{0,1} \left(\frac{Pr_f}{Pr_w}\right)^{0,25} \varepsilon_L \tag{4}$$

where: ε_L – correction factor for $L/d < 50$; L – length of the tube, d – inner diameter.

The correlation for turbulent flow was used for air flow, where Reynolds numbers were in the range from 22,000 to 105,000, depending on the configuration and the position of the section in the coil. The correlation for laminar flow was used for molten salt flowing in the annulus gap – Reynolds number varied in the range from 40 to 1140.

For the curved conduits correlation (3) or (4) must be multiplied by a correction factor ε_R [10]:

$$\varepsilon_R = 1 + 1{,}77\frac{d}{R} \tag{5}$$

where: R – radius of curvature, d – conduit's inner diameter.

For the characteristic numbers with index 'f' the value of arithmetic mean of inlet and outlet temperatures was used in the determination of the physical properties, whereas index 'w' stands for a wall temperature as a reference.

Convective heat transfer coefficient for the i-th section, α_i, was computed from the formula:

$$\alpha_i = \frac{Nu_i \lambda_i}{d_h} \tag{6}$$

Heat transfer coefficient related to the unit length for the i-th section, $k_{L,i}$, was computed from the formula [18]:

$$k_{L,i} = \left[\frac{1}{\pi}\left(\frac{1}{\alpha_i d_{i,i}} + \frac{\ln(d_{i,o}/d_{i,i})}{2\lambda} + \frac{1}{\alpha_o d_{i,o}}\right)\right]^{-1} \tag{7}$$

where α_i and α_o represent convective heat transfer coefficients at the inside and the outside of the inner tube; λ stands for thermal conductivity of the wall material.

The required length of the i-th section, L_i, was calculated from the formula [18]:

$$L_i = \frac{\dot{Q}}{\Delta T_{ln,i} k_{L,i}} = \frac{\dot{m}_a C_{a,i}(T_{a,in,i} - T_{a,out,i})}{\Delta T_{ln,i} k_{L,i}} \tag{8}$$

Overall heat transfer coefficient related to the outer surface of the inner tube for the i-th section, U_i, was computed from the formula [18]:

$$U_i = \frac{\dot{Q}}{\Delta T_{ln,i}\pi d_{i,o} L_i} = \frac{\dot{m}_a C_{a,i}(T_{a,in,i} - T_{a,out,i})}{\Delta T_{ln,i}\pi d_{i,o} L_i} \tag{9}$$

Results of the calculations are listed in Table 2, where arithmetic mean of the values of heat transfer coefficients related to the unit length ($k_{L,avg}$) and overall heat transfer coefficients related to the outer surface of the inner tube (U_{avg}) are given. Total length of the coil and minimum height of the helical coil for each case are the most important results of the analysis since they determine the size of the heat exchanger. Variations of an overall heat transfer coefficient along the pipe, in subsequent sections are shown in Fig. 4. The results presented in the table and in the Figure show an interesting relation between diameter of the coil and an overall heat transfer coefficient. With increasing of the diameter both $k_{L,avg}$ and U_{avg} decrease due to the decrease of convective heat transfer coeffi-cients, which is associated with decreasing of the veocity of the fluids. However, for the largest coil considered this trend has changed, which can be explained by non-line-arity of the relations of heat transfer coefficients and surface area as the functions of the coil diamater (as in the case of critical radius of insulation [18]).

Table 2. Arithmetic mean of the values of the heat transfer coefficients related to the unit length ($k_{L,avg}$) and overall heat transfer coefficients related to the outer surface of the inner tube (U_{avg}).

Configuration #	1	2	3	4
$k_{L,avg}$, W/(m·K)	6.38	5.55	5.29	5.56
U_{avg}, W/(m^2·K)	317.31	276.20	263.07	276.50
Total length, m	12.61	14.50	15.22	14.50
Min. height of the helical coil, m	0.59	0.72	0.71	0.67

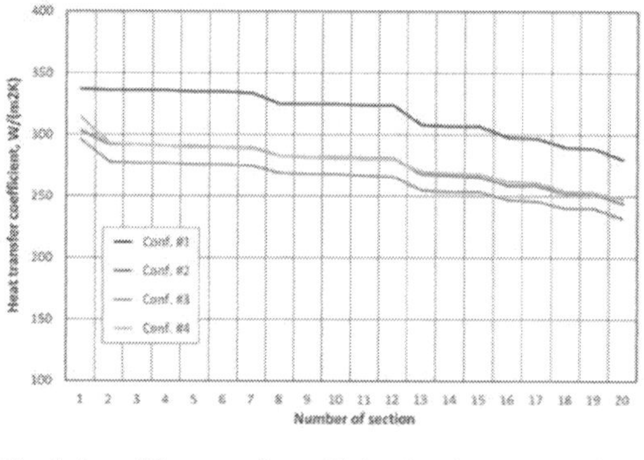

Fig. 4. Overall heat transfer coefficient in subsequent sections.

4 Pressure drop calculations

To choose the most suitable dimensions of the tubes an additional pressure drop analysis was conducted as it affects directly performance of the components following the TES, e.g. a turbine. The first step involved deter-mination of the friction factor for each of the 20 sections. Here, a formula proposed in [20] was used:

$$f = \frac{0.316}{Re^{0.25}} + 0.03\sqrt{\frac{d_{i,i}}{D}} \tag{10}$$

Subsequently, pressure drop at the i-th section, Δp_i, was calculated according to the formula (11) given by [18]. Table 3 illustrates the results of the calculations. Percentage pressure drop includes both phases of the operation, i.e. charging and discharging of TES unit.

$$\Delta p_i = f_i \frac{L_i}{d_{i,i}} \frac{\rho_i v_i^2}{2} \tag{11}$$

Table 3. Total and percentage pressure drop related to the air inlet pressure equal to 100 bar.

Configuration #	1	2	3	4
Total pressure drop, kPa	230.05	32.84	9.40	0.80
%	4.6	0.66	0.19	0.02

5 Results and discussion

It may be observed that the total heat transfer coefficient decreased slightly with growing dimensions of the tubes (configuration 1 to 4). However, in each of the cases it stayed at the same level of approximately 300 W/(m^2·K) in the first sections to decline below this value afterwards (Fig. 4). Hence, a simple heat transfer analysis did not determine the most suitable configuration.

On the other hand, differences in the results of pressure drop calculation are much more distinct. There is a clear tendency for the pressure drop to decrease as the inner diameter increases which is in agreement with the theory yielding $\Delta p_i \sim d^{-5}$ while keeping mass flow rate unchanged. Total pressure drop in the whole cycle charging/discharging related to the air inlet pressure equal to 100 bar experienced a significant decline along with dimensions growth. Starting at around 4.6% for the first configuration, it dropped to approximately 0.02% for the configuration #4. Keeping in mind that the overall heat transfer coefficients were roughly at the same level, configuration #4 was found the most suitable and thus, it was subjected to further analysis.

6 Numerical simulations of heat transfer and fluid flow in the heat exchanger

To verify the obtained results, an analysis based on numerical approach has been conducted with the use of ANSYS Fluent software.

To assure adequate accuracy of calculation, the computational domain for heat transfer and pressure drop analysis spreads only for the first two sections with the total length of 1080 mm. Grid system for the coils is shown in Fig. 5. It was composed of 453 768 elements and 468 318 nodes with max. element size of 1 mm within the boundary layer at the inner (air) side.

Solver type was set to pressure-based as within the analysed domain air density was assumed to be constant. Similar approach was adopted towards the rest of parameters of air, *HITEC* salt and AISI 304H stainless steel. Their values are listed in Table 4.

Table 4. Properties of air, *HITEC* salt and AISI 304H stainless steel used in numerical computation.

Parameter	Air	*HITEC*	AISI 304H ss
Density, kg/m^3	40.83	1740	8030
Specific heat, J/(kg·K)	1109	1567	500
Thermal conductivity, W/(m·K)	0.06	0.571	20
Dynamic viscosity, Pa·s	$3.851 \cdot 10^{-5}$	$1.3 \cdot 10^{-3}$	
Inlet temperature, °C	600	492	
Mass flow rate, kg/s	0.01	0.0074	

The realizable k-ε model of turbulence with enhanced wall treatment was adopted to simulate turbulent flow. The boundary conditions were set as follows: inlet temperature: air – 873 K, *HITEC* salt – 765 K; mass flow rate: air – 0.01 kg/s, *HITEC* salt – 0.0074 kg/s; outer wall heat flux – 0 W/m^2. To solve pressure-velocity coupling the SIMPLEC scheme was used as well as the second-order upwind scheme for the convection terms in the governing equations. Results of the simulation are shown as contours in Fig. 6 (inner tube, air) and Fig. 7 (annulus, *HITEC* salt).

Fig. 5. Grid system for helical coil tube-in-tube heat exchanger.

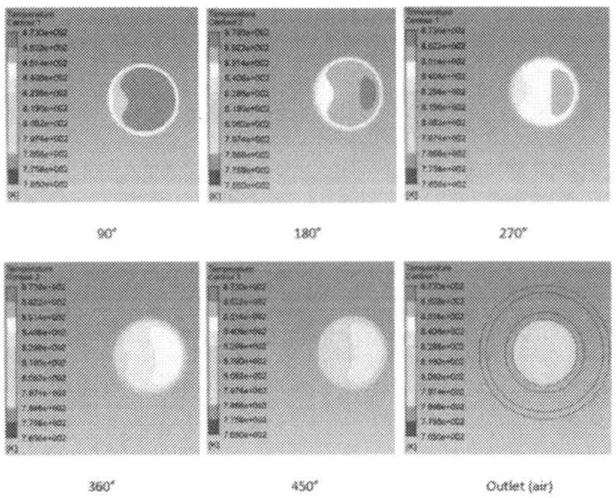

Fig. 6. Air temperature fields for inner tube; angle value is measured from the air inlet cross-section.

As it is depicted in Fig. 6, air outlet temperature is in the range of 819-829 K. Comparing these values to the one obtained by means of empirical correlations, that is 833 K, a slight discrepancy may be observed of roughly 4-14 K. The annulus region is characterised by larger differences in results as in the outlet (salt) temperature in a great part of the cross-section is in the range of 808-819 K. On the other hand, a significant part stays in the temperature area of 797-808 K. Here, the temperature obtained by empirical correlation was 803 K.

Concluding from Fig. 6 and Fig. 7, a clear influence of centrifugal force on the temperature distribution may be observed. As a result, secondary flows are induced and particles are forced to move from the inner to the outer part of the tube. Consequently, these radial flows cause more turbulence and so enhance heat transfer. The effects of centrifugal force are particulary visible in the inner region because of much higher velocities of air in comparison to those encountered in the annulus (approximately 1.3 m/s to 0.01 m/s).

A similar analysis was conducted for pressure drop in the first 2 sections. The results are depicted in Fig. 8. As it may be seen, pressure drop over these parts of the heat exchanger is in the range of 77-88 Pa. For comparison, formula (11) yielded 80 Pa. Thus, a relative error is of the order of a few per cents.

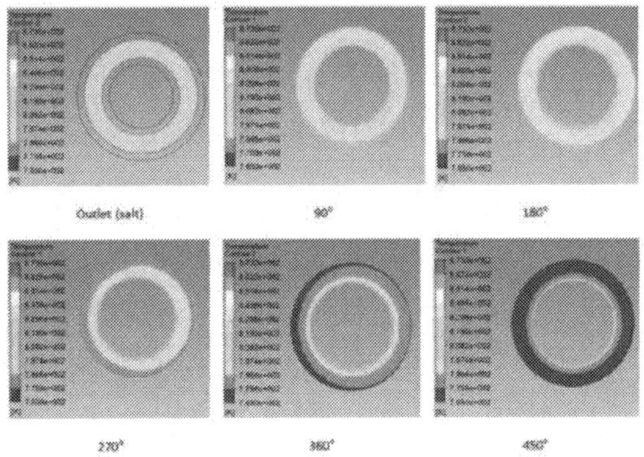

Fig. 7. *HITEC* salt temperature fields; angle value is measured from the air inlet cross-section.

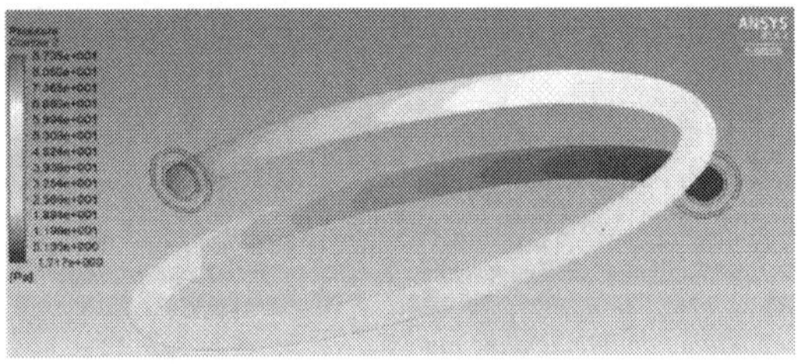

Fig. 8. Pressure drop over the first 2 sections expressed as an overpressure required to reach 0 Pa at the outlet cross-section.

6 Summary and conclusions

The final effect of the analysis performed in the study includes a design of a lab-scale prototype of a 17 kWh_{th} thermal energy storage unit for AA-CAES with molten *HITEC* salt used as a storing material and AISI 304H stainless steel as a wall material. Designed TES unit consists of a tube-in-tube helical coil heat exchanger, with tubes diameters 15.7, 19.1, 28.4, 31.8 and coil radius 127.2, pitch 37, height 670 and total length 14 500 (values in mm), located inside a high-temperature thermocline-type vessel with molten *HITEC* salt used as a storing material. The outer diameter and height of the vessel were set to be 400 and 800 mm, respectively. Air inlet pressure is equal to 100 bar. Its temperature drops from 600°C to 200°C while charging of the TES unit. On the other hand, *HITEC* temperature rises by approx. 370°C to reach 530°C at the outlet, operating at the pressure of 1.5 bar.

Considering the amount of salt contained only in the area surrounding the heat exchanger (around 106 kg), the capacity of the TES unit was estimated to be approx. 61.4 MJ/17 kWh. Given the salt mass flow rate of 0.0074 kg/s, the heat exchanger may operate for nearly 4 h at its design point. The proposed height and diameter of the vessel are the minimum ones to enclose the heat exchanger. However, by extending the dimensions of the vessel, it is possible to store more salt and so to lengthen the operational time while keeping the inner part, i.e. heat exchanger, unchanged. Such a flexibility broadens significantly the area of potential application.

It was proved that it is possible to obtain an effective cooling of compressed air in a CAES system by heat storage media using simple unit consisting of heat storage tank integrated with a heat exchanger.

References

1. M. Budt, D. Wolf, R. Span, J. Yan, Appl. Energ., **170**, 250 (2016).
2. M. Minutillo, L.A. Lubrano, E. Jannelli, J Energ. Storage, **4**, 135 (2015).
3. H. Xue, A. White, J. Energy Storage, **18**, 349 (2018).
4. G. Zanganeh, R. Khanna, C. Walser, A. Pedretti, A. Haselbacher, A. Steinfeld, Sol. Energy, **114**, 77 (2015)
5. G. Zanganeh, M. Commerford, A. Haselbacher, A. Pedretti, A. Steinfeld, App. Therm. Eng. **70**, 316 (2014).
6. L. Geissbühler et al., J. Energy Storage, **17**, 129 (2018).
7. V. Becattini et al., J. Energy Storage, **17**, 140 (2018).
8. A. Ferrara, G. Yenetchi, R. Haslett, R. Kosson, NASA Technical Reports 19800021356 (1977).
9. M.M. Kenisarin, Renew. Sust. Energ. Rev. **14**, 955-970 (2010).
10. J.E. Pacheco, et al., J. Sol. Energ-T ASME, **124**, 153 (2002).
11. Z. Yang, and S. V. Garimella, Appl. Energ., **103**, 256 (2013).
12. S.M. Flueckinger, S.V. Garimella, Appl. Energ., **116**, 278 (2014).
13. A. Pizzolato et al., Energy, **139**, 231 (2017).
14. A.K. Pathak, Y.K. Tembhurne, M. Gangwar, I.J. Cur. Res. Rev. **9**, 19, 40 (2017).
15. HITEC® Heat Transfer salt parameters; Retrieved from http://stoppingclimatechange.com (last access August 2018).
16. AISI 304H stainless steel parameters, Retrieved from https://www.steeleaglemalta.com (last access August 2018).
17. M. Szybiak, BSc Thesis, Warszawa (2018).
18. Y.A. Cengel, A.J. Ghajar, *Heat and Mass Transfer. Fundamentals and application.* (The McGraw-Hill Co., 2011).
19. R. Laskowski, A. Smyk, A. Rusowicz, A. Grzebielec, Entropy, **18**, 12, 444 (2016).
20. P. Mishra, S.N. Gupta, Ind. Eng. Chem. Process Des. Dev. **18**, 130 (1979).

Analysis of the operation characteristics of a selected wind power plant

Karol Tucki[1,*], *Anna* Bączyk[2], *Izabela* Wielewska[3], and *Michał* Piątkowski[1]

[1]Warsaw University of Life Sciences, Department of Production Organization and Engineering, Nowoursynowska Street 164, 02-787 Warsaw, Poland
[2]Warsaw University of Life Sciences, Department of Hydraulic Engineering, Nowoursynowska Street 159, 02-776 Warsaw, Poland
[3]UTP University of Science and Technology in Bydgoszcz, Department of Agronomy, Fordońska Street 430, 85-790 Bydgoszcz, Poland

Abstract. The purpose of this paper was to analyse the operation of a wind farm and the wind turbine yields with a total capacity of 46.5 MW. The analysis was divided into two stages: (1) the general characteristics of the facility, which included: a) distribution of wind speeds and the method of calculation and visualisation of the estimated electricity production, b) evaluation of the wind farm operation as per type of tested turbine and the wind farm terrain and (2) the impact of the facility on the quality of electricity in the distribution network.

1 Introduction

Wind energy in Poland is a dynamically developing sector of renewable energy sources [1-12]. Currently, the installed capacity in the wind farms alone is almost 6 GW, which is about 14% of the power installed by domestic power plants [13,14]. Wind conditions in the country have forced the development of investments mainly in the north-western part of the country. However, in many other regions, the landscape is also diversified by revolving rotors. The nearest future of investment in wind energy in Poland is shifting to the Baltic Sea, where large private and public companies plan to invest in MW offshore wind farms [15-20].

The most important advantage of wind energy is the fact that wind is a source of renewable energy that will never run out, unlike such raw materials as gas or coal, the ability to gain partial independence from electricity suppliers and a significant reduction of pollution. According to calculations, by replacing coal power plants with wind farms we could reduce the emission of carbon dioxide (700 g/kWh), nitrogen oxides (4.2 g/kWh), sulphur dioxide (5.5 g/kWh) and dust (49 g/kWh) [21]. However, wind power has a significant disadvantage, i.e. unpredictable electricity production as a consequence of wind speed variability.

The purpose of this paper was to analyse the operation of a wind farm and the wind turbine yields with a total capacity of 46.5 MW. The analysis was divided into two stages: (1) the general characteristics of the facility, which included: a) distribution of wind speeds and the method of calculation and visualisation of the estimated electricity production, b) evaluation of the wind farm operation as per type of tested turbine and the wind farm terrain and (2) the impact of the facility on the quality of electricity in the distribution network.

2 Materials and methods

This article is based on data originating from a company operating in the private sector. Accordingly, all data contained herein constitute a trade secret. Therefore, any information subject to industrial property rights and copyright, as well as any other information that could potentially lead to the identification the farm described herein has been intentionally omitted. At the same time, it should be emphasised that the paper is based on actual studies from different periods of operation of the farm, as well as data originating from project analyses. In this paper, the original name of the wind farm was replaced by the abbreviation WF, and the wind turbines composing the wind farm – the abbreviation WT(s), with the relevant index - A and B, for two types of generating units.

The wind farm in question covers an area of over 40 km^2 and is one of the largest facilities of this type in Poland. Its basic parameters are composed by (Fig. 1):

1) 12 turbines with a unit nominal power of 2 MW and a hub height of 105 m and a rotor diameter of 100 m,

2) 15 turbines with a unit nominal power of 1.5 MW and a hub height of 80 m and a rotor diameter of 82 m.

The total installed capacity of the units is 46.5 MW. Each turbine is equipped with a transformer increasing the voltage to 30 kV and connected to the main distribution point, R-30 kV. The final power output runs through a 115/30 kV block transformer over a 110 kV overhead line to the national power grid. The ratings of each turbine are shown in Tab. 1.

An important element of the WF (Wind Farm) equipment is a measuring mast with five wind speed sensors and two weather vanes placed at a height of about 30 to 60 m above the ground level. In addition, each wind turbine has an anemometer and a weather

* Corresponding author: karol_tucki@sggw.pl

vane installed at the top of the gondola. Weather analyses are carried out on a continuous basis during the operation of the wind farm. The expected service life of the facility is 20 years.

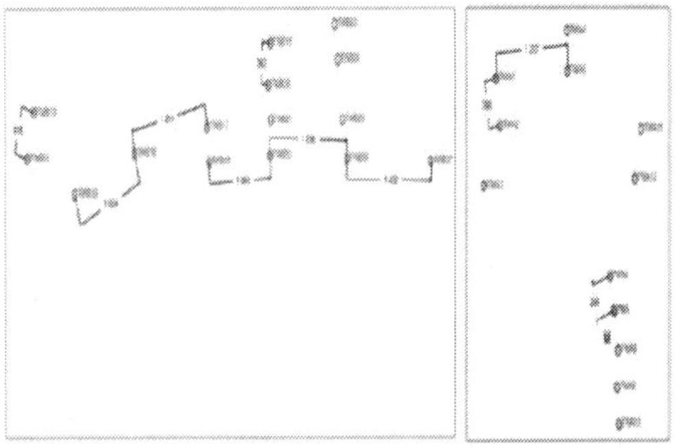

Fig. 1. Overview of a wind farm plan.

Table 1. Ratings of the most important elements of WTs.

Parameter	WT A	WT B
Number of turbines	12	15
Min. wind speed	3 m/s	3 m/s
Max. wind speed	22 m/s	25 m/s
Capacity	2.000 kW	1.500 kW
Operating temperature	-20°C to + 45°C	-30°C to + 40°C
Standard (wind class)	IEC IIb	IEC IIa
Diameter of the rotor	100 m	82 m
Rotation area	7.854 m^2	5.289 m^2
Type	Three degrees; two planetary segments and one screw segment	Three degrees; one planetary segment and two screw segments
Hub height	105 m	80 m
Frequency	50/60 Hz	50/60 Hz
Type	4-pole (50 Hz) / 6-pole (60 Hz) generator with dual power supply	6-pole (60 Hz) generator with dual power supply
Voltage	690 V	12.000 V
Rotation speed	9-17 rpm	16,7 rpm
Power [kV·A]	2100	1700, 350
Transmission [kV]	30/0.69	30/12
Weight	70 t	67.5 t
Hub height	5.4 m	4.1 m
Length	10.4 m	8.7 m
Width	3.5 m	3.4 m
Starting point	3 m/s	3 m/s
Switching point	22 m/s	25 m/s
Rated operating point	12.5 m/s	13.5 m/s

Analysing the distribution of domestic wind energy resources, the WF is located in the third (III) wind zone, on a five-point scale, and the conditions are regarded as favourable [18]. The average annual wind speed in the area of location of wind turbines is 2.8 m/s. The strongest winds occur in November reaching an average speed of 3.4 m/s, while the lowest winds occur in August, with an average speed of 2.2 m/s. The most frequent wind direction of the WF location is from the west (22%).

3 Results

Data analysis was carried out on the basis of measurements taken in the period from 1 February to 30 March 2011. The data from the C-1 and C-5 anemometers and the A-1 weather vane were used. The calculations were made for the whole wind power plant complex on the basis of the WF power curve (Fig. 2). The altitude was assumed for the gondola of WT_B, i.e. 80 m above the ground level.

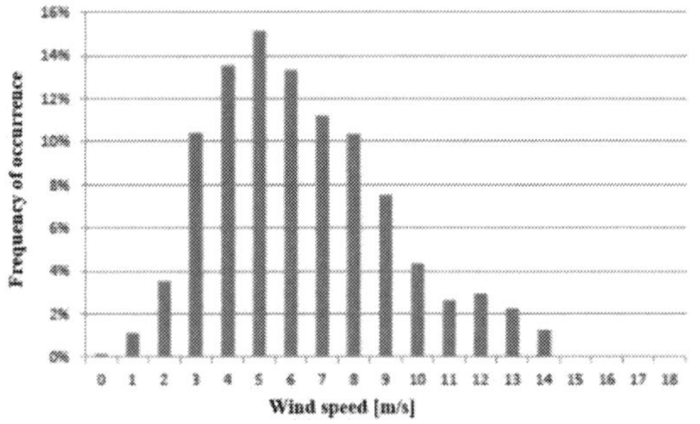

Fig. 2. Distribution of wind speed in the period from 1 February to 30 March 2011.

It can be observed that the prevailing speeds range from 3 to 8 m/s. They generate approx. 50% of the nominal gross power according to the information from the power curve of the WF (Fig. 4). Based on the power curve measurement data for the entire WF, the total amount of energy produced by the WF, on the assumption of continuous trouble-free operation of all units, would be 27.3 GW·h. The wind, most frequently blowing from the west, would generate almost 8 GW·h. If divided into individual types of wind farms, WT_A would generate about 15 GW·h, and WT_B - 12 GW·h (Fig. 3).

On the basis of the data of power production by individual turbines, an analysis of energy efficiency was carried out taking into account characteristic features of distribution in the field, the value of energy produced and the model of the turbine.

Wind farm production in May 2016 (Fig. 4) was not among the highest. You can observe relatively equal energy production by each WT of a given type, which indicates full availability of the turbines. The left side of the graph shows the values of energy produced by 12 WT_A type of turbines (WT_A - Wind Turbine type A). The discrepancies in results are due to the degree of production by the turbines of the same type, which can be understood when analysing the topography of the area and the configuration of the distribution of the generating units.

 https://doi.org/10.1051/e3sconf/20187001016

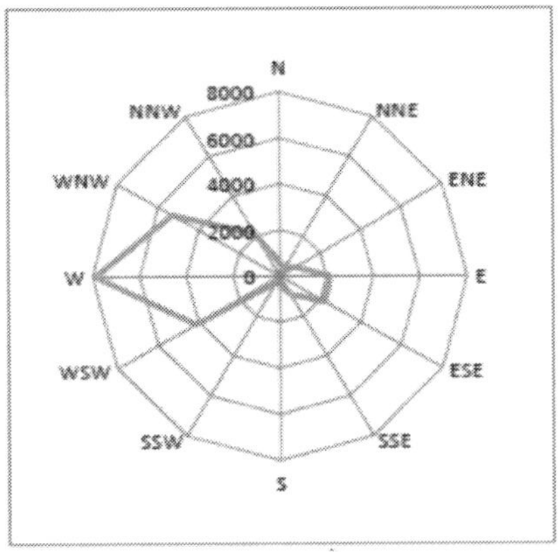

Fig. 3. Distribution of energy – Power rose.

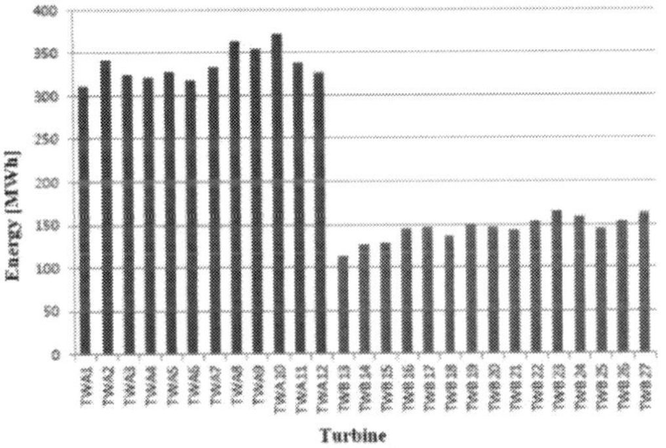

Fig. 4. Total amount of gross power generated by a single turbine in May 2016.

The first analysis was based on a test report verifying compliance with the parameters contained in the Distribution Grid Operating and Maintenance Manual (Grid Code - IRiESD). The manual describes in detail technical requirements for connecting the WF to the transmission network, and it is the operator's responsibility to present this report in the first year of operation of the WF.

The test aims to determine the power characteristics of the WF as a function of wind speed (WTs – Wind Turbine speed). The result will help to determine:

- the minimum wind speed at which WF and the individual WTs A and WTs B groups will start generating power,
- the minimum wind speed at which WF and the individual WTs A and WTs B groups will reach nominal power,
- the maximum wind speed at which WF and the individual WTs A and WTs B groups will shut down due to technical possibilities.

Observations were made for 320 days with full availability of turbines.

On the basis of the registered data, characteristics of the gross power capacity were determined for the whole wind farm (Fig. 5).

Based on the results of the data recorded during the observation of the WF, in the free operation mode, it is estimated that:

- the maximum gross power that the WF generates is approx. 47 MW,
- the minimum wind speed at which WF starts to operate is $v \approx 3.5$ m / s,
- the minimum wind speed at which WF reaches its nominal power is $v \approx 14$ m / s. Above this speed, the nominal power of the WF is maintained most accurately.

With the wind blowing at the speed of:

a) v=3.5-7.5 m/s , the characteristics change non-linearly from 0 to 21 MW,
b) v=7.5-9.5 m/s, a linear change in the characteristics was adopted, ranging between 21-38.5 MW,
c) v=9.5-13 m/s again, the characteristics change non-linearly from 38.5 to 47 MW,
d) v>13 m/s the characteristics curve is linear, with a constant power of 47 MW.

During the trial, turbines were not shut down due to any technical limitations caused by the maximum wind speed.

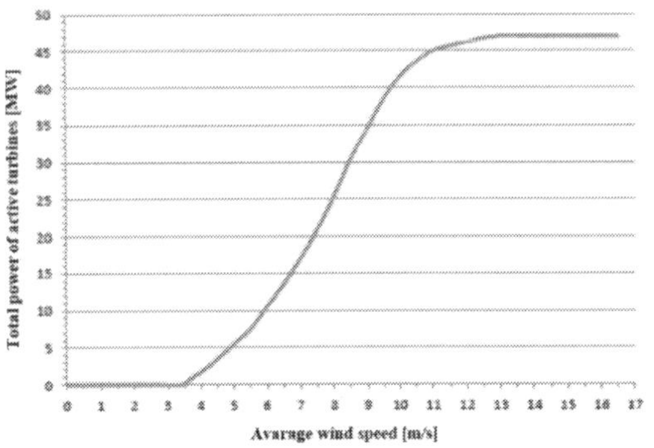

Fig. 5. Characteristics of the WF gross power as a function of wind speed.

As required by the DSO (Distribution System Operator), WFs must maintain power quality standards as a prerequisite for proper operation of wind farms. This is related to compliance with criteria defined in the Grid Code.

The quality of electricity was evaluated by analysing the measurements at the WF power output (GPZ R-110 kV). Reference was made to the following results: voltage, current, active and reactive power on the 110 kV line, asymmetry of voltages and currents, higher voltage harmonics and the resulting THD_U voltage distortion coefficient.

Based on an hour-long measurement of grid parameters, in which a free operational mode, shutdown, start-up and downtime of the WF were simulated, no sudden changes or surges in the voltage in excess of 3% at the WF connection point were observed. Neither were there any cyclical disturbances (Fig. 6). The maximum distortion of the THD_U coefficient (THD - Total Harmonic Distortion), over 99% of the observation time, did not exceed 2%. Accordingly, it may be concluded that the operation of the WF does not have a significant impact on voltage distortions.

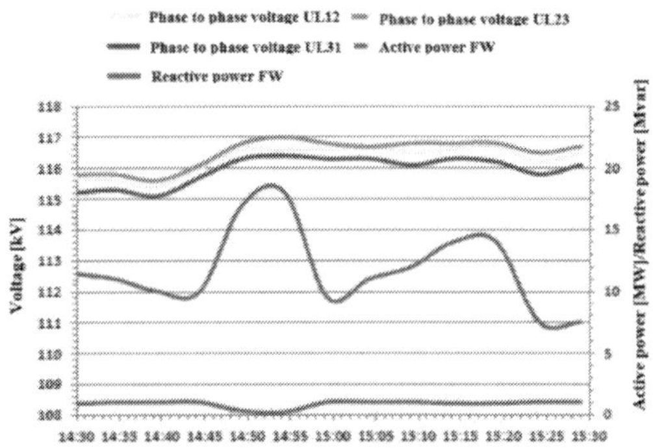

Fig. 6. GPZ R-110 kV. Effective voltage values as well as active and reactive power in the one-hour registration period.

The following characteristics (Fig. 7) shows the minimum voltage values due to a short circuit in the grid, for which the WF should remain in operation and allow for improvement of voltage conditions. In order to analyse the requirements, a reference must be made to the technical documentation for both types of turbines. When analysing the WTs technical documentation, it can be concluded that both types of turbines meet the Grid Code requirements in terms of occurrence of disturbances in the National Power Grid in the form of short-circuits.

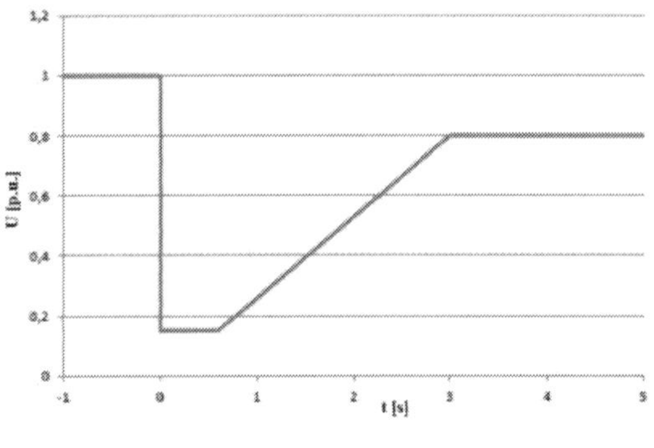

Fig. 7. Characteristics of the required WF operating range in case of network disturbances.

Another analysis was an attempt to verify changes in active power during the suspension of the free operating mode of the WF and the launch of the free operating mode of the WF. In addition, the WF''s demand for active and reactive power was verified after all turbines were stopped.

The values recorded during the test trial are shown in Fig. 8. A zero production value of active power PSP = 0 MW was set, which caused a sequential shutdown of the wind turbines. The setting was executed when the measured active power value at the connection point was determined (approx. 12.5 MW - on the axis of the graph - 60%). The 0 MW value was obtained after 4.5 minutes with an average gradient of 2.85 MW/min. During the programmed shutdown, the maximum active power consumption of the WF was PWF=-0.83 MW, while the reactive power fluctuated at ± 1 M_{var}. The behaviour of the WTs should be noted, which involved cyclic, temporary switching on and off of some of them, in order to compensate for the WF's active power demand and obtain the highest accuracy of limitation of the generation to the set active power PSP = 0 MW.

The launch consisted in restoring the WF to free operating mode. The generated power with an average gradient of 3.09 MW/min reached 18 MW, which was due to the then-current wind conditions.

Due to the inability, at the level of the SCADA system, to simultaneously discontinue the operation of all WTs, the WF demand for active power test was carried out along with the trial test of the WF's behaviour, i.e. reaction to frequency changes.

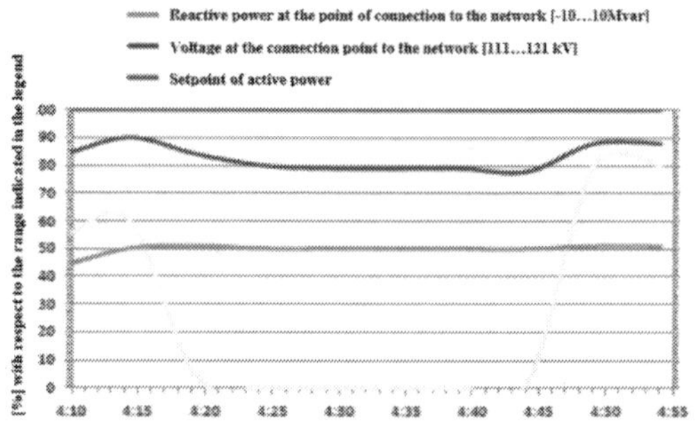

Fig. 8. Progress of the programmed trial of shutting down and starting the WF using the SCADA system.

What is significant importance about the test, is its option to discontinue the operation of the WF by the task of limiting the production of active power. Reactive power during the suspension of the operation of the WF at the point of connection to the National Power System was at the level of QWF = 0 M_{var}. The WT was switched on and off during the test in order to balance power for the WF's own needs. The Grid Code requires a maximum of 30% Pn/min in terms of the average gradient of power changes, which in the case of the analysed WF results in values below 13.95 MW/min. These requirements were met both for the suspension and start-up phase of the operation of the WF.

Following the trial tests, the case where the WF's power output was controlled during intervention work was examined (Fig. 9). Intervention work is a service provided by the WF to the National Power System in order to ensure real-time balance and management of active and reactive power flows. The purpose of such analysis is to verify the ability of the WF to maintain the set values of active power and to determine the rate of change of such set values.

The trial was initiated during free operation, with full availability of WTs. The wind speed provided good conditions for generating active power at about 17 MW. The active power generation limits were set in the following order:

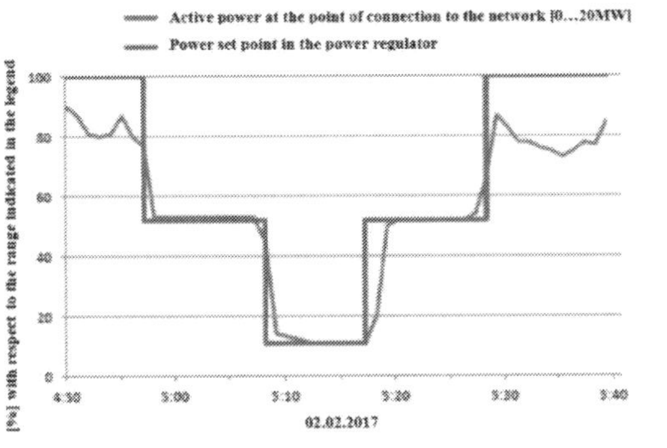

Fig. 9. Operation of the active power regulator of the WF during interventional work PSP = 46.5MW → 10.5MW→ 2.3MW→ 10.5MW→ 46.5MW.

The characteristics (Fig. 9) show slight deviations of active power values from the set value in the controller. The maximum deviations of the measured active power values were observed for the set PSP value = 2.3 MW and amounted to -7.4% P SP. The criterion of accuracy in maintaining the set value was 5%, which resulted in the excess of -55 kW. The average gradients of changes were as follows:

- 11.1 MW/min during intervention lowering of active power from 46.5MW to 10.5 MW,
- 8.59 MW/min during intervention lowering of active power from 10.5 MW to 2.3 MW.

Next, the WF's reaction to frequency changes during free operation were tested. Due to the limitations of the control system, the test was carried out independently for two types of turbines - WT_{A1-12} and WT_{B13-27}. During the test, the wind conditions enabled active power production outputs in the following ranges:

- WT_A: PFW ≈ 22-45% PN, i.e. 10-21 MW,
- WT_B: PFW ≈ 15-22% PN, i.e. 7 MW.

In the free operating mode, changes in the frequency received by the WT_A SCADA system were simulated, with the WT_B and the WT_B SCADA switched off, with the WT_A switched off. In both cases, the range of frequency changes was 47.5-51.6 Hz according to the following values:

f_{sym} = 47.5Hz→50 Hz→50.6Hz→51Hz→50.6Hz→51.6 Hz→51Hz→50,4Hz→ end of simulation

The reduction of power as a function of frequency is set in accordance with the parameters stated in the Grid Code. During free operation, the reference power is equal to the set power, and when the frequency increases above 50.4 Hz - in accordance with the applicable limits. Frequency changes in the fsym range = 47.5-50.5 Hz do not cause disturbances in the operation of the WTA turbines (Fig. 10) and they continue to generate power at a level resulting from wind conditions. An increase in frequency above 50.5 Hz results in a reduction of the power produced in proportion to the frequency obtained, until the turbines are completely shut down at the 51.5 Hz limit frequency. When assessing the rate of power change, the high accuracy of the response of the WT_A turbines can be observed. The turbines automatically resumed operation at a simulated frequency fsym = 51 Hz .

In case of WT_B turbines (Fig. 11), the response structure to changes in the simulated frequency value is similar to that of the WT_A turbines. However, the instantaneous power value, given the applicable limitations, is characterised by clear deviations.

Both types of turbines of the FW comply with the Grid Code requirements related to changes in the generated power in connection with frequency changes. It is assumed that the accuracy of power adjustment by the WT_B turbines could have been affected by unstable wind conditions during the test.

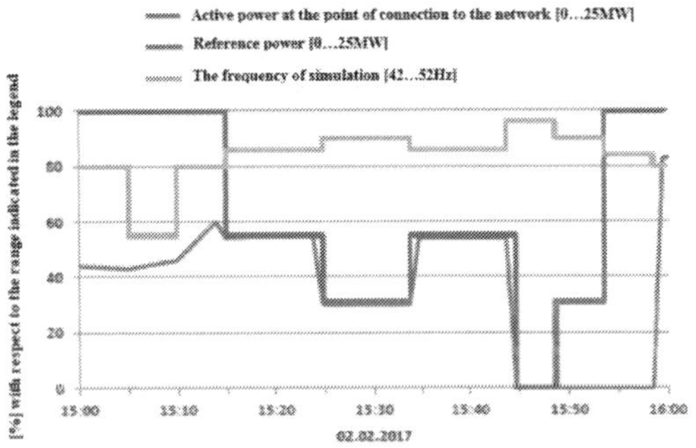

Fig. 10. Generation of active power by WT_A during the response test of the WF on frequency changes with inactive WT_B.

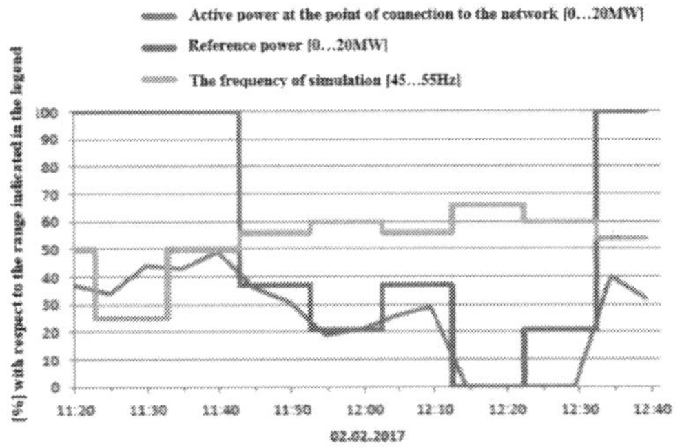

Fig. 11. Generation of active power by WT_B during the response test of the WF on frequency changes with inactive WT_A.

Another simulation consisted in verifying the dynamics and accuracy of the WF reactive power adjustment system and the evaluation of changes in the voltage value under the influence of reactive power adjustment. Changes in the reactive power value were set, and their values were read from, the recording device at the point of connection to the grid. The initial value of the set reactive power was 0 M_{var}, and then changed as follows:

QSP = +3 M_{var} → +4 M_{var} → -3 M_{var} → -4 M_{va}r → -4,5 M_{var} → +10,6 M_{var} → -10,6 M_{var} → 0 M_{var}

With the set power adjustment range, results with an accuracy not exceeding ± 5% were read from the recording device. As shown in Fig. 12, after the set point was achieved, the system was able to maintain the reactive power generation. The overshoot amplitude depends on the pitch of the change, e.g. ranging between +3 and +4 M_{var}, and -3 and -4 M_{var}, and is insignificant. Over time, these changes were from 1 minute for small pitches, and up to 3 minutes with pitches above 10 M_{var}.

The reactive power adjustment by the WF did not cause any significant voltage oscillations at the point of connection to the grid.

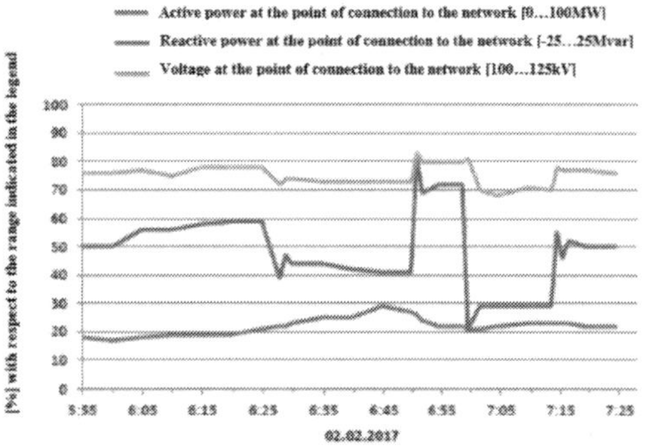

Fig. 12. Operation of the WF voltage regulation system operating in the reactive power adjustment mode.

The subsequent experiment consisted in verifying the dynamics and accuracy of the cosφ power coefficient of the adjustment system of the WF. In order to start the test, the cosφSP power coefficient adjustment mode was started. Next, in order to maintain the grid stability, the following operating steps were gradually introduced in the controller:

$$\cos\varphi SP = 1.0 \rightarrow 0.995i \rightarrow 0.985i \rightarrow 0.995p \rightarrow 1.0 \rightarrow 0.975p \rightarrow 0.975i \rightarrow 1.0$$

The quality of adjustment is not satisfactory though (Fig. 13). Considerable adjustment failures and poor accuracy of maintenance of the desired power coefficient were found. The immediate WF's response to the changes was evident. With regard to the reactive power adjustment, the WF's reaction, in the form of voltage at the grid connection point, is significant. However, the Grid Code guidelines do not define the operating ranges of the cosφ power coefficient adjustment systems.

Visible changes in the voltage at the connection point have a marginal effect on the operation and quality of the SE.

During the simulation of the reactive power production by the WF, maximum capacitive and inductive reactive power values were obtained. The wind conditions enabled generation at the production level of active power in the range of PWF = 32 -39 MW. The set points resulting from wind speed have been introduced to the reactive power controller, i.e.:

- Q_{SP1} = -15.3 M_{var}; during this time the active power generation was changed from 30 MW to 39 MW,
- Q_{SP2} = +10.8 M_{var}; during this time the active power generation was changed from 30 MW to 37.5 MW.

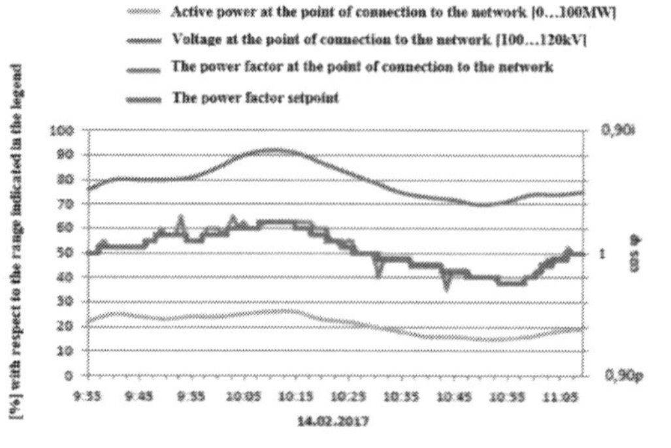

Fig. 13. Operation of the WF voltage regulation system operating in the power factor adjustment mode.

On the basis of the analysis, it was found that the WF, irrespective of generated active power, has the ability to transmit and receive reactive power due to the value of the power cosφ coefficient of 0.975 at the level of (Fig. 14):

- 10.3 M_{var}, with inductive grid load,
- -10.3 M_{var} with capacitive grid load.

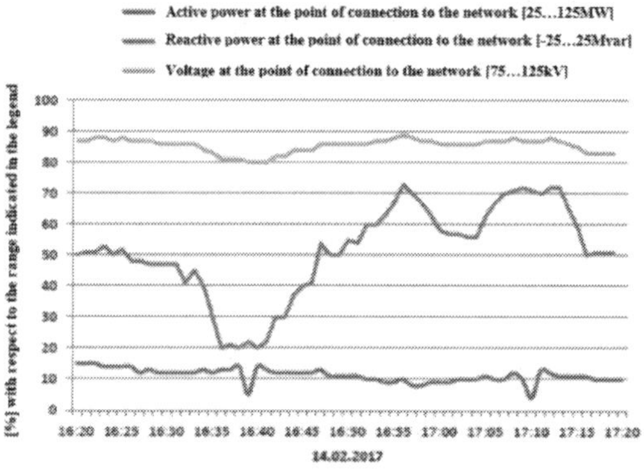

Fig. 14. Maximum reactive (capacitive and inductive) power generation process.

The values of reactive power at the point of grid connection during the test were maintained at +10,6 M_{var} and -15,3 M_{var}.

4 Conclusions

Based on the tests, conclusions can be drawn in three directions: (1) conclusions resulting from the analysis of atmospheric conditions at the wind farm, (2) those resulting from the wind turbines distribution and (3) those resulting from the interpretation of the parameters of the wind farm's impact on the quality of electricity. With regard to the first group of factors:

1) The incidence of the prevailing direction of the wind has a decisive impact on the distribution of wind farms in the field. The turbines were arranged optimally

with regard to the wind from the west and neighbouring directions.

2) The roughness of the terrain affects wind energy disturbances. Turbines were been placed in such a way as to minimise the impact of forestation, spatial development or natural unevenness of the area, on the generation of electricity.

3) Electricity production increases with increasing wind speed. For accurate calculations, the change in wind speed and altitude should be taken into account.

4) During the two-month wind measurement, the energy was calculated as GW·h. The share in the calculated generation of WT_A turbines was 56%, and the remaining , i.e. 44% was attributable to WT_B.

5) The analysis of energy measurements generated in May 2016 shows that the share of WT_A turbines of 65% in the total energy generation, and 35% in the case of WT_B, with power output of 6.18 GW·h. The highest production was obtained by the WT_{A10} turbine, generating 371 MW•h. The lowest value was obtained by the WT_{B14} – 112 turbine - 112 MW•h. It is estimated that such a poor result is influenced by the nearby forest which has an impact on the reduction of the mechanical wind energy quality.

Conclusions drawn from the distribution of the wind turbines:

1) Wind turbines are separated from the residential buildings at such a distance that the noise generated by the unit does not exceed 40 dB. In practice, no turbine is closer than 1 km from the residential buildings.

2) With respect to the line of the most frequent wind direction, the wind farms were arranged so that the neighbouring farm was not in line in front of the previous one. The reason is to avoid the effect of aerodynamic shading – the wake effect. The WT_{B14} and WT_{B15} turbines - 860 m are arranged closest to each other in the western wind line. They are arranged in such a way so that the WT_{B14c}turbine does not "shade" the WT_{B15} turbine.

Summing up the results related to the parameters characteristic of the wind farm's impact on the quality of electricity, it can be concluded that:

1) The wind farm begins its generation with wind blowing at a speed of about 3.5 m/s. The nominal power is obtained at an approximate speed of 14 m/s. These values are similar to those declared by the manufacturer.

2) During the simulation of suspension and start-up of the wind farm, there were no sudden voltage surges exceeding acceptable limits. No other interference was observed.

3). The wind farm meets the requirements of the IRiESD (Grid Code) related to the resistance (ability to continue operation) of the farm at the occurrence of a short-circuit in the grid. In addition, the farm installation is properly secured and allows continuous movement of the rotors in the event of this type of disturbance. During a 1-second fault, both types of turbines lose their ability to operate smoothly when the mains voltage drops to 0 in the lower limit, and over 120% of the nominal voltage, in the upper limit.

4) All regulatory tests are characterised by high accuracy and do not cause adverse changes to the grid connection point.

Taking into account the evaluation of the cases under consideration, it can be concluded that the WF in question does not affect the stability of the grid. The detected fluctuations were not related to the operation of the WF.

It should be emphasised here that the majority of tests consisted in performing simulations, not actual parameter changes. It is justified and reasonable, and related to the least interference in the system of power grids and simultaneous conduct of reliable research in order to mitigate the risk of failure.

When met, the conditions for the output power of the WF, do not allow for reflecting the full range of the analyses. The field tests allowed for the assessment of the WF adjustment capabilities as per the following variants: (1) start-up and discontinuation of the WF in free operating mode, (2) maintenance of the WF's operation in the event of a short circuit in the grid, (3) power adjustment in case of intervention work, (4) the WF's reaction to frequency changes, (5) voltage adjustment in the reactive power adjustment mode and the cosφ factor, (6) production capacity (transmission and consumption) of reactive power.

All the above operating modes and requirements for WFs have been assessed favourably. The wind farm has a beneficial effect on improving the quality of the grid. However, the essential parameter of the quality of operation of wind farms, that is their unpredictability, should be always considered.

References

1. R. Best, P.J. Burke, Energy Policy **118** (2018)
2. G. Dąbrowska, Elktro Info **1-2** (2018)
3. Z.Połecki, P. Kacejko, Rynek Energii **1** (128) (2017)
4. T. Popławski, M. Weżgowiec, Przegląd Elektrotechniczny **93** (2) (2017)
5. A. Augustyn, J.Kamiński, Rynek Energii **1** (134) (2018)
6. I.M. Łukasik, Rynek Energii **6** (97) (2011)
7. K. Badyda, H. Kapoń, Rynek Energii **3** (106) (2013)
8. A. Brzezińska-Rawa, J. Goździewicz-Biechońska, Renew and Sustain Energy Rev **38** (2014)
9. B. Igliński, R. Buczkowski, A. Iglińska, M. Cichosz, M. Plaskacz-Dziuba, Journal of Power Tech **95** (2) (2015)
10. J. Jurasz, P.B. Dąbek, B. Kaźmierczak, A. Kies, M. Wdowikowski, Energy 161 (2018)
11. J. Paska, T. Surma, Renew Energy **71** (2014)
12. A. Wędzik, T. Siewierski, M. Szypowski, Renewable and Sustainable Energy Reviews **75** (2017)
13. Mapa KSE. available at: www.pse.pl [07.09.2018]

14. Moc zainstalowana w instalacjach wykorzystujących energię wiatru (stan na dzień 31.03.2018). available at: www.ure.gov.pl [09.07.2018]
15. A. Dobrzycki, G. Wodnicki, Poznan Uni of Tech Acad Jour. Electr Eng **94** (2018)
16. K. Międlarz, P. Więcławski, 2018. Zagadnienia Aktualnie Poruszane Przez Młodych Naukowców 12. CREATIVETIME, Kraków
17. B. Igliński, A. Iglińska, G. Koziński, M. Skrzatek, R. Buczkowski, Renew and Sustain Energy Rev **64** (2016)
18. A. Ostrowska, Annales Universitatis Mariae Curie-Skłodowska. Sectio G. Ius **65** (1) (2018)
19. M.S. Maj, 2017. Offshore Wind Energy in Poland: Potential in Transformation of Polish Energy Sector. Engineer's Thesis. Warsaw University of Technology
20. M.M. Sokołowski, Energy Policy **106** (2017)
21. K. Kożuchowski, 2012 Meteorologia i klimatologia, Warszawa

Impact of key factors on expected development of onshore wind energy sector in Poland and development scenarios

Izabela Wielewska[3], *Karol* Tucki[1,*], *Anna* Bączyk[2], and *Magda* Trzaska[1]

[1]Warsaw University of Life Sciences, Department of Production Organization and Engineering, Nowoursynowska Street 164, 02-787 Warsaw, Poland
[2]Warsaw University of Life Sciences, Department of Hydraulic Engineering, Nowoursynowska Street 159, 02-776 Warsaw, Poland
[3]UTP University of Science and Technology in Bydgoszcz, Department of Agronomy, Fordońska Street 430, 85-790 Bydgoszcz, Poland

Abstract. The aim of the paper was to analyse the wind power market in Poland by reviewing the factors that shape and influence its current state and the possible development prospects. The paper was focused on legislative, environmental, manufacturing, sociocultural and economic factors. Barriers to the development of onshore wind power market and the expected development of wind energy in Poland in the years 2017-2020 were identified and measured based on a survey. The review of individual factors and the study performed present that legislative barriers and the introduction of the 'distance act' are factors with the biggest influence on the current stagnation of onshore wind energy sector. A review of the recommendations concerning the distance (from protected areas and housing) required to build wind farms set forth in literature shows that Poland is the only country with such harsh restrictions. With its good environmental conditions and technical capacities, Poland can become a European leader in the production of energy from wind. The only barrier is the legislative environment and political instability on the national level. Without improvements in this sector, there is no chance for new wind projects, as these factors are crucial for development of this type of energy.

1 Introduction

In the times of gradually diminishing resources of fossil fuels and a growing energy demand, renewable power seems to be a lifebelt for the global economy and the natural environment. Pursuant to directive of the European Parliament and of the Council no. 2009/28/CE [1-4], Poland was obligated to reach at least a 15% share of energy from renewable sources in gross final consumption of energy. Nevertheless, hard bituminous coal and lignite remain the main sources of energy in Poland, with the respective shares of 50% and 31.4% in the overall production of energy [5].

Wind energy is an important factor for Poland in the process of pursuing the EU's objective. Onshore wind power allows to obtain 5824 MW (69%) of the total of 8538 MW (as at 30.09.2017) [6] of installed capacity of RES. The economically and environmentally justified share of onshore wind power in the Polish energy market is estimated to amount to 10 GW, compared to today's 5.8 GW of installed capacity of onshore wind energy.

The onshore wind energy sector, developing dynamically since 2001, is now in stagnation caused by the implementation of new and amendments of the currently applicable legal deeds. The biggest of the barriers encountered so far is the Act of 20 May 2016 on investments in wind power stations, commonly known as the 'distance act'. The provisions that are especially important for the sectors apply to:

1) Restrictions concerning distance between wind turbines and occupied buildings/nature conservation areas,
2) Increase of the real property tax base,
3) Identification of wind farm location exclusively based on a local zoning plan.

The tragic situation of wind power is further enhanced by the current condition of the market of green certificates, i.e. negotiable certificates of origin confirming that electrical energy has been produced from renewable energy sources. The scale of oversupply estimated by PSEW (Polish Wind Energy Association) is the largest ever, with the price of green certificates being drastically low [7,8]. Trying to minimise or mitigate the problem, the Polish government introduced new amendments to the act on renewable sources of energy. The crucial changes in the renewed Act signed on 14 April 2017 apply to determination of the substitution fee, which now depends on market prices of certificates of origin [9]. Originally established at the level of PLN 300.03, it is now calculated based on 125% of the annual average weighted price of property rights in the previous calendar year. The act entered into force without prior consultations, and met with fierce criticism of representatives of the sector [10-15]. All these actions make the achievement of the objective set to Poland

* Corresponding author: karol_tucki@sggw.pl

impossible (a 15% share of energy from renewable sources in gross final consumption of energy) [16-20].

Poland has good wind conditions [21,22], but in order to be able to use them, it needs to create sustainable conditions for its development [23-25], in particular, by introducing stable legislation which is a standard in other countries [26-28].

The aim of the article is to present the current status of wind energy, indicate the significance of key factors for the prospects of its development and choose the most plausible development scenario.

2 Methodology

Two research techniques were used to achieve the objectives, i.e.: a survey conducted in February 2017 among employees directly employed in the wind energy field; and analysis, as part of which individual properties and incidents were identified and their relationship with the whole were described. Additionally, the research encompassed an overview of available scientific literature including legal deeds, technical articles, reports and studies. The survey, entitled "Analysis of development of wind energy in Poland", was sent to companies rendering services or delivering products for the purpose of wind energy production. 30 out of the 100 surveys sent were collected. The surveys were filled in by companies specialising in:

- environmental consultancy,
- noise measurements and monitoring for the existent and planned wind farms,
- electric power generation from renewable energy sources and provision of regulatory system services (PGE Energia Odnawialna S.A., a Polish state power company).

Relying on experts' knowledge and the literature available, 23 factors were distinguished with potential significance for the slowdown of wind energy in Poland [29-37]. The factors were divided into 5 groups (table 1) and the division was used as a basis for the Ishikawa diagram [38], with the aim to facilitate expert assessment of individual criteria. The groups of factors listed in table 1 were used as 2nd order objectives. Acting in accordance with the Ishikawa diagram principle, it was assumed that the impact of all groups of factors equals 100% (same as the impact of individual factors within the group). This principle was applied by the experts, who performed their assessment by breaking the 100% down into individual factors and showing their significance. The experts assessed the significance of individual factor groups and individual factors in the group [38]. Then, the consensus of expert opinions was analysed [38]. For this purpose, the concordance coefficient θ was calculated, based on formula (1):

$$\Theta = \frac{12\,S}{N_E^2\left(b^3 - b\right)} \qquad (1)$$

where:
S - sum of squares of factual deviations of rank values,
N_E - number of experts, b - number of factors assessed.

To find out whether the consensus of expert opinions was not accidental, the χ^2criterion was used, expressed by formula (2):

$$\chi^2 = \frac{S}{\frac{1}{12} N_E b(b+1) - \frac{1}{b-1}\sum_{i=1}^{N_E} T_i} \qquad (2)$$

where: T_i - similar rank coefficient.
If the χ^2 value was higher than the tabular value χ^{2tab}, and the concordance coefficient differed significantly from zero, the consensus of expert opinions was considered not accidental. Additionally, consensus of opinions was also assessed with the use of the coefficient of variation V, calculated based on formula (3):

$$V_j = \frac{g_j}{m_j} 100\% \qquad (2)$$

where: g_j - standard deviation, m_j - arithmetic mean of expert assessments.

Another method used in the study was the scenario method. The scenarios selected were supposed to facilitate expert assessment of the likelihood of application of individual wind power development prospects for the years 2017-2020. The respondents' task was to break down 100% of the 1st order elements into 2nd order factors with potential influence on the emergence and significance of 1st order barriers. The scenarios were as follows:

- significant cuts in all investments,
- increased number of turbines with less dynamic increase in installed capacities than before, despite the restrictions introduced by the distance act,
- increased number of household installations and offshore wind farms and reduced number of onshore wind farms with tall towers.

Table 1. 1st and 2nd order barriers which may potentially favour the freezing of wind power development in Poland.

Parameter symbol	Barriers
C21	Legal (legislative) barriers
C211	Restrictions introduced by legal regulations concerning the conditions, location and construction of wind power stations (including, among others, minimum distance between the wind power station and residential properties/combined use buildings/protected nature/ pilot forest complexes)
C212	Restrictions imposed by the local zoning plan; weak laws concerning spatial planning
C213	Stability of state and local government
C214	Lack of political motivation to support wind power; i.e. complicated procedures and long waiting time for permits required in order to build wind turbines, inadequate attention paid in the Polish power strategy and development concepts to the development of wind power
C215	Risks connected with the implementation of bidding system

Table 1: continuation.

C22	Economic barriers
C221	Relatively high costs of production of wind energy when compared to conventional energy
C222	High investment costs of implementation of new technologies that need to be incurred at the start
C223	Economic standing of the country (gross domestic product, inflation, unemployment rate, etc.)
C224	Interest rates and availability of credit facilities
C225	Inadequate possibilities to obtain subsidies from EU green funds
C23	Manufacturing and technological barriers
C231	Inadequately developed power grid for power reception from wind devices
C232	Obsolete grid infrastructure
C233	Difficulties pertaining to the planning and prediction of available capacity and volumes of power produced from wind power stations, resulting from dependency of the parameters on meteorological conditions
C234	Necessity to store some of the power generated at wind power stations as back-up power
C24	Sociocultural barriers
C241	Conflicts with ecological circles. No methods to avoid conflicts with ecological circles
C242	Ecological awareness of the society: wind power is perceived negatively - conflicts with local communities
C243	No interest in employment in wind power sectors
C244	Inadequate education on wind power at schools
C25	Environmental barriers
C251	Adverse wind conditions in areas available for the execution of wind energy projects
C252	Environmental risks related to wind power (farms' impact on birds and bats)
C253	Hazards to anthropological environment posed by wind power (noise emission, flicker effect, shadow cast by turbines, magnetic field emission, impact on value of the nearby properties)
C254	Limited area available to install wind power stations due to the existence of National Parks, Nature Reserves, Nature 2000 Areas, Ramsar (wetland) Areas, eco-corridors, Landscape Parks, Protected Landscape Areas, Natural and Landscape Complexes, ecological sites and natural monuments
C255	Negative perception of the impact of wind power stations on landscape

3 Results

Before the survey results were presented, the Act of 20 May 2016 on investments in wind power stations had been thoroughly analysed, as it has for some time had major impact on the condition of the sector. The document sets forth conditions and requirements of/for location of wind power stations within the vicinity of existent or planned residential premises. The Act introduced major restrictions, in particular, when it comes to minimum distance between wind turbines and occupied buildings and nature conservation areas, and the possibility to choose investment site exclusively based on local zoning plans (LZP). Poland is the only European state with such restrictive distance requirements [39-41]. Figure 1 and table 1 show selected EU states and their regulations concerning the restrictions and recommendations pertaining to distance between wind power stations and residential buildings.

Because of the restrictions introduced, there are currently no investors interested in the performance of new wind farm projects. In addition, all projects which are currently at the procedural stage needed to obtain a building permit have been withheld, which means that huge amounts of money spent on agreements with owners of the lands leased, tests carried out etc., have been lost. The already erected farms and previous investors are afraid that they may go bankrupt because of the raised taxes or decreasing prices of certificates of origin.

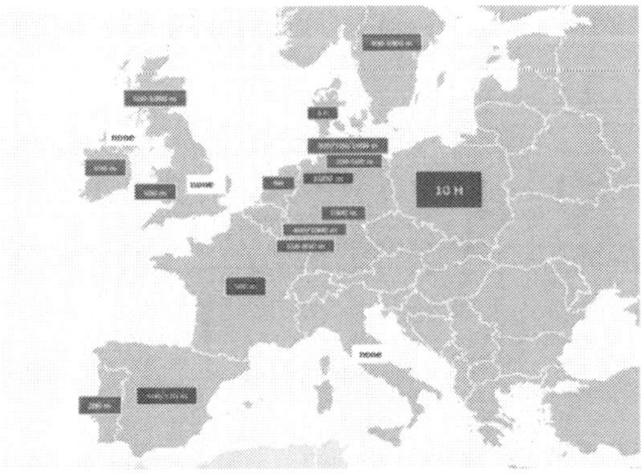

Fig. 1. Minimum distances of wind power stations from residential buildings in selected countries; blue: recommendations, red - requirements [author's own work based on [41].

Table 2. Minimum distances of wind power stations from residential buildings in selected countries [41].

Country	Distance
Poland	Distance **equal to or bigger than ten times the height of the wind power plant** measured from the ground level to the highest point of the structure, including technical elements, such as rotor and blades
Denmark	Distance from buildings must be **equal to or bigger than four times the height of the wind power station;** Distance from major roads and railroad rails must at least be equal to the overall height of the wind turbine. Additionally, the distance also depends on the noise level.
Germany	**Recommendations:** • Saara: **550-850 m** from residential properties, • Thuringia: **1000 m** from residential properties and recreational or historical sites,

Table 2: continuation.

Germany	• Rhineland-Palatinate: **400 m** from individual detached houses, 1000 m from residential areas, • Bremen: **200-500 m** from residential properties, • Schleswig-Holstein: **300 m** from areas with 1 to 4 residential properties, **500 m** from rural areas, **1000 m** from cities and holiday resorts. **Legal requirements:** • Lower Saxony/Saxony: 1000 m from residential properties • Hamburg: **300 m** from individual residential properties, **500 m** from residential areas, **50-100 m** from the nearest roads, railways, telecom networks, transformers, real estate boundaries, **200-500 m** from forests, wetlands, stand off areas of birds and bats and other protected areas.
England:	**No recommendations or legal requirements.**
Scotland:	**Recommendations:** Area within **2 km** from cities and villages; the assessment is performed *in concreto* depending on geographical conditions and visual factors, which is why the distance is usually smaller (**500-1000 m**), but can also be extended. Wind power stations **must not be built in the vicinity of national parks and landscape parks.**
Wales	**Recommendations:** • **500 m** from houses – the distance may vary, depending on background noise measurements and other neighbouring noise-generating investments.
Ireland	**Recommendations:** • **500 m**
Northern Ireland	**Lack of statutory regulations concerning the distance.** **Recommendations:** • Distance corresponding to the turbine height from base to the tip of the blade + 10 %
Italy	**No generally applicable regulations concerning the distance.**
Spain	Distance corresponding to **500 m** from residential properties, **120 m** from roads and power grids.
Portugal	Recommendations: • **200 m** from residential properties
Sweden	**Recommendations:** • **400-1000 m**
The Netherlands	**Recommendations:** • **four times the height of the wind turbine** [measured to the hub (tower)] from the neighbouring buildings
France	**500 m** from residential properties

The survey was prepared and presented to respondents 9 months after the effective date of the 'distance act'.

Respondents were of the opinion that the biggest negative impact on the development of wind power in Poland is exerted by legal factors (fig. 2). More than 1/2 of all respondents (62%), irrespective of seniority and work placement, indicated this factor as the most important [42]. Nearly 1/5 of respondents were of the opinion that the environmental factor plays major role in the development of wind power sector, with further 11% saying that sociocultural barriers may be significant for the development of wind energy. Only 3% of respondents considered manufacturing and technological barriers to hinder the growth of wind energy share in the country's power mix.

Fig. 2. Significance of 1st order barriers for the development of wind power in Poland.

Respondents indicated that among all legislative barriers, it was the introduction of the distance act that had the biggest impact on the hindered development of all wind projects. Significance of this factor was assessed as 58%. The runner-up to this factor, responsible (in the respondents 'opinion) for the poor condition of the market was (tab. 3) the lack of motivation to support wind power among legislators and complicated administrative procedures, including long waiting time for permits required in order to build wind turbines (18%), with successive places taken by: restrictions pertaining to poor coverage of the territory of Poland by local zoning plans (11%), stability of state and local government (7%) and risks connected with the implementation of bidding system (6%).

Table 3. Share of individual barriers in the development of Polish wind power.

Parameter symbol	Share, %	Coefficient of variability
C21		
C211	**58**	1.09
C212	11	
C213	7	
C214	18	
C215	6	
C22		
C221	31	1.55
C222	**36**	
C223	12	
C224	10	
C225	11	
C23		
C231	**44**	0.59
C232	29	
C233	14	
C234	13	
C24		
C241	36	0.75
C242	**45**	
C243	5	

Table 3: continuation.

C244	14	
C25		
C251	10	0.47
C252	24	
C253	15	
C254	**35**	
C255	18	

The second most important hazard to onshore wind power indicated by respondents was the environmental factor. The average percentage rate allocated to environmental barriers by respondents equalled 18. Among the environmental factors, the one related to the limited area available to carry out the investment, due to the presence of various nature conservation forms within a substantial portion of the country's area, was considered the most significant (35%). This restriction is closely connected with one of the legal factors, i.e. the requirement that the distance between wind power stations and residential buildings (or nature conservation areas) must not be smaller than 10 times the turbine length. 24% of respondents also listed the negative impact on birds and bats among the negative environmental effects of wind power stations. Another 18% were of the opinion that negative perception of the impact of wind power stations on landscape was an important barrier, while further 15% indicated the hazard to anthropological environment (noise emission, flicker effect, shadow cast by turbines, magnetic field emission, impact on value of the nearby properties). Only 10% was attributed by respondents to adverse wind conditions in areas available for the execution of wind energy projects. Sociocultural factors, among which are conflicts with local communities and ecological circles, received 11%, with further 6% being allocated to economic factors. Among the most significant hazards in 'economic' group, respondents enumerated high initial capital expenditures and relatively high costs of production of wind energy when compared to traditional energy.

In the opinion of respondents, the least important factors with the potentially inhibiting impact on wind energy in Poland were manufacturing and technological factors, which received 3% of votes. In this field, respondents pointed mainly to the inadequately developed power grid for power reception from wind devices.

Introduction of the act of 20 May 2016 on investments in wind power stations had a substantial limiting effect on the development of wind energy in Poland.

Out of the three scenarios proposed, the majority of respondents chose scenario 1, which stipulated significant cuts in investments in wind power. The likelihood that this option will be selected was assessed as 82% (tab. 4). 14% was attributed to scenario 3, assuming growth in the number of household systems and 5% to scenario 2, according to which wind power will keep thriving in Poland despite the distance act, although less dynamically than before. The coefficients analysed show that the consensus of expert opinions is not accidental.

Table 4. Scenarios of wind power development in Poland.

Scenario		Likelihood, %	Coefficient of variability
1	Between 2017 and 2020, there will be a significant cut down in investments in wind power, leading to a material decrease in the number of wind turbines, especially those with high towers.	**80**	0.55
2	In the years 2017-2020, despite the restrictions imposed by the act, the wind power sector will develop, which is going to translate into a bigger number of turbines, even if the development will not be as dynamic as it used to be.	3.33	3.55
3	In the years 2017-2020, despite restrictions imposed by the distance act, the number of wind turbines will grow, but their structure will change, i.e. turbines with low towers will outdistance turbines with tall towers and the number of offshore installations will grow.	16.67	2.00
Concordance coefficient		0.168	
Criterion $\chi 2$		10.067	

4 Conclusions

Analysis of the wind power market in Poland enabled assessment of the as-is state, the environmental and technical potential and the most plausible development scenario for the sector. Upon a review of factors significant for the past, presence and future of wind energy in Poland, their relevance for the shaping of the wind energy market was established. The following conclusions have been drawn from the study:

1) The review of legislative factors and an own study pointed to legislative barriers, including introduction of the distance act, as the most significant factor responsible for the current stagnation of the onshore wind energy sector. Available readings concerning recommendations and permitted distances between wind farms and residential buildings/nature conservation forms for different countries, show Poland as the only country with such harsh restrictions.

2) In the opinion of respondents, the second most important barriers hindering and preventing increase in the installed capacity of onshore wind power sector are environmental barriers, among which is a factor correlating with the distance act, i.e. limited space available for the installation of wind power stations due to the existence of nature conservation areas.

3) The analysis of sociocultural factors revealed benefits from the development of wind energy, mainly in terms of employment and additional funds for

communes/municipalities and the economy. In this phase, we found out that offshore wind energy is perceived positively by the local community and that there are no conflicts with fishing environments.

4) The share of individual manufacturing and technological factors demonstrated that barriers such as obsolete grid infrastructure and inadequately developed power grid can be eliminated by projects by PSE (Polish Power Grids) concerning expansion of the national transmission system in terms of power evacuation, connection of new energy generating sources and reconstruction of the central point.

5) Analysis of economic factors confirmed information about the dropping investment and operating costs due to expansion of the power market and technological progress.

6) With reference to the biggest barrier faced by the onshore wind energy sector, the most plausible prospect for the years 2017-2020 is a significant drop inthe number of wind turbines.

Analysis of the resources available and benefits of wind power showed that Poland has good environmental conditions and technical capacities and has the potential to become a European leader in wind power production. The only barrier here is the regulatory environment and political instability on the national level. Without improvements to this end, there is no chance for new wind projects to be implemented, as it is these very factors that make foundations for the development of wind energy.

References

1. Directive 2009/28/EC of the European Parliament and of the Council of 23 April 2009 on the promotion of the use of energy from renewable sources and amending and subsequently repealing Directives 2001/77/EC and 2003/30/EC.
2. R.L. Arantegui, A. Jäger-Waldau, Photovoltaics and wind status in the European Union after Paris Agreement. Renewable and Sustainable Energy Reviews 81 (2018), 2460-2471
3. C. Wustemeyer, R. Madlener, D.W. Bunn, A stakeholder analysis of divergent supply-chain trends for the European onshore and offshore wind installations. Energy Policy 80 (2015), 36-44
4. O. Ellabban, H. Abu-Rub, F. Blaaljerg, Renewable energy resources: current status, future prospects and their enabling technology. Renewable and Sustainable Energy Reviews 39 (2014), 748-764
5. http://wysokienapiecie.pl/oze/2307-w-2017-wiecej-pradu-z-wegla-mniej-z-oze [date of access: 12.01.2017]
6. https://www.ure.gov.pl/ [date of access : 18.01.2018]
7. http://wysokienapiecie.pl/oze/2586-cena-zielonych-certyfikatow-tge-zl-mwh [date of access: 12.01.2017]
8. http://gramwzielone.pl/trendy/26003/najwieksza-w-historii-nadpodaz-zielonych-certyfikatow [date of access: 12.01.2017]
9. G.M.J. Herbert, S Iniyan., D. Amutha, A review of technical issues on the development of wind farms. Renewable and Sustainable Energy Reviews 32 (2014), 619-641
10. http://wysokienapiecie.pl/prawo-energetyczne/1393-msz-ostro-krytykuje-ustawe-o-odleglosci-wiatrakow [date of access: 12.01.2017]
11. http://inwestycjeliniowe.pl/farmy-wiatrowe-nowe-zasady-lokalizacji-elektrowni-wiatrowej/ [date of access: 12.01.2017]
12. https://www.money.pl/gospodarka/wiadomosci/artykul/energetyka-wiatrowa-w-polsce-branza-walczy-o,34,0,2136354.html [date of access: 12.01.2017]
13. http://wysokienapiecie.pl/prawo-energetyczne/1388-ustawa-antywiatrakowa-dotknie-4-terytorium-polski-i-nieruchomosci-o-wartosci-7-mld-zl [date of access: 12.01.2017]
14. http://wysokienapiecie.pl/oze/1384-sad-najwyzszy-ustawa-antywiatrakowa-moze-byc-niekonstytucyjna [date of access: 12.01.2017]
15. http://wysokienapiecie.pl/prawo-energetyczne/1341-biuro-analiz-sejmu-ustawa-antywiatrakowa-zagrozi-realizacji-celu-oze-na-2020-r [date of access: 12.01.2017]
16. J.S. González, R. Lacal-Arántegui, A review of regulatory framework for wind energy in European Union countries: Current state and expected developments. Renewable and Sustainable Energy Reviews 56 (2016),588-602
17. Wykonanie celu OZE 2020. Analiza stanu obecnego i prognoza, red. Schnell R. SOLIVAN Adwokaci i Radcy Prawni
18. Wind Energy in Europe: Outlook to 2020. Wind Europe, 2017 Brussels
19. Wind Energy in Europe: Scenarios for 2030. Wind Europe, 2017 Brussels
20. The state of wind energy in Poland in 2016. PWEA 2017
21. R. McKenna, S. Hollnaicher, P. Ostman v.d. Leye, W. Fichtner, Cost-potentials for large onshore wind turbines in Europe. Energy 83 (2015), 217-229
22. B. Igliński, A. Iglińska, G. Koziński, M. Skrzatek, R. Buczkowski, Wind energy in Poland – history, current state, surveys, Renewable Energy Sources Act, SWOT analysis. Renewable and Sustainable Energy Reviews 64 (2016), 19-33
23. S. Ruggiero, V. Varho, P. Rikkonen, Transition to distributed energy generation in Finland: prospects and barriers. Energy Policy 86 (2015), 433-443
24. M. Garcia-Alvarez, L. Cabeza-Garcia, I. Soares, Analysis of the promotion of onshore wind energy in the EU: Feed-in tariff or renewable portfolio standard? Renewable Energy 111 (2017), 256-264

25. A. Brzezińska-Rawa, J. Goździewicz-Biechońska, Recent developments in the wind energy sector in Poland. Renewable and Sustainable Energy Reviews 38 (2014), 79-87
26. F. Hvelplund, P.A. Østergaard, N.I. Meyer, Incentives and barriers for wind power expansion and system integartion in Denmark. Energy Policy 107 (2017), 573-584
27. J. Nordensvärd, F. Urban, The stuttering energy transition in Germany: wind energy policy and feed-in tariff lock-in. Energy Policy 82 (2015), 156-165
28. L. Ho, Wind energy in Malaysia : past, present and future. Renewable and Sustainable Energy Reviews 53 (2016), 279-295
29. A. Darmani, E. Niesten, M. Hekkert, Characteristics of investors in onshore wind power in Sweden Environmental Innovation and Societal Transitions 24 (2017), 67-82
30. S. Höltinger, B. Salak, T. Schauppenlehrer, P. Scherhaufer, J. Schmidt, Austria's wind energy potential - a participatory modeling approach to assess socio-political and market acceptance. Energy Policy 98 (2016), 49-61
31. Elefheriadis I.M. and Anagnostopoulou E.A., Identyfying barriers in the diffusion of renewable energy sources. Energy Policy 80 (2015), 153-164
32. Z. Khan, A.A. Khan, Current barriers to renewavle energy development in Trinidad and Tobago. Startegic Planning for Energy and the Environment 36 (2017), 8-23
33. C. Ghisetti, S. Mancinelli, M. Mazzanti, M. Zoli, Financial barriers and environmental innovations: evidence from EU manufacturing firms. Climate Policy 17 (2017), 131-147
34. A. Tzouvelekis, A study of the potential for small wind turbines in Greece: current market status, barriers to growth and support policy options. Master thesis in Faculty of Technology, Policy and Management in Delft University of Technology
35. N. Mercer, G. Saba, A. Klinke, "Wind is not an issue for government": Barriers to wind energy development in Newfoundland and Labrador, Canada. Energy Policy 108 (2017), 673-683
36. S. Sen, S. Ganguly, Opportunities, barriers and issues with renewable Energy development – a discussion. Renewable and Sustainable Energy Reviews 69 (2017), 1170-1181
37. V. Mytilinou, A.J. Kolios, G. Di Lorenzo, A comparative multidisciplinary policy review in wind energy developments in Europe. International Journal of Sustainable Energy 36 (2017), 754-774
38. M. Krzywonos, J. Skudlarski, A. Kupczyk, J. Wojdalski, K. Tucki., 2015, Przem.Chem. 2015; 94/12
39. A.N. Commin, M.W.H. Davidson, N. Largey, P.P.J. Gaffney, D.W. Braidwood, S.W. Gibb, J. McClatchey, Spatial smoothing of onshore wind: implications for strategic development in Scotland. Energy Policy 109 (2017), 36-48
40. https://www.wind-watch.org/documents/european-setbacks-minimum-distance-between-wind-turbines-and-habitations/ [date of acess: 18.01.2018]
41. Legal opinion on the compliance of the act on wind farm investments with the community law, Warsaw 2016
42. D. Angelopoulos, R. Brückmann, F. Jirouš, I. Konstantinavičiūtė, P. Noothout, J. Psarras, L. Tesnière, B. Breitschopf, Risks and cost of capital for onshore wind energy investments in EU countries. Energy & Environment 2016, 27 (1), 82-104[http://journals.sagepub.com/doi/pdf/10.1177/0958305X16638573]

Innovative heat pipes and thermosyphons for renewable sources of energy application

Leonard Vasiliev[1,*], and *Alexander* Zhuravlyov[1]

[1]Luikov Heat and Mass Transfer Institute of the NAS of Belarus, Porous Media Laboratory, 15 P. Brovka str., 220072, Minsk, Belarus

Abstract. In the paper the authors present possibilities of use two-phase heat-conducting devices at cooling technologies for PV and PV/T solar cooling and air-conditioning. Three ways to improve heat conductive properties of heat pipes and thermosyphons with help of nanotechnologies are considered: use of nanofluids, application of nanocoatings on heat loaded surfaces and polymer composites with nanoparticles producing. A mini bubbles generation of the working fluid inside the transparent thermosyphon evaporator with nanocoating under the solar radiation influence was observed. Nanotechnologies allow to intensify heat transfer processes and reduce energy losses in original designs of heat pipes and thermosyphons.

Introduction

Issues in sustainable development and climate change are directly related to problems of heat and mass transfer. The field of heat and mass transfer holds some of the key phenomena of human activities and in many branches of industry: chemical, energy conversion and power generation, food-processing and pharmaceutics, material processing and metallurgy, transportation and space technology, electronics cooling, bioengineering and biomedical applications, indoor climate, and the care for the environment. Despite decades of development, the science and engineering of heat and mass transfer are continuously being confronted with new challenges in search of new and improved technologies. Some of the multiple facets of heat and mass transfer are still awaiting proper understanding and continue to be in worldwide research focus. Developments in computer simulations in synergy with sophisticated experimental and diagnostics techniques open new frontiers for discoveries and innovation in both conventional and emerging technologies. Recently, we have witnessed expansion of scales into micro and nanoscales, where some very new technologies are in the offing, and into mega- and giga-scales encountered in the atmosphere, oceans, interior of the Earth, and outer space. The need to collect and store solar energy has broadly resulted in two solar energy conversion methods being developed. The first one is photovoltaic and involves converting solar energy directly into electrical energy. The second is solar thermal and involves converting solar energy directly into thermal energy, or cold. Nanofluids, nanocoatings and nanocomposites inventions open a new niche in the heat pipe and thermosyphon design and use for solar energy application. The combination of nanofluid and small channels constitutes an innovating method providing effectiveness, compactness and low thermal resistance of heat pipes. In certain applications polymers composites reinforced by carbon nanowires and nanoparticles can successfully replace the metal envelope of heat pipes and thermosyphons.

The topic of this presentation is related to the field of cooling technologies for PV and PV/T solar cooling and air-conditioning. Solar energy is one of the most interesting solutions among renewable energy resources as it can be converted easily whether into heat, cold, or into electricity. The main problem when using such an energy source is its unfair time distribution which may cause mismatch between needs and availability. Long thermosyphons (vapordynamic thermosyphons, VDT, and flat polymeric loop thermosyphons, FPLT) are of great interest as components of heat exchangers for recuperation of energy of renewable sources (solar, ground) and upgrading their potential with the help of heat pumps. Transparent thermosyphons with nanofluids and nanocoated surface inside presented in this research program are a good tool to absorb solar radiation in the volume of the fluid flow. Vapordynamic thermosyphons cooling system is good solution for building-integrated photovoltaic technology. Development of the new environmentally – friendly and energy – efficient technologies will be vital to achieving investigation of a hybrid PHOTOVOLTAICS/THERMOCOLLECTOR (PV/T) systems, providing electricity and heat/cold simultaneously, represents an important step toward reducing dependency on fossil fuels.

Colloidal suspensions of nanosized particles in a fluid, have recently gained popularity as cooling fluids mainly due to their enhanced heat transfer capabilities [1].

Actually the micro and nanoparticles (NP) as nanocoating (NC) of the heat pipe evaporators and thermosyphons are suggested as means to increase the heat transfer coefficient with working fluid evaporation

* Corresponding author: Leonard.L.Vasiliev@gmail.com

 https://doi.org/10.1051/e3sconf/20187001018

and boiling on the heat loaded wall [2]. When NPs are properly dispersed, NFs can offer numerous benefits besides the anomalously high effective thermal conductivity, such as improved heat transfer and stability, microchannel cooling without clogging, the possibility of miniaturizing systems scaling, or reduction in pumping power, among others. NCs of the heat pipe evaporators and NPs based polymer composites design are considered as a mean to increase its effective thermal conductivity. The generation of engineered nanostructures represents a major breakthrough in material science and nanotechnology.

1 Nanofluids for heat pipes

In some cases it is interesting to make the minichannels heat exchanger and heat pipes envelope from transparent material (glass, plastic) and to heat the working fluid in volume by solar, or laser radiation, Fig. 1.

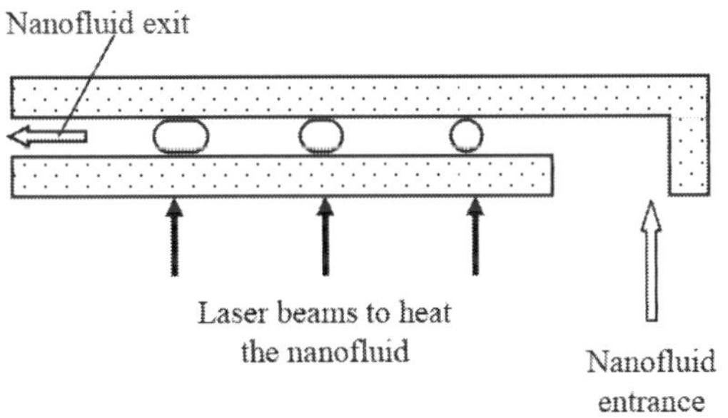

Fig. 1. Schematic of the flat minievaporator with transparent walls heated by impulse laser beam.

Bubbles generation phenomena in minivolumes filled with nanofluid due to the laser beam interaction with solid particles is the reason of NFs circulation in the mini heat pipe loop, or thermosyphon. This impulse is working as two-phase minipump, which initiate the fluid circulation inside minichannel, Fig. 2.

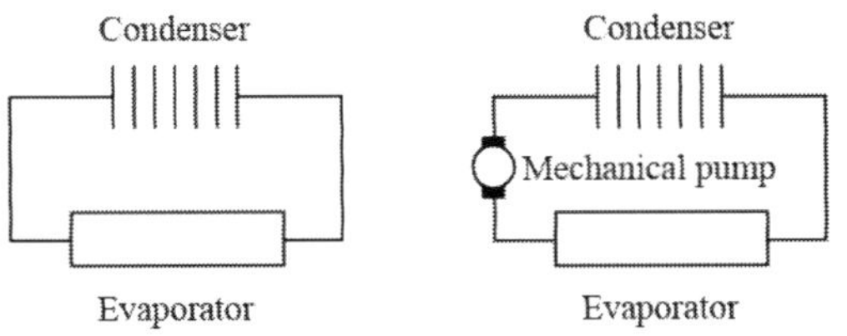

Fig. 2. Schematic of two-phase loop thermosyphon with nanofluid circulation (without mechanical pump – left, and with mechanical pump – right).

Bubbles are considered also as a motive force to organize the fluid circulation in pulsating heat pipes and loop thermosyphons. One of the major interesting topics is the investigation of the influence of metal oxide NPs (CuO, Al_2O_3 nanoparticles) immersed in the fluid (water) on the local bubbles generation and two-phase heat transfer intensification to compare with pure water.

It was shown [3, 4] that the heat flow generated by light-absorbing nanoparticles (gold, or Cu NPs) initiates more intense bubble generation. Transparent evaporators made from glass or plastic have a real practical interest for solar energy application (Figs. 3, 4). So, application of plasmonic nanofluids in photothermal use, for example in solar thermal receivers, is a strong subject of interest in current research.

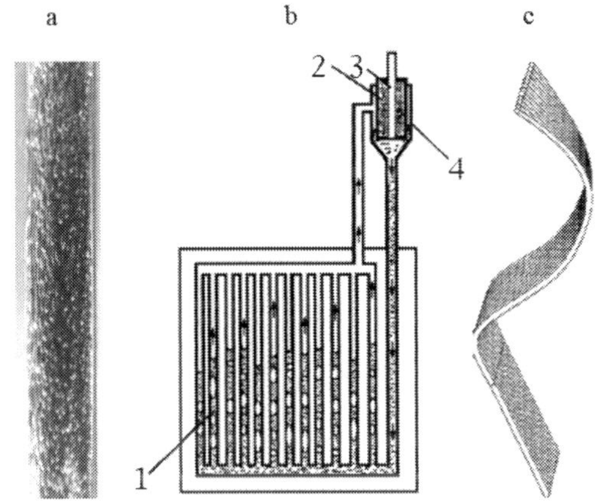

Fig. 3. Bubble generation in the evaporators of heat pipes and thermosyphons under solar radiation: a) nanofluid (Al_2O_3) absorbing solar radiation inside the transparent (glass) pipe; b) transparent pulsating heat pipes: 1 – multi evaporators, 2 – heat exchanger, 3 – cooler, 4 – condenser; c) polymeric transparent pulsating heat pipe [5].

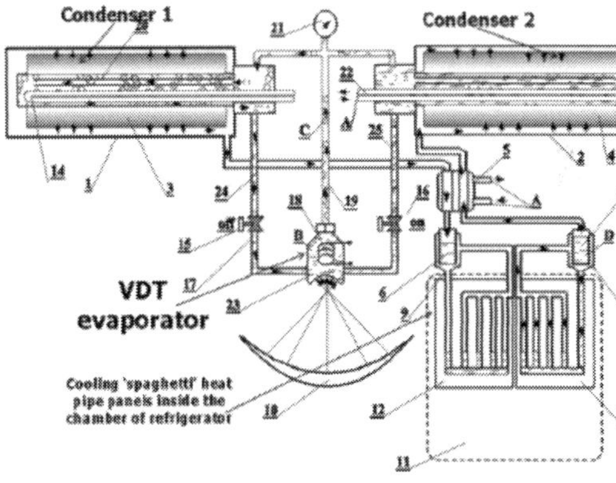

Fig. 4. VDT with two condensers as a cooler/heater for solar solid sorption refrigerator (nanofluid inside the evaporator) [4]: 1, 2 – adsorbers, 3, 4 – sorbent beds, 5 – liquid heat exchanger, 6, 7 – ammonia accumulators, 8, 9 – "spaghetti" heat pipe condensers, 10 – parabolic solar concentrator, 11 – refrigerator chamber,12 – "freezer", 13 – "spaghetti" evaporators , 14, 22 – cooling liquid heat exchanger, 15, 16 – valves, 17 – VDT liquid pipe, 18 – electric heater, 19 – VDT vapor pipe, 20 – adsorber vapor channels, 21 – pressure sensor, 23 – VDT evaporator (solar absorber), A – cooling liquid heat exchanger, B – VDT evaporator envelope, C – two phase flow of the working fluid, D – ammoniac low-temperature circuit.

2 Nanocoatings in heat pipes

In photovoltaic (PV) systems, incoming sunlight strikes a semiconductor material (crystalline silicon) and dislodges electrons in the material to cause the movement of electrons. The resulting electron flow generates the PV cells electrical output. However, only a small portion of incident solar energy (typically less than 20%) is transformed into electrical energy. The remaining incident energy increases the temperature of the PV cell and reduces its operational performance, Fig. 5.

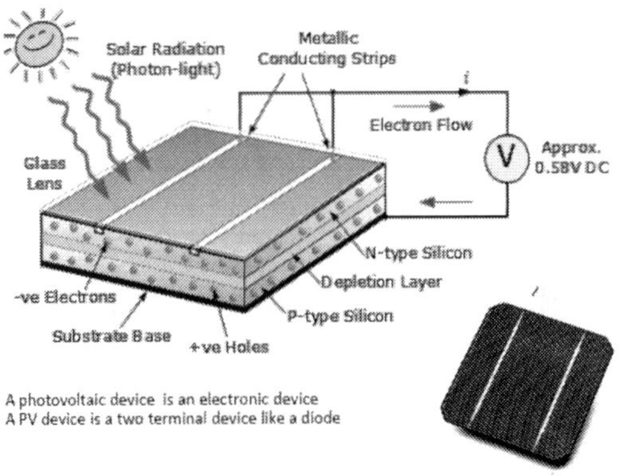

Fig. 5. Photovoltaic solar cells application efficiency depends on the silicon film temperature.

Good means to improve solar cells efficiency of electricity production is to cool solar cell near the ambient temperature using heat pipe heat exchanger (VDTs with long horizontal evaporator and condenser).

If the thermosyphon, or heat pipe envelope is made from the transparent material (glass), there is a possibility to organize the volumetric absorption of solar radiation inside the evaporator. Inside the annular evaporator of thermosyphon there is a metallic tubular solar receiver coated by nanoparticles such as CuO, or Al_2O_3 (Fig. 6). This receiver is partly immersed to the working fluid (water, propane). The temperature profile inside such thermosyphon is different to compare with conventional thermosyphon (Fig. 7).

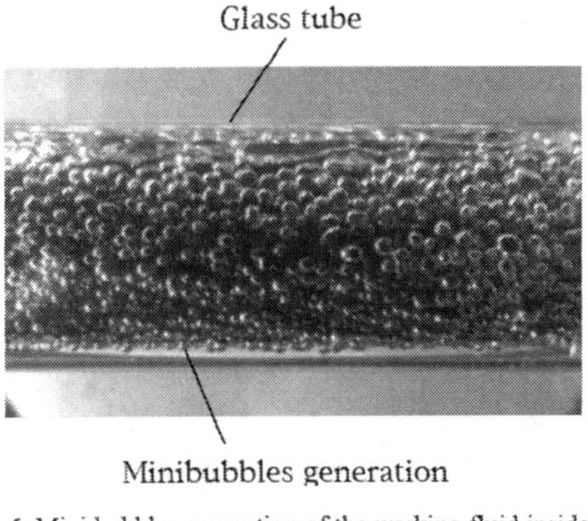

Fig. 6. Mini bubbles generation of the working fluid inside the transparent thermosyphon evaporator with nanocoating under the solar radiation influence. Working fluid – propane.

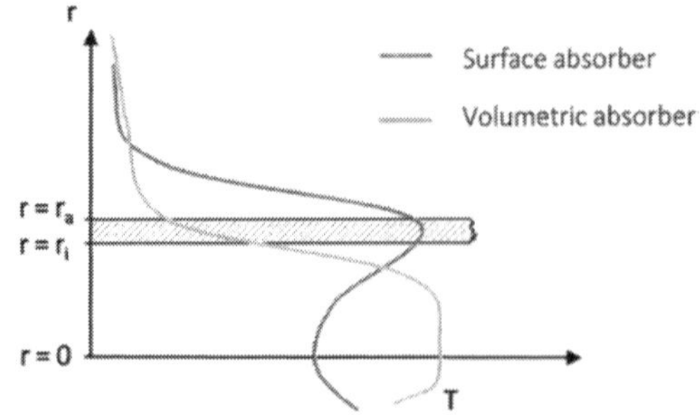

Fig. 7. Temperature field evolution (cross section) in copper tube solar thermal absorber with water (surface solar absorption), and transparent glass solar absorber with nanofluid (water + CuO nanoparticles; volumetric absorption of solar radiation) [6].

The copper loop thermosyphon has a thin and long horizontal evaporator (length 200 mm, diameter 10 mm) and liquid cooled cylindrical condenser [7], (Fig. 8). Miniature copper sintered powder porous coating less 1 mm thick is put inside the evaporator. Thermosyphon is made self-priming by carefully controlling condenser, vapor line and liquid line and the volume of the compensation chamber. The compensation chamber and fluid charge are set so that there is always fluid in the compensation chamber even if the condenser, vapor line and liquid line are completely filled with the liquid. Thermosyphon condenser is cooled by the liquid (water) from the cooling bath flowing through the annular liquid heat exchanger. The low temperature cooling liquid flow is maintained via a thermostat with its flow rate being measured by a flow meter. The temperature of the cooling liquid is regulated from 20 up to 80 °C to give a possibility to change the temperature of the vapor condensation inside thermosyphon.

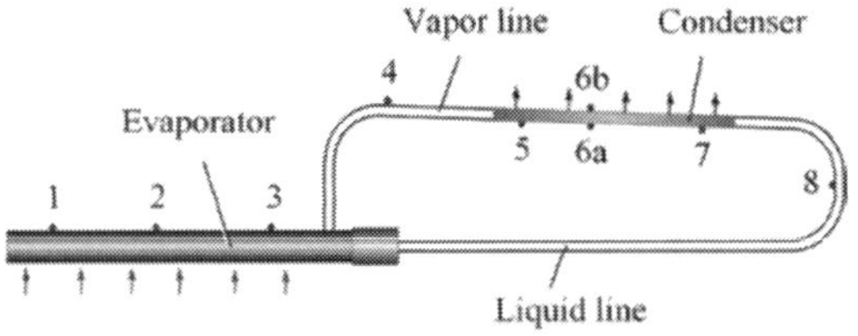

Fig. 8. Loop thermosyphon with cylindrical horizontally disposed evaporator and condenser; 1-8 are the thermocouples position on the thermosyphon surface.

The pressure drop during a working fluid circulation inside thermosyphon is calculated as

$$\Delta P_{cap} + \Delta P_g + \Delta P_{bg} \geq \Delta P_L + \Delta P_v + \Delta P_e + \Delta P_c \quad (1)$$

The notations ΔP_{cap} is the capillary pressure of the wick; ΔP_g is the gravitational pressure loss; ΔP_{bg} is the bubble generation pressure drop in the evaporator; ΔP_L is the pressure drop of the liquid flow; ΔP_v is the pressure drop of the vapor flow; and ΔP_e and ΔP_c are the preassure drop due to evaporation and condensation at the liquid-vapour interface respectively.

For the thermosyphon to function properly, the $\Delta P_{cap}+$ $+\Delta P_g$ $+\Delta P_{bg}$ ought to be larger or at least equal to the sum of all the pressure losses mentioned above. To increase the capillary pressure and permeability of the wick, the wick ought to have a heterogeneous porous structure (micro, mesa and macro pores). The wick will require a high liquid surface tension σ, low contact angle θ (i.e. high liquid wettability) and small powder size for low effective pore size radius R_{eff}.

The average temperature difference between the evaporator and condenser, ΔT_{avg}, is found by subtracting the average, front-end condenser temperature, T_c, from average front-end evaporator temperature, T_e. Both the evaporator and condenser temperatures were found using Eqs. (2), (3):

$$T_e = \frac{T_1 + T_2 + T_3}{3}, \tag{2}$$

$$T_{\tilde{n}} = \frac{T_5 + T_6 + T_7}{3}, \tag{3}$$

where T_1, T_2, T_3 and T_5, T_6, T_7 are the temperatures occurring at the respective thermocouples locations on the evaporator and condenser; T_4 and T_8 are the temperatures occurring at the respective thermocouples locations on the vapor and liquid lines of thermosyphon (Fig. 8). The cooling capacity of LTCE condenser is calculated from the following expression:

$$Q_{out} = m \cdot c_p (T_{out} - T_{in}), \tag{4}$$

where m, c_p, T_{out} and T_{in} represent the flow rate, specific heat at the constant pressure, outlet temperature and inlet temperature of chilled water, respectively.

$$R_{total} = \Delta T/Q = (T_e - T_c)/Q_{actual}, \tag{5}$$

where T_e and T_c are the average temperature of the evaporator and condenser; Q_{actual} is representing the average of the heat removal form of the condenser and supplied input power ($Q_{input} = (I \times V)$. The dimensions of thermosyphon charged with water are shown on the Table 1. The temperature of components and thermal resistance of thermosyphon via time and heat input is presented in Fig. 9 and in the Table 2.

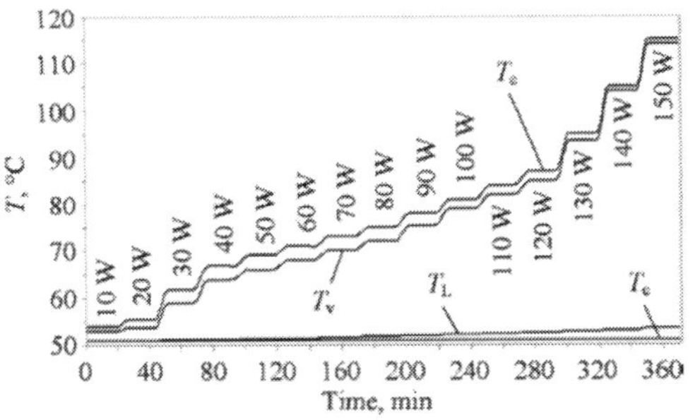

Fig. 9. Temperature of thermosyphon components via the time for different heat loads. Working fluid is water. T_e, T_v, T_c and T_L are the temperatures of evaporator, vapor, condenser and the working liquid of thermosyphon respectively.

The working fluid at the entrance of the evaporator is subcooled and its temperature is close to the temperature (T_{sink}) of the liquid inside the cooling bath.

Table 1. Dimensions of the thermosyphon, mm.

Dimension	**mm**
Length of the evaporator	200
Diameter/wall thickness of the evaporator	12/1
Length of the condenser	460
Diameter/wall thickness of the condenser	4/0.5
Length of the vapor line	300
Diameter/wall thickness of the vapor line	4/0.5
Length of the liquid line	220

Table 2. Thermal resistance R of the evaporator, condenser and total of thermosyphon (T_{sink} = 50 °C).

Heat load (W)	R_t **(K/W)**	R_e **(K/W)**	R_c **(K/W)**
10	0.3	0.1	0.2
20	0.22	0.085	0.135
30	0.36	0.1	0.26
40	0.39	0.075	0.32
50	0.35	0.06	0.29
60	0.32	0.05	0.27
70	0.3	0.043	0.26
80	0.3	0.029	0.26
90	0.28	0.027	0.26
100	0.29	0.01	0.28
110	0.29	0.01	0.28
120	0.29	0.008	0.28
130	0.4	0.015	0.38
140	0.4	0.007	0.4
150	0.4	0.007	0.4

Thermosyphon action was checked both with multiple increasing and decreasing of the heat load. Good performance characteristics of the copper-water LTCE are also confirmed by the almost complete absence of hysteresis phenomena (Fig. 10). The solid lines indicate the evaporator heat flux is increasing; the dotted lines indicate a decrease in the heat load.

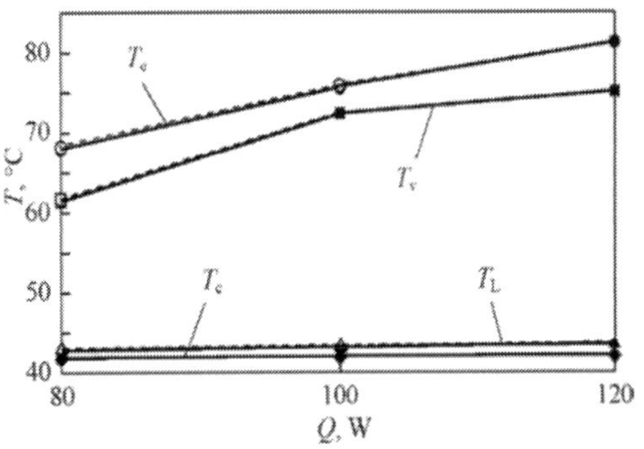

Fig. 10. Temperatures of thermosyphon via heat load Q with water as a working fluid: T_e, T_V, T_c and T_L are the temperatures of evaporator, vapor, condenser and working fluid (near the evaporator).

The solar PV collectors with thermosyphon coolers are shown on Fig. 11.They are cooled by the air (natural convection), or by the liquid forced convection. Evaporators of thermosyphons are inserted into the basement of the solar collector; finned condensers are installed on the back side of collector and are cooled by the natural convection of the air, or heat exchanger with water circulation

a

b

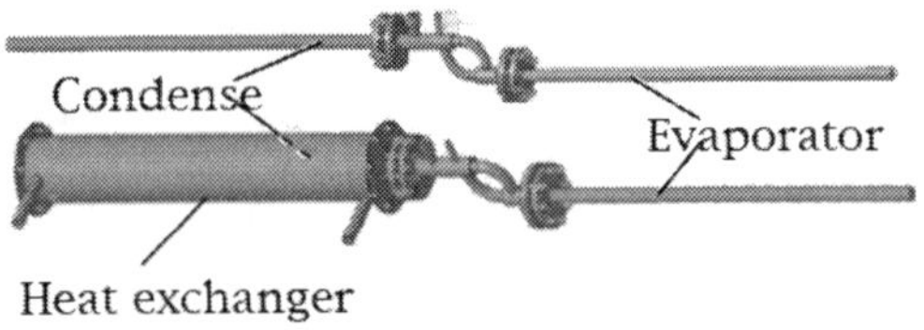

Fig. 11. Solar collectors (a) and VDT thermosyphons as its cooling components (b).

The VDT is less sensitive to the angle of its inclination (Fig. 12) to compare with conventional thermosyphon. VDT has low thermal resistance $R = 0.22$ K·W^{-1}. The conventional thermosyphon has its thermal resistance $R = 0.25\text{-}0.3$ K·W^{-1} for the near horizontal position of the evaporator, and $R > 0.5$ K·W^{-1} for its negative inclination to horizon. The heat flow $Q = 100$ W transferred by VDT is near 1.7 time more to compare with conventional thermosyphon $Q = 60$ W at the same mean temperature of the condenser.

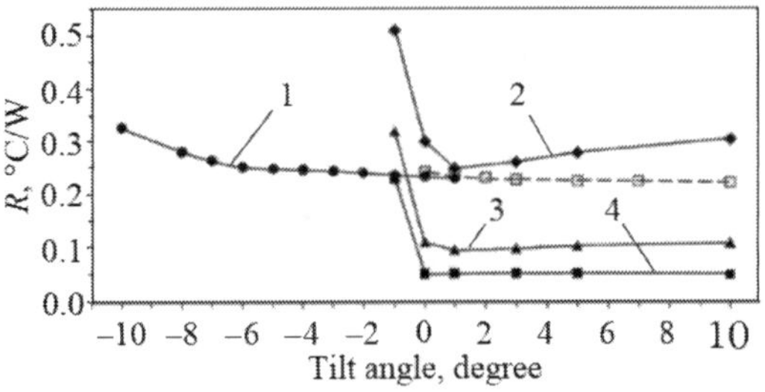

Fig. 12. Thermal resistance of VDT thermosyphon (1 – total) and conventional thermosyphon (2 – total, 3 – condenser, 4 – evaporator) as a function of the angle of thermosyphon inclination.

3 Polymer composites with nano-particles

Recently a new nanotechnology was developed to produce loop polymer thermosyphons capable of long-term operation without permeation by air or working fluids through their walls [8]. Polymer-metal composites are becoming an attractive subject due to their unique surface morphology. Nanocoating of the heat loaded surface of such polymer have a grand potential to increase the wettability and heat transfer intensity in small size heat transfer devices. Actually polymer heat transfer equipment is used in different devices. The application of horizontal flat polymer thermosyphons to apply the energy of alternative energy sources (solar and ground sources of energy generation), cooling and heating of the buildings and storehouses has a good future. Thermosyphon (Fig. 13) has a flat interface (for attaching heat-generating elements) and flexible transport lines (vapor and liquid pipes) between the flat evaporator and condenser. The working fluid of thermosyphon is isobutene (R 600).

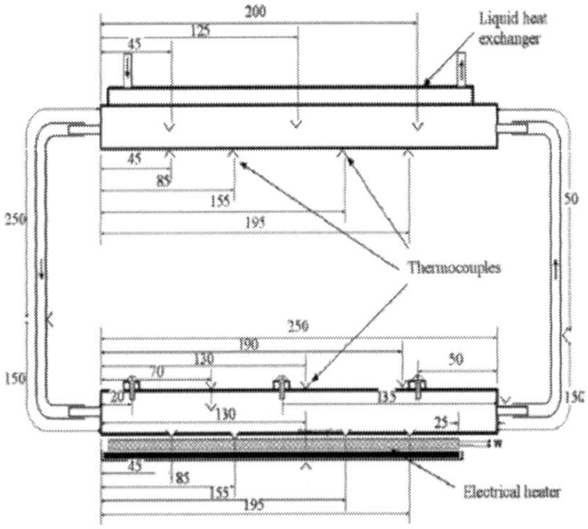

Fig. 13. Flat loop thermosyphon made from polymer composite (Polyamide reinforced by carbon nanofilaments and nano diamond particles).

The distance between the evaporator and condenser can constitute several meters. This thermosyphon operates autonomously under the action of the gravity forces and under the temperature drop between the heat source and heat sink. There are some rectangular minigrooves as a capillary structure inside the evaporator and condenser of the thermosyphon.

The envelope of thermosyphon is made of polyamide composite with nano carbon filaments and nano diamond particles to increase its effective thermal conductivity. The effective thermal conductivity of composite is equal to 11 W/(m·°C).

The width and length of thermosyphon is 50 and 250 mm, respectively. The width of the grooved surface inside the evaporator and condenser, where two-phase heat transfer occurs is 30 mm. The width and depth of the rectangular minigrooves inside the evaporator and condenser are 2.5 mm. The thickness of the evaporator and condenser is 10 mm. The width and thickness of the rectangular parts of frames are 10 mm, respectively.

The difference between the temperature of the external walls of the evaporator T_w and the saturated temperature of the adiabatic (vapor transport) zone T_{sat}, i. e., $T_w - T_{sat}$, was measured directly by four thermocouples, one junction of which was on the evaporator wall, while the other was placed in a thermally controlled liquid bath. The vapor saturation inside the thermosyphon was maintained using the liquid heat exchanger and thermostat. The total thermal resistance of the thermosyphon R_{TS} was calculated as

$$R_{TS} = \frac{T_e - T_c}{Q} \quad (6)$$

where T_e is the mean temperature of the evaporator (°C), T_c is the mean temperature of the condenser (°C), and Q is the heat flow (W).

The experimental technique was used to perform some necessary operations including temperature measurements (Agilent Data Acquisition Agilent Data Loger HP-34970A with a set of thermocouples connected to the computer), measurements of the heat flow Q from the evaporator to condenser and of the effective thermal conductivity of the thermosyphon envelope. To cool the thermosyphon condenser (liquid loop) a Joulabo F12 recirculation thermal bath with temperature-regulated accuracy of ±0.5 °C was used. The heat flow was supplied by an electric cartridge heater located on the bottom surface of the evaporator. The heat sink (liquid heat exchanger) was attached to the top of the condenser. All measurements were performed in a steady-state regime for the heat flow range of 50–100 W.

The temperature distribution along the evaporator, transport zone, and condenser as a function of the heat load is shown in Fig. 14.

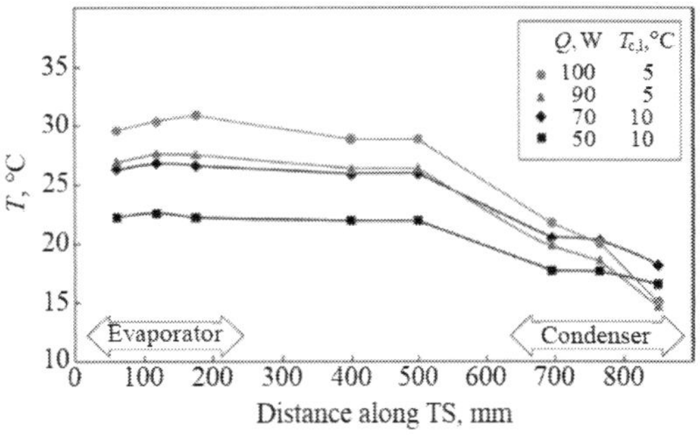

Fig. 14. Temperature distribution along the evaporator, transport zone and condenser of the thermosyphon as the function of heat input.

4 Conclusion

Recent advances in nanotechnology have allowed development of a new nanofluids, nanocoatings and nano polymer composites intensifying heat transfer processes and reducing energy losses in original designs of heat pipes and thermosyphons. Application of photo thermal microscopy allows visualizing individual bubbles generation in transparent mini-evaporator. Visualization of the hydrodynamic and heat transfer of two-phase flow in horizontal annular mini-channel allows analyzing the efficiency of porous coating influence on hydrodynamic and heat transfer in mini-channels and thermosyphon evaporators. Nanofluids and nanocoatings may be recommended in solar thermal applications due to very effective volumetric absorption and good solar radiation properties.

References

1. S. Kakaç, A. G. Yazicioglu. CONV-14, Int. Symp. Convect. Heat Mass Transf., 1 (2014).
2. F.P. Cabral, G. Ribatski. ENCIT 2010 – 13th Braz. Congr. Thermal Sci. and Eng. (2010).
3. D. Lapotko, E. Lukianova. Int. J. Heat Mass Trans. **48**, 227 (2005).
4. L. Vasiliev, D. Lapotko, E. Lukianova, A. Zhuravlyov, A. Shapovalov, L. Vasiliev Jr. 14th IHPC – 14th Int. Heat Pipes Conf. 138 (2007).
5. L. Vasiliev Jr. Heat exchange device made of polymeric material. Pat. US 2011/0067843 A1. 2011.
6. J. R. Eggers, S. Kabelak. Int. J., Thermophys. **36**, 2769 (2015).
7. L. Vasiliev, M. Rabetsky, L. Grakovich, V. Kulikousky, A. Zhuravlyov, M. Kuzmich, 19th International Heat Pipe Conf. and 13th Int. Heat Pipe Symp., Piza, Italy. 22 (2018),
8. L. P. Grakovich, M. I. Rabetskii, L. L. Vasiliev, L. L. Vasiliev Jr., S. P. Bogdanovich, S. S. Pesetskii. Heat Pipe Sci. and Techn. Int. J. **5**, 145 (2014).

Mathematical model of flat plate solar thermal collector and its validation

Michał Wasik[1,*], *Marcin A.* Bugaj[1], *Tomasz S.* Wiśniewski[1], *Maciej* Klein[2,3], *Patryk* Chaja[2], *Sebastian* Bykuć[2]

[1]Warsaw University of Technology, Faculty of Power and Aeronautical Engineering, Institute of Heat Engineering, ul. Nowowiejska 21/25, PL-00-665 Warsaw, Poland
[2] Institute Of Fluid-Flow Machinery Polish Academy Of Sciences, Fiszera st. 14, 80-231 Gdańsk, Poland.
[3]Gdańsk University of Technology, Faculty of Applied Physics and Mathematics, ul. Gabriela Narutowicza 11/12, PL-80-233 Gdańsk, Poland

Abstract. In order to predict solar thermal collector's performance and optimization of control algorithms, proper mathematical models are necessary. Computer calculation technique provides tools for determination of modern materials impact on improvement of heat transfer inside the collector and minimization of heat loss. Such analysis is impossible by using standard technical datasheet provided by producer or by using empirical formulas. In the paper the authors present a mathematical model of a flat plate solar thermal collector based on the Hottel-Whiller-Bliss equation and criterial formula. The iterative algorithm solved steady state heat transfer equations for a glazed and an unglazed collector. The validation experiment was conducted under class AAA sun simulator for different inlet temperature and solar irradiation values. The unglazed PV/T and glazed solar thermal collectors were tested. For PV/T the relative difference, between measured and computed outlet temperatures, was below 5% and the highest value was reached for the lowest inlet temperature. The validation study showed that the experimental results reached good agreement with simulation predictions. Presented computation algorithm enables to predict influence of geometry changes on collector performance. The model could be used for optimization of the construction without using CFD methods, which need large computation resources.

NOMENCLATURE

A – area, m^2,
a_s – thermal diffusivity, m^2/s,
c – specific heat at constant pressure, J/(kgK),
D – external diameter, m,
d– internal diameter, m,
F – fin efficiency,
F' – plate efficiency factor,
F_R – heat removal factor,
G – irradiance, $W/(m^2)$,
Gr– Grashof number, $(g\beta'\Delta TL^3/\nu^2)$,
h – heat transfer coefficient, $W/(m^2K)$,
k –thermal conductivity, W/(m·K),
L – characteristic dimension, m,
m – mass flow rate, kg/s,
N_c – number of glass cover,
Nu – Nusselt number,(hL/k)
Pr – Prandtl number, (a/k)
Q_u – useful heat rate, W,
Ra – Rayleigh number, (GrPr)
Re – Reynolds number, (wL/ν)
T – temperature, °C,
U – heat loss coefficient, $W/(m^2K)$,
W – distance between canals, m,
w – wind velocity, m/s,

Greek symbols

α – absorbance,
β – collector tilt angel from horizontal, deg,
β' – volumetric coefficient of expansion, 1/K,
δ – plate thickness, m,
ε – emissivity,
ν – kinematic viscosity, m^2/s
σ – Stefan-Boltzmann constant, $W/(m^2 K^4)$,

Subscripts

a – ambient,
c – convection,
f – fluid,
g – glass,
in – inlet,
l – total heat loss,
out – outlet,
p – plate,
r – radiation,
sky – sky,
t – top,
w – wind or wall,

* Corresponding author: mwasik@itc.pw.edu.pl

1 Introduction

Flat plate collectors are used widely for domestic hot water preparation and domestic heating. Mathematical models of solar thermal collector are used for simulation or optimization of system. Many researchers used the Hottel-Bliss-Whiller equation to compute collectors performance [1]–[4]. Some work applied the Hottel-Bliss-Whiller equation to multiflow numerical analyses [5], air collectors [6] and concentrated collectors [7] have been published so far. The process of the heat transfer between the absorber and an ambient is complicated, especially for the top losses. Heat is transferred by a convection and by a radiation. Empirical formula for the heat loss coefficient was proposed by Klein and further it was developed by others [4], [8]–[10]. The Klein formula has the following form [8], [9]:

$$U_t = 1/\{N_c\, T_p/C/[(T_p - T_a)/(N_c + f)]^e + 1/h_w\} +$$
$$+\sigma\,(T_p + T_a)(T_p + T_a)^2/[1/(\varepsilon_p + 0.00591\, N_c\, h_w) +$$
$$(N_c + f - 1 + 0.133\,\varepsilon_p)/\varepsilon_g - N_c] \quad (1)$$

$$f = (1 + 0.0889 h_w - 0.1166 h_w\,\varepsilon_p)(1 + 0.07866 N_c) \quad (2)$$

$$C = 520\,(1 - 0.000051\,\beta^2) \quad (3)$$

$$e = 0.43\,(1 - 100/T_p) \quad (4)$$

The Klein formula results in large errors due to calculation of convection and radiation parallel with unknown glass temperature [9]. A mathematical model of solar collector based on analogy between heat transfer and electrical circuit was proposed [2]. The model was developed and empirical equation for glass temperature was presented [9]. Many experiments of convection have been conducted and equations for heat transfer coefficient have been proposed [2], [6], [11]–[13]. The formulas compute heat transfer coefficient as linear function of air velocity, for example McAdams which is a popular formula [2], [6]:

$$h_w = 5.7 + 3.8\,w \quad (5)$$

During last years some research was conducted to optimize collectors area, storage tank volume and system pattern [4], [14]–[16]. Mathematical model of solar thermal collector used for optimization based on characteristic equation was presented in literature [14], [15]. The Hottel-Bliss-Whiller equation and empirical formula was also used in optimization for calculations of top heat loss coefficient [4]. Optimization of absorber plate geometry was done in reference [17], while complex optimization of geometry, physical properties of materials and system were not reported in the literature. Proposed mathematical model gives opportunity to analyse influence of material properties and geometry on collector efficiency. Moreover, it could be applied in order to reduced collector mass and production cost by using new materials.

2 Mathematical model

The mathematical model, presented by the authors, based on the Hottel-Whiller-Bliss equation has the following form [1]–[3]:

$$Q_u = AF_R[G(\alpha\tau) - U_l\,(T_{in} - T_a)] \quad (6)$$

Heat loss coefficient is a sum of bottom heat loss and top heat loss coefficients. The convection from the back of the collector was neglected. The back heat loss coefficient was computed from thermal resistance of the insulation. The top heat loss coefficient was computed from connection of thermal resistances of radiation and convection according to equation [2], [3], [9]:

$$U_t = 1/[1/(h_{c,g\text{-}p} + h_{r,g\text{-}p}) + 1/(h_{c,g\text{-}a} + h_{r,g\text{-}sky})] \quad (7)$$

Radiative heat transfer coefficient between plate and glass cover was computed by [2], [3], [9]:

$$h_{r,g\text{-}p} = \sigma\,(T_p^2 + T_g^2)(T_p + T_g)/(1/\varepsilon_p + 1/\varepsilon_g - 1) \quad (8)$$

For radiation from glass surface to the sky [2]:

$$h_{r,g\text{-}sky} = \sigma\,\varepsilon_g\,(T_{sky}^2 + T_g^2)(T_{sky} + T_g)/(T_g - T_a) \quad (9)$$

The convective heat transfer in the air gap was calculated from the Hollands equation [2], [3], [8]:

$$Nu = 1 +$$
$$1.44[1\text{-}1708/(Ra\cos\beta)]^+[1\text{-}1708(\sin\beta)^{1.6}/(Ra\cos\beta)] +$$
$$[(Ra\cos\beta/5830)^{1/3} - 1]^+ \quad (10)$$

From the glass surface heat is transferred to the ambient by a free or forced convection. For a free convection the Nusselt number was calculated by [2]:

$$Nu = 0.15\,Ra^{1/3} \quad (11)$$

The Nusselt number for a forced convection was computed by [2]:

$$Nu = 0.86\,Re^{1/2}\,Pr^{1/3} \quad (12)$$

The heat removal factor was calculated with a procedure applied by many researchers [2], [3], [18], [19]:

$$F_R = m\,c_f/(A\,U_l)\,\{1 - \exp[-A\,U_l\,F'/(m\,c_f)]\} \quad (13)$$

$$F' = 1/W/\{1/[D + (W - D)\,F] + U_l/(\pi\,d\,h_f)\} \quad (14)$$

where F is fin efficiency.

The heat transfer coefficient in plate channels was computed from the Michiejew equation [20]:

$$Nu = 0.15\,Re_f^{1/3}\,Gr_f^{0.1}\,Pr_f^{0.43}\,(Pr_f / Pr_w)^{0.25} \quad (15)$$

For a computation of the collector's outlet temperature inlet parameters are necessary. The model needs the meteorological and the flow data, such as: irradiance, wind speed, ambient and sky temperature, mass flow through the collector and inlet temperature.

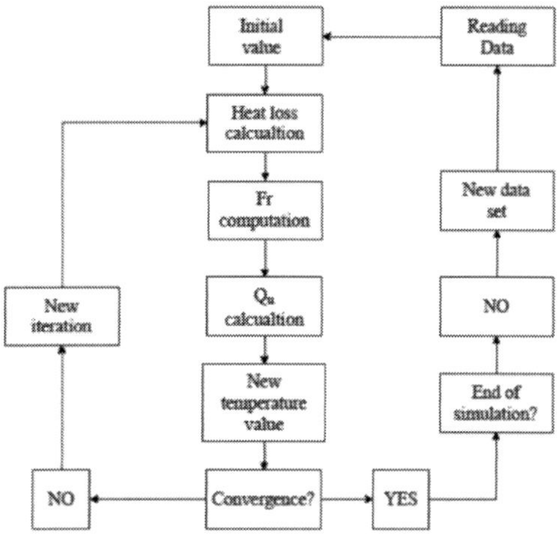

Fig. 1. Block diagram of computational algorithm.

2.1. Sensitivity analysis

The influence of the input parameters to the model was analysed by simulations. Testing data sets were prepared for each input parameter (solar irradiation, inlet temperature, ambient temperature, wind speed, mass flow rate). The outlet temperature was calculated for variation of one parameter when others were constant. Procedure was repeated with changed constant parameter. The influence was calculated from equation:

$$\Delta T_{out}/\Delta X = (T_{out}^{j+1} - T_{out}^{j-1}) X^{j} / T_{out}^{j} / (X^{j+1} - X^{j-1}) \quad (16)$$

where X is the examined parameter, j is the index of data. The results of analysis were presented in the figures below, except for the wind speed, which had the least influence.

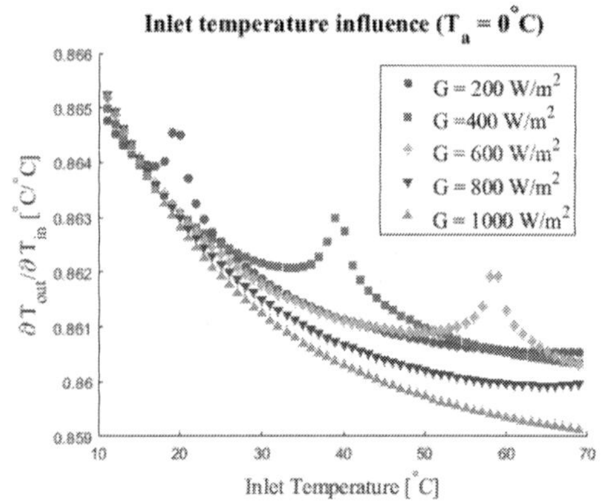

Fig. 2. Influence of inlet temperature on computed outlet temperature.

Influence of inlet temperature locally rising and then falling down. The rise is connected with the working point where efficiency of the collector is close to the zero value. After changing to the negative value influence of the inlet temperature is falling down (Fig. 2).

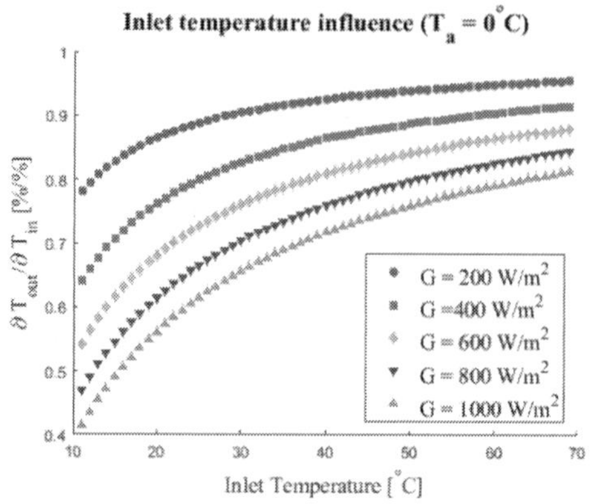

Fig. 3. Influence of inlet temperature on computed outlet temperature.

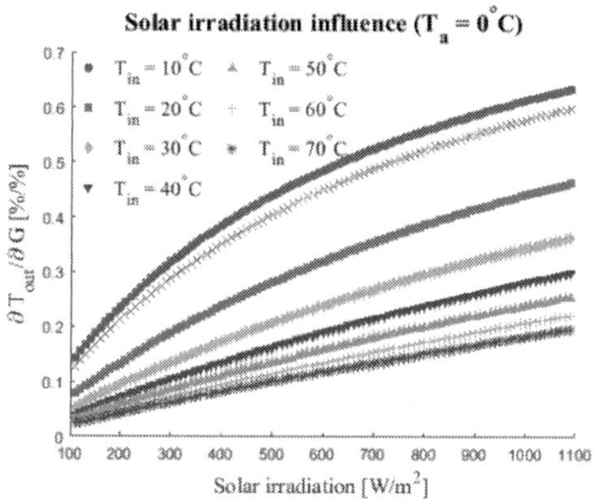

Fig. 4. Influence of solar irradiation on computed outlet temperature.

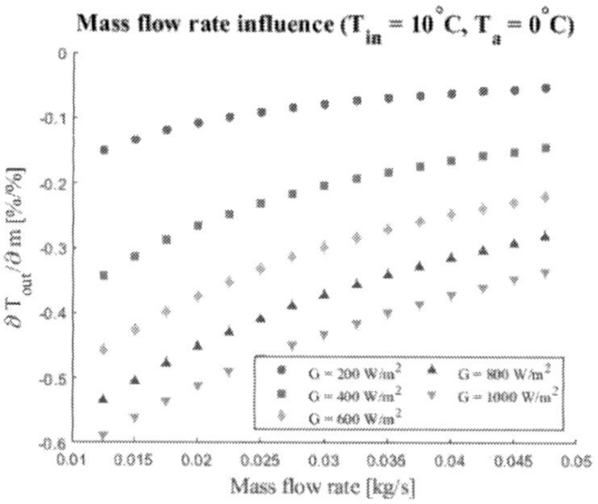

Fig. 5. Influence of mass flow rate on computed outlet temperature.

 https://doi.org/10.1051/e3sconf/20187001019

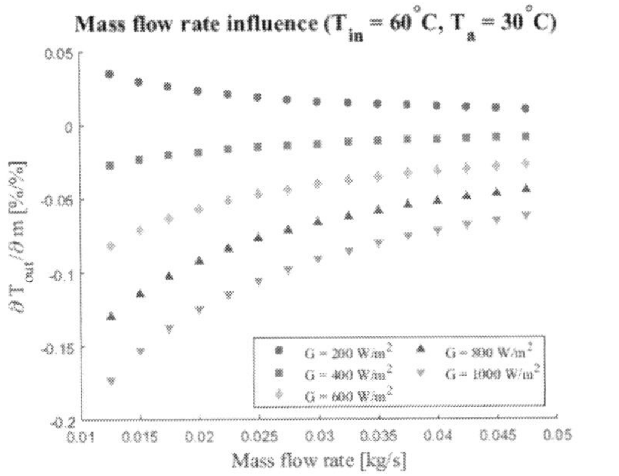

Fig. 6. Influence of mass flow rate on computed outlet temperature.

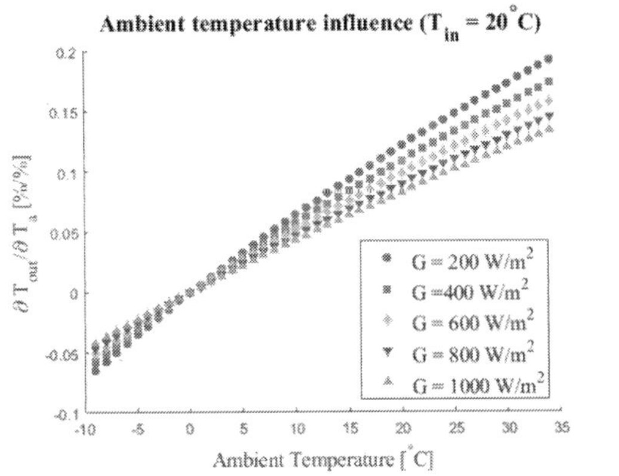

Fig. 7. Influence of ambient temperature on computed outlet temperature.

2.2 Convergence and iteration influence

The calculation algorithm had two convergence conditions:

• difference between new and old outlet temperature value lower than 0.001°K,

• conservation of the energy with an accuracy of 70 J.

The programme reached required convergence in about 10 iteration during sensitivity analysis. Results of convergence and number of iteration were presented in Fig. 8 and Fig. 9 for datasets used for inlet temperature and irradiation sensitivity analysis.

Number of iteration had approximately exponential influence on the convergence (Fig. 10).

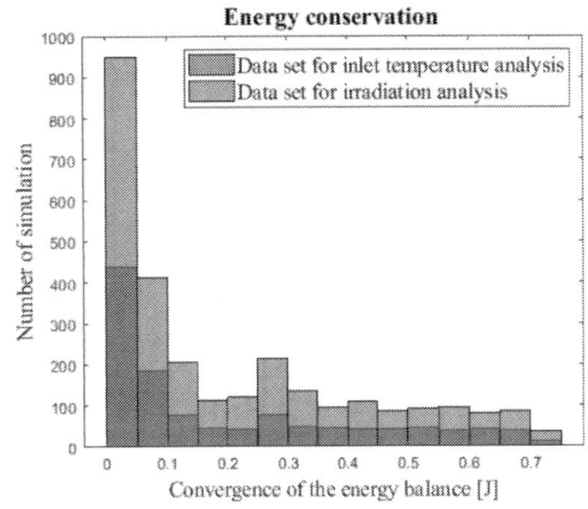

Fig. 8. Histogram of the energy covergence.

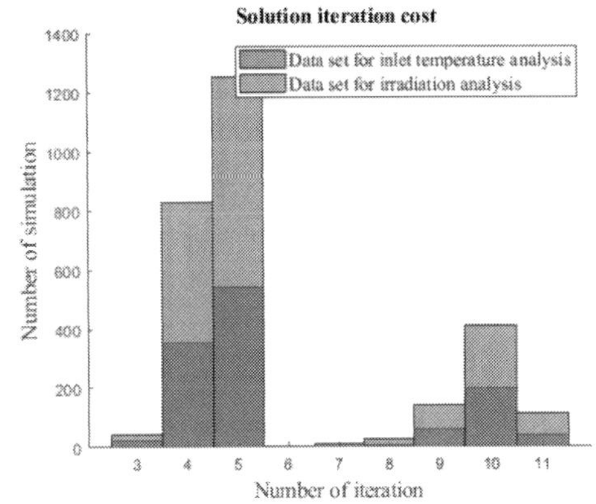

Fig. 9. Histogram of iteration required to reach convergence.

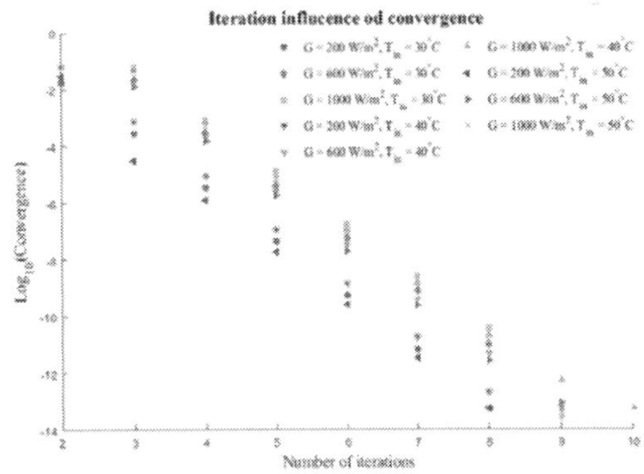

Fig. 10. Variation of convergence.

https://doi.org/10.1051/e3sconf/20187001019

3 Validation experiment

The experiment was conducted under class AAA solar simulator – LA150200 (Eternal Sun) in controllable conditions. The following parameters were measured during the experiment:

- irradiation, by SR11 Hukselflux pyrometer,
- infrared irradiation, by IR02 Hukselflux pyrgeometer,
- temperature, by PT100 A class resistant thermometers,
- mass flow, by Optimass 6400C flowmeter.

Weber Sol ECO 2,0 and Ensol 2,0 PV/T solar collectors were examined. Figure 11 presents deviation between measured and computed values of outlet temperature of the PV/T. The experiment was conducted for two different solar irradiations (820 and 1020 W/m^2) and for four inlet temperatures (14, 23, 32, 42°C). Mass flow rate during experiment was set about 1.2 l/min, Temperature in the experimental lab was set between 21 to 29°C. During each measurement ambient temperature was constant. Air velocity was constant and equal to 1 m/s. Experiment was conducted in steady state, the experiment's data were measured for at least 5 minutes. The uncertainty of the measured data was estimated based on standard deviation and meter accuracy. Temperature's uncertainty was estimated as 0.3°C, irradiation uncertainty was estimated as 16 W/m^2. Because of little influence on the computation results uncertainties of wind speed and mass flow rate were not taken into consideration.

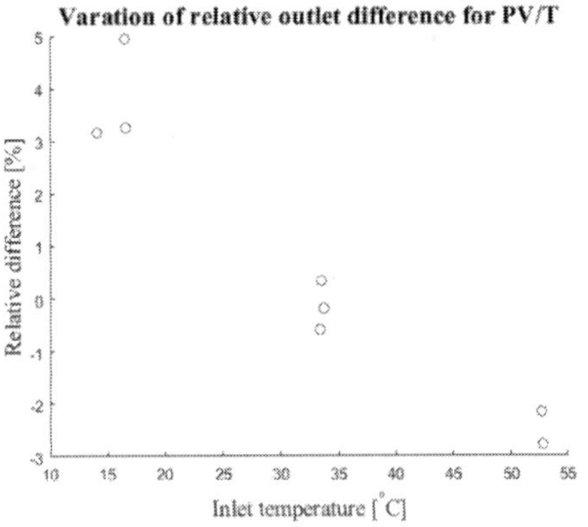

Fig. 11. Difference between measured and computed outlet temperatures of PV/T.

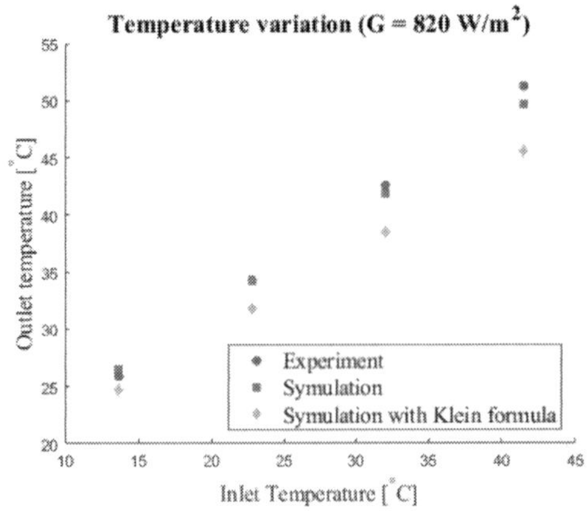

Fig. 12. Varation of outlet temperature mesured and simulated for irradiation of 820 W/m^2.

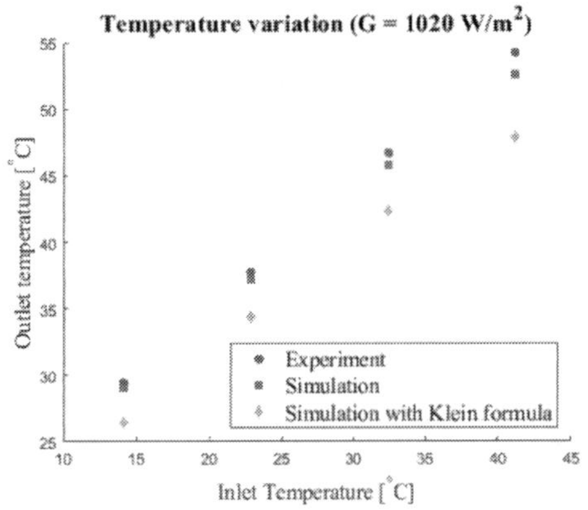

Fig. 13. Varation of outlet temperature mesured and simulated for irradiation of 1020 W/m^2.

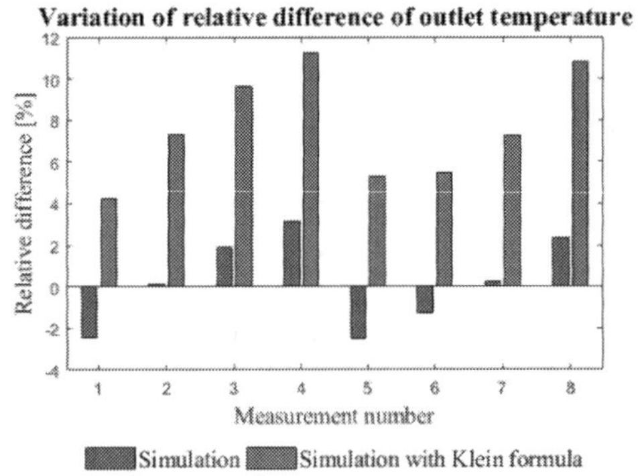

Fig. 14. Varation of difference beetwen mesured and simulated outlet temperature of presented model and for model, which used Klein formula.

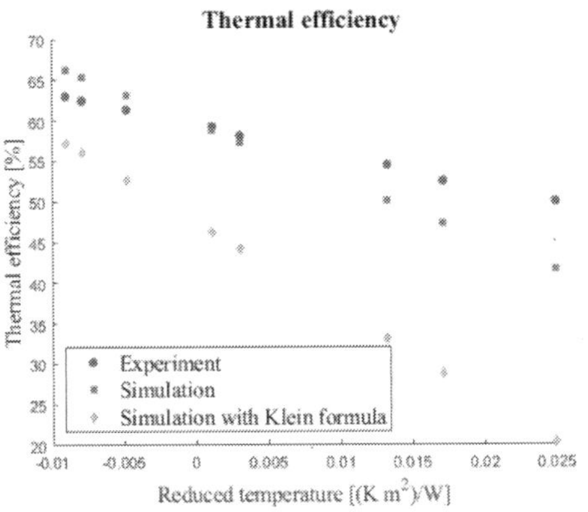

Fig. 15. Variation of Weber Sol ECO collector efficiency.

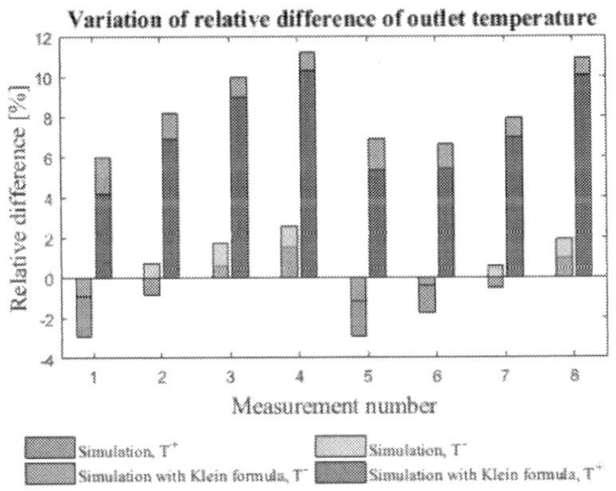

Fig. 16. Inluence of the inlet temperature uncertainty on outlet temperature.

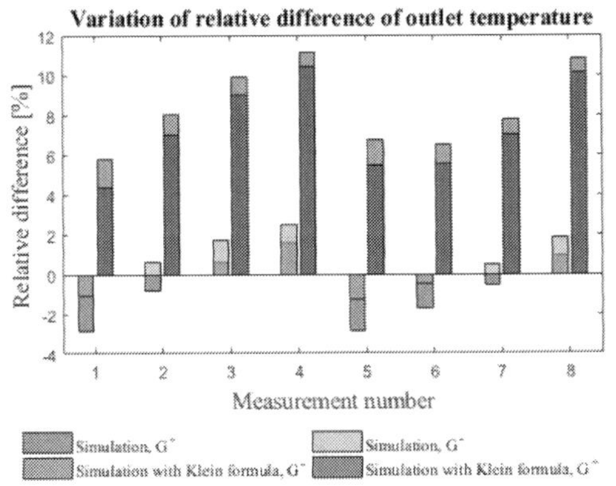

Fig. 17. Inluence of the solar irradiation uncertainty on outlet temperature.

4 Conclusions

The mathematical model of solar thermal collector was presented. The sensitivity analysis showed that inlet temperature value had significant influence on the computation results. 1% change in inlet temperature resulted in about 1% change in outlet temperature. Inlet temperature influence was presented in Fig. 2 and Fig. 3. Second factor with huge impact in sensitivity analysis was solar irradiation (Fig. 4). The influence of mass flow rate depended on the irradiation, ambient temperature and inlet temperature.

The solving algorithm was reaching required accuracy in about 10 iterations. Increasing precision of the solution did not increase significantly the number of iteration (Fig. 10).

The analysis of the comparison between the experiment and simulation leads to the following conclusions. First of all, the presented model achieved a good agreement with the experiment; a maximum difference about 3% was obtained (Fig. 14).

Uncertainty of the measured parameters had not significant influence on the computation results (Fig. 16 and Fig. 17).

The Klein formula in every simulation under-estimated collector efficiency and outlet temperature (Fig. 12, Fig. 13 and Fig 14). Moreover, every use of the Klein formula gave higher difference between experiments than the model. Computation costs of Klein formula algorithm and the model were similar. Both methods needed about 4 iterations during validation simulation. Simulation with Klein formula in most cases reached better convergence but results were not as closed to the experiment as the model's results. The presented model gave more accurate results than a model based on the Klein formula.

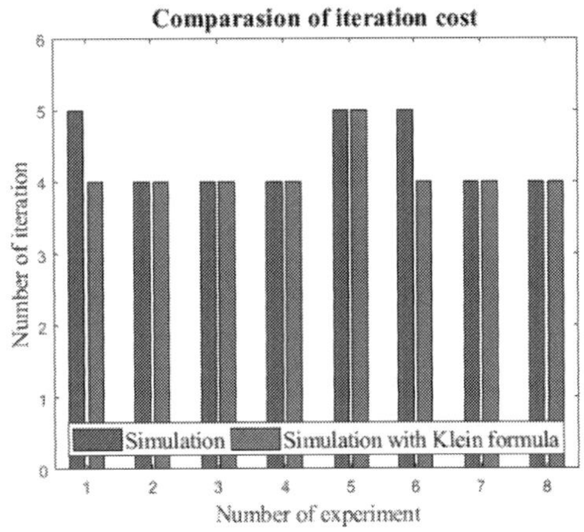

Fig. 18. Comparasion between models in respection to number of iteration.

According to the experiment for PV/T collector the model can be used in order to achieve 5% accuracy.

This work was supported by the own research fund of The Szewalski Institute of Fluid-Flow Machinery PASci (FBW27).

One of the authors (M. K.) was supported by the Foundation for Polish Science (FNP).

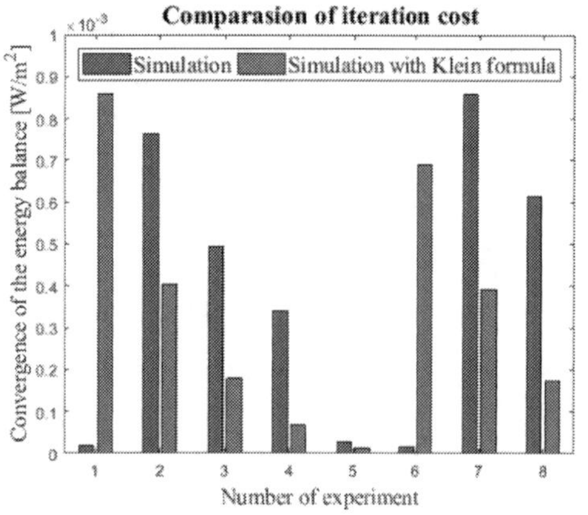

Fig. 19. Comparasion between models in respective to energy convergence.

References

1. V. Shemelin, T. Matuska, **2017** (2017)
2. J. A. Duffie, W. A. Beckman, W. M. Worek, *Solar Engineering of Thermal Processes, 4nd ed.*, **116**. New York: Wiley, 2003
3. Z. Pluta, *Podstawy teoretyczne fototermicznej konwersji energii słonecznej*. (Warszawa: Oficyna Wydawnicza Politechniki Warszaskiej, 2013)
4. D. M. Atia, F. H. Fahmy, N. M. Ahmed, H. T. Dorrah, Math. Comput. Model. **55** 1436–1449 (2012)
5. H. U. Helvaci, Z. A. Khan, Energy Convers. Manag. **106** 139–150 (2015)
6. F. Chabane, N. Moummi, D. Bensahal, A. Brima, Appl. Sol. Energy **50** 19–26 (2014)
7. W. T. Xie, Y. J. Dai, R. Z. Wang, Sol. Energy **91** 242–255 (2013)
8. N. Akhtar, S. C. Mullick, Energy **32** 1067–1074 (2007)
9. S. Kumar, S. C. Mullick, Ain Shams Eng. J. **3** 299–304 (2012)
10. Y. R. Sekhar, K. V Sharma, M. B. Rao, ARPN J. Eng. Appl. Sci. **4** 15–19 (2009)
11. M. Mattei, G. Notton, C. Cristofari, M. Muselli, P. Poggi, Renew. Energy **31** 553–567 (2006)
12. E. Skoplaki, A. G. Boudouvis, J. A. Palyvos, Sol. Energy Mater. Sol. Cells **92** 1393–1402 (2008)
13. C. Schwingshackl *et al.*, Energy Procedia **40** 77–86 (2013)
14. B. Durão, A. Joyce, J. F. Mendes, Sol. Energy **101** 160–166 (2014)
15. G. Fraisse, Y. Bai, N. Le Pierrès, T. Letz, Sol. Energy **83** 232–245 (2009)
16. N. Sharma, Varun, Siddhartha, Renew. Sustain. Energy Rev. **16** 1399–1411 (2012)
17. B. Kundu, Appl. Therm. Eng. **22** 999–1012 (2002)
18. J. Yazdanpanahi, F. Sarhaddi, M. Mahdavi Adeli, Sol. Energy **118** 197–208 (2015)
19. R. Chauhan, T. Singh, N. S. Thakur, N. Kumar, R. Kumar, A. Kumar, Renew. Sustain. Energy Rev. 0–1 (2017)
20. S. Wiśniewski, T. S. Wiśniewski, *Wymiana Ciepła*. (Warszawa: Wydawnictwo WNT, 2012)

Power increase of the power plant with single-circuit ORC plant with a wet medium, supplied with water in appropriate proportions from the Lower and Middle Triassic Reservoirs

Sławomir Wiśniewski[1,*], *Gabriela* Sołtysik[1], and *Władysław* Nowak[1]

[1]West Pomeranian University of Technology, Chair of Thermal Engineering, Al. Piastów 19, 70-310, Szczecin, Poland

Abstract. This article presents an analysis of the efficiency of the ORC power plant with a wet circulating medium. The power plant is supplied from two heat sources, which are geothermal water streams obtained from the Lower Triassic and Middle Triassic Reservoirs. In the ORC system, it was assumed that one of these streams is directed to the superheater and the other to the evaporator. In the work, two variants of ORC power supply were considered. The first one, in which the maximum stream of water from the Middle Triassic Reservoir $\dot{m}_{g2} = 55$ kg/s at a temperature of $T_{g1}^{2} = 104°C$ powers the evaporator, while the water stream from the Lower Triassic Reservoir $\dot{m}_{g1} = 25$ kg/s at a temperature of $T_{g1}^{1} = 128°C$ powers the superheater. In the second variant, the maximum water stream from the Lower Triassic Reservoir $\dot{m}_{g1} = 25$ kg/s and a temperature of $T_{g1}^{1} = 128°C$ feeds the evaporator, while the water stream from the Middle Triassic Reservoir $\dot{m}_{g2}$ at a temperature of $T_{g1}^{2} = 104°C$ supplies the superheater. In both variants, the geothermal water stream directed to the superheater is the size resulting from ORC plant operation and cannot exceed the maximum flow value for a given geothermal reservoir. The following working substances were taken into account in the analysis: cyclopropane, R143a, R32, propylene, R22, propane, R134a. The analysis carried out showed that in the ORC power plant with a two-source supply the maximum system power was obtained in the case of the first variant for the working medium R143a.

1 Introduction

In recent years, we have been dealing with a constant increase in energy consumption [1]. The need to cover the growing demand for energy forces the search for new sources of energy or increase the use of existing resources of fossil fuels and renewable energy carriers. Environmental degradation associated with the emission of pollutants from the energy sector using conventional energy carriers as well as international obligations [2] are forcing more and more interest in renewable energy sources. For this reason, in this paper an attempt to assess the efficiency of the ORC plant operation using geothermal resources as a heat source was made. In Poland, the resources of this energy are used to a small extent, and existing geothermal plants use it only for heating purposes [3]. Geothermal waters with a temperature of 100 - 130°C do not create the possibility of using these resources to generate electricity in a classical steam power plant (with geothermal water as a circulating fluid). In this temperature range of the heat carrier, it is real to use the ORC (Organic Rankine Cycle) power plant. In the literature, there are a lot of works on the ORC power plant. These are the works, among others, in which issues related to the evaluation of the efficiency of ORC systems supplied from different heat sources, their operation and the technical solutions used have been presented [4-7].

In the ORC plant, the circulating factor is a suitable organic substance, characterized by high saturation pressure of several MPa at low evaporation temperatures (below 100°C). Such substances are commonly used in refrigeration systems. The analysis of the properties of circulating fluids used in ORC systems was presented, among others, in [8-11]. These works clearly show that the right choice of a fluid has a significant impact on the ORC power plant efficiency.

Apart from the analysis of single-circuit ORC systems supplied from one source of heat, the literature also includes works on systems supplied from many heat sources. The works in which the ORC systems supplied from many sources were analyzed include, among others, [6, 13-15]. A characteristic feature of the systems analyzed in these works was that they consisted of many separate ORC circuits. The paper [16] presents an assessment of the effectiveness of a geothermal power plant with a single-circuit ORC power plant with a two-source power supply. The assessment of the efficiency of this plant, including the power value $N_{C\text{-}R}$ and the thermal efficiency of the Rankine cycle $\eta_{C\text{-}R}$, was carried out for the selected dry medium and wet agent. In the analyzed ORC system [16], as well as in this paper,

* Corresponding author: Slawomir.Wisniewski@zut.edu.pl

https://doi.org/10.1051/e3sconf/20187001020

geothermal water extracted from the Lower Triassic Reservoir and the Middle Triassic Reservoir were assumed as the heat sources.

It was assumed in the work [16] that in the case of the ORC plant with a dry working medium, the mass stream of geothermal water with higher temperature supplies the evaporator, and the mass stream of geothermal water with lower temperature feeds the preheater. In the case of a power plant with a wet working medium, the mass stream of geothermal water with a higher temperature feeds the superheater, and the stream of geothermal water with lower temperature powers the evaporator.

This paper presents the evaluation of ORC power plant efficiency with a wet circulating factor, supplied from two heat sources, i.e. geothermal water obtained from the Lower Triassic Reservoir and the Middle Triassic Reservoir, including two variants of this power supply.

In the first variant of feeding the ORC system, the maximum flow of geothermal water obtained from the Middle Triassic Reservoir $\dot{m}_{g2}$ at a temperature of T_{g1}^{2} feeds the evaporator. The size of mass stream of geothermal water from the Lower Triassic Reservoir $\dot{m}_{g1}$ with the supply temperature T_{g1}^{1} supplying the superheater was adjusted to the operating parameters of the analyzed ORC plant. The size of this stream was determined from the dependence (7) presented in the calculation methodology, however, it cannot exceed the maximum mass stream $\dot{m}_{g}$=25 kg/s.

In the second variant, the maximum stream of geothermal water from the Lower Triassic Reservoir with the stream $\dot{m}_{g1}$ and the temperature T_{g1}^{1} supplies the evaporator. The size of the mass stream of geothermal water from the Middle Triassic Reservoir, feeding the superheater, was adapted to the operating parameters of the analyzed ORC plant. The size of this stream of geothermal water $\dot{m}_{g2}$ was determined from dependence (17), however, it cannot exceed the maximum mass stream $\dot{m}_{g} = 55$ kg/s.

The exact description of the ORC system analyzed, along with the diagrams are presented in the next part of the work..

2 System description

The system under consideration is an ORC power plant with a two-source power supply. These sources of heat are two streams of geothermal water, the first extracted from the Lower Triassic Reservoir and the second extracted from the Middle Triassic Reservoir. The power plant comprises the preheater 5, evaporator 6, superheater 7, turbo-generator 2, condenser 3, circulation pump 4. It has been assumed that the circulating fluid in the power plant is a working medium from the group of so-called wet working fluids. The evaporation process of this factor in the ORC plant is carried out at a temperature of 67°C and the condensation process at a temperature of 29°C. The analysis includes seven working fluids: cyclopropane, R143a, R32, propylene, R22, propane, R134a. The analysis of the efficiency of the ORC plant operation with the above-mentioned factors was carried out for two variants of system supply from two different temperature heat sources. In the first variant, carried out in accordance with the diagram shown in Figure 1, the stream of water from the Lower Triassic Reservoir $\dot{m}_{g1}$ is directed to the superheater, whereas the water stream $\dot{m}_{g2}$ from the Middle Triassic Reservoir is directed to the evaporator. These streams, after leaving the superheater and evaporator are combined before the preheater. Therefore, a total stream of geothermal water $\dot{m}_{g3}$, with a temperature of 70°C, is directed to the preheater. After leaving the preheater, the cooled water returns to the Triassic Reservoirs. In each of the heat exchangers geothermal water gives energy to the organic medium. The working fluid first goes to the preheater 5, where it receives heat from the water stream and heats up from the condensation temperature to the evaporation temperature. Then the fluid goes to the evaporator 6, where it is evaporated, and then in the form of dry saturated steam is directed to the next heat exchanger, that is superheater 7. In the superheater 7, the temperature of the steam increases. Then the superheated steam goes to the turbo-generator 2, where it expands to the pressure corresponding to the condensing pressure. After leaving the turbo-generator 2 the steam goes to the condenser 3. As a result of cooling, the circulating medium in the condenser 3 is condensed. The condensate liquid by means of circulating pump 4 is conveyed into the preheater 5, where it is heated to the evaporation temperature and the cycle of particular transformations is repeated.

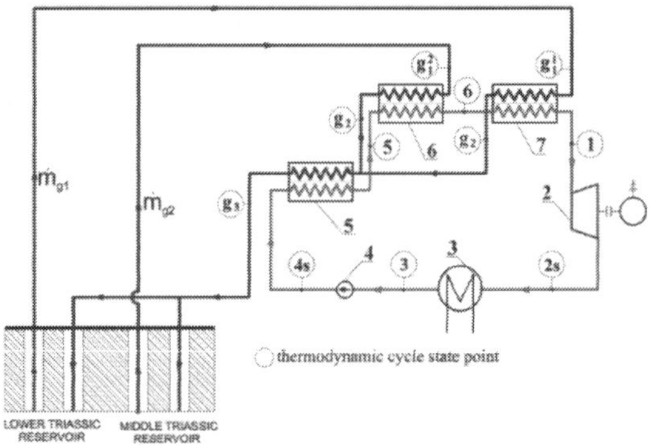

Fig. 1. ORC plant diagram with a two-source power supply for the first variant.

The second variant analyzed in the work is the ORC system operating in accordance with the scheme shown in Figure 2. The main difference between the principle of operation and the first variant is that in the second variant the geothermal water stream from the Lower Triassic Reservoir $\dot{m}_{g1}$ is directed to the evaporator, whereas the stream $\dot{m}_{g2}$ water from the Middle Triassic Reservoir is directed to the superheater.

The cycle of thermodynamic transformations carried out in individual ORC devices is presented in Figure 3.

The transformations shown in Figure 3 are carried out in the following devices:

1-2s: isentropic expansion of the steam turbine (to saturation state),
2s-3: condensation of steam in the condenser,
3-4s: isentropic increase of working fluid pressure in the pump,
4s-5: isobaric preheating of the working liquid in the preheater to the saturation state,
5-6: isobaric-isothermal evaporation of the liquid in the evaporator,
6-1: isobaric superheated steam in the superheater.

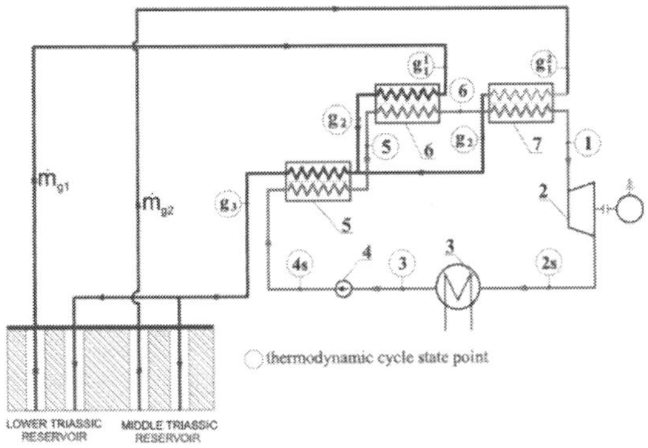

Fig. 2. ORC plant diagram with a two-source power supply for the second variant.

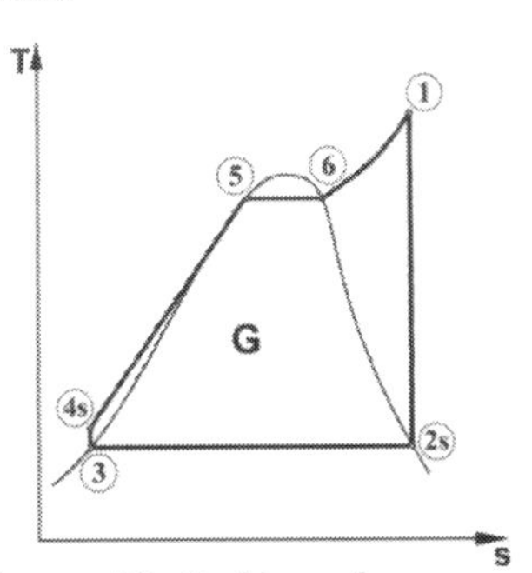

Fig. 3. T-s diagram of the Rankine cycle.

3 Methodology of calculations

The paper presents the assessment of the efficiency of the power plant operation with a single-circuit ORC power plant with a two-source power supply. As a source of heat, the Lower Triassic Reservoir and the Middle Triassic Reservoir were adopted. For calculations it was assumed that in the case of the Lower Triassic Reservoir the maximum mass flow of geothermal water is $\dot{m}_{g1}$=25 kg/s, and the temperature of the extracted water is T_{g1}^1=128°C. In the case of the Middle Triassic Reservoir, the maximum mass flow of geothermal water is $\dot{m}_{g2}$=55 kg/s, and the temperature of the extracted water is T_{g1}^2=104°C. It has been assumed that the geothermal water temperature at the outlet from the superheater and evaporator is T_{g2}=70°C.
For the above-mentioned temperatures, the appropriate specific enthalpies appearing in the energy balance equations of the geothermal water as an energy carrier have been identified $h_{g1}^1(T_{g1}^1)$, $h_{g2}(T_{g2})$, $h_{g3}(T_{g3})$.

After adopting the type of working medium on the basis on the specific enthalpy in the characteristic points of the Rankine cycle, the proper enthalpies of the overheating Δh_{superh}, evaporation Δh_{evap}, preheating Δh_{preh} were determined from the following dependencies:

$$\Delta h_{superh} = h_1 - h_6 \tag{1}$$

$$\Delta h_{evap} = h_6 - h_5 \tag{2}$$

$$\Delta h_{preh} = h_5 - h_{4s} \tag{3}$$

where: h - specific enthalpy of the organic factor at the characteristic points of the Rankine cycle.
In the first variant, the ORC power plant is supplied with a stream of geothermal water $\dot{m}_{g1}$ from the Lower Triassic Reservoir with a temperature T_{g1}^1, which is directed to the superheater, and the geothermal water stream $\dot{m}_{g2}$ from the Middle Triassic Reservoir with temperature T_{g1}^2, which feeds the evaporator. In order to perform thermal-flow calculations for the first variant, the energy balance equations of three heat exchangers in the following form have been used:
for superheater:

$$\dot{m}_{g1}\left[h_{g1}^1(T_{g1}^1) - h_{g2}(T_{g2})\right] = \dot{m}_n \Delta h_{superh} \tag{4}$$

for evaporator:

$$\dot{m}_{g2}\left[h_{g1}^2(T_{g1}^2) - h_{g2}(T_{g2})\right] = \dot{m}_n \Delta h_{evap} \tag{5}$$

for preheater:

$$\dot{m}_{g3}\left[h_{g2}(T_{g2}) - h_{g3}(T_{g3})\right] = \dot{m}_n \Delta h_{preh} \tag{6}$$

In order to determine the mass flow of geothermal water $\dot{m}_{g1}$ from the Lower Triassic Reservoir feeding the superheater, the dependences (4) and (5) were divided on two sides. After performing the appropriate transformations, the expression for stream $\dot{m}_{g1}$ supplying the superheater has got the following form (7):

$$\dot{m}_{g1} = \left(\frac{(h_{g1}^2 - h_{g2})}{(h_{g1}^1 - h_{g2})}\frac{\Delta h_{superh}}{\Delta h_{evap}}\right)\dot{m}_{g2} \tag{7}$$

Next, after adding dependencies (4) and (5) respectively and performing the relevant transformations, the dependence on the mass stream of the working medium of the power plan $\dot{m}_n$ was determined:

$$\dot{m}_n = \frac{\dot{Q}_{superh} + \dot{Q}_{evap}}{(\Delta h_{superh} + \Delta h_{evap})} \tag{8}$$

The heat flows supplied to the superheater $\dot{Q}_{superh}$ and the evaporator $\dot{Q}_{evap}$ which appeared in the equation (8) were determined after using the left sides of dependences (4) and (5).
The water temperature T_{g3} leaving the preheater is higher than the temperature of the medium at the point T_{4s} by ΔT=3K.

$$T_{g3} = T_{4s} + \Delta T \tag{9}$$

The specific enthalpy of water h_{g3} on the outflow from the preheater was read from the Refprop [17] program on the basis of temperature T_{g3}.
The mass flow of water $\dot{m}_{g3}$ leaving the preheater was determined on the basis of the balance equation for the preheater (6):

$$\dot{m}_{g3}=\frac{\dot{m}_n \Delta h_{preh}}{[h_{g2}(T_{g2})-h_{g3}(T_{g3})]} \tag{10}$$

The heat flow supplied to the power plant $\dot{Q}_d$, is the sum of heat fluxes that are fed to the superheater $\dot{Q}_{superh}$, the evaporator $\dot{Q}_{par}$ and the preheater $\dot{Q}_{preh}$:

$$\dot{Q}_d=\dot{Q}_{superh}+\dot{Q}_{evap}+\dot{Q}_{preh} \tag{11}$$

The dependence on the power plant capacity N_{C-R} takes the following form:

$$N_{C-R}=l_{C-R}\,\dot{m}_n \tag{12}$$

where: $l_{C-R}=l_{turbine}-l_{pump}$ – unit work of the Rankine cycle.
The power plant efficiency η_{C-R} was determined from the following formula:

$$\eta_{C-R}=\frac{N_{C-R}}{\dot{Q}_d} \tag{13}$$

In the second variant, the ORC power plant is also supplied with geothermal water from the Lower Triassic Reservoir with mass flow $\dot{m}_{g1}$ and temperature T^1_{g1}, with the difference that this stream is directed to the evaporator, and geothermal water from the Middle Triassic Reservoir with the stream $\dot{m}_{g2}$ and temperature $T^2_{g1}=104°C$ is directed to the superheater.
In order to perform thermal and flow calculations for the second variant, energy balance equations of three heat exchangers are used, which take the form:
for superheater:

$$\dot{m}_{g2}\left[h^2_{g1}\left(T^2_{g1}\right)-h_{g2}\left(T_{g2}\right)\right]=\dot{m}_n\Delta h_{superh} \tag{14}$$

for evaporator:

$$\dot{m}_{g1}\left[h^1_{g1}\left(T^1_{g1}\right)-h_{g2}\left(T_{g2}\right)\right]=\dot{m}_n\Delta h_{evap} \tag{15}$$

for preheater:

$$\dot{m}_{g3}\left[h_{g2}\left(T_{g2}\right)-h_{g3}\left(T_{g3}\right)\right]=\dot{m}_n\Delta h_{preh} \tag{16}$$

First, to determine the stream of geothermal water feeding the superheater, the dependence (14) should be divided by (15). After the transformation of these equations, the dependence on the stream of geothermal water $\dot{m}_{g2}$ feeding the superheater is as follows:

$$\dot{m}_{g2}=\left(\frac{\left(h^1_{g1}-h_{g2}\right)}{\left(h^2_{g1}-h_{g2}\right)}\cdot\frac{\Delta h_{superh}}{\Delta h_{evap}}\right)\dot{m}_{g1} \tag{17}$$

The further calculation algorithm is the same as in the calculation for the first variant.

4 Calculation results

All thermal and caloric parameters of circulating factors as well as geothermal water were determined using the basis of organic factors RefProp [17]. Corresponding statements of these values for circulating factors of the ORC power plant and geothermal water circuits are presented in Tables 1 and 2.

Table 1. Comparison of values of temperatures and specific enthalpies in the ORC power plant at characteristic points of circulation for two selected factors for the first and second variant.

Working fluid		Circulation point					
		1	2s	3	4s	5	6
R143a	T [°C]	75,16	29,00	29,00	30,89	67,00	67,00
	h [kJ/kg]	417,7	399,8	245,9	248,00	232,1	392,2
R32	T [°C]	91,55	29,00	29,00	31,09	67,00	67,00
	h [kJ/kg]	549,3	515,9	253,4	256,2	341,9	486,3

Table 2. Comparison of the values of temperatures and specific enthalpies of geothermal water at individual points of the power plant for the first and second variant.

Working fluid		Circulation point			
		g^1_1	g^2_1	g2	g3
R143a	T [°C]	128	104	70	33,89
	h [kJ/kg]	537,9	436,05	293,07	142,00
R32	T [°C]	128	104	70	34,09
	h [kJ/kg]	537,9	436,05	293,07	142,85

The obtained calculation results characterizing the ORC power plant efficiency for circulating factors R143a and R22 for the first and the second variant are presented in Tables 3, 4, 5 and 6.

Table 3. Comparison of the results of calculations of the quantities characterizing operation of the ORC power plant of the first variant with the working medium R143a.

Δh_{superh}	Δh_{evap}	Δh_{preh}	Δh_{cond}	Δh_c
h_1-h_6	h_6-h_5	h_5-h_{4s}	$h_{2s}-h_3$	h_1-h_{4s}
25,57	69,08	75,10	153,92	169,75
$l_{turbine}$ [kW]	l_{pump} [kW]	l_{C-R} [kW]	$\dot{m}_{g1}$ [kg/s]	
17,94	2,11	15,83	11,89	
$\dot{Q}_{superh}$ [kW]	$\dot{Q}_{evap}$ [kW]		$\dot{m}_n$ [kg/s]	
2910,83	7863,90		113,84	
T_{g3}[°C]	h_{g3}		$\dot{m}_{g3}$	
33,89	142,00		56,59	
$\dot{Q}_{preh}$ [kW]	$\dot{Q}_d$ [kW]		N_{C-R} [kW]	η_{C-R} [%]
8549,90	19324,63		1801,84	9,32

In order to demonstrate the legitimacy of the proposed ORC plant solution supplied from two heat sources, the power capacity of the ORC power plant supplied from one heat source was calculated. Determining the power of the power plant supplied from one heat source, it was assumed that the maximum available stream of geothermal water from the Lower Triassic Reservoir or the Middle Triassic Reservoir is used. The results of these calculations for all analyzed factors are presented in Table 7.

Table 4. Comparison of the results of calculations of the quantities characterizing operation of the ORC power plant of the first variant with the R32 working medium.

Δh_{superh}	Δh_{evap}	Δh_{preh}	Δh_{cond}	Δh_c
$h_1 - h_6$	$h_6 - h_5$	$h_5 - h_{4s}$	$h_{2s} - h_3$	$h_1 - h_{4s}$
62,93	144,48	85,65	262,56	293,06
$l_{turbine}$ [kW]	l_{pump} [kW]	$l_{C\text{-}R}$ [kW]	$\dot{m}_{g1}$ [kg/s]	
33,34	2,84	30,50	13,99	
$\dot{Q}_{superh}$ [kW]	$\dot{Q}_{evap}$ [kW]		$\dot{m}_n$ [kg/s]	
3424,99	7863,90		54,43	
T_{g3}[°C]	h_{g3}		$\dot{m}_{g3}$	
34,09	142,85		31,03	
$\dot{Q}_{preh}$ [kW]	$\dot{Q}_d$ [kW]		$N_{C\text{-}R}$ [kW]	$\eta_{C\text{-}R}$ [%]
4662,00	15950,89		1659,98	10,41

Table 5. Comparison of the results of calculations of the quantities characterizing operation of the ORC plant of the second variant with the R143a working medium.

Δh_{superh}	Δh_{evap}	Δh_{preh}	Δh_{cond}	Δh_c
$h_1 - h_6$	$h_6 - h_5$	$h_5 - h_{4s}$	$h_{2s} - h_3$	$h_1 - h_{4s}$
25,57	69,08	75,10	153,92	169,75
$l_{turbine}$ [kW]	l_{pump} [kW]	$l_{C\text{-}R}$ [kW]	$\dot{m}_{g1}$ [kg/s]	
17,94	2,11	15,83	15,84	
$\dot{Q}_{superh}$ [kW]	$\dot{Q}_{evap}$ [kW]		$\dot{m}_n$ [kg/s]	
6119,50	2265,14		88,59	
T_{g3}[°C]	h_{g3}		$\dot{m}_{g3}$	
33,89	142,00		44,04	
$\dot{Q}_{preh}$ [kW]	$\dot{Q}_d$ [kW]		$N_{C\text{-}R}$ [kW]	$\eta_{C\text{-}R}$ [%]
6653,33	15037,97		1402,15	9,32

Table 6. Comparison of the results of calculations of the quantities characterizing operation of the ORC power plant of the second variant with the R32 working medium.

Δh_{superh}	Δh_{evap}	Δh_{preh}	Δh_{cond}	Δh_c
$h_1 - h_6$	$h_6 - h_5$	$h_5 - h_{4s}$	$h_{2s} - h_3$	$h_1 - h_{4s}$
62,93	144,48	85,65	262,56	293,06
$l_{turbine}$ [kW]	l_{pump} [kW]	$l_{C\text{-}R}$ [kW]	$\dot{m}_{g1}$ [kg/s]	
33,34	2,84	30,50	18,64	
$\dot{Q}_{superh}$ [kW]	$\dot{Q}_{evap}$ [kW]		$\dot{m}_n$ [kg/s]	
6119,50	2665,25		42,36	
T_{g3}[°C]	h_{g3} [kW]		$\dot{m}_{g3}$ [kg/s]	
34,09	142,85		24,15	
$\dot{Q}_{preh}$ [kW]	$\dot{Q}_d$ [kW]		$N_{C\text{-}R}$ [kW]	$\eta_{C\text{-}R}$ [%]
3627,86	12412,60		1291,76	10,41

Table 7. Comparison of the results of calculations of the quantities characterizing operation of the ORC power plant of the second variant with the R32 working medium.

Power variant	W I	W II	Single-source power supply $T^1_{g1} = 128°C$ (25 kg/s)	Single-source power supply $T^2_{g1} = 104°C$ (55 kg/s)
Working fluid	N_{C-R} [kW]	N_{C-R} [kW]	N_{C-R} [kW]	N_{C-R} [kW]
Cyclopropane	1074,5	836,2	785,1	1008,9
R143a	1801,8	1402,2	1023,4	1315,1
R32	1660,0	1291,8	899,8	1156,4
Propylen	1230,1	957,2	860,4	1105,7
R22	1223,2	951,8	831,0	1067,9
Propan	1157,6	900,8	855,2	1099,0
R134a	1134,3	882,7	844,8	1085,7

The comparison of ORC plant efficiency for the R143a and R32 factors for the first and second variant is shown by means of the bar graph in Figure 4.
The comparison of the power of the ORC power plant for the R143a and R32 factors for the first and second variant is presented by means of a bar graph in Figure 5.

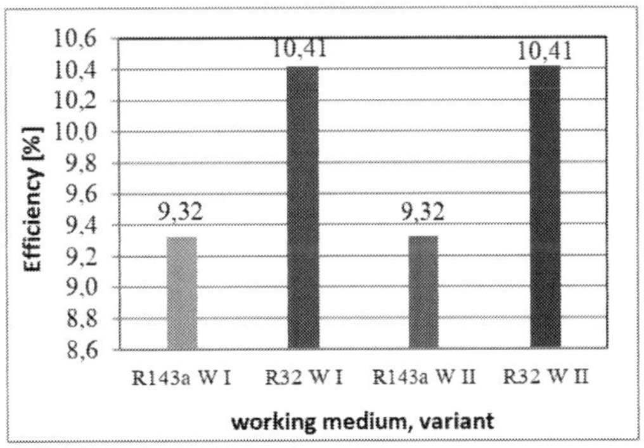

Fig. 4. Efficiency of considered variants.

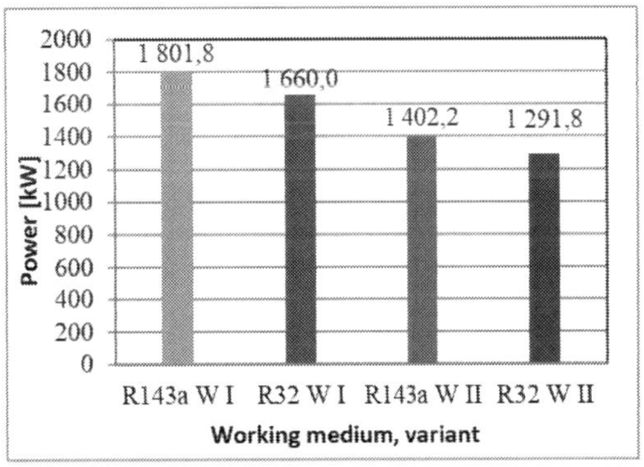

Fig. 5. The power of considered variants.

5 Conclusion

In the paper, it was shown that among the two analyzed variants of ORC power station supply with two heat sources, the first power option is more advantageous. Therefore, having at the disposal heat sources similar to those analyzed in this work, the heat carrier with lower temperature and higher mass stream should be directed to the ORC evaporator, and the heat carrier with a higher temperature to the superheater.
The above conclusion is also confirmed by the fact that the obtained power quantities for: cyclopropane, R143a, R32, propylene, R22, propane and R 134a for the first variant were obtained with incomplete use of the geothermal water stream from the Lower Triassic Reservoir, whose maximum value was assumed at 25 kg/s. The values of this geothermal water stream for the analyzed working factors, according to the order in which they are listed, are: $\dot{m}_{g1}$=2,09 kg/s, $\dot{m}_{g1}$= =11,89 kg/s, $\dot{m}_{g1}$=13,99 kg/s, $\dot{m}_{g1}$=3,61 kg/s, $\dot{m}_{g1}$= =4,67 kg/s, $\dot{m}_{g1}$=1,71 kg/s, $\dot{m}_{g1}$=1,44 kg/s using the maximum flow of geothermal water from the Middle

Triassic Reservoir $\dot{m}_{g2} = 55$ kg/s directed to the ORC system evaporator. From the above, it also follows that by feeding the ORC system with a heat carrier stream with a lower temperature (geothermal water from the Middle Triassic Reservoir), the required heat carrier stream with a higher temperature delivered to the ORC superheater is much smaller. This makes it possible to take ORC systems supplied from two heat sources into consideration, in which one heat source is, for example, waste heat, and the second heat source is, for example, a biomass-fired boiler (or other fuel). Therefore, the possibility of using such a solution is not limited to using geothermal water as a source of heat.

Analyzing the obtained results of calculations regarding the ORC system with a two-source power supply, it is clearly visible that the type of circulating medium used significantly affects the power of the system. This results directly from the data presented in Table 7. In the case of the first variant, the highest system power, amounting to 1802,8 kW, was obtained using the circulating medium R143a, whereas the lowest one - 1074,5 kW for the cyclopropane. Using the same heat sources, the power obtained in the analyzed ORC plant with the R143a factor (the first variant) is about 68% greater than the power obtained in the ORC power plant with the cyclopropane factor. This also confirms the conclusions from many previous works indicating the appropriate choice of working medium in the ORC system. It is suitable not only in terms of temperature parameters of the heat source, but also in terms of maximizing the ORC system power for a given technological solution.

Summarizing the above, it can be concluded that further testing of the ORC power plant with a two-source power supply, aimed at increasing the power, should be continued using only the first power variant.

References

1. BPStatistical Review of World Energy, June 2018, 67th edition; https://www.bp.com/content/dam/bp/en/corporate/pdf/energy-economics/statistical-review/bp-stats-review-2018-full-report.pdf
2. DIRECTIVE 2009/28/EC OF THE EUROPEAN PARLIAMENT AND OF THE COUNCIL of 23 April 2009 on the promotion of the use of energy from renewable sources and amending and subsequently repealing Directives 2001/77/EC and 2003/30/EC
3. B. Kepińska, A review of geothermal energy uses in Poland in 2016–2018, Geological Exploration Technology, Geothermal Energy, Sustainable Development, Vol. 57, (261), pp. 11-27, (in Polish), (2018)
4. Zhang Shengjun, Wang Huaixin, Guo Tao, Performance comparison and parametric optimization of subcritical Organic Rankine Cycle (ORC) and transcritical power cycle system for low-temperature geothermal power generation, Applied Energy 88, pp. 2740–2754, (2011)
5. H. Ghasemi, M. Paci, A. Tizzanini, A. Mitsos, Modeling and optimization of a binary geothermal power plant, Energy 50, pp 412-428, (2013)
6. A.A. Stachel, S. Wiśniewski, Influence of the type of working fluid in the lower cycle and superheated steam parameters in the upper cycle on effectiveness of operation of binary power plant, Archives of Thermodynamics, Vol. 36, no. 1, pp 111-123, (2015)
7. D. Moyaa, C. Aldásd, P. Kaparajue, Geothermal energy: Power plant technology and direct heat applications, Renewable and Sustainable Energy Reviews 94, pp 889–901, (2018)
8. A. Borsukiewicz-Gozdur, W. Nowak, Comparative analysis of natural and synthetic refrigerants in application to low temperature Clausius–Rankine cycle, Energy 32, 344-352, (2007)
9. O. Badr, S.D. Probert, W.P. O'Callaghan, Selecting a working fluid for a Rankine-cycle engine, Appl. Energ. 21, 1–42, (1985)
10. A.A. Lakew, O. Bolland, Working fluids for low-temperature heat source, Applied Thermal Engineering 30, pp 1262-1268, (2010)
11. B. Saleh, G. Koglbauer, M. Wendland, J. Fischer, Working fluids for low-temperature organic Rankine cycles, Energy 32, pp 1210–1221, (2007)
12. H. Chen, D.Y. Goswami, E.K. Stefanakos, A review of thermodynamic cycles and working fluids for the conversion of low-grade heat, Renewable and Sustainable Energy Reviews, 14 (9), pp. 3059-3067, (2010)
13. W. Nowak, S. Wiśniewski, A concept to fuel the ORC power plant by heat sources of different temperatures, Thermodynamics in Science and Technology Proceedings of the 1-st International Congress on Thermodynamics, Part 2 L. Bogusławski (Editor) Poznań, pp. 877-886, (2011)
14. A. Borsukiewicz-Gozdur, Dual-fluid-hybrid power plant co-powered by low-temperature geothermal water. Geothermics 39, 170-176, (2010)
15. W. Nowak, S. Wiśniewski, A. Borsukiewicz-Gozdur, Siłownia parowa z wieloźródłowym zasilaniem, Archiwum energetyki, Vol. 42, no. 2,pp 103—112, (in Polish), (2012)
16. S. Wiśniewski, G. Sołtysik, W. Nowak, Evaluation of the effectiveness of one-cycle ORC power plant supplied by geothermal water from lower and middle Triassic reservoirs, using a wet and dry working fluids, Geological Exploration Technology, Geothermal Energy, Sustainable Development, Vol. 57, (261), pp. 37-45, (in Polish) (2018)
17. NISIT. RefProp 9.0, Standard Reference Database 23, Version 9.0, , Reference Fluid Therodynamic and Transport Properties. National Institute of Standards and Technology, Gaithersburg, MD, USA. (2010)

Simplified methodology to estimate the emissivity for roof covers

Zbigniew Zapałowicz[1,*]

[1]Department of Heat Engineering, Faculty of Mechanical Engineering and Mechatronics, West Pomeranian University of Technology, Szczecin, al. Piastów 17, 70-310 Szczecin, Poland

Abstract. In a number of countries, producers of roof covers are not obliged by law to estimate the reflectivity and emissivity of their products and make them generally available. However, information concerning these parameters that characterize the surface of roof cover is relevant for building's heat calculations. Adequate choice of kind and color of the roof surface affects the amount of heat transferred between the interior and the environment. Simplified research methodology has been proposed in the paper. This methodology requires that experimental research is carried out and characteristic temperatures of roof and of the air are determined. It is then possible to estimate the emissivity for roof cover from the model of energy balance.

Nomenclature

G – irradiance, W/m^2
h – convective heat transfer coefficient, W/(m^{2}K)
q – heat flux, W/m^2
r – reflectivity
R – heat resistance, (m^{2}K)/W
t – temperature, °C
T – absolute temperature, K
U – overall heat transfer coefficient, W/(m^{2}K)
w – wind velocity, m/s

Greek symbols

δ – thickness, m
ε – emissivity
λ – thermal conductivity, W/(m·K)
σ – Stephan-Boltzmann constant, $5.67 \cdot 10^{-8}$ W/m^{2}K^{-4}

Indexes

a – absorption
A – ambient
C – roof covering (roofing paper)
e – equivalent
H – sky (horizon)
r – irradiance
R – interior
S – sun
SI – styrofoam on the side of roofing paper
SII – styrofoam on the side of interior
β – incidence angle of solar beams
λ – conductivity

1 Introduction

Effective energy consumption is one of fundamental strategic targets of governments in many countries [1]. Decrease of energy consumption causes in each economy that the so far used sources are sufficient and advancement of the energy sector can be less dynamic.

A significant consumer of heat and electricity is the sector of construction [2-5]. One of ways to diminish the heat energy consumption in constructing is adequate choice of applied building materials [6-8]. The issue has been recognized also by producers and scientists of roof covers. Appropriate choice of kind and color of roofing affects meaningfully the energy balance of a building [9-12]. The amount of energy transferred between the building's interior and its environment can be significantly reduced by application of the right roofing material with right reflectivity and emissivity of roofing outer layers [13-15]. In some countries, producers of roof covers have created organizations to promote the so called "cool roofing" [16]. They publish information concerning a.o. reflectivity and emissivity of outer layers of roof covers.

Unfortunately in too many countries, producers of roof covers are not associated and there are not juridical regulations that would oblige the producers to include reflectivity and emissivity in technical data concerning their products.

The present paper proposes simplified estimate methodology of emissivity for roof covers. The methodology requires, as a matter of fact, experimental research to be carried out, however, its implementation is not difficult.

* Corresponding author: zbigniew.zapalowicz@zut.edu.pl

2 Model of heat transfer for roof

In papers [17, 18] the model of heat transfer for roof proposed by author is being presented. It is assumed for the model that the roof is in a quasi-state condition. Fig.1 shows energy fluxes that are transferred among the roof and particular media.

Energy balance equation written for 1m^2 of roof has the general shape:

$$G_{Sa} = q_{C-A} + q_{C-H} + q_{\lambda}, \tag{1}$$

On the left side of equation (1), there is irradiance absorbed by the roof cover G_{Sa}. The irradiance is part of total irradiance, $G_{S\beta}$, that reaches the surface of roofing. The remaining part of irradiance, G_{Sr}, is reflected by the outer surface of the cover. On the right side of the equation (1), there are three heat fluxes, that is, the flux transferred between the roofing and the surrounding free air, q_{C-A}, the flux radiated from the outer layer of the cover to the sky (horizon), q_{C-H}, and the flux conducted by the construction of roofing and transferred into the interior of the building, q_{λ}. It is assumed for equation (1) that only transverse heat fluxes are analyzed. The roofing, as well as materials used for roof constructing are characterized with relatively low thickness which allows to neglect the heat transfer in the longitudinal direction. Particular energy fluxes can be determined from the given below equations:

$$G_{Sa} = (1 - r_C)G_{S\beta}, \tag{2}$$

$$q_{C-A} = h_{C-A}(T_C - T_A), \tag{3}$$

$$q_{C-H} = \varepsilon_C \sigma (T_C^4 - T_H^4), \tag{4}$$

$$q_{\lambda} = U_e(T_C - T_R). \tag{5}$$

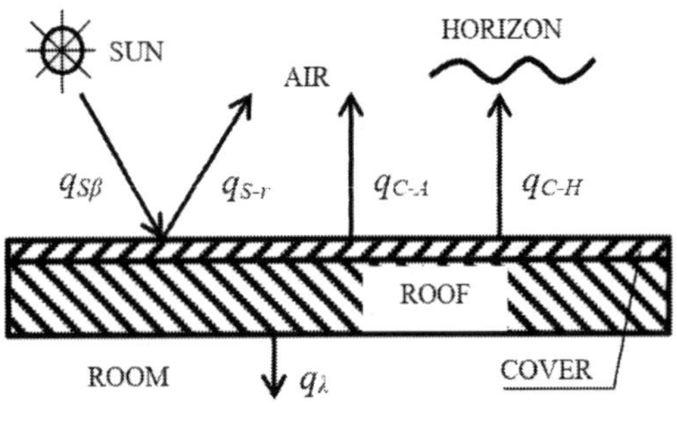

Fig. 1. Energy fluxes for roof [17, 18].

After substitution of relation (2) ÷ (5) in equation (1), the following equation has been obtained:

$$(1 - r_C)G_{S\beta} = h_{C-A}(T_C - T_A) +$$

$$+\varepsilon_C \sigma (T_C^4 - T_H^4) + U_e(T_C - T_R). \tag{6}$$

In equation (6), two parameters characterizing properties of roofing are present, that is, reflectivity r_C and emissivity ε_C. Heat properties of roof are related by parameter U_e, called equivalent heat transfer coefficient. Its value depends on resistance of heat conduction by particular layers of roof R_i, and on resistance of heat transfer between the ceiling and the inside of interior R_R. Hence:

$$U_e = 1/R_e = 1/(\textstyle\sum_{i=1}^{N} R_i + R_R), \tag{7}$$

Heat resistances can be calculated from the equations:

$$R_i = \delta_i/\lambda_i, \tag{8}$$

$$R_R = 1/h_R. \tag{9}$$

Heat resistance of a flat layer of material depends on its thickness δ_i and on coefficient of heat conduction λ_i. Heat transfer between free air and roof cover depends on coefficient of heat transfer h_{A-C}, and heat transfer between the ceiling and the interior depends on coefficient h_R.

Thermodynamic stimulants for heat transfer are the temperature gradients. In heat balance equation (6) for roof, mean temperatures of particular media are applied. In order to calculate the sky temperature, the simplified Swinbank formula has been applied [19]:

$$T_H = 0.0553 \cdot T_A^{1.5}. \tag{10}$$

Application of equation (6) for calculations of parameters that characterize optic properties of roof covers requires the knowledge of all the remaining parameters present in equations (6) – (10). Besides, in equation (6) both reflectivity r_C, as well as emissivity ε_C are unknown. So, value of one of the above parameters should be assumed, or an additional equation to define the relation between them should be written.

However, if energy balance for roof cover is valid only for the night time, equation (6) is simplified to the shape:

$$0 = h_{C-A}(T_C - T_A) + \varepsilon_C \sigma (T_C^4 - T_H^4) +$$

$$+U_e(T_C - T_R). \tag{11}$$

Emissivity of roof cover can be determined from the above equation:

$$\varepsilon = \frac{h_{C-A}T_A + U_e T_R - (h_{C-A} + U_e)T_C}{\sigma(T_C^4 - T_H^4)}. \tag{12}$$

3 Experimental stand

In order to determine temperatures of particular media, an experimental stand was built. The stand is presented in Fig. 2. Rectangular prism with dimensions: height – 0.88m, length – 1.14m, and width – 0.84 m was a model of a building with a flat roof. The stand was placed on the roof of the building of Department of Heat Engineering, West Pomeranian University of Technology, Szczecin (Poland), on its south-west side. Walls of the stand were made of plywood with thickness of 0.014m and insulated with rock wool with thickness of 0.005m at the inner sides. Also plywood was used to build the roof. The roof was insulated with styrofoam with thickness of 0.1m. The outer surface of styrofoam was covered by the black roofing paper with unknown properties. The floor of the stand was insulated with a styrofoam plate with thickness of 0.1m.

In order to measure temperatures, five sensors of type Pt100 were used. They were connected with data recorder of type APAR206. The temperature sensors were placed:

- in the centre of roof cover surface (roofing paper),
- in the centre of styrofoam surface on the side of roofing,
- in the centre of styrofoam surface on the side of stand's interior,
- in the centre of inner space of the stand,
- behind the curtain, in order to measure the ambient temperature.

a)

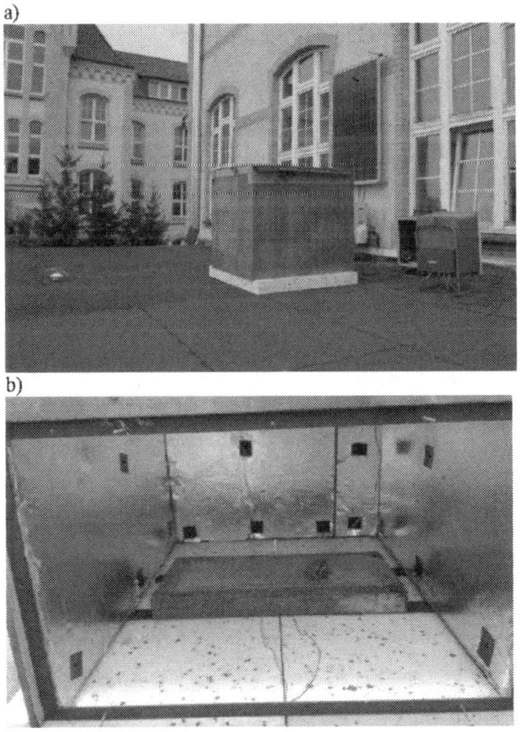

b)

Fig. 2. Experimental stand: a) view b) interior (photo A.Walterowicz [20]).

Temperatures of particular layers of roof, of environment, and in of interior were measured every 3 minutes.

4 Research results

Experimental research was carried out in the time from 25.08.2017 to 14.09.2017. Initial analysis of obtained results is presented in paper [20]. Fig.3. shows characteristic temperatures distributions for media in particular night periods within the time under research.

Times of sunrise and sunset were determined on the basis of a computer program from the website [21]. It can be concluded from the analysis of measurements results that the lowest are temperatures of outer and inner surfaces of roofing paper. Air temperatures in the interior and at the inner surface of styrofoam have the highest values. Ambient temperatures have values higher than temperature of outer surface of roofing paper, but lower than values of air temperature in the interior. Thus, on the basis of Fig.3, it can be stated that heat is transferred from the outer air to the roof cover at night. The roof cover receives also the heat from the inner air. On the other hand, roof cover loses heat as a result of irradiance to the sky.

It was assumed for calculations that coefficient of heat transfer from the free air to the roof cover was known. Value of this coefficient was calculated with assumption that wind velocity was equal 1m/s [22]:

$$h_{A-C} = 2.8 + 3.0w. \quad (13)$$

The coefficient equals $h_{A-C} = 5.8$ W/(m^{2}K). In turn, equivalent heat transfer coefficient equals $U_e = 0.18$ W/(m^{2}K) [17]. The above value was calculated because heat resistance of styrofoam was known and the coefficient of heat transfer from styrofoam to inner air was assumed.

Fig.4 shows calculations results for emissivity for black roofing paper determined for the time of night. Values of emissivity change in a wide range. It may be caused by application of simplified Swinbank formula in order to calculate equivalent temperature of sky in the proposed methodology. Temperature of sky depends on content of water vapor in atmosphere and on cloudiness. It depends also on the time of the year and on the time of the day [23]. Choice of an adequate relation to calculate the temperature of sky, as Pluta [24] claims, is particularly relevant when the operation of solar installations in the night time is analyzed. Thus, if the sky is cloudy, temperature of sky calculated from simplified Swinbank formula (10) requires to be revised, or only cloudless (or nearly cloudless) nights in time under research can be analyzed.

Because of the above fact, analysis of irradiance distribution in the investigated time (Fig.5) was made. The end of August 2017 was the time in which it was stated that the sky was very little cloudy, as irradiance distribution was regular and time temperatures changes were mild. Thus, three August nights were chosen for further analysis, i.e. 25/26, 27/28, and 28/29 of August. Results of temperatures measurements and calculated values of roofing paper emissivity are shown in Figs 6 and 7. It results from analysis of temperatures distributions for chosen nights that the mildest changes occurred at night 25/26 of August 2017. Roofing paper emissivity calculated from equation (12) shows tendency to oscillate around the mean value of this parameter equal 0.63. So, it can be assumed that the black roofing paper under research is characterized by emissivity close to this value.

 https://doi.org/10.1051/e3sconf/20187001021

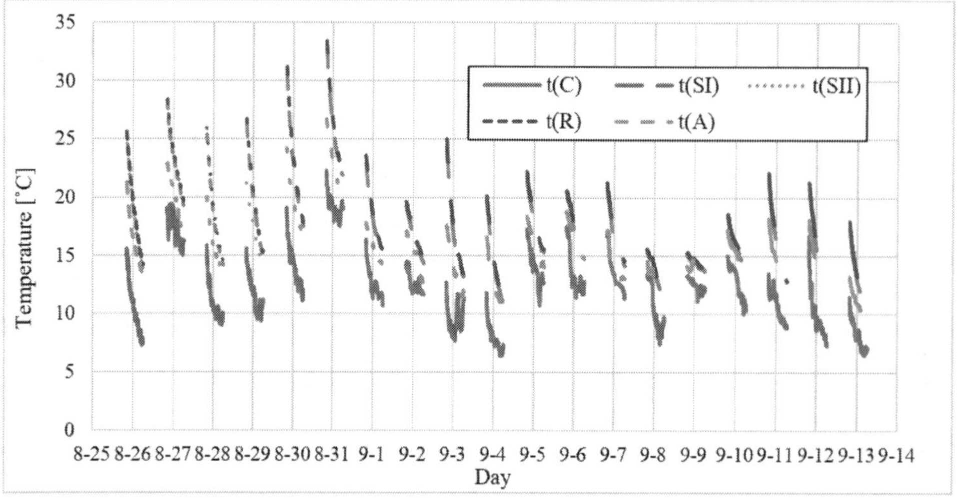

Fig. 3. Temperatures distributions in particular nights measured in the time from 25.08.2017 to 13.09.2017.

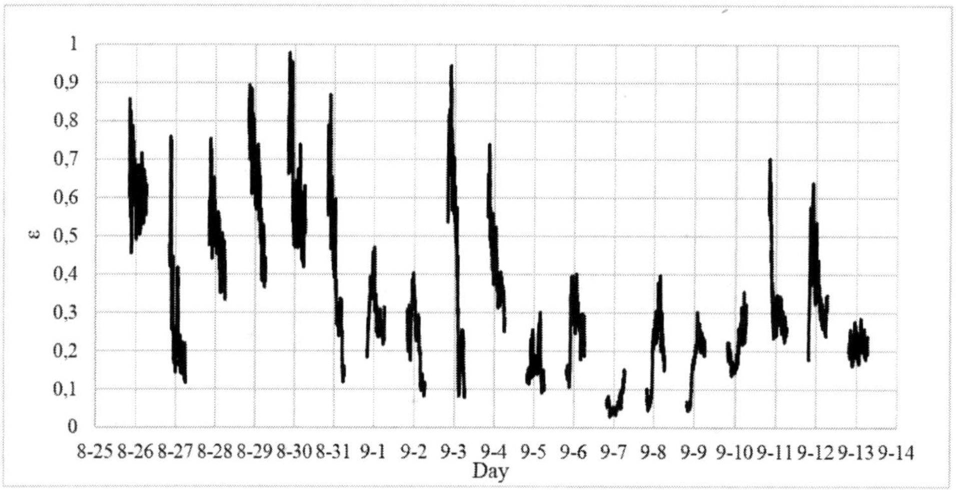

Fig. 4. Emissivity calculated for the night time in the period from 25.08.2017 to 13.09.2017.

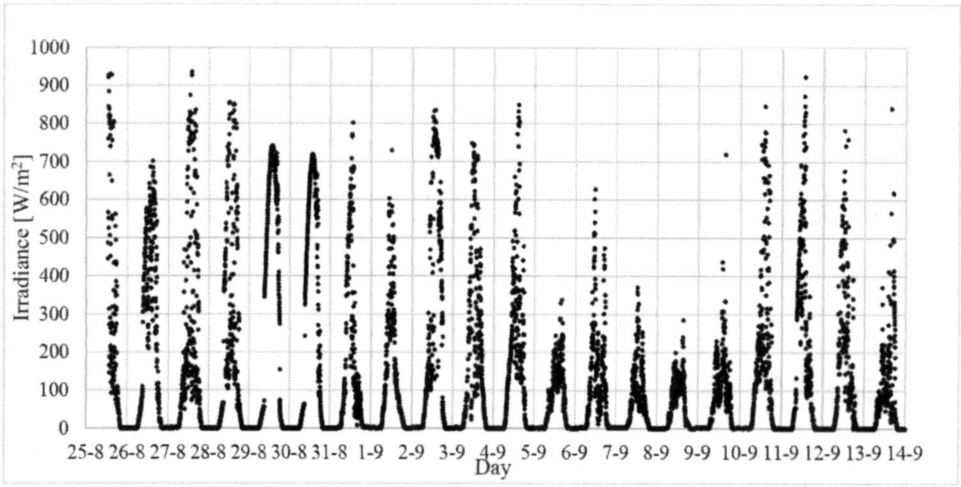

Fig. 5. Irradiance in the period from 25.08.2017 to 14.09.2017.

However, the proposed methodology of determination the emissivity requires further modifications in order to a.o. more explicitly determine the value of temperature of sky.

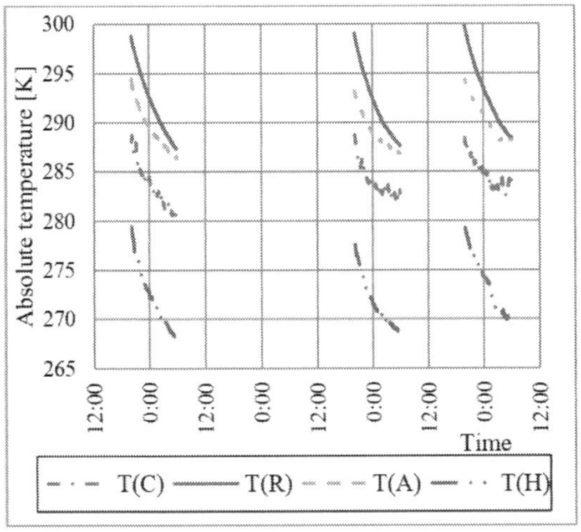

Fig.6. Characteristic absolute temperatures of media; nights 25/26, 27/28, and 28/29 of August 2017.

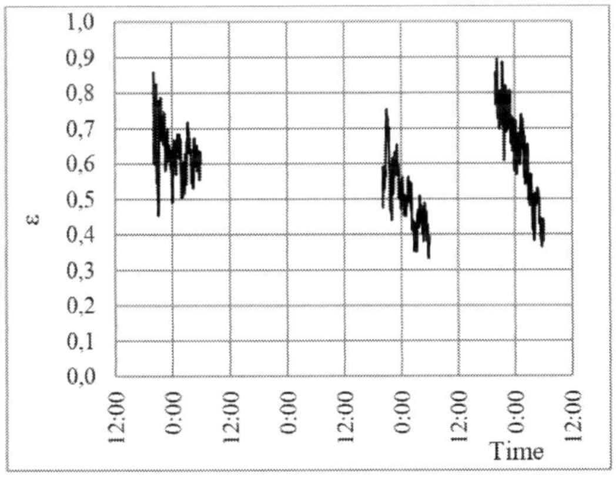

Fig. 7. Emissivity for roofing paper under research; nights 25/26, 27/28, and 28/29 of August 2017.

5 Conclusions

On the basis of carried out analysis of research results, the following conclusions can be proposed:

a) the methodology of research proposed in the paper allows to estimate the value of emissivity for roof cover,
b) the methodology requires experimental research to be carried out where determination of mean temperatures of particular media are needed; temperatures for a cloudless and nearly windless night should be chosen for calculations,
c) equation describing the temperature of sky should be modified or replaced by another one in order to increase accuracy of proposed methodology,
d) for the black roofing paper under research, the emissivity equals ca 0.63.

References

1. Directive 2012/27/EU of the European Parliament and of the Council of 25 October 2012 on energy efficiency, amending Directives 2009/125/EC and 2010/30/EU and repealing Directives 2004/8/EC and 2006/32/EC, Official Journal of the European Union, L 315/1, (2012)
2. L. Pérez-Lombard, J. Ortiz, Ch. Pout, Energ. Buildings, **40**, 394-398 (2008)
3. H. Zhao, F. Magoulès, Renewable and Sustainable Energy Reviews, **16**, 6, 3586-3592 (2012)
4. N. Fumo, Ren.Sust. Energ. Rev., **31**, 53-60 (2014)
5. P. Kivimaa, M. Martiskainen, Energ. Effic., **11**, 1, 31-51 (2018)
6. Y. Anand, A. Gupta, A. Maini, Avi Gupta, A. Sharma, A. Khajuria, S. Gupta, S. Sharma, S. Anand, S.K. Tyagi, J. Eng., Article ID 685640, 9 pages (2014)
7. E. Di Giuseppea, M. Pergolinib, F. Stazib, Energ. Procedia, **134**, 404–413, (2017)
8. O. Sandin, J. Nordin, M. Jonsson, J. Coat. Technol. Res., **14**, 4, 817–821 (2017)
9. https://energy.gov/energysaver/energy-efficient-home design/cool-roofs
10. https://heatisland.lbl.gov/coolscience/cool-roofs
11. https://www.epa.gov/heat-islands/using-cool-roofs-reduce-heat-islands
12. J. Šabíková, L. Krčmárová, J. Sustain. Construct. Mater. Technol., **1**, 2, 60–63 (2016)
13. H. Suehrckea, E.L. Petersonb, N. Selbyd, Energ. Buildings, **40**, 2224–2235 (2008)
14. R. Levinson, H. Akbari, Energ. Effic., **3**, 1, 53-109 (2010)
15. A.L. Piselloa, F. Cotanaa, L. Brinchia, Energ. Procedia, **45**, 453 – 462 (2014)
16. http://coolroofs.org
17. Z. Zapałowicz, J. Mech. Energ. Eng., **41**, 1, 107-112 (2017)
18. Z. Zapałowicz, 24 Symp. Nutzung Regenerative Energiequellen und Wasserstofftechnik, 9. – 11.11.2017 Stralsund, Germany, 167-176 (2017)
19. J.A. Duffie, W.A. Beckman, *Solar Energy Thermal Processes*. John Wiley & Sons, New York, (1991)
20. A. Walterowicz, Lic. Work, West Pomeranian University of Technology, Szczecin (2018), (in Polish)
21. http://darekk.com/sun/sunrise-sunset-calculator?lat=53.42444444&lon=14.56027778&z=10&tz=+1&loc=Szczecin
22. J. Cieśliński, J. Mikielewicz. *Nonconventional Sources of Energy*. Wyd. Politechniki Gdańskiej, Gdańsk p.59, (1996), (in Polish)
23. H. Nowak, Infrared Physics, **29** (2-4), 231-232 (1989)
24. Z. Pluta, *Podstawy teoretyczne fototermicznej konwersji energii słonecznej*. Oficyna Wyd. Politechniki Warszawskiej, Warszawa (2000) (in Polish)

Influence of desorption temperature on the thermodynamic performance of adsorption heat pump

Katarzyna Zwarycz-Makles[1,*]

[1]West Pomeranian University of Technology, Department of Heating, Ventilation and Heat Engineering, al. Piastow 50, 70-311 Szczecin, Poland

Abstract. In the paper an analysis of the desorption temperature effect on the thermodynamic efficiency of the adsorption heat pumps is presented. The thermodynamic performance of heat pump is determined by Coefficient of Performance (COP) as well as exergetic efficiency coefficient (η_{ex}) at the adsorption equilibrium conditions and compared to the performance at heat of evaporation of the working fluid conditions. Possible estimation of reduced efficiency of adsorption silica gel/water heat pump, as distinct from the equilibrium efficiency in realistic technical system is presented.

1 Introduction

In order to utilize the waste heat from industrial processes and low grade heat, such as solar energy or geothermal energy, adsorption systems are used in refrigeration and air conditioning systems as an alternative to vapor compression ones. Adsorption heat pumps utilize heat to drive thermodynamic cycles that transfer heat from a lower temperature to a higher temperature for applications such as space cooling and refrigeration or for improving energy efficiency of space heating or water heating even integrated with district heating network [1-3].

Activated carbon/methanol, activated carbon/ammonia, zeolite/water and silica gel/water are the adsorbent/adsorbate pairs commonly used in practical realised adsorption cycles. Adsorption heat pumps do not require moving parts for circulation of working fluid. They can work with low temperature driving energy sources, even it can be thermal energy storage. But they are characterised by many disadvantages such as low COP value, intermittently working principles, require advanced technology and special designs to maintain high vacuum [4-6].

The thermodynamic performance of heat pump is determined by COP as well as exergetic efficiency at the adsorption equilibrium conditions and compared to the performance at heat of evaporation of the working fluid conditions. Possible estimation of reduced efficiency of adsorption silica gel/water heat pump, as distinct from the equilibrium efficiency in practical technical system is presented.

2 Adsorption heat pump cycle

The operation of adsorption heat pumps is based on the ability of porous adsorbent to adsorb vapor (adsorbate) at low temperature, and to desorb it when heated. The performance of an adsorption heat pump depends on many parameters, such as adsorbent and adsorbate properties, system design and operating conditions. The adsorption heat pump under consideration consists of an evaporator, an adsorber/desorber and a condenser (Fig. 1). The adsorber/desorber is encapsulated; it contains tubes with deposited adsorbent.

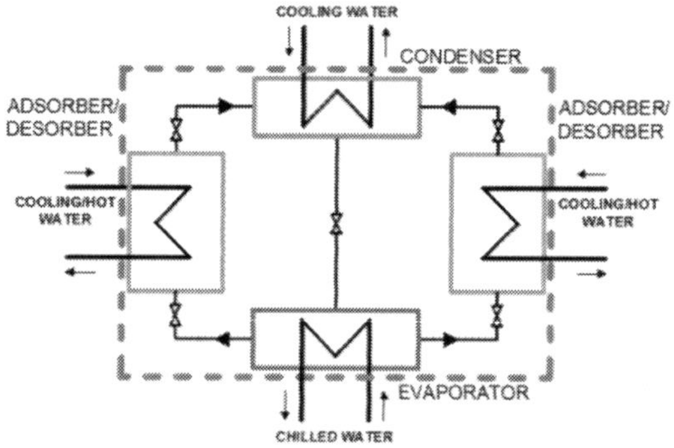

Fig. 1. Scheme of considered conventional single stage adsorption heat pump cycle.

The adsorption process is followed by preheating process to lift up the temperature and to lift up the pressure of the sorption reactor up to the condenser pressure. The desorption process starts afterward, then the temperature of sorption reactor approaches to the heat source temperature over time. Next precooling process starts, it cools down the adsorber/desorber to have low pressure equal to the evaporator pressure. When refrigerant from the evaporator is transported to the condenser and returns back to the evaporator then

* Corresponding author: kzwarycz@zut.edu.pl

one cycle is completed and restarts circulation again [2, 7].

General equation of energy balance is written for all elements of the adsorption heat pump model:

$$\dot{Q}_{ac} = \dot{Q}_s - \dot{Q}_r + \dot{Q}_g \quad (1)$$

where $\dot{Q}_{ac}$ is the heat flow accumulated in the process, $\dot{Q}_s$ is the heat flow supplied to the process, $\dot{Q}_g$ is the heat flow generated and $\dot{Q}_r$ is the heat flow discharged from the process. Assessment of the exergy efficiency of the adsorption heat pump requires determining the organization of the adsorption cycle carried out in each of the absorbers included in its composition.

The operation mechanism of heat pump includes a number of cyclic transient processes. These processes run at different temperature and pressure levels in following stages. A simple adsorption cycle indicating all the heat transfer for a complete cycle is shown in Fig. 2.

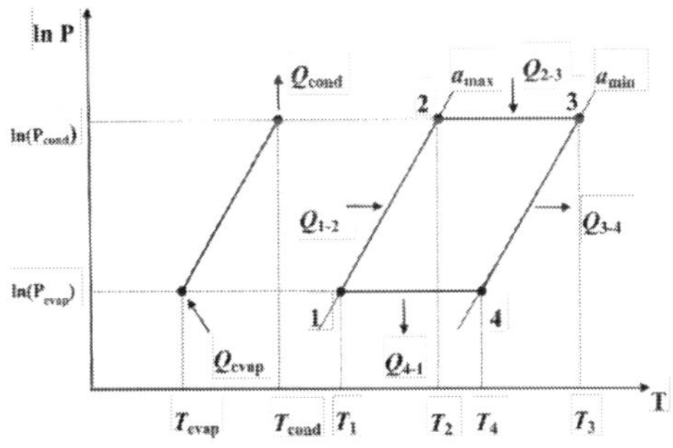

Fig. 2. Clapeyron diagram of the adsorption heat pump cycle.

(1-2) Isosteric heating of the saturated adsorbent along the isoster from temperature T_1 to the minimal desorption temperature T_2. The temperature of adsorbent bed is increased from T_1 to T_2 by heating the adsorbent bed without desorption. The amount of heat, which should be transferred to the adsorbent bed from external source of temperature T_H is given by the relation:

$$Q_{1-2} = \int_{T_1}^{T_2} \left[m_a\left(c_{p,a} + a\, c_{p,v}\right) + m_t c_{p,t}\right] dT \quad (2)$$

(2-3) Isobaric desorption of the sorbate due to adsorbent heating from T_2 to the temperature $T_3 = T_{des}$, where T_{des} is the external heat source temperature. Subsequent condensation process in the condenser, where heat Q_{cond} is released to the outside of temperature T_{ot}. The pressure of the process remains constant. Part of the heat, which is transferred to the adsorbent bed, increases the temperature of adsorbate-adsorbent pair and adsorbent bed, while the other part causes the desorption process:

$$Q_{2-3} = \int_{T_2}^{T_3} \left[m_a\left(c_{p,a} + a\, c_{p,v}\right) + m_t c_{p,t}\right] dT + \int_2^3 m_a \Delta H_a \, da \quad (3)$$

(3-4) Isosteric cooling of the adsorbent along the isoster down from maximum temperature of the cycle T_3 to temperature T_4. During this process, both the pressure and the temperature of the adsorbent bed decrease to the evaporator's values. Adsorbate concentration in adsorbent is constant. The amount of released heat is given by:

$$Q_{3-4} = \int_{T_3}^{T_4} \left[m_a\left(c_{p,a} + a\, c_{p,v}\right) + m_t c_{p,t}\right] dT \quad (4)$$

(4-1) Isobaric adsorption driven by adsorbent cooling down to $T_1 = T_{evap}$. Vaporisation of the adsorbate in the evaporator is started due to heat Q_{evap} transferred from external source of temperature T_L to the evaporator. Adsorbate concentration in adsorbent is increasing. During adsorbing of the adsorbate in the adsorbent heat is released due to heat of adsorption:

$$Q_{4-1} = \int_{T_4}^{T_1} \left[m_a\left(c_{p,a} + a\, c_{p,v}\right) + m_t c_{p,t}\right] dT + \int_4^1 m_a \Delta H_a \, da \quad (5)$$

The heat accumulated in the condenser can be determined based on the adsorbate mass desorbed by the adsorbent:

$$Q_{cond} = m_a \Delta a \Delta H_{evap} \quad (6)$$

where Q_{cond} is heat of condensation at the condenser temperature, and Δa is mass of adsorbate condensed at condenser. The amount of heat extracted in the evaporator Q_{evap}:

$$Q_{evap} = m_a \Delta a \Delta H_{evap} + \int_{T_{cond}}^{T_{evap}} m_a \Delta a \, c_{p,v} \, dT \quad (7)$$

where Q_{evap} is heat of evaporation and ΔH_{evap} is enthalpy of evaporation.

The heat of evaporation causes cooling effect, which can be used for cooling purposes. Heat of condensation can be employed for heating purposes.

The total amount of heat required to execute bed regeneration Q_{des} could be equal to heat supplied by the hot water from external heat source (generator) during the adsorbent regeneration (desorption) and affects the desorption of the adsorbate by increasing the temperature of the tube with the deposited adsorbent. Its value can be calculated from the following relationship:

$$Q_{des} = Q_{1-2} + Q_{2-3} \quad (8)$$

The total amount of heat released during stages 3 and 4 is:

$$Q_{ads} = Q_{3-4} + Q_{4-1} \quad (9)$$

3 Adsorption heat pump efficiency

Energy efficiency of adsorption heat pumps is reduced by thermodynamic losses due to the temperature differences between the cycle and ambient or external sources. Designed temperatures of adsorption and desorption are the key factors of affecting efficiency.

The cooling effect in cycle occurs during the isobaric adsorption process (4-1) when the adsorbate is evaporated by gathering heat from external heat source (outside environment) [5, 8]. The heating effect appears during the isobaric desorption process (2-3) when the adsorbate is condensed by releasing heat to external source. In addition to the isobaric desorption process, the adsorbent bed is also cooled during (3-4) and (4-1) processes. The heat released during these processes can also be utilized for heating purposes in any process [6, 9].

In this paper thermodynamic analysis of an ideal cycle of single stage adsorption heat pump with reference to desorption temperature is performed. The efficiency of the working cycle is characterised by the Coefficient of Performance (COP), the ratio of the amount of produced cold (heat) to the amount of supplied heat [4, 6, 10]:

for cooling

$$\mathrm{COP}_{\mathrm{cool}} = \frac{Q_{evap}}{Q_{1-2}+Q_{2-3}} = \frac{Q_{evap}}{Q_{des}} \tag{10}$$

for heating

$$\mathrm{COP}_{\mathrm{heat}} = \frac{Q_{cond}+Q_{3-4}+Q_{4-1}}{Q_{1-2}+Q_{2-3}} = \frac{Q_{cond}+Q_{ads}}{Q_{des}} \tag{11}$$

For every heat pumping machine the maximal theoretical Coefficient of Performance is the Carnot COP [11-14]. In adsorption heat pumps, with assumption of reversibility of all processes and for the equilibrium cycle, the entropy generation is equal to zero $\Delta S = 0$ [15-17]. For an ideal cycle with assumption of zero thermal masses, during isobaric desorption $T_3 = T_{des}$ and during adsorption phase $T_1 = T_{ads}$, there is no temperature difference between the adsorber (or desorber) and the external heat source, so the entropy generation is zero, the energy balance (the first law) can be written:

$$Q_{cond} - Q_{evap} - Q_{des} + Q_{ads} = 0 \tag{12}$$

and the entropy balance (the second law):

$$-\frac{Q_{cond}}{T_{cond}} + \frac{Q_{evap}}{T_{evap}} + \frac{Q_{des}}{T_{des}} - \frac{Q_{ads}}{T_{ads}} = \Delta S \geq 0 \tag{13}$$

Then Carnot COP is based on the three temperature levels: evaporation temperature, desorption temperature and condensation temperature:

for cooling

$$\mathrm{COP}_{\mathrm{Carnot-cool}} = \frac{\left(\frac{1}{T_{cond}}\right)-\left(\frac{1}{T_{des}}\right)}{\left(\frac{1}{T_{evap}}\right)-\left(\frac{1}{T_{cond}}\right)} = \frac{\left(1-\frac{T_{cond}}{T_{des}}\right)}{\left(\frac{T_{cond}}{T_{ads}}-1\right)} \tag{14}$$

for heating

$$\mathrm{COP}_{\mathrm{Carnot-heat}} = 1 + \mathrm{COP}_{\mathrm{Carnot-cool}} = \frac{\left(\frac{1}{T_{evap}}\right)-\left(\frac{1}{T_{des}}\right)}{\left(\frac{1}{T_{evap}}\right)-\left(\frac{1}{T_{cond}}\right)} = 1 + \frac{\left(1-\frac{T_{cond}}{T_{des}}\right)}{\left(\frac{T_{cond}}{T_{ads}}-1\right)} \tag{15}$$

However experiments and calculations described by Zelenko, Heifets [18] and Sharonov, Aristov [19] proved that equilibrium $\mathrm{COP}_{\mathrm{Carnot}}$ cannot be achieved by the improving the heat exchange process in elements of the heat pump. The losses due to cooling the working fluid in the evaporator are negligible as compared to the heat of evaporation ΔH. Therefore Sharonov and Aristov [19] presented definition of COP for cooling and heating processes (below respectively) as ratio of enthalpy:

for cooling

$$\mathrm{COP}_{\mathrm{c-c\,cool}} = \frac{\Delta H_{evap}}{\Delta H_{ads}} \tag{16}$$

for heating

$$\mathrm{COP}_{\mathrm{c-c\,heat}} = 1 + \mathrm{COP}_{\mathrm{c-c\,cool}} = \frac{\Delta H_{cond}+\Delta H_{ads}}{\Delta H_{ads}} \tag{17}$$

where ΔH_{evap} is the enthalpy of evaporation, ΔH_{ads} is the average enthalpy of adsorption in the heat pump and ΔH_{cond} is enthalpy of condensation.

At temperatures along isobaric stages the entropy is generated due to the heat supply at $T_1 - T_{ot} > 0$ or the heat removal at $T_{\mathrm{ot}} - T_{cond} > 0$, that reduces the cycle efficiency below the Carnot maximum values.

According to Clausius-Clapeyron equation for the gas pressure and assumption of Trouton's rule [19, 20] of intersection of adsorption isoster:

$$T_2 = \frac{{T_{cond}}^2}{T_{evap}} \tag{18}$$

relations (16-17) leads to relationships:

for cooling

$$\mathrm{COP}_{\mathrm{c-c\,cool}} = \frac{T_{cond}}{T_{des}} \tag{19}$$

for heating

$$\mathrm{COP}_{\mathrm{c-c\,heat}} = 1 + \frac{T_{cond}}{T_{des}} \tag{20}$$

There is degradation of efficiency due to entropy generation caused by external thermal coupling of the isothermal heat sources to the temperature varying adsorber, thus both $\mathrm{COP}_{\mathrm{c-c\,cool}}$ and $\mathrm{COP}_{\mathrm{c-c\,heat}}$ are always lower than Carnot COP.

The thermodynamic efficiency of adsorption heat pump is determined not only by Coefficient of Performance (COP) but also by thermodynamic (exergetic) efficiency coefficient (η_{ex}) [6, 11, 15] designated for the adsorption equilibrium conditions. The thermodynamic efficiency (exergetic efficiency) relates to actual performance of a system to its expected maximum performance as determined from ideal Carnot cycle and designated for the adsorption equilibrium conditions:

for cooling

$$\eta_{\mathrm{ex-cool}} = \frac{\left(\frac{T_{ot}}{T_L}-1\right)Q_{evap}}{\left(1-\frac{T_{ot}}{T_H}\right)Q_{des}} \tag{21}$$

$$\eta_{ex-cool} = \frac{COP_{cool}}{COP_{Carnot-cool}} \quad (22)$$

for heating

$$\eta_{ex-heat} = \frac{\left(\frac{T_{ot}}{T_L}-1\right)(Q_{cond}+Q_{ads})}{\left(1-\frac{T_{ot}}{T_H}\right)Q_{des}} \quad (23)$$

$$\eta_{ex-heat} = \frac{COP_{heat}}{COP_{Carnot-heat}} \quad (24)$$

Compared to the reduced efficiency due to effect of enthalpy of evaporation, condensation and average enthalpy of adsorption, exergetic efficiency could be expressed by:
for cooling

$$\eta_{ex-cool\ cc} = \frac{COP_{cool}}{COP_{c-c}} \quad (25)$$

for heating

$$\eta_{ex-heat\ cc} = \frac{COP_{heat}}{COP_{c-c}} \quad (26)$$

Equations (25-26) represent possible estimation of reduced efficiency of adsorption silica gel/water heat pump, as distinct from the equilibrium efficiency in realistic technical system.

4 Results and conclusions

Using transient numerical model, developed and described in detail in previous Author's articles [21-23] to simulate silica gel/water adsorption heat pump system numerical analysis was done. The results of computer simulation were used to calculate the coefficient of performance for heating and cooling purposes.

To describe the adsorption characteristics the adsorption equilibrium equation (Dubinin-Astakhov) with appropriate coefficients was used (references [21-23]). Mathematical model is established comprising the set of heat and mass balance partial differential equations (PDEs), together with the initial and boundary conditions and is solved by the numerical method of lines (NMOL). The PDEs are reduced to a set of ordinary differential equations (ODEs) and is solved using Runge-Kutta 4th order numerical method. The main calculation element (adsorption/desorption column) contains three sub-elements that describe the component of the adsorber/desorber heat exchanger: heating/cooling water, metal pipe (tube of column), working bed (silica gel). For each calculation sub-element the heat and mass balance partial differential equations (PDEs) were formulated.

Performance is conducted for the two-bed adsorption heat pump (adsorber/desorber) with silica gel adsorbent and water as adsorbate. The adsorption equilibria of selected adsorbent-adsorbate pair was sourced from available literature [24-27]. Numerical simulation was conducted for Dubinin-Astakhov adsorption equilibria, for 1,5m length of sorption column, and temperatures $T_{cond} = 303\ K$, $T_{evap} = 283\ K$, $T_{ads} = 313\ K$. For the calculation, more parameters that are necessary are according to paper [21]. Results of efficiency for cooling and heating purposes of silica gel/water adsorption heat pump in function of desorption temperature are presented in Figures 3-6 and Tables 1-4.

Table 1 and Fig. 3 shows the effect of desorption temperature on the COP coefficients including reduced COP for analysed adsorption heat pump in case of cooling applications. Maximum value of COP (Carnot) is presented as $COP_{Carnot-cool}$. As can be seen, the increase in desorption temperature slightly increases COP_{cool}, however it is lower than $COP_{c-c\ cool}$ due to cycle limitations of evaporation heat. Coefficient $COP_{c-c\ cool}$ estimates reduced efficiency of adsorption silica gel/water heat pump, as distinct from the equilibrium efficiency in practical technical system. Similar reduction in thermodynamic efficiency is presented in Table 2 and Fig. 4.

Table 1. Comparison of COP coefficients (for cooling).

T_{des}	$COP_{Carnot-cool}$	$COP_{c-c\ cool}$	COP_{cool}
$[K]$	[-]	[-]	[-]
343	1,65	0,88	0,071
353	2,00	0,86	0,154
358	2,17	0,85	0,179
363	2,34	0,83	0,198
368	2,50	0,82	0,211

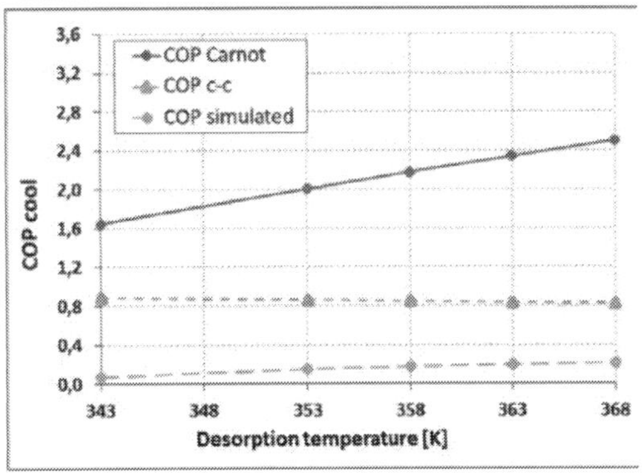

Fig. 3. Simulated COP (for cooling purposes) compare to reduced COP of silica gel/water adsorption heat pump in function of desorption temperature.

Table 2. Comparison of efficiency coefficients (for cooling).

T_{des}	$\eta_{ex-cool}$	$\eta_{ex-cool\ cc}$
$[K]$	[-]	[-]
343	0,043	0,080
353	0,077	0,179
358	0,082	0,212
363	0,085	0,237
368	0,085	0,257

For heating applications of heat pump, comparison of COP coefficients in function of desorption temperature is shown in Table 3 and Fig. 5, and thermodynamic efficiency variation in Table 4 and Fig. 6. Carnot boundary value of COP is presented as $COP_{Carnot-heat}$.

The efficiency of heat pump is higher with lower temperature of the heat source as generator. Difference between basic COP and $COP_{c-c\,heat}$ obtained according to limitations of heat of evaporation still indicates possibilities of improving efficiency in the realistic technical systems.

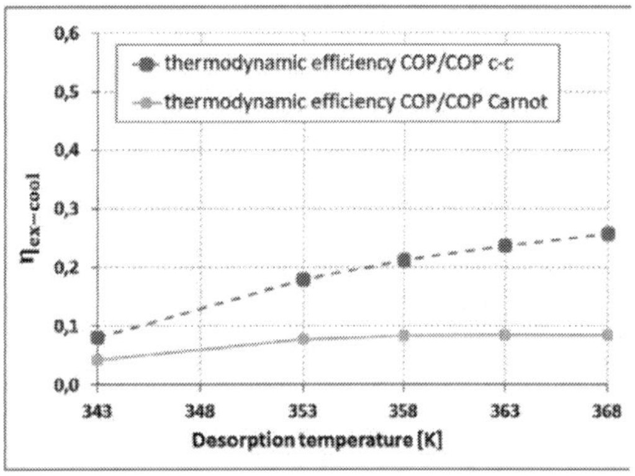

Fig. 4. Thermodynamic efficiency (for cooling purposes) of silica gel/water adsorption heat pump in function of desorption temperature.

Table 3. Comparison of COP coefficients (for heating).

T_{des}	$COP_{Carnot-heat}$	$COP_{c-c\,heat}$	COP_{heat}
$[K]$	[-]	[-]	[-]
343	2,65	1,88	1,780
353	3,00	1,86	1,619
358	3,17	1,85	1,375
363	3,34	1,83	1,216
368	3,50	1,82	1,105

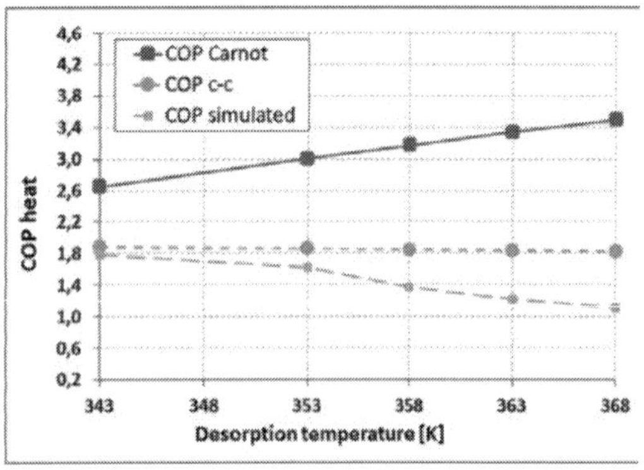

Fig. 5. Simulated COP (for heating purposes) compare to reduced COP of silica gel/water adsorption heat pump in function of desorption temperature.

Energy efficiency of adsorption heat pumps is reduced by thermodynamic losses due to the temperature differences between the cycle and ambient or external sources. Designed temperatures of adsorption and desorption are the key factors of affecting efficiency.

The considerable difference between the equilibrium efficiency and the theoretical efficiency is evidence that dissipative structure characteristic for the adsorption heat pump are present in the working cycle. Consideration of real nonequilibrium vapor heat transfer makes possible to estimate the upper limit of the heat pump attainable with the heat of evaporation.

Moreover, efficiency of equilibrium in adsorption heat pump is always lower than the Carnot one due to the thermal entropy production caused by the external thermal coupling. Difference between basic COP and COP_{c-c} obtained according to limitations of heat of evaporation still indicates possibilities of improving efficiency in the realistic technical systems. The results presented in the cited articles [18-20] refer to various known working pairs as adsorbate and adsorbent, however for the silica gel/water working pair in the simulated adsorption heat pump performance is demonstrated in presented paper.

Coefficient COP_{c-c} estimates reduced efficiency of adsorption silica gel/water heat pump, as distinct from the equilibrium efficiency in practical technical system. This limit considerably exceeds the simulated efficiency, thus confirming that there is potential for improving performance of adsorption heat pumps in technical systems.

Table 4. Comparison of efficiency coefficients (for heating).

T_{des}	$\eta_{ex-heat}$	$\eta_{ex-heat\ cc}$
$[K]$	[-]	[-]
343	0,672	0,945
353	0,539	0,871
358	0,433	0,745
363	0,364	0,663
368	0,316	0,606

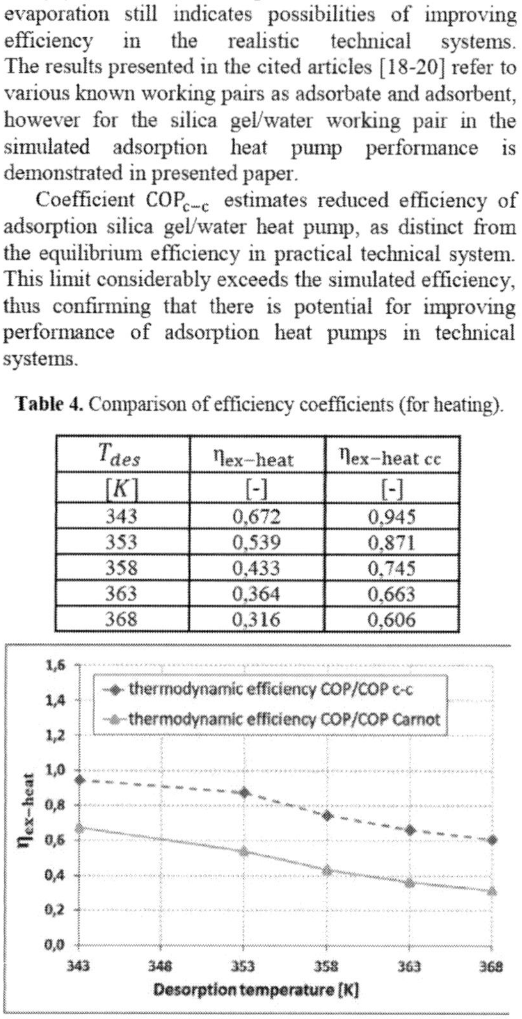

Fig. 6. Thermodynamic efficiency (for heating purposes) of silica gel/water adsorption heat pump in function of desorption temperature.

The COP and thermodynamic efficiency η_{ex} depend on generation (desorption) temperature T_{des}. It is connected with mass of generated adsorbate, thus with type of used adsorbent/adsorbate pair and its characteristic heat of evaporation.

In the application of analysed silica gel/water adsorption heat pump for refrigeration/cooling processes, the COP and efficiency slightly increases with increasing desorption temperature.

For heating purposes, when the interest is in utilising low grade heat sources, with analysed silica gel as adsorbent and water as working fluid and with characteristic evaporation heat, the efficiency of heat pump is higher with lower temperature of the heat source as generator.

It can be observed that chosen silica gel/water working pair generates higher efficiency for heating than cooling utilisation of heat pump. Therefore, the type of adsorbent/adsorbate pair for the purpose of adsorption heat pump should be adapted. Some pairs will contribute to greater efficiency in refrigeration applications and other in heating applications.

Nomenclature

a - mass/amount adsorbed/desorbed on adsorbent at time τ,[kg kg^{-1}] (uptake)
c_p - specific heat,[J kg^{-1} K^{-1}]
Ex - exergy (heat exergy) [kJ kg^{-1}]
ΔH_a - isosteric heat of adsorption, [J mol^{-1}]
Ex - exergy (heat exergy) [kJ kg^{-1}]
ΔH - enthalpy of evaporation, [kJ kg^{-1}]
m - mass/amount, [kg mol^{-1}]
p - pressure [Pa]
Q - heat, [J]
$\dot{Q}$ - heat flow, [W]
S - entropy [kJ kg^{-1}K^{-1}]
T - temperature, [K]
P - pressure, [Pa]

Greek letters

η - efficiency [-]

Subscripts

a - adsorbent
ads - adsorption
cc - refers to ratio of enthalpy
cond - condenser
cool - cooling
des - desorption
ex - exergetic
evap - evaporator
f - fluid (water)
heat – heating
H - high (warm source)
L - low (cold source)
ot - surrounding/environment
t - tube/pipe
v - adsorbate (water vapor)

References

1. T. Nunez, W. Mittelbach, H.M. Henning, Development of an adsorption chiller and heat pump for domestic heating and air-conditioning applications. Appl. Therm. Eng. **27**, 13, pp. 2205–12, (2007)
2. A. Akisawa, T, Miyazaki, Mutli-bed adsorption heat pump cycles and their optimal operation. *Advances in Adsorption Technology*, (Nova Science Publishers, 2014)
3. D. Mugnier, V. Goetz, Energy storage comparison of sorption systems for cooling and refrigeration, Sol. Energy **71**, 1, pp. 47–55, (2001)
4. F. Meunier, Second law analysis of a solid adsorption heat pump operating on reversible cascade cycles: application to the zeolite–water pair, Heat Recovery Syst. **5**, 133, pp. 141, (1985)
5. H. Demir, M. Mobedi, S. Ülkü, A review on adsorption heat pump: Problems and solutions, Renew. Sust. Energ. Rev. **12**, 9, pp. 2381-2403, (2008)
6. M. Pons, F. Meunier, G. Cacciola, R. E. Critoph, M. Groll, L. Puigjaner, B. Spinner, F. Ziegler, Thermodynamic based comparison of sorption systems for cooling and heat pumping. Int. J. Refrig. **22**, 1. pp. 5-17, (1999)
7. H. T. Chua, K.C. Ng, et al.: Transient modeling of a two-bed silica gel–water adsorption chiller. Int. J. Heat Mass Transfer **47**, pp. 659–669, (2004)
8. J. Aman, DS-K. Ting, P. Henshaw, Residential solar air conditioning: Energy and exergy analyses of an ammonia–water absorption cooling system, Appl. Therm. Eng. **62**, 2, pp. 424–432, (2014)
9. Y.B. Gui, R.Z. Wang, W. Wang, J.Y. Wu, Y.X. Xu, Performance modelling and testing on a heat-regenerative adsorptive reversible heat pump, Appl. Therm. Eng. **22,** pp. 309-320, (2002)
10. N. Douss, F. Meunier, Effect of operating temperatures on the coefficient of performance of active carbon-methanol systems, Heat Recovery Systems and CHP **8**, 5, pp. 383–392, (1988)
11. J. Szargut, R. Petela, *Egzergia*, (PWN, Warszawa, 1965)
12. S. de Oliveira, *Exergy Production, Cost and Renewability*, (Springer, London, 2013)
13. R. Pohorecki, S. Wroński, *Kinetyka i termodynamika procesów inżynierii chemicznej*, (WNT, Warszawa, 1979)
14. E. E. Vasilescu, R. Boussehain, M. Feidt, , A. Dobrovicescu, Energy and exergy optimization of the adsorption refrigeration machines with simple and double effect, Termotehnica **1-2**, pp. 80-86, (2007)
15. A. Hepbasli, A key review on exergetic analysis and assessment of renewable energy resources for a sustainable future, Renew. Sust. Energ. Rev. **12**, 3, pp. 593-661, (2008)
16. H. Gunerhan, A. Hepbasli, Exergetic modeling and performance evaluation of solar water heating systems for building applications, Energ. Buildings **39**, 5, pp. 509-516, (2007)
17. E. Hürdoğan, , O. Buyükalaca, , A. Hepbasli, , T. Yılmaz, Exergetic modeling and experimental performance assessment of a novel desiccant cooling system, Energ. Buildings **43**, 6, pp. 1489-1498, (2011)

18. V.L. Zelenko, L.I. Heifets, Limiting efficiency of adsorption heat pump, Moscow University Chemistry Bulletin **62**, 1, pp. 9–12, (2007)
19. V.E. Sharonov, Yu.I. Aristov, Chemical and adsorption heat pumps: Comments on the second law efficiency, Chem. Eng. J. **136**, pp. 419–424, (2008)
20. R.E. Critoph, Performance limitation of adsorption cycles for solar cooling, Sol. Energy **41**, 1, pp. 21–31, (1988)
21. K. Zwarycz-Makles, W. Szaflik, Comparison of Analytical and Numerical Models of Adsorber/desorber of Silica Gel-water Adsorption Heat Pump, Journal of Sustainable Development of Energy, Water and Environment Systems **5**, 1, pp. 69-88, (2017)
22. K. Zwarycz-Makles, D. Majorkowska-Mech, Numerical Gear's and Runge-Kutta Discretization methods in differential equations of adsorption in adsorption heat pump, CPOTE 2016: 4th International Conference on Contemporary Problems of Thermal Engineering, (Silesian University of Technology, 2016)
23. K. Zwarycz-Makles, K. Kuczynski, Model and simulation of six-bed silica gel-water adsorption heat pump, CPOTE 2016: 4th International Conference on Contemporary Problems of Thermal Engineering, (Silesian University of Technology, 2016)
24. S.U. Rege, R.T. Yang, M.A. Buzanowski, Sorbents for air prepurification in air separation, Chem. Eng. Sci. **55**, 21, pp. 4827–4838, (2000)
25. Y. Wang, M.D. LeVan, Adsorption Equilibrium of Carbon Dioxide and Water Vapor on Zeolites 5A and 13X and Silica Gel Pure Components, Chem. Eng. J. **54**, 10, pp. 2839–2844, (2009)
26. J.S. Lee, J.H. Kim, J.T. Kim, J.K. Suh, J.M. Lee, C.H. Lee, Adsorption equilibria of CO2 on zeolite 13X and zeolite X/Activated carbon composite, Chem. Eng. J **47**, 5, pp. 1237–1242, (2002)
27. I. Park, K.S. Knaebel, Adsorption breakthrough behaviour: unusual effects and possible causes, American Institute of Chemical Engineers Journal **38**, 5, pp. 660–670, (1992)

An investigation of heat transfer coefficient during refrigerants condensation in vertical pipe minichannels

Tadeusz Bohdal[1], *Henryk* Charun[1], *Marcin* Kruzel[1,*], and *Małgorzata* Sikora[1]

[1]Koszalin University of Technology, Department of Energetics, ul. Racławicka 15-17, PL - 75-620 Koszalin, Poland

Abstract. This article presents the results of experimental research of heat transfer coefficient in air cooled vertical pipe minichannels during R404A, R407C and R410A high-pressure refrigerants condensation. The study determined local and average heat transfer coefficient values in full range of vapour quality, $x = 0 \div 1$. On the basis of experimental investigations, the dependence of heat transfer on the vapour quality x, mass flux density G and minichannel's internal diameter d_w was obtained.

1 Introduction

The dynamic expansion in global development forces the unceasing need for innovative solutions in design and construction of equipment. These solutions aim at intensifying utility parameters such as power, efficiency and reducing their size while lowering production costs [1]. Researchers are therefore looking for the best solutions from a technical and economic point of view. Increase of efficiency factor results in considerable usage of equipment and that generates increase in production costs [3]. One of the main causes of wear is the inefficient cooling. Due to the limited size of conventional HE's, micro- condensers fits perfectly into the needs of the market. But we still have to be aware of ecological considerations. The smaller the size of the system, the less risk to the environment - in case of leak of the refrigerant to the environment.

Small refrigeration systems are commonly used, for example, in personal computers or smartphones. However, miniature heat exchangers will be gradually displaced on the market by "micro" heat exchangers [2]. Growing demands and requirements for heat exchangers constructors forces new solutions. The authors assume that the efficiency of the micro heat exchangers is mostly influenced by: the nature of the refrigerant flow, the type of refrigerant used, and the influence of the surface tension forces. Not meaningless stays the type of heat reception and channels orientation.

There is still a small number of heat exchangers designed with an air-cooled microchannels. A similar situation occurs with vertical channels.

The article [19] describes study on HTC in horizontal rectangular multiport minichannel with and without fins using R134a. Authors fabricated a new experimental apparatus in order to obtain explicit local condensation heat transfer coefficient measurements over a range of test conditions. The test section was an 852 mm long horizontal rectangular multiport minichannel with and without fins having 20 channels with a hydraulic diameter of 0.64 mm and 0.81 mm. The measurements were done over a range of saturation temperature from 30 to 35 °C with mass fluxes ranging from 50 to 200 $kg \cdot m^{-2} \cdot s^{-1}$. The effects of vapour quality, mass fluxes, channel geometry, and saturation temperature on the heat transfer coefficient have been clarified and analyzed. The experimental results were compared with the well-known condensation heat transfer models available in the open literature. All of the existing correlations were failed to capture the present experimental heat transfer coefficient with a high degree of accuracy. As a consequence, a new condensation heat transfer correlation was developed based on the present experimental data and validated with 750 data points collected from the available journal.

Article by Chen et. al. [2] describes FC-72 refrigerant's condensation model in rectangular channel with hydraulic diameter d_h = 1mm. The model was confronted with research results available in the world literature. It has been noted that the vapor velocity of the medium is significantly higher than that of its condensate. There was also found that an increase in the size of the bubbles goes along with the increase in the mass flux density.

An experimental investigation was carried out in paper [20] for the benefit of refrigerant mass flux, local vapor quality, coolant flow rate and inlet coolant temperature on the local condensation heat transfer coefficient. Flow visualization was also carried out during the flow of condensation using a high-speed camera integrated with a microscope. HFE-7100, a dielectric and eco-friendly refrigerant was used in rectangular multimicrochannels with a hydraulic diameter of 0.57 mm. Experiments were performed at a saturation temperature of 60 °C, mass flux range 48-126 kg / (m2 s), coolant flow rate range 0.5-1.1 L / min and

* Corresponding author: marcin.kruzel@tu.koszalin.pl

inlet coolant temperature range 20-40 °C. The results showed that the local condensation heat transfer coefficient increases with increasing mass flux and decreases with decreasing local vapor quality. A negligible effect of the coolant side conditions, saturation-to-wall temperature difference, on the local condensation heat transfer coefficient was found. The main flow regime was an annular flow, while slug and bubbly flow were found at some operating conditions. The experimental results were correlated for heat transfer rates. also, two existing flow pattern maps, for conventional and mini / microchannels, were used to compare the current flow pattern results.

In paper [21], the successive evaporation and condensation of water-ethanol and water-methanol liquid film mixtures for cooling purpose was investigated. The simulations were achieved using a finite difference model to describe the heat and mass transfer during this process. The studied process relied on the absorption of energy at the evaporator and then the release of this energy at the condenser. The evaporation and heat exchange abilities of the mixtures composition ware analysed to identify the appropriate mixture for the new proposed cooling approach. The results indicated that more energy equilibrium occurs with high concentration of water in the mixture with the advantage of having low bulk and interface temperature.

A new, universal approach to predict the heat transfer coefficient in minichannels and microchannels were presented by Kim and Mudawar [5]. The concept applies to a number of fluids with significantly different thermophysical properties, as well as for variable channel geometry and flow parameters. The authors collected a database of 28 different sources, including over 4000 measurement points. The data concern condensation of 17 types of working fluids in single channels and multiports, and their hydraulic diameters are in the range: d_h = 0.424 - 6.22 mm. Mass flux density G = 53 - 1403 $kg{\cdot}m^{-2}{\cdot}s^{-1}$. Two new correlations were proposed: the first for the annular flow, the second for plug flow and bubbly flow.

Fronk and Garimella [4] investigated the heat transfer coefficient and pressure drop during carbon dioxide (CO2) condensation in rectangular minichannels. The channels hydraulic diameters were d_h = 0.1 mm and 0.16 mm. Channels were made of copper by electroforming. Thermo-flow parameters were measured at mass flux density G = 400, 600 and 800 $kg{\cdot}m^{-2}{\cdot}s^{-1}$ and the full range of refrigerant's vapor quality x. Authors compared the results with the investigations data from several existing correlations. The best agreement for was obtained with Garimella correlation [5]. For the heat transfer coefficient, the best agreement was for authors [6,7,8,9] data.

2 Subject of research

The subject of the research were vertical pipe minchannels, made of stainless steel with an inner diameter d_w = 0.5 ÷ 2.5 mm and length L = 1000 mm. The refrigerants R404A, R407C and R410A were used for the tests. The section of the minichannel was cooled with air at a flow velocity of = 17 $m{\cdot}s^{-1}$, and the refrigerant was supplied with set and timed thermal and flow parameters. Minichannels were placed in an air channel measuring 60x120mm. Experimental research has been carried out in the following range of thermal and flow parameters:.

Table 1. Range of parameters.

Parameter	Range
mass flux density	G = 50 ÷ 200 $kg{\cdot}m^{-2}{\cdot}s^{-1}$
heat flux density	q = 0 ÷ 10 $kW{\cdot}m^{-2}$
vapor quality	x = 0 ÷ 1
saturation temperature	t_s = 20 ÷ 40 °C

The construction of the experimental facility enabled the measurement of the temperature and pressure of the refrigerant, the temperature of the channel wall and cooling air over the length of the measuring section. The flow rate of the refrigerant and coolant were also measured.

3 Experimental facility

The main element of the test stand was the tube minichannel with an internal diameter d_w and an overall length of 1000 mm (on the active length of 950 mm, the heat transfer coefficient was measured).

Figure 1 shows the actual view of the measurement facility, while in Figure 2 the schematic diagram of the facility.

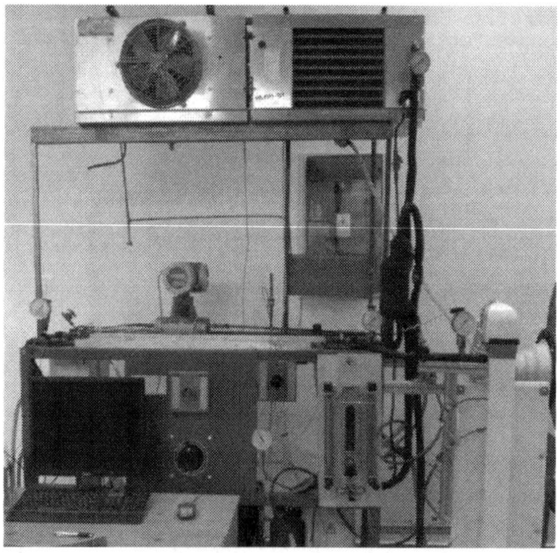

Fig. 1. Overall view of test stand.

The refrigerant condensation process during refrigerants flow took place in pipe minichannels. The superheated vapour of the refrigerant was directed by a control valve

to the measuring section after compressing in the reciprocating compressor.
A water-cooled heat exchanger was installed before the refrigerant at the inlet of the measuring section. On the basis of the thermal balance of the exchanger, the vapour quality was calculated. Exchanger received the heat of overheated refrigerant. The intensity of heat transfer was controlled by the change of the water flow rate cooling the exchanger. So parameterized refrigerant vapour supplied the measuring section.

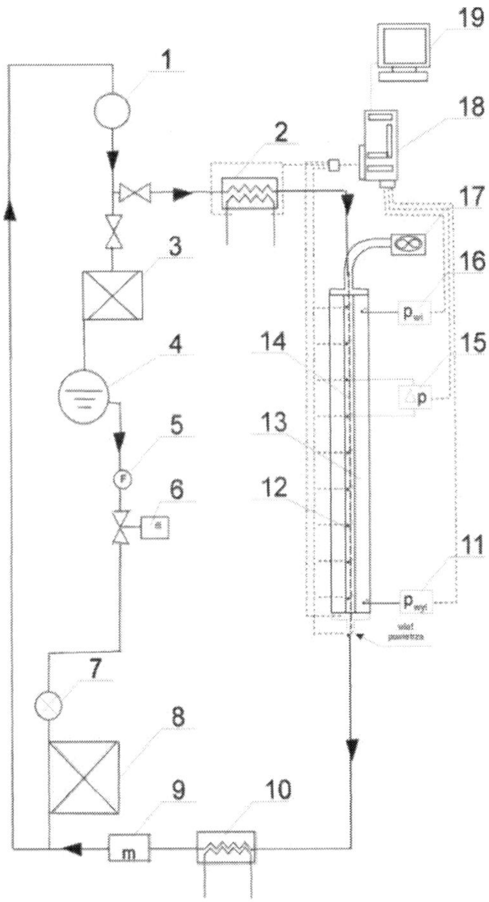

Fig. 2. Schematic diagram of test stand: 1) compressor, 2) pre heat exchanger, 3) condenser, 4) refrigerant liquid tank, 5) filter, 6) electromagnetic valve, 7) expansion valve, 8) fan refrigerator, 9) refrigerant mass flow meter, 10) heat exchanger, 11) inlet pressure sensor, 12) thermocouples, 13) air channel, 14) pipe minichannel, 15) differential pressure measurement, 16) outlet pressure sensor, 17) fan, 18) data acquisition system, 19) computer.

4 Test results

Condensation process analysis requires the acquaintance of the distribution of local heat transfer coefficient values at L length of the refrigerant flow. Figure 3 shows the dependence of the local heat transfer coefficient α_x on the change of mass flux density G for inner diameter d_w = 0,5 mm.

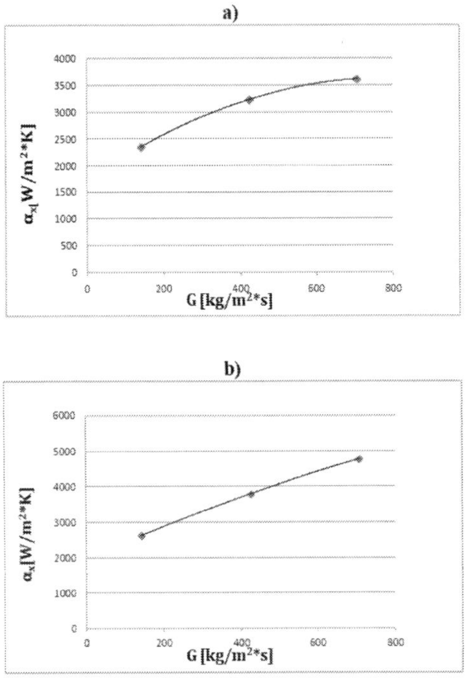

Fig. 3. Experimental investigation results of local heat transfer coefficient during condensation of a) R410A, b) R404A refrigerants in vertical pipe minichannel with internal diameter d = 0.5 mm.

As it is shown on figure 3 the local heat transfer values increases along with the mass flux density G rises. A similar relationship occurs for both examined refrigerants. Figure 4 shows the dependence of local heat transfer values on the inner diameter d_w.

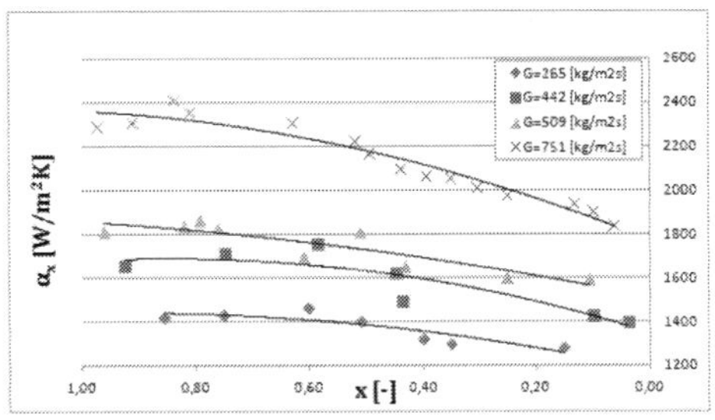

Fig. 4. Experimental investigation results of local heat transfer coefficient dependence on vapour quality x during R404A condensation in pipe minichanel with d_w = 2.0 mm inner diameter.

The dependence of local heat transfer coefficient on vapour quality is presented on figure 4. The increase of vapour quality x results in HTC values increase. Highest values were noticed for x = (0.6 ÷ 0.9) which is the proper condensing area, in which the heat transfer is most efficient.

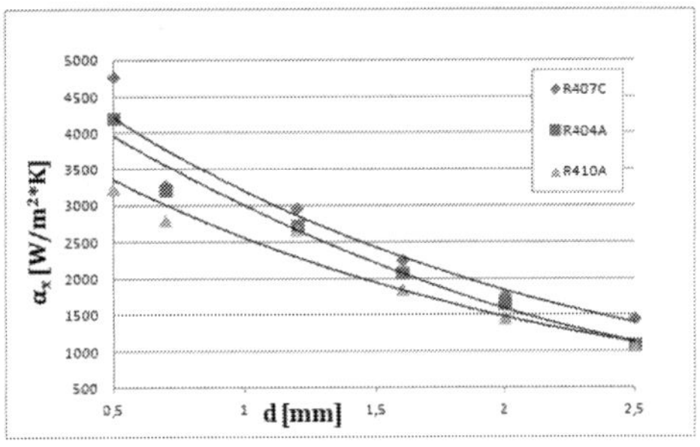

Fig. 5. Experimental investigation results of local heat transfer coefficient dependence on inner diameter during condensation of all three examined refrigerants.

Figure 5 shows that efficiency of heat transfer depends on the size of the inner diameter d_w. Values of local heat transfer coefficient rise with the reduction of inner diameter. Highest HTC values were noticed during condensation of R407C refrigerant in d_w = 0.5 mm wide minichannel. A similar trend for all examined refrigerants.

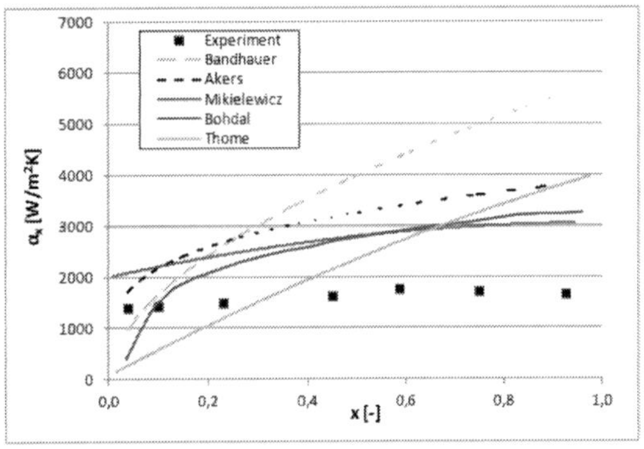

Fig. 6. Experimental investigation results of local heat transfer coefficient during condensation of a) R410A, b) R404A refrigerants in vertical pipe minichannel with internal diameter d = 0.5 mm.

The results of own empirical studies were compared with the results of calculations according to the correlations of other authors [10-14].

The graphical comparison shows the adjustment of existing heat exchange prediction models to the results of the experiment.

Figure 6 shows that the results of calculations according to the analyzed models do not fit into the course of the experiment results. A different course of calculation results may be caused by different (much lower) values of heat flow q that characterize heat exchange in air-cooled condensers. The calculations indicated best agreement with experimental results for Thome's correlation (figure 7).

But the results were not satisfactory. Mean absolute percentage error MAPE was 32% for R404A, 35% for R407C and 47% for R410A. The results of the comparison with the correlations according to the other authors were subject to a significant error (MAPE) of more than 50%. Such a large discrepancy may result from different ranges of applicability of particular correlations.

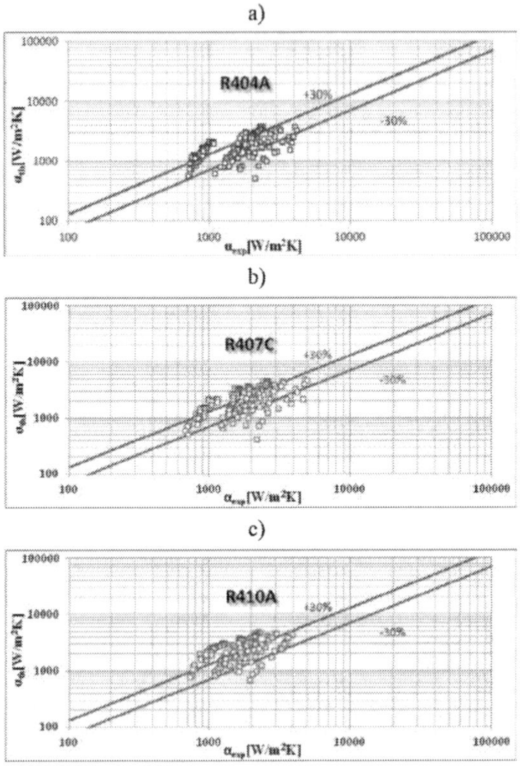

Fig. 7. Comparison of experimental data of the local heat transfer coefficient with the results of calculations according to the Thome's model for a) R404A, b) R407C and c) R410A refrigerants condensation.

Heat exchange during the refrigerants two-phase flow through a minichannel is a very complicated process. Therefore, quantifying this process should be considered both the nature of the refrigerant flow and its properties. With this in mind, a model was new model was developed. The model is intended for prediction of heat transfer coefficient values in a vertical air-cooled minichannels. Based on the analysis of the results of own experimental research data of R407C, R410A and R404A refrigerants condensation process in minichanels, correlation of various authors and dimensional analysis a new correlation for local heat transfer coefficient and local pressure drop was obtained. Correlation parameters were determined using nonlinear regression model [18] estimated by *Levenberg – Marquardt* method (in Statistica's standard pacage).

Local heat transfer coefficient α_x. was calculated by:

$$Nu_x = 0.63 \cdot {Re_l}^{0.35} \cdot {p_r}^{-0.49} \cdot {Pr_l}^{0.79} \cdot \left(\frac{x}{1-x}\right)^{0.22} \quad (1)$$

$$\alpha_x = \frac{Nu \cdot \lambda_l}{d_w} \quad (2)$$

The usefulness of the above correlation was confirmed experimentally in the field of operation of air-cooled minichannels, i.e. for low values of heat flux density q = =500 ÷ 5000 W·m^{-2}.

In order to verify the scope of applicability of the existing model describing heat exchange during refrigerants condensation in mini- and microchannels, the method of residual analysis using the error MAPE (mean absolute percentage error) was applied:

$$MAPE = \frac{1}{n}\sum_{i=1}^{n}\left[\frac{|Nu_{th}-Nu_{exp}|}{Nu_{exp}}\right]\cdot 100\% \qquad (3)$$

Nu_{th} means the result of calculations according to the correlation of individual authors, whereas Nu_{exp} is the value of a given quantity determined experimentally, n - is the number of measurements made. MAPE shows the average error that should be expected when anticipating the behavior of a given model in subsequent studies.

In the case of the Nusselt number, the MAPEs were calculated for the Thome [12], Akers [13], Bandhauer [14], Shah [15], Tang [16], and Cavallini [] models. The results were compared with the author's correlation (Table 2).

Table 2. Comparison of HTC prediction model MAPE for R407C refrigeration.

	MAPE %						
d_w [mm]	Thome	Akers	Shah	Tang	Bandhauer	Cavallini	autor
2,5	55	190	498	601	152	147	11
2	43	158	399	408	135	78	6
1,6	26	131	264	334	113	83	6
1,2	21	83	198	185	99	64	8
0,7	25	61	162	111	85	68	5
0,5	41	33	99	96	58	71	10
Mean value	35	109	270	289	107	85	8

Figure 8 contains a quantitative graphic comparison for MAPE errors of particular correlations included in the study.

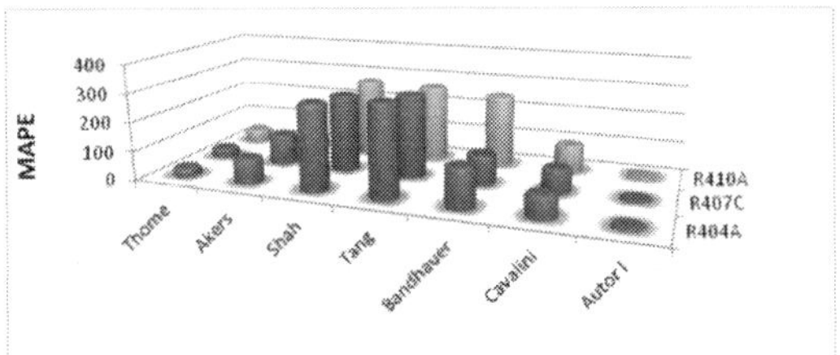

Fig. 8. Comparison of MAPE errors for the calculation of the local Nusselt number according to different Nu_{th} authors with the results of experimental Nu_{exp}.

The results of the local Nusselt number comparison for all examined internal diameters are presented below. Figure 9 shows the diffrences bettween theoretical and experimental results. for condensation in vertical pipe minichannel.

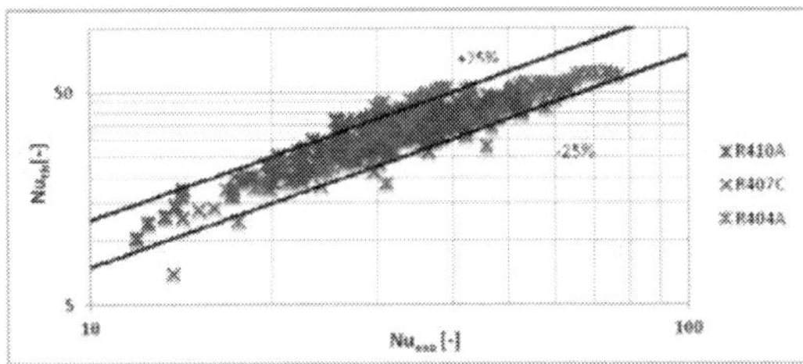

Fig. 9. Comparison of experimental results with the results of calculations according to the authors model (1) for the local Nusselt number.

Discrepancies between experimental values and calculated values are within ± 25%. The correlation describes the results of experimental studies with an average percentage error of 8%. For a two-phase flow, this result was considered satisfactory.

4 Conclusions

1) Efficiency of heat transfer increases along with increasing mass flux density G of the refrigerant in the minichannel.

2) Local values of the heat transfer coefficient α_x increases with the increase in the vapor quality x of the refrigerant, in the range of about $x = 0 \div 0.8$, and then decrease slightly.

3) Correlations currently available in literature describe the heat transfer during condensation in small diameter water cooled minichannels. They do not have unfortunately analogous use in the case of condensation in air-cooled minichannels. The condensation process in air-cooled minichannel runs at significantly lower heat flux density values.

4) Appliance of existing correlations describing condensation in water-cooled minichannels used for air-cooling was considered unsatisfactory, because the experiment's discrepancy range is greater than ± 50% for heat exchange.

5) To calculate the local values of heat transfer coefficient during refrigerants condensation in air cooled minichannels (for low heat flux densities q = 500 ÷ 5000 W·m^{-2} it is proposed to use dependencies determined on the basis of author's experimental investigations according to the dependence (1).

Nomenclature

Symbols

α_x – local heat transfer coefficient, [W·m^{-2}·K^{-1}]
d_w – internal diameter, [mm]
d_h – hydraulic diameter, [mm]
G – mass flux density, kg/[m^2·s],
p_r – reduced pressure [-]
p_s – saturation pressure [kPa]
p_k – critical pressure [kPa]
q – heat flux, [kW·m^{-2}]
v – velocity [m·s^{-1}]

Acronyms

HE – Heat Exchanger
HTC – Heat Transfer Coefficient
MAPE – Mean Absolute Percentage Error

References

1. T. Bohdal, M. Kruzel, M. Sikora, Journal of Mechanical and Energy Engineering, vol. 1(41), no. 2. (2017).
2. A. H. Al-Zaidi, M. M. Mahmoud, T. G. Karayiannis, Experimental Thermal and Fluid Science 90 (2018) 153-173.
3. M. Kruzel, Wydawnictwo Uczelniane Politechniki Koszalińskiej, (2012) 95-104.
4. M. B. Fronk, S. Garimella: Internationoal Journal of Heat and Mass Transfer 100 (2016) 150-164.
5. S.M. Kim, I. Mudawar, International Journal of Heat and Mass Transfer Vol. 55 (11-12). - 2012. 3246-3261.
6. A. Agarwal, T.M. Bandhauer, S. Garimella, Int. J. Refrig. 33 (6) (2010) 1169–1179.
7. M.K. Dobson, J.C. Chato, Condensation in smooth horizontal tubes, J. Heat Transfer 120 (1) (1998) 193–213.
8. W. W. Wang, T.D. Radcliff, R.N. Christensen, Exp. Thermal Fluid Sci. 26 (5) (2002) 473–485.
9. D.P. Traviss, W.M. Rohsenow, A.B. Baron, ASHRAE Trans. 79 (Part 1) (1973) 157–165.
10. T. Bohdal, H. Charun, M. Sikora, International Journal of Refrigeration, vol. 59. - 2015, pp. 210-223.
11. D. Mikielewicz, Heat Transfer Engineering, Taylor & Francis, 2016, pp. 1158-1171.
12. J.R. Thome, J. El Hajal, A. Cavallini, Int. J. Heat Mass Transf. 46, (2003), 3365-3387.
13. W. W. Akers, H. F. Rosson, Chemical Eng. Progr. Symposium 1960, Series 50. - 1960, pp. 145-149.
14. T.M. Bandhauer, S. Agarwal, S. Garimella J. Heat Transfer, vol. 128, (2006), 1050-1059.
15. M.M. Shah, Int. Journal of Refrigeration 67 (2016), 22-41.
16. L.Tang Ph.D., College Park, MD University of Maryland, (1997).
17. A. Cavallini, G. Censi, D. Del Col, L. Doretti, A. Longo, A. Rossetto, Heat Transfer Eng. 27 (8), (2006), pp. 31-38.
18. M. Dobosz, Akademicka Oficyna Wydawnicza EXIT, Warszawa (2003), 233-276.
19. M. Mostaqur Rahman, K. Kariya, A. Miyara, Int. Journal of Heat and Mass Transfer 116, (2018), pp. 50-60.
20. Ali H. Al-Zaidia, M.M. Mahmouda, T.G. Karayianniss, Experimental Thermal and Fluid Science, 90, (2018), pp. 153-173.
21. M. Najim, M. Feddaoui, A. Nait Alla, A. Charef, A.E. Kabeel, International Journal of Heat and Mass Transfer 122, (2018), pp. 895–912.

Condensation of Novec 649 refrigerant in pipe minichannels

Tadeusz Bohdal[1], *Henryk* Charun[1], and *Małgorzata* Sikora[1,*]

[1]Koszalin University of Technology, Department of Energy, ul. Racławicka 15-17, PL -75-620 Koszalin, Poland

Abstract. The paper presents the results of experimental investigation of Novec 649 refrigerant condensation in tube minichannels. This is a low-pressure refrigerant. This investigations are basis for flow structures visualization during condensation in pipe minichannels. The local and the average values of pressure drop ($\Delta p/L$) and heat transfer coefficient α in the whole range of the changes of vapour quality ($x = 1 \div 0$) were calculated. On the basis of the obtained test results there was illustrated the influence of the vapour quality x, the mass flux density G and the inner diameter of channel d changes on the studied parameters. These results were compared with the calculation results based on the dependencies of other authors.

1 Introduction

Due to the development of technologies and miniaturization of power equipment (while increasing their efficiency), compact heat exchangers are very important area of studies around the world. Exchangers of this type are built from mini and microchannels, which results in an increase of heat transfer intensity by a small surface. The minichannels are those channels which hydraulic diameter d_h < 3mm. Especially important in this case are two-phase diabatic flows, with accompanying phase change like condensation or boiling [11].

Two-phase flow phenomena in conventional channels in diabatic conditions are completely different than those received in minichannels. Therefore, the transfer of the results obtained in the field of two-phase flow in conventional channels, to mini- and microchannels is incorrect and causes errors in interpretation of thermal and flow phenomena. It is known that in the small diameter channels significant impact on the mechanism of momentum and energy transport have viscosity and surface tension, while gravity and inertia forces are less important [14]. In the work of Coleman and Garimella [7, 9] was found that for channels with a hydraulic diameter greater than 10 mm, the channel dimension and surface tension didn't significantly influence on flow and thermal parameters. This proves that the experimental investigations of refrigerants condensation in mini and microchannels are very important. Many scientists from around the world conduct research in this area for years.

For example, Fronk and Garimella [8] presented a comparison of modern methods for calculating the heat transfer coefficient and flow resistance in the condensation process in conventional and minichannels. They pointed to the usefulness of computational methods of authors such as: Silver [19], Bell and Ghaly [3], Thome [23] and Cavallini et al. [5,6]. The proposed methods are based on the analysis of the condensation mechanism in the flow, but their weakness is the degree of complexity of equations.

Zhang et al. [28] in his study presented the results of R22, R410A and R407C refrigerant condensation in single circular minichannels with internal diameters of 1.088 and 1.289 mm, with parameters in the ranges of: saturation temperature T_s = 30 - 40 °C, mass flux density G = 300 - 600 kg/m^{2}s and vapour quality x = 0.9-0.1. As expected, it was found that pressure drops increase when the mass flux density, G, and vapour quality, x, increase, but in the range of $x > 0.8$, this influence is much smaller. What is more the influence of the channel diameter and refrigerant type on pressure drops was shown.

Son and Oh [20] presented experimental results of pure R22, R134a and R410A refrigerants pressure drops during condensation in a vertical circular channel with an inner diameter of 1.77 mm and a length of 1.22 m for the ranges of mass flux density G = 450 - 1050 kg/(m^{2}s), and saturation temperature T_s = 40°C. The experimental results illustrated the dependence of the vapour quality x, on the pressure drop for R22, R134a and R410A refrigerants in pipe minichannels with an internal diameter of 1.77 mm. It was shown that the pressure drops for R410A are the lowest in the same mass flux density.

Not only internal diameter of the channel have important influence on condensation phenomena. Second very important thing is kind of refrigerant. Rajan [17] showed in his work that R410A refrigerant is more effective than R22 because the heat transfer coefficient for R410A is higher by approximately 30 % at boiling, and, in the same time, the pressure drop is lower by approximately 40%.

Due to the destructive impact of refrigerants from CFC and HCFC groups on the ozone layer and the

*Corresponding author: malgorzata.sikora@tu.koszlin.pl

https://doi.org/10.1051/e3sconf/20187002002

intensification of global warming, radical actions were taken in order to eliminate derived chlorinated substances from a common use. The first international agreement which was aimed to protect the environment was the Montreal Protocol [16]. The consequence of this document is the Act on substances that deplete the ozone layer [25] passed by the European Union and ratified by the Polish Parliament. Removed refrigerants are successively replaced with ecological refrigerants having similar physicochemical properties Most of new refrigerants are mixtures. As a replacements of R22 refrigerant are used zeotropic mixtures as: R404A, R407C and R410A [4, 12]. The most popular homogenous refrigerant R134a is removed now because of very high global warming potential (GWPR134a = 1430), and companies are looking for new substitutes as a R1234yf or R1234ze [1, 26, 13]. According to the Directive of the European Parliament and the Council of Europe No. 2006/40/EC the use of fluorinated greenhouse gases for which GWP value is greater than 150 is limited in all newly manufactured refrigeration and MAC (Mobile Air Conditioning) systems [1, 24, 26].

What is more, now is a lot of new mixtures as a working fluids and scientists investigate two-phase flow phenomenon with they use. For example, Zhuang and all [29] made experimental investigation of methane condensation in a horizontal smooth tube in range of mass flux density G=99-255 kg/m^{2}s. Methane is natural refrigerant with came back to use. Wen and all [27] made numerical investigations on R1234ze(E) and propane condensation in channels with internal diameter d = 1 - 4.9 mm. Results of heat transfer and frictional pressure parameters calculations were compared with results for R134a refrigerant. Ghim and Lee in work [10] made an experimental investigations of condensation in ORC of R245fa, Novec 649, HFE7000 refrigerants. Investigations was conducted in straight horizontal channel with internal diameter d = 7.75mm in range of mass flux density G = 150 - 700 kg/m^{2}s, vapour quality x = 0.95-0.05. Authors showed, that heat transfer coefficient α during condensation is highest for R245fa then HFE 7000 and the lowest for Novec 649.

This work concerns condensation of Novec 649 refrigerant in pipe minichannels. This working fluid is environmentally friendly, its ODP is 0, GWP 1 and its atmospheric lifetime is about 5 days [1]. This is low pressure refrigerant and is one of most reference fluids for mobile applications [15].

From the given publication review, it can be concluded that the number of publications concerning the condensation of new, environmentally friendly refrigerants in minichannels is not bid enough. There is no clearly preferred method of heat transfer coefficient and pressure drop calculation. There is need of further study of the condensation process of these refrigerants, especially in mini and microchannel. Very interesting will be a investigation of two-phase flow structures for new refrigerants.

2 Experimental study

Subject of study was condensation of Novec 649 refrigerant in tubular minichannels made from stainless steel with an internal diameter d = 0.5, 0.7 and 1.2 mm Fig.1shows the view of this kind of minichannels. On fig.2 are showed example properties of refrigerant Novec 649.

Fig. 1. View of used stainless steel minichannels.

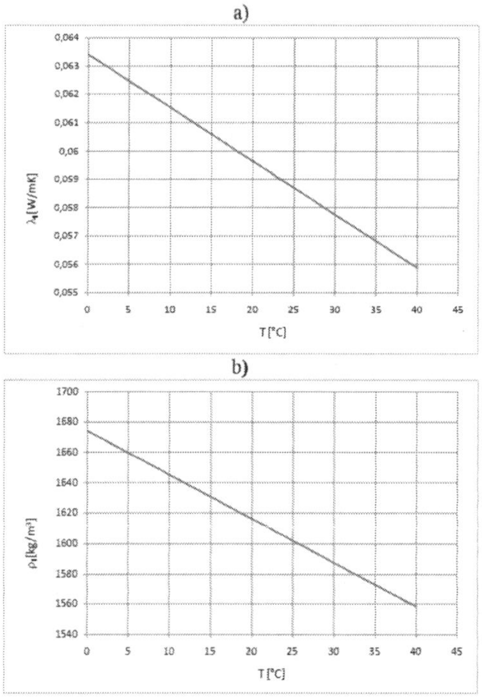

Fig. 2. Dependence of example properties on refrigerant Novec 649 temperature; a) liquid conductivity λ, b) liquid density ρ [30].

The heat and flow investigations were conducted in terms of parameters:

- inner diameter of minichannel: d = 0,5 - 1,2 mm,
- mass flux density: G = 790 ÷ 5525 kg/m^{2}s,
- heat flux density: q = 0 ÷ 200 kW/m^2,
- saturation temperature: t_s = 30 ÷ 70°C,
- vapor quality: x = 1÷ 0.

2.1 Test stand

Schematic diagram of the test stand for thermal – flow and structure visualization researches of two-phase flow during condensation of Novec 649 refrigerant is shown in fig. 3. On the schematic diagram there are listed two sections, named as A and B. Section A were used for thermal and flow experimental investigations of refrigerants condensation in pipe minichannels. The investigation allowed experimental determining of the average and local values of heat transfer coefficient α and flow resistance Δp/l during condensation.

Section B can be used to visualisation of flow structures during condensation of refrigerants in minichannels.

In fig. 4 is shown real view of heat and flow measurement section, with was cooling by water. The working length of the channel was L = 0.15 m. For measure temperature of channel wall and cooling water were used K-type thermocouples.

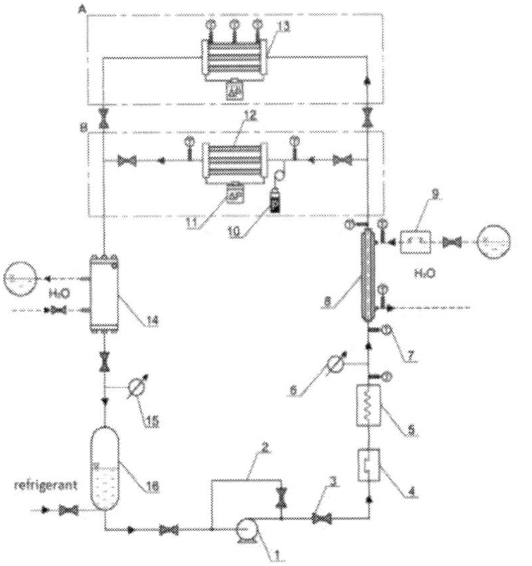

Fig. 3. Schematic diagram of the test stand A – Section for thermal and flow investigations, B – Section for experimental visualization of two-phase flow structures; 1 – ceramic pump, 2 – bypass for the flow control, 3 – valve, 4 – flowmeter for refrigerant, 5 – evaporator, 6 – manometer, 7 – K-type thermocouples, 8 – pre-heat exchanger "pipe in pipe" type , 9 – flowmeter for cooling water, 10 – pressure sensor, 11 – differential pressure sensor, 12 – horizontal section for visualization, 13 – horizontal section of a heat-flow measurement, 14 – heat exchanger, 15 – manometer, 16 – tank of liquid refrigerant.

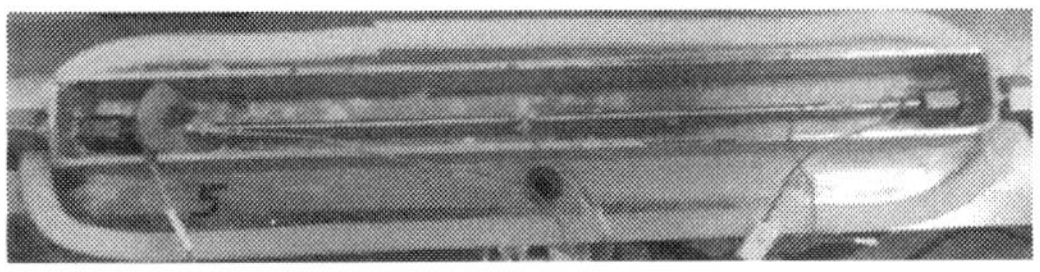

Fig. 4. Real view of heat and flow measurement section.

2.2 Investigation results

Fig. 5 and 6 show the thermal and flow characteristics of the Novec 649 refrigerant condensations, for example, in a pipe minichannel with an internal diameter d =0.5, 0.7 and 1.2 mm. The provided characteristics illustrate a dependence of the local heat transfer coefficient, α_x and local pressure drops, $(\Delta p/L)_x$ on the vapour quality, x and the mass flux density, G.

a)

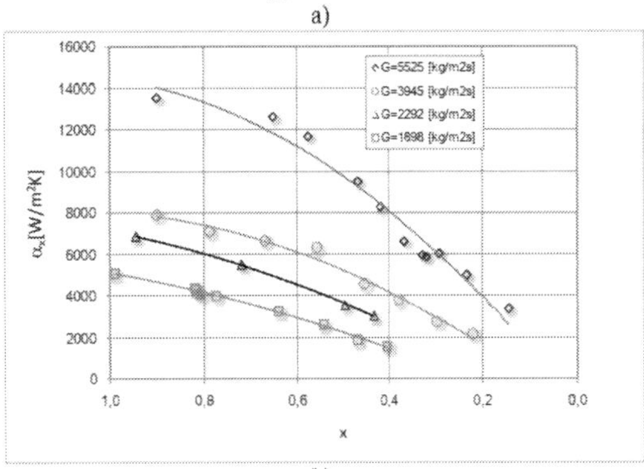

b)

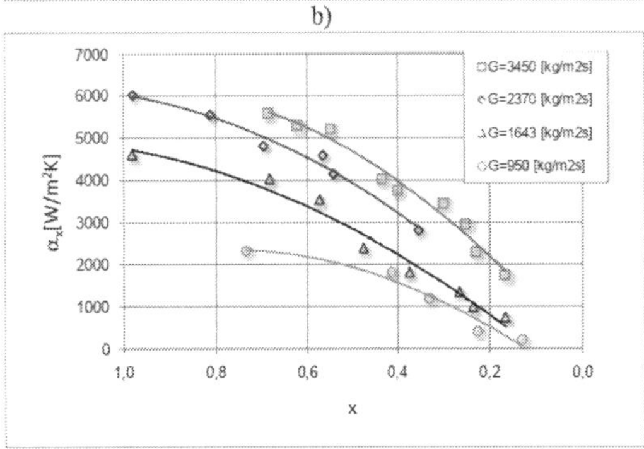

c)

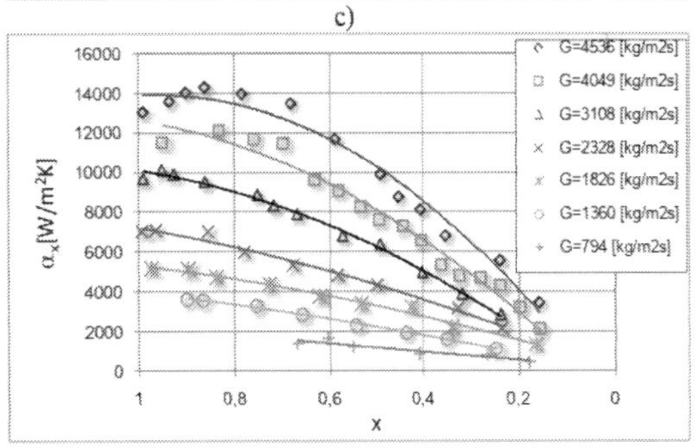

Fig. 5. Local heat characteristics of Novec 649 refrigerant condensation in pipe minichannels: a) d=0.5mm, b) d=0.7mm, c) d=1.2mm.

The characteristics of the thermal and flow changes were made for the whole range of vapour quality, from x = 1 to x = 0. The value of the local heat transfer coefficient for the condensation process initially increases, reaches a maximum value near to x = 0.8 and then follows. The reduction of the vapour quality during condensation affect increases the amount of saturated liquid in the two-phase mixture. For the liquid phase, the heat transfer coefficient is smaller than for the two-phase mixture, which justifies the decrease in the value of the heat transfer α_x coefficient. Reduction of the inner diameter, d of the pipe minichannel enhances the heat transfer in the condensation process.

In fig. 6 are shown similar flow characteristics. They show changes of the local pressure drops, $(\Delta p / L)_x$, with a decrease of the vapour quality, x of the two-phase mixture. Pressure drops depends on vapour quality like the heat transfer coefficient α. Intensification of the heat transfer during condensation, implemented to reduce the minichannel internal diameter, requires an increase in the driving energy of the flowing refrigerant mixture. Proof of this is the increase of pressure drops $(\Delta p/L)_x$ with a reduction of the minichannel internal diameter.

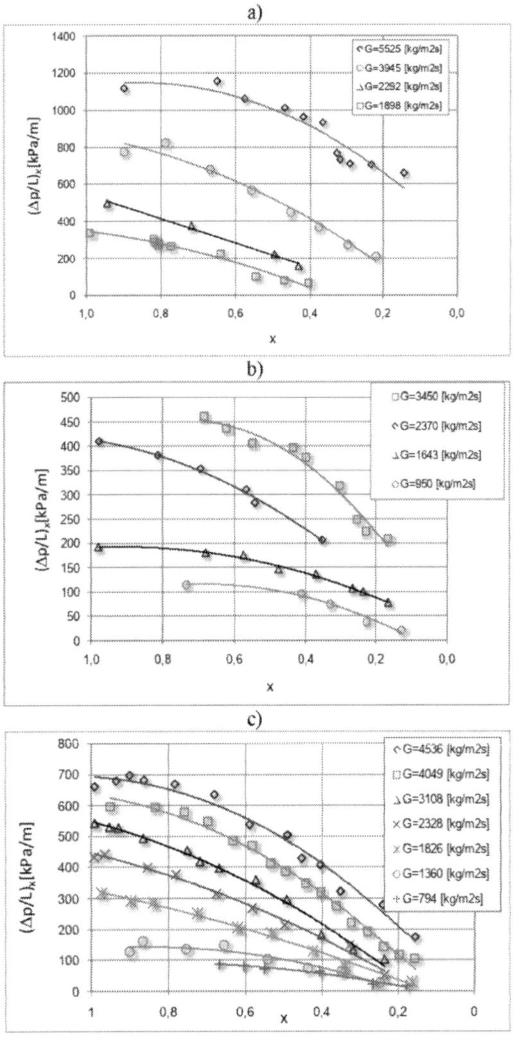

Fig. 6. Local flow characteristics of Novec 649 refrigerant condensation in pipe minichannels a) d=0.5mm, b) d=0.7mm, c) d=1.2mm.

Fig. 7 presents a example dependences of average heat transfer coefficient α_a and average pressure drops $(\Delta p/L)_x$ on mass flux density for the constant vapour quality x=0.6 during condensation Novec 649 in minichannel with internal diameter d=0.7 mm. It showed very brightly dependence of this two heat and flow parameters on the mass flux density G.

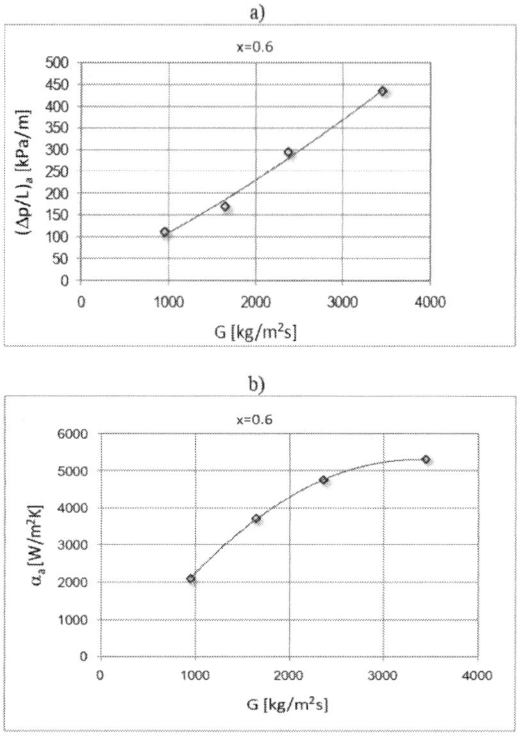

Fig. 7. Average a) flow, b) heat characteristic of Novec 649 refrigerant condensation in minichanal with d=0.7 mm.

Own experimental results of the local heat transfer coefficient α_x were compared with the calculation results of the correlation of other authors: Shah [18], Akers [2], Thome [22] and Tang end all [21]. Results of this comparison are showed in fig.8 and fig.9.

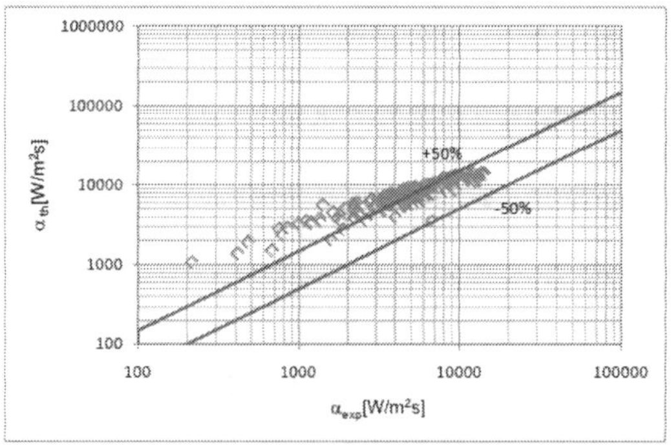

Fig. 8. Comparison of experimental heat transfer coefficient with calculations from correlations of Shah [18].

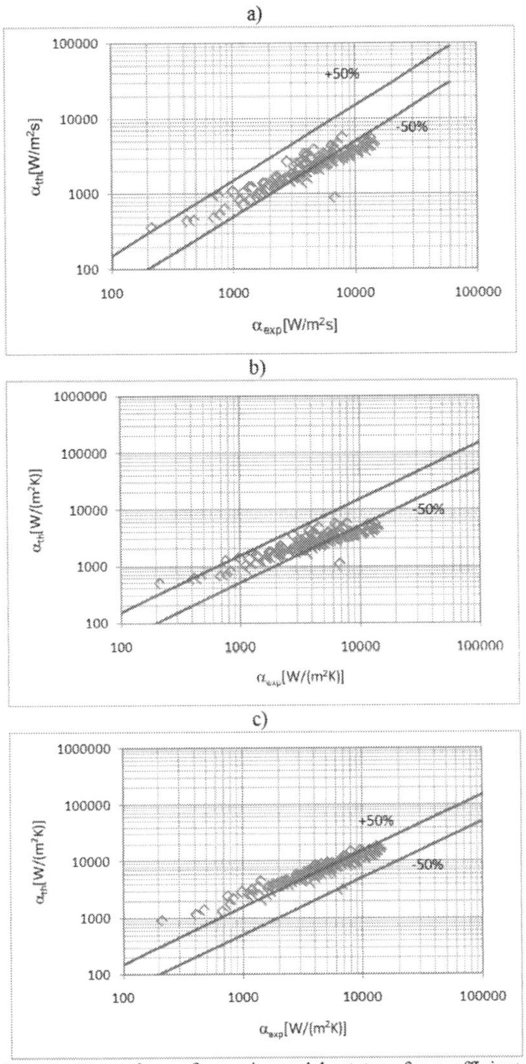

Fig. 9. Comparison of experimental heat transfer coefficient with calculations from correlations of: a) Akers [2], b) Thome [22], c) Tang [21].

The best accuracy of experimental investigation of new refrigerant Novec 649 condensation in minichannels can be observed to the Thome correlation. Unfortunately, in this case the deviation is 50% which is not a satisfactory result. This may be due to the properties of the refrigerants or the small diameter of the tested channels. This means that more research should be done on the Novec 649 refrigerant, in a larger number of small diameter channels, and then determine the relationship describing the condensation process of this and other similar refrigerants in minichannels.

3 Conclusions

1) Pipe mini-channels can be used in the construction of compact refrigeration condensers. In this way heat transfer process during condensation is intensified by a high degree of structural compactness. The heat transfer coefficient is the measure of this intensification compared with conventional channels. The determination of the value of the heat transfer coefficient and the pressure drop during this process constitutes the primary problem for designers.

2) Thermal and flow experimental investigations of Novec 649 refrigerant condensation in pipe minichannels in terms of the channel internal diameter changes is very important. This is the new environmentally friendly refrigerant with low GWP and ODP.

3) Thermal characteristics describing condensing of refrigerants were done, with a description of dependences of local and average heat transfer coefficient, depending on the vapour quality and mass flow density. It is found that the heat transfer coefficient depends on parameters such as thermal parameters of condensation process (saturation temperature), the internal diameter of the tubular minichannel, mass flow density and vapour quality.

4) The thermal characteristics of the tested refrigerant were compared with the results of calculations prepared by other authors. The discrepancies between the experimental results and the calculation results exceeded the range of ± 50 % , which gives unsatisfactory results and made a new correlation is necessary.

5) There is a need of extension of research with new refrigerants and observation of flow structures, with affect the thermal and flow parameters of the process.

References

1. R. Akasaka., K. Tanaka, Y. Higashi, *Thermodynamic property modeling for 2,3,3,3-tetrafluoropropene (HFO-1234yf)*. International Journal of Refrigeration, vol. 33, 52 – 60, (2010)
2. W. Akers, O.K. Deans, O.K. Crosser, *Condensation heat transfer within horizontal tubes*. XChem. Eng. Progr., vol. 54, 89-90, (1985)
3. K.J. Bell, M.A. Ghany, *An approximate generalized design method for multicomponent partial condenser*, AIChE Symposium Series, vol. 69, , pp. 72-79, (1973)
4. T. Bohdal H. Charun, M. Sikora, *Eksperymentalne charakterystyki cieplno – przepływowe skraplania roztworów zeotropowych w kanałach konwencjonalnych i minikanałach*. Chłodnictwo, vol. 6, 14 – 19, (2013)
5. A. Cavallini, , D. Del Col, L. Doretti, M. Matkovic, L. Rossetto, *Condensation in horizontal smooth tubes: A new heat transfer model for heat exchanger design*, Heat Transfer Engineering. vol. 27(8), pp. 31-38. (2006)
6. A. Cavallini, G. Censi, D. Del Col, L. Doretti L. Rossetto, *Heat transfer coefficient HFC refrigerants during condensation at high temperature inside a enhanced tube*, Proc. Int. Refrigeration and Air Conditioning Conference, Pardue University, paper 563, R8-2, (2002)
7. J. Coleman, S. Garimella, *Two-Phase Flow Regimes in Round, Square and Rectangular Tubes During*

Condensation of Refrigerant R134a, Int. J. of Refrigeration, vol. 26 (1), pp. 117-128, (2003)

8. B.M. Fronk, S. Garimella, *In-tube condensation of zeotropic fluid mixtures: A review*, Int. Journal of Refrigeration, vol. 36, pp. 534-561, (2013)
9. S. Garimella, *Condensation flow mechanisms in microchannels: basis for pressure drop and heat transfer models,* Heat Trans. Eng., vol. 25(3), pp. 104-116, (2004)
10. G. Ghim, J. Lee, *Condensation heat transfer of low GWP ORC working fluids in a horizontal smooth tube*, Int. J. of Heat and Mass Transfer, vol. 104, pp. 718-728, (2017)
11. W. Kuczyński, A. Denis, *Modeling the instabilities of the condensation process of the R134a and R404A refrigerants in pipe minichannels in the conditions of periodic hydrodynamic disturbances*. Journal of Mechanical and Energy Engineering, Vol. 1(41) No. 2, November, pp. 179 – 188, (2017)
12. W. Kuczynski, H. Charun, *Modeling of a two-phase region length of the condensation of R134a and R404A refrigerants in pipe minichannels with periodic hydrodynamic instabilities*. Heat Transfer Engineering, Vol. 35, Issue 9, pp. 850 – 862,
13. W. Kuczyński, *Modeling of the propagation of a pressure wave during the condensation process of R134a refrigerant in a pipe minichannel under the periodic conditions of hydrodynamic disturbances* Int. J. of Heat and Mass Transfer, vol. 56, Issue 1-2, pp. 715 – 723
14. D. Mikielewicz, R. Andrzejczyk, *Comparative study of flow condensation in conventional and small diameter tubes*, Archives of Thermodynamics, vol. 33 (2), pp. 67-83, (2012)
15. M. Preißinger, J.A.H. Schwöbel, A. Klamt, D. Brüggemann, *Multi-criteria evaluation of several million working fluids for waste heat recovery by means of Organic Rankine Cycle in passenger cars and heavy-duty trucks*, Applied Energy, 206, pp. 887-899, (2017)
16. Montreal Protocol on Substances that Deplete the Ozone Layer. drawn in Montreal on September 16, 1987. Dz. U. of December 23, 1992.
17. S. Rajan, *Comparison of Refrigerants R410A and R404A for use in low temperature applications: a computer model study*, Ph.D. Thesiss, Mechanical Engineering in the Graduate College of the University of Illionois at Urbaba-Champaign, (2011)
18. M.M Shah, *A general correlation for heat transfer during film condensation inside pipes*, Int. J. of Heat and Mass Transfer, vol. 22, pp. 547-556, (1979)
19. L. Silver, *Gas cooling with aqueous condensation*, Trans. Inst. Chem. *Eng.,vol. 25, 1947, pp. 30-42.*
20. C.H. Son, H.K. Oh, *Condensation pressure drop of R22, R134a and R410A in a single circular microtube*, Int. J. of Heat Mass Transfer, vol. 48(8). 1437-1450, (2012)
21. L. Tang, *Empirical study of new refrigerant flow condensation inside horizontal smooth and micro-fin tubes*, Univeity of Maryland at College Park, Ph.D. Thesis, p. 251, (1997)
22. J.R. Thome, J. El Hajal, A. Cavallini: *Condensation in horizontal tubes. New heat transfer model based on flow regimes.* Int. J. of Heat and Mass Transfer, vol. 46, p. 3365 – 3387, (2003),
23. J.R. Thome, *Condensation in plain horizontal tubes: Recent advances in modeling of the transfer to pure fluids and mixture*, J. of the Braz. Soc. Of Mech. Sci.&Eng.,vol. 27, no. 1, pp. 23-30, (2005)
24. The National Archives. Regulation (EC) No 842/2006 of the European Parliament and of the Council of on certain fluorinated greenhouse gases.
25. Ustawa o postępowaniu z substancjami zubożającymi warstwę ozonową. Dz. U. nr 52, poz. 537 z dnia 3 marca 2001 r.
26. Ch. Wang, *System performance of R-1234yf refrigerant in air-conditioning and heat pump system – An overview of current status.* Applied Thermal Engineering, vol. 73, 1 – 9, (2014)
27. J. Wen, X. Gu, S. Wang, Y. Li, J. Tu, *The comparison of condensation heat transfer and frictional pressure drop of R1234ze(E), propane and R134a in a horizontal mini-channel*, Int. J. of Refrigeration, (2018), Accepted Manuscript, DOI 10.1016/j.ijrefrig.2018.03.006
28. H.Y Zhang, J.M. Li, N. Liu, B.X. Wang, *Experimental investigation of heat transfer and pressure drop of R22, R410A and R407C in mini-tubes*, Int. J. of Heat and Mass Transfer. Vol.55. pp. 3522-3532, (2012)
29. Zhu X.R. Zhuang, G. F. Chen, X. Zou, Q. L. Song, M.Q.Gong, *Experimental investigation on flow condensation of methane in a horizontal smooth tube*, Int. J. of Refrigeration, 78, pp.193-214, (2017)
30. www.multimedia.3m.com/mws/media/5698650/3mtm-novectm-649-engeneered-fluid.pdf

Experimental investigation of two-phase thermosyphon heat exchanger charged with acetone

Janusz T. Cieśliński[1,*], and *Maciej* Fabrykiewicz[1]

[1]Gdańsk University of technology, Department of energy and Industrial Apparatus, 80-233 Gdańsk, Poland

Abstract. This paper presents thermal characteristics of prototype of a two-phase thermosyphon heat exchanger (TPTHEx) charged with acetone as a working fluid. The TPTHEx consists of two horizontal cylindrical vessels connected by two risers and a downcomer. Tube bundles placed in the lower and upper cylinders work as an evaporator and a condenser, respectively. The tested TPTHEx operates in a vacuum. Therefore, the working liquid is boiled in temperatures ranging from 33°C to 62°C. The overall heat transfer coefficient (OHTC) of the tested TPTHEx was estimated by the use of the Wilson method and the modified Peclet equation. The results obtained indicate a superiority of water over acetone as a working fluid. Moreover, it was shown that having a lower pressure in the shell-side of TPTHEx results in a higher overall heat transfer coefficient. The Wilson method and the modified Peclet equation predict OHTC with satisfactory agreement.

1 Introduction

A two-phase thermosyphon heat exchanger (TPTHEx) is a special type of heat exchanger that is used in the food industry or chemical installations [1,2]. The mechanisms governing the heat transfer process in such heat exchangers with shell-side boiling and condensation are far from being completely understood [3]. Two-phase thermosyphons can be divided into two main groups: a thermosyphon tube with a countercurrent flow of liquid and vapor [4-6] and a two-phase loop where the evaporator is connected to the condenser by a riser and downcomer [7,8]. Different aspects of the performance of the two-phase thermosyphon have been investigated and the state-of-the-art design is presented in [9]. The purpose of the present study was to examine acetone as a working fluid as well as the influence of the operating pressure inside the shell on the overall performance of a two-phase thermosyphon heat exchanger. Distilled water as a reference working fluid was also used. A Wilson method was used to reduce the experimental data. Moreover, the modified Peclet equation was applied to estimate the overall heat transfer coefficient of the tested TPTHEx.

2 Experimental setup

The test stand consists of three main systems: the prototype TPTHEx, heating water loops and cooling water loops. The test facility is capable of determining the overall heat transfer coefficient of TPTHEx. A schematic diagram of the test stand is shown in Fig. 1. Heating and cooling water loops each contain a centrifugal pump, a flowmeter and a vent tank. A district heating network and a cooling tower are used as a heat source and heat sink, respectively. Heating and cooling water flow rates are controlled by regulating valves and are measured by magnetic flowmeters Danfoss MAG 3100 which are accurate within ±0.25% of the measured value. Pressure inside shell is measured by Trafag NA/8891 pressure transducer accurate to ±0.3% of the measured value. The average temperature of the heating and cooling water at the inlet and outlet of the evaporator and condenser tube bundles of TPTHEx was measured using resistance temperature devices Pt100 with an accuracy of ±0.1K.

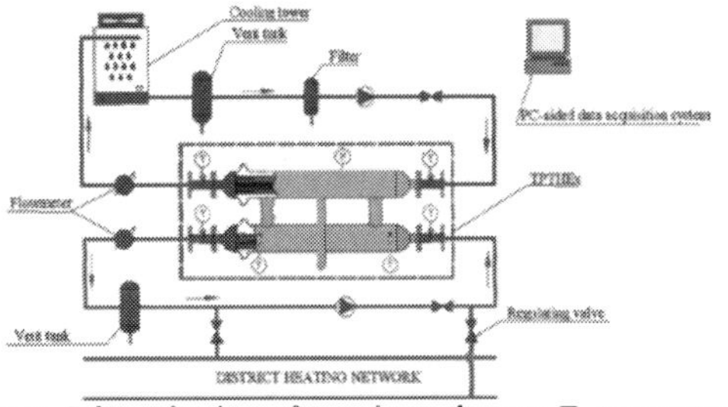

Fig. 1. Schematic view of experimental setup; T- temperature sensors and P- pressure sensors.

2.1 Prototype two-phase thermosyphon heat exchanger

The prototype two-phase thermosyphon is a shell and tube, horizontal heat exchanger in the form of a welded 1.4404 stainless steel construction. The shell consists of two cylindrical vessels, measuring 159 mm in diameter and 1 m in length, which are connected by two risers and

* Corresponding author: jcieslin@pg.gda.pl

a downcomer. The evaporator is designed as a tube bundle consisting of 19 smooth, tubes with a triangular arrangement and a pitch equal to s/d=2.0. The condenser is designed as a tube bundle consisting of 31 smooth stainless steel tubes (outside diameter was 10 mm) with a triangular arrangement and pitch equal to s/d=1.8.

2.2 Experimental procedure

Before the tankage of the TPTHEx with working fluid, an absolute pressure of 5 kPa was created inside the shell. During the tests the absolute pressure inside the shell ranged from 8 to 91 kPa, which corresponds to the operating temperatures from 33.7°C to 61.7°C. The heating as well as cooling water mass flow rates ranged from 0.2 to 3.03 kg/s. The monitoring of the temperature and pressure readings was facilitated by the use of a PC-aided data acquisition system. All data readings have been performed during steady-states. More details concerning the experimental setup and procedure are presented in [10].

3 Data reduction

3.1 Evaporator heat flux

Heat flux transferred in the evaporator was calculated using the formula

$$\dot{q}_e = \frac{\dot{Q}_e}{A_e} \tag{1}$$

where the heat transfer rate $\dot{Q}_e$ was estimated using the measured volume flow rate and the measured hot water temperatures at the inlet $t_{hot,1}$ and outlet $t_{hot,2}$:

$$\dot{Q}_e = \dot{V}_{hot} \cdot \rho_{e,hot} \cdot c_{p_{e,hot}} \cdot (t_{hot,1} - t_{hot,2}) \tag{2}$$

3.2 Overall heat transfer coefficient according to the Wilson method

The overall heat transfer coefficient according to the Wilson method was calculated as:

$$k_{o,W} = \frac{1}{\frac{1}{k_{e,W}} + \frac{1}{k_{c,W}}} \tag{3}$$

The overall heat transfer coefficient of evaporator $k_{e,W}$ was estimated as:

$$k_{e,W} = \frac{1}{\frac{1}{\alpha_{e,iW}} + \frac{\delta_t}{\lambda_t} + \frac{1}{\alpha_{e,oW}}} \tag{4}$$

The overall heat transfer coefficient of condenser $k_{c,W}$ was estimated as:

$$k_{c,W} = \frac{1}{\frac{1}{\alpha_{c,iW}} + \frac{\delta_t}{\lambda_t} + \frac{1}{\alpha_{c,oW}}} \tag{5}$$

Following the Wilson method, the heat transfer coefficient inside the tube can be estimated as :
- for $\dot{m}_{f,hot}$ = const. and $\dot{m}_{f,cold}$ = var.

$$\alpha_{e,i,W} = \frac{1}{C_3 - 2 \cdot \frac{\delta_W}{\lambda_W} - \frac{1}{\alpha_{boi}} - \frac{1}{\alpha_{con}}} \tag{6}$$

$$\alpha_{c,i,W} = \frac{1}{C_2 - w_{cold}^p} \tag{7}$$

- for $\dot{m}_{f,cold}$ = const. and $\dot{m}_{f,hot}$ = var

$$\alpha_{c,i,W} = \frac{1}{C_3 - 2 \cdot \frac{\delta_W}{\lambda_W} - \frac{1}{\alpha_{boi}} - \frac{1}{\alpha_{con}}} \tag{8}$$

$$\alpha_{e,i,W} = \frac{1}{C_2 - w_{hot}^p} \tag{9}$$

C_2 and C_3 are Wilson method constants and were determined by the use of linear regression [11]. The pool boiling heat transfer coefficient α_{boi} was predicted by the use of Cooper correlation [12]. The condensation heat transfer coefficient α_{con} was calculated from the Gutkowski correlation [13].

3.3 Overall heat transfer coefficient according to the modified Peclet equation

Following the approach proposed by Cieśliński and Fiuk in [10] the overall heat transfer coefficient for a two-phase thermosyphon heat exchanger can be calculated as

$$k_{o,P} = \frac{1}{\frac{A_e \cdot \Delta T_e}{\dot{Q}_e} + \frac{A_c \cdot \Delta T_c}{\dot{Q}_c}} \tag{10}$$

where the heat transfer rate transferred in evaporator $\dot{Q}_e$ is given by Eq. 2 and the heat transfer rate transferred in condenser $\dot{Q}_c$ is estimated using the formula:

$$\dot{Q}_c = \dot{V}_{cold} \cdot \rho_{c,cold} \cdot c_{p_{c,cold}} \cdot (t_{cold,1} - t_{cold,2}) \tag{11}$$

where the volume flow rate of cold water $\dot{V}_{cold}$, the temperature of the inlet $t_{cold,1}$ and the temperature of the outlet $t_{cold,2}$ of the condenser were measured values. The mean evaporator fluid-to-wall temperature difference, ΔT_e, reads

$$\Delta T_e = \bar{t}_{hot} - \bar{t}_f \tag{12}$$

The mean condenser fluid-to-wall temperature difference, ΔT_c, reads

$$\Delta T_c = \bar{t}_f - \bar{t}_{cold} \tag{13}$$

https://doi.org/10.1051/e3sconf/20187002003

where $\bar{t}_{hot}$ and $\bar{t}_{cold}$ are the arithmetic means of the measured inlet and outlet temperatures of hot and cold water, respectively, and $\bar{t}_f$ is the arithmetic mean of measured working fluid temperature. The detail of the calculation procedure is given in [10].

4 Results

In order to validate the experimental apparatus and procedure, the present results obtained for distilled water as the working fluid were compared to the data reported by Cieśliński and Fiuk [10] for the same configuration of the TPTHEx (Fig. 2). The overall heat transfer coefficient was determined using a modified Peclet equation (Eq. 10). Keeping in mind the complexity of the heat transfer process occurring in TPTHEx, the agreement between the present results and the data published in [10] is satisfactory. These results confirm that the current facility can be used to provide highly repeatable data.

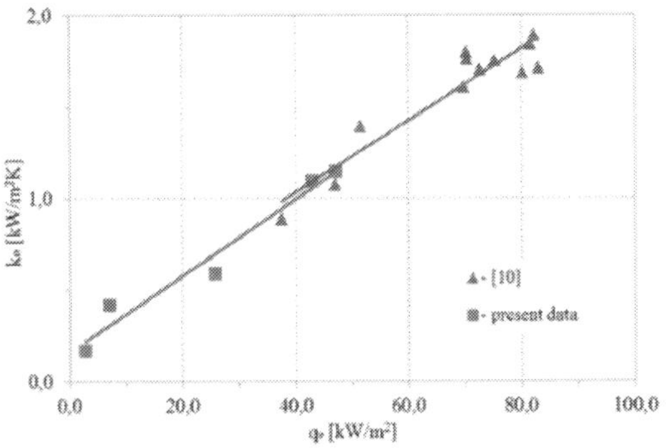

Fig. 2. Overall heat transfer coefficient vs. heat flux of evaporator for distilled water as the working fluid at an operating pressure inside the shell of about 13 kPa.

Figure 3 shows boiling curves of the TPTHEx evaporator for two tested working fluids, i.e. distilled water and acetone. The boiling curve for acetone is shifted right - towards higher wall superheat. It means that there was heat transfer deterioration. The obtained results confirm the tendency observed in the literature that water displays superiority over acetone during pool boiling heat transfer [14].

Figure 4 shows the influence of the operating pressure on TPTHEx evaporator performance with acetone as a working fluid. A decrease of the absolute pressure inside a shell from about 75 to 52 kPa results in heat transfer deterioration – the boiling curve shifts left towards higher wall superheats. Again, the obtained results confirm the tendency observed in the literature, that a decrease in the operating pressure results in heat transfer deterioration during pool boiling [15,16].

Figure 5 shows the overall heat transfer coefficient against the evaporator heat flux for water and acetone estimated by two methods, i.e. Wilson approach and modified Peclet equation. It is important that the Wilson method underestimated the OHTC for water – the difference is about 25% and overestimated the OHTC for acetone - the difference is about 33%.

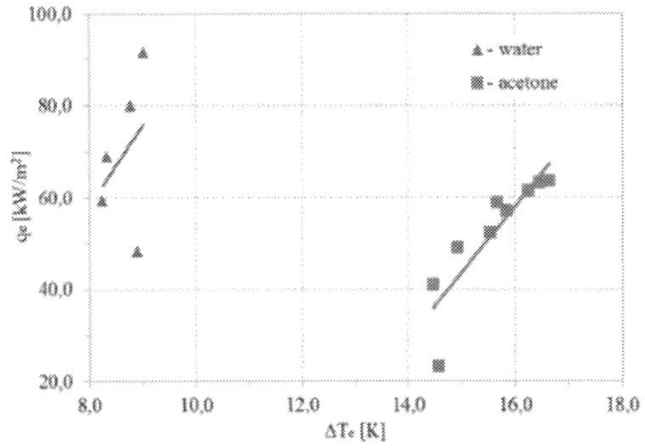

Fig. 3. Boiling curves of the TPTHEx evaporator for water and acetone.

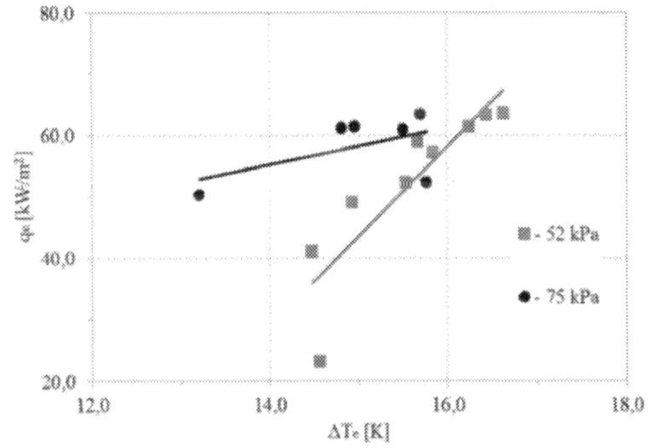

Fig. 4. Boiling curves of the TPTHEx evaporator for different operating pressure inside a shell.

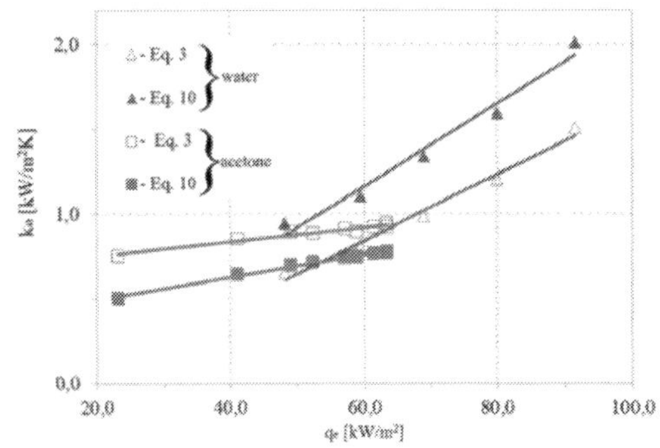

Fig. 5. Overall heat transfer coefficient vs. evaporator heat flux.

5 Conclusions

- It was shown that water displays superiority over acetone as a working fluid in the tested two-phase thermosyphon heat exchanger.
- The data obtained shows that an increase in the operating pressure results in performance improvement of the evaporator of the tested two-phase thermosyphon heat exchanger.

• The Wilson method and the approach proposed by Cieśliński and Fiuk, which is based on the modified Peclet equation, predict the overall heat transfer coefficient of the tested two-phase thermosyphon heat exchanger with satisfactory agreement.

References

1. L.L. Vasiliev, Applied Thermal Engineering, **25**, 1-19 (2005)
2. Ł. Vasiljew, Ł. Grakowicz, D.K. Chrustaljew, Teplowyje *truby w sistemach s wozobnowliajemnymi istocznikami energii, Nauka i Technika*, Minsk (1988)
3. L.S. Pioro, I.L. Pioro, *Industrial Two-Phase Thermosyphons, Begell House Inc.*, New York, Wallingford (UK) (1997)
4. S. Khandekar, M. Groll, *Insights into the performance models of closed loop pulsating heat pipes and some design hints, Proc. 18th National & 7th ISHMT-ASME Heat and Mass Transfer Conference*, IIT Guwahati, India, Paper No: G-440 (2006)
5. M. Bezrodny, J. Goscik, Refrigeration Air-Conditioning Eng, **1**, 5–10 (1997) (in Polish)
6. D. Chisholm, *Heat Exchangers Design Handbook, Part 3, Thermal and Hydraulic Design of Heat Exchangers, Begell House Inc.*, New York, Wallingford (UK), Chapter 3.10.1 (1998)
7. N. Gavotti, F. Polášek, *Proc. of Eurotherm Seminar No. 6, Single and Two-Phase Natural Circulation*, Genoa, Italy, 229–238 (1999)
8. R. Khodabandeh, B. Palm, Proc. of the 5th ExHFT World Conference, Thessaloniki, Greece, 191–196 (2001)
9. J.T. Cieśliński, Archives of Thermodynamics, **37**, 23–40 (2016)
10. J. T. Cieśliński, A. Fiuk, Appl. Therm. Eng, **51**, 112-118 (2013)
11. J.T. Cieśliński, J. Rubalewski, Technika Chłodnicza i Klimatyzacyjna, **195**, 206–211 (2012) (in Polish)
12. M.G. Cooper, Adv. Heat Transfer, **16**, 157-239 (1984)
13. K. Gutkowski, Refrigeration - Selected Numerical Problems, WNT Warsaw (1972) (in Polish)
14. S. Fazel, M. Jamialahmadi, A. Safekordi, Iran. J. Chem. Chem. Eng., **27**, 135-150 (2008)
15. J.T. Cieśliński, K. Krasowski, J Enhanced Heat Transfer, **20**, 165-177 (2013)
16. J.T. Cieśliński, T. Kaczmarczyk, Chemical and Process Engineering, **32**, 321-332 (2011)

Nomenclature

A	– heat transfer area, m^2
C_2, C_3	– constants of Wilson method
c_p	– specific heat, J/(kgK)
d	– diameter, m
k	– overall heat transfer coefficient, W/(m^2K)
$\dot{m}$	– mass flow rate, kg/ s
$\dot{Q}$	– heat transfer rate, W
$\dot{q}$	– heat flux, W/m^2
p	– pressure, Pa
t	– temperature, K
ΔT	– wall-to-fluid temperature difference, K
w	– velocity, m/s
$\dot{V}$	– volume flow rate, m^3/s

Greek symbols

α	– heat transfer coefficient, W/(m^2K)
δ	– thickness, m
λ	– thermal conductivity, W/(mK)
ρ	– density, kg/m^3

Subscripts

boi	– boiling
c	– condenser
cold	– cold water
con	– condensation
e	– evaporator
f	– working fluid
hot	– hot water
i	– inside
o	– outside
P	– Peclet equation
t	– tube
w	– wall
W	– Wilson method
1	– inlet
2	– outlet

Experimental investigation of forced convection of water/EG-Al_2O_3 nanofluids inside horizontal tube

Janusz T. Cieśliński[1], and *Przemysław* Kozak[1,*]

[1]Gdansk Uniwersity of Technology, Narutowicza 11/12, 80-233 Gdańsk, Poland

Abstract. In this paper, forced convection of water/EG-Al_2O_3 nanofluids inside horizontal stainless steel tube is studied experimentally. As base fluid served water/EG mixture of two ratios (60:50 and 50:50). Nanoparticle mass concentrations was 0.1% or and 1%. Transition and turbulent flow regimes were tested. Average heat transfer coefficient and pressure drop values were determined for nanofluids and compared to base fluids. Deterioration of heat transfer for all tested nanofluids has been recorded compared to base fluid. Negligible increase of pressure drop for nanofluids compared to base fluids has been observed.

1 Introduction

Simultaneously with research aimed at enchancing heat transfer by modifying the construction of a thermal apparatus (change of surface roughness, surface development [1], use of flow turbulence elements) or using additional elements in the installation causing the liquid to pulsate or vibrate, numerous works are carried out to change thermophysical properties of conventional working fluids or proposing new products that can replace traditional liquids. In last two decades strong interest in fluids with suspended nanosize solid particles (called nanofluids) results in large number of experimental and numerical studies. So far carried out experiments dealing with forced convection of nanofluids in both laminar and turbulent conditions and conducted under the same thermal-hydraulic conditions, i.e., for the same Reynolds number and temperature of the process showed that the degree of intensification of heat transfer is higher than can be expected from an increase in thermal conductivity [2]. Some of the reserchears shown that adding Al_2O_3 or TiO_2 nanoparticles to distilled water have considerable impact on thermal conductivity of base fluid [3]. Azmi et al. [4] show that for given operating temperatures heat transfer coefficients can be almost similar for nanofluids with different thermal conductivity. Changes in fluid internal shear stress and weakening of the inter-particle and inter-molecular adhesion forces with decreasing of the nanofluid viscosity can be the main reason of this behavior. Higher thermal conductivity of nanofluids compared to the base fluid can be caused by several mechanisms proposed by Khurana et al. [5]. These are: Brownian motion (random motion of the nanoparticles), interfacial layer (liquid layer close to nanoparticle), aggregation (formation of a linear assembly of nanoparticle chain), thermophoresis (nanoparticle motion inducted by thermal gradients) and intensification of turbulence (small solid particles in nanofluid develop turbulent eddies). All of this facts lead to the conclusion that the thermal conductivity is not the only reason for the increase of heat transfer coefficient in forced convection conditions. Liquid properties, such as viscosity, density and specific heat, also have impact on heat transfer coefficient during forced convection. From the other hand, geometry of the system, i.e. the diameter and the length of a tube, also change conditions of the heat transfer process.

Many groups of investigators conducted experiments to determine heat transfer coefficients for different types of nanofluids. Noghrehabadi and Pourrajab [6] studied heat transfer inside circular tubes under uniform constant heat flux boundary condition. Three volume concentrations, i.e. 0.1%, 0.3% and 0.9% of water/γ-Al_2O_3 were tested under laminar flow regime. Comparison with distilled water shows that average heat transfer coefficient increases of about 16.8% at 0.9% volume concentration and Re=2070. Enhancement of heat exchange was particularly significant in the entrance region of the test section and decreases with axial distance. Sahin et al. [7] collate experimental data for water/CuO nanofluid tested inside horizontal aluminium pipe. They concluded that the particle volume concentrations higher than 1% vol. were not appropriate with respect to the heat transfer performance of the water/CuO nanofluid. The highest heat transfer enhancement was achieved for 0,5% volume concentration and Re= 16.000. At higher than 1% particle volume fraction, the Nusselt number decreases with increasing particles loading. Another, widely tested type of nanofluid is liquid based on water/EG mixture, as an often used working fluid for car engine cooling, solar collector systems etc. Azmi et al. [8] investigated forced convection of Al_2O_3 nanofluids in 60:40, 50:50, 40:60 water and ethylene glicol mixture ratio. Reynolds number was varied between 3000 to 25000. Maximum

* Corresponding author: przkozak@pg.edu.pl

heat transfer enhancement was noticed at a working temperature of 70^0C for 60:40 water/EG base fluid and 1% volume concentration of nanoparticles, and riched 24,6%. Autors also mentioned that for all mixtures friction factor increases slightly as particle concentration increased. Kulkarni et al. [9] showed that not only particle loading, but also size of this particles have substantial influence on heat transfer process. They tested nanofluids consisted of SiO_2 nanoparticles and water/ethylene mixture (40:60 % by weight). Nanoparticle of 20 nm, 50 nm and 100 nm diameters were used. Published results shown that the heat transfer increases with the particle size increase. According to Kulkarni et al. larger diameter particles results in lower viscosity of nanofluids, what cause that the Reynolds number increases. What is more, for larger particles also particle Peclet number increases. Combination of this two effects is a reason for the increasing of the heat transfer coefficient with increasing of the particle size.

Also more complicated geometries of tubes and inserts were tested with the different types of nanofluids. Mohhamadiun et al. [10] combined in their study three heat transfer enhancing methods, i.e. corrugation, inserts and nanofluids. They investigated fluid flow of the EG/ Al_2O_3 nanofluid under turbulent flow conditions. Nanofluids of 0.5, 1 and 1.5% volume concentrations of Al_2O_3 were prepared and tested with three twist ratios of twisted tape. Maximum intensification of heat transfer was found with the use of 0.5% particle loading and lowest twisted tape ratio and was defined as maximum thermal performance factor equal to 4,2. Some of researchers go further and conducted experiments to identify application possibilities of nanofluids. Cieśliński et al. [11] used water-Al_2O_3 nanofluids instead of traditional working fluid in flat solar collector. The tests were conducted for three mass concentration of nanoparticles: 0.1%, 0.2% and 0.4%, for two values of radiation: 1050 W/m² and 400 W/m², and volume flow rates respectively 60 l/h, 120 l/h and 170 l/h. The efficiency was calculated according to the recommendations of the Polish Standard. The results obtained show that the application of nanofluids instead of pure water improves efficiency of the tested flat plate solar collector for given range of thermal radiation. Independent of the volume flow rate of the nanofluid, the maximum efficiency was recorded for nanoparticle mass concentration equal to 0.2%. Kaya et al. [12] also carried out the experiment for solar collector. In this case, evacuated U-tube solar collector was used. Water/EG base fluid in 50:50 proportion was used to prepare nanofluids with volume concentrations of ZnO nanoparticles of 1, 2, 3, 4%. Maximum growth of solar collector efficiency was noticed for 3,0% nanoparticle loading and working fluid mass flow rate of 0,045 kg/s and it was calculated as almost 27% increase comparing to pure base fluid. Sidik et al. [13] presented comprehensive review on the application of nanofluids for cooling system of the traditional car engines. From rich database they come to the conclusion that heat transfer coefficient in coolant radiator can be improved up to 50% compared to the base coolant. From the other hand, problem is that with adding nanoparticles pressure drop is also increasing, so the optimum performance of cooling system can be reached for lower concentration of nanoparticles (<1% by vol).

This paper shows an experimental study on heat transfer potential of water-ethylene glycol/Al_2O_3 nanofluids during forced convection inside horizontal stainless steel tube.

2 Experimental setup

Scheme of the experimental test loop used in this experiment is shown in Fig. 1.

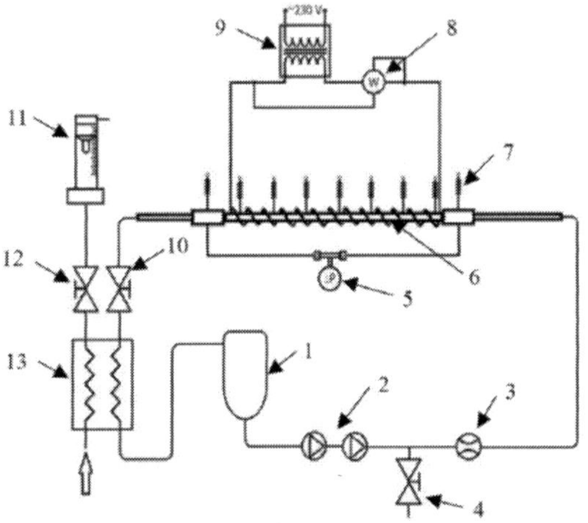

Fig. 1. Scheme of the test rig: 1 – reservoir tank, 2 – rotary pumps, 3 – flow meter, 4 – drain valve, 5 – differential pressure transducer, 6 – test tube, 7 – temperature sensors, 8 – wattmeter, 9 – autotransformer, 10 – valve, 11 – rotameter, 12 – cooling water control valve, 13 – plate heat exchanger.

Nanofluid circulates in the test loop by use of the group of rotary pumps (2). Reservoir tank (1) is made from glass in order to observe the potential settlement of the nanoparticles during the test. To measure the mass flow rate of the nanofluid magnetic flow meter (3) is used. Control valve (10) and different pump operation sequences are used to change mass flow rate. Test tube (6) with 8 mm internal diameter, 1 mm wall thickness and 2000 mm length is made from stainless steel. Tube is connected to the autotransformer (9) and works as a direct heater. Heat flux is estimated due to the measurement of electrical power by wattmeter (8) assuming no heat loss to the environment. The hydrodynamic entry section is long enough to accomplish fully developed region at the entrance of the heat transfer test section. Plastic (PTFE) fittings are attached at both ends of the test section to thermally isolate the connection. Two types of temperature sensors (7) are used: two thermocouples of J-type to measure the nanofluid temperature at the inlet and the outlet of the test section and six Pt100 sensors (mounted to the surface of the tube) are used to measure the axial wall temperature distribution of the test tube. The inside wall temperature is calculated assuming radial heat conduction within the tube wall. Cooling section is responsible for keeping temperature of tested fluid on required value. Tap water flowing through the plate heat exchanger (11) cools hot nanofluid. Flow rate of cooling water is controlled by valve (12) and measured by rotameter (13).

3 Preparation and characterization of the tested nanofluids

In the present study Al_2O_3 nanoparticles were used while mixture of distilled, deionized water and ethylene glycol was applied as a base fluid. Two volume proportions of this mixture (W:EG) were used, i.e. 60:40 and 50:50. Nanofluids with 0,1 and 1% nanoparticle mass concentrations were tested during the experiment. Alumina (Al_2O_3) nanoparticles have mean diameter estimated to be lover than 50 nm according to the manufacturer (Sigma-Aldrich Co.). The tested nanofluids were prepared by use of the two-step method. Nanoparticles of the required amount and base liquid were mixed together in ultrasonic bath. There wasn't used any dispersant to stabilize the suspension.

4 Data reduction and uncertainty estimation

The Nusselt number is calculated as:

$$Nu = \overline{\alpha} d_i / \lambda_{nf} \quad (1)$$

Thermal conductivity of the tested nanofluids λ_{nf} is calculated from the Maxwell model [14]:

$$\lambda_{nf} = \lambda_f + \frac{3(\lambda_n - \lambda_f)\Phi}{\left(\frac{\lambda_n}{\lambda_f} + 2\right) - \left(\frac{\lambda_n}{\lambda_f} - 1\right)\Phi} \quad (2)$$

where λ_n is taken from the producer of nanoparticles (Sigma Aldrich Ltd.) and λ_f is mesured by KD2 (Decagon Devices Inc.) thermal properties analizer.
Average heat transfer coefficient is calculated as:

$$\overline{\alpha} = \dot{Q}/A_i(\overline{T}_w - \overline{T}_f) = P/A_i(\overline{T}_w - \overline{T}_f) \quad (3)$$

where heat transfer area is calculated as:

$$A_i = \pi d_i L \quad (4)$$

and average nanofluid temperature $\overline{T}_f$ reads:

$$\overline{T}_f = (T_{in} + T_{out})/2 \quad (5)$$

Average inside wall temperature of the tube $\overline{T}_w$ is calculated as [15]:

$$\overline{T}_w = \overline{T}_o - \frac{P}{2\pi\lambda_t L} \ln(d_o/d_i) \quad (6)$$

where $\overline{T}_o$ is calculated as an arithmetic mean of measured outside wall temperatures in six positions along the tube - Fig.1:

$$\overline{T}_o = \sum_{j=1}^{j=6} T_{oj}/6 \quad (7)$$

Reynolds number is calculated as:

$$Re = w d_i / \nu_{nf} \quad (8)$$

where fluid velocity w is calculated as:

$$w = \dot{V}/A_{cs} \quad (9)$$

Kinematic viscosity of the tested nanofluids is calculated as:

$$\nu_{nf} = \mu_{nf}/\rho_{nf} \quad (10)$$

Dynamic viscosity of the tested nanofluids is calculated from Brinkman formula [16]:

$$\mu_{nf} = \frac{\mu_f}{(1-\Phi)^{2.5}} \quad (11)$$

where μ_f is mesured by Rheotest LK 2.2 (Medingen GmbH) viscometer.
Density of the nanofluids is calculated as [17]:

$$\rho_{nf} = (1-\Phi)\cdot\rho_f + \Phi\cdot\rho_n \quad (12)$$

The uncertainties of the measured and calculated parameters are estimated by mean-square method. Because heat flux is calculated from the formula

$$q = P / \pi d_i L \quad (13)$$

the experimental uncertainty of the heat flux is estimated as follows:

$$\Delta q = \sqrt{\left(\frac{\partial q}{\partial P}\cdot\Delta P\right)^2 + \left(\frac{\partial q}{\partial d_i}\cdot\Delta d_i\right)^2 + \left(\frac{\partial q}{\partial L}\cdot\Delta L\right)^2} \quad (14)$$

where the absolute measurement errors of the electrical power ΔP, inside tube diameter Δd_i and active length of the tube ΔL are 10 W, 0.02 mm, and 0.2 mm, respectively. So, the maximum overall experimental limits of error for heat flux extended from ± 1.4% for maximum heat flux up to ±3% for minimum heat flux.

Because average heat transfer coefficient is calculated from the formula

$$\overline{\alpha} = q/(\overline{T}_w - \overline{T}_f) = q/\Delta T \quad (15)$$

the experimental uncertainty for the average heat transfer coefficient is calculated as:

$$\Delta\overline{\alpha} = \sqrt{\left(\frac{\partial\overline{\alpha}}{\partial q}\Delta q\right)^2 + \left(\frac{\partial\overline{\alpha}}{\partial\Delta T}\delta\Delta T\right)^2} \quad (16)$$

where the absolute measurement error of the temperature difference, $\delta\Delta T$, estimated from the systematic error analysis equals $\pm$ 1 K. The maximum error for average heat transfer coefficient was estimated to $\pm$7.0%.

5 Results

Figure 2 shows average Nusselt number versus Reynolds number for tested base fluids and prepared nanofluids for selected heat flux equal to 20 kW/m^2. It is seen that addition of nanoparticles results in Nusselt number decrease for given Reynolds number. Nusselt number (heat transfer coefficient) decreases with the nanoparticle loading increase with maximum decrease of about 25% for water:EG ratio equal to 50:50 and 1% of Al_2O_3 nanoparticle concentration. Pressure drop was presented in Fig. 3. Negligible pressure drop for all tested nanofluids has been recorded.

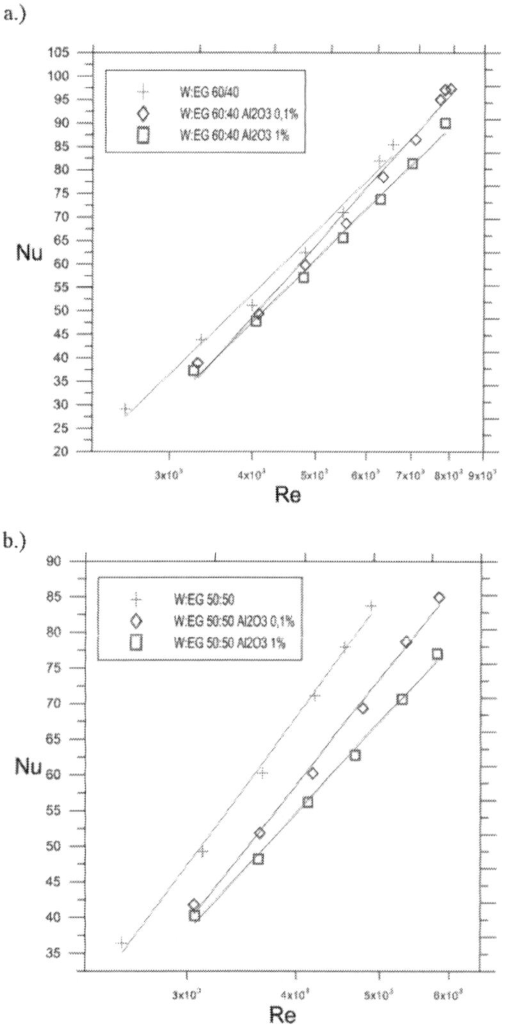

Fig. 2. Average Nusselt number for water/EG-Al_2O_3 nanofluids: a) water:EG (60:40), b) water:EG (50:50).

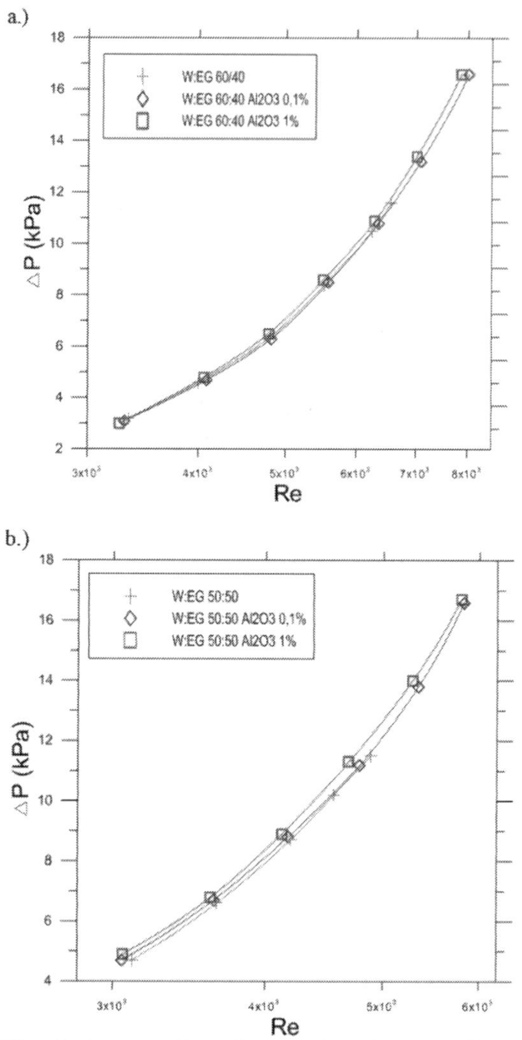

Fig. 3. Pressure drop for water/EG-Al_2O_3 nanofluids: a) water:EG (60:40), b) water:EG (50:50).

6 Conclusions

Obtained results revealed:

- degradation of heat transfer for water/EG-Al_2O_3 nanofluids with 0,1% and 1% nanoparticle mass concentrations within tested range of Reynolds number compared to base fluids,
- negligible increase of pressure drop for all tested nanofluids compared to base fluids.

References

1. R. Andrzejczyk, T. Muszyński, Arch. of Therm. **37**, 137-159 (2016)
2. R. Saidur, K.Y. Leong, H.A. Mohammad, Ren. & Sust. En. Rev. **15**, 1646–1668 (2011)

3. J.T. Cieśliński, K. Krygier, S. Smoleń, Of. Wyd. Pol. Rzeszowskiej, *Termodyn. i wym. ciepła w bad. proc. cieplno-przepł.*, 87-108 2014
4. W.H Azmi, A.K Hamid, N.A. Usri, R. Mamat, M.S. Mohamad, Intern. Comm. in Heat & Mass Transf. **76**, 24-32 (2016)
5. D. Khurana, R. Choudhary, S. Subudhi, Heat Mass Transf., (2016)
6. A. Noghrehabadi, R. Pourrajab, J. of Mech. Sc. & Tech. **30**, 943-952 (2016)
7. B. Sahin, E. Manay, E. Akyurek, J. of Nanomat., (2015)
8. W.H. Azmi, N.A. Usri, R. Mamat, K.V. Sharma, M.M. Noor, App. Therm. Eng. **112**, 707-719 (2017)
9. D.P. Kulkarni, P.K. Namburu, H. Ed Bargar, D.K. Das, Heat Transf. Eng. **29**, 1027-1035 (2008)
10. H. Mohammadium, M. Mohammadium, M. Hazbehian, H. Maddah, Heat Mass Transf. **52**, 141-151 (2016)
11. J.T. Cieśliński, B. Dawidowicz, A. Popakul, App. Mech. & Mat. **831**, 181-187 (2016)
12. H. Kaya, K. Arslan, N. Eltugral, Ren. En. **122**, 329-338 (2018)
13. A. Sidik, M. Yazid, R. Mamat, Ren. & Sust. En. Rev. **75**, 137-144 (2017)
14. W. Duangthongsuk, S. Wongwises, Exp. Therm. and Fluid Sc. **33**, 706–714 (2009)
15. W. Pudlik, Heat transfer and heat exchangers, D.L. of GUT, (2012)
16. D. Khurana, R. Choudhary, S. Subudhi, Heat & Mass Transf., (2016)
17. A.T. Utomo, H. Poth, P.T. Robbins, A.W. Pacek, Intern. J. of Heat & Mass Transf. **55**, 7772-7781 (2012)

Nomenclature

A - area, m^2,
A_t - heat transfer area, m^2,
d - diameter, m,
ED - ethylene glycol,
L - active tube length, m,
P - electrical power, W,
$\dot{Q}$ - heat flow rate, W,
T - temperature, °C,
$\overline{T}$ - average temperature, °C,
$\dot{V}$ - volume flow rate, m^3/s,
w - fluid velocity, m/s,

Greek symbols

$\overline{\alpha}$ - average heat transfer coefficient, W/(m^{2}K),
λ - thermal conductivity, W/(m·K),
μ - dynamic viscosity mPa·s
ν - kinematic viscosity, m^2/s,
ρ - density, kg/m^3,
Φ - nanoparticle volumetric fraction,

Subscripts

cs - cross section
f - fluid
i - inside
in - inlet
n - nanoparticle
nf - nanofluid
o - outside
out - outlet
t - tube
w - inside wall

Experimental investigation on influence of microcapsules with PCM on propylene glycol rheological properties

Krzysztof Dutkowski[1], and *Jacek Jan* Fiuk[1,*]

[1]Koszalin University of Technology, Faculty of Mechanical Engineering, Raclawicka 15-17, 75-620 Koszalin, Poland

Abstract. This paper details an experimental study that was performed to investigate rheological properties of microencapsulated phase change slurry - mPCM (Micronal® DS 5039 X and ERGOLID EKO®- an aqueous solution of propylene glycol). Seven samples of mPCM slurry were prepared with different mass ratio of the Micronal® to the ERGOLID EKO®: 30:70, 40:60, 50:50, 60:40, 70:30, 80:20 and 90:10. The apparent viscosity-shear rate curves were obtained for spindle speed from 0.01 to 100 rpm (shear rate 0.0132 to 132.00 s^{-1} respectively). The steady state measurement of viscosity was carried out when the slurry reached constant temperatures, namely: 10.0; 15.0, 17.5, 20.0, 22.5, 25.0, 27.5, 30.0, 40.0 and 50.0^{0}C. The apparent viscosity of slurries increases with the mPCM concentration in dispersion rises. Only the sample of 30% Micronal® may be considered as a Newtonian fluid within the test range. Increasing the shear rate ultimately causes apparent viscosity to decrease down to the Newtonian plateau, where it seems to be constant. The variation of the viscosity as a function of temperature and microcapsules concentration was accurately represented by a modified Vogel-Tamman-Fulcher equation at a mean absolute error (MAE) of 10.76%.

1 Introduction

Technological advancements have made mass production of capsules with a diameter of the order of micrometers available today. Hence, flow is possible for a whole dispersion of base fluid and microcapsules filled with phase-changing material. Intensive research has been undertaken to determine the thermal properties and flow characteristics of microencapsulated Phase Change Material Slurries (mPCM slurry), which are classified as Latent Functional Thermal Fluid (LFTF)[1]. LTFT is a two-phase working fluid used in heat exchange systems and is characterized by a greater heat transfer capacity (by latent heat in the phase transition temperature range) compared to conventional single-phase fluids.

MPCM slurries offer many advantages and can be used either as thermal storage materials or heat transfer fluids due to: 1) their high storage capacity during phase change; 2) the possibility of using the same medium either to transport or to store energy, as these slurries are pumpable (thus reducing heat transfer losses); 3) heat transfer at an approximately constant temperature; 4) a high heat transfer rate due to the elevated ratio surface/volume; 5) a lower pumping power, as a consequence of the reduction in mass flow rate due to the higher heat capacity; 6) a better heat exchange than conventional heat transfer fluids, due to the decrease in fluid temperature as a consequence of the higher heat capacity. Furthermore, these novel fluids have a more advantageous thermal energy storage density than conventional systems of sensible heat storage in water, and can compete with macroencapsulated PCM tanks. Moreover, the response time may be shorter using these PCM emulsions or mPCM slurries as a storage material than with macro- encapsulated PCM. The tanks will be simpler as there is no need to macroencapsulate, thus conventional tanks can be used [2].

Viscosity plays a significant role in determining the pressure drop and characteristics of fluid flow described by Reynolds and Grashoff number (higher viscosity leads to the lower turbulence of the slurry, so the heat transfer coefficient may be lowered).

2 Previous study on mPCM slurry viscosity

Zhang and Zao [3] experimentally investigated the thermal and rheological properties of a series of prepared mPCM slurries fabricated by dispersing micro-encapsulated PCM into water with an appropriate amount of surfactant. The mass ratio of mPCM to water and surfactant was 10:90:1, 25:75:1, 35:65:1, respectively. The mPCM slurry can be considered as Newtonian fluid when the shear rate is higher than 200 s^{-1} and the PCM microcapsules' mass fraction lower than 35 wt. %. The viscosity was higher for larger particle slurries.

Delgado et al. [2] experimentally investigated the microencapsulated PCM slurry with three different PCM mass fractions (14, 20 and 30%). The studied mPCM slurry consisted of microcapsules of paraffin coated by a

* Corresponding author: jacek.fiuk@tu.koszalin.pl

polymer and dispersed in water through detergents. These PCM microcapsules had a diameter range from 1 to 20 μm. The viscosity - shear rate curves using a control stress rheometer were obtained through a shear sweep from 0.001 to 1000 s^{-1} and at a constant temperature (T = 27°C). The equation that best predicts the shape of the flow curve for the three mPCM slurries is the non-Newtonian model, namely the Carreau model.

Allouche et al. [4] investigated the 45% w/w aqueous dispersion of RT15 paraffin microcapsules supplied by CIBA chemicals (UK) while a rheological rotational rheometer (UM/MC 100 PHYSICA) was used. Experiments were performed applying constant shear rates from 10 to 500 s^{-1}. Tests were repeated in a temperature range of 7–20°C. The rheological study revealed a non-Newtonian fluid behaviour of the PCM for different temperature settings.

Kong et al. [5] used mPCM slurries produced by Thies Technology Inc. as heat transfer fluids. The phase change material (PCM) used in the study was butyl stearate. Each microcapsule was made by encapsulating the PCM using polyurea as a shell material, which made up about 35% of the total volume of each microcapsule. The mPCM particle size range used in the experiments was between 5 and 10 μm in diameter. The apparent viscosity of mPCM slurries was measured using a coaxial rotating drum viscometer from Brookfield Engineering Laboratories Inc. Microencapsulated PCM slurries with 2.1, 5.9, 8.3 and 10.9 wt. %, respectively, were tested in a temperature range of 5–35°C. The viscosity was measured at three different rotational speeds of 50, 60 and 100 rpm. The experimental results showed that the viscosity of mPCM slurry is independent of the shear rate and behaves as a Newtonian fluid.

In a paper by Wang et al. [6] experimental research on the natural convective heat transfer with microencapsulated phase change material suspensions is described. The microencapsulated PCM slurry was formulated with PCM powder with N-hexadecane as the core material and a diameter range of 10 ÷ 40 μm. The water–propanol mixture was the base fluid along with the addition of dispersants. The viscosity of 10 ÷ 30 wt. % mPCM suspensions was measured with the TADHR-2 rheometer at a room temperature of 20 °C and shear rate of up to 100 s^{-1}. The 10 wt. % and 20 wt. % mPCM slurry behave as Newtonian fluids while the 30 wt. % mPCM slurry shows the shear thinning non-Newtonian characteristic.

Fu et al. [7] conducted experiments on thermo-physical properties of n-tetradecane&polystyrene-silica composite nanoencapsulated phase change material slurry. The mean particle size of Tet@PS-SiO2 mPCM was 151.3 nm. The mPCM water slurry contained 5% PCM and 15 wt. % amount of antifreeze. The viscosity of the slurry was investigated by a Brookfield DV-II + Pro rotation viscometer with S61 rotor type at 100 rpm in a temperature range from 5 to 25°C. It was found that the viscosities of all slurries were slightly higher than that of the base fluid.

Generally, the mPCM slurry is considered as a Newtonian fluid when the mass fraction of mPCM is up to 30% [4–6,8,9] or when the shear rate is large enough [3]. After adding additives such as surfactant agents, the mPCM slurry stopped behaving as a non-Newtonian fluid and started behaving as a Newtonian fluid [10,11]. The viscosities of slurries rose with the increase of the mass fraction of microcapsules up to several dozen times [5–8,12–15].

The purpose of this study is to investigate the rheological properties of mPCM slurry. In this paper, an mPCM slurry made of a new product, the Micronal® DS 5039 X and ERGOLID EKO® (an aqueous solution of propylene glycol) - as a base fluid, with seven different mass fractions was obtained. The study examines the effect of Micronal® DS 5039 X mass fraction on slurry viscosity along with the effect of temperature increase (10-50°C) on slurry viscosity. Detailed steady state condition characteristics and a predictive model for this property were obtained.

3 Experimental setup and procedure

3.1 Slurry properties

Micronal® DS 5039 X (BASF) is an aqueous dispersion of microencapsulated paraffin mixture with highly cross-linked polymethyl methacrylate (PMMA) shell. According to the manufacturer the product contains: 1,2-benzisothiazol-3(2H)-one (22 ppm); 2-methyl-2H-isothiazol-3-one (22 ppm) and a (3:1) mixture of 5-chloro-2-methyl-4-isothiazolin-3-one and 2-methyl-4-isothiazolin-3-one (13.5 ppm); and a range of particle size of 1 - 5 μm. Micronal® DS 5039 X contains 42.0-44.0% of solids in water. The main peak melting temperature of the product is around 25°C, the enthalpy of fusion (liquid) 41 kJ/kg, the latent heat (solid content) 100 kJ/kg, and the total thermal storage capacity (solid content, integral 10–30°C) 142 kJ/kg. The advantages of the special microencapsulation process that is used to manufacture Micronal® DS 5039 X are that the product is free of formaldehyde, has high mechanical stability and resistance to high temperatures, and that the polymer capsule is able to resist damage. Mechanical stability should be understood as consistency stability, colour and uniformity of the concentrate structure. A mechanically stable product should maintain the abovementioned features even after prolonged storage (several months). Micronal® DS 5039 X can be employed passively or in combination with an active cooling system. Some other Micronal® products (e.g. Micronal® DS 5001) has been incorporated in mortar, concrete or plasterboard as a passive system, while it is also used in active systems such as slurries [16].

Seven samples of mPCM slurries by mechanical mixing with the following mass ratios of Micronal® DS 5039 X to ERGOLID EKO®: 30:70, 40:60, 50:50, 60:40, 70:30, 80:20 and 90:10, were prepared respectively. The ERGOLID EKO®, applied as a base fluid, is a recommended in climatic conditions of Poland commercial product for usage in solar collector installations. The ERGOLID EKO® contents of 37%

pure propylene glycol and 63% of water with minor fraction of additives. Because Micronal® DS 5039 X contains 42.0-44.0% of solids in water, the actual mass fraction of microcapsules in the tested suspension was respectively: 12.9, 17.2, 21.5, 25.8, 30.1, 34.4 and 38.7%.

3.2 Experimental set-up

The viscosity of the slurry was investigated by a Brookfield DV-II+ Pro rotation viscometer (Brookfield Co., Ltd., USA) with a small sample adapter, consisting of a cylindrical sample chamber and spindle coupled with a PolyScience circulating water bath for temperature control (Fig. 1). The small sample adapter provides a defined geometry system for accurate viscosity measurements of small sample volumes at precise shear rates. As the SC4-18 spindle type was used for this viscometer to execute experiments, the sample injected volume was 6.7 ml. The small sample adapter's rheologically correct cylindrical geometry provides extremely accurate viscosity measurements and shear rate determinations. The Brookfield viscometer measures viscosity by measuring the force required to rotate a spindle in a fluid at a specified temperature in relation to its shear rate. On the DV+II viscometer type, the viscosity can be read directly or by using the Rheocalc program installed on a computer connected to the viscometer.

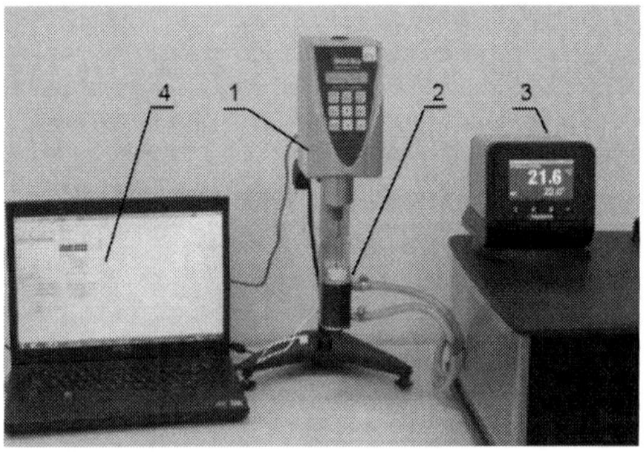

Fig. 1. Experimental set-up: 1 - Brookfield DV-II + Pro viscometer, 2 – small sample adapter, 3 – PolyScience circulating water bath, 4 – computer with the Rheocalc program.

The sample chamber fits into a water jacket so that precise temperature control can be achieved when a circulating water bath is used. The stirring action of the rotating spindle, plus the small sample volume, helps one to keep the temperature gradient across the sample to a minimum. A direct readout of the sample temperature is provided using sample chambers with an embedded RTD sensor connected to the viscometer.

3.3 Test conditions and procedure

In this experiment, rotational tests were carried out. These tests entail applying torque (or stress) and measuring the strain, in order to obtain viscosity values. The apparent viscosity-shear rate curves were obtained through a spindle speed from 0.01 to 100 rpm (shear rate 0.0132 to 132.00 s^{-1} respectively). The measurement of viscosity was carried out when the slurry reached a constant temperature, namely: 10.0; 15.0, 17.5, 20.0, 22.5, 25.0, 27.5, 30.0, 40.0 and 50.0°C. At a constant temperature the value of viscosity at eighteen spindle speeds was collected by the Rheocalc program after five minutes of waiting. As a tests for each temperature were repeated three times, mean value of three readings are presented in the figures. Although the Brookfield temperature probe accuracy is ±1°C, the precision of the bath temperature measurement is ±0.1°C. As the viscosity accuracy is ±1.0% of the full scale range (according to spindle type and rotational speed), the allowable error during the experiment is ±3 mPas (for maximum spindle speed).

4 Results and discussion

Fig. 2 shows the example relationship between the apparent viscosity and shear rate of Micronal® DS 5039 X slurry.

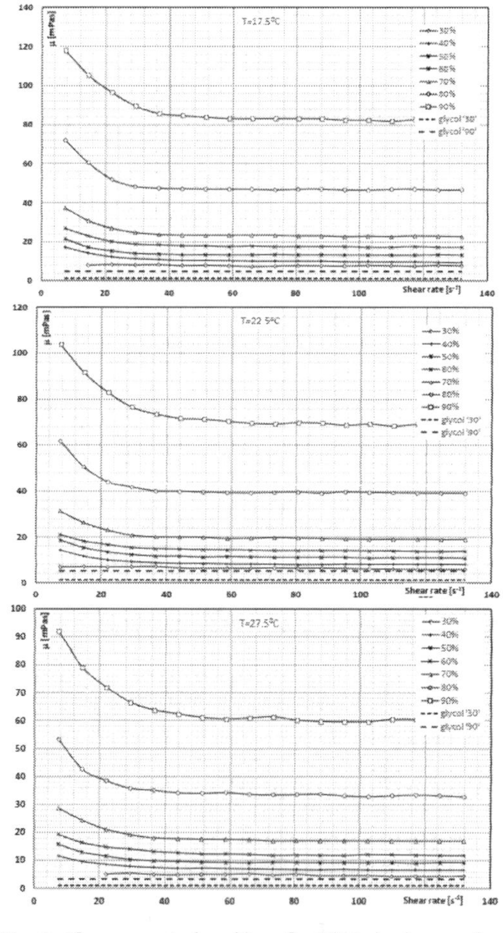

Fig. 2. The apparent viscosities of mPCM slurries vs. shear rates at different concentrations.

Each chart represents viscosity results for a different temperature. In each plot, there are several curves corresponding to different mass ratios of Micronal® DS 5039 X to ERGOLID EKO®, namely 30%, 40%, 50%, 60% 70%, 80% and 90%, respectively. Two specific fluids and its viscosities are given as a baseline for the purpose of comparison with obtained results. In Fig. 2 they are displayed with dashed line. First viscosity is the viscosity of a base fluid, which is propylene glycol-water mixture with glycol to water quantity at the same level as in solution of Micronal® DS 5039 X / ERGOLID EKO® with ratio 30:70, it is designated for convenience as glycol '30'. Second viscosity is by analogy, that of base fluid containing the same amount of glycol to water as solution Micronal® DS 5039 X / ERGOLID EKO® with ratio 90:10.

It was observed that, the apparent viscosity of 90% Micronal® DS 5039 X slurry is approximately several hundred times that of solution of propylene glycol with water at 10°C to several dozen times that of solution of propylene glycol with water at 50°C (not presented here).

It can be seen from Fig. 2 that, for each temperature of the experiment, the apparent viscosity increases exponentially with the concentration of PCM microcapsules. It is observed that when increasing the shear rate, the apparent viscosity decreases to the Newtonian plateau, where the viscosity seems to be constant. This shear-thinning behaviour has also been found in the majority of the works in this field found in the literature [5–7,12,13].

Fig. 3 presents the results of one's own research and the results of tests according to [3]. This research study [3] was carried out using another BASF Micronal® DS 5008 X product. The viscosity of the 25% aqueous suspension was measured at 15°C. According to the comparison of test results, Micronal® DS 5008 X's apparent viscosity characteristics are faster, while the viscosity of this suspension is comparable to the 70% viscosity of the Micronal® DS 5039 X suspension.

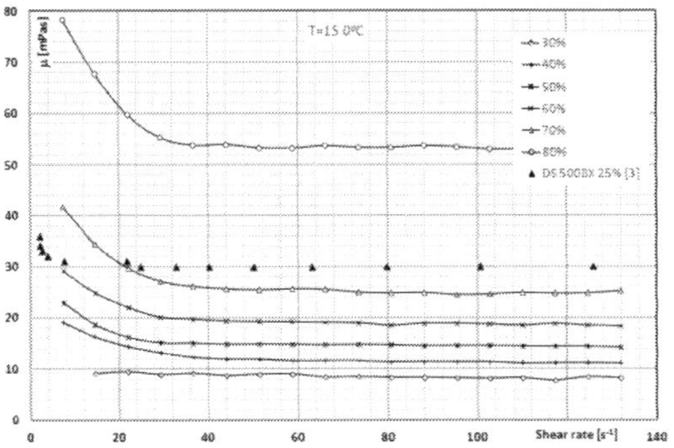

Fig. 3. View of apparent viscosity characteristics for mPCM slurries at different concentrations.

The results of one's own research indicate that shear-thinning behaviour is valid for every case except for the slurry with 30% concentration. In this particular case, mPCM slurry may be considered as a Newtonian fluid in the whole range of shear rate. This fact is confirmed by a detailed visual inspection of the characteristics obtained for other temperature values (not presented here).

Another regularity can be observed. With an increasing shear rate, apparent viscosity drops rapidly at first, then becomes lower until the value of apparent viscosity does not change much. The slurries under examination in the experiment were in the range of pseudoplastic fluids. This shear thinning behavior can be explained through the spatial layout of the microcapsules in suspension. When a slurry is stable and at rest, the particles are dispersed randomly in the continuous phase. When the slurry begins to be sheared, there is no cooperative motion between the microcapsules. The capsules move in an unorganized manner, not necessarily in accordance with the bulk flow direction and therefore the apparent viscosity is high. When the slurry is sheared at high shear rates, the microcapsules start to move in layers that form in parallel to the main flow. Due to this, the average distance between microcapsules decreases in the direction of flow and increases in the direction perpendicular to it. This change of the spatial layout makes the motion much easier, which can be observed as a decrease in viscosity [2].

It may be seen from Figs. 2 and 3 that after exceeding a certain value of shear rate, the decrease in apparent viscosity becomes less prominent and apparent viscosity drops in a linear fashion or becomes constant. The higher the slurry temperature, the lower this threshold shear rate. When the sample temperature was 50°C, a linear characteristic is displayed after exceeding a shear rate of 30÷40 s^{-1}, and for a sample at 20°C, this threshold shear rate is between 40÷60 s^{-1}. This shear thinning behaviour has been explained by Delgado et al. [2] in relation to microcapsules. It is stated that when microcapsules are randomly dispersed in a base fluid in a stagnation condition, the apparent viscosity is high. When approaching higher shear rates, the particles form layers with a low distance between particles in the flow direction and high distance between various layers. This effectively lowers friction between layers and, therefore, apparent viscosity. The phenomenon observed in Figs. 2 and 3 shows that with the higher temperature of sample, the transition to the Newtonian plateau is quicker as the formation of the abovementioned layers occurs much earlier. This may be caused by lower viscosity of the carrier fluid itself at higher temperatures.

5 Apparent viscosity of slurry – predictive model

There are many formulas that serve to describe the viscosity of non-Newtonian fluids. Some of them, in their original or modified form, take into account the influence of concentration (equations based on Einstein's Brownian motion theory [17,18]), the influence of temperature (equations originating from Arrhenius's law [19–23]), or the impact of the shear rate (as the Ostwald equation [24–26] also known as the Power-law model). It is known that all the above-mentioned factors highly influence viscosity.

In present work, the viscosity of slurry created from Micronal® DS 5039 X product and an aqueous solution of propylene glycol was calculated according to a modified Vogel-Tamman-Fulcher equation:

$$\mu = exp\left(B_1 + \frac{B_2}{T} + \frac{B_3}{T^2}\right), \quad (1)$$

where μ is the viscosity of fluid, T [K] its temperature and $B_{i=1,2,3}$ dimensionless fitting coefficients. On the one hand, a second-order term was added [20] to obtain a better fit between the experimental and calculated viscosity data. On the other hand, several researchers [21,22,27,28] have used a fitting coefficients B_i as a function of the components concentration ϕ:

$$B_i = b_{i1} + b_{i2} \cdot \emptyset + b_{i3} \cdot \emptyset^2, \quad (2)$$

where b_{i1}, b_{i2}, b_{i3} are empirical constants (fitting parameters).

The values of dimensionless parameters for the model of viscosity were acquired by least-squares regression method. The regressed values are summarized as below:

$$\begin{cases} B_1 = 0.47 - 1{,}97 \cdot \emptyset + 22.5 \cdot \emptyset^2 \\ B_2 = 38.9 - 13.62 \cdot \emptyset - 5.48 \cdot \emptyset^2 \\ B_3 = -125 - 640 \cdot \emptyset + 1500 \cdot \emptyset^2 \end{cases} . \quad (3)$$

In addition to the obtained values of b_{ij} coefficients, the value of mean absolute error MAE was calculated. The MAE describes average model performance error

$$MAE\ \% = \frac{100}{N} \cdot \sum_1^N \left(\frac{|\mu_{th} - \mu_{exp}|}{\mu_{exp}}\right); \quad (4)$$

where N - sample count. The mean absolute error for investigated cases is MAE = 10.76%.

Fig. 4 presents a comparison of the calculation results of dynamic viscosity according to the dependences (1) and (3) with the results of experimental research. An analysis of over 1250 cases shows that 85.1% of the results were calculated with a deviation of less than ± 20%.

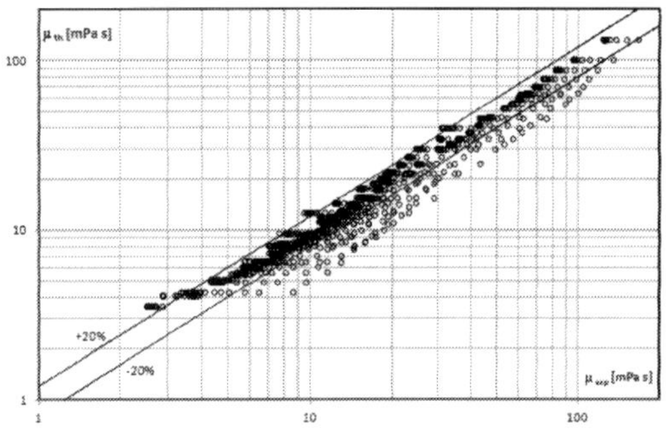

Fig. 4. The comparison of the calculation results of viscosity with the results of experimental research.

6 Conclusions

Experimental investigations were carried out to investigate the microencapsulated phase change slurries (Micronal® DS 5039 X – aqueous solution of propylene glycol) in terms of their rheological properties. The Brookfield DV-II+ Pro rotation viscometer (Brookfield Co., Ltd., USA) with the small sample adapter, consisting of a cylindrical sample chamber and spindle coupled with circulating water bath was used for measuring viscosity equipment. Experiments were conducted on seven types of mPCM slurry with the following mass ratios of Micronal® DS 5039 X to aqueous solution of propylene glycol: 30:70, 40:60, 50:50, 60:40, 70:30, 80:20 and 90:10. The apparent viscosity-shear rate curves were obtained through a shear rate from 0.0132 s^{-1} up to 132.00 s^{-1}. The following conclusions of these experimental investigations may be drawn:

1) The apparent viscosity of Micronal® DS 5039 X – aqueous solution of propylene glycol slurries increased with a rising mPCM concentration.
2) With an increasing shear rate, the apparent viscosity of slurry decreases to the Newtonian plateau, where the viscosity seems to be constant. The higher the concentration, the lower the shear rate value becomes a threshold for linear behaviour. The higher the temperature of the slurry, the lower the shear rate value, after which the apparent viscosity characteristic becomes linear or constant.
3) Only the 30% Micronal® DS 5039 X slurry (12.9 wt% solid of mPCM in liquid) can be considered as a Newtonian fluid within the whole test region (shear rate 0.0132 to 132.00 s^{-1}).
4) The measured values of viscosity were used to estimate the optimal values of fitting coefficients of modified Vogel-Tamman-Fulcher equation.
5) It was found that 85.1% of the results were correctly calculated using the proposed values of the modified Vogel-Tamman-Fulcher equation coefficients.

The authors are thankful to the BASF, Germany, for supplying samples of Micronal® DS 5039 X.

We would also like to thank Dr. Paul McNamara for conducting the linguistic verification of this article.

References

1. R. Yang, H. Xu, Y. Zhang, Sol. Energy Mater. Sol. Cells. **80,** 405–416 (2003)
2. M. Delgado, A. Lázaro, C. Peñalosa, B. Zalba, Appl. Therm. Eng. **63,** 11–22 (2014)
3. G.H. Zhang, C.Y. Zhao, Renew. Energy. **36,** 2959–2966 (2011)
4. Y. Allouche, S. Varga, C. Bouden, A.C. Oliveira, Energy Convers. Manag. **94,** 275–285 (2015).
5. M. Kong, J.L. Alvarado, W. Terrell, C. Thies, Int. J.

Heat Mass Transf. **101**, 901–914 (2016)

6. L. Wang, J. Zhang, Y. Wang, X. Lin, N. Xie, H. Chen, Appl. Therm. Eng. **99**, 583–590 (2016)
7. W. Fu, X. Liang, H. Xie, S. Wang, X. Gao, Z. Zhang, Y. Fang, Energy Build. **136**, 26–32 (2017)
8. L. Chen, T. Wang, Y. Zhao, X.R. Zhang, Energy Convers. Manag. **79**, 317–333 (2014)
9. M. Jurkowska, I. Szczygieł, Appl. Therm. Eng. **98**, 365–373 (2016)
10. Y. Yamagishi, H. Takeuchi, A. Pyatenko, N. Kayukawa, Am. Inst. Chem. Eng. **45**, 696–707 (1999)
11. M. Delgado, A. Lázaro, J. Mazo, B. Zalba, Renew. Sustain. Energy Rev. **16**, 253–273 (2012)
12. C. Liu, Z. Ma, J. Wang, Y. Li, Z. Rao, Int. J. Heat Mass Transf. **115**, 737–742 (2017)
13. L.Y. Li, D. Zou, X.F. Ma, X.S. Liu, Z.G. Hu, J.R. Guo, Y.Y. Zhu, Energy Convers. Manag. **135**, 170–177 (2017)
14. Z. Qiu, X. Ma, P. Li, X. Zhao, A. Wright, Renew. Sustain. Energy Rev. **77**, 246–262 (2017)
15. L. Liu, G. Alva, X. Huang, G. Fang, Renew. Sustain. Energy Rev. **66**, 399–414 (2016)
16. J. Giro-Paloma, G. Oncins, C. Barreneche, M. Martínez, A.I. Fernández, L.F. Cabeza, Appl. Energy. **109**, 441–448 (2013)
17. P. Zhang, Z.W. Ma, Z.Y. Bai, J. Ye, Energy. **106**, 63–72 (2016)
18. S. Mossaz, J.A. Gruss, S. Ferrouillat, J. Skrzypski, D. Getto, O. Poncelet, P. Berne, Appl. Therm. Eng. **81**, 388–398 (2015)
19. D. V. Boger, An Introduction to Rheology, 1989.
20. D.I. Sagdeev, M.G. Fomina, G.K. Mukhamedzyanov, I.M. Abdulagatov, J. Chem. Thermodyn. **43**, 1824–1843 (2011)
21. L.A. Tirona, R.B. Leron, A.N. Soriano, M.H. Li, J. Chem. Thermodyn. **77**, 116–122 (2014)
22. C.P. Liao, R.B. Leron, M.H. Li, Fluid Phase Equilib. **363**, 180–188 (2014)
23. D.I. Sagdeev, M.G. Fomina, I.M. Abdulagatov, Fluid Phase Equilib. **450**, 99–111 (2017)
24. A.K. Pandey, M.S. Hossain, V. V. Tyagi, N. Abd Rahim, J.A.L. Selvaraj, A. Sari, Renew. Sustain. Energy Rev. **82**, 281–323 (2018)
25. L. Huang, M. Petermann, Int. J. Heat Mass Transf. **83**, 479–486 (2015)
26. B. Muñoz-Sánchez, J. Nieto-Maestre, E. Veca, R. Liberatore, S. Sau, H. Navarro, Y. Ding, N. Navarrete, J.E. Juliá, Á.G. Fernández, A. García-Romero, Sol. Energy Mater. Sol. Cells. **176**, 357–373 (2018)
27. S. Garg, G. Murshid, F.S. Mjalli, A. Ali, W. Ahmad, Chem. Eng. Res. Des. **118**, 121–130 (2017)
28. G. Murshid, H. Ghaedi, M. Ayoub, S. Garg, W. Ahmad, J. Mol. Liq. **250**, 162–170 (2018)

Effect of air parameters, water temperature, and number of pool occupants on moisture gains

Agnieszka Garnysz-Rachtan[1,*], and *Zbigniew* Zapałowicz[1]

[1]West Pomeranian University of Technology, Szczecin; Department of Heat Engineering, al. Piastów 17, PL-70-310 Szczecin, Poland

Abstract. Thermal calculations for indoor swimming pools require that amount of water evaporated from the pool's water surface, as well as water evaporated from the floor surrounding the pool and from the bodies of the occupants are to be determined. It means in practice that amount of vapor transferred to the air in the pool hall depends most of all on physical parameters of water and air. One of more important factors that affect water evaporation is also the way the pool is operated. The present article shows only chosen relations applied to determine the values of evaporation rate from occupied pool. The relations also account for the mode of pool's operation. The aim of the paper is to analyse the effect of changes of temperature and of relative air humidity in the hall, of water temperature and of air velocity above the water surface, as well as of the number of occupants on moisture gains in the hall. The above data let choose the right relation to be applied in calculations for the water evaporation rate.

1 Introduction

One of the steps in constructing ventilation systems or air-conditioning systems for a pool hall is determination of the amount of evaporated water in the object [1, 2]. The literature gives a number of relations that allow to determine the evaporation rate for indoor swimming pools [1–8]. This paper presents only methodologies to determine moisture gains in occupied pools. Persons present in the pool hall cause increase of evaporation intensity in comparison with the same parameters at evaporation from a calm water surface. The additional gains are caused by increase of water surface as a result of its roughness by water evaporation from the floor around the pool, and by wet bodies of occupants, as well as by emission of vapor during their breathing. Authors of so far proposed relations to determine the evaporation rate account differently for the additional moisture gains. The case will be discussed further in the paper (section 2).

Indoor swimming pools can be divided according to the operation mode as [2, 9]: residential pools (private), recreational pools (of general purpose), school pools, competition pools, and therapeutic pools. These objects are characterized with various requirements concerning water temperature, temperature and relative humidity of air in the hall, as well as air velocity over the water surface [2, 8, 9]. Table 1, elaborated on the basis of [2, 9], shows the most frequent ranges of changes for these parameters. A number of authors advise to assume the air temperature in the hall to be by 2÷4 K higher than the water temperature. Air temperatures ranges in the hall given in Table 1 are determined by assumption that the air temperature is by 2 K higher than water temperature. The tally of parameters neglects the ones that are valid for residential and therapeutic pools. Besides, the last column in Table 1 presents water surface area required per each occupant.

The aim of the paper is to analyse the effect of changes of temperature and relative air humidity in the hall, water temperature and air velocity above the water surface, as well as the number of occupants on moisture gains in the hall. The above data let choose the right relation to be applied in calculations for the water evaporation rate.

Table 1. Required parameters of water and air for chosen types of swimming pools [2, 9].

pool type	water temperature	air temperature	air humidity	air velocity	water surface area per occupant
	°C	°C	%	m/s	m²/person
recreational	24÷32	26÷34	40÷70	0,05÷0,3	3÷3,5
school	26÷30	28÷32			5÷6
competition	22÷28	24÷30			8÷10

* Corresponding author: agnieszka.garnysz@zut.edu.pl

2 Relations that determine evaporation rate in pools with occupants

The literature gives a number of relations that allow to determine the evaporation rate for indoor swimming pools. The present article is confined only to presentation of the relations which relate to occupied pools [1, 6, 10–15]. Chosen relations used to determine evaporation rate in pools during their operation are presented in Table 2.

According to Dalton's law, the amount of evaporated water is proportional to difference of vapor partial pressures in the thin layer of air direct above the water surface (in the boundary layer) and in the air around the pool (Δp), as well as to evaporation coefficient (ε):

$$E = \varepsilon \cdot \Delta p \tag{1a}$$

Having regard to the relation between vapor partial pressure and content of humidity in the air, difference of partial pressures Δp present in equation (1a) can be replaced by the difference of humidity content Δx, then equation (1a) has the form:

$$E = \sigma \cdot \Delta x \tag{1b}$$

Common feature of relations given in Table 2 and applied to calculate the evaporation rate is that coefficients ε, σ are composite functions. Therefore, some of researchers propose only modifications of relations that apply for calm water surface, in order to take into consideration the effect of additional moisture gains caused by presence of occupants.

Labohm [1], in equation (7), took into account the evaporation coefficient which could be calculated from Sprenger relation ($\sigma = 25 + 19 \cdot w$). According to this author, the value of evaporation rate obtained by means of Sprenger formula should be increased by 200-250%, by application of correction factor $K_L = 3 \div 3{,}5$. Then, correction factor takes into account the additional water evaporation from the wet surface of occupants' bodies, from wet floors, and as a result of rough water in the pool. ***Kappler*** [12] refers also to Sprenger relation. However, in order to take into account additional gains of vapor emitted by occupants, and as a result of evaporating of the water film wetting their bodies, he gives different values of correction factor in equation (6) – coefficient K_K. According to ***Besler*** [10], the following factors should be related in calculations concerning occupied pool: temperature rise on water surface, evaporation from the wet surfaces of occupants' bodies, and significant increase of water surface through its turbulence. It results from Besler's considerations that the relation of Ferencowicz [10], applied to determine evaporation rate from calm water, can be used for calculations, however on condition that this parameter will be increased by 180% – equation (3).

Table 2. Relations applied to calculate evaporation rates in occupied swimming pools.

researcher, source	relation	remarks	number of equation
ASHRAE, [14]	$E = F_A \cdot \left[3{,}6 \cdot \frac{(0{,}089 + 0{,}0782 \cdot w) \cdot \Delta p}{Y}\right]$	$F_A = 0{,}5$ residential pools $F_A = 0{,}65$ condominium and therapeutic pools $F_A = 0{,}8$ hotel pools $F_A = 1{,}0$ public, school, competition pools	(2)
Besler, [10]	$E = 2{,}8 \cdot (A_F + 17{,}4 \cdot w) \cdot 10^{-3} \cdot \frac{760}{p_b} \cdot \Delta p$	$A_F = 23$ for $t_w < 30°C$; $A_F = 28$ for $30°C < t_w < 40°C$	(3)
Biasin&Krumme, [11]	$E = \left[0{,}118 + 0{,}01995 \cdot \frac{\Delta p}{133{,}3} \cdot \varepsilon_{BK}\right]$	$\varepsilon_{BK} = \frac{N}{A \cdot 0{,}167} \approx 6 \cdot N^*$; $\varepsilon_{BK} \leq 0{,}75$	(4)
Hens, [6]	$E = 0{,}0000409 \cdot (1 + 8{,}46N^*) \cdot \Delta p$		(5)
Kappler, [12]	$E = K_K \cdot (25 + 19 \cdot w) \cdot \Delta x$	$K_K = 1{,}1$ private indoor pools $K_K = 1{,}1 \div 1{,}15$ general purpose pools (recreational pools) $K_K = 1{,}2$ hotel and school pools	(6)
Labohm, [1]	$E = K_L \cdot (25 + 19 \cdot w) \cdot \Delta x$	$K_L = 3 \div 3{,}5$	(7)
Shah, [6]	$E = \left(1{,}9 - 21(\rho_a - \rho''_{wg}) + 5{,}3N^*\right) \cdot E_o$ where: $E_o = \max\begin{cases} 35\rho''_{wg}(\rho_a - \rho''_{wg})^{1/3} \cdot \Delta x \\ 0{,}00005 \cdot \Delta p \end{cases}$	Equation is applicable to $N^* \geq 0{,}05$; For $N^* < 0{,}05 \rightarrow$ to perform linear interpolation between E/E_o at $N^* = 0{,}05$ and $E/E_o = 1$ at $N^* = 0$. For $(\rho_a - \rho''_{wg}) < 0$, use $(\rho_a - \rho''_{wg}) = 0$	(8)
Smith et al., [15]	$E = 1{,}26 \cdot \left[3{,}6 \cdot \frac{(0{,}089 + 0{,}0782 \cdot w) \cdot \Delta p}{Y}\right]$		(9)
VDI/Recknagel, [13]	$E = \frac{\beta_b}{461{,}52 \cdot [0{,}5 \cdot (t_w + t_a) + 273{,}15]} \cdot \Delta p$	$\beta_b = 21$ private pool $\beta_b = 28$ swimming/recreational pool; with depth less than 1,35 m $\beta_b = 40$ swimming/recreational pool; with depth more than 1,35 m $\beta_b = 50$ wave pool during work	(10)

Smith, Löf and Jones [15] carried out experimental research for indoors and outdoors pools, both for calm water surface, and during the activity of occupants. The quantity of evaporating water from indoor pool was estimated on the basis of energy balance for water in the pool, by means of water temperature measurements results. The authors compared the obtained experimental data with calculation results obtained from Carrier equation recommended at that time in the ASHRAE handbook. They came to the conclusion that evaporation rate determined from Carrier equation should be increased by 26%. Smith, Löf and Jones proposed equation (9) for calculations. In the meantime, as it was given in [14], experimental research showed that evaporation rate determined from Carrier equation had too high value. The problem was noticed also by ***ASHARE*** which introduced coefficient F_A, called activity coefficient, in equation (2) in 2007. Values of F_A were not determined on the basis of experimental research, but they were stated arbitrarily on the meeting of members of ASHRAE [14].

The next group of relations applied for calculations of moisture gains are equations that require determination of occupants' number. ***Hens*** [6] presented relation (5), obtained on the basis of measurement results carried out at university pool where one of parameters was the number of occupants per pool surface area (parameter N^*). ***Biasin & Krumme*** [11] determined the quantity of evaporating humidity in a real object, on the basis of the quantity of condensate collected by the air conditioning system operating in the pool hall, changing rooms, and shower rooms. On the basis of additional measurements concerning the operation of showers, Biasin & Krumme came to the conclusion that moisture present in shower rooms does not affect meaningfully the water evaporation rate. According to the carried out regression analysis of measurement data, they proposed the relation (4). ***Shah*** proposed relation (8) in [6]. He elaborated the relation on the basis of measurement data analysis in experimental research for occupied indoor pools, however, he obtained the data from publications of other authors. Shah took into regard the effect of occupants' number and of difference of air density in the hall and over the evaporation surface. Evaporation rate E_o, present in equation (8) concerns evaporation from unoccupied pool and is discussed by Shah in [14].

The newest issue of ***Recknagel et al.*** [13] presents relation (10), given in updated guidelines of Association of German Engineers ***VDI 2089-1:2006-2009.*** According to [16], equation (10) is a formula resulting from the equation of mass conservation and value of mass transfer coefficient β_b was determined by recalculation of water evaporation coefficients present in the former version of VDI 2089 guidelines. Given values of mass transfer coefficient β_b take into account evaporation from overflow gutter, wet clothes and surfaces of occupants' bodies, and water vapor included in the air breathed out by occupants.

3 Research results and discussion

3.1 Basic assumptions

Relations given in Table 2 were applied to calculate the rate of evaporation of water.

In order to determine parameters of boundary layer, it is necessary to know the temperature of this layer. One of ways to determine the boundary layer temperature is to assume for calculations the temperature of water surface. It can be determined from the relation [1, 2]:

$$t_{WS} = t_w - 0{,}125 \cdot (t_w - t_{WB}) \qquad (11)$$

Papers [1, 2] point out that temperature of boundary layer is assumed as equal with water temperature in order to simplify calculations, and respective corrections caused by the above assumption are taken into regard by correction factors. A principle has been adopted in the present paper, that if the reference does not point out clearly that water temperature should be applied for calculations, then water surface temperature is used. On the above basis, water temperature was applied in calculations for equations (3), (4), (6), (9), (10), and water surface temperature calculated from equation (11) was applied in equations (2), (5), (7), (8)

Among relations presented in Table 2, three of them require previous determination of values of correction factors with regard to the kind of pool under consideration. In case of relation proposed by ASHRAE (2), pools analysed in the paper can be regarded as the ones for which activity factor F_A equals 1,0. Kappler formula (6) does not take into regard the correction factors for competition pools, and it distinguishes various values of correction factor for school and public pools. Because of the above fact, mean value of correction factor (K_K = 1,18) was assumed for calculations for school and public pools. In formula proposed by VDI/Recknagel (10), two variants of correction factors were assumed for public and recreational pools, depending on the depth of the pool. There is a lack of data concerning the correction factors for competition pools. The present paper presents calculations for two values of this factor, that is β_b = 28, and β_b = 40. In case when formula of Labohm (7) is used, it is unknown for which types of pools the values of correction factor are given. Labohm has only given the changes range for this factor. Therefore, mean value of correction factor, K_L = 3,3, was assumed for the calculations. In turn, formulas given by Biasin & Krumme (4), Hens (5), and Shah (8) require the number of occupants per pool surface area (parameter N^*). Thus, on the basis of required area of water surface per one occupant (Table 1), mean occupants' number per 1 m² of surface was determined for chosen type of pool. Values of parameter N^* equalled respectively: 0,31 occupants/m² for recreational (public) pools, 0,18 occupants/m² for school pools, and 0,11 occupants/m² for competition pools.

3.2 Moisture gains calculation in swimming pool hall

It was assumed for calculations that water temperature in the pool equalled 27°C, and temperature and relative humidity of the air in the hall equalled 29°C and 55%, respectively (mean values of particular parameters, according to Table 1). Requirements of thermal comfort for occupants state that air velocity over the water surface should not exceed 0,15 m/s [4, 14, 17], so the above value was assumed for calculations. The number of occupants per pool's surface area (parameter N^*) was determined with regard to limitations of application of the Biasin & Krumme formula (4), that is $\varepsilon_{BK} \leq 0{,}75$. Limit value of coefficient ε_{BK} appeared when the number of occupants per 1 m² of water surface equalled about 0,13. Besides, it was assumed that atmospheric pressure equalled 1013,25 hPa.

Calculations results for evaporation rate obtained from analysed relations are presented in Fig. 1.

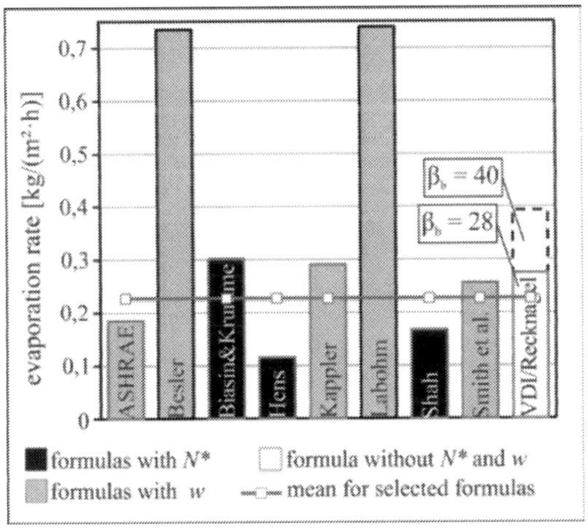

Fig. 1. Comparison of evaporation rate calculated by means of relations given in Table 2 (N^* = 0,13 occupants/m²; t_w = 27°C; t_a = 29°C; φ = 55%; w = 0,15 m/s).

Values of evaporation rate calculated on the basis of equations proposed by Besler (3) and Labohm (7) are distinctly higher than the ones determined by means of the remaining relations (in the extreme case the difference equalled 543%). For this reason, these relations were neglected in further analysis. After they were rejected from the analysis, difference of values of obtained results reached 162%.

In case of relation advised by VDI/Recknagel (10), assumption of higher value of coefficient $\beta_b = 40$ caused that calculated value of evaporation rate was by 43% higher than the one calculated for the lower value of the coefficient $\beta_b = 28$. Thus, value of evaporation rate calculated for the lower value of β_b is closer to the results obtained by means of remaining relations. Therefore, for further analysis, $\beta_b = 28$ was assumed in relation (10) advised by VDI/Recknagel.

Fig. 1. shows also the mean value of evaporation rates calculated for the parameter in question, obtained after application of equations proposed by ASHRAE (2), Biasin & Krumme (4), Hens (5), Kappler (6), Shah (8), Smith et al. (9), and VDI/Recknagel (10) – when $\beta_b = 28$. On the basis of the above, it was stated that values obtained from equations of Smith et al. (9), VDI/Recknagel (10), ASHRAE (2) were at closest to the mean value of evaporation rate. Whereas similar values of absolute differences in comparison with the mean value of evaporation rate were obtained by means of equations of Kappler (6), and Shah (8). The difference was positive when equation of Kappler (6) was applied and negative when equation of Shah (8) was applied. The biggest discrepancies among calculations for evaporation rate appeared when equation of Hens (5) was applied.

3.3 Effect of conditions in swimming pool hall on moisture gains

Conditions that are present inside the object affect relevantly the evaporation rate. This section of the paper presents results of research on effect of changes of air temperature and relative humidity of the air in the hall, of water temperature, and of air velocity over the water surface on evaporation rate. Ranges of changes of the above parameters given in Table 2 were assumed for calculations. Whereas in case of analysis concerning effect of the number of occupants on moisture gains, the range of changes of parameter $N^* = 0{,}05 \div 0{,}33$ was assumed. The lower value of the above range is the limit value of application of the relation of Shah (8), and the upper value results from the minimal water surface area per one occupant (Table 1). The remaining parameters are compatible with assumptions made in sections 3.1 and 3.2.

Effect of temperature and relative air humidity in the swimming pool hall

Values of evaporation rate in function of temperature and relative air humidity in the hall, calculated from relations given in Table 2, are presented in Fig. 2. Additionally, the chart shows runs of value changes of evaporation rate for three values of parameter N* characteristic for three types of pools.

On the basis of analysis of calculations results given in Fig.2a, it can be observed that evaporation rate in the object rises with increase of air temperature in the hall, at simultaneous rise of water temperature (that is when relation $t_w = t_a - 2$ is kept). In turn, rise of relative air humidity in the hall (Fig. 2b) decreases the rate of evaporation water.

Intensity of changes of evaporation rate with the change of temperature (Fig. 2a) and of relative air humidity (Fig. 2b) is for most of relations nearly the same. For example, decrease of relative air humidity from 70% to times 40% causes increase of evaporation rate by about 2,5 times. Only evaporation rate calculated from Biasin & Krumme relation (4) changes to a smaller degree.

In case of equations where the number of occupants is not required, that is, when calculations are carried out

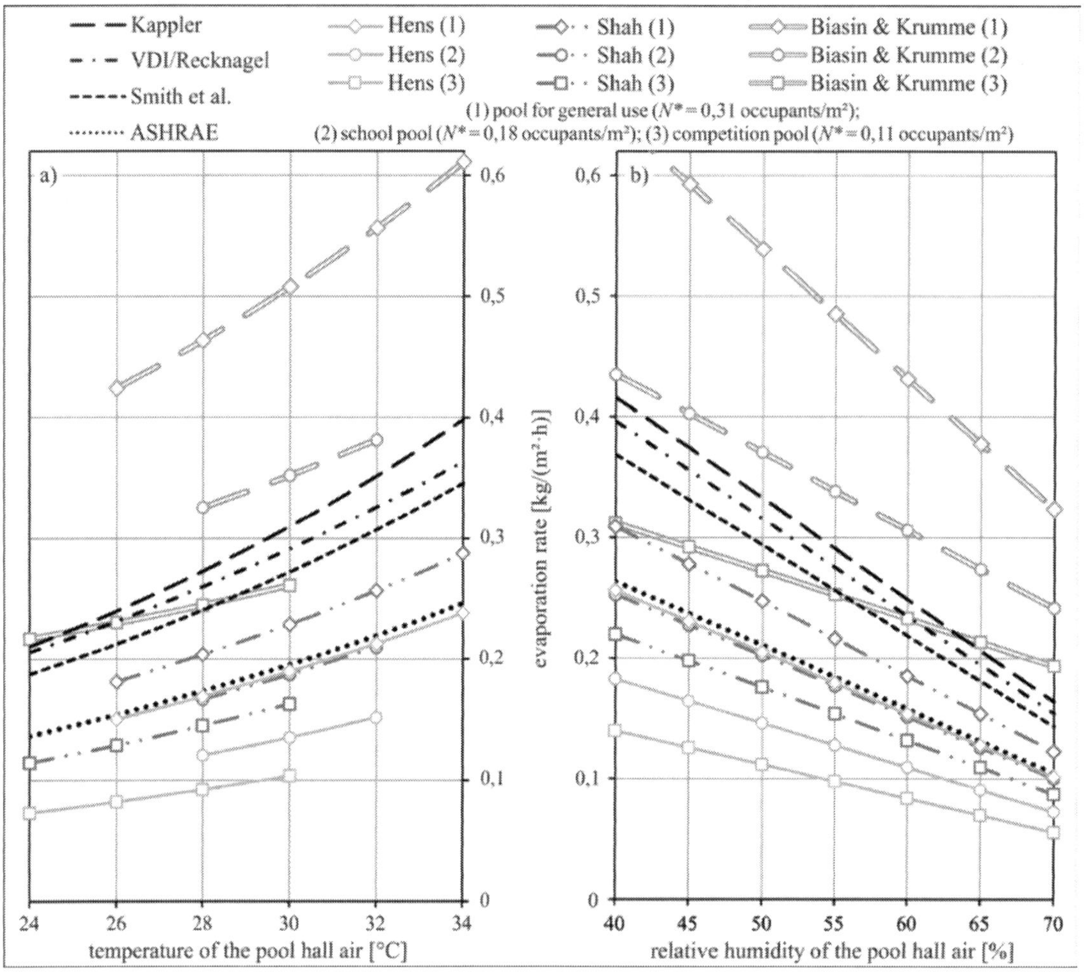

Fig. 2. Evaporation rate in function of: a) air temperature in the swimming pool hall ($t_w = t_a - 2$; $\varphi = 55\%$, $w = 0{,}15$ m/s); b) relative air humidity in the swimming pool hall ($t_a = 29°C$, $t_w = 27$; $w = 0{,}15$ m/s).

by means of equations of Kappler (6), VDI/Recknagel (10), and Smith et al. (9), evaporation rate values are similar. Differences among values of this parameter equal from 12% to 14%, depending on temperature and relative air humidity in the hall. Evaporation rate runs are also similar. If effect of air temperature in the hall (Fig. 2a) is analysed, then it is stated that by lower temperatures evaporation rate values obtained from equation VDI/Recknagel (10) are closer to the ones obtained from Kappler equation (6), and at higher air temperature, evaporation rate values are closer to the ones obtained from the equation of Smith et al. (9). On the other hand, when relative air humidity (Fig. 2b) changes, runs illustrating the changes of evaporation rate have the same character. Distinctly lower values of evaporation rate are obtained from equation of ASHRAE (2) in comparison to values calculated by means of equations of Kappler (6), VDI/Recknagel (10), and Smith et al. (9). For each of considered pool types, the lowest evaporation rate values are obtained by means of Hens relation (5). In turn, evaporation rates obtained from the Shah equation (8) are close to values calculated from equation of ASHRAE (2). In case of a pool with the highest number of occupants (public pool), values of evaporation rate are higher. They are at lower for the pool with lower occupants density (competition pool) and they are nearly equal for a school pool.

For school pools and recreational pools, evaporation rates, calculated by means of Biasin & Krumme (4) relation, are distinctly higher than the one obtained by means of remaining equations. However, it should be kept on mind that the results are obtained for the coefficient $\varepsilon_{BK} > 0{,}75$.

Effect of the water temperature in the pool

Fig. 2 presents calculation results for evaporation rate against change of air temperature in the hall, by assumption that relation $t_w = t_a - 2$ is valid. So, it can be concluded that analysis concerning effect of water temperature on evaporation rate is analogical as in the case of effect of air temperature in the hall on the evaporation rate.

Effect of air velocity over the water surface

Calculation results of the evaporation rate calculated from relations presented in Table 2, for selected values of air velocity over water surface are presented in Fig. 3.

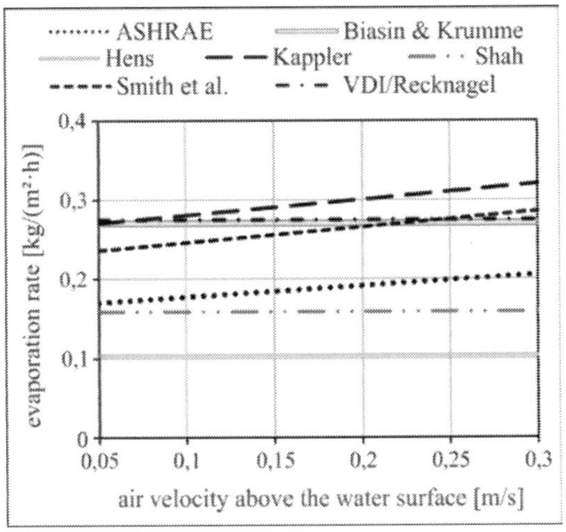

Fig. 3. Evaporation rate in the function of air velocity over the water surface (N^* = 0,13 occupants/m²; t_w = 27°C; t_a = 29°C; φ = 55%).

The maximum values of the evaporation rate are obtained by application of relation of Kappler (6), whereas the minimum values are obtained by application of Hens (5). Differences in obtained results of evaporation rate equal from 165% (for w = 0,05 m/s) to 214% (for w = 0,3 m/s).

Only three formulas given in Table 2 take into regard the variation of moisture gains depending on the change of air velocity – relations: ASHRAE (2), Kappler (6), and Smith et al. (9). For these relations, rise of air velocity from the value of w = 0,05 m/s to the value of w = 0,3 m/s, that is by 500% results only in 20% rise of evaporation rate. It can be then concluded that air velocity change over the water surface does not essentially affect the evaporation rate. Discrepancies in evaporation rate result most of all from the choice of particular relation, and not directly from the assumed value of air velocity.

Effect of occupants' number

Results of calculations for evaporation rate in the function of the number of occupants per 1 m² of water surface are presented in Fig. 4. The chart presents also the mean value of evaporation rate calculated only on the basis of the relations where the number of occupants was not taken into regard (relations: ASHRAE (2), Kappler (6), Smith et al. (9), and VDI/Recknagel (10)) – mean ◄.

Values of evaporation rate calculated from equations advised by Kappler (6), Smith et al. (9), and VDI/Recknagel (10) are higher than the mean ◄. On the other hand, evaporation rate determined from equation of ASHRAE (2) is lower than the mean ◄.

Relation proposed by Biasin & Krumme (4) takes into regard the number of occupants and is valid when the values of parameter N^* are lower than 0,13 occupants/m². Additionally, Biasin & Krumme state that no significant increase of evaporation rate is observed above this value, which means that evaporation rate remains unchanged despite the rise of occupants' number – see run B1 in Fig. 4. However, as it is presented in a number of publications, this limitation is rejected and thus the calculated evaporation rates can be overestimated for the values of N^* > 0,13 occupants/m². In this case, changes of evaporation rate run are illustrated by the run B2 in Fig. 4. Values of evaporation rate obtained by means of equation of Biasin & Krumme (4), with a small number of occupants (0,05÷0,1 occupants/m²) are lower than the mean ◄.

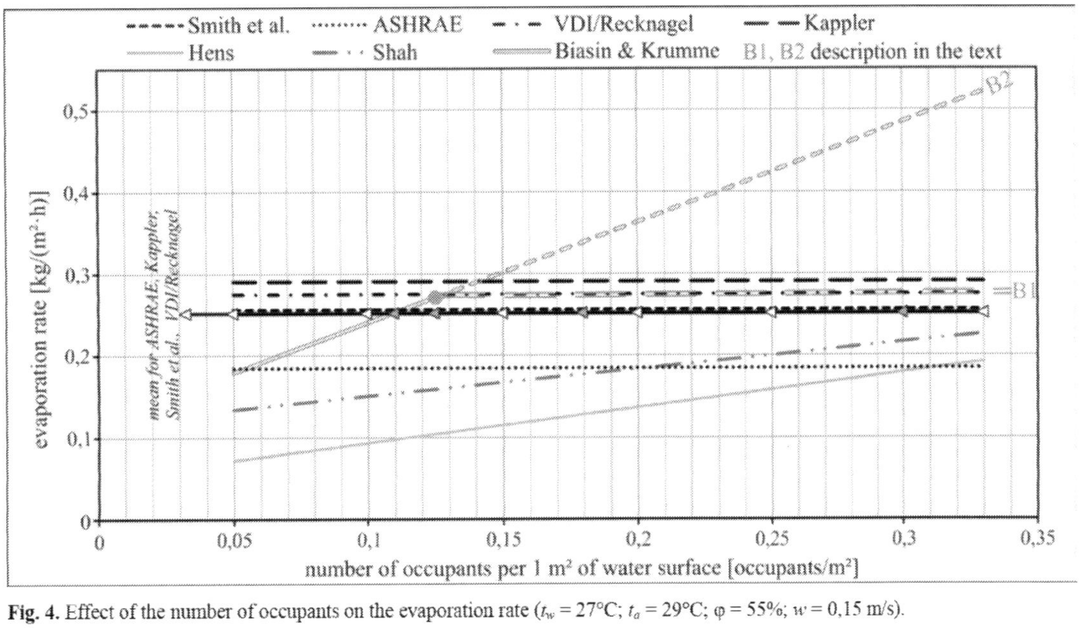

Fig. 4. Effect of the number of occupants on the evaporation rate (t_w = 27°C; t_a = 29°C; φ = 55%; w = 0,15 m/s).

However, when the number of occupants exceeds the value of 0,1 occupant/m², values of evaporation rate are very close to the mean ◄.

Values of evaporation rate calculated from equations of Shah (8) and Hens (5) rise with the rise of occupants' number. Still, these values are always lower than the mean ◄.

4 Conclusion

On the basis of the carried analysis, it can be stated that:

a) values of evaporation rate calculated according to relations proposed by Besler (3) and Labohm (7) differ significantly from values obtained with application of the remaining relations;
b) values of evaporation rate calculated according to relations proposed by Kappler (6), Smith et al. (9), and VDI/Recknagel (10) are close to each other and to the mean value of evaporation rate calculated only on the basis of the relations where the number of occupants is not taken into regard;
c) effect of air velocity over the water surface on moisture gains are properly described by relations proposed by Smith et al. (9) and by Kappler (6);
d) relation proposed by Biasin & Krumme (4) considers properly the effect of occupants' number on evaporation rate; however, it should not be applied for the value of $N^* > 0{,}13$ occupants/m²;
e) values of evaporation rate calculated from relations advised by Hens (5) and Shah (8), and taking into regard the effect of number of occupants, are significantly lower than values of the parameter calculated in case when the number of occupants is not taken into regard.

Symbols

A_F coefficient in equation (3),
A area of pool surface, m²,
E evaporation rate from occupied pool, kg/(m²·h),
E_o evaporation rate from unoccupied pool, kg/(m²·h),
F_A activity factor in equation (2),
K_K correction factor in equation (6),
K_L correction factor in equation (7),
N number of pool occupants,
N^* number of occupants per pool area, $N^* = N/A$, occupants/m²,
p''_{vwg} vapor partial pressure in the boundary layer, Pa,
p_{va} vapor partial pressure at the temperature and relative humidity of the swimming pool hall air, Pa,
p_b atmospheric pressure, Pa,
t_{WB} wet-bulb temperature, °C,
t_a temperature of the swimming pool hall air, °C,
t_{WS} water surface temperature, °C,
t_w pool water temperature, °C,
x''_{wg} specific humidity of air in boundary layer, kg/kg,
x_a specific humidity of air at the temperature and relative humidity of the swimming pool hall air, kg/kg,
w air velocity above the water surface, m/s,
Y latent heat of vaporization of water, kJ/kg,
β_b water evaporation mass transfer coefficient in equation (10), m/h,
Δp $\Delta p = p''_{vwg} - p_{va}$, Pa
Δx $\Delta x = x''_{wg} - x_a$, kg/kg,
ε evaporation coefficient (related to pressure), kg/(m²·h·Pa),
ρ''_{wg} air density in boundary layer, kg/m³,
ρ_a air density in the swimming pool hall, kg/m³,
σ evaporation coefficient, kg/(m²·h).

References

1. M. Jaskólski, Z. Micewicz, *Wentylacja i klimatyzacja hal krytych pływalni* (IPPU MASTA, Gdańsk, 2000)
2. H.G. Sabiniak, M. Pietras, *Klimatyzacja obiektów basenowych* (Wydawnictwo Politechniki Łódzkiej, Łódź, 2010)
3. A. Isańska-Ćwiek, A. Napiórkowska, Instal, **3**, 46–48, (2007)
4. K. Ratajczak, E. Szczechowiak, Ciepłownictwo, Ogrzewnictwo, Wentylacja, **4**, 141–148, (2010)
5. M.M. Shah, Energy and Buildings, **35**(7), 707–713, (2003)
6. M.M. Shah, ASHRAE Transactions, **119**(2), 450–455, (2013)
7. M.M. Shah, ASHRAE Transactions, **120**(2), (2014)
8. F. Asdrubali, Energy and Buildings, **41**(3), 311–319, (2009)
9. A. Napiórkowska, Ciepłownictwo, Ogrzewnictwo, Wentylacja, **5**, 27–31, (2002)
10. G. Besler, Ciepłownictwo, Ogrzewnictwo, Wentylacja, **5**, 142–147, (1972)
11. K. Biasin, W. Krumme, Electrowärme International, **32**, 115–129, (1974)
12. H.P. Kappler, *Baseny kąpielowe* (Arkady, Warszawa, 1977)
13. H. Recknagel, E. Sprenger, R.E. Schramek, *Kompendium ogrzewnictwa i klimatyzacji łącznie z zagadnieniami przygotowania ciepłej wody i techniki chłodniczej* (OMNI SCALA, Wrocław, 2008)
14. M.M. Shah, ASHRAE Transactions, **118(2)**, 460–466, (2012)
15. C.C. Smith, G. Löf, R. Jones, ASHRAE Transactions, **104**, 514–523, (1998)
16. K. Więcek, Ciepłownictwo, Ogrzewnictwo, Wentylacja, **3**, 29–31, (2009)
17. H.G. Sabiniak, M. Pietras, Ciepłownictwo, Ogrzewnictwo, Wentylacja, **1**, 28–33, (2005)

Boiling of a refrigerant of low GWP on the surface with copper microchannels

Robert Kaniowski[1,*], and *Robert* Pastuszko[1]

[1]Kielce University of Technology, Faculty of Mechatronics and Mechanical Engineering, al. 1000-lecia Państwa Polskiego 7, PL-25-314 Kielce, Poland

Abstract. The boiling curves and heat transfer coefficients between the heating surface and fluid were investigated in the paper. Copper samples with horizontal microchannels of rectangular cross-section, variable depth and width were the objects of the study. The following geometrical parameters have been used: microchannel width 0.2; 0.3 and 0.4 mm, depth between 0.2 and 0.5 mm (change every 0.1 mm). Boiling refrigerant was Novec-649 (GWP = 1), and the experiment was performed at atmospheric pressure. Geometrical parameters impact, within a given range of heat flux 3 – 130 kW/m^2, on the heat transfer process was determined.

1 Introduction

Digital systems, nuclear reactors, engines, gas turbines are devices generating significant heat fluxes. Miniaturization and increase of power of electronic systems necessitate searching for compact cooling units of high efficiency.

Great demand for effective heat dissipation, mostly from microelectronic chips, determines development of small heat exchangers with the phase change. Passive method is one of the most effective forms of thermal energy transfer with simultaneous maintenance of relatively fixed temperature of the heating surface. Designers of heat exchangers strive to reduce their volume and weight.

In available publications, there are no analytical dependencies which would allow for determining heat transfer coefficient (HTC) at pool boiling for various geometries of structural surfaces. The available correlations concern mostly smooth flat or cylindrical surfaces.

Jaikomar and Kandlikar [1] performed studies on the heat transfer on samples of copper surface in form of open microchannels with porous fin tops (channel widths: 300 – 762 μm, depths: 200 – 400 μm) at atmospheric pressure using FC-87 as boiling fluid. The highest HTC of 20 kW/m^{2}K was obtained for the channel with the width and depth of 400 μm. Gheitaghy et al. [2] analysed copper surfaces with microchannels inclined at 45° in relation to horizontal plane, where the microchannels' parameters were as follows: widths: 0.5–0.7 mm, depths: 0.5–1 mm. The experiment was performed using water as working fluid. The highest HTC (about 120 kW/m^{2}K) was obtained for the channel width of 0.5 mm and depth of 1.4 mm. In turn, Walunj and Sathyabhama [3] investigated rectangular, parabolic and stepped microchannels fabricated on the 10 mm diameter copper rod with channel widths: 250–800 μm, height of 500 μm. They also used water as the working fluid. At the heat flux of 100 kW/m^2, the HTC increased by 21–35%. In the paper [4], the subject of the study was pool boiling of Novec-7300 fluid and self-rewetting water-heptanol mixtures on bare copper surface and a copper surface coated with copper-plated nanofibers. Authors have found that the heat transfer coefficients were greater for surfaces with cooper-plated nanofibre.

Studies on pool boiling heat transfer with water, ethanol, FC–72 and Novec-649 from tunnel structures, microfins, microchannels, microcavities [5–8] and boiling heat transfer with FC-72 flowing in narrow channels [9–11] have been conducted at the Kielce University of Technology for more a decade.

Due to unfavourable impact of traditional refrigerants and working fluids on the climate, it is necessary to replace these fluids of high global warming potential (GWP), such as HFCs (hydrofluorocarbons) and PFCs (perfluorocarbons) with more environmentally friendly alternatives, for instance a new class of fluoroketone fluids, including Novec 649. These fluids were being explored for their use in direct and indirect heat transfer systems and ORC (Organic Rankine Cycle) systems.

The objectives of the present paper are to determine the effect of microchannel width on the HTC for Novec-649 boiling.

2 Experimental setup

Heat transfer measurements were performed using the experimental stand presented in Fig. 1 and Fig. 2 [5–8]. The main module allows for measuring the temperature difference between fluid and the heating surface

* Corresponding author: kaniowski@tu.kielce.pl

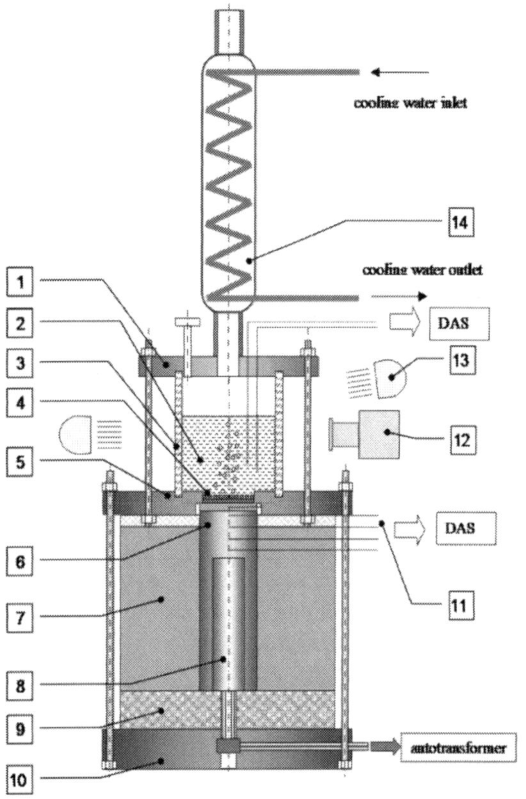

Fig. 1. Main module; 1 – top flange, 2 – boiling liquid, 3 – glass vessel, 4 – sample, 5 – Teflon annular flange, 6 – copper bar, 7 – insulation, 8 – cartridge heater, 9 – bottom insulation, 10 – retaining flange, 11 – thermocouples, 12 – digital camera, 13 – lights, 14 – condenser, DAS – Data Acquisition System.

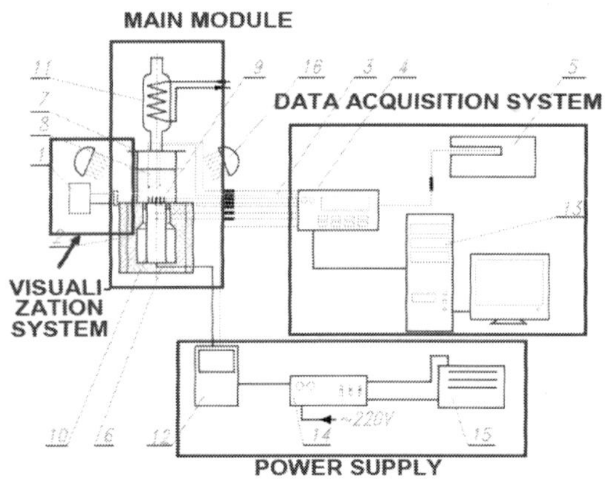

Fig. 2. Pool boiling measurement stand 1 – digital camera, 2 – Teflon casing, 3 – compensating leads, 4 – data logger, 5 – dry-well calibrator, 6 – insulation, 7 – glass vessel, 8 – boiling liquid, 9 – investigated sample, 10 – copper bar with cartridge heater, 11 – condenser, 12 – wattmeter, 13 – PC, 14 – power supply and fuses, 15 – autotransformer, 16 – lights.

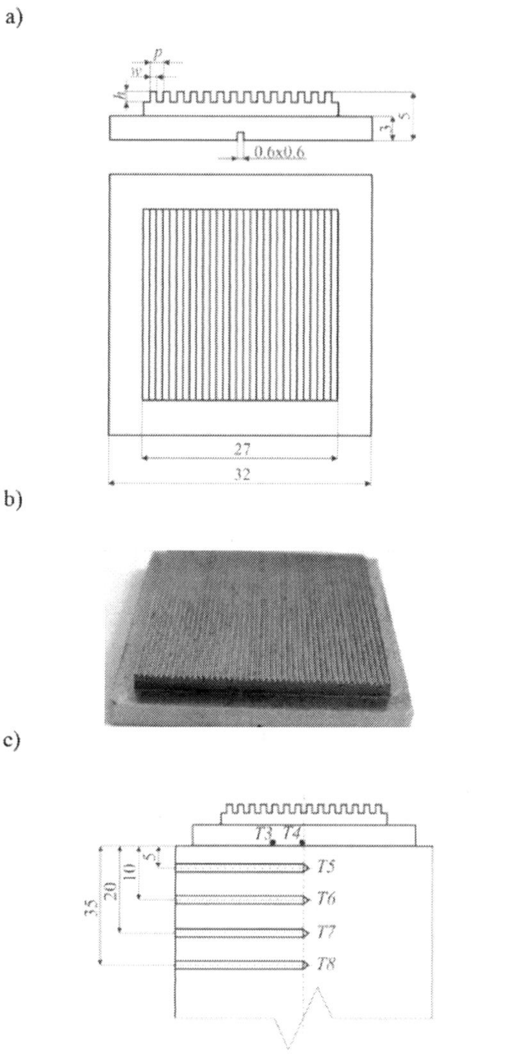

Fig. 3. a) dimension symbols MC, b) view of the specimen MC, c) arrangement of thermocouples.

(superheating) and determining the heat flux. These parameters are necessary to determine the boiling curves.

The cylinder diameter corresponds to the diagonal of the sample base. A 1000 W electric cartridge heater, 16 mm in diameter and 100 mm in length, was installed inside the heating bar.

The estimated uncertainties were as follows: low heat flux (2 kW/m^2): heat flux ±35%, heat transfer coefficient ±40%, high heat flux (120 kW/m^2): heat flux ±28%, heat transfer coefficient ±32%.

Measurement of temperature has been performed using K type thermocouples (NiCr-NiAl) of diameter 0.5 mm; their arrangement is presented in Fig. 3c. FLUKE Hydra series II data acquisition system was applied. In order to fix the sample on the heating cylinder, a layer of tin of thickness δ_{Sn} = 0.1 mm and thermal conductivity λ_{Sn} = 66.5 W/(mK) was used. Dimensions and photo of the sample are presented in

Fig. 3a, b, respectively. Linearity of temperature profile read from the thermocouples *T3, T4, T5, T6, T7* and *T8*, Fig. 4, was checked, where the determination coefficient R^2 reached the value above 0.99. This proved correctness of the assumption concerning one-dimensional heat transfer in the top section of the heating cylinder.

The specimens with test surfaces were made of copper and had parallel grooves with a constant pitch, made with an end mill of 0.2 – 0.4 mm in diameter (CNC machining process). The test section consisted of a 32 x 32 mm^2 square copper specimen with a 27 x 27 mm^2 boiling region. The mean roughness of the microchannel surface was 0.15 μm. Table 1 compiles the surface codes and specifications according to Fig. 3a.

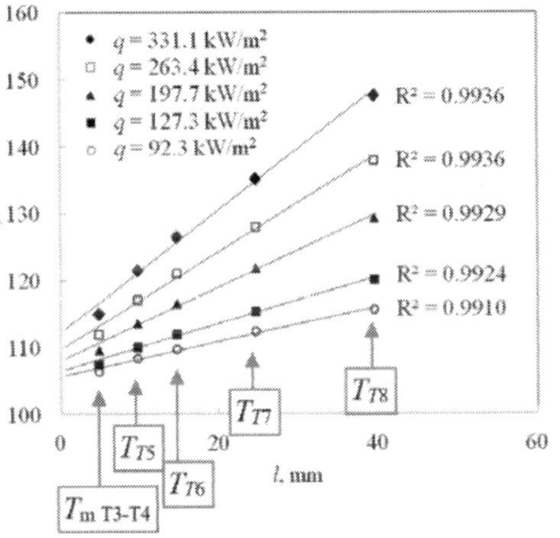

Fig. 4. Temperature vs. thermocouple location in the heating bar, water.

Table 1. MC surface codes and specifications.

Sample code	w mm	h mm	p mm
MC-0.2-0.4-0.4	0.2	0.4	0.4
MC-0.2-0.3-0.4	0.2	0.3	0.4
MC-0.2-0.2-0.4	0.2	0.2	0.4
MC-0.3-0.5-0.6	0.3	0.5	0.6
MC-0.3-0.4-0.6	0.3	0.4	0.6
MC-0.3-0.3-0.6	0.3	0.3	0.6
MC-0.3-0.2-0.6	0.3	0.2	0.6
MC-0.4-0.5-0.8	0.4	0.5	0.8
MC-0.4-0.4-0.8	0.4	0.4	0.8
MC-0.4-0.3-0.8	0.4	0.3	0.8
MC-0.4-0.2-0.8	0.4	0.2	0.8

With the assumption that the heat transfer is one-dimensional, the heat flux was calculated on the basis of the temperature gradient in the copper bar from:

$$q = \frac{\lambda_{Cu}\left(T_{T8} - T_{T5}\right)}{l_{T5-T8}}, \qquad (1)$$

where: $l_{T5\text{-}T8}$ = 30 mm is the distance between thermocouples *T5* and *T8*.

The heat transfer coefficient is expressed as:

$$\alpha = \frac{q}{\Delta T}, \qquad (2)$$

The difference between temperatures of the heated surface and liquid ΔT (superheat) is shown by the following equation, according to Fig. 3:

$$\Delta T = \frac{T_{T3} + T_{T4}}{2} - \frac{T_{T1} + T_{T2}}{2} - q\left(\frac{\delta_s}{\lambda_{Cu}} + \frac{\delta_{Sn}}{\lambda_{Sn}}\right), \qquad (3)$$

where δ_s is distance between microchannel bottom (base) and thermocouples *T3* and *T4*.

3 Results

Before the actual measurements, the fluid was degassed on the experimental stand for about 15 minutes by boiling it. The measurement data was recorded after stabilization of all temperature values within the cooper bar. Further measurements were performed at increasing heat flux until the beginning of the boiling crisis was observed.

Figures 5 and 6 show the dependence of the heat transfer coefficient on the heat flux, at the growing heat flux on MC surfaces for boiling liquid Novec-649. In case of the investigated samples with microchannels, significant intensification of heat transfer was noticed comparing to smooth surface, however the heat transfer coefficient depends on the height and width of the microchannel. Among the investigated samples, the greatest HTC within the range of heat flux 3 – 90 kW/m^2 was obtained for the sample MC-0.2-0.4-0.4 where the heat transfer coefficient was α = 8.5 W/m^{2}K, for the top limit of the given range of heat flux.

4 Conclusion

The analysis of the graphs shows the following conclusions:

1) Comparison of boiling curves of Novec-649 shows the impact of height and width of the microchannel on the heat transfer coefficient.

2) Analyzed surfaces with microchannels allow to obtain heat transfer coefficients within the range of 4.5 – 9 kW/m^2, which in relation to the flat surface gives a 2 – 4-fold increase in HTC.

3) The greatest HTC in the range of 3 – 90 kW/m^2 can be obtained using microchannels of width 0.2 mm and depth 0.3 – 0.4 mm. The small width is associated with a small pitch of microchannels, which in turn makes it possible to obtain the largest number of microchannels on the width of the sample.

4) Surfaces with the widest channels (0.4 mm) allow for obtaining low superheating – initiation of boiling takes place at superheating of 1 – 1.5 K, which results in great HTC value at small heat fluxes.

5) The main reason for the increase in the heat transfer coefficient when increasing the heat flux was the growing number of active nucleation sites at the bottom of microchannels and its side surfaces.

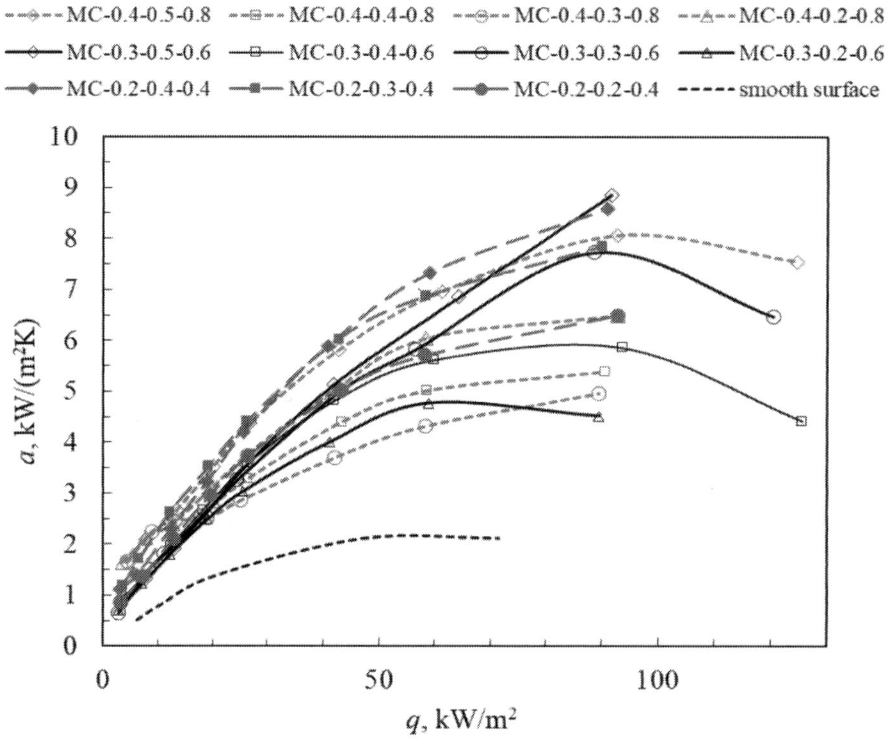

Fig. 5. Boiling curves for Novec-649, heat transfer coefficient vs. heat flux.

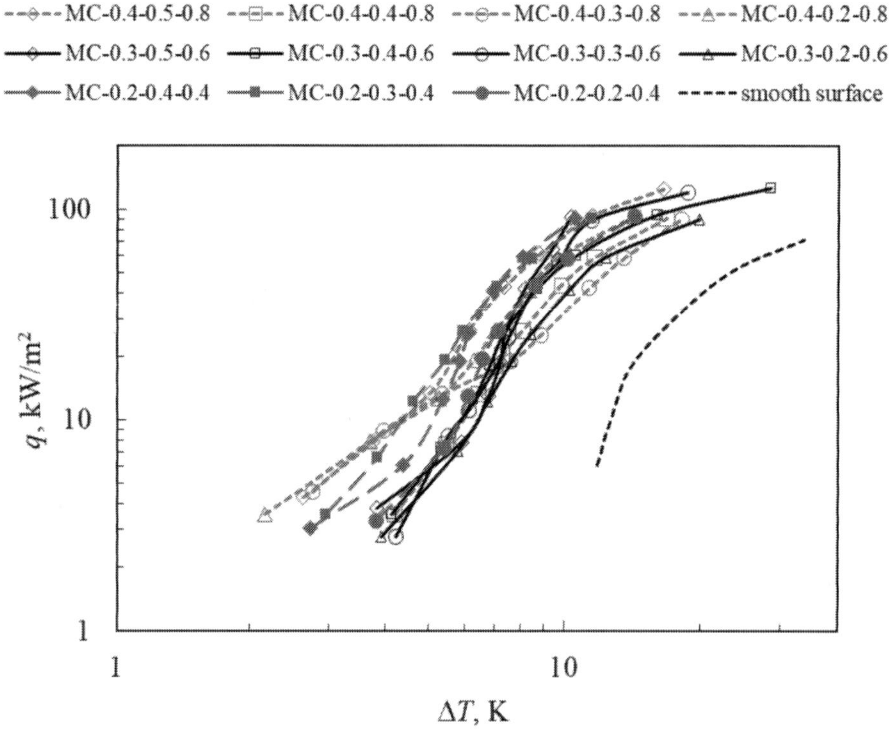

Fig. 6. Boiling curves for Novec-649, heat flux vs. temperature difference.

Nomenclature

GWP – global warming potential,
h – depth, m,
HTC – heat transfer coefficient,
l – distance between thermocouples, m,
MC – microchannel,
p – pitch, m,
q – heat flux, kW/m^2,
T – temperature, K,
w – width, m,

Greek symbols

α – heat transfer coefficient, $W/(m^2K)$,
δ – thickness, mm,
λ – thermal conductivity, $W/(m{\cdot}K)$,
ΔT – difference of temperature, K,

Subscripts

Cu – copper,
m – mean,
s – sample,
Sn – tin.

References

1. A. Jaikumar, S. Kandlikar, App. Therm. Eng. **91**, 426–433 (2015).
2. A.M. Gheitaghy, A. Samimi, H. Saffari, App. Therm. Eng. **126**, 892–902 (2017).
3. A. Walunj, A. Sathyabhama, App. Therm. Eng. 128, 672–683 (2018)
4. R. P. Sahu, S. Sinha-Ray, S. Sinha-Ray, A. L., Int. J. Heat & Mass Transfer, **95**, 83–93 (2016).
5. R. Kaniowski, R. Pastuszko, Ł. Nowakowski, EPJ Web of Conferences **143**, 02049 (2017).
6. R. Kaniowski, R. Pastuszko, EPJ Web of Conferences **180**, 02041 (2018).
7. R. Kaniowski, R. Pastuszko, EPJ Web of Conferences **180**, 02042 (2018).
8. R. Pastuszko, Int. J. Therm. Sci., **125** 197–209 (2018).
9. K. Strąk, M. Piasecka, B. Maciejewska, Int. J. Heat & Mass Transfer, **117**, 375-387 (2017).
10. B. Maciejewska, K. Strąk, M. Piasecka, Int. J. Num. Meth. for Heat & Fluid Flow, **28**, 1, 206-219 (2017).
11. M. Piasecka, K. Strąk, B. Maciejewska, Heat Transfer Eng., **38**, 3, 332-346 (2016).

Bubble departure diameter determination for pool boiling on surface with microchannels

Robert Kaniowski[1], *Robert* Pastuszko[1,*], *Joanna* Kowalczyk[1] and *Łukasz* Nowakowski[1]

[1]Kielce University of Technology, Faculty of Mechatronics and Mechanical Engineering, al. 1000-lecia Państwa Polskiego 7, PL-25-314 Kielce, Poland

Abstract. The paper presents visualization investigations into pool boiling heat transfer for open microchannel surfaces. The experiments were carried out with saturated water, ethanol, FC-72 and Novec-649 at atmospheric pressure. Parallel microchannels fabricated by machining copper sample were about 0.2 to 0.5 mm wide and 0.2 to 0.5 mm deep. The diameter of departing bubble was calculated for the microchannel surface on the basis of buoyancy force and surface tension force balance. The visualization carried out was aimed at determining the diameters of the departing bubbles at various heat fluxes for four working fluids.

1 Introduction

Modern technology determines the development of devices that are dimensionally smaller and have an increasing demand for energy. The leading problem of modern technology is the reception of high heat flux. This is an extremely significant issue from an economic, as well as technological point of view, and in the context of environmental protection. Devices that generate high-value heat fluxes include e.g. digital systems, nuclear reactors, combustion engines, gas turbines. The achievement of high values of heat transfer coefficients reduces the dimensions and consumption of materials needed for the construction of heat exchangers, and also reduces energy consumption during technological processes. Miniaturization and increased power of electronic circuits have given rise to the need for more efficient cooling units. A tendency has been observed in the design of heat exchangers to reduce their volume and weight. A passive cooling method comprises heat transfer between the refrigerant and the heat-emitting surface at boiling. The boiling heat transfer is very intensive, but in the case of technically smooth surfaces it does not provide sufficient dissipation of high heat fluxes. Therefore, there is a need to develop methods to increase the heat transfer surface. These include the use of extended heating microsurfaces, e.g. by creating parallel incisions in the form of grooves. The progress of experimental and theoretical research has led to the development of a variety of surfaces. A series of studies related to active heat exchange for FC-72 during the flow of liquid through channels has been described in [1-3].

The diameter of a departing vapor bubble is one of the basic parameters used in the description of pool boiling, hence the issue has been analyzed by various authors. The available literature provides many formulas for the bubble departure diameter, but most of them relate to pool boiling on flat, smooth surfaces. In relation to boiling on surfaces such as tunnel-pore or subsurface tunnels created in the fins or minifins [4-6], the authors usually present a complicated methodology of calculations.

Ramaswamy et al. [7] have specified 6 forces, the combination of which has enabled the calculation of the departing bubble diameter: (1) unsteady growth force, (2) buoyancy force, (3) surface tension force, (4) lift force, (5) bubble inertia force and (6) liquid inertia force. The authors have assumed that in the case of inertia driven growth, the velocity of the bubble front is constant, hence there was no inertia force.

Pastuszko [8] proposed calculating the departing bubble diameter for the microfins with wire mesh surface on the basis of buoyancy force and surface tension balance referred to the pore diameter. Cooke and Kandlikar [9] used the high speed video to show that the bubbles attached to each side of the microchannel. They assumed that the bubbles were spherical and there was liquid in the microchannel below the bubble. Walunj and Sathyabhama [10] used force balance similar to given by Ramaswamy et al. [7]. Due to the use of microchannels with different geometries, the authors gave their own dependence on surface tension force, including channel base and top width as well as channel depth. Their bubble diameter model predicted the experimental data with mean absolute error of abort 5.6%.

2 Investigated enhanced surfaces

The specimens with test surfaces were made of copper and had parallel grooves with a constant pitch, made

* Corresponding author: tmprp@tu.kielce.pl

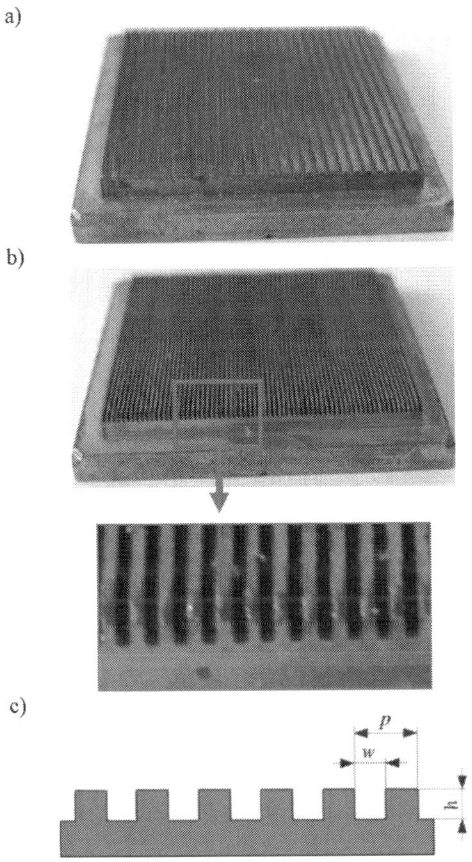

Fig. 1. View of specimen (a) MC 0.5-0.5-1.0 (b) MC 0.2-0.5-0.4 with magnification, (c) MC dimension symbols.

with an end mill of 0.2 to 0.5 mm in diameter (CNC machining process), Fig.1 a, b. The test section consists of a 32 x 32 mm^2 square copper specimen with a 27 x 27 mm^2 boiling region. Figure 1b shows a magnification of the microchannels and Fig. 1c dimension symbols.

The static contact angle was measured using *Biolin Scientific's Attension Theta Tensiometer*. The test consisted of placing a drop of liquid of approximately 4 μl on the surface of a metal sample. An image was recorded with a fast monochrome camera, starting from the moment the drop touched the surface of the disc for about 10 seconds. After the completion of research, an analysis of the images was performed. Results obtained after approximately 0.2 seconds from the moment of application of drops onto the surface of the tested sample were selected. An average of the achieved results was calculated. Tests were performed using distilled water, ethanol, FC-72 and Novec-649. Droplets were applied onto a copper sample.

Figure 2 shows the contact angle between a copper specimen and water. The smallest contact angle, therefore the best surface moistening results, were achieved with ethanol, while the worst wetting results were observed with water and amounted to 99.1°, as in Fig. 3. This means that water does not moisten copper, as the achieved wetting angle is greater than 90°.

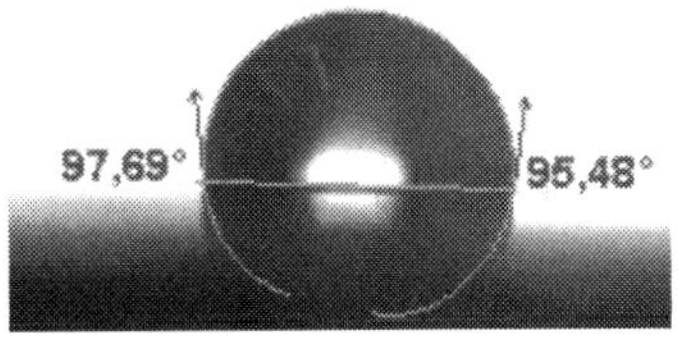

Fig. 2. Contact angle between a copper sample and water.

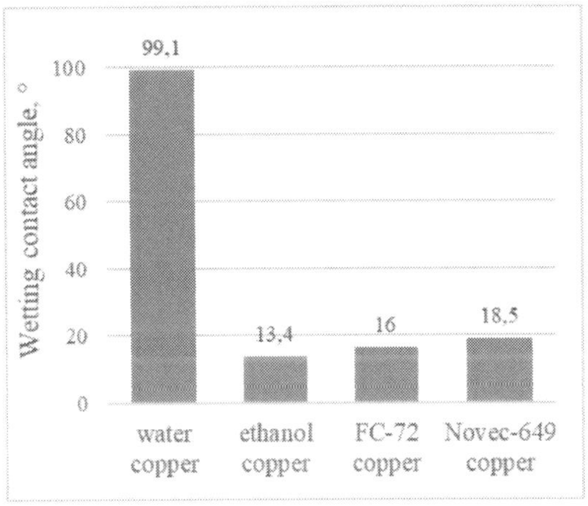

Fig. 3. Contact angle between a copper sample and working fluid.

Figure 4 shows the view of the MC 0.2-0.5-0.4 specimen and its roughness profile, Fig. 5, which was determined using a *Talysurf CCI – Lite non–contact 3D Profiler*, a measurement system for optical, non-contact analysis of geometric structures of surfaces [11]. Table 1 compiles the surface codes and specifications with mean roughness on the top of microfins and the bottom of selected microchannels.

Table 1. MC surface codes and specifications.

Sample code	*w* mm	*h* mm	*p* mm	Ra_t μm	Ra_{bt} μm
MC-0.2-0.5-0.4	0.2	0.5	0.4	24	30
MC-0.2-0.4-0.4	0.2	0.4	0.4	14	25
MC-0.2-0.3-0.4	0.2	0.3	0.4	20	14
MC-0.2-0.2-0.4	0.2	0.2	0.4	30	30
MC-0.3-0.5-0.6	0.3	0.5	0.6	14	25
MC-0.3-0.4-0.6	0.3	0.4	0.6	18	18
MC-0.3-0.3-0.6	0.3	0.3	0.6	20	12
MC-0.3-0.2-0.6	0.3	0.2	0.6	18	12
MC-0.4-0.5-0.8	0.4	0.5	0.8	25	21
MC-0.4-0.4-0.8	0.4	0.4	0.8	31	14
MC-0.4-0.3-0.8	0.4	0.3	0.8	25	15
MC-0.4-0.2-0.8	0.4	0.2	0.8	22	12
MC-0.5-0.5-1.0	0.5	0.5	0.8	31	41
MC-0.5-0.4-1.0	0.5	0.4	0.8	36	28
MC-0.5-0.3-1.0	0.5	0.3	0.8	22	15
MC-0.5-0.2-1.0	0.5	0.2	0.8	20	13

https://doi.org/10.1051/e3sconf/20187002008

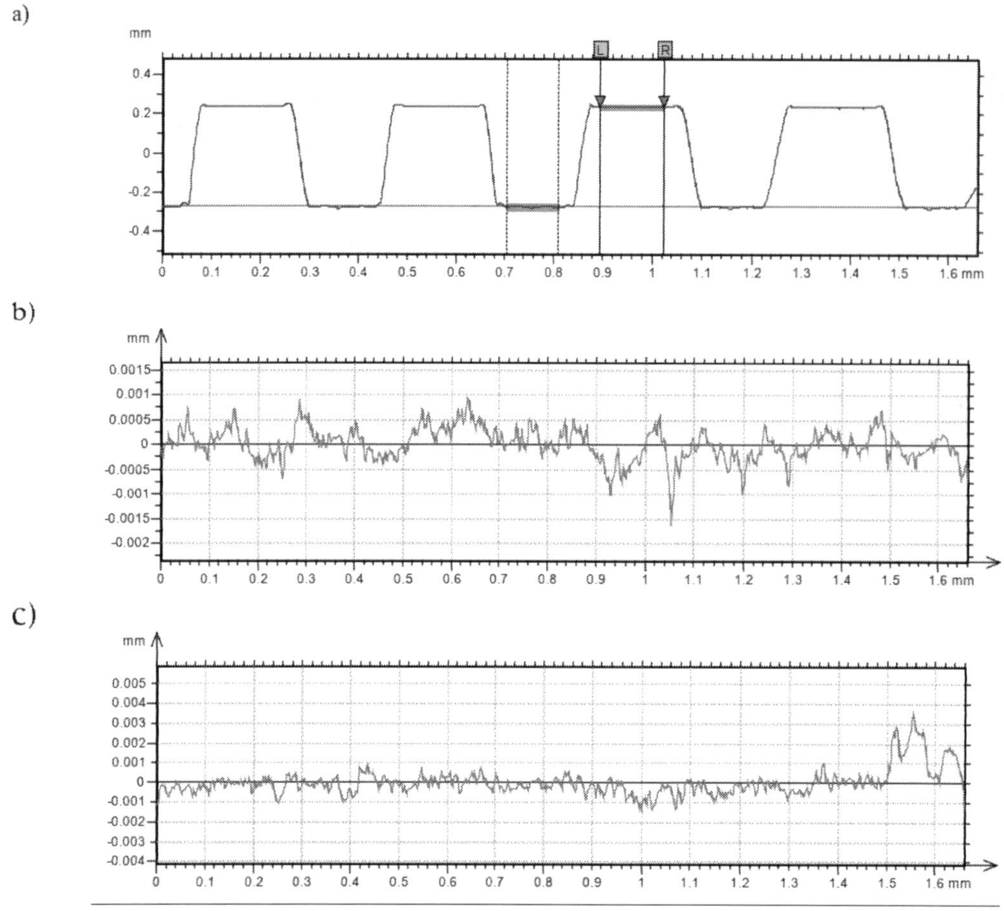

Fig. 4. MC 0.2-0.5-0.4 (a) microchannel outline, (b) microchannel top, (c) microchannel bottom.

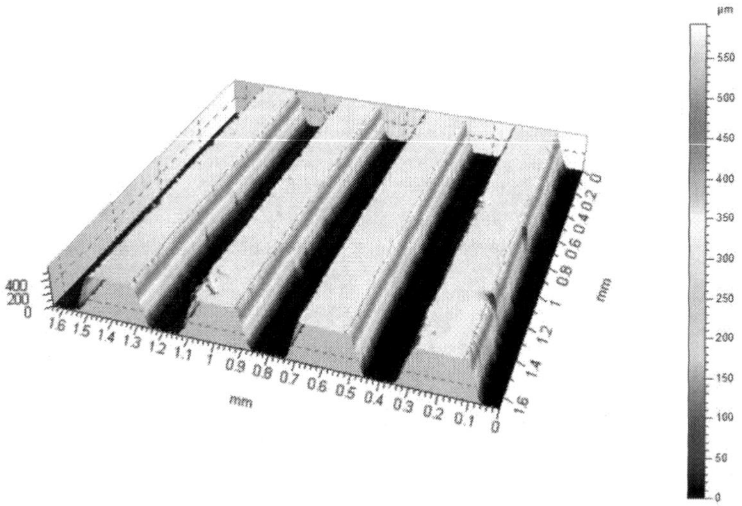

Fig. 5. Profile of the surface roughness of the microchannel, MC 0.2-0.5-0.4.

3 Experimental setup

The diagram of the measuring unit is presented in Fig. 6 [8]. The following modules constituted the experimental setup:

A. main module (items 1—6);
B. vapor cooling and condensate recovery (item 7);
C. temperature measurement and data acquisition module (items 8—10);
D. power supply module (item 11);
E. visualization module (digital camera/digital video camera, item 12).

The visualization images were obtained using a high speed digital monochrome camera MV-D1024-160-CL (*Photonfocus*) with resolution of 500x250 pixels at 458 fps as well as with photo camera Casio EX-FH20.

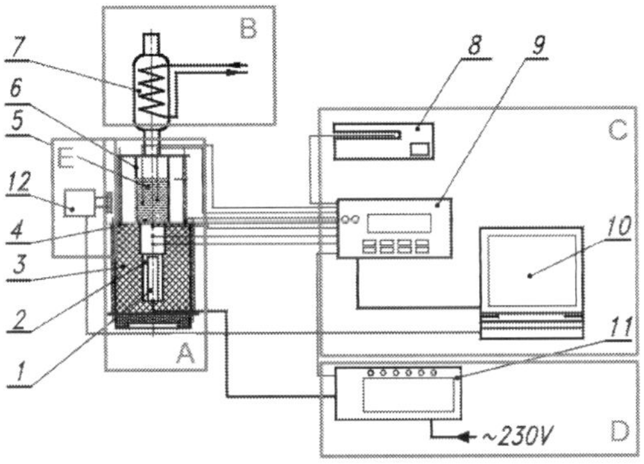

Fig. 6. Schematic diagram of the test system: 1 – cartridge heater; 2 – copper bar; 3 – insulation; 4 – investigated structure; 5 – boiling liquid; 6 – glass vessel; 7 – condenser; 8 – dry-well calibrator; 9 – data logger; 10 – PC; 11 – power regulator; 12 – digital camera/video camera.

4 Results

4.1. Boiling visualization

Figure 7 illustrates boiling of Novec-649 on the surface MC-0.3-0.2-0.6 at atmospheric pressure. Most bubbles are generated in the microchannel space, but we can also observe bubbles in contact with the upper surface of the microfins, as well as bubbles departing at the tops of the microfins.

The largest bubbles, with a diameter corresponding to 2 – 3 pitches of microchannels, were formed as a result of coalescing.

4.2 Bubble departure diameter determination

The authors proposed the calculation of the diameter of departing bubbles for the analyzed surface on the basis of buoyancy force and surface tension force balance. This is a rough estimate that does not take into account the inertia force during bubble growth. Buoyancy force can be calculated from:

$$F_{bu} = \frac{\pi d_b^3}{6} g(\rho_l - \rho_v) \tag{1}$$

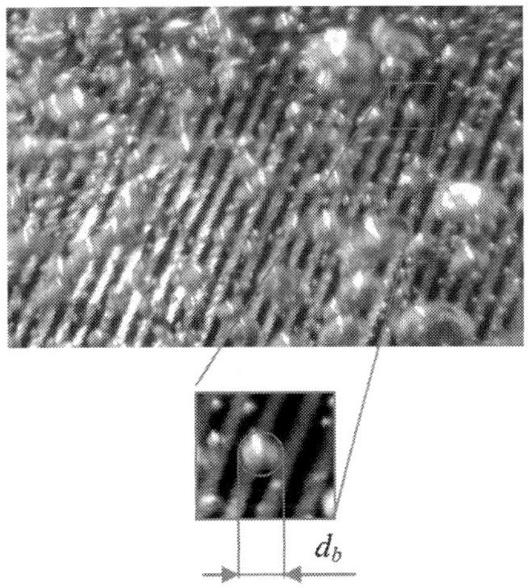

Fig. 7. Visualization observations of pool boiling of Novec-649 on the microchannel surface MC-0.3-0.2-0.6, q=2.8 kW/m^2.

Surface tension force acting on the bubble at departing is determined by assuming that, just before detachment, the bubble base (neck) is in contact with the upper edges of microfins along the length equal to the width of the microchannel (Fig. 8). The length of the contact line is therefore L_c=2w, and dependence on the surface tension force is defined as:

$$F_s = \sigma L_c \sin\Theta \tag{2}$$

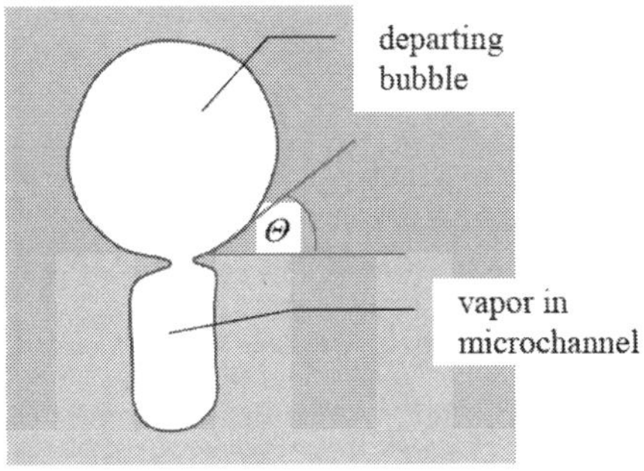

Fig. 8. Bubble departing at microfins tops.

The contact angle θ is shown in Fig 8. Table 2 also shows for comparison the capillary length values, where this parameter is defined as:

$$L_{cap} = \sqrt{\frac{\sigma}{g(\rho_l - \rho_v)}} \tag{3}$$

For the analyzed microchannel widths, the calculated diameter of departing bubbles is smaller than the capillary length. According [9] the surface between liquid and vapor has a constant radius if the channel width is less than the capillary length.

Table 2. Selected parameters and calculated values of the departing bubbles diameters for four working fluids.

Parameters at p = 101.3 kPa		Water	Ethanol	FC-72	Novec-649
T_{sat}, ^{0}C		100	78.3	56.4	49.0
ρ_l, kg/m^3		959	757	1602	1513
ρ_v, kg/m^3		0.597	1.43	13.24	13.42
i_{lv}, kJ/kg		2251	963	94.9	88.0
λ_l, W/mK		0.68	0.169	0.055	0.059
σ_l, N/m		0.0589	0.0177	0.0081	0.0108
d_b, mm	w = 0.2 mm	1.68	0.75	0.48	0.56
	w = 0.3 mm	1.92	0.86	0.55	0.64
	w = 0.4 mm	2.11	0.95	0.6	0.71
L_{cap}, mm		2.5	1.55	0.72	0.86

Assuming that F_{bu}=F_s, on the basis of formulas (1) and (2) we can establish the bubble departing diameter as:

$$d_b = \left(\frac{12\sigma w \sin\Theta}{\pi g(\rho_l - \rho_v)} \right)^{1/3} \tag{4}$$

It should be noted that the contact angles determined by the authors were defined in static conditions, but with the increase of bubble size, their values are subject to change. The diameters of departing bubbles for contact angle values provided in Fig. 1 have been specified in Table 2.

Figure 9 shows influence of heat flux on departing bubble diameter for four working fluids. At small heat fluxes (q < 25 kW/m^2 for water and ethanol and q < 7 kW/m^2 for FC-72 and Novec-649), averaged diameters are approximate in value to the predicted ones in equation 4. At higher heat fluxes there are significant discrepancies between experimental measurements and computational values, which can be explained by the tendency to coalesce.

It was assumed a modified suction-evaporation boiling mode in microchannels (Fig. 10). The whole process of bubble formation consists of a waiting period, a growth period and a liquid intake period. A vapor bubble forms in the corner at the microchannel base and grows as a result of liquid evaporation in the microchannel space. The regions of vapor outflow and bubble generation are determined. The inside of the microchannel is filled with vapor, except for a thin film of liquid at the walls and in menisci.

5 Conclusions

The experimental studies and theoretical calculations discussed in this paper led to the following conclusions:

- Determined experimentally static contact angles were used to determine the diameter of bubbles depart from the surface of micro-channels. These angles were then

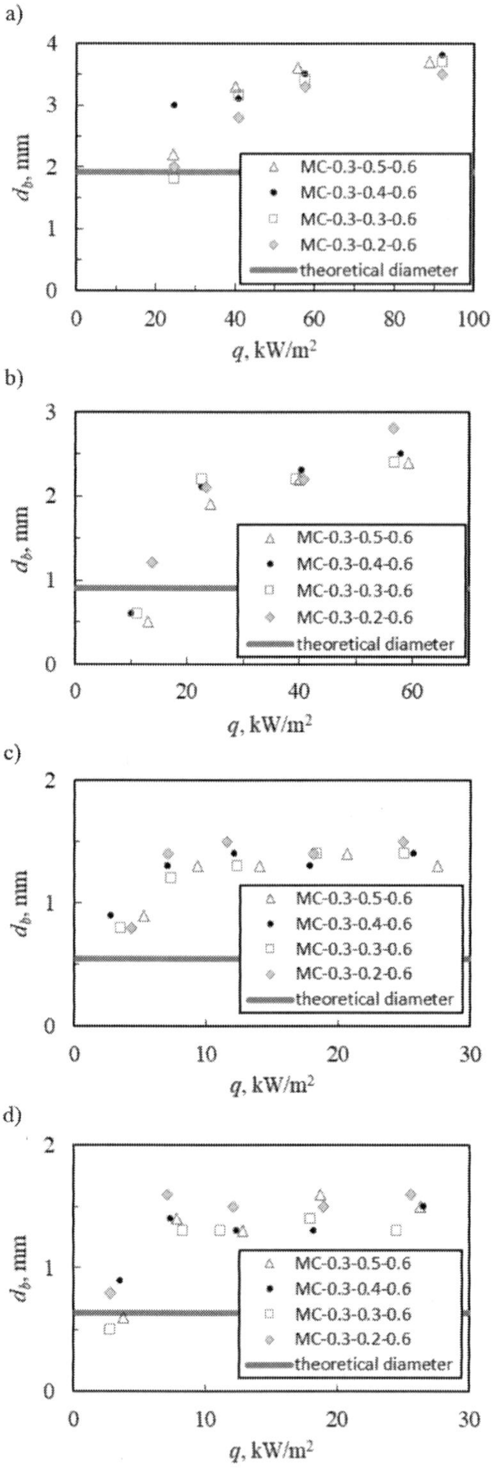

Fig. 9. Averaged departure bubble diameters for (a) water, (b)ethanol, (c) FC-72, (d) Novec-649, theoretical value d_b was calculated from eq. 4.

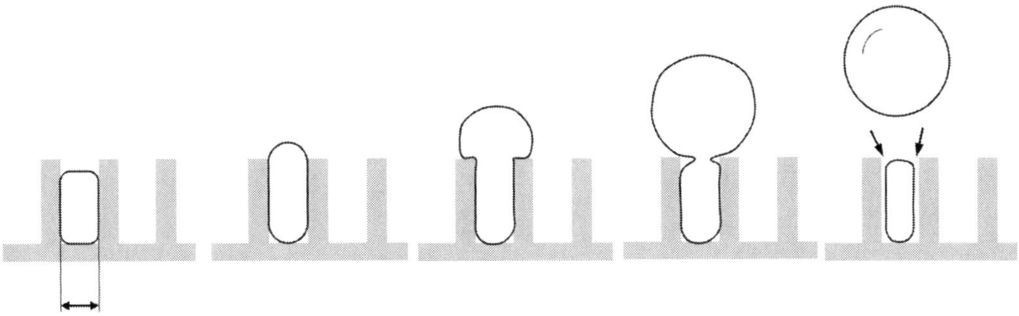

Fig. 10. Bubble growing and departing cycle in microchannel.

related to the cycle of formation of vapor in micro-channels, with the assumption that the microchannel was completely filled with vapor.

- In the case of small superheats between the heating surface and the liquid, vapor bubbles take spherical shapes and do not coalesce.
- Bubble diameters became stable after reaching the limit of about 40 kW/m^2 for water, 20 kW/m^2 for ethanol and 7 kW/m^2 for FC-72 and Novec-649.
- The contact angle for water is 5.5 – 7.4 times greater than the contact angle for ethanol, FC-72 and Novec.
- Theoretically determined bubble departing diameters are consistent with experimental data for low superheats and heat fluxes. Even three times larger diameters determined experimentally at higher heat fluxes compared to computational values can be explained by changes in the contact angle at boiling temperatures and the tendency of bubbles to coalesce.

Nomenclature

d	– diameter, m,
F	– force, N,
i_{lv}	– heat of evaporation, J/kg
g	– gravity, m/s^2,
h	– microchannel depth, m,
L	– length, m
MC	– micro-channel,
p	– pitch, m,
p	– pressure, Pa,
q	– heat flux, W/m^2,
Ra	– mean roughness, μm,'
T	– temperature, K
w	– width, m.

Greek symbols

Θ	– contact angle, °,
λ	– thermal conductivity, W/(m·K),
ρ	– density kg/m^3,
σ	– surface tension, N/m,

Subscripts

b	– bouble,
bt	– bottom surface of microfin,
bu	– buoyancy
c	– contact
cap	– capillary,
l	– liquid,
s	– surface tension,
sat	saturation,
t	– top surface of microfin,
v	– vapor.

References

1. M. Piasecka, Heat Tran. Eng. **35** (10), 903–912 (2014)
2. B. Maciejewska, M. Piasecka, Heat Mass Trans. **53**, 1211–1224 (2017)
3. B. Maciejewska, M. Piasecka, Int. J. Heat Mass Tran. **107**, 925–933 (2017)
4. R. Pastuszko, R. Kaniowski, EPJ Web of Conferences **25**, 02019 (2012)
5. R. Pastuszko, EPJ Web of Conferences 45, 01020 (2013)
6. R. Kaniowski, R. Pastuszko, EPJ Web of Conferences **143**, 02050 (2017)
7. C. Ramaswamy, Y. Joshi, W. Nakayama, W. B. Johnson, Int. J. Heat Mass Transfer **43**, 4257-4269 (2003)
8. R. Pastuszko, Int. J. Therm. Sci. **125**, 197–209 (2018)
9. D. Cooke, S. G. Kandlikar, Int. J. Heat Mass Transfer **55** (2012)
10. A. Walunj, A. Sathyabhama, App. Therm. Eng. 128, 672–683 (2018)
11. R. Kaniowski, R. Pastuszko, L. Nowakowski, EPJ Web of Conferences **143**, 02049 (2017).

Selected thermal and flow issues in a reversed thermosiphon with a steam liquid lifter

Michał Klugmann[1,*], *Paweł* Dąbrowski[1], and *Dariusz* Mikielewicz[1]

[1]Gdańsk University of Technology, Faculty of Mechanical Engineering, Department of Energy and Industrial Apparatus, Narutowicza 11/12, 80-233 Gdańsk, Poland

Abstract. The work presents initial studies of a reversed thermosiphon with a steam liquid lifter and a minigap evaporator. The main interest of the authors is the evaporator, especially the heat exchange efficiency and flow issues, such as maldistribution and pressure drop. However, because the thermosiphon circuit of such a construction is characterized by cyclic working conditions, which are quite difficult in analysis the authors focused on this specificity first. The article presents the concept of a test circuit and preliminary results for distilled water and for a minigap evaporator with a smooth heat exchange surface 100 x 200 mm and minigap thickness of 1 mm. Visualizations of the separator's work and the minigap boiling structures are presented.

1 Introduction

All technology sectors are endeavouring to miniaturization with simultaneous performance improvement. That causes an increase of heat flux which is necessary to dissipate from the surface. For example, the average value of heat flux for computer processors is from 2 to 4.5 MW/m^2 (up to 45 MW/m^2 in small areas); for insulated gate bipolar transistors (IGBT) - from 6.5 to 50 MW/m^2. Hence, there is a need to focus on modern kinds of highly compacted and high efficiency heat exchangers design and to maximize the heat transfer. Over the past decades, many scientists have focused on mini and microchannel geometries. It have been well described in the literature. There were also concepts to apply such geometries to the construction of highly efficient, compact heat exchangers. Much less attention was paid to the minigaps. Meanwhile, the minigap geometry seems to be – in some cases - even more interesting because of technically easier construction and very high heat transfer coefficients which have already been observed.

In addition, in recent times, the renaissance of all types of passive heat and mass transfer techniques has been observed. For example, in the water supply technique the gravitational systems based on the natural differences in terrain levels return to the favours, which were commonly used at the turn of the 19th and 20th century. Noticeable attention is paid to passive methods of heat exchange intensification. Thermosiphon circuits, a few decades ago widely used in heating systems, now focus the researchers' attention again. However, these are not only just simple systems that use single-phase convective flow from the lower to the upper heat source but also reversed systems, giving the heat flow from top to bottom, using two-phase flows and microscale geometries. This results in completely new research issues, both in the field of basic research as well as operational problems. One of such issues is cyclic/pulsatile work characteristic for some designs of reversed thermosiphons. This is a specific type of working conditions, from the borderline of a steady and transient state, which has rarely been investigated by researchers yet. One of the issues that takes on a key meaning in this type of work is a maldistribution phenomenon which can occur wherever there is a necessity of the working medium distribution within a complex flow system [1]. Examples of such systems may be minichannel and minigap plate heat exchangers.

The aim of the present paper is the research of reversed thermosiphon with a steam liquid lifter regarding the way the system works and the impact of basic parameters on its operation. The working fluid is distilled water operating under sub-atmospheric pressures.

2 State of the art

Different aspects of two-phase thermosiphons performance have been investigated over the years. Filippeschi [2] analysed Closed Loop Two Phase Thermosiphon (CLTPT). The purpose of this study was to find a correlation between mass flow rate and heat flow rate. Authors have been running experiments with water and ethanol as a working fluid, where heat flow rate was up to 1.7 kW. The mass flow rate were measured at three different modes: a continuous mode, an integral mode and an indirect mode. They reported that the quantitative prediction of the performance of CLTPT appears to be quite difficult due to the individual

*Corresponding author: garyb@pg.gda.pl

and the inter-correlative effects of the various variables(operating pressure, filling ratio) but the maximum mass flow rate always can be found with the increase of the heat load.

Buschmann and Franzke [3] conducted experiment on vertical two-phase thermosiphon utilizing deionized water and water based nanofluids with titanium dioxide and gold particles in different concentrations. Their results show a decrease of the thermal resistance with increasing concentration of nanoparticles. Authors achieved a maximal thermal resistance reduction of about 24%.

The employment of nanoparticles in thermosiphon performance was also an aim of study conducted by Cieśliński [4]. The results obtained with a nanofluid water-Al2O3 solution as a working fluid were compared with a distilled water. Study showed that addition of Al2O3 nanoparticles results in a higher overall heat transfer coefficient but increase in nanoparticle concentration from 0.01% to 0.1% causes a decrease of the overall heat transfer coefficient.

The air to air two phase closed thermosiphon based heat exchangers were investigated experimentally by Danielewicz et al. [5]. Authors found out that effectiveness of heat exchanger increased as the ratio between both massflow rates increased. They validated the computational model that was used to predict the heat transfer coefficient, effectiveness, pressure drop and heat extraction duty of featured heat exchanger.

Zhu and Hrnjak [6] presented experimental evaluation of a two-phase thermosiphon loop in two modes: conventional and with separation. Microchannel heat exchanger was used as an evaporator. Authors measured temperature distribution over the evaporator surface in time. It seems that thermosiphon with separation of vapor and liquid phase stabilizes in shorter time than conventional one. Moreover separation causes lower mass flow rate oscillations at start-up. Besides, at low heat load in steady state operation, the total pressure drop in two phase line of the system with separation is lower than that of the system without separation.

Another utilization of microchannels in passive, gravitational systems has been realized by Panse and Kandlikar [7]. They used Open Microchannel Manifold (OMM) with taper configuration as an evaporator in thermosiphon loop. Ethanol was used as a working fluid. Results shown that OMM provides superior heat transfer while maintaining a low pressure drop performance. Authors prepared the boiling curves for 4% taper and 6% taper manifolds. A maximum heat flux of 136 W/cm^2 was recorded at a wall superheat of 42°C resulting in a highest heat transfer coefficient of 34.1 kW/(m^{2}K) with 6% taper manifold.

Khodabandeh and Furberg [8] conducted research on single rectangular microchannel working as an evaporator in a two-phase thermosiphon loop. Authors enhanced the heat transfer surface of the evaporator with a copper nano- and micro-porous structure. Then the heat transfer of the enhanced evaporator was compared to a smooth surface evaporator. The main observation was that the enhanced structure surface decreased the oscillations at the entire range of heat fluxes and enhanced the heat transfer coefficient.

Bieliński and Mikielewicz [9] developed a one-dimensional separated two-phase flow and heat transfer model in a closed thermosyphon loop with minichannels and a minipump. Studies indicated that the presented model is a power tool for increasing the effectiveness of computer cooling. If the mass flow rate is not high enough to circulate the necessary fluid to transport heat from evaporator to condenser, the minipump can be used to promote natural circulation. Another paper by Bieliński [10] validated mentioned above model by experimental measurements of volumetric flow rate. Author claimed that the steady-state conditions played a dominant role during the two-phase flow in the thermosiphon loop.

As shown above, there are some works that are trying to develop a microchannel geometries in a gravity-driven loops, i.e. thermosiphons. But there are hardly any research that conducts experiment on a microgap heat exchanger in application to passive systems (thermosiphons or heat pipes). However, there are some works on microgaps that are not necessarily related to passive systems but can be useful.

Nishikawara and Nagano [11] developed a mathematical model to investigate the heat transfer characteristics of a capillary evaporator in a loop heat pipe with a microgap between the case and the wick. Calculations were compared with measured results obtained for a polytetrafluoroethyleneethanol loop heat pipe. Authors found out that the heat-transfer coefficient have a local maximum against the gap distance.

Tamanna and Lee [12] dealt with various configurations of microgaps with expanding depth towards the outlet. A microgap with a constant depth of 200 µm was used as a reference one. The gaps with increasing depth had the same initial depth of 200 µm (inlet) and the final depth of 300 µm and 460 µm (outlet). Water with different mass fluxes from G=400 kg/m^{2}s to G=1000 kg/m^{2}s was used as the working fluid. A variable heat flux of 0 W/cm^2 ÷ 80 W/cm^2 was supplied to water. Authors claimed that inlet pressure oscillations increase with increasing mass flow rate for a specific heat flux. This oscillations also increase with increasing heat flux for a predetermined mass flow rate. Pressure oscillations at the inlet to the exchanger can be lowered, using microgaps with expanding depth from 200 µm to 300 µm. However, it should be emphasized that further increasing of depth to 460 µm does not give positive results. A similar situation occurs when it comes to temperature fluctuations.

Alam et al. [13] carried out tests to determine the heat transfer coefficient. They used microgaps at three different depths: 190 μm, 285 μm and 381 μm, through which water ran at an inlet temperature of 86°C. The influence of the mass flow rate and the heat flux supplied to the heat exchanger on heat transfer coefficient and pressure drops was investigated. They found out that smaller the size of the microgap, the higher the local heat transfer coefficient. In addition, reducing the mass flow of the flowing medium also results in an increase in the heat transfer coefficient.

Later Alam, Lee, and Yap [14] decided to carry out research on the same experimental rig as before, but this time they put the main emphasis on the surface roughness of microgaps. Flow visualization showed that the surface roughness has a significant impact on the formation of bubbles. The bubble nucleation density increases with increasing roughness. It is worth mentioning that this effect is greater for larger gap depths. The most important positive effect that can be achieved by using rough surfaces is to improve the heat transfer coefficient. The authors did not observe increased pressure drops when using rough surfaces. However, higher roughness results in increased fluctuations in pressure at the inlet to the heat exchanger.

As can be seen from the literature review, thermosiphons are very promising heat transfer devices. The lack of external source of working fluid circulation results in higher system efficiency and better reliability. There are some applications of passive methods with microchannels that provides high performance and miniaturization. Unfortunately, there is hardly any study that combines minigaps or microgaps in passive systems, and those structures ensures very good performances of heat transport. Current study aims to fill this gap in literature. Authors built the specially designed apparatus of reversed thermosiphon with minigap evaporator. The evaporator is above the condenser, which is justified in solar applications.

3 Experimental setup

Experimental rig, Fig.1., like any inverted thermosiphon has an evaporator located above the condenser. In the discussed system, this height difference, and thus, the lifting height is about 1.35 m. The evaporator (4) is a plate heat exchanger composed of a single smooth plate with a heat exchange surface of 100 x 200 mm. Between the plate and the cover of the exchanger, a 1 mm thick minigap is arranged.

The heat for the evaporator can be supplied in two ways: a) radiative, through a transparent cover, similar to a solar collector; b) using a water circuit, where the temperature and flow of the water can by regulated in the thermostat. The assumption is to obtain the boiling structures in the minigap which can push the liquid upward into the separator (5) located above the evaporator. It is a liquid transport mechanism similar to those found in the mammoth pump.

In the separator, a liquid part of the working medium flows to the condenser through a pipeline (2) in the bottom of the separator, while the vapour part goes to the pipeline (3) at the top of the separator and further goes to the condenser through the second tube.

The condenser (6) is a coil immersed in a thermostat which allows precise temperature control and, thereby, the amount of heat removed from the system.

In steady state with an equal thermal balance, the system works in a characteristic cyclical way, throwing the portions of liquid and vapour into the separator at regular time intervals.

Evaporator and separator are equipped with sight-glasses To observe and register the minigap boiling structures and the way of the separators work.

The system is equipped with temperature sensors at locations of interest and also at additional points on the length of pipelines to accurately analyze the temperature distribution in the entire circuit. The pressure difference is measured at the inlet and outlet of the evaporator. In addition, system pressure is measured.

The flowrate of the working medium is measured at the inlet to the evaporator (7) and at the outlet of the liquid pipeline from the separator (8). The flow rate in the vapour pipeline must be estimated from the mass balance.

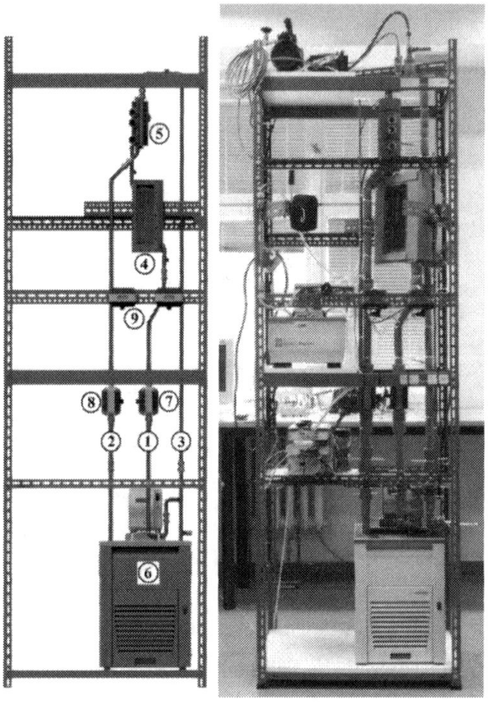

Fig. 1. Diagram and picture of experimental rig: 1- cold (input) pipeline, 2 – hot (output) liquid pipeline, 3 – hot (output) vapour pipeline, 4 – minigap evaporator, 5 – separator, 6 – condenser, 7 – flowmeter for cold pipeline, 8 – flowmeter for hot liquid pipeline, 9 – display.

4 Experimental results

4.1 System operation over the time

The following is a comparison of the way the system works when supplying heat by means of a water circuit at two different mass flow rates. This results in an approximately twofold difference in the heat flux density supplied in the evaporator to the working medium. Registrations were carried out for the time of 100 seconds.

https://doi.org/10.1051/e3sconf/20187002009

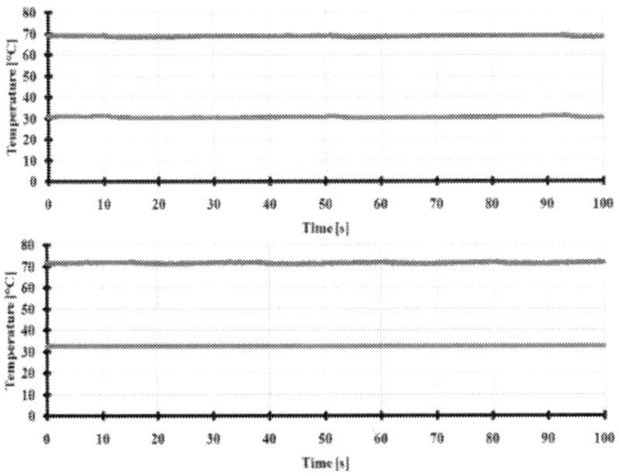

Fig. 2. Temperatures in evaporator (blue – input, red – output) for average heat flux supplied to minigap: top = 7.6 kW/m^2, bottom = 14 kW/m^2.

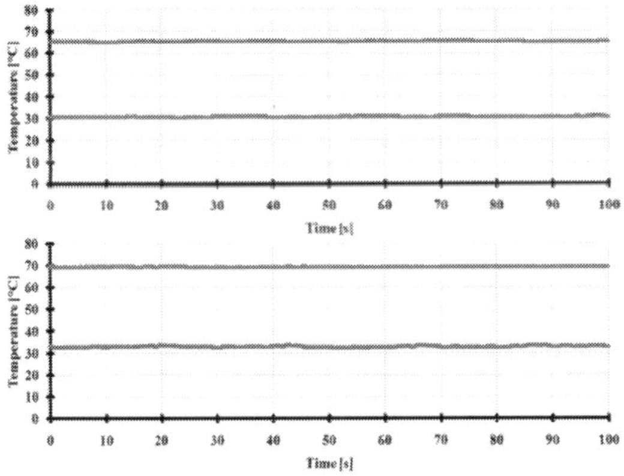

Fig. 3. Temperatures in condenser (blue – input, red – output) for average heat flux supplied to minigap: top = 7.6 kW/m^2, bottom = 14 kW/m^2.

Equal temperature distributions in time in the evaporator and condenser, as well as the fact that the working fluid is being cooled in the condenser to its initial temperature prove that the system, in both cases, works in even heat balance conditions and attains the equilibrium conditions.

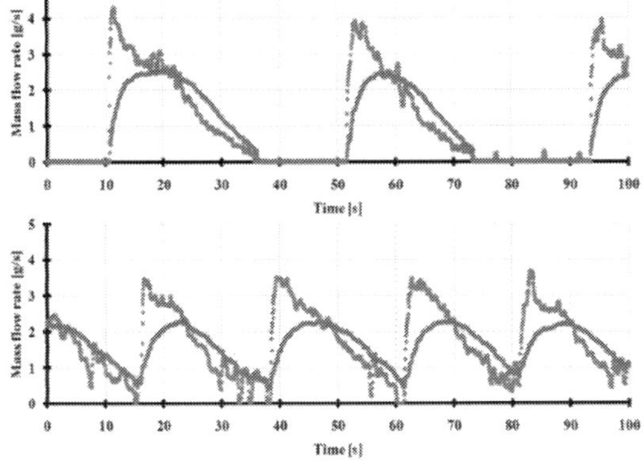

Fig. 4. Mass flowrate: blue – at evaporator input, red – at separator output (liquid pipeline) for average heat flux supplied to minigap: top = 7.6 kW/m^2, bottom = 14 kW/m^2.

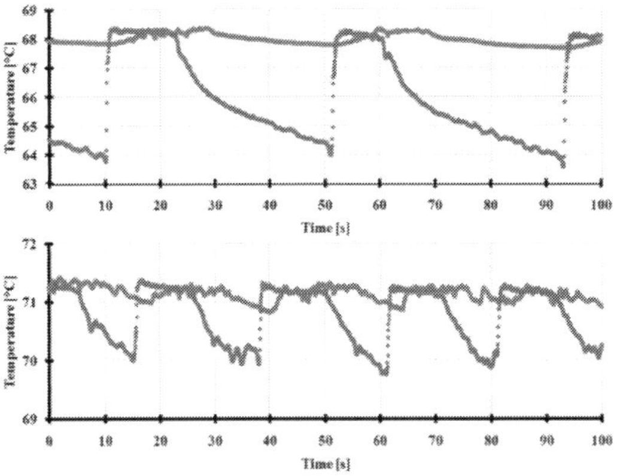

Fig. 5. Temperatures on separator's ouputs: blue – liquid pipeline, red – vapour pipeline for average heat flux supplied to minigap: top = 7.6 kW/m^2, bottom = 14 kW/m^2.

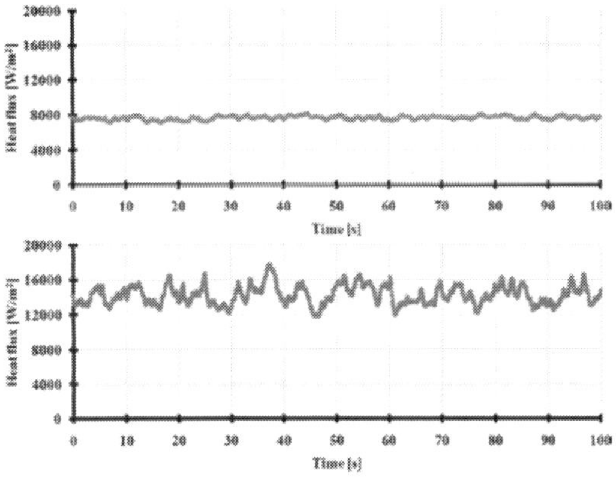

Fig. 6. Heat flux supplied to minigap: top = 7.6 kW/m^2, bottom = 14 kW/m^2.

In the mass flow rate charts there appears characteristic pulsating nature of the system operation. It can be seen, that for almost twice heat flux delivered the almost twice density of work cycles occurs.

Considering that - in both cases - the temperature differences in exchangers are of the same order of magnitude, we can conclude about how the system increases the amount of transferred heat. This is confirmed by the Fig. 7. where more heat flux density points are included.

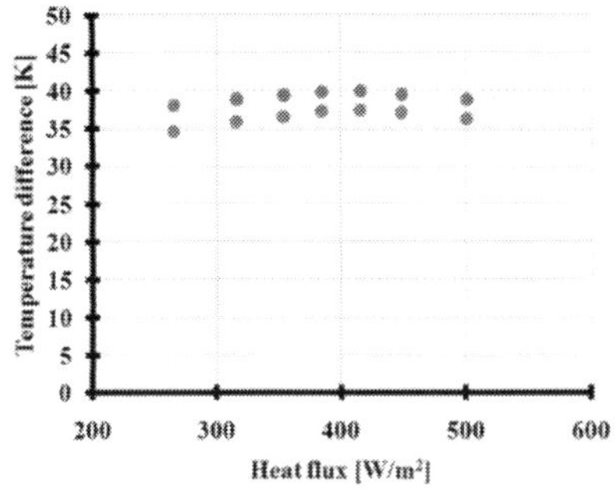

Fig. 7. Temperature differences as a function of heat flux supplied to minigap; red – evaporator, blue – condenser.

Qualitatively, the shape of the mass flux curves is similar in both cases. For a higher heat flux, the maximum values are even smaller. On the other hand, for a smaller heat flux the periods of no flow appears, what indicates that it's not an optimal work point for this system.

For heat flux equal to 14 kW/m^2 a much smaller temperature gradient of vapour at the exit from the separator can be seen (less than 1.5 K vs 4 K for 7.6 kW/m^2). This can also testify the more even system operation.

4.2 Visualisation of work of the key system components

Because one of the walls of the evaporator is made of heat-resistant glass it is possible to register flow structures in parallel with the collection of measurement data. So, by observing the work of the evaporator it can be determined whether the given parameters are optimal or not.

The visualization of the separator's work is carried out using its special design made of plexiglass. For testing more aggressive fluids the brass version is used, as can be seen on the experimental rig diagram, Fig. 1.

The preferred way of work occurs when the boiling appears on the entire surface of the exchanger's plate. This can happen when the temperature of the working medium at the inlet to the evaporator is close to the saturation temperature. The final boiling structure at the evaporator outlet also matters because it is responsible for mass transport to the separator. For this reason, too high vapour quality is not preferred because most of the mass is transported by a less effective vapour phase.

Fig. 9. Presents a fairly favourable situation with an inlet temperature of about 10 K below the saturation temperature. There is a churn structure at the outlet of the evaporator what means that such a structure appears in the separator (Fig. 11.)

The situation of Fig. 10. Occurs when a bubbly structure is obtained at the outlet of the evaporator.

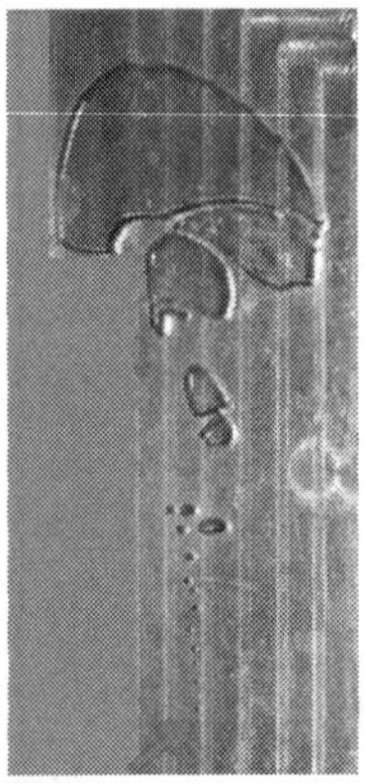

Fig. 8. Formation and growth of vapour bubbles; the arrow indicates nucleation point.

Fig. 9. The development of flow boiling on a vertically oriented plate of minigap evaporator (minigap thickness = 1 mm); visible flow structures (from the bottom to top): bubbly, slug and churn.

Fig. 10. Registration of single working cycle of separator ejecting boiling liquid of a bubbly structure (duration = 1 s, time interval = 80 ms, exposure time = 0.5 ms).

https://doi.org/10.1051/e3sconf/20187002009

Fig. 11. Throw of a churn flow structure in separator (exposure time = 0.5 ms).

5 Experimental uncertainty analysis

The system experimental uncertainty was determined using the sequential perturbation method of error analysis [15]. This method allows determining of the total experimental error by including errors originating from individual sources into general database and averaging it using a root sum square method (RSS). According to this method, part of measurements made using measuring instruments shows a relatively high accuracy and the resulting inaccuracies should not exceed several percent. A detailed list can be found in the Table 1.

Table 1. Partial experimental uncertainties.

Measured number	Measuring range	Absolute error	Relative error
Temperature	90 K	0.5 K	0.6 %
Evaporator pressure difference	5 kPa	10 Pa	0.2 %
System pressure	100 kPa	1.8 kPa	1.8 %
Mass flowrate (working medium)	5 g/s	0.125 g/s	2.5 %
Mass flowrate (heating circuit)	8.33 g/s	0.172 g/s	2.1 %
Metric dimensions	0.2 m	0.00005 m	0,025 %
Minigap thickness	0.001 m	0.00005 m	5%

Using the above mentioned method, it was estimated that the thermal balance of the entire system was determined with an error of 5%.

6 Conclusions

1) The system shows a cyclic way of operation, however ensuring stable way of heat transfer in the evaporator and condenser.
2) The experiment proved that a highly effective heat exchange is possible in the exchanger of a relatively small heat exchange surface. However, from the point of view of planned solar applications, it can be seen that the achieved heat flux densities are much too large. It follows, that the distilled water is not suitable as a working medium for such an applications due to the very large heat capacity. The necessity to work in a deep underpressure is also a serious disadvantage.
3) A very advantageous feature of the tested system is it's dynamic adjusting to the amount of supplied heat by changing the number of working cycles in the time, while the temperatures in exchangers remains unchanged.

Acknowledgement

The work presented in the paper was funded from the National Science Centre research project No. 2015/19/D/ST8/03201 in years 2016–2019.

References

1. P. Dąbrowski, M. Klugmann, and D. Mikielewicz, AoT, **38/3**, 135 (2017)
2. A. Franco and S. Filippeschi, Exp. Therm. Fluid Sci., **51**, 302 (2013)
3. M. H. Buschmann and U. Franzke, Int. J. Refrig., **40**, 416 (2014)
4. J. T. Cieslinski, AoT, **37/2**, 23 (2016)
5. J. Danielewicz, M. A. Sayegh, B. Śniechowska, M. Szulgowska-Zgrzywa, H. Jouhara, Energy, 77, 82 (2014)
6. L. Zhu, P. Hrnjak, Appl. Therm. Eng., **121**, 879 (2017)
7. S. S. Panse, S. G. Kandlikar, Int. J. Heat Mass Transf., **106**, 546 (2017)
8. R. Khodabandeh, R. Furberg, Int. J. Therm. Sci., **49**/7, 1183 (2010)
9. H. Bieliński, J. Mikielewicz, AoT, **37/1**, 3 (2016)
10. H. Bieliński, AoT, **37/3**, 109 (2016)
11. M. Nishikawara, H. Nagano, Int. J. Therm. Sci., **102**, 39 (2016)
12. A. Tamanna, P. S. Lee, Int. J. Heat Mass Transf., **89**, 390 (2015)
13. T. Alam, P. S. Lee, C. R. Yap, L. Jin, Int. J. Multiph. Flow, **42**, 164 (2012)
14. T. Alam, P. S. Lee, C. R. Yap, J. Heat Mass Transf., **64**, 28 (2013)
15. R. J. Moffat, Exp. Therm. Fluid Sci., **1**, 3 (1988)

Experimental research on hydrodynamic instabilities during condensation of the pro-ecological refrigerant R1234yf in tubular minichannels

Waldemar Kuczyński[1,*], and *Aleksander* Denis[1]

[1]Technical University of Koszalin, Department of Energy, av. Raclawicka 15-17, PL 75-620 Koszalin, Poland

Abstract. The following paper presents the results of preliminary experimental research on the influence of instabilities of a hydrodynamic type on the condensation phase change process in tubular minichannels. The research was focused on a new pro-ecological refrigerant, R1234yf, intended as a substitute for R134a that currently is being phased out. The flow condensation phase change process was investigated for both steady and un-steady conditions in singular tubular minichannels with an internal diameter d = {1,44; 2,30; 3,30} mm. The scope of the analysis of the experimental data covered an estimation of propagation velocities for both pressure and temperature instabilities as well as the shrinkage of the condensation zone. The results were also compared with the previous results obtained for the flow condensation phase change of R134a refrigerant in tubular minichannels with the same internal diameters.

1 Introduction

The refrigerant is a substance or a blend that circulates within a closed loop. It absorbs heat while evaporating at a low temperature and pressure, and it rejects heat while condensing at a high temperature and pressure. Refrigerants are used in fridges, freezers, air conditioners, heat pumps and Organic Rankine Cycles. Due to an unequivocal impact on the environment, refrigerants are subjected to many international regulations, such as:

- The Montreal Protocol on Substances that Deplete the Ozone Layer (1987),
- The United Nation Framework Convention on Climate Change (1992),
- The Regulation (EU) No 517/2014 of the European Parliament and of the Council of 16 April 2014 on Fluorinated Green-House Gases.

The first agreement introduced the Ozone Depletion Potential (ODP) indicator that illustrates the negative impact of each refrigerant on the ozone layer. As a result of this agreement, all chlorine based refrigerants (ODP>0) were banned from use. The second agreement introduced the Global Warming Potential (GWP) indicator that illustrates a contribution of each refrigerant to the global warming effect with carbon dioxide as a reference pollutant ($GWP_{CO2}=1$). The EU Regulation No 517/2014, however, sets a roadmap for the phase-out of the so-known as fluorinated green-house gases – in particular hydro-fluoro-hydrocarbons (HFC), per-fluoro-hydrocarbons (PFC), sulphur hexafluoride and other fluorinated refrigerants. It aims at a 66% reduction in emission of fluorinated green-house gases (base level 2010). One of those green-house gases is the R134a refrigerant that is currently commonly used in vehicle air conditioning systems and heat pumps. According to the roadmap, it has already been banned from use in brand new installations since January the 1st 2017. Currently, the replacement for the R134a refrigerant are R1234yf for vehicle air conditioning systems and R1234ze(E) for heat pumps. The new refrigerants have significantly lower GWP indicators, $GWP_{R1234yf}=4$, $GWP_{R1234ze(E)}=7$, they are non-toxic hydro-fluoro-olefins (HFO), but contrary to R134a, they are officially considered as slightly flammable [1,2,3,4,5].

All in all, from the engineering perspective, the main conclusion of those regulations is that nowadays the priorities in the refrigerant slection process look as follow:

- environmental impact,
- operational safety,
- thermodynamic efficiency,
- other technical aspects (like the corresponding type of oil, etc.),
- economics.

The following research was done using minichannels. According to Kandlikar [6], they have an ID (internal diameter) between 0.2 and 3 mm, and according to Mehendale [7] between 1 and 6 mm. Wider channels are known asknown as conventional ones, and smaller are known asknown as meso, micro or nano channels depending on their ID.

The flow through conventional channels is governed by inertia forces. In minichannels, however, the flow is governed by capillary forces like viscosity and surface tension [8,9].

Minichannels allow one to shrink heat exchangers and to reduce the coolant mass in the system (in other

* Corresponding author: waldemar.kuczynski@tu.koszalin.pl

 https://doi.org/10.1051/e3sconf/20187002010

words, F-gas emission in the case of leakage). Moreover, much higher heat flux density enables a rapid development of high power density electronics. Although it comes at the expense of higher hydraulic losses, only in the case of a high flow rate the pressure penalty exceeds the heat transfer gain [10]. The heat transfer can be further enhanced using the microfins [11], parallel channels configuration [12] or by operating in an unstable two-phase region [13].

Instabilities are inherently connected with phase changes, evaporation and condensation, and the two-phase flow [14]. The first discovered instability was the Ledinegg's flow excursion [15] – the pressure drop versus flow rate (or the N-curve) for heated channels may have up to 3 different points of operation. The occurrence of instabilities depends on the flow rate, heat flux and medium conditions at the inlet (like quality, superheating or subcooling) [16] and system characteristics (like the presence of internal compressible volumes) [17]. The instabilities mainly affect flow distribution in parallel channels [18], flow pattern [19] and pressure drop [20].

In conclusion, most researchers focus on flow boiling instabilities due to more severe potential consequences connected with an uncontrolled temperature rise (like burn-out or meltdown). Some researchers focus on flow condensation instabilities, some on the thermal-hydraulics of mini and microchannels, but only few have investigated flow condensation instabilities in mini-channels [21].

The experimental research on the influence of hydrodynamic instabilities on the flow condensation process in minichannels has already been conducted for those refrigerants that are currently being phased-out [22 - 33]. However, the influence of this type of instabilities on the phase change of new pro-ecological refrigerants is not known. It is well known that during the flow condensation of the currently used refrigerants under hydrodynamic instabilities, the instabilities propagate in a wave form. The two-phase mediums are especially susceptible for this phenomenon.

Hydrodynamic instabilities occurring in two- and multi-phase mediums during flow condensation inside minichannels may have a different origin [8]:

- onset of phase-change,
- density variation,
- flow variation,
- pressure variation,
- temperature variation,
- quality variation,
- shock-wave,
- Kelvin-Helmholtz instability due to the interference of surface waves.

Among different instabilities connected to mass flux oscillations, the acoustic wave and condensation front are of a prime importance. Each one of them has a different characteristics, especially the propagation velocity within the medium. Both of them are considered as dissipative and dispersive. Dissipative means that the process is irreversible (generates entropy), thus it does not self-sustain, but decays in time. Dispersive means that the propagation velocity strongly depends on the frequency of induced instabilities.

The condensation front manifests itself as a displacement of a sudden drop of wall temperature, in other words, a shrinkage of the high efficiency condensation zone.

A typical condenser during a steady state operation is supplied with a superheated gas and discharges a sub-cooled liquid. Therefore, it can be divided into 3 characteristic zones: cooldown of superheated gas, two-phase condensation and sub-cooling of liquid. The size of the condenser is constant; thus, any transient, hydrodynamic instabilities in particular, affects\ the distribution of these zones. As a result the performance of the condenser drops, and in the worst case scenario, the whole system may be damaged.

All the wave phenomena are strongly coupled, in other words, an onset of one of them usually triggers an onset of all of them. For example, a dynamic change of the flow rate triggers variations of density, pressure, temperature and quality. This paper aims to characterize the hydrodynamic instabilities present during condensation of the pro-ecological refrigerant R1234yf, both qualitatively and quantitatively.

2 Apparatus

The diagram of the apparatus used for the experimental part of the research is shown in FigureFigure 1. Refrigerant leaves the tank in liquid state at a high pressure and room temperature. Then it decreases pressure thanks to expansion valve and evaporates while absorbing ambient heat in the evaporator. The compressor increases the pressure and temperature of the gaseous refrigerant. While leaving the compressor, the refrigerant's stream splits in two: most of the flow returns to the tank via the condenser, where it turns back into a liquid while rejecting heat, and the remaining part (the exact amount is adjusted using the regulation valve) is directed to the test section. At the entrance, there is a cut-valve that induces the instabilities. Its timings are controlled via a controller.

The test section (Figure 2) is a minichannel armed with type K thermocouples and pressure sensors that is immersed in water bed. To maintain the constant temperature of the cooling water, a water pre-conditioner is used. A residual gaseous refrigerant is condensed in the after cooler. The liquefied refrigerant flows through a flow-meter and returns to the main stream. Analog voltage-based signals from the thermocouples, the flow meter and the pressure sensors are converted to digital ones using A/D adapters and stored using a PC.

Figure 3 shows a sample of a change of flow in time registered using the apparatus, there is a slight delay between the pressure wave and the flow recovery. Just before the cut-valve re-opening, the minichannel is being re-flooded by liquid and after the re-opening the incoming gas encounters a "cork" of stagnant liquid. Moreover, the maximum flow rate after the cut-valve re-opening exceeds the steady state flow rate (~1.85 kg/h) multiple times, depending on the frequency.

 https://doi.org/10.1051/e3sconf/20187002010

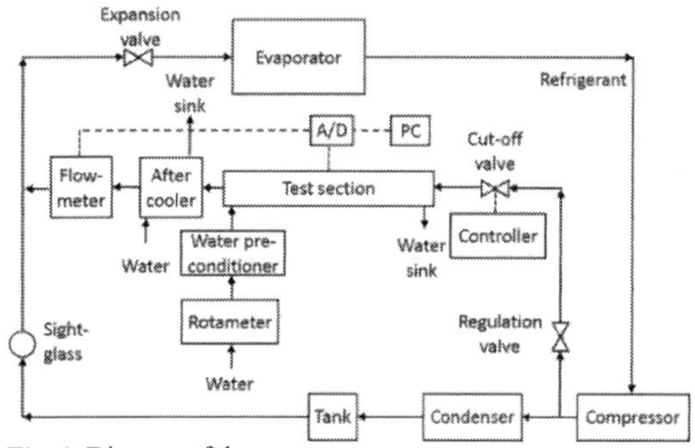

Fig. 1. Diagram of the apparatus.

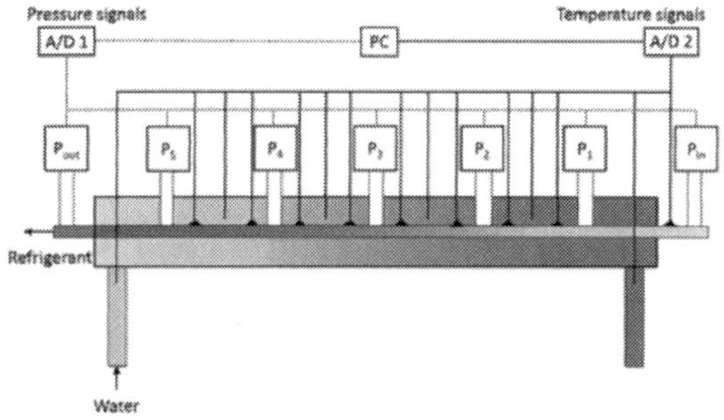

Fig. 2. Diagram of the test section.

Figure 4 represents an evolution of pressure in time registered by multiple sensors installed along the minichannel's length. Right after the cut-valve closure, the pressure suddenly drops, next it recovers to a plateau and after the valve re-opening it increases very rapidly. At first glance, they seem identical, but due to pressure losses the signals are slightly shifted vertically and horizontally due to the pressure wave travelling through the system.

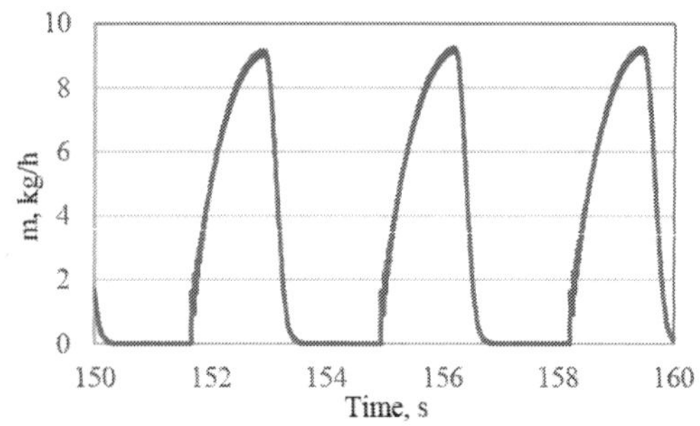

Fig. 3. Sample of flow oscillations.

Figure 5 illustrates temperature oscillations registered by multiple sensors distributed along the channel. The thermal phenomena are much slower than the pressure wave, thus the differences in signals are much easier to observe. Again, the vertical shift is due to the heat transfer between the refrigerant and cooling water, and the horizontal shift is due to the condensation front travelling through the system.

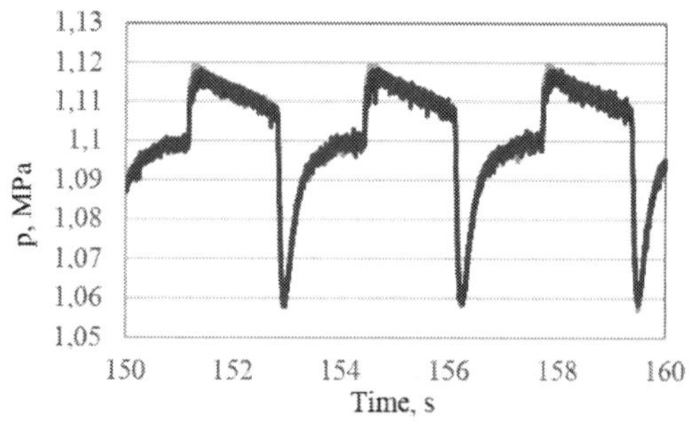

Fig. 4. Sample of pressure oscillations for multiple sensors

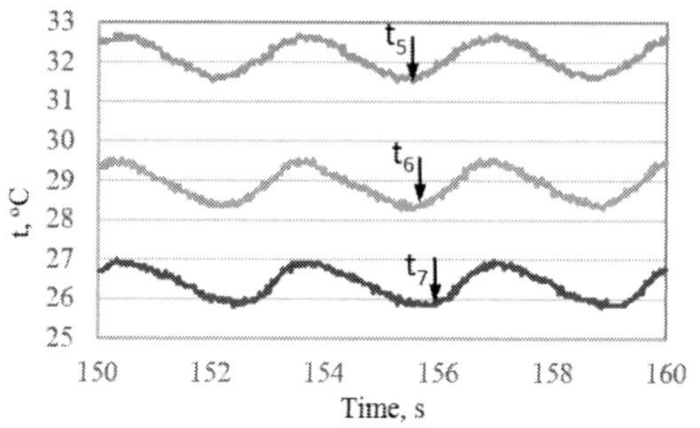

Fig. 5. Sample of temperature oscillations for multiple sensors.

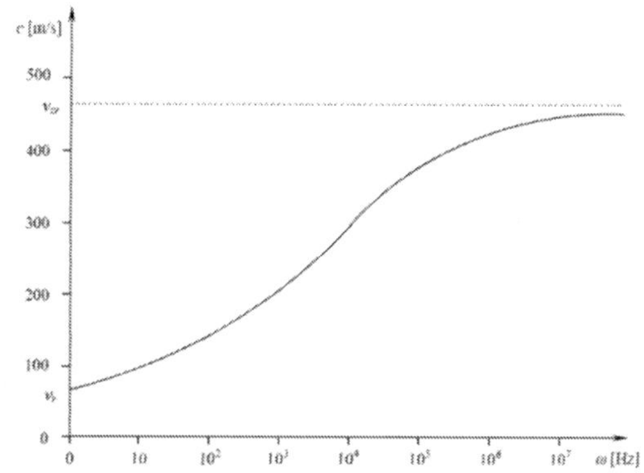

Fig. 6. Pressure wave velocity as a function of frequency for conventional channels [25].

3 Methodology

The literature provides a basic theory on the dispersive character of the periodic instabilities, or in other words how the frequency of those disturbances affects the their propagation velocity. The curve (shown in Figure 6) has 2 characteristic asymptotes. The first one is known asknown as an equilibrium velocity:

$$v_E = v(\omega \to 0) \quad (1)$$

and it represents a situation when the wave period is longer than the relaxation time (time of re-establishing of the thermodynamic equilibrium) and the pressure wave propagates slower. The second asymptote is known as a frozen velocity:

$$v_F = v(\omega \to \infty) \quad (2)$$

and it represents a situation when the wave period is shorter than the relaxation time and the pressure wave propagates much faster.

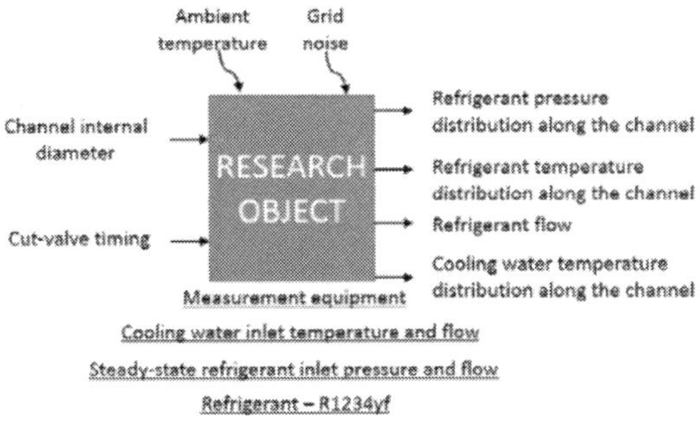

Fig. 7. Research object.

The research object for the experimental part is shown in Figure 7. The internal diameter of the channel and cut-valve timings were the input values. Flow, and pressure and temperature distributions were the output values. The constants of the model were the equipment type, cooling water inlet conditions and steady state refrigerant inlet conditions. Known influences that could not be eliminated were ambient temperature variation during the long lasting experiments and noises coming the complex power supply grid that supplies electricity to the laboratory and noticeably affected the operation of the measurement system. The experimental data was acquired using the DasyLab software.

The methodology of the experimental research part was analogical to the other researchers of this subject, like Teng, Lee, Kuczynski or Bohdal. Signals from multiple sensors (examples shown in Figures 8 and 9 for pressure and temperature, respectively) were compared in order to find a time shift between their responses to the cut-valve operation.

For each frequency, multiple time shifts were averaged in order to minimize the uncertainty of the average value. The distance between sensors was well known, and the final results were obtained using this simple formula – distance over time shift:

$$\bar{v} = \Delta l / \Delta \bar{\tau} \quad (3)$$

where:
$\Delta\tau$ – time shift,
Δl – distance between sensors.

An estimation of time shifts (Figures 8 and 9) "by hand" could be arbitrary. Therefore, to eliminate the human factor, a curve fitting procedure was employed. For pressure signals the 4 parameter sigmoid was selected as a reference curve:

$$y(\tau) = \frac{a}{1+exp(-k*(\tau-\tau_0))} + c \quad (4)$$

where:
a – gain,
k – shape parameter,
τ_0 – inflection point,
c – initial value.

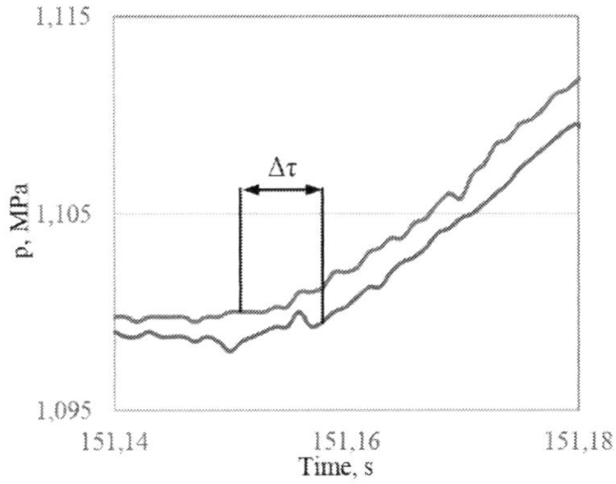

Fig. 8. Time shift between pressure sensors.

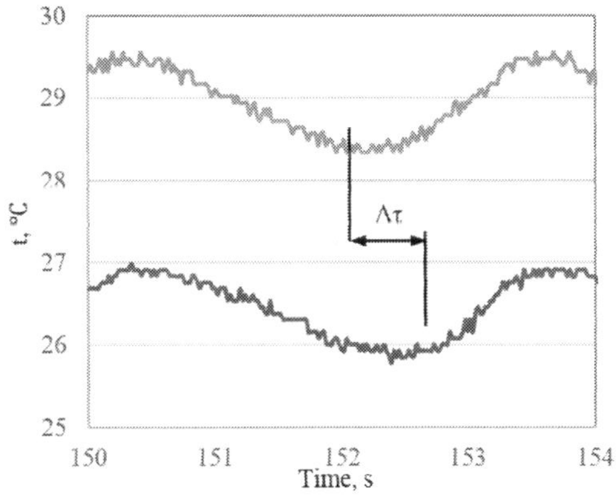

Fig. 9. Time shift between temperature sensors.

For temperature signals, however, the resized and shifted sinusoid was selected as a reference curve:

$$y(\tau) = a * sin(\tau + \tau_0) + c \quad (5)$$

where:
a – resize parameter,
τ_0 – phase,
c – average value.

The curve fitting procedure was done using a combination of Python's libraries such as numpy, pandas, pylab and scipy.

4 Results

The scope of the preliminary research covered:

- One refrigerant – R1234yf,
- Three internal diameters ID={3.3; 2.3; 1.44 mm},
- Single channel configuration,
- Refrigerant mass flux G=60 ÷ 316 kg/(m^2*s),
- Refrigerant inlet pressure p_{in}=1.09 ÷ 1.17 MPa (saturation temperature t_{sat}= 42.6 ÷ 45.5 °C),
- Steady state and periodic hydrodynamic instabilities induced by a cut-valve for 40 different

frequencies (the wave period between 0.1 and 4 seconds with 0.1 increment).

Figure 10 shows the correlation between the velocity of the pressure wave and the frequency of periodic instabilities for the R1234yf refrigerant and three different internal diameters of the minichannel. The shape of the curves agrees with the one for conventional channels shown in Figure 6. The frozen velocity in minichannels, however, is reached for much smaller frequencies. Both frozen and equilibrium velocities decrease with the internal diameter of the channel. There are many similarities between the R1234yf and R134a refrigerants in terms of pressure instabilities. The values of frozen velocities for both refrigerants are comparable, nevertheless, the equilibrium velocities for the R1234yf are substantially higher. For ID=3.3mm, the frozen velocity reaches 350 m/s, which is comparable with the speed of sound for normal conditions. Therefore, it is reasonable to regard this size as a transition between mini and conventional channels.

Figure 12 illustrates a correlation between the velocity of the condensation front and the frequency of periodic instabilities for the R1234yf refrigerant and the three different internal diameters of the minichannel. The curves also demonstrate similarities with one shown in Figure 6. Again the frozen velocity is reached for much smaller frequency and both asymptotes decrease with the internal diameter. However, their values are much closer to the ones for R134a shown in Figure 13.

Figure 14 represents the time-averaged values of wall temperature along the channel during periodic instabilities for the R1234yf refrigerants. The plateau that is clearly visible for the highest frequency represents the stable two-phase condensation zone. For smaller frequencies, the plateau is not present at all. This means that the high efficiency condensation zone was constantly moving along the channel. In the case of R134a (Figure 15), the position of the high efficiency condensation zone was more stable.

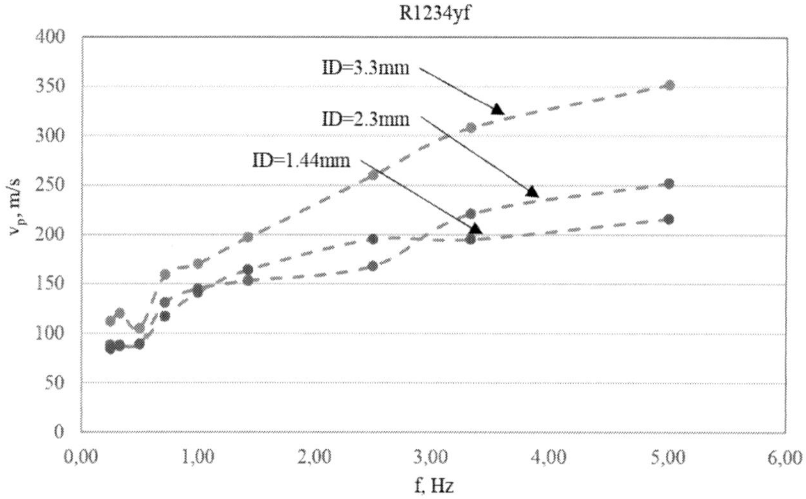

Fig. 10. Propagation velocity of pressure instability for R1234yf refrigerant and different IDs.

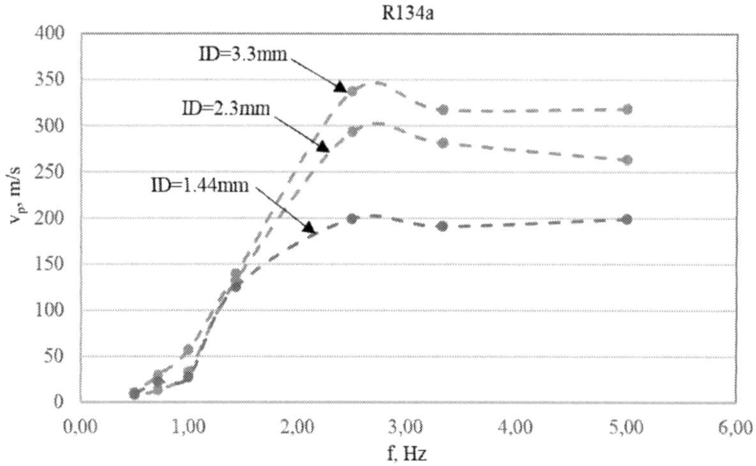

Fig. 11. Propagation velocity of pressure instability for R134a refrigerant and different IDs [8].

 https://doi.org/10.1051/e3sconf/20187002010

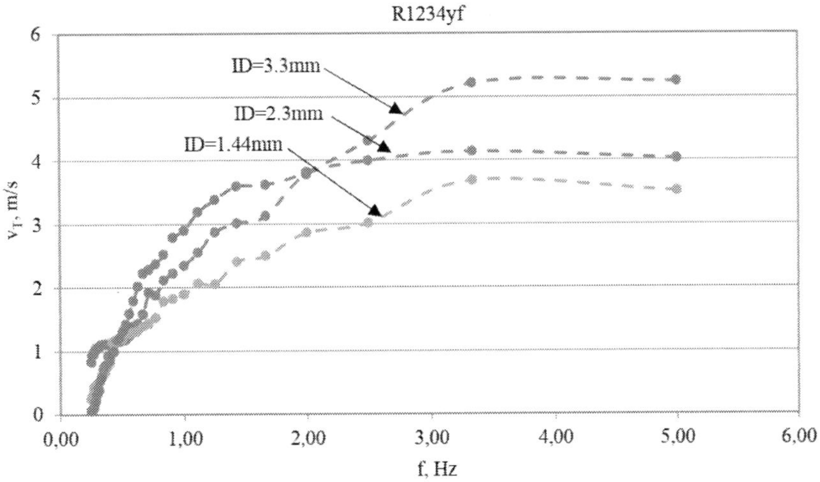

Fig. 12. Propagation velocity of temperature instability for R1234yf refrigerant and different IDs.

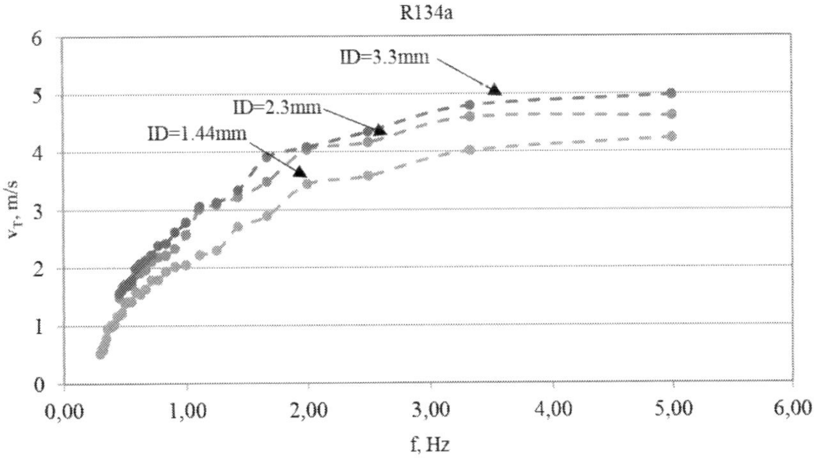

Fig. 13. Propagation velocity of temperature instability for R134a refrigerant and different IDs [8].

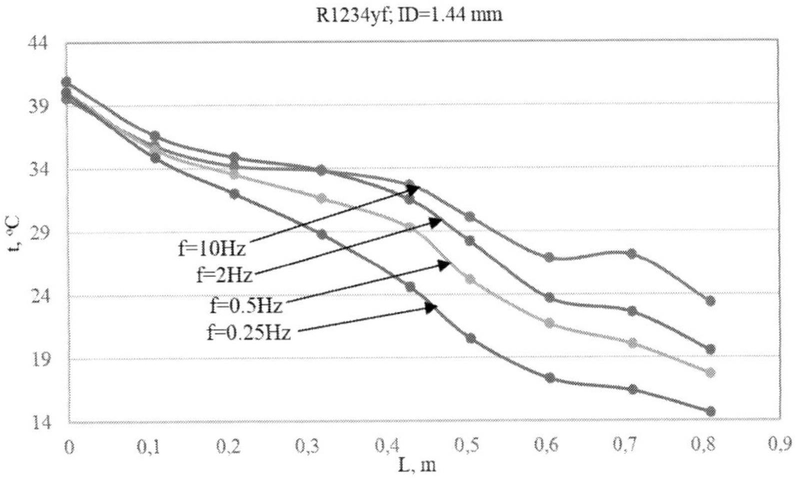

Fig. 14. Temperature distribution along the minichannel for R1234yf refrigerant (ID=1.44 mm) and different frequencies of instabilities.

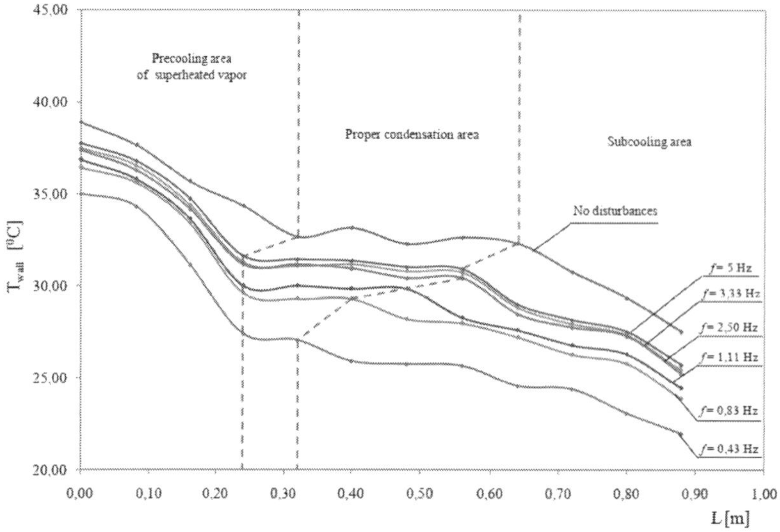

Fig. 14 Results of the experimental investigations concerning the influence of periodically generated instabilities on the length of the areas of superheat removal, proper condensation and subcooling of the R134a refrigerant depending on the frequency *f* of the generated disturbances in minichannel with the internal diameter $d = 1.44$ mm [8].

4 Conclusions

An analysis of premilimary experimental research on hydrodynamic instabilities during condensation of the pro-ecological refrigerant R1234yf in tubular minichannels showed that:

1) For pressure instabilities, shape of the functions obtained for the R1234yf refrigerant (Figure 10) agree with the one for conventional channels (Figure 6). However, frozen velocity in minichannels is obtained for much smaller frequencies. The asymptotes tend to decrease for smaller IDs like in the previous studies made for the R134a refrigerant (Figure 11). Frozen velocities for the R1234yf refrigerant are comparable with those for R134a found in the literature, equilibrium velocities; however, they are significantly higher. ID=3.3 mm is transitional between conventional and minichannels, thus the pressure wave velocity is comparable with the speed of sound for normal conditions.

2) For temperature instabilities, the shape of the functions obtained for the R1234yf refrigerant (Figure 12) agrees with the one for conventional channels (Figure 6). However, the frozen velocity in minichannels is obtained for much smaller frequencies. The asymptotes tend to decrease for smaller IDs like in the previous studies made for the R134a refrigerant (Figure 13). Both frozen and equilibrium velocities for the R1234yf refrigerant are comparable with those for R134a found in the literature.

3) The high efficiency condensation zone shrank with a decrease of the frequency of the instabilities as expected (Figure 14). However, the plateau that was clearly visible during the investigation of the R134a refrigerant (Figure 15) did not occur due to a significant movement of the zone.

References

1. T. Bohdal, Bubble boiling in flow of refrigerating media. Journal of Mechanical and Energy Engineering, Vol. 1 No. 1 (41), June 2017, pp. 57 – 64.
2. T. Bohdal, M. Kruzel, M. Sikora, Analisys of heat transfer coefficient during refrigerant condensation in vertical pipe minichannel. Journal of Mechanical and Energy Engineering, Vol. 1 No. 1 (41), June 2017, pp. 65 – 70
3. European Standard, PN-EN 14511:2009 *Air conditioners, liquid chilling packages and heat pumps with electrically driven compressor for space heating and cooling*
4. Polish Act, *Act on the renewable Energy sources of 20th February 2015*, Dz.U z 2015 r., poz. 478 (in Polish)
5. Report, Bitzer - Refrigerants, edition no. 19, 2016 (in Polish)
6. S. Kandlikar, S. Garimella, D. Li, S. Colin, M.R. King, *Heat transfer and fluid flow in minichannels and microchannels* (2006)
7. S.S. Mehendale, A.M. Jacobi, R.K. Shah, *Fluid flow and heat transfer at micro- and meso-scales with application to heat exchanger design*, Applied Mechanics Rev. 53(2000), 7, pp. 175–193
8. W. Kuczyński, *Hydrodynamic instabilities during condensation of refrigerants in minichannels*, 2013 (in Polish)
9. K.A. Triplett, S.M. Ghiaasiaan, S.I. Abdel-Khalik, A. LeMouel, B.N. McCord, *Gas–liquid two-phase flow in microchannels Part II: void fraction and*

pressure drop, Int. J. Multiphas Flow 25 (1999) pp. 395–410

10. Y.Y. Yan, T.F. Lin, *Condensation heat transfer and pressure drop of refrigerant R-134a in a small pipe*, Int. J. of Heat and Mass Transfer (1999) pp. 697 – 708
11. T. Nulboonrueng, J. Kaewon. S. Wongwises, *Two-phase condensation heat transfer coefficients of HFC-134a at high mass flux in smooth and micro-fin tubes*, Int. Comm. Heat Mass Transfer 30 no. 4 (2003) pp. 577 – 590
12. S.G. Kandlikar, *Fundamental issue related to flow boiling in minichannels and micro-channels*, Experimental Thermal and Fluid Mechanics and Thermodynamics, 2002, vol. 26, p. 129-146
13. H. Yuncu, O.T. Yildirim, S. Kakac, *Two-phase flow instabilities in a horizontal single boiling channel* Appl. Sci. Res. 48 (1991) pp. 83 – 104
14. J. Bouré, A. Bergles, L. Tong, *Review of two-phase flow instabilities*, Nucl. Eng. Des. 25 (1973) pp. 165–192
15. M. Ledinegg, *Instability of flow during natural and forced circulation*, Die Wärme 61 (8) (1938) pp. 891–898.
16. Y. Ding, S. Kakaç, *Dynamic instabilities of boiling two-phase flow in a single horizontal channel*, Experimental Thermal and Fluid Science (1995), pp. 327 – 342
17. J. Maulbetsch, P. Griffith, *A study of system-induced instabilities in forced convection flows with subcooled boiling*, MIT engineering projects Lab Report 5382-35, 1965
18. K. Akagawa, T. Sakaguchi, *Study on distribution of flow rates and flow stabilities in parallel long evaporates*, Bull. JSME 14 (1971) pp. 837–848
19. Y. Buyevich, I. Natalukha, *Self-oscillating regimes of nucleate, transition and film boiling*, Int. J. Heat Mass Transfer 38 (2) (1996) pp. 2363–2373
20. A. Stenning, *Instabilities in the flow of a boiling liquid*, J. Basic Eng. (1964) pp. 213–217
21. T. Bohdal, M. Łomiak, *Condensation of refrigeration medium under condition of unit disturbances,* An International Journal of Turbulence, vol. 11, 2005, pp. 221 - 224
22. W. Kuczyński, T. Bohdal, H. Charun, *Impact of periodically generated hydrodynamic disturbances on the condensation efficiency of R134a refrigerant in pipe mini-channel,* Experimental Heat Transfer 26 (1), 2013, pp. 64 - 84.
23. W. Kuczyński, *Modeling of the propagation of a pressure wave during the condensation process of R134a refrigerant in a pipe minichannel under the periodic conditions of hydrodynamic disturbances,* International Journal of Heat and Mass Transfer Volume 56, Issues 1–2, 1 January 2013, pp. 715-723.
24. W. Kuczyński, T. Bohdal, H. Charun, *Influence of hydrodynamic instability on the heat transfer coefficient during condensation of R134a and R404A refrigerants in pipe mini-channels,* International Journal of Heat and Mass Transfer Volume 55, Issues 4, 1 January 2013, pp. 1083-1094.
25. Z. Bilicki, P. Downar-Zapolski, *Wave phenomena related to two fluid and homogenous two-phase flow models*, Prace INP PAN Gdańsk, no. 94, 1992, pp. 19 – 42 (in Polish)
26. R. Laskowski, M. Jaworski, *Maximum entropy generation rate in a heat exchanger at constant inlet parameters.* Journal of Mechanical and Energy Engineering, Vol. 1 No. 1 (41), June 2017, pp. 65 – 70.
27. M. Sikora, *Flow structures during refrigerants condensation.* Journal of Mechanical and Energy Engineering, Vol. 1 No. 1 (41), June 2017, pp. 101 – 106.
28. T. Bohdal, M. Kruzel, M. Sikora, *An investigation of heat transfer coefficient during refrigerants condensation in vertical pipe microchannel.* Journal of Mechanical and Energy Engineering, Vol. 1(41) No. 2, November 2017, pp. 163 – 170.
29. W. Kuczynski, A. Denis, *Modeling the instabilities of the condensation process of the R134a and R404A refrigerants in pipe minichannels in the conditions of periodic hydrodynamic disturbances.* Journal of Mechanical and Energy Engineering, Vol. 1(41) No. 2, November 2017, pp. 179 – 188.
30. T. Bohdal, H. Charun, M. Sikora, *Pressure drop during condensation of refrigerants in pipe minichannels*, Archives of Thermodynamics (2012) 33(1), 87-106
31. T. Bohdal, K. Widomska, M. Sikora, *The analysis of thermal and flow characteristics of the condensation of refrigerant zeotropic mixtures in minichannels*, Archives of Thermodynamics (2016) 37(2), 41-69
32. T. Bohdal, H. Charun, M. Sikora, *Heat transfer during condensation of refrigerants in tubular minichannels*, Archives of Thermodynamics (2012) 33(2), 3-22
33. T. Bohdal, H. Charun, M. Sikora, *Empirical study of heterogeneous refrigerant condensation in pipe minichannels*, International Journal of Refrigeration (2015) 59, 210-223

Modelling of a two-phase zone during condensation of R1234yf refrigerant in tubular minichannels under periodic dynamic disturbances

Waldemar Kuczyński[1,*], and *Aleksander* Denis[2]

[1]Technical University of Koszalin, Department of Energy, av. Raclawicka 15-17, PL 75-620 Koszalin, Poland

Abstract. The paper presents an experimental analysis of the influence of dynamic disturbances on the size of the condensation zone of the appropriate new R1234yf environment-friendly refrigerant in circular mini-channels. Experimental tests were conducted for single minichanells placed horizontally, with an inner diameter $d = \{1.44; 2.30; 3.30\}$ mm. A range of frequencies from f = 0.2 - 5 Hz of periodically generated disturbances showed an explicit and unfavourable influence on the decrease of the length of the proper condensation region. This influence reduced the efficiency of the process. A computational model was developed to calculate the length of the two-phase region under disturbance conditions. The model proposed was verified by the results of the experiment, and compliance was obtained in the range of ±25%.

1 Introduction

A heat exchanger known as the condenser is one of the basic components of a power system that performs a reversed thermodynamic cycle. This component is crucial in terms of the overall efficiency of refrigeration systems or heat pumps. The basic input criteria for optimisation of the operation of such systems is the available heat transfer area of the condenser. This applies both to the heat rejection by the working fluid (main medium) and heat absorption by the coolant (auxiliary medium). The hydraulic losses also play an essential role in the process. Both heat transfer processes (heat rejection and heat absorption by the working fluid) affect the heat transfer rate in a different way. Therefore, it is very important to provide proper conditions that intensify the overall heat transfer coefficient of the condenser [1 - 7]. Modern trends in the design of power machinery are focused on the energy efficiency. As a consequence, the power systems are becoming smaller and smaller owing to the employment of mini and microchannels as well as new pro-ecological refrigerants [8]. According to the 2020 forecast, the share of miniaturized devices should reach up to 25% of all refrigeration, power, medical, IT, aviation and aerospace equipment produced [8].

The manufacturing process of the above-mentioned systems require the employment of heat exchangers (condensers and evaporators) based on channels with a hydraulic diameter significantly smaller than the so-called conventional ones. The classification of those systems has already been developed. This classification is based on the size of the internal diameter of the channel –channels with internal diameter below the range 3 – 6 mm are known as minichannels [9].

Such systems, as well as those conventional ones, require properly selected working fluid. So far, one of the most commonly used refrigerants were the so-called fluorinated greenhouse gases, especially: hydrofluorohydrocarbons, perfluorohydrocarbons, sulphur hexafluoride and other fluids that contain fluorine. The fluorinated greenhouse gases group includes the R134a refrigerant. Although this refrigerant is commonly used in currently operated systems, it has been phased-out since 01.01.2017 and the process should be completed by 31.12.2020. The scope of the applications of this refrigerant covers air conditioning systems in mobile applications and heat pumps. The suggested substitute for the R134a refrigerant is the chemical compound labelled R1234yf. Its GWP index (Global Warming Potential) is very small, just 4, comparing to 1430 in case of its predecessor R134a [10,11].

During the condensation phase change in minichannel based systems, as well as in conventional systems, there are plenty of thermal-hydraulic, structural and operational problems. Those problems include the possibility of the occurrence of unsteady working conditions induced by instabilities of internal (static) and dynamic (external) origin. This paper pays special attention on phenomena connected with the onset and propagation of disturbances in the systems that perform condensation phase change of refrigerants [12 - 16].

The static instabilities are inherently connected with the change of the state parameters of the system. Basing on the record of state parameters evolution in time, it is possible to predict when the static instability will occur

* Corresponding author: waldemar.kuczynski@tu.koszalin.pl

https://doi.org/10.1051/e3sconf/20187002011

(the system is considered deterministic). As a result, the system will reach an another steady state or will oscillate around this state. A perfect example of static instabilities is the flow excursion (also known as the Ledinegg instability) [17,18]. It involves pressure drop due to hydraulic losses during phase change. This instability may occur in a specific state of the system when the total pressure drop decreases due to increase of flow rate.

When the system is under influence of thermal or hydraulic forces that introduce inertia, the flow becomes unstable and the dynamic instabilities occur. The exact state of the system becomes probabilistic or stochastic. The flow instabilities in two-phase media (liquid-gas) propagate in a form of two phenomena: a pressure wave (acoustic wave travelling with the speed of sound), flow rate fluctuations (due to the oscillations of void fraction) and temperature instability also known as the evaporation front (for the evaporation phase change) or condensation front (for the condensation phase change) [12-16, 19-23].

Those phenomena have a wave character described with different propagation velocities. The acoustic waves are characterized by high frequency, while flow rate and temperature oscillations are characterized by much smaller frequency.

Previous experimental research on influence of instabilities on flow phase change of refrigerants in minichannels allowed partial recognition of those phenomena [1-6, 12-16]. However, there are no publications that concern those phenomena regarding new pro-ecological refrigerants that are meant as a substitutes for those being phased out.

2 Computational model

During the condensation process in tubular minichannels and conventional condensers of compressor based refrigeration systems, it is possible to distinguish three characteristic sub-zones inside the condenser: pre-cooling of superheat vapour, proper condensation (two-phase) and sub-cooling of the liquid. Figure 1 shows the temperature distribution of the refrigerant and the coolant along the condenser, including the above-mentioned sub-zones.

In the first sub-zone, the superheated vapour is being cooled-down until it reaches the condensation temperature (saturated steam with quality x=1.0). From this point on, provided that there is a temperature gradient at the wall, the two-phase proper condensation sub-zone begins. The flow condensation process is determined by process parameters and takes place until the saturated liquid state (x=0.0) is reached. From this point on, the third sub-zone begins. It is responsible for the cool-down of the liquid. The most intense heat transfer mode is the two-phase flow condensation that takes place inside the proper condensation sub-zone. Single phase cool-down and sub-cooling sub-zones are characterized by much lower heat transfer coefficients. The onset of instabilities, both static and dynamic, may cause the disturbance of distribution of sub-zones inside the condenser, including a shrinkage of the active area of proper condensation characterized by the highest heat transfer coefficient [12].

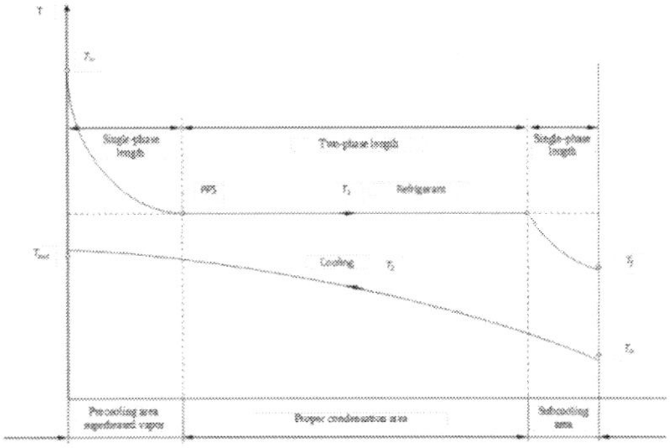

Fig. 1. Diagram of the distribution of the temperature of refrigerant and cooling medium during condensation inside the horizontal pipe of condenser; T_p – superheated vapour temperature, T_s – temperature of condensation (saturation), T_f – final temperature of the process, T_{in}, T_{out} - temperature of cooling medium, at the beginning and end of process respectively, PPS – starting point of condensation [12].

The author of the paper [7] suggested a theoretical model that allows one to determine the size of each of the sub-zones of the condenser under steady-state operation. However, in order to determine the size of the proper condensation sub-zone for the R1234yf refrigerant for flow condensation inside minichannels under periodic dynamic instabilities, a novel model developed for the R134a and R404A refrigerants was used [12].

2.1 Apparatus

The experimental research is aimed at a verification the presented mathematical model. At the same time, it allowed the Authors to determine the influence of periodically generated hydrodynamic disturbances on the size of each of the heat transfer sub-zones during flow condensation of the R1234yf refrigerant inside tubular minichannels.

During experimental research, the tubular minichannels made of stainless steel with internal diameters d={1.44; 2.30; 3.30 mm} were used. The research was conducted using an apparatus that was specially designed for this experiment. The same apparatus was previously used for the investigation of the R134a refrigerant. Minor changes were introduced in order to adjust the installation to the new pro-ecological refrigerant R1234yf. The Diagramdiagram of the apparatus in presented in Figure 2. Figure 3 shows the test section in details.

The instrumentation of the apparatus allowed the Authors to measure the following parameters:

- pressure p_s of the refrigerant during flow along the tubular minichannel,
- wall temperature along the tubular minichannel,
- pressure of the refrigerants at the inlet and outlet of the test section p_{out}, p_{in},

https://doi.org/10.1051/e3sconf/20187002011

- temperature of the refrigerant at the inlet and outlet of the test section T_{wej}, T_{wyj},
- mass flow rates of the refrigerant $\dot{m}_{R1234a}$ and the cooling water $\dot{m}_{H_2O}$.

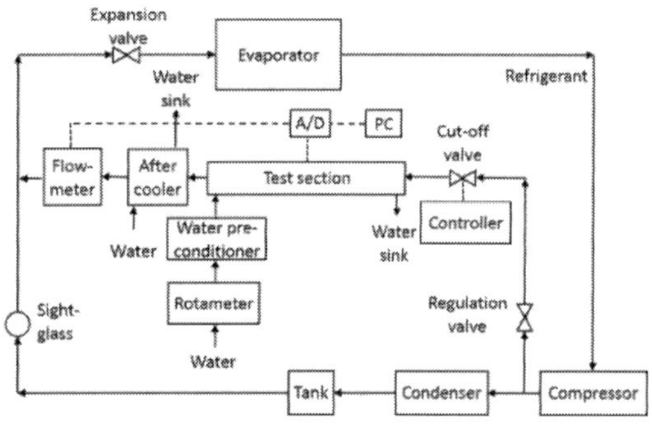

Fig. 2. Diagram of the apparatus.

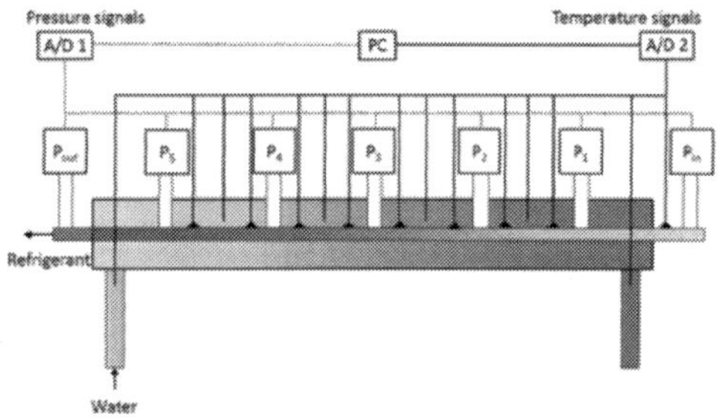

Fig. 3. Diagram of the test section.

The measurements were done for the following intervals of the parameters:

- mass flux of the refrigerant G = 100 ÷ 4100 kg/(m^2·s),
- saturation temperature of the refrigerant T_s = 20 ÷ 55°C,
- frequency of the instabilities f = 0,20 ÷ 5 Hz.

The research was done for a constant value of mass flow rate of the refrigerant during the steady state that preceded the onset of the instabilities. The test section with channel length L = 880 mm was armed with piezoelectric pressure sensors and K-type thermocouples maintaining a constant distance between the sensors. The voltage based signals collected from the pressure and temperature sensors, and the flow meter were measured using:

- thermoelectric temperature sensors made of *NiCr-Ni* (each sensor was individually calibrated prior to the research) coupled with A/D adapter *USB PersonalDaq/3000* with *PDQ30* extension module,
- piezoresistive pressure sensor (*Cerabar TPMP131* type) coupled with A/D adapter,
- flow meter *Coriolisa Promass 80A*.

All the equipment was connected to a PC with data acquisition and processing software.

2.2 Mathematical model

According to the temperature distribution of the refrigerant flowing along the horizontal minichannel, shown in Figure 1, during the condensation of the R1234yf refrigerant, it is clearly visible that the proper condensation sub-zone onsets for both steady and unsteady conditions. If at the inlet that supplies superheated vapour T_p to the minichannel, there is an electromagnetic regulation valve, then it is possible to generate periodic hydrodynamic instabilities simply by alternating opening and closure of this valve. The frequency of those instabilities can be calculated using the following formula [24]:

$$f = \frac{1}{\tau_o + \tau_z} \tag{1}$$

where: τ_o is the opening time and a τ_c is the closure time of the valve that supplies the refrigerant to the test section.

The possibility of determining the length of the two-phase sub-zone L for the R1234yf under previously stated conditions was also taken into account. During the modelling process, the following assumptions were made:

- inside the straight, axisymmetric tubular minichannel with the length L and constant internal diameter d, there is a mass flow rate of refrigerant $\dot{m}$;
- the refrigerant condenses during the flow along the tubular minichannel until it reaches the saturated liquid state (quality x=0.0) and then it is being sub-cooled;
- periodic instabilities with constant frequency are generated by the opening and closure of the cut-valve at the inlet of the refrigerant to the tubular minichannel;
- the mass flow rate $\dot{m}$ of the refrigerant is under oscillatory fluctuations;
- the instabilities propagate along the axis of the horizontal channel of the length L in a form of pressure wave with velocity v_p that is a functions of the amplitude and frequency.

From the condensation heat balance for elementary length of tubular minichannel in the two-phase sub-zone, it results:

$$d\dot{Q}_i = q_i \cdot \pi \cdot D \cdot dL_{TFP}, \tag{2}$$

where:

d – internal diameter of the minichannel,
q_i – local heat flux,
dL_{TFP} – elementary length of the minichannel.

The heat flux removed along within the two-phase sub-zone of the minichannel is described by the following formula:

$$\dot{Q} = \dot{m} \cdot (h_G - h_L) = \dot{m} \cdot r, \tag{3}$$

where:

$\dot{m}$ – average mass flow rate under steady conditions,
h_G – specific enthalpy of the refrigerant at the inlet to the two-phase sub-zone (saturated vapour with quality x = 1),

h_L – specific enthalpy of the refrigerant at the outlet of the two-phase sub-zone (saturated liquid with quality x = 0).

Combining Equations (2) and (3), it can be written as follow:

$$d\dot{Q} = \dot{m} \cdot dh \,. \tag{4}$$

By introducing a constant C = π · D into Equation (2), Equation (4) can be rewritten:

$$C \cdot q_i \cdot dL_{TFP} = \dot{m} \cdot dh \,. \tag{5}$$

The heat flux qi from Equation (5) should be regarded as a local parameter that depends both on time and the distance from the channel's inlet up to the beginning of the two-phase sub-zone: $q_i = q_i(L, t)$, d_h however is stands for the local elementary drop of the specific enthalpy in this cross-section.
The change (oscillations) of the mass flow rate of refrigerant caused by the periodically generated hydrodynamic instabilities can be treated as a propagation of a small amplitude and frequency monochromatic wave along the axis of the horizontal channel. Based on the following studies [25,26], the oscillatory variation of the mass flow rate at the minichannel's inlet can be determined using the following formula:

$$\dot{m}_{inst.} = \dot{m} \cdot [1 + A \cdot \sin(2 \cdot \pi \cdot f \cdot \tau) \tag{6}$$

where $\dot{m}$, A, f and τ are mass flow rate before the onset of oscillations, amplitude of the oscillations, frequency and time, respectively. The maximum amplitude of the oscillations of mass flow rate A equals to the undisturbed radius of internal surface of the condensate film. This amplitude can be calculated using the following formula:

$$A = A_0 \cdot \exp^{\omega k} \,. \tag{7}$$

The minimal thickness of the undisturbed condensate film equals to: $\delta = \frac{r_0}{2}$, where r_0 is the internal diameter of the minichannel [12,19]. The A_0 parameter stands for the initial amplitude of the instabilities. Its value depends on the vapour inlet pressure and equals $A_0 = -p_G$. The ω parameter stands for an increase of the amplitude A, and k stands for the wave number [19]:

$$k = \frac{2\pi}{\lambda} = \frac{\omega}{v_p} \,, \tag{8}$$

where:
λ – wave length,
v_p – propagation velocity of the instability, $v_p = \frac{l}{\Delta\tau}$,
l – distance between the pressure sensors,
$\Delta\tau$ – delay of the pressure signal,
ω – pulsation (angular frequency) of the base parameter, $\omega = \frac{2\pi}{T}$ [21, 23],
T – wave period ($T = t = t_o + t_c$).

By combing Equations (5) and (6), the following is obtained:

$$\dot{m}_{inst.} \cdot dh = C \cdot q_i \cdot dL_{TFP}, \tag{9}$$

and after integration:

$$\int_{h_1}^{h_2} \dot{m}_{inst.} \cdot dh = \int_0^L C \cdot q \cdot dL_{TFP} \,, \tag{10}$$

and further:

$$\dot{m}_{inst.} \cdot \Delta h = C \cdot \int_0^L q \cdot dL_{TFP} \,. \tag{11}$$

Equation (11) was employed to determine the length of the proper condensation sub-zone of the tubular minichannel. It can be solved using numerical methods, provided that the distribution of heat flux $q_i = q_i(L,t)$ is known. Assuming that the length of the two-phase sub-zone depends on the frequency of periodically generated instabilities f, the problem is simplified.

Calculations of the average heat flux $\overline{q}$ along the length of the proper condensation sub-zone L_{TFP} inside the minichannel were made following the method presented in [7]. According to the authors of this paper, the average heat flux $\overline{q}$ along the length of the proper condensation sub-zone L equals to:

$$\overline{q} = \frac{1}{L}\int_0^L q(L) dL_{TFP} \text{ thus } \int_0^L q(L) dL = \overline{q} \cdot L_{TFP} \tag{12}$$

and finally:

$$\dot{m}_{inst.} \cdot \Delta h = C \cdot \overline{q} \cdot L_{TFP} \,, \tag{13}$$

where:

$$\dot{m}_{inst.} = (G_{inst.} \cdot F). \tag{14}$$

Assuming that the internal cross-section of the channel is constant $dA/dz = 0$, the length of the two-phase proper condensation sub-zone L_{TPF} can be calculated using the following formula:

$$L_{TPF} = F \cdot \frac{G_{inst.} \cdot \Delta h}{C \cdot \overline{q}} \tag{15}$$

For the minichannel with a constant internal cross-section area $F = \frac{\pi \cdot d^2}{4}$ and the perimeter $C = \pi \cdot d$, Equation (15) takes the following form:

$$L = d \cdot \frac{G_{inst.} \cdot \Delta h}{4 \cdot \overline{q}} \,, \tag{16}$$

or

$$L = d \cdot \frac{G_{inst.} \cdot r \cdot (1-x)}{4 \cdot \overline{q}}, \qquad (17)$$

where r is the latent heat of condensation, and $\overline{x}$ is the average quality of the refrigerant.

Based on Formula (17), the length of the two-phase condensation sub-zone can be calculated, provided that all the assumptions of the model are satisfied.

Taking into account the fact that the external instabilities are induced by the valve, the model included the valve's characteristics – relative valve flow coefficient k_v. This coefficient determines the relation between the valve position and the flow rate. Depending on pressure drop at the valve, there are two main characteristics: internal and operational. The value of k_v is determined as follow [27]:

$$k_v = \frac{K_v}{K_{v100}}, \qquad (18)$$

where:

K_v – computational valve flow coefficient according to [27] - this is the characteristic factory size of the valve, m^3/h,

K_{v100} – nominal (catalogue) value of valve flow coefficient according to the manufacturer, m^3/h[27].

The coefficient K_{vw} is presented, according to [27], in the following form:

$$K_{vw} = 10 \cdot \dot{m} \cdot \sqrt{\frac{v_2}{\Delta p}}, \qquad (19)$$

Provided that: $p_2 \geq \frac{p_1}{2}$,

where:

$\dot{m}$ – mass flow rate of the refrigerant,

Δp – pressure drop at the valve,

v_2 – specific volume of vapour for p_2 and T_1,

p_1 – pressure at the inlet,

p_2 – pressure at the outlet,

T_1 – temperature of the refrigerant before the valve,

Equation (17), after substitution of Equation (18), takes the following form:

$$L = k_v \cdot \frac{d \cdot G_{inst} \cdot r \cdot (1-\overline{x})}{4 \cdot \overline{q}}. \qquad (19)$$

3 Experimental results

The dynamic external instabilities during condensation of the R1234yf refrigerant inside tubular minichannel manifested themselves in the form of fluctuations of the mass flow rate and, consequently, the fluctuations of the pressure and p_s and the saturation temperature T_s. Figures 4 - 6 show the examples of oscillations of the mass flow rate, pressure p_s and wall temperature T_{wall}.

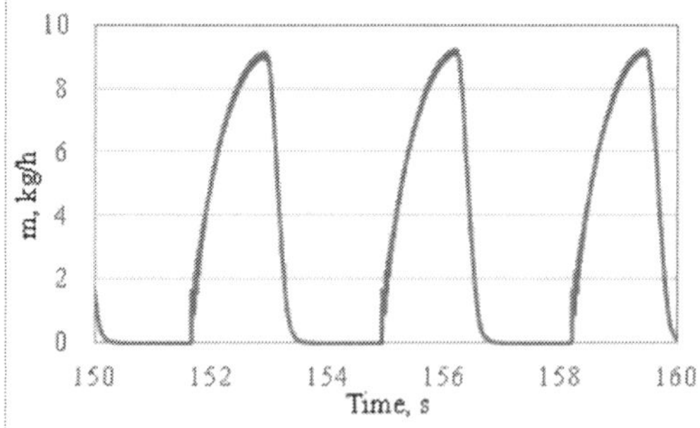

Fig. 4. Sample of flow oscillations.

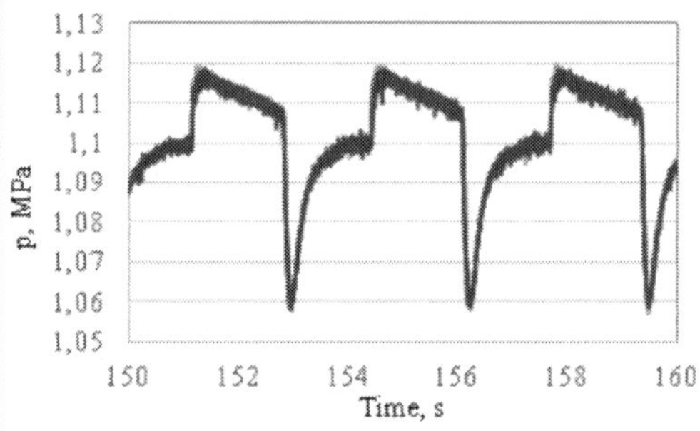

Fig. 5. Sample of pressure oscillations for multiple sensors.

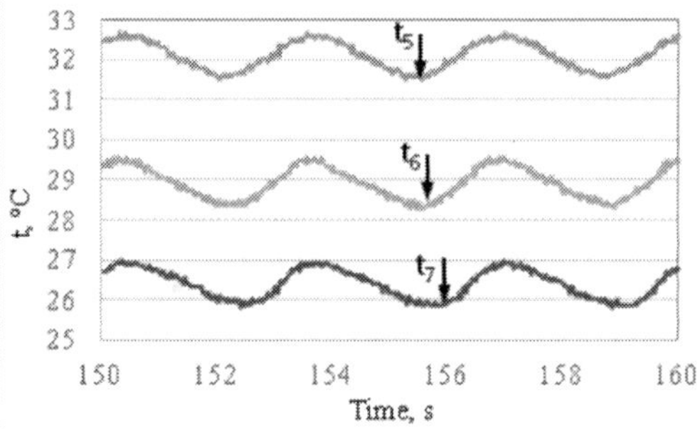

Fig. 6. Sample of temperature oscillations for multiple sensors.

Based on the obtained experimental results, the temperature distribution of the refrigerants along the investigated minichannel was obtained. Figure 7 shows a summary that includes the distribution of sub-zones depending on the frequency of the periodic instabilities for minichannel with internal diameter d = 1.44 mm.

The summary presented in Figure 7 allows one to visualize the influence of hydrodynamic instabilities on condensation inside the tubular minichannels. The external instabilities were generated through a periodic sudden closure of the inflowing refrigerant. As a result, the length of both superheat removal and proper condensation zones contracted, while the sub-cooling section expanded. The higher the closure time is, the smaller superheat removal and proper condensation zones are. The change in sub-zones distribution is significant, thus the thermal performance of the condenser drops. An increase of the valve closure time translates into a lower frequency of the instabilities *f*, see Equation (1).

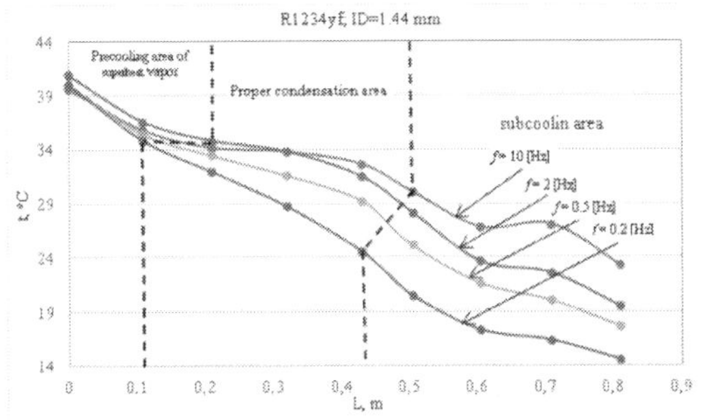

Fig. 7. Results of the experimental investigations concerning the influence of periodically generated instabilities on the length of the of the superheat removal, proper condensation and sub-cooling sub-zones for the R1234yf refrigerant depending of the frequency *f* of the generated disturbances in mini-channels with internal diameter d = 1.44 mm.

4 Computational results

The computational results of the length of the two-phase condensation sub-zone L_{cal} were compared with the experimental results for the R1234yf refrigerant for minichannels with internal diameters d = {1.44; 2.30; 3.30} mm.

Within the scope of the statistical analysis that aimed to compare numerical results obtained using the presented model and the experimental results, the relative mean absolute error (RMAE) criterion was used [24]:

$$RMAE = \frac{1}{n} \cdot \sum_{i=1}^{n} \frac{(\hat{y}_i - y_i)}{y_i} \cdot 100\% , \qquad (22)$$

where:

$\hat{y}_i$ – simulated value (numerical value obtained using a verified model),

y_i – empirical value (experimental),

n – number of measurement points.

Based on the calculations, it was found that the relative mean absolute error in the case of the two-phase proper condensation sub-zone does not exceed ±25%. It was confirmed by comparing the experimental results with calculations using Formula (19). The results are shown in Figure 8. The value or error is higher than in the case of an estimation of the proper condensation sub-zone size for the R134a refrigerant (±15%). This may be due to the properties of the R1234yf refrigerant and requires further investigations.

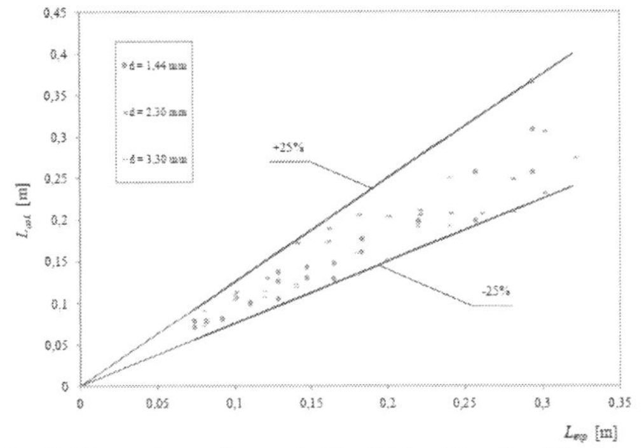

Fig. 8. Comparison of numerically obtained length of the two-phase proper condensation sub-zone Lcal according to Formula (19) with experimental results L_{exp}.

5 Conclusions

1) Periodically generated external dynamic instabilities during condensation of the R1234yf refrigerant in minichannels with internal diameter d = 1.44; 2.30 and 3.30 mm, just like in the case R134a refrigerant, cause fluctuations of the mass flow rate of the working fluid that leads to further wave changes of the pressure and wall temperature.

2) Periodically generated dynamic instabilities within the frequency range f = 0,25 - 5 Hz have an influence on the size of the sub-zones inside the minichannel based condenser, especially, the size of the proper condensation sub-zone. Due to the highest heat transfer coefficient, the size of this zone determines the thermal performance of the condenser. The frequency of the generated instabilities *f* has a negative influence on the thermal performance of the condenser.

3) The model presented in this paper was compared with the experimental results. The difference between experimental and numerical results did not exceed ±25% range. The value of this error is significantly higher than in the case of the estimation of the proper condensation sub-zone for the R134a refrigerant (±15%).

4) Nevertheless, despite the higher error of the model for the R1234yf refrigerant, Formula (19) is still applicable.

References

1. T. Bohdal, *Bubble boiling in flow of refrigerating media*. Journal of Mechanical and Energy Engineering, Vol. 1 No. **1 (41)**, June 2017, pp. 57 – 64.
2. T. Bohdal, M. Kruzel, M. Sikora, *Analisys of heat transfer coefficient during refrigerant condensation in vertical pipe minichannel*. Journal of Mechanical and Energy Engineering, Vol. 1 No. **1 (41),** June 2017, pp. 65 – 70.
3. R. Laskowski, M. Jaworski, *Maximum entropy generation rate in a heat exchanger at constant inlet parameters*. Journal of Mechanical and Energy

Engineering, Vol. 1 No. **1 (41),** June 2017, pp. 65 – 70.

4. M. Sikora, *Flow structures during refrigerants condensation*. Journal of Mechanical and Energy Engineering, Vol. 1 No. **1 (41),** June 2017, pp. 101 – 106.
5. T. Bohdal, M. Kruzel, M. Sikora, *An investigation of heat transfer coefficient during refrigerants condensation in vertical pipe microchannel*. Journal of Mechanical and Energy Engineering, Vol. **1(41)**, No. 2, November 2017, pp. 163 – 170.
6. W. Kuczynski, A. Denis, *Modeling the instabilities of the condensation process of the R134a and R404A refrigerants in pipe minichannels in the conditions of periodic hydrodynamic disturbances*. Journal of Mechanical and Energy Engineering, Vol. **1(41)** No. 2, November 2017, pp. 179 – 188.
7. A. Paliwoda, *Calculation of evaporators and condensers of refrigeration equipment. Vol. I Basic equations and dependencies*. Vol. II. Length of pipes and flow resistance of refrigerants. Refrigeration Year IX. Notebook 3, 1974, pp. 1 - 6. in Polish
8. Z. Gnutek, A. Nemś, *Trends in the development of machinery and power equipment in the era of miniaturization*. Materials of the XXX Congress of Thermodynamics, Wroclaw 2008, volume I, 318 - 324. in Polish
9. S.M. Ghiaasiaan, *Two-phase flow, boiling, and condensation in conventional and miniature systems*. Cambridge University Press 2008, pp. 68 - 76.
10. D. Colorado, J.A. Hernandez, W. Rivera, *Comparative study of a cascade cycle for simultaneous refrigeration and heating operating with ammonia, R134a, butane, propane and CO_2 as working fluids*. International Journal of Sustainable energy 2011, 06.
11. D.H. Kim, H.S. Park, M.S. Kim, *Optimal temperature between high and low stage cycles for R134a/R410A cascade heat pump water heater system*. Experimental Thermal and Fluid Sci 2013, vol. **47**, s. 172-179.
12. W. Kuczynski, H. Charun, *Modeling of a Two-Phase Region Length of the Condensation of R134a and R404A Refrigerants in Pipe Minichannels With Periodic Hydrodynamic Instabilities*. Heat Transfer Engineering Volume **35**, Issue 9, pp. 850 – 862.
13. W. Kuczynski, *Modeling of the propagation of a pressure wave during the condensation process of R134a refrigerant in a pipe minichannel under the periodic conditions of hydrodynamic disturbances*. International Journal of Heat And Mass Transfer Volume **56**, Issue 1-2, pp. 715 – 723.
14. W. Kuczynski, T. Bohdal, H. Charun, *Impact of Periodically Generated Hydrodynamic Disturbances on the Condensation Efficiency of R134a Refrigerant in Pipe Mini-Channels*. Experimental Heat Transfer, Vol. **26**, Issue 1, pp. 64 – 84.
15. W. Kuczynski, *Pressure wave propagation during the condensation process of the R404A refrigerant in mini-condenser under periodic hydrodynamic disturbances*. International Journal of Heat And Mass Transfer Volume **67**, pp.404-415.
16. W. Kuczynski, H. Charun, T. Bohdal, *Modeling of temperature instabilities during condensation of R134a refrigerant in pipe minichannels*. International Journal of Heat And Mass Transfer. Volume **111**, pp.83-93.
17. S. Kakac, B. Bon, *Areview of two-phase flow dynamic instabilities in tube boiling systems*. International Journal of Heat and Mass Transfer **51** (3–4), 2008, pp. 399 – 433.
18. T. Zhang, T. Tong, J-Y. Chang, Y. Peles, R. Prasher M.K Jensen, J. Wenf, P. Phelan, *Ledinegg instability in microchannels*. International Journal of Heat and Mass Transfer **52**, 2009, pp. 5661 - 5674.
19. K.H. Ardron, R.B. Duffey, *Acoustic wave propagation in a flowing liquid-vapor mixture*. International Journal Multiphase Flow **4**, 1977, pp. 303 - 322.
20. L. Cao, S., Kakaç, H.T. Liu, P.K. Sarma, *The effects of thermal non-equilibrium and inlet temperature on two-phase flow pressure drop type instabilities in an upflow boiling system*. Int. J. Therm. Sci. **39**, 2000, pp. 886 - 895.
21. O. Comakli, S. Karsli, M. Yilmaz, *Experimental investigation of two phase flow instabilities in a horizonatl in-tube boiling system*. Energy Conversion and Management **43**, 2002, pp. 249 - 268.
22. Y. Ding, S. Kakaç, *Dynamic instabilities of boiling two-phase flow in a single horizontal channel*. Experimental Thermal and Fluid Science, 1995, pp. 327 - 342.
23. A.N. Pavlenko, V.V. Lel', *Model of self-maintaining evaporation front for superheated liquids*. Proc. 3th Int. Conference on Multiphase Flow, ICMF'98, Lyon, France **4**, 1998, pp. 3-5.
24. P. Kartaschoff, Frequency and Time, Publishing House of Communications and Communications Warsaw 1995. in Polish
25. T.N. Veziroglu, *Two-phase flow instabilities, final report. NSF Project CME 79-20018*, Clean Energy Research Inst. Coral Gables, FL 1983.
26. G.B. Whitham, Linear and nonlinear waves. Wiley, New York 1974.
27. International Standard IEC 60534-1. 1987., 15

The effect of reduced pressure on carbon dioxide flow boiling heat transfer in minichannels

Dariusz Mikielewicz[1,*], and *Blanka* Jakubowska[1]

[1]Gdańsk University of Technology, Faculty of Mechanical Engineering, ul. Narutowicza 11/12, 80-233 Gdańsk, Poland

Abstract. In the paper presented are the results of the study on the effect of reduced pressure on flow boiling heat transfer data in minichannels as well as conventional ones. That effect renders that most of heat transfer correlations fail to return appropriate results of predictions. Mostly they have been developed for the reduced pressures from the range 0.1-0.3. The special correction has been postulated to the in-house model of flow boiling and condensation which modifies the two-phase flow multiplier as well as the temperature gradient in pool boiling. Four two-phase flow multiplier models were tested for this purpose, i.e. due to Friedel, Tran, Müller-Steinhagen and Heck and finally its in-house modification for applicability to minichannels. The model has been tested against a large selection of experimental data collected from various researchers to investigate the sensitivity of the in-house developed model. The collected experimental data came from various studies from literature and were conducted for the full range of quality variation and a wide range of mass velocity and saturation temperatures. In the work are presented the results of calculations obtained using the in-house developed semi empirical model on selected experimental flow boiling data related to carbon dioxide.

1 Introduction

Nowadays, there is an increasing interest in refrigerants featuring low Global Warming Potential (GWP). The reason of this concern can be attributed to the growing number of regulations and laws prohibiting the use of some of synthetic refrigerants. According to these regulations the new fluids used in e.g. air-conditioning and refrigeration applications cannot be manufactured with fluorinated greenhouse gases having GWP greater than 150 [1]. Within that document most of the substances used in refrigeration system have been regulated due to its Ozone Depletion Potential (ODP). Consequently one of the working fluids most extensively used in medium evaporation temperatures, such as for example R134a with GWP=1430 needs to be replaced by more environmentally friendly fluids. Previous studies have considered R152a [2] and the natural refrigerant CO_2 [3] as possible replacements for R134a. However, R152a is a flammable refrigerant and for the CO_2 system, the working pressure is significantly higher than for R134a system. Apart from that drawback CO_2, as compared to other contemporary fluids is a relatively safe one. It is non-toxic, non-flammable, non-explosive, inexpensive and can be coupled with most metals and plastics. Design of evaporators for use of the CO_2 requires the exact determination of the heat transfer coefficient during flow boiling. Available in the literature empirical correlations give different results as compared to the results obtained experimentally. There is hardly any robust and recommended correlation for the purpose of calculation of carbon dioxide two-phase heat transfer, despite some devoted contributions [3]. There are many available experimental investigations into flow boiling heat transfer of all mentioned above working fluids in the literature. Unfortunately, the results published for some of these fluids are still inconsistent and difficult in mathematical modelling [4]. In the literature there are many empirical correlations for modelling of flow boiling heat transfer. As mentioned earlier, in case of CO_2 most of well-known models does not exhibit a good consistency with experimental data. It was authors intention to show the performance of their own approach in predicting of flow boiling heat transfer coefficient on the example of the data collected from literature using the in-house developed model [5]–[8]. Based on the evidence of comparisons with experimental data two corrections incorporating the effect of reduced pressure have been postulated to the authors own model. The first one modifies the temperature gradient in flow boiling and is applicable only to that situation. The second correction relates to modification of the two-phase flow multiplier and influences both the flow boiling or flow condensation. Presented model provides improved consistently of the predictions with the experimental data for the case of carbon dioxide data in a wide range of reduced pressures.

2 The modelling

The versatile semi-empirical model for calculations of flow boiling and flow condensation originally due to J.

* Corresponding author: Dariusz.Mikielewicz@pg.edu.pl

 https://doi.org/10.1051/e3sconf/20187002012

Mikielewicz [5] and the final version due to D. Mikielewicz et al. [6]–[8] has been tested for a significant number of experimental data and has returned satisfactory results for the case of the flow boiling process for numerous fluids. The fundamental hypothesis of the model is the fact that heat transfer during flow boiling with bubble generation can be modeled as a sum of two contributions constituting the total energy dissipation in the flow, namely the energy dissipation due to the shearing flow without the bubbles and dissipation resulting from the bubble generation. The final version of the model reads:

$$\frac{\alpha_{TPB}}{\alpha_{LO}} = \sqrt{R_{MS}{}^{n} + \frac{C}{1+P}\left(\frac{\alpha_{Pb}}{\alpha_{LO}}\right)^{2}}, \qquad (1)$$

In equation (1) $C = 1$ for flow boiling and $C = 0$ for flow condensation, Equation (1) also includes the empirical correction P defined by equation (2). Occurring in the equation (1) the two-phase multiplier is raised to the power n (n=0.76 for turbulent, n=2 for laminar flow).

$$P = 2{,}53\cdot 10^{-3}\cdot \mathrm{Re}^{1.17}\cdot Bo^{0.6}\cdot\left[\left(\phi^{2}_{LO}\right)_{MS}-1\right]^{-0.65}, \qquad (2)$$

The two-phase flow multiplier present in eq. (1) has been recommended thus far as a Muller-Steinhagen and Heck [9] model or its in-house modifications (described in Table 1), but it must be stressed that any definition of the multiplier can be used for that modelling provided that it captures the convective flow boiling character properly.

In calculations presented below the sensitivity of the developed model to the selection of the two-phase flow multiplier has also been examined. For that purpose four models of two-phase flow multipliers were introduced into equation (1), namely modified correlation due to Friedel [10], Tran [11], Muller-Steinhagen and Heck [9], and the latter taking into account of the non-adiabatic effect [12]. Relationships, which describe the two-phase multiplier using during this analysis have been shown in Table 1.

Table 1. Two-phase multiplier models.

Autor(s)	Equation
Muller-Steinhagen and Heck [9]	$R_{MS} = \left[1+2\left(\frac{1}{f_1}-1\right)\cdot x\cdot Con^{m}\right]$ (3) $(1-x)^{1/3}+\frac{1}{f_{1z}}x$ Turbulent flow $f_1=(\rho_L/\rho_G)(\mu_L/\mu_G)^{0.25}$, $f_{1z}=(\mu_G/\mu_L)(\lambda_L/\lambda_G)^{1.5}(c_{pL}/c_{pG})$ Laminar flow $f_1=(\rho_L/\rho_G)(\mu_L/\mu_G)$; $f_{1z}=(\lambda_G/\lambda_L)$ m=0 for conventional channels m = -1 for flow in minichannels $\mathrm{Con} = \sqrt{\frac{\sigma}{g(\rho_l-\rho_g)\cdot d_h^2}}$
Friedel [10]	$R_F = E + \frac{3.24\cdot F\cdot H}{Fr^{0.045}\cdot We^{0.035}}$ (4) $E=(1-x)^2+x^2\left(\frac{\rho_l f_v}{\rho_v f_l}\right)$ $F = x^{0.78}(1-x)^{0.2224}$ $H=\left(\frac{\rho_l}{\rho_v}\right)^{0.91}\left(\frac{\mu_v}{\mu_l}\right)^{0.19}\left(1-\frac{\mu_v}{\mu_l}\right)^{0.7}$
Tran [11]	(5) $R_T = 1+\left[4{,}3\cdot\frac{(dp/dz)_{GO}}{(dp/dz)_{LO}}\right]\times$ $\left[N_{conf}\cdot x^{0,875}\cdot(1-x)^{0,875}+x^{1,75}\right]$ $N_{conf}=\sqrt{\frac{\sigma}{g(\rho_l-\rho_g)\cdot d_h^2}}$
Muller-Steinhagen and Heck with non-adiabatic effects [12]	(6) $R_B = \begin{cases} R_{MS}\cdot\left(1-\frac{B}{2}\right) & for\ 0.1<x\le 1 \\ R_{MS}\sqrt{1+\left(\frac{8\cdot\alpha_{Pb}\cdot d_w}{\lambda_l\cdot \mathrm{Re}\cdot \mathrm{Pr}\cdot\xi_0\cdot R_{MS}}\right)^2} & for\ 0\le x\le (\end{cases}$ $B=\frac{2\cdot q\cdot\frac{\rho_l}{\rho_v}}{f_r\cdot G(s-1)\cdot r}$, $s=\sqrt[3]{\frac{\rho_l}{\rho_v}}$
Zhang and Webb [13]	(7) $R_{ZW}=(1-x)^2+\frac{2.87\cdot x^2}{P_r}+1.68\cdot x^{0.8}\cdot(1-x)^{0.25}\cdot P_r^{-1}$

The model adopted for consideration may be used both for predicting heat transfer coefficients during boiling and condensation. It was expected that the accuracy of model predictions could be improved by some modifications to the empirical correction P, here by incorporation of the reduced pressure effect. Taking into account the reduced pressure in the empirical correction P in case of modeling the condensation process in the flow will not affect the obtained results of the calculations. This is due to the fact that the empirical correction P, and thus the reduced pressure included in the calculations, are considered only in the part associated with the generation of bubbles, which is non-existent when modeling the condensation process. The condensation process can only be influenced if the reduced pressure effect is applied to the model of two-phase flow multiplier.

As mentioned above it was expected that the accuracy of model predictions could be improved by some modifications to the empirical correction P, here by incorporation of the reduced pressure effect. The modified in the present work empirical correction P yields:

$$P=\left(\frac{p_{sat}}{p_{kr}}\right)^{a} 2{,}53\cdot 10^{-3}\,\mathrm{Re}^{1.17}\,Bo^{0.6}\left[\left(\phi^{2}_{LO}\right)_{MS}-1\right]^{-0.65} \qquad (8)$$

The convective part of equation (1) can also be modified to include the effect of reduced pressure in the form:

$$R_N = R_{MS} \cdot \left[1 - \left(\frac{P_{sat}}{P_{crit}} \right)^{a} \right] + 1 \qquad (9)$$

Therefore the general form of the model (1) with account of reduced pressure in the convection term and bubble generation term reads:

$$\frac{\alpha_{TBP}}{\alpha_{LO}} = \sqrt{\left\{ R_{MS} \cdot \left[1 - \left(\frac{P_{sat}}{P_{crit}} \right) \right]^{a_1} + 1 \right\}^{a} + \frac{C}{1 + \left(\frac{P_{sat}}{P_{crit}} \right)^{a_2} \cdot 2.53 \cdot 10^{-3} \cdot \mathrm{Re}^{1.17} \cdot Bo^{0.6} \cdot \left[R_{MS} - 1 \right]^{-0.65}} \left(\frac{\alpha_{Pb}}{\alpha_{LO}} \right)^2} \qquad (10)$$

The exponents in eq. (9) have been derived using the multiple regression method and are equal to a_1=1.61 and a_2=-1.There is additionally another term in eq. (9) which should account for the fact that it may be prone to the reduced pressure. That is the pool boiling heat transfer coefficient. In the considered model the generalized Cooper model is used. That model is featuring the reduced pressure as one of the independent parameters. For that reason no amendments are required to that issue. The final form of the generalized dependence is described

$$\alpha_{Pb} = A \cdot q^{\frac{2}{3}} \cdot M^{-0,5} \cdot P_r^{0.12} \cdot \left(-\lg P_r \right)^{-0.55} \qquad (11)$$

The term A in the formula (11) is a constant dependent on the type of refrigerant and in the case of freons this value is 55.

Specification of tested data is presented in Table 2. Some of the data fall into the range of minichannels, some of them to conventional ones. The value of the confinement number *Con* is used to distinguish between these data. Additionally in Table 2 presented are values of reduced pressure. Using the Kew and Cornwell [14] criterion, the available data bank was divided into conventional size channels and minichannels.

Table 2. List of experimental data.

Author(s)	Fluid	*Con*	P_{sat}/P_{kr}
Docoloumbier et al. [15]		**1.611** – **1.412**	0.359 – 0.472
Pamitran et al. [16]		0.241 – 0.483	0.485 – 0.61
Mastrullo et al. [17]	CO_2	0.113 – 0.138	0.382 – 0.548
Yoon et al. [18]		0.083 – 0.099	0.472 – 0.61
Choi et al. [19]		0.415	0.61
Oh et al. [20]		0.071 – 0.274	0.413 – 0.777
Dang et al. [21]		0.274	0.69
Kim et al. [22]		0.096 – 0.16	0.413 – 0.777
Wu et al. [23]		**0.526** – **0.78**	0.136 – 0.472
Cho et al. [24]		0.048 – 0.149	0.472 – 0.777
Zhao et al. [25]		0.223	0.19

3 Results

It can be concluded that the transition from conventional size channels to minichannels takes place on average at channel diameter smaller than 1.5 mm, i.e. *Con* > 0.5. It should be also noted that the analysis of the parameters from Table 2, indicates the fact that collected for scrutiny experimental research covers a full range of quality and a relatively wide range of mass velocity.

In the first instance the attention will be drawn to the issue how the Muller-Steinhagen and Heck model is peforming at different reduced pressures in comparison to the Zhang and Webb model [25]. The latter has been especially developed to model pressure drop in minichannels and has a capability of differentiating between different values of reduced pressure. The comparison of the diagrams presented in Figures 1 shows that distributions of the pressure drop component for the same channel diameter are very similar for the case of the Zhang and Webb model and the model proposed here (9). Comparisons of the pressure drop determined with these two models at constant mass velocity G and different channel diameters show some differences, but they are smaller than 20% between simulations.

In Fig. 2 presented are the results of predicting the two-phase pressure drop in tubes using different models of two-phase flow multipliers. The data from Docoloumbier et al. [23] has been taken. Apparently, the model using formulation (9) performs best of other ones considered.

Belyaev et al. [25] tried to confirm the hypothesis that in case of high value of reduced pressure, the two-phase flow structures in small diameter channels are similar to those occurring in conventional size diameter. Based on their study they observed that when reduced pressure is greater than 0.4 then is no differences between heat transfer during flow boiling in minichannels and conventional channels. Mauro [26] reports that at the same value of reduced pressure, the thermodynamic properties of refrigerants are very similar, wherein in case of transported properties, these properties are more divergent.

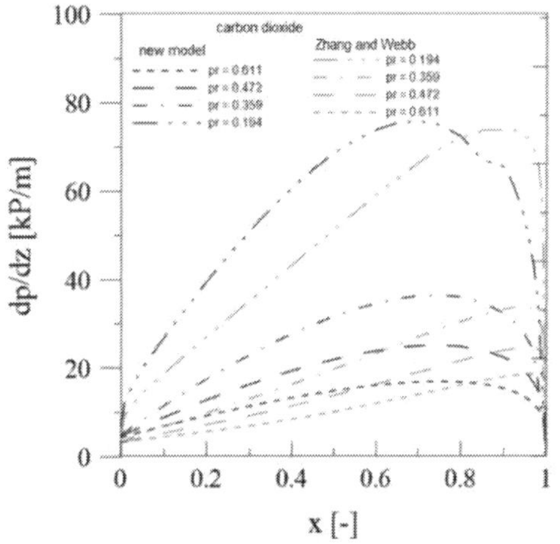

Fig. 1. Pressure drop distribution in function of quality for CO_2 at different values of mass velocity; G=400 kg/(m²s), T_{sat}=0°C, d = 1mm, comparison of modified Muller-Steinhagen and Heck model (3) and Zhang and Webb.

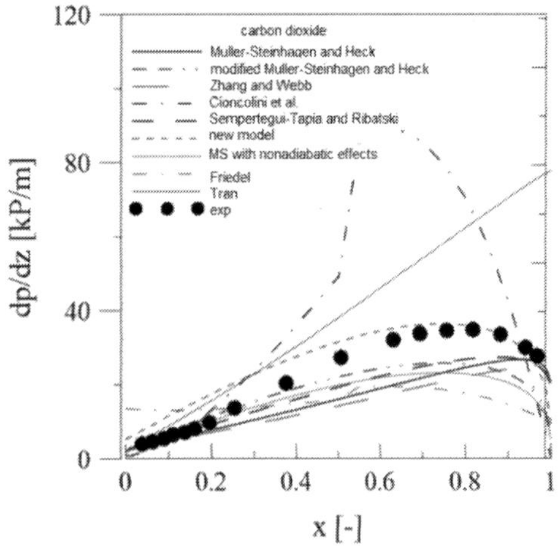

Fig. 2. Pressure drop distribution in function of quality for CO_2 at different values of mass velocity; G=400 kg/(m²s), T_{sat} = -10°C, d = 1.42 mm, comparison of different two-phase multiplier models [15].

In the following part, the basic model and its subsequent modifications, which have been selected for discussion, will be analyzed with respect to predictions of the heat transfer coefficient. Calculation are carried out using the model described by equation (5), where different two-phase multiplier definitions have been assumed. The results of calculations are presented in Fig. 3.

In Figures 4-6 presented are the results of calculations of heat transfer coefficient using the model presented pressure drops related to boiling in flow using the relationships specified in Table 1. The presented calculations show that the models proposed by eq. (10) gave the very good match with the experimental data.

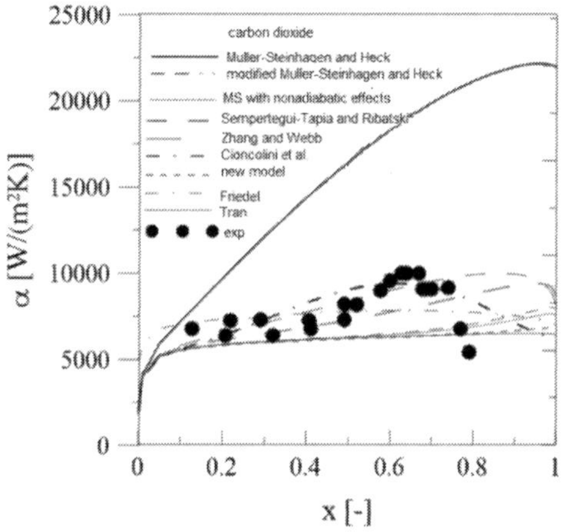

Fig. 3. Pressure drop distribution in function of quality for CO_2 at different values of mass velocity; d = 0,53 mm, G = 200 kg/m²s, q = 10 kW/m², T_{sat} = 0°C, comparison of different two-phase multiplier models [15].

The results of the analysis include all the collected experimental data specified in Table 2. According to the adopted systemization of channels, experimental data of Docoloumbier and al. [15] (CO_2), Wu and others [23] (CO_2), Con ≥ 0.5 was obtained while the remaining experimental database collected corresponds to the flow through a channel with conventional diameter. In performed calculations MAE=36.22%.

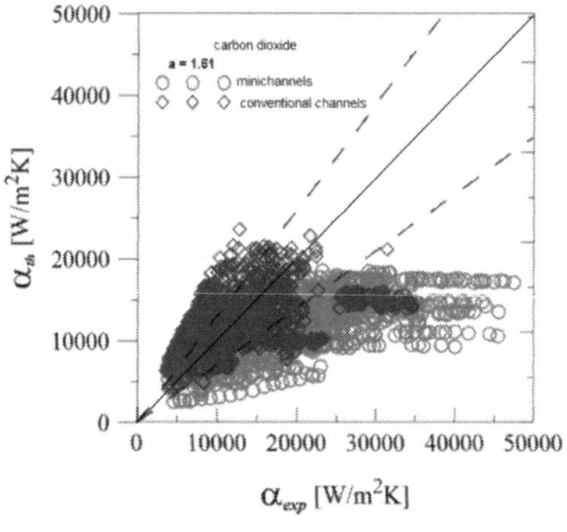

Fig. 4. Correlation of theoretical and empirical heat transfer data for CO_2 at a=1.61.

The graphs show that the correlation equations adopted for analysis reflect the experimental data in a satisfactory manner.

In addition, it should be added that to reduce the average relative error and increase the coefficient of determination in the linear regression model, the amount of experimental data falling within the error limits of ± 30% in relation to the original form of the model adopted for analysis also increased. For the case of considered carbon dioxide data it amounts to 52%. The histogram of deviations is presented in Fig. 6.

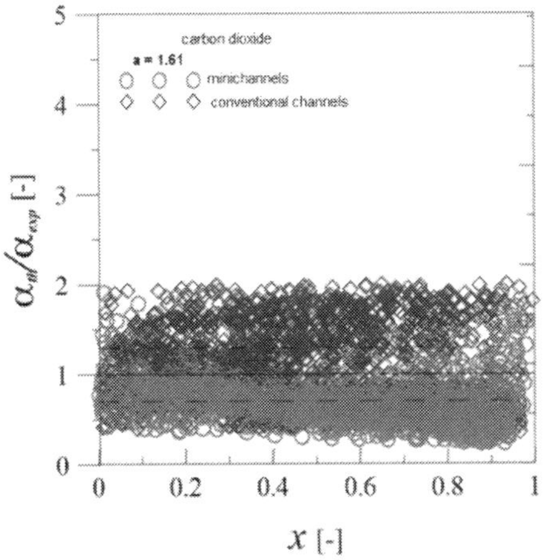

Fig. 5. Correllation of theoretical to empirical heat transfer ratio data in function of quality for CO_2 at a=1.61.

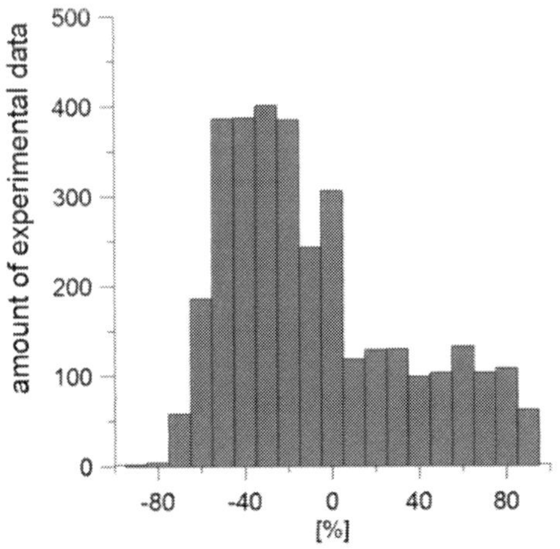

Fig. 6. Histogram of deviations of the calculateds results α_{th}.

4 Conclusions

The paper presents the analysis of the results of flow boiling calculations using an authors' own model to predict heat transfer coefficient in a wide range of reduced pressures for the case of carbon dioxide. The special correction has been postulated to the in-house model of flow boiling and condensation in which modified was the two-phase flow multiplier as well as the temperature gradient in pool boiling. Four two-phase flow multiplier models were tested for this purpose, i.e. due to Friedel, Tran, Müller-Steinhagen and Heck and finally its in-house modification for applicability to minichannels. The latter model has been tested against a large selection of experimental data collected from various researchers to investigate the sensitivity of the in-house developed model. The collected experimental data were conducted for the full range of quality variation and a wide range of mass velocity and saturation temperatures. The results show that change of the model which describes the two-phase multiplier is significant in case of CO_2, where the best compliance with experimental data obtained using the Tran correlation. The results also show that taking into account appropriate two-phase multiplier model and reduced pressure effects can significantly contribute to the convergence with experimental data compared to original model. In authors' opinion, the proposed method to calculate the heat transfer coefficient is a reliable tool in engineering calculations e.g. designing heat exchangers [28-30].

Results presented in the paper have been carried out within the project 2017/25/B/ST8/00755 funded by the National Science Centre, Poland in years 2018-2021.

References

1. United Nations Environment Program (UNEP) Montreal Protocol on Substances that Deplete the Ozone Layer, 1997, (Final Act. United Nations, New York, USA).
2. M. Ghodbane, *An investigation of R152a and hydrocarbon refrigerants in mobile air conditioning*, Int. Proceedings of the international congress and exposition, pp. no. 1999–01–0874.
3. L. Cheng, G. Ribatski, and J. R. Thome, *New prediction methods for CO2 evaporation inside tubes: Part II-An updated general flow boiling heat transfer model based on flow patterns*, Int. J. Heat Mass Transf., vol. **51**, no. 1–2, pp. 125–135, 2008.
4. D. Mikielewicz, B. Jakubowska, *Prediction of flow boiling heat transfer coefficient for carbon dioxide in minichannels and conventional channels*, Archives of Thermodynamic, vol. **37**, no. 2, pp. 89–106, 2016.
5. J. Mikielewicz, *Semi-empirical method of determining the heat transfer coefficient for subcooled staurated boiling in a channel*, Int. J. Heat Transfer, vol. **17**, pp. 1129–1134, 1973.
6. D. Mikielewicz, J. Mikielewicz, and J. Tesmar, *Improved semi-empirical method for determination of heat transfer coefficient in flow boiling in conventional and small diameter tubes*, Int. J. Heat Mass Transfer, vol. **50**, no. 19–20, pp. 3949–3956, 2007.
7. D. Mikielewicz and J. Mikielewicz, *A Common Method for Calculation of Flow Boiling and Flow Condensation Heat Transfer Coefficients in Minichannels With Account of Nonadiabatic Effects*, Heat Transfer Engineering, vol. **32**, no. 13–14, pp. 1173–1181, 2011.
8. D. Mikielewicz, R. Andrzejczyk, B. Jakubowska, J. Mikielewicz, *Analytical Model with Nonadiabatic Effects for Pressure Drop and Heat Transfer During Boiling and Condensation Flows in Conventional*

Channels and Minichannels, Heat Transfer Eng., vol. **37**, no. 13–14, pp. 1158–1171, 2016.

9. H. Müller-Steinhagen, K. Heck, *A simple friction pressure drop correlation for two-phase flow in pipes*, Chem. Eng. Process, vol. **20**, pp. 197–308, 1986.
10. L. Friedel, *Improved friction pressure drop correlations for horizontal and vertical two-phase pipe flow*, European Two-Phase Flow Group Meeting, 1979, p. Paper E2.
11. T. N. Tran, M. C. Chyu, M. W. Wambsganss, *Two-phase pressure drop of refrigerants during flow boiling in small channels: an experimental investigation and correlation development,* Int. J. Refrigeration, vol. **26**, pp. 1739–1754, 2000.
12. D. Mikielewicz, J. Wajs, R. Andrzejczyk, M. Klugmann, *Pressure drop of HFE7000 and HFE7100 during flow condensation in minichannels*. International Journal of Refrigeration, vol. **68**, 226-241, 2016.
13. M. Zhang and R. L. Webb, *Correlation of two-phase friction for refrigerants in small-diameter tubes*, Exp. Thermal and Fluid Science, vol. **25**, no. 3–4, pp. 131–139, 2001.
14. P. A. Kew, K. Cornwell, *Correlations for the prediction of boiling heat transfer in small diameter channels*, Appl. Therm. Eng., vol. **17**, pp. 705–715, 1997.
15. M. Docoulombier, S. Colasson, B. J., P. Haberschill, *Carbon dioxide flow boiling in a single microchannel – Part II: Heat transfer*, Exp. Therm. Fluid Sci., vol. **35**, pp. 597 – 611, 2011.
16. A. S. Pamitran, K.-I. Choi, J.-T. Oh, Nasruddin, *Evaporation heat transfer coefficient in single circular small tubes for flow natural refrigerants of* C_3H_8, NH_3 *and* CO_2, Int. Multiph. Flow, vol. **37**, pp. 794–801, 2011.
17. R. Mastrullo, A. W. Mauro, A. Rosato, G. P. Vanoli, *Carbon dioxide heat transfer coefficients and pressure drops during flow boiling: Assessment of predictive methods*, Int. J. Refrig., vol. **33**, no. 6, pp. 1068–1085, 2010.
18. S. H. Yoon, E. S. Cho, Y. W. Hwang, M. S. Kim, K. Min, Y. Kim, *Characteristics of evaporative heat transfer and pressure drop of carbon dioxide correlation development*, Int. J. Refrig., vol. **27**, no. 2, pp. 111–119, 2004.
19. K.-I. Choi, A. S. Pamitran, C.-Y. Oh, J.-T. Oh, *Boiling heat transfer of R22, R134a and* CO_2, Int. J. Refrig., vol. **30**, no. 8, pp. 1336–1346, 2007.
20. H.-K. Oh, C. H. Son, *Flow boiling heat transfer and pressure drop characteristics of* CO_2 *in horizontal tube of 4.57 mm inner diameter*, Appl. Therm. Eng., vol. **31**, no. 2–3, pp. 163–172, 2011.
21. C. Dang, N. Haraguchi, E. Hihara, *Flow boiling heat transfer of carbon dioxide inside a small-sized microfin tube*, Int. J. Refrig., vol. **33**, no. 4, pp. 655–663, 2010.
22. Y. J. Kim, J. M. Cho, M. Kim, *Experimental study on the evaporative heat trans- fer and pressure drop of CO 2 flowing upward in vertical smooth and microfin tubes with diameter of 5 mm*, Int. J. Refrig., vol. **31**, no. 5, pp. 771–779, 2008.
23. J. Wu, T. Koettig, C. Franke, D. Helmer, T. Eisel, F. Haug, J. Bremer, *Investigation of heat transfer and pressure drop of CO2 two-phase flow in a horizontal mini-channel*, Int. J. Heat Mass Transf., vol. 54, pp. 2154–2162, 2011.
24. J. M. Cho, M. S. Kim, *Experimental studies on the evaporative heat transfer and pressure drop CO2 in smooth and micro-fin tubes of the diameters of 5 and 9.52 mm*, Int. J. Refrig., vol. **30**, no. 6, pp. 986–994, 2007.
25. X. Zhao, P. K. Bansal, *Flow boiling heat transfer characteristic of CO 2 at low temperatures*, Int. J. Refrig., vol. **30**, pp. 937–945, 2007.
26. A. V Belyaev, A. N. Varava, A. V Dedov, A. T. Komov, An experimental study of flow boiling in minichannels at high reduced pressure, *Int. J. Heat Mass Transf.,* vol. 110, pp. 360–373, 2017.
27. A. W. Mauro, Flow boiling of refrigerants: from low to high reduced pressure, in *UIT Summer School*, 2017.
28. J. Wajs, D. Mikielewicz, Influence of metallic porous microlayer on pressure drop and heat transfer of stainless steel plate heat exchanger, *Applied Thermal Engineering*, **93**, 1337–1346, 2016.
29. J. Wajs, D. Mikielewicz, E. Fornalik-Wajs, Thermal performance of a prototype plate heat exchanger with minichannels under boiling conditions, *Journal of Physics Conference Series* **745**(2016), No. 032063. DOI:10.1088/1742-6596/745/3/032063
30. J. Wajs, D. Mikielewicz, B. Jakubowska, Performance of the domestic micro ORC equipped with the shell-and-tube condenser with minichannels, *Energy*, **157**(2018) 853-861.

Thermal calculation of heat exchangers with simplified consideration of axial wall heat conduction

Wilfried Roetzel[1,*], and *Chakkrit* Na Ranong[2]

[1]Helmut Schmidt University/University of the Federal Armed Forces Hamburg, Institute of Thermodynamics, Hamburg, Germany
[2]Technische Hochschule Nürnberg Georg Simon Ohm, Faculty of Process Engineering, Nuremberg, Germany

Abstract. The usual thermal design and rating methods [1, 2] for heat exchangers neglect axial wall heat conduction in the separating walls and external shells of recuperators and in the solid matrix of regenerators. This may lead to undesirable undersizing. In this paper a simplified model is developed for the fast estimation of axial wall conduction effects in counterflow, parallel flow and mixed-mixed cross-flow recuperators. The dispersion model [3] is used to describe the performance deterioration of the exchanger with an effective fluid dispersion Peclet number for the correction of the heat transfer coefficients or mean temperature difference. The method is tested against analytical and numerical calculations for counterflow and parallel flow with good results. It is also shown how the method can be adapted to thermal regenerators and the related thermal calculation methods [1, 2]. An alternative approach is suggested for the consideration of lateral heat conduction resistance in the solid matrix.

1 Introduction

The usual simple heat exchanger design and rating methods [1, 2] neglect the negative effect of axial heat conduction in the separating wall and the outer shells as well as in the solid matrix of regenerators. This may lead to undesirable underdesign. Kroeger [4] was the first to present an analytical closed form solution for balanced and unbalanced counterflow with axial conduction in the separating wall with adiabatic ends. Recently Aminuddin and Zubair [5] published several analytical solutions and calculated results on the performance of counterflow heat exchangers with consideration of axial heat conduction in the separating wall and heat losses from the shell to the vicinity. They assumed various realistic boundary conditions at the wall ends which are useful for the application of the solutions to the cell method or to series connections of exchangers.

In this paper a simple approximation is developed for the fast estimation of the performance deterioration due to axial heat conduction not only in the separating wall but also in the outer shells. Heat losses to the vicinity are neglected. Since axial wall heat conduction has a similar effect as axial fluid dispersion, the dispersion model [3] is used in the approximation method. First counterflow and parallel flow recuperators are considered.

2 System of governing differential equations

The steady state heat transfer process in a counterflow or parallel flow heat exchanger with adiabatic outside surface and axial wall heat conduction can be described with the following system of ordinary differential equations:

$$\frac{dT_1}{dx} - N_1(T_w - T_1) - N_{a1}(T_{wa1} - T_1) = 0 \quad (1)$$

$$\frac{dT_2}{dx} \pm N_2(T_w - T_2) \pm N_{a2}(T_{wa2} - T_1) = 0 \quad (2)$$

$$\frac{d^2T_w}{dx^2} + N_1\mathrm{Pe}_{w1}(T_1 - T_w) + N_2\mathrm{Pe}_{w2}(T_2 - T_w) = 0 \quad (3)$$

$$\frac{d^2T_{wa1}}{dx^2} + N_{a1}\mathrm{Pe}_{wa1}(T_1 - T_{wa1}) = 0 \quad (4)$$

$$\frac{d^2T_{wa2}}{dx^2} + N_{a2}\mathrm{Pe}_{wa2}(T_2 - T_{wa2}) = 0 \quad (5)$$

The number of transfer units N_i of fluids $i = 1,2$

$$N_i = \frac{\alpha_i A_i}{\dot{W}_i}, \quad N_{ai} = \frac{\alpha_{ai} A_{ai}}{\dot{W}_i} \quad (6)$$

are formed with the individual heat transfer coefficients alone, where $\alpha_i A_i$ describes the heat exchange of fluid i with the separating wall and $\alpha_{ai} A_{ai}$ the exchange with the inside surface of the outer wall or shell.

The wall Peclet numbers Pe_w are defined for the separating wall and the outer wall as

$$\mathrm{Pe}_{w,i} = \frac{\dot{W}_i L_i}{\lambda_{w,i} A_{qw,i}}, \quad \mathrm{Pe}_{wa,i} = \frac{\dot{W}_i L_i}{\lambda_{wa,i} A_{qwa,i}}. \quad (7)$$

* Corresponding author: roetzel-suelfeld@t-online.de

For counterflow and parallel flow $L_1 = L_2 = L$ and $\lambda_{w1} A_{qw1} = \lambda_{w2} A_{qw2} = \lambda_w A_{qw}$. In eq (2) the positive sign is for counterflow and the negative sign for parallel flow.

The boundary conditions and further conditions in brackets are:

Counterflow

$$\begin{aligned} x=0: T_1=1 \quad & (T_2=P_2) \\ x=1: T_2=0 \quad & (T_1=1-P_1) \end{aligned} \tag{8}$$

Parallel flow

$$\begin{aligned} & x=0: T_1=1, T_2=0 \\ & (x=1: T_1=1-P_1, T_2=P_2) \end{aligned} \tag{9}$$

Counterflow and parallel flow

$$x=0, x=1: \frac{dT_w}{dx} = \frac{dT_{wa1}}{dx} = \frac{dT_{wa2}}{dx} = 0 \tag{10}$$

The system of five ordinary differential equations of first and second order together with the required eight boundary conditions are implemented into the numeric computing environment of the mathematical software Maple for counterflow and parallel flow, respectively. The resulting real-valued two-point ($x = 0$ and $x = 1$) boundary value problems are solved by a finite difference method based on the trapezoidal scheme with Richardson extrapolation enhancing the accuracy [6]. The numerical solutions and analytical solutions of Kroeger [4] for special cases are used for the test of the approximation method developed in this paper.

3 Infinite thermal conductivities of the walls

First the most disadvantageous case of infinite thermal conductivities of the walls is considered. The wall Peclet numbers become zero and the system of differential equations, eqs (1) – (4), reduces to eqs (1) and (2). The wall temperatures t_w, t_{wa1} and t_{wa2} take constant values. The following equation has been derived for the calculation of $P_1 = P_{1,0}$ with index 0 for zero wall Peclet numbers.

$$\begin{aligned} \frac{1}{P_{1,0}} = & \frac{N_{a1}}{N_1(N_1+N_{a1})} + \frac{1}{1-\exp[-(N_1+N_{a1})]} + \\ & R_1\left[\frac{N_{a2}}{N_2(N_2+N_{a2})} + \frac{1}{1-\exp[-(N_2+N_{a2})]}\right] \end{aligned} \tag{11}$$

This equation is valid for counterflow, parallel flow and any flow arrangement. The indices 1 and 2 can be exchanged. In the limiting case $\dot{W}_2 = \infty$, $R_1 = N_2 = N_{a2} = 0$ the right hand side term in eq (11) turns to $R_1/N_2 = \dot{W}_1/(\alpha_2 A_2)$. For $N_{a1} = N_{a2} = \infty$, eq (11) applies to the two-sided stirred tank [1, 2].

4 Weak wall heat conduction effects

The limiting case of very low thermal wall conductivities ($\lambda_w \to 0$, $\lambda_{wa1} \to 0$, $\lambda_{wa2} \to 0$) is considered in which fluid and wall temperatures do not change remarkably. The wall temperatures of the outside walls assume the temperature of the adjacent fluid $i = 1,2$, $T_{wa,i} = T_i$, as the outside surface is adiabatic. So the local axial conductive heat flows in the outer walls $a,i = a,1$ and $a,2$ are

$$\dot{Q}_{wa,i} = -\lambda_{wa,i} A_{qwa,i} \frac{dT_{wa,i}}{dx} = -\frac{\dot{W}_i L}{\mathrm{Pe}_{wa,i}} \frac{dT_i}{dx}. \tag{12}$$

With the local temperature of the separating wall

$$T_w = T_1 \frac{\alpha_1 A_1}{\alpha_1 A_1 + \alpha_2 A_2} + T_2 \frac{\alpha_2 A_2}{\alpha_1 A_1 + \alpha_2 A_2} \tag{13}$$

the local conductive heat flow in this wall can be expressed as

$$\begin{aligned} & \dot{Q}_w = -\lambda_w A_{qw} \frac{dT_w}{dx} = \\ & -\frac{\dot{W}_1 L}{\mathrm{Pe}_{w1}} \left(\frac{\alpha_1 A_1}{\alpha_1 A_1 + \alpha_2 A_2} \frac{dT_1}{dx} + \frac{\alpha_2 A_2}{\alpha_1 A_1 + \alpha_2 A_2} \frac{dT_2}{dx} \right). \end{aligned} \tag{14}$$

The substitutive effective axial dispersive heat flows in fluids $i = 1,2$ are

$$\dot{Q}_{d,i} = -\lambda_{d,i} A_{q,i} \frac{dT_i}{dx} = -\frac{\dot{W}_i L}{\mathrm{Pe}_i} \frac{dT_i}{dx}. \tag{15}$$

The basic idea of this model is to shift the conductive axial heat flows from the walls into the adjacent fluids. This yields effective Peclet numbers for the dispersion model. Since the separating wall is in contact with both fluids, its heat flow has to be distributed according to the (unknown) fractions φ_1 for fluid 1 and $\varphi_2 = 1 - \varphi_1$ for fluid 2. The substitutive dispersive heat flows are accordingly expressed as

$$\dot{Q}_{d,1} = \varphi_1 \dot{Q}_w + \dot{Q}_{wa,1}, \tag{16}$$

$$\dot{Q}_{d,2} = \varphi_2 \dot{Q}_w + \dot{Q}_{wa,2}. \tag{17}$$

The eqs (16) and (17) represent the basic concept of the model for weak conductive effects. Substituting eqs (12), (14) and (15) into eqs (16) and (17) and rearranging yields

$$\begin{aligned} & \frac{1}{\mathrm{Pe}_1} = \\ & \frac{\varphi_1}{\mathrm{Pe}_{w1}} \left(\frac{\alpha_1 A_1}{\alpha_1 A_1 + \alpha_2 A_2} \pm R_1 \frac{\alpha_2 A_2}{\alpha_1 A_1 + \alpha_2 A_2} \right) + \frac{1}{\mathrm{Pe}_{wa,1}}, \end{aligned} \tag{18}$$

$$\begin{aligned} & \frac{1}{\mathrm{Pe}_2} = \\ & \pm \frac{\varphi_2}{\mathrm{Pe}_{w1}} \left(\frac{\alpha_1 A_1}{\alpha_1 A_1 + \alpha_2 A_2} \pm R_1 \frac{\alpha_2 A_2}{\alpha_1 A_1 + \alpha_2 A_2} \right) + \frac{1}{\mathrm{Pe}_{wa,2}}. \end{aligned} \tag{19}$$

The positive sign is for counterflow and the negative sign for parallel flow. The negative sign occurs as for parallel flow $dT_2/dT_1 \leq 0$. The Peclet numbers Pe_1 and Pe_2 could be used if the fractions φ_1 and φ_2 were known. They are determined in the following.

The overall conductive effect can be expressed with the dimensionless dispersive mean temperature difference [3]

$$\Theta_d = \Theta_{\mathrm{lg}} - \left(\frac{P_1}{\mathrm{Pe}_1} + \frac{P_2}{\mathrm{Pe}_2}\right) = \Theta_{\mathrm{lg}} - \frac{P_1 + P_2}{\mathrm{Pe}_{1,2}}. \quad (20)$$

This equation defines also $\mathrm{Pe} = \mathrm{Pe}_{1,2}$ which is equal for both the fluids. The deteriorative conduction effect should yield positive values $\mathrm{Pe} = \mathrm{Pe}_{1,2} > 0$ for all cases of counterflow and parallel flow.

Multiplying eqs (18) and (19) by P_1 and P_2, respectively, summation of both equations and dividing by P_1 yields with $P_2/P_1 = R_1$

$$\frac{1+R_1}{\mathrm{Pe}} = \frac{1}{\mathrm{Pe}_{w1}}(\varphi_1 \pm R_1\varphi_2)\left(\frac{\alpha_1 A_1}{\alpha_1 A_1 + \alpha_2 A_2} \pm R_1 \frac{\alpha_2 A_2}{\alpha_1 A_1 + \alpha_2 A_2}\right) + \frac{1}{\mathrm{Pe}_{wa,1}} + \frac{R_1}{\mathrm{Pe}_{wa,2}}. \quad (21)$$

This equation reveals that the condition $\mathrm{Pe} > 0$ can only be fulfilled throughout if

$$\varphi_1 = \frac{\alpha_1 A_1}{\alpha_1 A_1 + \alpha_2 A_2}, \quad \varphi_2 = \frac{\alpha_2 A_2}{\alpha_1 A_1 + \alpha_2 A_2}. \quad (22)$$

Substituting eq (22) into eq (21) finally yields

$$\frac{1+R_1}{\mathrm{Pe}} = \frac{1}{\mathrm{Pe}_{w1}}\left(\frac{\alpha_1 A_1}{\alpha_1 A_1 + \alpha_2 A_2} \pm R_1 \frac{\alpha_2 A_2}{\alpha_1 A_1 + \alpha_2 A_2}\right)^2 + \frac{1}{\mathrm{Pe}_{wa,1}} + \frac{R_1}{\mathrm{Pe}_{wa,2}}. \quad (23)$$

The eq (23) remains valid if the indices 1 and 2 are exchanged. The positive sign is for counterflow and the negative sign is for parallel flow. The dispersive Peclet number Pe has to be applied to both fluids.

Substituting eq (22) into eqs (18) and (19) yields formulas for Pe_1 and Pe_2 which have to be applied to fluid 1 and fluid 2, respectively.

With counterflow $\mathrm{Pe}_1 > 0$ and $\mathrm{Pe}_2 > 0$, which means that, owing to wall heat conduction, for fixed inlet and outlet temperatures the mean temperature difference $\theta_m = T_{1m} - T_{2m}$ decreases, the mean temperature T_{1m} decreases and T_{2m} increases.

With parallel flow one of the Peclet numbers may be negative. If e.g. $\mathrm{Pe}_{wa,1} = \mathrm{Pe}_{wa,2} = \infty$ (no outer wall effect) and $N_1 > N_2$, then $\mathrm{Pe}_1 > 0$ and $\mathrm{Pe}_2 < 0$. This means that in this case $\theta_m = T_{1m} - T_{2m}$ decreases, T_{1m} decreases and T_{2m} decreases as well, but less than T_{1m}. This is in qualitative accordance with the real process.

The application of Pe_1 and Pe_2 from eqs (18), (19) and (22) gives the same results as Pe from eq (23). The two Peclet numbers Pe_1 and Pe_2 yield more insight in the process, applying only one Peclet number Pe according to eq (23) is more appropriate for the approximation method developed in the following.

5 Approximation equation for Pe

The relative error of Pe from eq (23) decreases with increasing values of the wall Peclet numbers and the resulting Pe for both fluids. For lower values of Pe eq (23) provides too low values of Pe. In the limiting case $\mathrm{Pe}_w = \mathrm{Pe}_{wa,1} = \mathrm{Pe}_{wa,2} = 0$ eq (23) gives the wrong result $\mathrm{Pe} = 0$. This case has been investigated in section 3 and the correct value of $\mathrm{Pe} = \mathrm{Pe}_0$ can be calculated from eq (20). Substituting in eq (20) Θ_d according to the well known relationship for any flow arrangement with or without dispersion

$$\Theta_d = \frac{P_1}{\mathrm{NTU}_1} = P_1\left(\frac{1}{N_1} + \frac{R_1}{N_2}\right) \quad (24)$$

yields the formula for the calculation of $\mathrm{Pe} = \mathrm{Pe}_0$:

$$\frac{1+R_1}{\mathrm{Pe}_0} = \frac{\Theta_{\mathrm{lg}}}{P_{1,0}} - \left(\frac{1}{N_1} + \frac{R_1}{N_2}\right). \quad (25)$$

In eq (25) $\mathrm{Pe}_{1,0}$ is determined from eq (11) and Θ_{lg} is the dimensionless logarithmic mean temperature difference for counterflow or parallel flow, respectively, calculated with $P_{1,0}$ and R_1.

The eqs (20) and (25) are also exactly valid for the pure cross-flow and are approximations for other one-pass flow arrangements, if Θ_{lg} is replaced by the correct mean temperature difference for the flow arrangement under consideration.

With the limiting Peclet number Pe_0 from eq (25) and the Peclet number from eq (23) which is now denoted with Pe_∞ the approximation equation is formed

$$\mathrm{Pe} = \left(\mathrm{Pe}_0^m + \mathrm{Pe}_\infty^m\right)^{\frac{1}{m}} \quad (26)$$

which yields exact values of Pe for $\mathrm{Pe}_\infty = \infty$ and for $\mathrm{Pe}_\infty = 0$.

The comparison of eq (26) with analytical and numerical results from eqs (1) – (10) provides a value of $m = 0.87$. In the range $5 \leq \mathrm{Pe} \leq 23$ for counterflow and $9 \leq \mathrm{Pe} \leq 140$ for parallel flow the mean relative error of Pe is about 5 % and the mean relative error of P_1 or P_2, calculated with Pe, is about 0.5 %.

The temperature change P_1 can be calculated by correcting the overall heat transfer coefficient (kA) according to

$$\frac{1}{(kA)_\lambda} = \frac{1}{(kA)} + \frac{1}{\mathrm{Pe}}\left(\frac{1}{\dot{W}_1} + \frac{1}{\dot{W}_2}\right) \quad (27)$$

https://doi.org/10.1051/e3sconf/20187002013

and applying the corrected value $(kA)_\lambda$ to the known formulas for counterflow and parallel flow. Other ways are possible as shown in [3].

6 Application of the model to mixed-mixed cross-flow

The eqs (11) and (24) – (26) can directly be applied to mixed-mixed cross-flow. Only the eq (23) has to be replaced by a new equation and in eq (25) Θ_{1g} has to be replaced by the mean temperature difference for mixed-mixed cross-flow.

Now in the separating wall axial heat conduction takes place in the two flow directions and both the flow lengths L_i may be different. So in the wall Peclet numbers usually $L_1 \neq L_2$ and $\lambda_{w1} A_{qw1} \neq \lambda_{w2} A_{qw2}$ (see eq (7)).

Applying again the principle of shifting the conductive heat flows into the fluids leads to the following equation of $\mathrm{Pe} = \mathrm{Pe}_\infty$:

$$\psi_1 = \mathrm{Pe}_{w1}^{1/2}(1+R_1)^{1/2}\left(1+\frac{\alpha_2 A_2}{\alpha_1 A_1}\right) \tag{28}$$

$$\psi_2 = \mathrm{Pe}_{w2}^{1/2}(1+R_2)^{1/2}\left(1+\frac{\alpha_1 A_1}{\alpha_2 A_2}\right) \tag{29}$$

$$\frac{1}{\mathrm{Pe}_\infty} = \left(\frac{1}{\psi_1}\right)^2 + \left(\frac{1}{\psi_2}\right)^2 + \frac{1}{\mathrm{Pe}_{wa1}(1+R_1)} + \frac{1}{\mathrm{Pe}_{wa2}(1+R_2)}. \tag{30}$$

Using eq (28) and (29) the eq (23) for counterflow (+) and parallel flow (-) can be rearranged to a similar form:

$$\frac{1}{\mathrm{Pe}_\infty} = \left(\frac{1}{\psi_1} \pm \frac{1}{\psi_2}\right)^2 + \frac{1}{\mathrm{Pe}_{wa1}(1+R_1)} + \frac{1}{\mathrm{Pe}_{wa2}(1+R_2)}. \tag{31}$$

In the limiting cases of one constant fluid temperature, $\psi_1 = \infty$ or $\psi_2 = \infty$, the eqs (30) and (31) become identical. Since both equations have been derived according to the same model they should reach approximately the same accuracy, also in the general case $\psi_1 < \infty$ and $\psi_2 < \infty$.

Using the Peclet numbers $\mathrm{Pe}_{(30)}$ from eq (30) and $\mathrm{Pe}_{(31,\pm)}$ from eq (31) enables the general presentation:

$$\frac{1}{\mathrm{Pe}_\infty} = \frac{a}{\mathrm{Pe}_{(31,+)}} + \frac{1-a}{\mathrm{Pe}_{(31,-)}} \tag{32}$$

with

$a = 0$: Parallel flow
$a = 1/2$: Mixed-mixed cross-flow
$a = 1$: Counterflow.

Other one-pass symmetrical flow arrangements may be described with individual values of $0 \leq a \leq 1$.

7 Application of the model to regenerators under steady periodic operation

The method derived for counterflow and parallel flow recuperators can also be adapted and applied to fixed-bed and rotary regenerators. The outer walls are considered as part of the solid matrix and need not be considered separately.

7.1 Infinite axial thermal conductivity of the solid matrix

This limiting case corresponds to the steady state heat transfer process in two mixed-unmixed cross-flow heat exchangers coupled by a circulating transversely mixed flow stream [7, 8]. The temperature change P_1

$$\frac{1}{P_1} = \frac{R_{1s}}{1-\exp\left[-R_{1s}\left(1-e^{-N_1}\right)\right]} + \frac{R_{1s}}{1-\exp\left[-R_{2s}\left(1-e^{-N_2}\right)\right]} \tag{33}$$

with the fluid to solid matrix capacity ratios:

For the fixed-bed regenerator

$$R_{i,s} = \frac{\dot{W}_i \tau_i}{C_{s,i}} \tag{34}$$

and the rotary regenerator

$$R_{i,s} = \frac{\dot{W}_i(\tau_1+\tau_2)}{C_{s,t}} \tag{35}$$

The capacity $C_{s,i}$ in eq (34) is the capacity of one matrix "i" and $C_{s,t}$ in eq (35) is the total capacity of the rotating matrix. The time τ_i is the duration of period "i" and the sum $(\tau_1 + \tau_2)$ in eq (35) is the duration of one revolution of the matrix.

The numbers of transfer units N_1 and N_2 in eq (33) are formed with the corrected heat transfer coefficients $\alpha_{i,o}$ according to

$$\frac{1}{\alpha_{i,o}} = \frac{1}{\alpha_i} + \frac{1}{\alpha_s} = \frac{1}{\alpha_i} + \phi\frac{\delta}{\lambda_s}. \tag{36}$$

They take the lateral heat conduction resistance inside the solid matrix into account. The estimation of the additional resistance $1/\alpha_s = \phi\,\delta/\lambda_s$ is subject to Hausen's theory of regenerators [9, 1, 2] and will be discussed later in this paper.

Once the temperature change P_1 is determined from eq (33), the Peclet number for $\lambda_s = \infty$ can be calculated. For regenerators the previous eq (25) has to be replaced by

$$\frac{1+R_1}{\text{Pe}} = \frac{\Theta_{\text{lg}}}{P_1^{(33)}} - \left(\frac{1}{N_1} + \frac{R_1}{N_2}\right)\frac{k_0}{k}. \quad (37)$$

For rotary regenerators:

$$R_1 = \frac{\dot{W}_1}{\dot{W}_2} = \frac{1}{R_2} \quad (38)$$

For fixed-bed regenerators:

$$R_1 = \frac{\dot{W}_1\tau_1}{\dot{W}_2\tau_2} = \frac{1}{R_2} \quad (39)$$

The correction factor k/k_0, which is in reality a correction of the mean temperature difference, takes the deviations of the time averaged longitudinal temperature profiles in the regenerator from those in the equivalent counterflow recuperator into account. This correction has been and still is discussed and calculated by many researchers [9, 10]. It is calculated for a solid matrix with infinite lateral and zero axial thermal conductivity. Results are given in the relevant literature [9, 10].

7.2 Low axial thermal conductivity of the solid matrix

The previous eqs (23) or (31), (28) and (29) can directly be applied using the following definitions of the wall Peclet numbers.

Fixed-bed regenerator:

$$\text{Pe}_{wi} = \text{Pe}_{s,i} = \frac{\dot{W}_i L \tau_i}{\lambda_s A_{qs,i}(\tau_1 + \tau_2)} \quad (40)$$

Rotary regenerator:

$$\text{Pe}_{wi} = \text{Pe}_{s,i} = \frac{\dot{W}_i L}{\lambda_s A_{qs,i}} \quad (41)$$

The cross-section for heat conduction $A_{qs,i}$ in eq (40) belongs to the one solid matrix „i" under consideration. The cross-section $A_{qs,i}$ in eq (41) is the total axial cross-section of the rotating solid matrix.

The products $\alpha_i A_i$ in eqs (23), (28) and (29) have to be substituted according to

Fixed-bed regenerator:

$$\alpha_i A_i = \alpha_{i,0} A_i \quad (42)$$

Rotary regenerator:

$$\alpha_i A_i = \alpha_{i,0} \tau_i \quad (43)$$

7.3 Final correction for axial solid matrix conduction

With Pe_0 from eq (37), considering eqs (33) – (39), and with Pe_∞ from eqs (23) and eqs (40) – (43) the final Peclet number can be calculated using eq (26). Then eq (27) is applied to the correction of the overall heat transfer coefficient (kA).

For the rotary regenerator

$$\frac{1}{kA} = \left(\frac{1}{\alpha_{1,0}A_1} + \frac{1}{\alpha_{2,0}A_2}\right)\frac{k_0}{k} \quad (44)$$

and eq (27) yields the corrected value $(kA)_\lambda$. In eq (44) A_1 and A_2 are the actual heat transfer surfaces in the hot and cold section.

For the fixed-bed regenerator eq (27) has to be adapted to the definitions in Hausen's theory [9, 1, 2] leading to

$$\frac{1}{(kA)_\lambda(\tau_1+\tau_2)} = \frac{1}{(kA)(\tau_1+\tau_2)} + \frac{1}{\text{Pe}}\left(\frac{1}{\dot{W}_1\tau_1} + \frac{1}{\dot{W}_2\tau_2}\right) \quad (45)$$

The surface A in eq (45) is the heat transfer surface of one solid matrix.

If inlet and mean outlet temperatures are given, eq (20) can be used for the correction of the mean temperature difference instead of eq (27) [3].

7.4 Consideration of lateral conductive resistance in solid matrix

7.4.1 Equivalent wall thickness

The function ϕ in eq (36) can be calculated according to Hausen [9, 1, 2] for the plane wall of thickness δ, for the cylinder of diameter δ and the sphere of diameter δ. For other geometries Hausen introduced the equivalent wall thickness

$$\delta_{eq} = \frac{\delta}{2} + \frac{V_s}{A} \quad (46)$$

to be used in the equations for the plane wall.

A new formula for the equivalent wall thickness is proposed which gives better agreement with exact calculations for spherical elements than eq (46)

$$\frac{1}{\delta_{eq}} = \frac{2}{3}\frac{1}{\delta} + \frac{1}{6}\frac{A}{V_s}. \quad (47)$$

For the cylinder eqs (46) and (47) yield the same results.

7.4.2 Alternative calculation model

In the theory on thermal regenerators [9, 10] the transient process is calculated under the assumption of zero axial and infinite lateral thermal conductivity or zero wall thickness, $\delta_{eq}/\lambda_s = 0$. If $\delta_{eq}/\lambda_s > 0$ the additional resistance $1/\alpha_s$ from eq (36) is added.

According to previous investigations on temperature oscillation heat transfer processes [11], not only the heat transfer coefficient has to be corrected (eq (36)) but also the wall thickness or the capacity of the elements. (Index s in this paper is index w in ref [11].) If $\delta_{eq}/\lambda_s > 0$ the reduced effective capacity $\widetilde{C}_s \le C_s$ should be used for the determination of Hausen's reduced period Π [9, 10]. The reduction $\widetilde{C}_s/C_s = \widetilde{X}/X$ and the related heat transfer resistance $1/\alpha_s$ can be calculated according to the temperature oscillation model [11]. The relevant equations are given in the appendix. The resulting values of α_s are equal to or a few percent lower than those of Hausen's approach. The required capacity reduction yields another minor or negligible correction on the safe side. The limiting case of infinite storage capacity does no longer occur.

8 Conclusions

1. The effect of axial heat conduction in the walls of recuperators and in the solid matrix of regenerators can be described and estimated with the axial dispersion model and an effective dispersive Peclet number.
2. The derived approximation model can be applied to counterflow, parallel flow and mixed-mixed cross-flow recuperators as well as counterflow and parallel flow regenerators.
3. The effect of lateral heat conduction in the solid matrix of regenerators can be described with a temperature oscillation model, developed earlier for the evaluation of temperature oscillation experiments [11]. A new formula for Hausen's equivalent wall thickness [1, 2, 9] can improve and simplify the calculation methods for regenerators.

References

1. VDI Wärmeatlas, Edition 11, VDI-GVC (ed.), Springer 2013
2. VDI Heat Atlas, VDI-GVC (ed.), Springer-Verlag Berlin Heidelberg 2010
3. Roetzel W., Na Ranong Ch., Fieg, G.: New axial dispersion model for heat exchanger design. *Heat and Mass Transfer*, 47 (2011), pp. 1009–1017
4. Kroeger P.G.: Performance deterioration in high effectiveness heat exchangers due to axial heat conduction effects. *Adv. Cryog. Eng.*, 12 (1967), pp. 363–372.
5. Aminuddin M., Zubair S.M.: Analytical solution to counter-flow heat exchanger subject to external heat flux and axial conduction. *Int. Journal of Refrigeration*, 74 (2017), pp. 20–35.
6. Ascher U.M., Mattheij R.M.M., Russell R.D.: *Numerical solution of boundary value problems for ordinary differential equations*. Classics in applied mathematics 13, Society for Industrial and Applied Mathematics, Philadelphia 1995.
7. Martin H.: *Wärmeübertrager*. Georg Thieme Verlag, Stuttgart New York, 1988.
8. Na Ranong Ch., Roetzel W.: Stationäres und instationäres Verhalten von zwei gekoppelten Kreuzstromwärmeübertragern mit quervermischtem Umlaufstrom. In: Wärmeaustausch und erneuerbare Energiequellen, Tagungsband *VIII. Internationales Symposium Szczecin-Leba* 18.-20.9.2000, 269 – 276, Herausg. W. Nowak, TU Szczecin, 2000, ISBN 83-87423-14-9.
9. Hausen H.: *Wärmeübertragung im Gegenstrom, Gleichstrom und Kreuzstrom*. 2. Auflage, Springer-Verlag, Berlin Heidelberg New York 1976. English translation: McGraw-Hill, New York, 1983.
10. Baclic B.S., Dragutinovic G.D.: *Operation of counterflow regenerators*. Computational Mechanics Publications, Southampton, UK and Boston, USA, 1998.
11. Roetzel W., Na Ranong Ch.: Evaluation of temperature oscillation experiment for the determination of heat transfer coefficient and dispersive Peclet number. In Proceedings HTRSE 2016, pp. 229 – 241. Extended version: *Archives of Thermodynamics*, Vol. 39 (2018), No. 1, 91–110, DOI: 10.1515/aoter-2018-0005.

Appendix

1 Calculation of α_s in eq (36) according to the oscillation model [11]

Equivalent wall thickness

$$\delta_{eq} = 2\left(\frac{V_s}{A}\right)_{eq} \tag{A1}$$

Substitution into eq (47)

$$\left(\frac{A}{V_s}\right)_{eq} = \frac{4}{3}\frac{1}{\delta} + \frac{1}{3}\left(\frac{A}{V_s}\right) \tag{A2}$$

Dimensionless equivalent wall thickness

$$X_{eq} = \left(\frac{V_s}{A}\right)_{eq}\sqrt{\frac{\pi\rho c}{(\tau_1+\tau_2)\lambda}}, \quad \left(\Omega = \frac{2\pi}{\tau_1+\tau_2}\right) \tag{A3}$$

Equivalent Nußelt number

$$\mathrm{Nu}_{s,eq} = \frac{\alpha_s}{\lambda}\left(\frac{V_s}{A}\right)_{eq}, \quad \mathrm{Nu}_s = \widetilde{\mathrm{Nu}}_s X/\widetilde{X} \tag{A4}$$

$$\mathrm{Nu}_{s,eq} = \frac{X\left(\sinh^2 2X + \sin^2 2X\right)}{\left(\sinh 2X - \sin 2X\right)\left(\sinh^2 X + \cos^2 X\right)} \tag{A5}$$

Application of eqs (A2), (A3), (A5) and (A4) yields α_w as function of δ and (V_s/A). For the cylinder $\delta_{cyl} = 4(V_s/A)_{cyl}$. For the sphere $\delta_{sph} = 6(V_s/A)_{sph}$.

https://doi.org/10.1051/e3sconf/20187002013

2 Calculation of the reduced effective capacity $\widetilde{C}_s$ according to ref [11]

Effective capacity ratio

$$\frac{\widetilde{C}_s}{C_s} = \frac{\widetilde{V}_s}{V_s} = \frac{\widetilde{X}}{X} \tag{A6}$$

For the plane wall

$$\frac{\widetilde{X}}{X} = \frac{\sinh^2 2X + \sin^2 2X}{2X(\sinh 2X + \sin 2X)(\sinh^2 2X + \cos^2 X)} \tag{A7}$$

Instead of eq (A2) the following eq (A8) is recommended for the application to eq (A7)

$$\left(\frac{A}{V_s}\right)^*_{eq} = \frac{0.87}{0.87 + X}\frac{2}{\delta} + \frac{X}{0.87 + X}\left(\frac{A}{V_s}\right) \tag{A8}$$

For precise calculations of cylinder and sphere (α_w and $\widetilde{C}_s$) use the equations (A3) and (A4) in the appendix of ref [11].

Nomenclature

A – area, m^2,
a – variable, eq (32),
C – heat capacity, J/K,
k – overall heat transfer coefficient, $W/(m^2K)$,
L, l – flow length, m,
m – exponent, eq (26)
N – number of transfer units, formed with heat transfer coefficient, $N = \alpha A/\dot{W}$,
Nu – Nußelt number, $\mathrm{Nu} = \frac{\alpha}{\lambda}\left(\frac{V}{A}\right)$, eq (A4),
NTU – number of transfer units, $kA/\dot{W}$,
P_1 – dimensionless temperature change of flow stream 1, $P_1 = \frac{T_{1,in} - T_{1,out}}{T_{1,in} - T_{2,in}}$,
P_2 – dimensionless temperature change of flow stream 2, $P_2 = \frac{T_{2,out} - T_{2,in}}{T_{1,in} - T_{2,in}}$,
Pe – Peclet number for fluid dispersion, $\mathrm{Pe} = \dot{W}L/(\lambda_d A_q)$,
Pe_w – wall Peclet number, $\mathrm{Pe}_w = \dot{W}L/(\lambda_w A_{qw})$,
$\dot{Q}$ – heat flow stream, W,
R – capacity ratio, $R_1 = \dot{W}_1/\dot{W}_2 = 1/R_2$,
for fixed bed regenerators $R_1 = \frac{\dot{W}_1\tau_1}{\dot{W}_2\tau_2} = \frac{1}{R_2}$,
T – temperature, K,
V – volume, m^3,
$\dot{W}$ – heat capacity rate, W/K,
X – dimensionless wall thickness, eq (A3),
x – dimensionless flow length, $x = l/L$

Greek symbols

α – heat transfer coefficient, $W/(m^2K)$,
δ – wall thickness or diameter, m,
Θ – dimensionless mean temperature difference, $\Theta = \Delta T_M/(T_{1,in} - T_{2,in})$,
λ – thermal conductivity, W/(m·K),
Π – reduced period of regenerator,
τ – duration or period, s,
ϕ – function, eq (36),
φ – fraction, eqs (16) and (17),
ψ – dummy variable, eqs (28) and (29)

Subscripts and superscripts

a – outside,
cyl – cylinder,
d – dispersive,
eq – equivalent,
i – counter,
lg – logarithmic,
m – mean value,
q – cross-section,
s – solid matrix,
sph – sphere,
t – total,
w – wall,
λ – corrected for heat conduction,
0 – corrected for conductive resistance, or zero wall Peclet number,
∞ – infinite wall Peclet number,
1, 2 – fluid 1, 2,
~ – effective reduced value

Research on flow boiling heat transfer for vertical upward and downward flows along a minichannel with a smooth heated surface

Kinga Strąk[1,*], and *Magdalena* Piasecka[1]

[1]Kielce University of Technology, Faculty of Mechatronics and Mechanical Engineering, al. 1000-lecia P.P. 7, 25-314 Kielce, Poland

Abstract. The paper reports results for flow boiling heat transfer in a 1.7 mm deep minichannel vertically-oriented with upward and downward flow. The heated element for HFE-649 flowing upward or downward in a channel was a smooth plate. Infrared thermography allowed determining changes in temperature on the outer plate side. Two-phase flow structures were recorded through a glass pane at the other side of the channel being in contact with the fluid. Analysis of the results was performed on the basis of experimental series obtained for the same heat flux for upward and downward flows and two mass flow velocities. The results are presented as relationships between the heat transfer coefficient or the plate temperature and the channel length, boiling curves, and between the heat flux and the heat transfer coefficient and two-phase flow structure images. The impact of mass flow velocity on the heat transfer coefficient and two-phase flow structures for vertical upward and downward flows were discussed.

1 Introduction

Heat transfer in small channels has been studied intensively over the last few years. The main application of small channels was cooling electronic systems and their components. Mini heat exchangers will reduce their sizes and costs, while the consumed power is identical in comparison to conventional sized heat exchangers. They are used to provide higher cooling capability for new technologies. Owing to the change of the state which accompanies flow boiling in small channels, it is feasible to meet contradictory needs simultaneously, i.e. obtain a heat flux as large as possible at small temperature difference between the heated surface and the saturated liquid while retaining small dimensions of heat transfer systems.

Extensive studies are conducted to understand the phenomena with change of phase (boiling or condensation during flow in minichannels [1-22]: and 'pool boiling' [23-26]). Research reported in the literature finds that different orientation of test sections is an important factor in flow boiling heat transfer models in mini spaces. In [8], the researchers examine complex trends of flow boiling critical heat flux (CHF) in a rectangular channel at eight angular positions (0°, 45°, 90°, 135°, 180°, 225°, 270°, 315°) in both microgravity and Earth gravity. The authors reported that CHF sensitivity to orientation was lower for upflow and upward-facing heated wall positions (0°–90°) and higher for downward flow and downward-facing wall positions (180° – 270°). The CHF mechanism in Earth gravity is shown to be highly sensitive to flow orientation at very low velocities.

In [9], while questioning the effect of gravity on CHF mechanisms, the authors show that the influence of orientation can be negated when the influence of gravity perpendicular and parallel to heating walls is overcome and when the sufficient length of the heated wall is provided to ensure liquid contact.

In [10], subcooled flow boiling critical heat flux measurements are taken over a range of flow positions: 0°, 45°, 90°, 135°, 180°, 225°, 270°, and 315° with respect to gravity. The findings exhibit an increase in CHF with increasing velocity and a considerable consolidation of data at different positions at the highest velocity.

The results of the Kielce University of Technology study on flow boiling heat transfer with FC-72 in a minichannel positions: 0°, 45°, 90°, 135° and 180° with respect to the horizontal plane are described in [11]. In their previous paper [12], the authors report the results of flow boiling heat transfer experiment with several refrigerants (FC-72, HFE-7100 or HFE-7000) in an asymmetrically heated rectangular minichannel inclined at seven angles (0°, 30°, 60°, 90°, 120°, 150°, and 180°) relative to the horizontal plane. The highest values of local heat transfer coefficients were observed at the inclination angle of 60° with HFE-7000 and at 180° with FC-72 (the subcooled boiling region) and all minichannel positions with FC-72 (the saturated boiling region).

* Corresponding author: kzietala@tu.kielce.pl

2 Experimental stand

The experimental setup comprises the main loop with refrigerant HFE-649 (3M), the data and image acquisition system, the supply and control system and the lighting system (Fig. 1a).

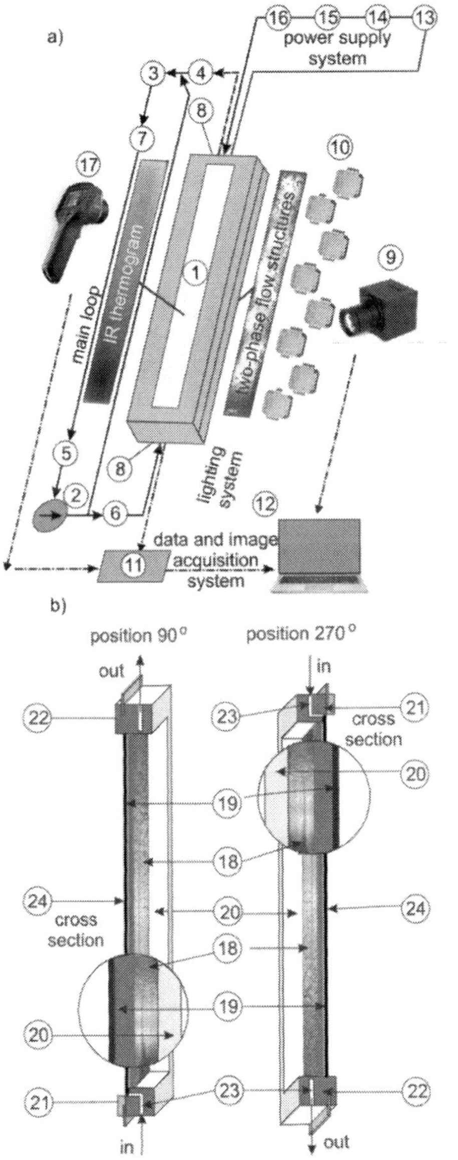

Fig. 1. a) The schematic diagrams of: a) the main systems of the experimental setup, 1-test section with a minichannel; 2-gear pump, 3-compensating tank/pressure regulator, 4-tube-type heat exchanger, 5-filter, 6-mass flowmeter, 7-deaerator, 8-pressure meter, 9-fast camera, 10-high power LEDs (400 W), 11-data acquisition station, 12-pc computer, 13-inverter welder, 14-shunt, 15-ammeter, 16-voltmeter, 17-infrared camera; b) the test section for vertical upward and downward flows along the minichannel: 18-minichannel, 19-heated plate, 20-glass panel, 21-channel body, 22-front cover, 23-thermocouple, 24-black paint layer.

The vertically-oriented rectangular minichannel with the upward flow (denoted as position 90°) and the downward flow (denoted as position 270°)°, was an essential element of the setup (Fig. 1b).

The minichannel had a depth of 1.7 mm, a width of 16 mm and a length of 180 mm. Its heated wall was a smooth plate of thickness of about 0.45 mm (7, Fig. 2a). This plate was made of Haynes-230 alloy manufactured by Haynes Int. Inc. (USA). In addition to excellent high-temperature strength and oxidation resistance, the alloys exhibit superior long-term stability, fabricability [27] and has good weldability [28] and allows you to create diffusion joints - possibilities of combining metal sheet with metallic foams using a sintering powder mixture in dissociated ammonia [29].

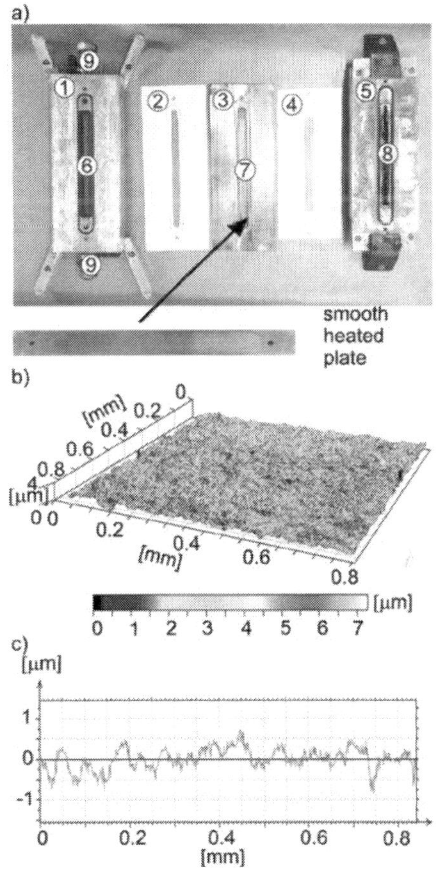

Fig. 2. a) Image of main elements of the test section. 1-front cover, 2,4-teflon plate, 3-graphite plate, 5-channel body, 6-glass panel, 7-heated plate, 8-micanite panels, 9-thermocouple; b,c) analysis of a fragment of the plate's smooth surface: 3D topography (b) and roughness profile (c).

Figures 2b-c show the 3D topography of fragment of the plate's smooth surface (b) and a roughness profile (c). The surface parameters roughness were: arithmetic mean deviation of the roughness profile Ra = 0.175 μm and arithmetic mean height Sa = 0.346 μm. Image of the test section with the minichannel was shown on Fig. 2a. The channel depth is provided by a Teflon plate of adequate thickness (2). The outer smooth side of the heated plate of the minichannel was strengthened on both sides with narrow micanite panels (8) to prevent the plate surface from deforming.

In the experimental series was measure the surface temperature of the outer side of the plate in the central, axially symmetric part of the channel, using a contactless method using an E60 *FLIR* infrared camera. The accuracy of the infrared camera temperature measurement is ± 1 °C or ± 1% within the temperature range of 0 – 120 °C [13]. The two-phase flow structures were observed simultaneously through a glass panel at the other side of the minichannel by an SP-5000M-CXP2 fast camera (manufactured by *JAI*). K-type thermocouples and pressure meters were placed at the inlet and outlet of the minichannel. The power supply and control system consists of: an inverter welder, a shunt, an ammeter and a voltmeter. The lighting system comprises high power LEDs (400 W) to provide light for the two-phase flow structures.

During experiments, after the desired values of the pressure and flow rate were reached, the electric power supplied to the heated plate was increased gradually to achieve an increase in the heat flux transferred to the fluid. It caused a change in the heat transfer between the plate and the fluid from single phase convection to nucleate boiling during refrigerant flow along the minichannel.

3 Calculations of the local heat transfer coefficient

The authors of [11, 15-18] presented various methods of determining the local heat transfer coefficient in the two-dimensional approaches employing Trefftz or Beck methods obtained from solving the inverse problem. Trefftz functions are used for various heat transfer calculations [30-33].

In this paper, one-dimensional approach is used for calculations, taking into account the heat flow direction, which is perpendicular to the direction of the fluid flow in the minichannel. Local heat transfer coefficients in the subcooled boiling region are determined from the following equation:

$$\alpha(x) = \frac{\frac{I \cdot \Delta U}{A} - q_{w,loss}}{T_P(x,\delta) - T_f(x) - \left(\frac{I \cdot \Delta U}{A} - q_{w,loss}\right)\frac{\delta_P}{\lambda_P}} \quad (1)$$

where x – distance from the minichannel inlet along the direction of the flow, T_P – the plate temperature measured by infrared thermography, T_f – fluid temperature determined on the basis of the linear distribution of fluid temperature along the length of the minichannel from the inlet to the outlet, λ_P – the coefficient of thermal conductivity of the plate, δ_P – the plate thickness, I – current, ΔU – voltage drop, A – surface area of the heated plate, $q_{w,loss}$ – heat loss to the surroundings [13].

4 Results

All measurements were recorded for the assumed heat flux supplied to the minichannel's heated plate, in stable similar thermal and flow conditions, thus the experimental results relate to the steady state. Data from experiments conducted using vertical orientation of the test section with a minichannel (position 90° - upward flow and position 270°- downward flow, see Fig. 1b) and two different mass flow velocities ($Q_m = 6 \cdot 10^{-3}$ kg/s and $Q_m = 12 \cdot 10^{-3}$ kg/s), at the subcooled boiling region, were analyzed. The results are presented as relationships between the heated plate temperature and the distance from the minichannel inlet (Fig. 3), the heat transfer coefficient and the distance from the minichannel inlet (Fig. 4), boiling curves (Fig. 5a), the heat flux vs. the heat transfer coefficient (Fig. 5b) and images of two-phase flow structures (Fig. 6).

Analysis of temperature distributions on the heated surface, shown in Fig. 3, indicated certain similarities and differences depending on upward and downward flows and mass flow velocities. The highest plate temperatures were achieved at position 270° and lower mass flow velocity ($Q_m = 6 \cdot 10^{-3}$ kg/s) for the highest heat flux (48 kW/m^2), Fig. 3c. For lower heat fluxes (23 and 35 kW/m^2), the highest temperature distributions was obtained at position 90° and at the same mass flow velocity (Fig. 3). The lowest plate temperatures were observed at position 270° and higher mass flow velocity ($Q_m = 12 \cdot 10^{-3}$ kg/s) for all heat fluxes. Moreover, at position 270° and lower mass flow velocity locally low plate temperatures were also noticed, see Fig. 3.

The highest values of the local heat transfer coefficients were achieved at position 270° and lower mass flow velocity and 35 kW/m^2 heat flux, Fig. 4b. For other heat fluxes, the higher heat transfer coefficient values were obtained at position 270° and higher mass flow velocity, Figs. 4a,c. The lowest heat transfer coefficient were observed at position 270° and lower mass flow velocity and at the highest heat flux, Fig. 4c. Moreover, it can be observed that local heat transfer coefficients at position 90° and all mass flow velocities reached similar values in the central part of the minichannel.

Boiling curves shown in Fig. 5a were constructed for upward and downward flows along the minichannel and two mass flow velocities. These curves were plotted as the heat flux against the difference T_P-T_f at 0.135 m distance from the minichannel inlet. The boiling curves are similar, except one obtained at position 270° and lower mass flow velocity with visible temperature difference drop accompanying the onset of boiling. The plots of the other boiling curves do not exhibit such a temperature drop. The relationships between the heat flux and the heat transfer coefficient in Fig. 5b confirm the findings from the analysis of the previous figures. The highest difference between the curve plots is observed at the position 270° and lower mass flow velocity.

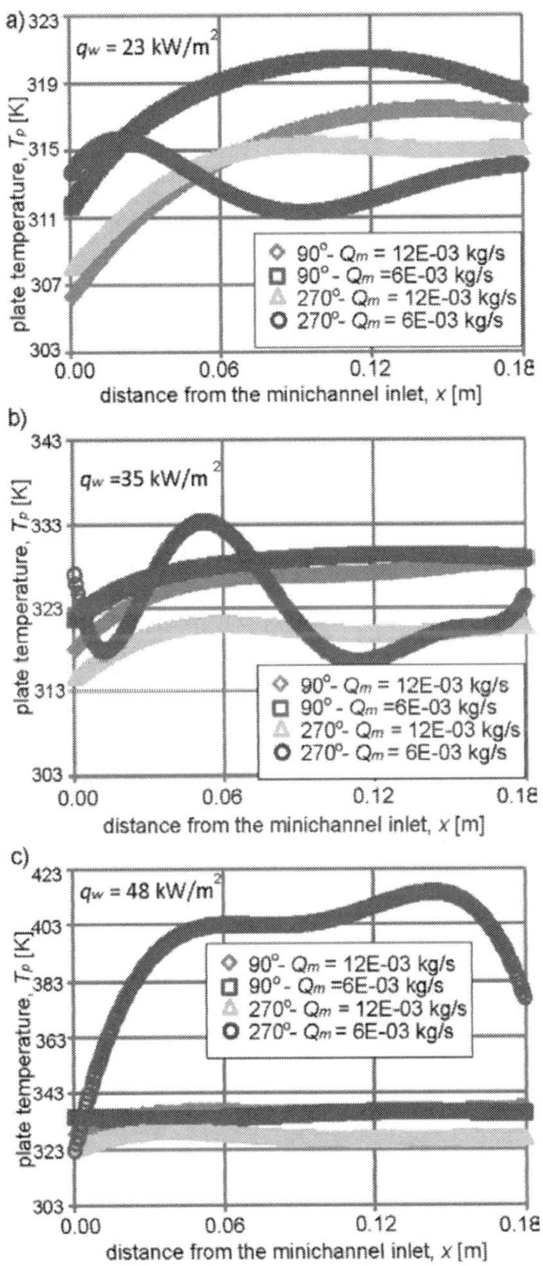

Fig. 3. The plate temperature vs. the channel length, for upward and downward flows along the minichannel and two selected mass flow velocities, at the subcooled boiling region.

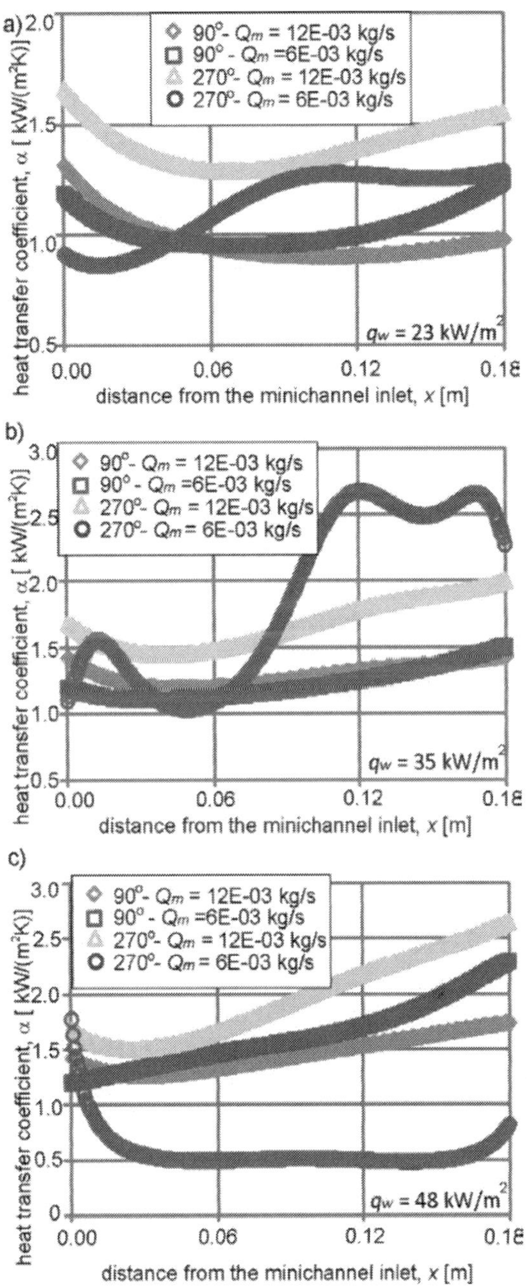

Fig. 4. The heat transfer coefficient vs. the channel length, for upward and downward flows along the minichannel and two selected mass flow velocities, at the subcooled boiling region.

https://doi.org/10.1051/e3sconf/20187002014

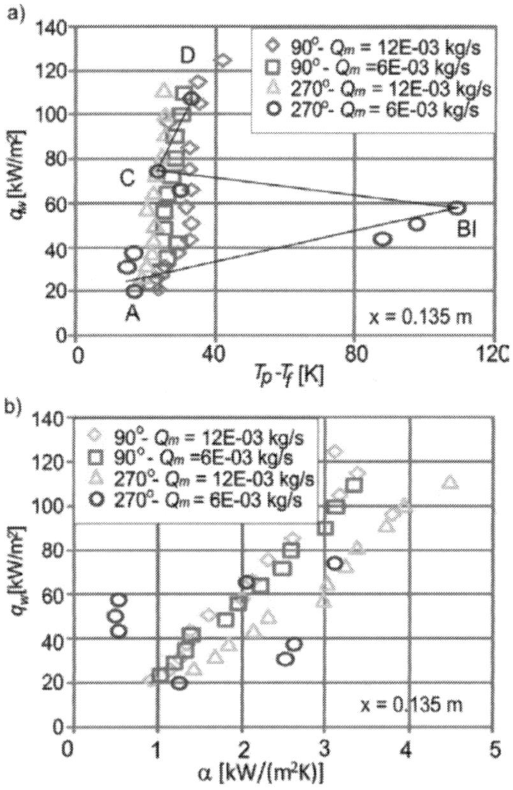

Fig. 5. a) Boiling curves, b) the heat flux vs. the heat transfer coefficient, constructed for 0.135 m distance from the minichannel inlet, BI – boiling incipience.

Images of selected HFE-649 two-phase structures, recorded for selected measurement series in the minichannel for the entire boiling region are shown in Fig. 6. In vertical upflow (position 90°), Figs. 6a-b, only tiny, single vapour bubbles were observed along the whole length of the minichannel and more bubbles were observed near outlet (Fig. 6a, A). With the aid of the dominant buoyancy force, the bubbles moved upward in the direction of the outlet (where the fluid was most heated). Compared to position 90°, more gaseous phase was observed during boiling at the position 270° (Figs. 6c-d). In vertical downflow (position 270°) relative to gravity (dominant with respect to the buoyancy force). The developed annular structures formed at the lowest heat flux at the inlet of the minichannel (Figs. 6c-d, I). The increasing heat flux "pushed" the annular structure upwards, thereby forming vapour locks along the minichannel length (Figs. 6c-d, II). At lower mass flow velocity at the position 270° (Fig. 6c, B), the annular structures forming in the inlet are larger compared with the case of the same channel position and higher mass flow velocity (Fig. 6d, C).

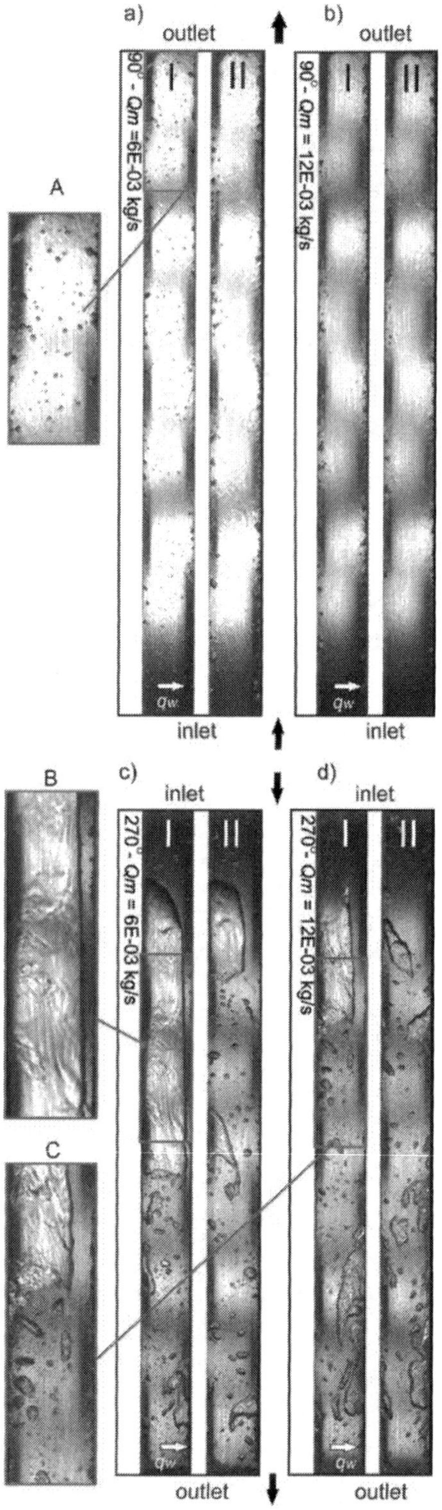

Fig. 6. Images of two-phase flow structures for upward and downward flows: position 90° (a, b) and position 270° (c, d), at mass flow velocities: $Q_m = 6\cdot10^{-3}$ kg/s (a, c) and $Q_m = 12\cdot10^{-3}$ kg/s (b, d).

 https://doi.org/10.1051/e3sconf/20187002014

5 Conclusions

This paper presents the results of HFE-649 flow boiling heat transfer in a rectangular minichannel vertically-oriented for upward and downward flows and two mass flow velocities ($6\cdot10^{-3}$ kg/s and $12\cdot10^{-3}$ kg/s). Infrared thermography allowed to determine changes in the temperature on the outer side of the channel. The two-phase flow structures were recorded through a glass pane at the other side of the minichannel being in contact with the fluid. The results are shown as relationships between: the heat transfer coefficient or the plate temperature and the distance along the minichannel length, boiling curves, the heat flux versus the heat transfer coefficient, and in the form of two-phase flow structure images.

The analysis indicated that the highest plate temperature were obtained for the downflow case at lower mass flow velocity and for at the highest heat flux. The highest values of the local heat transfer coefficients were achieved under downflow conditions and lower mass flow velocity and the average heat flux. For other heat fluxes, the higher heat transfer coefficient values were obtained at the downflow orientation and higher mass flow velocity. The lowest heat transfer coefficient were observed under downflow conditions and lower mass flow velocity and at the highest heat flux.

The constructed boiling curves were similar, except one obtained for downward flow and lower mass flow velocity, with visible temperature difference drop accompanying the onset of boiling. In vertical upflow, only tiny, single vapour bubbles were observed along the whole length of the minichannel. With the aid of the dominant buoyancy force, the bubbles moved upward in the direction of the outlet (where the fluid was most heated). At the downward flow relative to gravity, the two-phase flow structures generated during boiling behaved differently - more bubbles were observed at the inlet and the developed annular structures formed at the lowest heat flux (dominant gravity with respect to the buoyancy force).

The research reported here in was supported partially by a grant from the National Science Centre, Poland (No. DEC-2016/23/N/ST8/01247).

References

1. M. Piasecka, Heat Transf. Eng. **35** (10), 903-912 (2014)
2. T. Bohdal, H. Charun, M. Sikora, Int. J. Refrig. **59**, 210-223 (2015)
3. T. Bohdal, W. Kuczynski, Heat Transf. Eng. **32**(5), 359-368 (2011)
4. T. Bohdal, K. Dutkowski, Annu. Set Environ. Prot. **13**, 409-423 (2011) (in Polish)
5. D. Mikielewicz, J. Wajs, R. Andrzejczyk, et al. Int. J. Refrig. **68**, 226-241 (2016)
6. D. Mikielewicz, M. Klugmann, J. Wajs, Int. J. Therm. Sci. **65**, 79-91 (2013)
7. R. Kaniowski, M.E. Poniewski, Arch. Thermodyn. **34**(2), 3-22 (2013)
8. C.R. Kharangate, C. Konishi, I. Mudawar, Int. J. Heat Mass Transf. **92**, 467–482 (2016)
9. C. Konishi, I. Mudawar, M.M. Hasan, Int. J. Heat Mass Transf. **65**, 203–218 (2013)
10. J.S. Brower, J.F. Klausner, *Proc. of HT-FED2004 Westin Charlotte & Convection Center Charlotte*, Charlotte, North Carolina, USA, 1–6 (2004)
11. M. Piasecka, B. Maciejewska, Exp. Therm. Fluid Sci. **68**, 459–467 (2015)
12. K. Strąk, M. Piasecka, B. Maciejewska, Int. J. Heat Mass Transf. **117**, 375–387(2018)
13. M. Piasecka K. Strąk, B. Maciejewska, Heat Transf. Eng. **38**(3), 332–346 (2017)
14. B. Maciejewska, K. Strąk, M. Piasecka, Procedia Eng. **157**, 82–88 (2016)
15. B. Maciejewska, M. Piasecka, Int. J. Heat Mass Transf. **107**, 925–933 (2017)
16. B. Maciejewska, M. Piasecka, Heat Mass Transf. **53**(4), 1211–1224 (2017)
17. S. Hozejowska, M. Piasecka, Heat Mass Transf. **50**(3), 1053-1063 (2014)
18. B. Maciejewska, K. Strąk, M. Piasecka, Int. J. Numer. Methods Heat Fluid Flow **28**(1), 206-219 (2018)
19. M. Piasecka, K. Strąk, Heat Transf. Eng., **40** (13-14) (2019)
20. M. Piasecka, K. Strak, B. Grabas, Arch. Metall. Mater. **62** (4), 1983-1990 (2017)
21. K. Strąk, M. Piasecka, EPJ Web Conf. **143**, 02116 (2017)
22. K. Strąk, M. Piasecka, EPJ Web Conf. **180**, 02098 (2018)
23. R. Pastuszko, EPJ Web Conf. **45**, 01020 (2013)
24. R. Pastuszko, R. Kaniowski, EPJ Web Conf. **25**, 02019 (2012)
25. L. J Orman, J Enhanc Heat Transf. **23**(2), 137-153 (2016)
26. L. Dąbek, A. Kapjor, L. J. Orman, MATEC Web Conf. **168**, 07001 (2018)
27. www.haynesintl.com/pdf/h3009.pdf
28. P. Młynarczyk, S. Spadło, W. Depczyński, et al., *Proc. Met. 2015 24th Int. Conf. Metall. Mater.*,792-797 (2015)
29. W. Depczynski, *Proc. Met. 2016 25th Int. Conf. Metall. Mater.*, 682-687 (2016)
30. K. Grysa, A. Maciag, Int. J. Heat Mass Transf. **100**, 627-633 (2016)
31. A. Maciag, J. Theor App Mech, **49**(1), 97-116 (2011)
32. M. J. Cialkowski, A. Frackowiak, K. Grysa, Int. J. Heat Mass Transf. **50** (11-12), 2170-2181 (2007)
33. S. Blasiak, A. Pawinska, Int. J. Heat Mass Transf. **90**, 710-718 (2015)

Numerical modelling of air-cooler using simplified methods

Kamil Śmierciew[1], *Jerzy* Gagan[1], and *Dariusz* Butrymowicz[1,*]

[1]Bialystok University of Technology, Wiejska 45C, 15351 Bialystok, Poland

Abstract. The fin-and-tube air coolers have been extensively used in refrigeration systems applied to cold storage chambers. The performance of the heat exchanger affects the efficiency of the systems and makes the study of heat exchanger becomes important. Prediction of the temperature, humidity, as well as velocity distribution in cold storage chamber requires accurate prediction of operation of the finned air cooler. The presence of the air cooler unit is usually taken into account by the investigators, but with very simplified geometry and physics. Results of numerical modelling using the computational fluid dynamics (CFD) software ANSYS FLUENT of fin-and-tube air cooler is presented in the paper. Two different approaches were used: the Dual Cell Model and porous media model.

1 Introduction

Plate fin-and-tube heat exchangers are widely used in several areas such as HVACR systems. Many experimental studies available in the open literature investigate the air-side heat transfer performance of several types of fins used in finned tube heat exchangers. Among the numerical modelling studies, most of them are focused on performance of heat exchanger, mostly heat transfer coefficient at different geometries or operation conditions [1,2]. Numerical studies on performance of the fin-and-tube heat exchanger under frost condition can also be found [3,4,5,6]. However, in some cases, heat exchanger is a part of a large system, where performance of this systems is the most interesting issue. This is also the case of the cold storage chamber. The prediction of the distribution of temperature, humidity, and velocity of air is required in order to ensure the appropriate storage conditions for the stored goods. In such a case an air cooler is a part of the system so that operation of the heat exchanger has to be included in numerical model of the cold storage chamber. However, numerical modelling of operation of the air cooler in a cold storage chamber may be thought as still open problem which is the motivation of the present paper.

The presence of the air cooler unit is usually taken into account in various approaches of modelling of cold storage chamber but with very simplified geometry and physics. The fans forcing the air circulation are mostly represented as infinitely thin plates with assumed pressure jump. The heat exchanger is modelled by means of various approaches. In the work of Chorausia and Goswami [7] the square shaped fins of coils of evaporator in 2D simulation were taken at constant temperature. Hoang et al. [8] modelled the cooler as a block interacting with the surrounding air flow through the momentum sources computed from the pressure rise in fans and pressure drop due to resistance of the cooler tubes. Delele et al. [9] treated the cooler as porous medium with dominated inertial resistance, calculated by taking into account losses due to wall friction, entrance and exit, and acceleration and deceleration effects. Heat loss to the cooler and the mass source due to condensation/evaporation of water vapour on the cooler was also incorporated in the model.

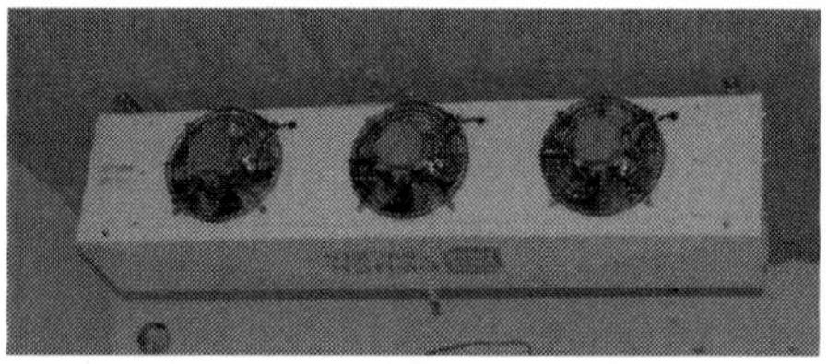

Fig. 1. Front view of the analysed fan-air cooler.

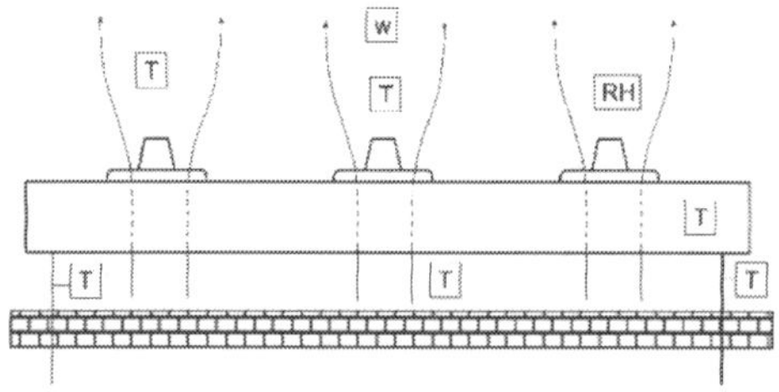

Fig. 2. Location of the measurement sensors.

Overall heat transfer coefficient through the inside surface is calculated from known classical formulae. The amount of transferred heat from air is estimated by using a substitute overall heat transfer coefficient, which value changes within time. Values of the substitute overall heat transfer coefficient may be estimated on the basis of the formulae presented in known publications [1].

* Corresponding author: d.butrymowicz@pb.edu.pl

 https://doi.org/10.1051/e3sconf/20187002015

Location of the measurement sensors is shown in Fig. 2. During experiments the temperature of glycol at air cooler inlet and outlet was measured. Temperature of air in front of the air cooler and behind the cooler, air velocity and relative humidity were also measured.

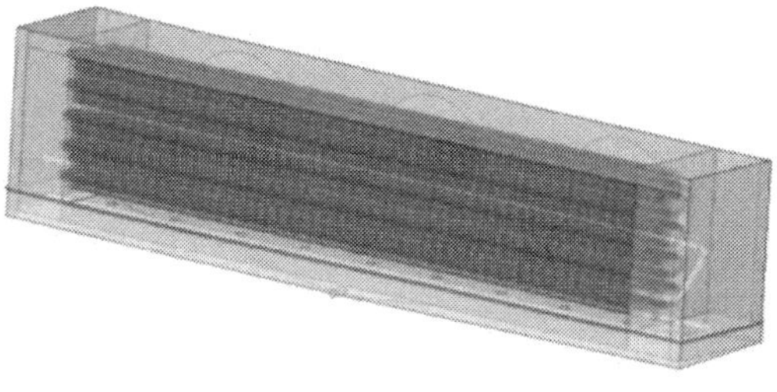

Fig. 3. View on analysed fin-and-tube heat exchanger (air-cooler), isometric view at rear side.

2 Analysis and modelling

The heat exchanger presented in Fig. 1 and Fig. 3 was taken into consideration. Tested heat exchanger consists of three fans with motors, 18 cooling pipes, 2 heating pipes for defrosting and 150 fins. For numerical simulation some assumption has been made in order to simplify geometry. It was assumed that fans will be represented by plane, and motor will be skipped. Only 1/6 of the heat exchanger is modelled with vertical symmetrical plane. The modelled part contains 25 fins. All of the external walls are assumed to be adiabatic.
The main assumption in modelling was application of porous material as the heat exchanger core. The porous zone is described by source term added to the momentum equation:

$$\frac{\partial}{\partial t}(\rho\vec{u})+\nabla\cdot(\rho\vec{u}\vec{u})=-\nabla p+\nabla\cdot(\bar{\bar{\tau}})+p\vec{g}+\vec{F} \quad (1)$$

where

$$\bar{\bar{\tau}}=\mu\left[\left(\nabla\vec{u}+\nabla\vec{u}^{T}\right)-\frac{2}{3}\nabla\cdot\vec{u}I\right]. \quad (2)$$

In the source term in the momentum equation, S_i, the viscosity losses (Darcy's law) – first term on the right hand side of eq. (3) and internal losses – second term on the right hand side of eq. (3) are included:

$$S_i=-\left(\sum_{j=1}^{3}D_{ij}\mu u_j+\sum_{j=1}^{3}C_{ij}\frac{1}{2}\rho|u|u_j\right) \quad (3)$$

where: S_i is the source term in momentum equation, $(i=x,y \text{ or } z)$, $|u|$ is velocity module and D and C are given matrices. The momentum loss is connect with pressure gradient in porous media. Pressure loss is proportional to velocity of flow. For homogenious porous materials it is:

$$S_i=-\left(\frac{\mu}{\alpha}u_i+C_2\frac{1}{2}\rho|u|u_i\right) \quad (4)$$

where α is permeability, C_2 is coefficient of internal resistance per unit thickness, and D and C are diagonal matrices with $1/\alpha$ and C_2 respectively, on main diagonal and zeros for rest entries. In laminar flow through porous material pressure drop is directly proportional to velocity and C_2 can be omit. Ignoring convective and diffusive phenomena, model of flow in porous material is reduced to Darcy's law:

$$\nabla p=-\frac{\mu}{\alpha}\vec{u} \quad (5)$$

For higher flow velocity constant C_2 depends on losses generated by the internal resistance. This coefficient can be considered as loss coefficient in reference to length in flow direction. In the case of a substitute for porous material the permeability term might be omitted and term representing internal resistance only can be used. This leads to:

$$\nabla p=-\sum_{j=1}^{3}C_{2ij}\left(\frac{1}{2}\rho u_j|u|\right) \quad (6)$$

or considering all directions x, y, z:

$$\Delta p_x\approx\sum_{j=1}^{3}C_{2xj}\Delta n_x\frac{1}{2}\rho u_j|u|;$$

$$\Delta p_y\approx\sum_{j=1}^{3}C_{2yj}\Delta n_y\frac{1}{2}\rho u_j|u|;$$

$$\Delta p_z\approx\sum_{j=1}^{3}C_{2zj}\Delta n_z\frac{1}{2}\rho u_j|u|$$

The main idea for substitution of the heat exchanger with porous material is to find performance characteristic of porous material as dependence of the pressure drop on flow velocity. In CFD code used in present simulations porous material is described by three terms: porosity P (scalar), permeability α, and coefficient of internal resistance C_2 (both are vectors). Pressure drop across the heat exchanger was calculated as follows:

$$\frac{\Delta p}{\Delta l}=\frac{\mu}{\alpha}u+\frac{1}{2}C_2u^2 \quad (7)$$

Table 1 represents the predicted pressure drop across the tested heat exchanger at different velocities for air. It should be noted that for the investigated heat exchanger there are not any information concerning pressure loss, since it was made for operation at specific conditions in the cold store chamber. According to the data sheets given by the heat exchanger producer the pressure drop for dry air-cooler is 6.7 Pa.
Results were interpolated with second order polynomial with the following function

$$Y=5.0765X^2+24.0975X-9.0636, \quad (8)$$

where $X=u$ i $Y=\Delta p/n$.

Table 1. Results of numerical calculation.

No.	Velocity u	Pressure drop Δp
	(m/s)	(Pa)
1	0.974	1.73
2	2.0	5.349
3	3.22	10.904

From eq. (8) the coefficients C_2 and α were calculated: $C_2 = 92.46812$ and $\alpha = 6.64799 \cdot 10^{-8}$, therefore, $\frac{1}{\alpha} = 1.50421 \cdot 10^{7}$, taking $a_2 = \begin{bmatrix} 5.0765 \\ 24.0975 \\ -9.0636 \end{bmatrix}$ and density $\rho = 1.22$ kg/m^3 and kinematic viscosity $\mu = 1.78 \cdot 10^{-5}$ Pa·s. Last required parameter is porosity *P*. For modelled domain consisting 25 fins and 25 areas between fins the porosity *P* = 0.96296.

For numerical simulation of the heat exchanger with porous material as heat exchanger core the volumetric heat source for heater or heat sink for cooler should be defined. Under real operation conditions coolant is heated by air and coolant temperature is rising. In simulation conditions there is no refrigerant, so the heat taken from air must be reject by heat sink. Based on experimental results, the heat sink per cubic meter was calculated:

$$\dot{Q}_{sink} = \frac{\dot{Q}}{V}, \tag{9}$$

where V is heat exchanger volume [m^3], $\dot{Q}$ is cooling capacity [kW].

3 Results and discussion

3.1 Results of numerical simulation of an air-cooler with porous material core

ANSYS Fluent 16.0 was used in modelling. Fins and spaces between fins in geometrical model of the heat exchanger were replaced by uniform brick with holes for tubes. The coolant flow through the tubes was omitted in calculation, and volumetric heat sink of the specified value was used instead. All external walls were assumed as adiabatic. Simulation were set as steady state. In numerical simulation velocity inlet was set as boundary condition. The average velocity and temperature at heat exchanger outlet were calculated. Also, the colour contour plots of temperature and velocity on various planes across the heat exchanger were generated. Values of inlets and parameters are shown in Table 2. Due to limited volume of the paper the details concerning the solver setup have been skipped. It should be noted that due to technical conditions in the cold storage chamber, the measurements during experiments were taken about 30 cm before and behind heat exchanger.

Table 2. Experimental and numerical results.

Inlet parameter	Experimental outlet		Numerical outlet	
T_{in} = 275.0 K	T_{out}	274.2 K	T_{out}	274.3 K
u_{in} = 0.94 m/s	u_{out}	2.57 m/s	u_{out}	2.82 m/s

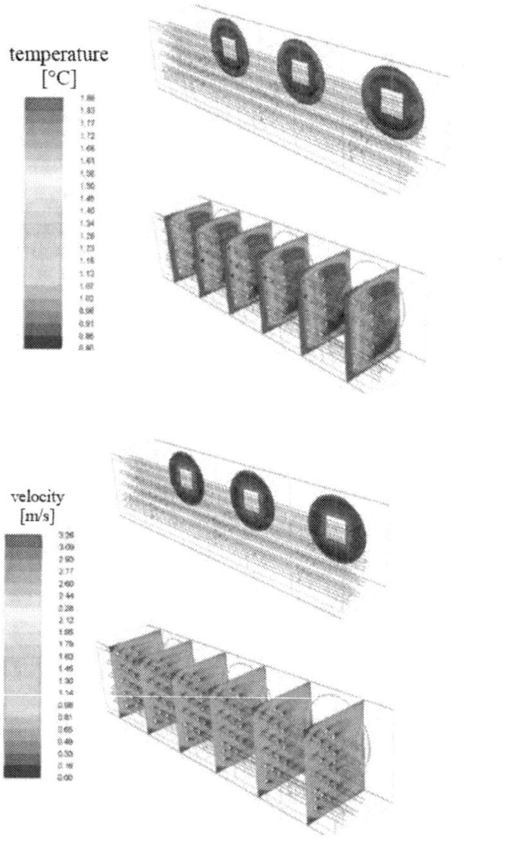

Fig. 4. Temperature field (upper part of the figure) and velocity field (lower part of the figure) at heat exchanger exemplary control plane.

Temperature contour plots at different control planes are presented in Fig. 4. It can be seen that in upper and lower part of the heat exchanger air temperature is higher than inside the core. Also, higher temperature of air can be seen in upper and lower part at the outlet section (see upper part of Fig. 6). This is the result of the existence of large spaces between heat exchanger cover and heat exchanger core (fins surfaces). These empty spaces of the air-cooler are free from flow resistance; therefore air easily passes thru these spaces with no particular change of temperature.

Velocity plots at the same control planes are also presented in Fig. 4. Maximum value u_{max} = 3.26 m/s exists at the exit plane. As it was mentioned earlier the mean exit velocity is $u_{out,EXP}$ = 2.57 m/s while in CFD simulation the mean value was $u_{out,CFD}$ = 2.82 m/s.

Note that the mean velocity taken for comparison not lies exactly at the same plane due to physical limitation of the measurement which was pointed out above. Moreover, outlet velocity resulting from the inlet velocity and flow resistances. Inlet velocity was calculated from the balance equation, and pressure drop due to flow resistance of porous material was found indirectly. It should be noted that there is one physical phenomenon not included in numerical simulations, namely operation of the air-cooler under frost conditions. Growth of frost has additional impact on flow resistance and pressure drop. Nevertheless, it can be concluded in general that modelling of the heat exchanger with porous material predicts outlet temperature and velocity with high accuracy if the pressure drop and heat sink (or source) are known.

3.2 Results of numerical simulation of an air-cooler with dual cell heat exchanger model

Modelling of the heat exchanger with dual cell model may be thought as a bit different approach in comparison with modelling using porous material core. Dual cell model assumes that modelled domain contains separate co-located meshes and couples the two flows only through heat transfer at the heat exchanger core. Also, if the heat exchanger has passes or sections, domain for each section/pass should be divided. Figure 5 show the analysed heat exchanger with section division (see also right-hand side of Figure 3). Each of 9 sections has to be divided into two passes. Therefore, entire heat exchanger would have 19 overlapping zones (1 brick for air which was set as a primary fluid and 18 smaller bricks for glycol which were set as an auxiliary fluid – 2 passes × 9 sections). Note that tubes for glycol are represented by entire core of domain. Every zone of glycol, i.e. inlets and outlets for next pass/section should be thermally coupled. Such complicated domain might be very hard for preprocessing, solver setup and converging and controlling of simulation. It should be noted that literature dealing with application of dual cell heat exchanger model are limited to theoretical construction and limitation of this model provided by ANSYS in ANSYS FLUENT manual [10]. Due to numerous limitations of this model and time-consuming preprocessing procedure in reference to results offered by dual cell model only one section (contains 2 pass) was numerically investigated. Detail schematic of modelled section is shown in Fig. 5.

The procedure for geometry preparation for dual cell model is proposed in ANSYS FLUENT manual [9]. Glycol was the cooling fluid and air was a cooled gas. Operating parameters were found from experimental investigations. Two cases were taken into consideration, before defrost and after defrost.

On the basis of experimental investigation, it was found that air-cooler capacity before defrost was $\dot{Q}_{exp1} \cong 370$ W and $\dot{Q}_{exp2} \cong 1180$ W after defrost. Since only part of heat exchanger was taken into consideration (1/6 × 1/9 = 1/54) the cooling capacity have to be recalculated. Finally, capacities $\dot{Q}_{CFD1} \cong 7$ W and $\dot{Q}_{CFD2} \cong 22$ W before and after defrost, respectively were set. In the dual cell heat exchanger model some input parameters have to be known, e.g. desired heat capacity, then experimental data that defines how heat transfer relates to the fluid flow rates have to be set. As a result user can get outlet temperature of fluids for desired heat capacity. Procedure for solver setup are given in ANSYS FLUENT manual [9].

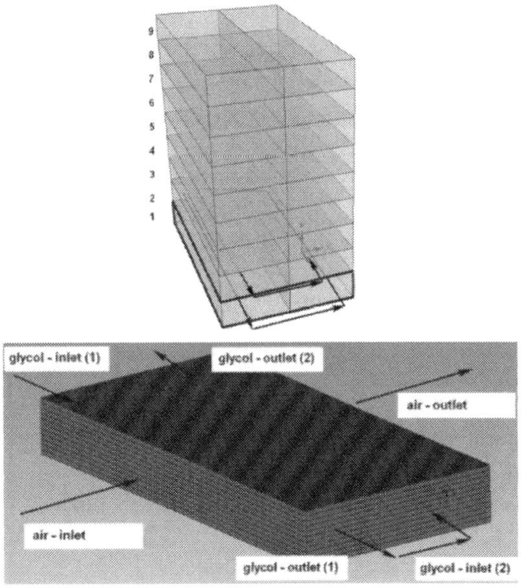

Fig. 5. Numerical domain in original (upper part of the figure) and simplified with boundary description (lower part of the figure).

Boundary conditions were as follows:

- for operation conditions before defrost: $\dot{m}_{in,air}$ = 0.01619 kg/s, $\dot{m}_{in,glycol}$ = 0.09392 kg/s, $T_{in,air}$ = 275.3 K, $T_{in,glycol}$ = 271.1 K;
- for operation conditions after defrost: $\dot{m}_{in,air}$ = 0.01291 kg/s, $\dot{m}_{in,glycol}$ = 0.09611 kg/s, $T_{in,air}$ = 275.5 K, $T_{in,glycol}$ = 270.6 K.

Comparison of numerical and experimental results for operation of air-cooler before and after defrost is shown in Table 3.

Table 3. Experimental and numerical results for operation before and after defrost.

parameter	Experimental	Numerical
	operation before defrost	
thermal capacity Q	6.895 W	6.888 W
outlet air temperature tout	274.1 K	274.9 K
parameter	Experimental	Numerical
	operation after defrost	
thermal capacity Q	22.01 W	22.04 W
outlet air temperature tout	274.1 K	273.85

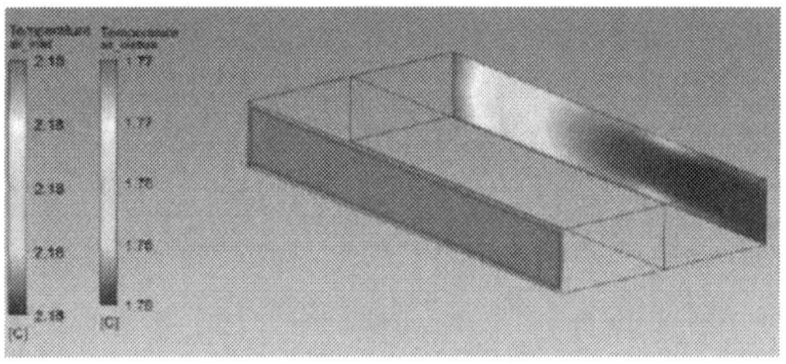

Fig. 6. Air temperature field at inlet and outlet.

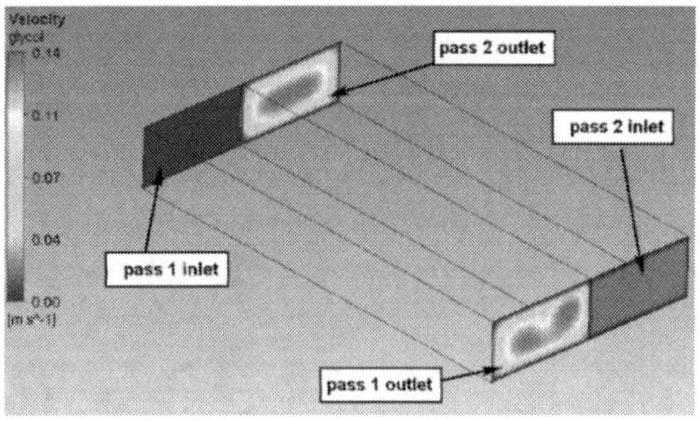

Fig. 7. Glycol velocity field at inlet and outlet.

Temperature field of air for operation conditions before defrost at inlet and outlet plane is resented in Fig. 6. The figure shows that air temperature at the outlet is not uniform. However, the temperature difference between left and right side of the outlet plane is 0.01 K

Velocity field of auxiliary fluid (glycol) at inlets and outlets of both passes is presented in Fig. 7. This figure clearly shows not physical conditions at the transition between 1st pass outlet and 2nd pass inlet. This is an effect of limitation of the discussed dual cell model. First pass outlet velocity is calculated during simulations, while inlet to second pass is defined as boundary conditions (normal to the plane with given value – velocity or mass flow rate) which by default is uniform. Both passes are thermally coupled only, therefore information about velocity profile at the second run inlet should be provided by user, e.g. using UDF function.

4 Conclusions

Results presented in this paper leads to the following conclusions:

1) Both applied approaches of the heat exchanger modelling have some advantages and disadvantages and limitations.
2) Both approaches require some experimental data necessary for information about heat exchanger performance. Both methods are dedicated to simulation of the heat exchanger as a part of large domain, e.g. storage chamber, cooling room etc. where the heat exchanger is not playing major role, where operation of heat exchanger is not required in details but in general, where the heat exchanger cannot be omitted but it can be simplified.
3) Both approaches predict operation of the heat exchanger with good or acceptance agreement with experiments.

The research results presented in the paper were completed within the statutory activities S/WM/1/2018 and financed by Ministry of Science and Higher Education and supported by the National Centre of Research and Development, Contract No. PBS1/A8/7/2012.

References

1. A. Mecke, I. Lee, J.R. Baker jr., M.M. Banaszak Holl, B.G. Orr, Eur. Phys. J. E **14**, 7 (2004)
2. M. Ben Rabha, M.F. Boujmil, M. Saadoun, B. Bessaïs, Eur. Phys. J. Appl. Phys. (to be published)
3. H. Bilirgen, S. Dunbar, E.K. Levy, 2013, Numerical modeling of finned heat exchangers, Applied Thermal Engineering 61, pp. 278-288
4. B. Selma, M. Désilets, P. Proulx, 2014, Optimization of an industrial heat exchanger using an open-source CFD code, Applied Thermal Engineering 69, pp. 241-250
5. J. Cui, W.Z. Li, Y. Liu, Y.S. Zhao, 2011a, A new model for predicting performance of fin-and-tube heat exchanger under frost condition, International Journal of Heat and Fluid Flow 32, pp. 249–260.
6. J. Cui, W.Z. Li, Y. Liu, Z.Y. Jiang, 2011b, A new time- and space-dependent model for predicting frost formation, Applied Thermal Engineering 31, pp. 447-457
7. M.K. Chourasia, T.K. Goswami, 2007, Steady state CFD modeling of airflow, heat transfer and moisture loss in a commercial potato cold store, Int. J. Refrig. 30: 672-689.
8. M.L. Hoang, P.Verboven, J. De Baerdemaeker, B.M. Nicolai, 2000, Analysis of the air flow in a cold store by means of computational fluid dynamics, Int. J. Refrig. 23: 127-140.
9. M.A. Delele, A. Schenk, E. Tijskens, H. Ramon, B.M. Nicolai, P. Verboven, 2009a, Optimization of the humidification of cold stores by pressurized water atomizers based on a multiscale CFD model, J. Food Eng. 91,pp. 228-239
10. ANSYS FLUENT 16.0, 2015, *Theory Guide*.

Preliminary research on mass/heat transfer in mini heat exchanger

Joanna Wilk[1*], *Sebastian* Grosicki[1], and *Krzysztof* Kiedrzyński[1]

[1]Thermodynamics Department, the Faculty of Mechanical Engineering and Aeronautics, Rzeszów University of Technology, al. Powstańców Warszawy 12, 35-959 Rzeszów, Poland

Abstract. In the paper the authors present the facility for model investigations of heat/mass transfer in the exchanger characterised by small dimensions. Determination of heat transfer coefficients is an important issue in the design of mini heat exchangers. The built facility enables measurements of mass transfer coefficients with the use of limiting current technique. The coefficients received from the experiment are converted into heat transfer coefficients basing on the analogy between mass and heat transfer. The exchanger considered consists of nine parallel minichannels with a square cross-section of 2mm. In real conditions during the laminar flow through the minichannels the convective heat transfer occurs. Analogous conditions are maintained during the model mass transfer experiment. The paper presents the experimental facility and the preliminary results of measurements in the form of voltammograms. The voltammograms show the limiting currents being the base of mass transfer coefficient calculations.

1 Introduction

Heat transfer in miniscale is an important subject of study in science and technology. Understanding the phenomena occurring in miniscale is essential for designing devices with small dimensions. To explore heat transfer and fluid flow in minichannels many factors should be taken into account. Determination of heat transfer coefficients is the base of the research and design of the mini heat exchangers. In practice, measurements of heat transfer coefficients are made using the thermal balance method. However, in some cases heat transfer experiments can be difficult to carry out. In the case of minichannels the surface and fluid temperature measurements present the significant problem due to small size of the channel. Thermal measurements in minichannels are often difficult because of the small dimensions of the channel which often makes it impossible to measure the basic thermal parameters such as temperature or heat flux. Mass transfer methods can be used as alternative techniques, heat transfer coefficients then are being calculated from a mass/heat transfer analogy. Mass transfer experiments are widely applied to determine heat transfer coefficients on the basis of empirical correlations for mass transfer.

In the paper the authors present the facility for model investigations of heat/mass transfer in the exchanger characterised by small dimensions. The facility enables measurements of mass transfer coefficients with the use of limiting current technique. The heat exchanger examined consists of the parallel minichannels with a square cross-section.

Some authors have used the limiting current technique in the research on mass transfer in minichannels. Papers [1-4] deal with the investigation of mass transfer in circular micro- and minichannels. In [1] and [5, 6] the authors present the results of mass transfer measurements in minichannels of high-speed heat regenerator under rotation conditions. In turn the works [7, 8] contain investigation of rectangular minichannels using the limiting current technique. The paper [9] presents the applicability of the limiting current technique for the research on mass/heat transfer in minichannels and provides an overview of results of mass transfer measurements. The present work and the planned further research may be the supplement to the current available results on mass/heat transfer in minichannels.

2 Research method

2.1 Limiting current technique

The limiting current technique is well known method used in mass transfer investigations. Some researchers use this technique in their study on mass/heat transfer. In addition to the papers presented above the following latest works on mass/heat transfer in nanofluids can also be cited: [10-15].

Limiting current technique involves observing controlled ion diffusion at one of the electrodes, usually the cathode. Once an external voltage is applied to the electrodes which are immersed in the electrolyte, anionic reduction occurs at the cathode and oxidation at the

* Corresponding author: joanwilk@prz.edu.pl

anode. As a result, electric current arises in the external circuit. The current is proportional to the number of ions reacting at the electrode per unit time. According to Faraday's law, the magnitude, I, of the current generated is given by

$$I=nFAN \qquad (1)$$

where: n – number of electrons consumed in electrode reaction, F - Faraday constant, F=96493×10^3 As/kmol, A - surface area of the cathode, N - molar flux density.
If only the diffusion process of ion transport is taken into account, according to Fick's law and the Nernst model (linear dependence of ion concentration vs the distance from the electrode surface in the diffusion layer) [16], one may write

$$I=nFAh_D(C_b\text{-}C_w) \qquad (2)$$

where: h_D - mass transfer coefficient, C_b, C_w – the bulk and at the wall ion concentration respectively.
If the anode surface is much bigger than the cathode surface, a further increase in the applied voltage will not lead to increasing of current intensity. The limiting current occurs. Under these conditions the ion concentration at the cathode surface C_w approaches zero. Based on the measurement of the limiting current I_p, the mass transfer coefficient can be calculated from the equation

$$h_D= I_p/(nFAC_b). \qquad (3)$$

2.2 Principle of mass/heat transfer analogy

There exists an analogy between mass and heat transfer processes. In mass transfer by means of diffusion a fundamental role is played by the gradient of the concentration of the molecules or ions exchanged, whereas in heat transfer it is the gradient of the fluid temperature. Similar terminology and mathematical models are used to describe both exchange processes. Heat transfer results may be converted from those of mass transfer under equivalent experimental conditions. It arises from the similarity of the equations governing the mass and heat transport processes. Sherwood and Nusselt numbers are analogous dimensionless numbers which represent mass and heat transfer intensity. Results of the experiments on mass transfer during the flow through the channel are expressed in the well-known empirical equation

$$\mathrm{Sh}= C\cdot \mathrm{Re}^p \mathrm{Sc}^q \qquad (4)$$

where: Re – Reynolds number, Sc– Schmidt number, Sh– Sherwood number, C, p, q – empirical constants.
According to the analogy of mass and heat transfer processes the results for heat transfer can be expressed in the form

$$\mathrm{Nu}= C\cdot \mathrm{Re}^p \mathrm{Pr}^q \qquad (5)$$

where: Nu – Nusselt number, Pr– Prandtl number. Dimensionless numbers are given by

$$\mathrm{Nu}= h\cdot d_H/k \quad , \quad \mathrm{Sh}= h_D\cdot d_H/D, \qquad (6)$$

$$\mathrm{Pr}= \nu/a, \quad \mathrm{Sc}= \nu/D, \qquad (7)$$

$$\text{and} \quad \mathrm{Re}=w\cdot d_H/\nu, \qquad (8)$$

where: h – heat transfer coefficient, d_H – hydraulic diameter of the minichannel, k – thermal conductivity, a – thermal diffusivity, D – diffusion coefficient, ν – kinematic viscosity, w – mean fluid velocity.

It is difficult to fulfil all the mass and heat transfer analogy conditions. The requirement of fluid properties similarity expressed by equal Schmidt and Prandtl numbers is especially difficult to meet. If Sc and Pr are not equal, the ratio of equations (5) to (4) indicates that the results of the mass transfer experiment may be converted to the corresponding heat transfer formula

$$\mathrm{Nu/Sh}= (\mathrm{Pr/Sc})^q \qquad (9)$$

The exponent q is the constant determined from empirical investigations. Basing on the available data, the analogy proposed by Chilton and Colburn [17] seems to be most useful when applied in internal flow through minichannels. In the Chilton-Colburn relation the value of the exponent q was found to be 1/3.

2.3 Electrochemical system

A classical system for measurements using the limiting current technique is the reduction of ferricyanide ions at the cathode and oxidation of ferrocyanide ions at the anode. A solution of $NaOH$ or KOH is used as the background electrolyte. The oxidation-reduction process under convective-diffusion controlled conditions is written as:

$$Fe(CN)_6^{-4} \leftrightarrow Fe(CN)_6^{-3} + e, \qquad (10)$$

in this case n=1.
In the present work an aqueous solution of equimolar (0.01 molar) quantities of $K_3Fe(CN)_6$ and $K_4Fe(CN)_6$ and molar solution of $NaOH$ as the background electrolyte were used. Electrodes applied have been made from nickel. To achieve the process under diffusion controlled conditions written by eq.(10) the removing of oxygen from the electrolyte is required. In the present study oxygen was removed by washing the electrolyte with nitrogen.

3 Experimental investigations

3.1 Research stand

Mass transfer measurements were performed on the research stand built specifically for the mini heat exchanger investigations. The general scheme of the measuring stand is shown in Fig. 1. Elements of the facility which had contact with the electrolyte were made of chemically resistant materials. Oxygen was released from the electrolyte by nitrogen bubbling. Because the decomposition in concentration of the reacting ions occurs under sunlight [18] the electrolyte had no contact with sunlight within the whole measuring system. The anode bigger than cathode was located behind the cathode (see Fig. 1) in the direction of the electrolyte flow. Test section used for the electrochemical experiment is shown in Fig. 2.

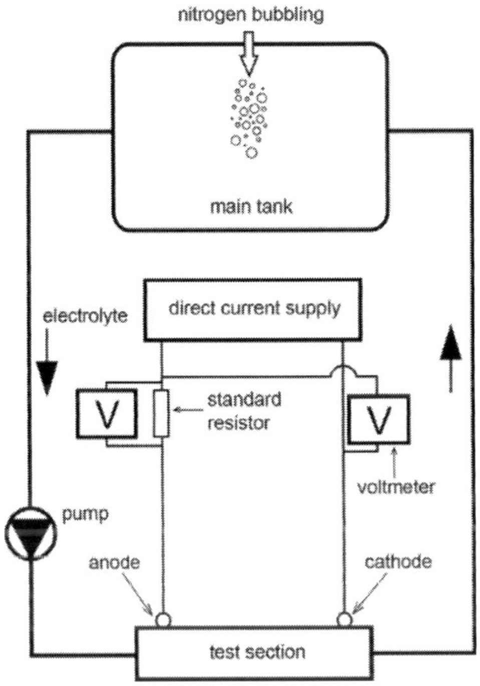

Fig. 1. Scheme of the experimental facility.

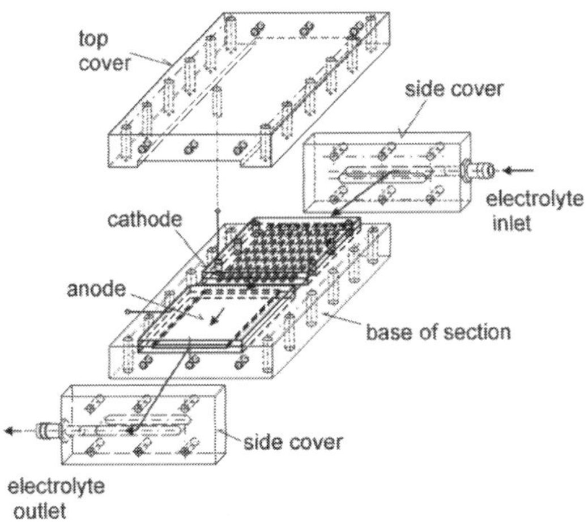

Fig. 2. Test section.

The part of test section was the model of the mini heat exchanger considered. Made from nickel the model of the exchanger consisted of nine parallel minichannels with a square cross-section of 2 mm. The inner surface of the minichannels worked as the cathode in the electrochemical experiment. Thus, the cathode surface was modelling the heat transfer surface in real conditions. Fig. 3 presents the geometry of the mini heat exchanger tested. Dimensions are given in mm.

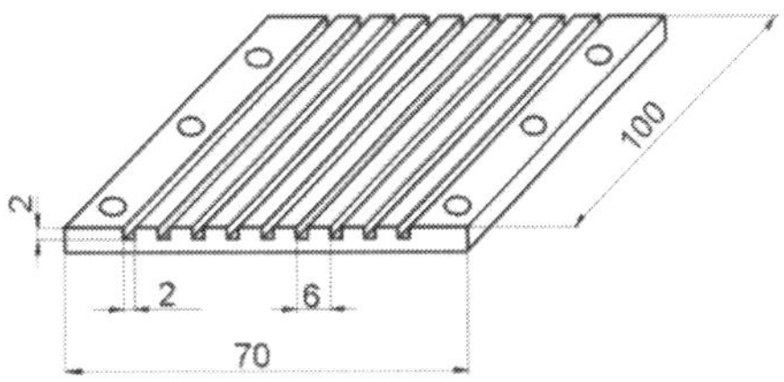

Fig. 3. Geometry of the mini heat exchanger.

Before the experiment the surfaces of electrodes were specially prepared. They were polished with the use of greasy diamond abrasive compound. The electrode polishing has reduced the surface roughness and the same the most suitable measurements has been achieved [18].

3.2. Measurement procedure

The base parameter measured on the research stand was the limiting current representing controlled ion diffusion at the cathode. The increasing of the applied voltage induced the increase in current until reaching the I_p. Current parameters were controlled and measured using voltmeters, direct current supply and standard 1 Ω resistor.

Temperature conditions of the experiment were regulated using the air conditioning system. In this way, the temperature was maintained at 22°C. Constant temperature of the electrolyte allowed the determination of temperature dependent thermophysical properties of the electrolyte: density ρ, dynamic viscosity μ and diffusion coefficient D. The properties have been calculated according to the formulas [19, 20]

$$\rho=1137{,}9-0{,}33T \tag{11}$$

$$\mu=1{,}12\cdot10^{-6}\exp(2056{,}8T^{-1}) \tag{12}$$

$$D=1{,}772\cdot10^{-6}\exp(-2348{,}9T^{-1}) \tag{13}$$

where T is the electrolyte temperature in K.

Values of the electrolyte density, dynamic viscosity and diffusion coefficient are summarized in Table 1.

Table 1. Electrolyte properties at 22°C.

ρ, kg/m^3	μ, N·s/m^2	D, m^2/s	C_b, kmol/m^3
1039,65	$1194{,}56\cdot10^{-6}$	$6{,}17\cdot10^{-10}$	$29{,}96\cdot10^{-4}$

Before the experimental investigations the ion concentration has been measured. Iodometric titration technique has been used [15]. Because of minichannels of the exchanger examined are characterized by a large ratio L/d_H (where L is a channel length), the decrease of ion concentration in axial direction may occur.

According to the scheme presented in Fig.4 the difference between the ion concentration at the electrolyte inlet C_{in} and at outlet C_{out} exists. Basing on the balance of the electrolyte molar flux [9] and taking into account eq.(3) the ion concentration at the minichannel outlet is given by

$$C_{out} = C_{in} - I_p/(nF\dot{V}), \quad (14)$$

where $\dot{V}$ is the electrolyte volumetric flow.

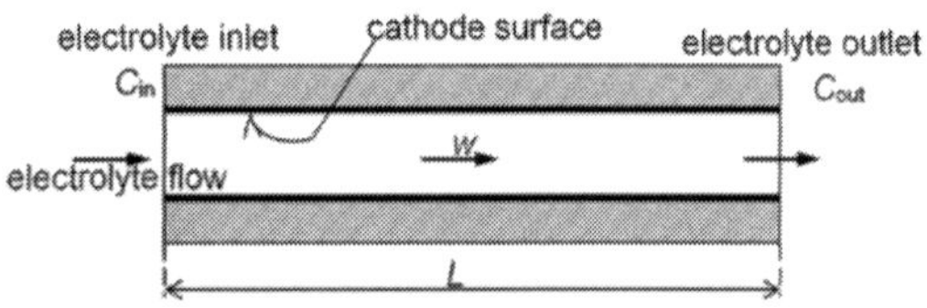

Fig. 4. Electrolyte flow through a long minichannel.

The value of C_{in} was measured by iodometric titration and the bulk ion concentration C_b necessary for mass transfer coefficients calculations was calculated from

$$C_b = \frac{C_{in} - C_{out}}{\ln(C_{in}/C_{out})}. \quad (15)$$

The obtained value of C_b is in Table 1.

The next parameter necessary for development of the research results was the volumetric flow of the electrolyte $\dot{V}$. Values of $\dot{V}$ were determined for five configurations of the pump. The volumetric method has been applied. The flow time of a given electrolyte volume has been measured. On the basis of $\dot{V}$ the mean velocity of the electrolyte flowing through the minichannels of the exchanger has been obtained.

4 Results and discussion

4.1. Measurement results

As a primary results of the measurements plots of the current passing through the cathode against applied voltage have been obtained. The plots named voltammograms are shown in Fig. 5.

As can be seen in Fig. 5 the voltammograms are characterised by the appearance of a flattened section in the curve. It represents the limiting current value. The most representative flat section is in the case of the lowest electrolyte velocity. Reynolds number for this case is about 1430 which indicates a laminar flow. In other cases, there is a transitional and turbulent flow what may be the reason for shortening the flat section of voltammograms representing limiting current.

Mean mass transfer coefficient at the inner surface of the square minichannel of the exchanger has been calculated for laminar flow conditions. The parameters received: mean velocity of the electrolyte, inner surface of minichannels – cathode surface, limiting current and mass transfer coefficient are summarised in Table 2.

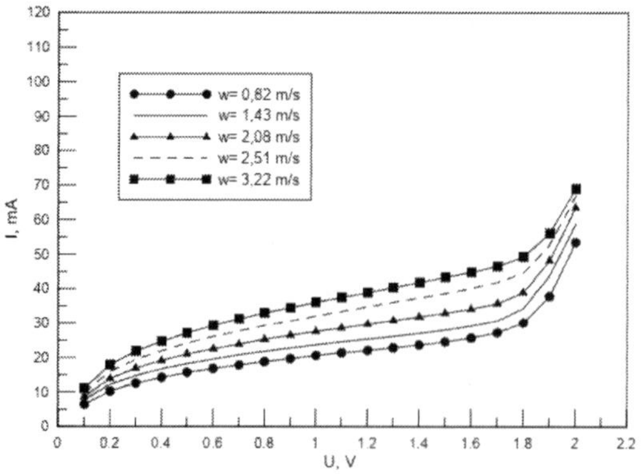

Fig. 5. Voltammograms for the reduction of ferricyanide ions at the cathode being the inner surface of the nine square minichannels of the exchanger.

Table 2. Results of measurements of basic parameters

w, m/s	A, m^2	I_p, A	h_D, m/s
0,82	$7{,}2\cdot10^{-3}$	$23\cdot10^{-3}$	$11{,}05\cdot10^{-6}$

The average relative uncertainty of mass transfer coefficient measurements with the use of limiting current technique is given by

$$\frac{\Delta h_D}{h_D} = \left(\left(\frac{\Delta I_p}{I_p}\right)^2 + \left(\frac{\Delta A}{A}\right)^2 + \left(\frac{\Delta C_b}{C_b}\right)^2\right)^{1/2} \quad (16)$$

where: ΔI_p, ΔA, ΔC_b – are the mean uncertainties of the partial measurements of limiting current, cathode surface and ion concentration, respectively. Basing on the previous investigations of circular minichannels [2, 3, 15] the $\Delta h_D/h_D$ value was estimated to be about 2,5%.

4.2 Mass/heat transfer analogy results

In order to preliminary validation of the results received calculations based on the analogy between mass and heat transfer have been performed. In the first step dimensionless numbers for mass transfer have been calculated. Next, Chilton-Colburn analogy in the form of eq.(6) has been applied. Prandtl number equal 3,25 was adopted for calculations. Values of dimensionless numbers are included in Table 3.

Table 3. Dimensionless numbers for mass and heat transfer during the laminar flow through the square minichannel.

Re	Sh	Sc	Nu	Pr
1430	35,82	1862,24	4,31	3,25

Dirichlet boundary condition of constant temperature at the wall for heat transfer corresponds to the situation of constant concentration at the wall for mass transfer. The mass transfer investigations carried out by limiting current technique are characterised by constant ion concentration at the cathode. Thus, the analogical condition of constant temperature is correct. The literature data about fully developed laminar flow in square channels give the Nusselt number to be 2.98 [21] under a constant wall temperature boundary condition. In turn the another conventional correlation from the literature [22] for the prediction of Nusselt number during the fully developed laminar flow through the rectangular duct give Nu to be 3,55 in the case of square channel. As a most of the literature data on heat transfer in minichannels give the results of an enhancement in heat transfer in comparison to conventional channels, the results obtained in current study may be considered correct.

5 Conclusions

Preliminary experimental investigations on mass/heat transfer in square minichannels of the model of mini heat exchanger have allowed to formulate some conclusions from the obtained results.

Under laminar flow conditions the voltammograms received were characterised by the longer and more pronounced flat section representing the limiting current than in transitional and turbulent regime. Thus, more accurate results of mass transfer measurements with the use of limiting current technique can be expected under laminar flow conditions.

As the channel considered was characterized by the high value of L/d_H, changes of ion concentration along the channel length were occurred. Therefore, it is important to take into account the mean value of concentration in the calculation of mass transfer coefficient.

The use of the analogy of mass and heat transfer gave a higher values of heat transfer coefficient than those in conventional channels. However, further research for obtaining mass/heat transfer processes in square minichannels of a model heat exchanger are necessary.

References

1. J. Wilk, Int. J. of Heat and Mass Transfer **47**, 1979-1988 (2004)
2. J. Wilk, Exp. Thermal and Fluid Sci. **33**, 267-272 (2009)
3. J. Wilk, Exp. Thermal and Fluid Sci. **38**, 107-114 (2012)
4. O. N. Sara, O. B. Ergu, M. E. Arzutug, S. Yapici, Int. J. of Thermal Sci. **48**, 1894-1900 (2009)
5. B. Bieniasz, Int. J. of Heat and Mass Transfer **53**, 3166-3174 (2010)
6. B. Bieniasz, Heat Mass Transfer **50**, 1211-1223 (2014)
7. R. E. Acosta, R. H. Muller, C. W. Tobias, AIChE J. **31**, 473-482 (1985)
8. O. B. Ergu, O. N. Sara, S. Yapici, M. E. Arzutug, Int. Commun. Heat Mass Transfer **36**, 618-623 (2009)
9. J. Wilk, Exp. Thermal and Fluid Sci. **57**, 242-249 (2014)
10. H. Beiki, M. N. Esfahany, N. Etesami, Microfluid Nanofluid **15**, 501-508 (2013)
11. H. Beiki, M. N. Esfahany, N. Etesami, Int. J. of Thermal Sci. **64**, 251-256 (2013)
12. N. Keshishian, M. N. Esfahany, N. Etesami, Int. Commun. Heat Mass Transfer **46**, 148-153 (2013)
13. C. Liu, H. Lee, Y. H. Chang, S. P. Feng, J. Colloid Interface Sci. **469**, 17-24 (2016)
14. J. Wilk, S. Grosicki, J. of Physics **745**, 032084 (2016)
15. J. Wilk, S. Grosicki, Int. J. of Thermal Sci. **129**, 280-289 (2018)
16. A. J. Bard, L. R. Faulkner, *Electrochemical methods* (John Wiley & Sons 2005)
17. T. H. Chilton, A. P. Colburn, Ind. And Engineering Chemistry **26**, 1183-1187 (1934)
18. D. A. Szanto, S. Cleghorn, C. Ponce-de-Leon, F. C. Walsh, AIChE J. **54**, 802-810 (2008)
19. M. Hopkowicz, Z. Pietrzyk, Inż. Chem. **4**, 843-853 (1977)
20. B. Bieniasz, J. Wilk, Int. J. of Heat and Mass Transfer **38**, 1823-1830 (1995)
21. S. G. Kandlikar, S. Garimella, D. Li, S. Colin, M. R. King, *Heat transfer and fluid flow in minichannels and microchannels*, (Elsevier Ltd. 2006)
22. P.-S. Lee, S. V. Garimella, D. Liu, Int. J. of Heat and Mass Transfer **48**, 1688-1704 (2005)

The study on the heat recovery from air compressors

Mariusz Broniszewski[1], and *Sebastian* Werle[2,*]

[1]Bulten Polska S.A., Bukietowa Street 60, 43-300 Bielsko-Biała, Poland
[2]Silesian University of Technology, Institute of Thermal Technology, Konarskiego Street 22, 44-100 Gliwice, Poland

Abstract. Effective use of utilities in production plants is an issue that is becoming increasingly significant in energy policy of Europe. Production of compressed air consumes 3% of the total electricity consumption in Europe. In order to produce compressed air, approximately 10-20% of electricity transferred to compressor is used, the rest is lost due to lack of tightness and heat losses. The aim of this work is to evaluate the possibility to install recovery systems on air compressors to recover the lost waste heat and its management for the needs of heating office buildings/production halls and analysis of investment profitability. The investment will be supported by co-financing in the form of energy efficiency certificates (so-called white certificates), which the production plant will be able to sell after receiving them from Polish Energy Regulatory Office in return for completing an investment that consists recovering waste heat from air compressors.

1 Introduction

In the present world, increasing importance is attached to conducting activity in harmony with sustainable development of a company. It means that while planning company's operations, all aspects that may have an influence on the surrounded environment have to be taken into consideration. One of the aspects that have the biggest impact on the environment, which commonly occurs, is the utilities consumption, including the electricity consumption. Figure 1 shows the amount of the electricity is consumed in the world within different types of sectors. The CAS (compressed air systems) according to Diego Vittorini and Roberto Cipollone [1] is responsible for up to 20% industry electricity consumption. As we can see, there is a great potential of energy we can save/recover from compressed air systems. The CAS, which is energy demand subprocess for many companies, has also a great potential for improvements under both the energy and CO_2 emissions point of view.

The structure of electricity generation sources in Poland is based on coal. The electricity generation using this fuel is characterized by a high index of emission. Due to this fact, the effectiveness of the energy use by the enterprises have to increase. The standard emission index of the electricity generated in Poland equals to 1.191 t CO_2/MWh, whereas the European average equals to 0.460 t CO_2/MWh [2]. European standard emission indexes for particular countries are presented in Figure 2.

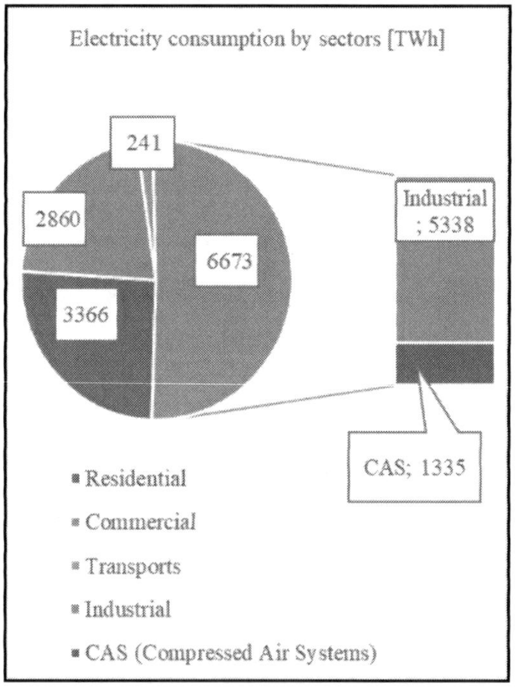

Fig. 1. Electricity consumption by sectors [1].

* Corresponding author: sebastian.werle@polsl.pl

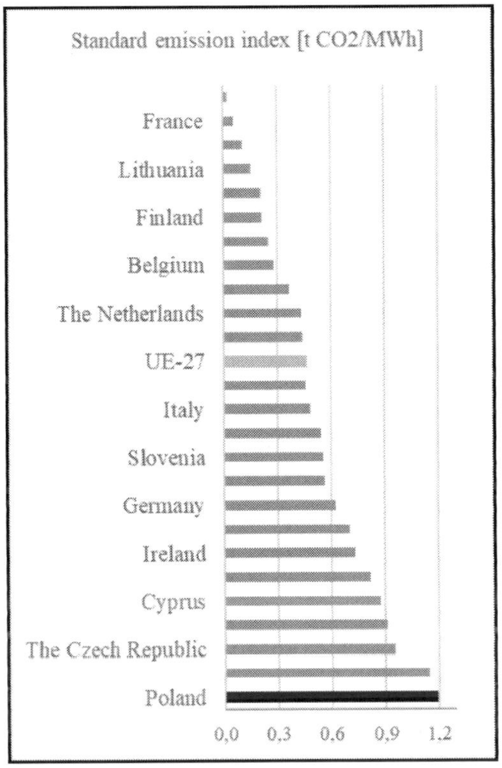

Fig. 2. National and European emission indexes related to electricity consumption [2].

As was mentioned above, the electricity in Poland is mostly produced from coal. In 2009-2012, more than 93% of total electricity produced in Poland was based on the coal combustion. Due to increasingly strict European Union Directives, Poland had to start to limit the use of this raw material, at the same time increase production of energy from renewable energy sources (RES). Not only European guidelines cause that Polish coal has to be replaced. Another reason is the fact, that nowadays coal production from the coal mine is becoming more difficult. Coal must be mined from deeper layers, so the mines are becoming unprofitable. Moreover, deposits of coal are slowly running out. For example, PGE Group (Polish Energy Group) [2] which manages the power plant in Bełchatów, estimates that deposits in the nearby mine will run out within next three years. Consequently, Poland is facing the challenge. The alternative fuels have to be found and energy production from RES have to be increased [3]. Additionally, the new solutions aiming at maximizing effectiveness of electricity use have to be proposed.

The aim of the work is to analyse profitability of installing recovering heat systems on compressors in Bulten Poland S.A. and estimate simple payback time (SPBT). This will help in order to being independent from heat supplier and will lead to reuse waste heat for their purposes, reducing simultaneously CO_2 emission and increasing energy efficiency in the company. Compressed air is significant utility for many of industrial companies, and because of that fact, there are variety possibilities of implementing this solution into industry.

Additional benefit that can be achieved from investments within improving energy efficiency in company can be possibility for obtaining energy efficiency certificates (so-called white certificates) which can be sold on the market bringing the company money and shortening simple payback time of the investment. This study will analyse simple payback time of proposed solution with and without taking into account energy efficiency certificates to show the big value of the motivation in the form of white certificates.

2 Present energy efficiency issues

Observing changes in energy consumption in the EU in the years 1990-2012, we can notice opposing tendencies in various sectors. While in agriculture and forestry, or in industry, there was a decrease in energy consumption respectively by 24.6% and 23.1%, the increase was noticed in the services sector by 36.6% and in transportation by 23.8%. These changes are partly the result of actions taken to increase energy efficiency [4]. The greenhouse gas emissions from the industrial sector have also changed [5]:

- for industrial processes: 462 million tonnes CO_2 in 1990, 394 million tonnes in 2000, 403 million tonnes for 2005, 321 million tonnes in 2012,
- for manufacturing and construction industries: 861 million tonnes in 1990, 707 million tonnes in 2000, 663 million tonnes in 2005, 533 million tonnes in 2012.

In 2008-2012, retail electricity prices for industry increased in the EU by an average of 3.5% per year and gas prices by 1%. It is estimated that electricity prices for industry in the EU are twice as high as in the US and Russia and 20% higher than in China [4].

Measures aiming at increasing energy efficiency may be somehow independent, may also be linked to environmental audits, and in larger enterprises may be a valuable, mandatory element of the management. The initial step to assess the potential benefits of using measures to improve energy efficiency in the company is energy audit. According to Directive 2012/27 / EU [6], an energy audit means a systematic procedure aimed at gaining adequate knowledge about the profile of existing energy consumption of a given activity or industrial plant, determining how and in what quantity it is possible to obtain cost-effective energy savings and to inform about the results of its.Proposed solution in the study was preceded by the proper energy audit which consisted analysis of potential energy savings investments in the company. From several proposals one was chosen as the most efficient and economically reasonable- recovering heat form compressors.

3 Compressed air in industry

The aspect of compressed air that is very often ignored in production plants, is the cost of its generation (electricity consumption) which in the significant manner may unnecessarily overstate production costs (pressure losses at pipelines, leakages). One of methods for rational use of electricity transferred to an air compressor is heat collection, which is produced as a by-product at compressed air production. Thus, electricity is directly transformed into compressor operation and indirectly into heat, which can be introduce to the central heating system and, as a result, provide appropriate temperature in offices and production halls. Then the heat consumption from grid, which in Poland is also mostly produced from coal, is limited. Additionally, the carbon footprint of an enterprise (carbon dioxide emission index, that is the amount of emitted CO_2 and other substances, calculated for CO_2 equivalent that an enterprise emits to the atmosphere as a result of its activity) is limited [6].

Compressed air is considered to be cheap, or even free source of energy. However, this is a wrong approach. Production of compressed air consumes 3% of the total electricity consumption in Europe. About 10-20% of electricity transferred to the compressor is used for production of compressed air, the rest is lost due to lack of tightness and heat losses [7,8].

For the needs of European Union countries, proposals of savings in compressed air systems were developed, with determination of savings potential according to index of possibility to apply, given as percentage values. The following operations were highlighted:

- Reduction of air leaks,
- Reduction of pressure losses that result from friction,
- Appropriate designing of new compressed air systems,
- More frequent changes of filters,
- Compressors upgrading,
- Variable rotary drives,
- High performance electric motors,
- Application of advanced Control systems,
- Waste heat recovery,
- Improvement of cooling, drying and filtration systems,
- Adjusting appropriate final devices (receiving) [8,9]

Energy costs during the utilization time of a compressor may reach more than 75% of the total cost during life cycle (as shown in Figure 3). It has been claimed that improvements within energy efficiency of compressors can bring energy savings in amount of 20-50% energy used for their running [10,11].

The cost of compressed air, depending on electricity price and devices efficiency in Bulten Poland SA is estimated for approximately 0.04 PLN/m^3. In one of the industrial plants that produces connecting materials for automotive industry, calculations were made at the price of 0.04 PLN/m^3 and monthly compressed air consumption of 1 774 632 m^3, monthly costs of compressed air production (without taking into consideration exploitation costs) amounted to PLN 70 985.28. During compressed air production, waste heat is generated as by-product, which the company decided to develop. Because of high maintenance costs of producing compressed air in industrial area solutions aimed at reducing costs and increasing energy efficiency are required.

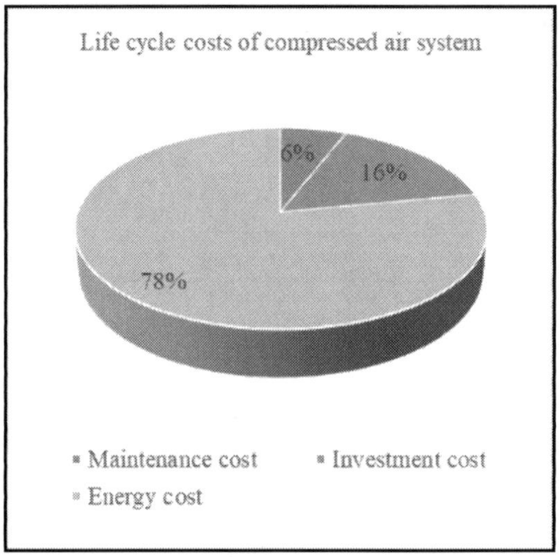

Fig. 3. Life cycle costs of compressed air system [10].

4 The possibilities of recovering waste heat from air compressors, the example of Bulten Poland S.A.

Bulten Poland SA is a factory producing fasteners for the automotive industry, which means that in the production process, the company uses a significant amount of compressed air, e.g. at the stage of pressing and rolling of products. The compressed air is essential utility for such a production, therefore proper compressed air system is very important for maintaining high production standards.

Generation of compressed air is always related to the need of cooling it. Compressed air or a factor mediating in heat exchange (e.g. oil) is cooled by a forced ambient air flow (smaller compressors) or by cooling water (bigger compressors). In both cases, it is possible to restore this heat and use it for useful purposes. Heat restore from positive displacement compressors (e.g. screw), consists of cooling oil (oil as coolant and greasing agent) in a heat exchanger, where on the secondary side heating water flows. Oil circulation is characterized by relatively high temperature (up to 90°C), so factor collecting heat (water) may be heated to usefully high temperatures.

In heating substation rooms, there is an isolated compressor room with screw compressors, oil compressors cooled by air, produced by BOGE Kompressoren Otto Boge GmbH & Co. Due to a small distance between heating substation devices from the planned heat recovery systems, there are very favourable conditions for effective connecting both systems and waste heat recovery from compressors.

For the purposes of heat recovery from warm oil that cools compressors, an external heat recovery system "Duotherm" is envisaged, delivered by compressors producer BOGE. The system is based on a plate heat exchanger, to which oil circulation is channelled from the compressor and water circulation from recovery

 https://doi.org/10.1051/e3sconf/20187003001

installation. The amount of heat collected in the exchanger is regulated by an integrated three-way valve with a thermostatic actuator. Water flow in compressors circulations is forced by operation of variable rotary pumps, controlled by an analogue signal in order to keep constant water temperature, measured behind recovery exchangers. The flow of heated water is directed towards the heat buffer that at the same time serves as a hydraulic clutch. In order to measure the amount of obtained heat, it is planned to install a calorimeter LC1 [12].

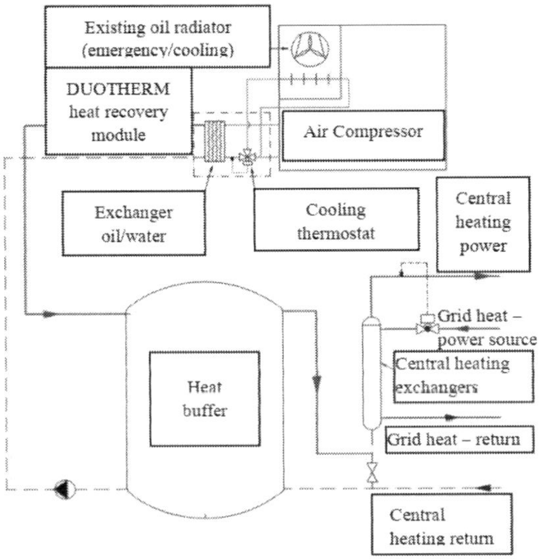

Fig. 4. Heat recovery scheme.

The planned installation of heat recovery from compressors, that is presented as block diagram at Figure 4 will contain such main devices as:

- Heat recovery module "DUOTHERM" for each compressor:

  Duotherm 150 for compressor SD-150

  2 pcs. Duotherm 100 for compressor SD-100/SD-90

  2 pcs. Duotherm 60 for compressors SD-60

- Heat buffer(s) of the total volume of about 10 m^3 for balancing variability of heat recovery and heat collection and to hydraulic compression of systems.
- Circulating pumps of the heat recovery system. Depending on the accepted design solution, each module may have its own pump or it is also possible to apply one pump and a direct-run distribution system with automatic shut-off valves.
- Automatics controlling the heat recovery system and contributing to maximization of heat recovery (control cabinets, programmable controller, automation devices).
- Calorimeter for counting the amount of the heat obtained from the compressors.
- Piping and installation fittings necessary for connecting systems and appropriate operation of the installation.

A compressor room, due to its constant operation nature and a relatively regular power, may be a very stable source of waste heat. Compressors operate in a cascade, depending on demand for compressed air. Usually, all compressors are in motion, whereas one of them (mostly SD-150) serves a "driving in" function, i.e. depending on the pressure in the compressed air grid, it operates in load/lighten mode (neutral gear). Some of the Company Departments Plant work 7 days a week (e.g. Department of Heat Treatment), the other - 5-6 days a week. For each compressor, powers of the main drive, the fan and the total nominal power were determined and calculated energy consumption. Additionally, average power input and load index was determined. The data is presented in Table 1.

Table 1. Energy powers and consumption by particular air compressors [12].

No.	Compressor type	Main drive power [kW]	Power of radiator fan [kW]	Total nominal power [kW]	Energy consumption – in total [kWh]	Energy consumption – main drive [kWh]	Average intake power [kW]
1.	SD-150	110	4.0	114	760 510	733 825	100.3
2.	SD-100	75	2.2	77.2	582 644	566 040	76.8
3.	SD-90*	65	2.2	67.2	484 163	468 312	63.8
4.	SD-60*	45	1.5	46.5	335 023	324 216	44.2
5.	SD-60*	45	1.5	46.5	335 023	324 216	44.2
In total		**340**	**11.4**	**351.4**	**2 497 363**	**2 416 609**	**329.3**

*Metering of energy consumption is only at compressors SD-150 and SD-100. Energy consumption by remaining compressors was estimated on the basis of work time and power of drives.

The possibilities to recover heat from compressors are mostly based on heat collection from warmed-up oil, which circulates in internal compressor circulation as a coolant and a greasing agent. This heat is received by the heat exchanger, where mostly 65÷75% of energy transferred to the compressor drive may be recovered. The compressor value declared by the producer, i.e. 72% was assumed for calculations.

Monthly demand for heat with waste heat potential possible to recover is presented in Table 2. The amount of heat obtained from the recovery installation and usefully used for the needs of central heating was estimated.

Demand for heat was determined on the basis of 'degree-days', through which heat consumption was calculated, expressed as the standard season, taking real data from years 2014 and 2015 for calculation; the amounts were compared to average monthly temperatures from the past 10 years.

In order to calculate profitability of the investment and payback time, one should compare the heating costs incurred before modernization (Figure 5 and Figure 7) and the heating costs after modernization (Figure 6 and Figure 8) and use of waste heat recovered from air compressors for this purpose. Total costs of heat before and after modernization have been shown in Figure 9.

Fixed and variable costs have been calculated with following formulas:
Fixed cost:

$$K_f = 12 \cdot \Sigma_{Os} \cdot q_0 . \quad (1)$$

Variable costs:

$$K_v = Q_{real} \cdot \Sigma_{Oz} . \quad (2)$$

where:

Os - sum of fixed charges for ordered power [PLN / MW / month]

Σ_{Oz} - the sum of variable fees for heat consumption, for distribution services, etc. [PLN / MW]

q_0 - ordered power [MW]

Q_{real} - the amount of heat consumption [MW]

Table 2. Monthly demand for heat with the waste heat potential possible to recover [12].

Month	Demand for heat [GJ]	Work time [h]	Energy consumption [kWh]	Maximum recovery [GJ]	Transmission efficiency [%]	Application efficiency [%]	Used waste heat [GJ]
I	1530.1	696	229 188	574.8	95	98	535.2
II	1428.9	624	205 479	515.4	95	98	479.8
III	959.5	696	229 188	574.8	95	94	513.3
IV	669.3	672	221 285	555.0	95	90	474.5
V	50.2	696	229 188	574.8	95	9	50.2
VI	0.0	672	221 285	555.0	95	0	0.0
VII	0.0	696	229 188	574.8	95	0	0.0
VIII	0.0	384	126 449	317.2	95	0	0.0
IX	12.5	696	229 188	574.8	95	2	12.5
X	613.8	672	221 285	555.0	95	90	474.5
XI	970.4	696	229 188	574.8	95	94	513.3
XII	1374.5	384	126 449	317.2	95	98	295.3
Σ	**7 609.1**	**7 584**	**2 497 363**	**6 264.0**	–	–	**3 348.8**

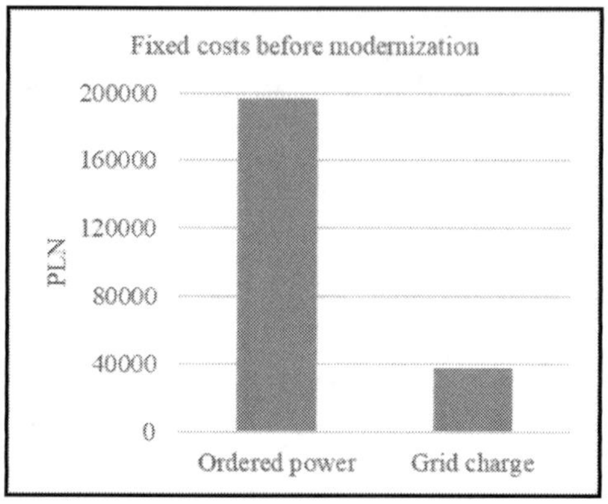

Fig. 5. Fixed costs before modernization [13].

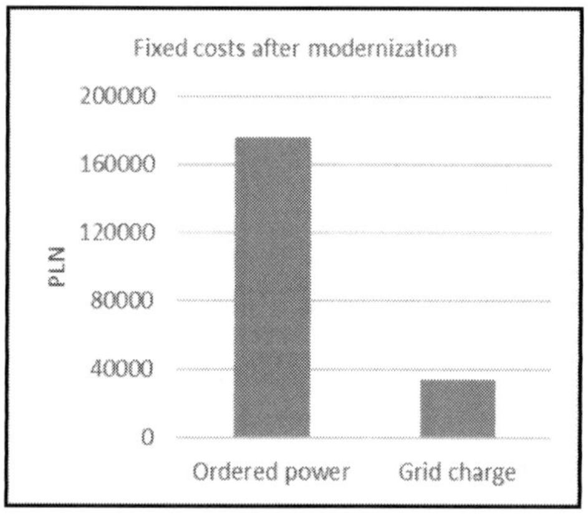

Fig. 6. Fixed costs after modernization [13].

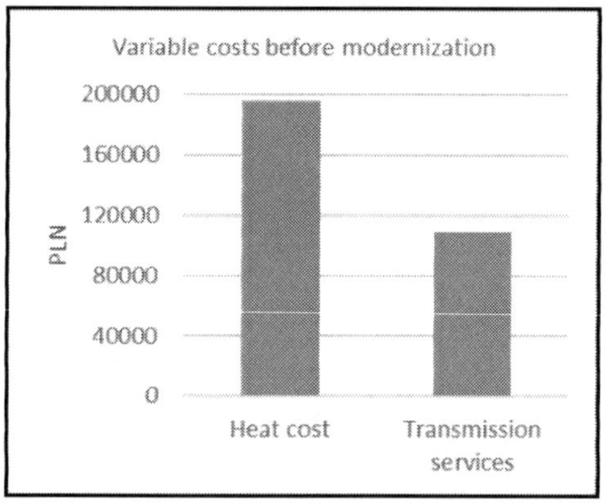

Fig. 7. Variable costs before modernization [13]

Table 3 shows consumption calculation costs of additional electricity devices, estimated modernization costs expenditures and simple investment payback time (SPBT)

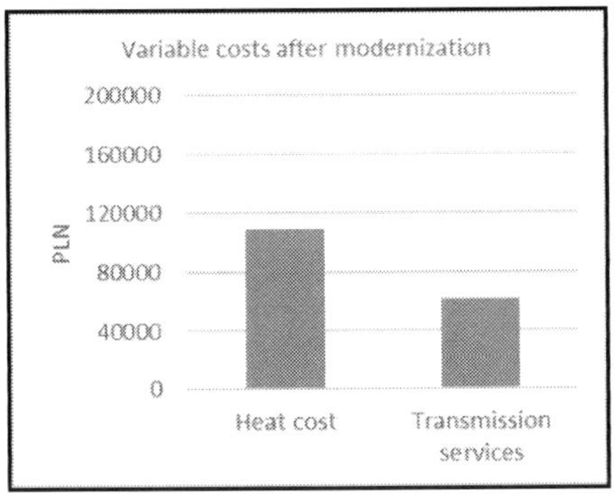

Fig. 8. Variable costs after modernization [13]

Table 3. Costs of additional electricity devices and estimated costs of modernization with SPBT.

Additional electricity devices	Electricity consumption to drive additional devices	4,800 kWh
	Average unit variable cost of electricity	65 EUR /MWh
	Cost of consumption of additional electricity after modernization	315 EUR /year
Costs before/after modernization	Exploitation cost before modernization	125 313 EUR /year
	Exploitation cost after modernization	88 741 EUR /year
	Annual savings for heat recovery	157 255 PLN/year
Estimated costs of modernization	Installation design and trade-specific agreements	2 325 EUR
	Installation of 5 "DUOTHERM" heat recovery modules	23 255 EUR
	Heat buffers	3 256 EUR
	Automatics driver elements (valves and pumps)	5 814 EUR
	Control automatics (driver, sensors, cables)	4 651 EUR
	Fittings, pipelines, refractory shapes, insulation	3 488 EUR
	Installation works and start-up	8 372 EUR
	IN TOTAL (investment costs related to concept)	51 161 EUR
SPBT	Simple investment payback time	1.4 (years)

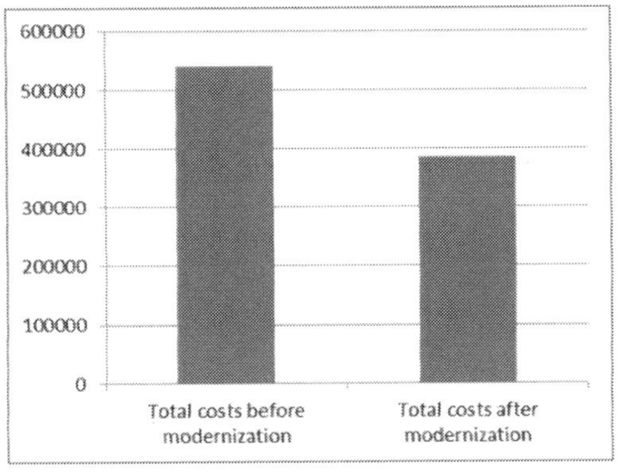

Fig. 9. Total costs of heat before and after modernization [13].

5 White certificates

The system of energy efficiency certificates, so-called "white certificates", uses market mechanisms to support increasing energy efficiency. For entrepreneurs, this is an opportunity to receive additional cash for modernization that resulting in energy savings in the enterprise. However, it should be remembered that the system can only be used for new investments. The exception is the transitional period, valid until 30 September 2017, when latecomers can still apply to the President of URE (Energy Regulatory Authority) for white certificates for projects that were completed in the period from January 1, 2014 to September 30, 2016. Each entrepreneur can benefit from the white certificates system if meets the conditions imposed by the Energy Efficiency Act, conducts an energy efficiency audit, and the application submitted will be considered positively by the President of the Energy Regulatory Authority.

When talking about the tender, one need to be aware that formally we dealt with tender procedures for projects of three categories:

- Increasing electricity saving by end-users,
- Increasing electricity saving by devices for own needs,
- Lowering electricity, heat or natural gas losses in transmission or distribution [14].

Bulten Polska S.A. took part in the tender procedure for white certificates no. 1/2015, announced on 29 December 2015. The proposed solution name was: Waste heat recovery from the compressors for the needs of plant heating. The concept involved increasing electricity saving at end-user by useful application of waste heat from cooling the compressors. Submitting documents for the tender procedure was preceded by a profound analysis of profitability and calculation of the size of electricity saving resulting from implementation of the abovementioned solution. This value is expressed as *toe* (equivalent amount of one ton of mineral oil of net calorific value equal to 41.868 GJ or 11.63 MWh) [15]. URE (Energy Regulatory Authority) granted to Bulten Polska S.A. white certificates of the value equal to 53.00 *toe* for implementation of proposed solution that comprises installation of recovery systems of waste heat from air compressors, which after sale generated profit of 11 660 EUR (average price for *1 toe* was 220 EUR). The profit from sale of white certificates significantly lowered anyways profitable payback time.

Tab. 5 shows how much decrease the payback time taking into account benefits from white certificates.

Table 4. Simple investment payback time after deduction benefit from white certificates.

IN TOTAL (investment costs related to concept)	51 161 EUR
Profit from white certificates	11 660 EUR
IN TOTAL (after deduction benefit from white certificates)	39 501 EUR
Simple investment payback time	1.1 (years)

6 Conclusions

Modern European politics strongly emphasises promoting sustainable development in enterprises. That is why, we mostly need to aim at increasing entrepreneurs' awareness in rational use of utilities, such as electric current or heating. The result of increasing awareness should be planning activities that would aim at increasing energy efficiency in production plants (and not only) by analyses (starting from even the simplest) of profitability of investment in recovery systems of waste heat from technological processes of production plants and secondary processes (e.g. such as production of compressed air). The companies that have to produce compressed air in their activity, should cautiously look at the entire compressed air system, it turns out that this utility, even though often ignored, proves great potential for application in heat recovery systems and of using it for own needs.

Above presented results show, that recovering heat form air compressors are profitable and can be successfully implemented in industrial area, in companies that use compressed air in their production processes. Additional advantage of implementing this kind of solution for improving energy efficiency is possibility for gaining white certificates, which can significantly shorten payback time of the investment.

References

1. D. Vittorini, R. Cipollone. Energy **102**, p. 502-515 (2016). *Energy savings potencial in existing industrial compressors.*
2. Technical enclosure to instruction for filling SEAP template: Emission index, http://www.soglasheniemerov.eu/IMG/pdf/technical_annex_pl.pdf, (on-line 08.11.2017).
3. Electricity production in Poland, https://energiadirect.pl/poradniki/produkcja-energii-elektrycznej-w-polsce, (on-line 08.11.2017).
4. T. Skoczkowski, S. Bielecki, Energetyka **1**, p. 9-14 (2016). *Measures to improve Energy efficiency in industry and their evaluation.*
5. Smarter, greener, more inclusive? Indicators to support the EU. Europe 2020 strategy. Eurostat Statistical Books, (2015).
6. Directive 2012/27/EU of the european parliament and of the council of 25 october 2012 on energy efficiency
7. Conscious choices of consumers – product's carbon footprint, http://www.akademiaodpadowa.pl/445,a,12-swiadome-wybory-konsumenckie-slad-weglowy-produktu.htm, (on-line 08.11.2017).
8. Compressed air as form of energy, http://www.silvent.com/pl/w-czym-mozemy-pomoc/optymalizacja-energii/sprezone-powietrze-jako-forma-energii/, (on-line 08.11.2017).
9. Energy efficiency of compressed air systems, http://enspol.eu/sites/default/files/Efektywno%C5%9B%C4%87%20Energetyczna%20System%C3%B3w%20Spr%C4%99%C5%BConego%20Powietrza.pdf, (on-line 08.11.2017).
10. S. Mousavi, S. Kara, B. Kornfeld, Procedia CIPR **15**, p. 313-318 (2014). *Energy Efficiency of Compressed Air Systems.*
11. How to save energy in compressed air systems, http://docplayer.pl/3287990-Jak-oszczedzac-energie-w-systemach-sprezonego-powietrza.html, (on-line 06.11.2017).
12. Assessment of possibilities for reusing waste heat to cover needs for heating in Bulten Poland S.A., https://www.researchgate.net/publication/320585127_ASSESSMENT_OF_POSSIBILITIES_FOR_REUSING_WASTE_HEAT_TO_COVER_NEEDS_FOR_HEATING_IN_BULTEN_POLAND_SA_OCENA_MOZLIWOSCI_WYKORZYSTANIA_CIEPLA_ODPADOWEGO_W_CELU_ZASPOKOJENIA_ZAPOTRZEBOWANIA_NA_MOC_GRZEWCZA_W, (13.11.2017).
13. Report from audit of energy efficiency of a project: Waste heat recovery from compressors for the needs of heating plant in Bulten Polska S.A.
14. Experiences from the first tender for white certificates, http://www.bialecertyfikaty.com.pl/artykuly/doswiadczenia-z-pierwszego-przetargu-na-biale-certyfikaty, (on-line 15.11.2017).
15. M. Robakiewicz. Rynek Instalacyjny **1-2**, p. 69-72 (2017). *System of White Certificates after the amendment to the Act.*

Investigations of prototype ejection refrigeration system driven by low grade heat

Dariusz Butrymowicz[1,*], *Jerzy* Gagan[1], *Kamil* Śmierciew[1], *Michał* Łukaszuk[1], *Adam* Dudar[1], *Andrzej* Pawluczuk[1], Adam Łapiński[1], and *Adam* Kuryłowicz[1]

[1]Bialystok University of Technology, Wiejska 45C, 15351 Bialystok, Poland

Abstract. One of possibilities of reduction of F-gas emission is application of low grade heat to drive the refrigeration systems as well as application of natural or low warming impact working fluids. The own experimental investigation of the ejection refrigeration system operating with refrigerant R-1234zeE are presented and discussed. The system is driven with low grade heat source of temperature below 70°C and thermal capacity approximately 90 kW. The experiments covered the effect of condensation, evaporation and generation temperatures on the capacity and thermal efficiency of the ejection refrigeration system operating for the air-conditioning purposes. Obtained results demonstrated that the proposed system may be thought as the promising heat driven refrigeration system with application of low grade motive heat sources.

1 Introduction

Buildings are responsible for about 30–45% of the global energy demand. The use of air conditioners and electric fans accounts for about a fifth of the total electricity in buildings around the world – or 10% of all global electricity consumption [1]. Over the next three decades, the use of ACs is set to soar, becoming one of the top drivers of global electricity demand [1]. Therefore, this sector has become an important target for reducing carbon emissions and energy consumption, because of its relevance in all these important areas [2].

Cogeneration is one of the possible way to increase the efficiency of electric energy production. While the production of electricity and heat can be conducted in an efficient manner during the heating season due to sale of the heat for the heating systems, utilization of combined production of electricity and heat decreases in summer due to the low demand for heat. This means that the solution for increasing the share of electricity produced in cogeneration is to develop effective ways of managing of the district heat during the summer. This heat is used for domestic tap water heating mainly, which involves not only the small amount of heat that is supplied to the heating system, but also the low or even very low parameters. These conditions may be thought as potentially suitable for the application of the ejection refrigeration system.

In ejection refrigeration system (Fig. 1) instead of pressurising the refrigerant by a mechanical compressor, an ejector compresses refrigerant vapour flowing from the evaporator and discharges it to the condenser. The motive vapour is generated in the vapour generator which is heated by low-temperature heat source. This paper presents development of refrigeration system equipped with vapour ejector operating with environmentally friendly new refrigerant HFO-1234zeE. Selection of this fluid was based on the comprehensive analysis of the working fluids for the ejector system with respect to Regulation of the European Parliament and the EU Council No. 517/2014 enacted on April 16th, 2014 [3,4]. It was shown by several researchers that ejection system can effectively operate with various natural and low warming impact synthetic refrigerants [5-8].

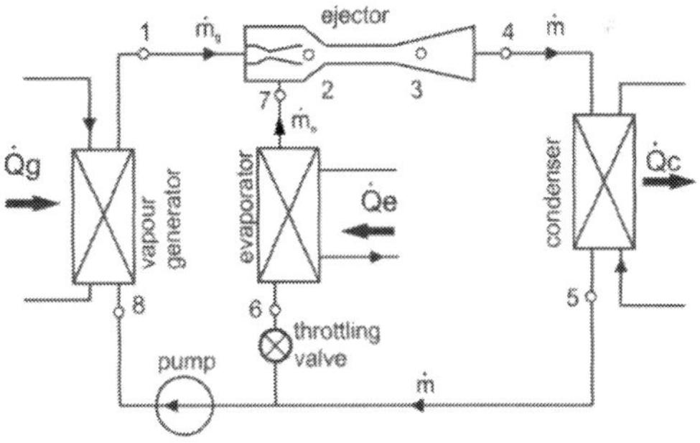

Fig. 1. Schematic of the ejector refrigeration system: $\dot{Q}_g$ - motive heat, $\dot{Q}_e$ - cooling capacity, $\dot{Q}_c$ - condenser capacity, $\dot{m}_g$ - primary (motive) mass flow rate, $\dot{m}_e$ - secondary (sucked in by ejector) mass flow rate, m - sum of $\dot{m}_g$ and $\dot{m}_e$, 1,2,...- thermodynamic state of working fluid.

Ejection refrigeration systems have been extensively investigated numerically and experimentally in recent years. The most comprehensive state-of-the-art study recent on ejection technology was presented by Chen et al. [9]. Besagni et al. [10] presented review on ejection refrigeration. These reviews gave all detail information

* Corresponding author: d.butrymowicz@pb.edu.pl

on ejector working principles, selection of the working fluids and ejection refrigeration technology as well. Elbel and Lawrence [11] presented review focused on ejectors applied for work recovery.

Butrymowicz et al. [12] presented the first experimental results of the ejector refrigeration system operating with isobutane driven by low grade heat source. Results presented in [12] showed that it is possible to use the ejection refrigeration system driven by heat gained from the district heat distribution network. During summer season the temperature of hot water is below 70°C, and sometimes even below 65°C. This means that application of the ejection refrigeration system may be thought as an effective solution for the production of chilled water for the air-conditioning with cooperation of the district heating system. The motive heat for the ejection system can be gained directly from the district heating system. This is much more attractive way for production of chilled water in terms of application because it allows for utilisation of low temperature heat during summer season for the air conditioning purposes.

In this paper the own investigation of the ejection refrigeration system operating with HFO group refrigerant R-1234zeE have been presented and discussed. It was shown by the authors in [13] that the ejection air-conditioning system driven by solar heat can effectively operate with low warming impact refrigerant R-600a (isobutane). However, isobutane is classified as explosive, therefore it might be considered as dangerous in large commercial refrigeration and air-conditioning. Here, ejection-based refrigeration system with safe R-1234zeE driven by low temperature heat source of app. 90 kW capacity is presented.

2 Methodology and apparatus

Experiments covered various operation parameters. For all of the runs the generator and the evaporator temperatures were kept constant, and condensation temperature was a variable parameter. The average operation parameters with deviations for the system are listed in Table 1, and separately for the ejector in Table 2.

The experimental rig presented in Fig. 2 was build and instrumented for operation with R1234zeE. The main elements of the rig are listed in the figure caption. The geometry of the test ejector is presented in Fig. 3. The testing stand was equipped with the temperature sensors and pressure transducers marked in Fig. 3 as (RTD) and (P), respectively, installed at the critical locations and other locations of interest. Pressure transducers of accuracy 0.25% of the full range were used during the experiments. The pressure transducer of 0-2.5 MPa scale was used to measure the motive stream parameters and 0-1 MPa were used at the other locations. The RTD temperature sensors (Pt100) of the accuracy 0.20% of measured value were applied for the experiments. The Coriolis mass flow meters of the accuracy 0.15% of measured value were used. COP measurement error was calculated by means of the total derivative approach.

The test rig was equipped with additional loops: the first for the thermal load for the evaporator and the second one for the condenser cooling. These systems allow for adjustment of the refrigerant flow rates as well as for changes of operation parameters in the wide range. The condenser cooling system was equipped with an automatically controlled dry cooler. The thermal load system was equipped with automatically controlled electrical heater. Both systems are fully instrumented with transducers for measurement of temperatures, pressures and flow rates with high accuracy. The rig is equipped with control valves enabling the adjustment of the operation parameters of the motive vapour at the inlet to the motive nozzle of the ejector.

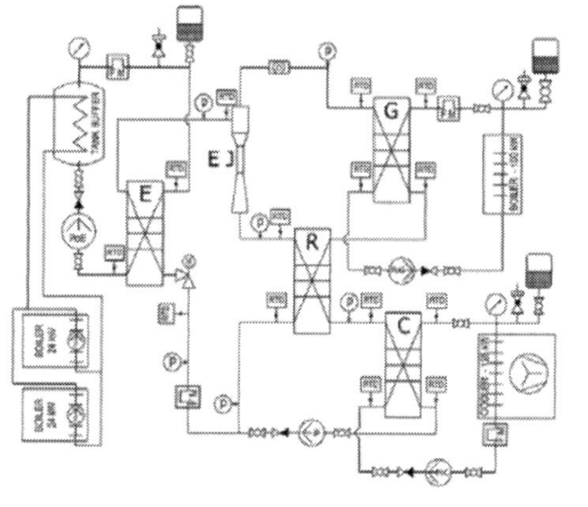

Fig. 2. Schematic of the testing stand: vapour generator (G); condenser (C); evaporator (E); ejector (EJ); refrigerant pump (P); regenerative heat exchanger (R); throttling valve (M); RTD – temperature sensor, p – pressure transducer, FM – flow meter.

During the experiments the electrical heaters thermal capacity was controlled to maintain the required saturation condition in the vapour generator. The stand is also equipped with control valves enabling the adjustment of the operation parameters of the motive vapour at the inlet to the motive nozzle of the ejector. The diaphragm pump HYDRA-CELL with 3 kW/700 RPM motor was used as the liquid refrigerant pump.

The ejector was designed and manufactured for the nominal operation conditions assumed as $t_{g,sat}$ = 60°C, $t_{e,sat}$ = 0°C. The operation conditions of the first test runs (marked in tables as run No. 1) are very close to the nominal operation conditions, i.e. $t_{g1,sat}$ = 58°C, $t_{e1,sat}$ = 0°C. For the second run (run No. 2) the evaporation temperature was increased $t_{e2,sat}$ = 6°C, $t_{g2,sat}$ = 58°C, and the third run (run No. 3) corresponds to the ultra-low motive parameters, i.e. $t_{g3,sat}$ = 48°C and $t_{e3,sat}$ = 7°C. For all of the runs an average superheating of the motive vapour was 5 - 6 K, therefore, the motive temperature was lower than 65°C, and it was even below 55°C for run No. 3. Such low motive temperature makes this system unrivalled in the use of low temperature heat sources.

https://doi.org/10.1051/e3sconf/20187003002

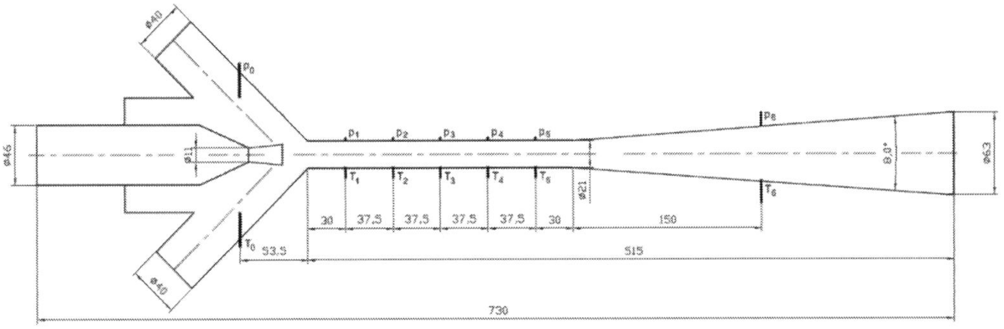

Fig. 3. Schematic of the tested ejector with dimensions in mm.

Under this range of the heat sources the motive temperature the ejection cycles can be considered to be genuinely competitive with absorption and adsorption refrigeration systems. The condensation temperature and corresponding saturation pressure was varied by means of control of the cooling conditions of the condenser, i.e., the mass flow rate of the cooling water was changed for every measurement point. For most the cases of the operation of the refrigeration system the motive source temperature and evaporation temperature are fixed. The condensation temperature is a function of the cooling fluid temperature which can vary in time, and therefore the assumed type of settings of the operation conditions may be thought as adequate for the actual operation of the ejection system.

For the ejector applications, the entrainment ratio is the most important parameter used to describe the ejector performance. Commonly, the entrainment ratio is defined as:

$$U = \frac{\dot{m}_e}{\dot{m}_g}, \tag{1}$$

where: $\dot{m}_g, \dot{m}_e$ are the primary (motive) and the secondary fluid mass flow rates, respectively.

The performance of the ejector system may be also presented by the relationship between the compression ratio versus mass entrainment ratio, Π. The compression ratio in this paper is thought as ratio of compression produced by ejector, p_c-p_e, and the motive pressure difference p_g-p_e. Therefore, compression ratio is defined as:

$$\Pi = \frac{p_c - p_e}{p_g - p_e}, \tag{2}$$

where p_c is condensation pressure, p_g – vapour generation pressure, p_e – evaporation pressure.

Another parameter describing the ejection system efficiency is the coefficient of performance COP. The COP is calculated as the ratio of the cooling capacity to the motive energy delivered to the system. In the ejection systems the motive power is the sum of the motive heat and electrical power consumed by the liquid pump Pp:

$$COP = \frac{\dot{Q}_e}{\dot{Q}_g + P_p}, \tag{3}$$

where thermal capacities of heat exchangers were calculated as the product of the mass flow rate of refrigerant and the change of the specific enthalpy between the both sides of the heat exchangers:

$$\dot{Q}_e = \dot{m}_e \left(h_{e,o} - h_{e,i} \right), \tag{4}$$

$$\dot{Q}_g = \dot{m}_g \left(h_{g,o} - h_{g,i} \right). \tag{5}$$

Table 1. Operation parameters of the tested ejection refrigeration system.

Generator saturation cond.		Evaporator saturation cond.		Condenser saturation cond.	
$p_{g,sat}$	$t_{g,sat}$	$p_{e,sat}$	$t_{e,sat}$	$p_{c,sat}$	$t_{c,sat}$
MPa	°C	MPa	°C	MPa	°C
run no.1					
1.23 ±0.02	58.3 ±0.3	0.22 ±0.03	0.2 ±1.0	Vary in range 0.416 and 0.600	Vary in range 19.2 and 31.3
run no.2					
1.211 ±0.03	57.8 ±0.4	0.275 ±0.01	6.3 ±0.8	Vary in range 0.430 and 0.569	Vary in range 20.2 and 29.4
run no.3					
0.950 ±0.04	48.0 ±1.6	0.276 ±0.01	6.8 ±0.9	Vary in range 0.425 and 0.490	Vary in range 19.8 and 24.5

In this paper the power consumed by the pump was not measured directly, but according to the manufacturer data sheet for the pump, the maximum motor input power is 3 kW. For the purpose of the analysis presented here the power consumed by pump was calculated on the basis of electric current measurement. The average value I = 4.5 A was taken to calculate the power consumed by mechanical pump. Here, it is:

$$P_p = \sqrt{3} \cdot U \cdot I \cdot \cos\varphi = 2.06 \pm 0.15 \text{ kW},$$

* Corresponding author: d.butrymowicz@pb.edu.pl

where $\cos\varphi = 0.66$ was taken from the motor plate.

The last parameter analysed in this paper is the efficiency of the ejector. Relation proposed by Dvořák and Vit [14] was used for this purpose:

$$\eta = \frac{\dot{m}_e}{\dot{m}_g} \frac{1 - \left(\frac{p_e}{p_c}\right)^{\frac{\kappa-1}{\kappa}}}{\left(\frac{p_g}{p_c}\right)^{\frac{\kappa-1}{\kappa}} - 1}, \tag{6}$$

where κ was calculated from equation of state [15].

Table 2. Operation parameters of the ejector.

ejector inlet motive vapour		ejector inlet secondary vapour		ejector outlet discharge	
p_g	t_g	p_e	t_e	p_d	t_d
MPa	°C	MPa	°C	MPa	°C
run no.1					
1.179 ±0.02	63.7 ±1.2	0.218 ±0.03	8.4 ±2.0	varied in range 0.420 and 0.612	varied in range 42.6 and 54.6
run no.2					
1.160 ±0.02	64.7 ±1.1	0.272 ±0.02	11.0 ±1.5	varied in range 0.441 and 0.579	varied in range 30.6 and 50.8
run no.3					
0.950 ±0.04	54.2 ±1.3	0.276 ±0.01	9.6± 2.2	Varied in range 0.425 and 0.490	varied in range 36.1 and 43.7

The ejection refrigeration systems generally may be thought as low efficient devices, therefore there is a clear need for further development of simple and low-cost method for improving of COP. The internal heat transfer by means of the internal heat exchanger which is located at the ejector discharge line/ liquid line may be thought as the most simple and effective approach. As it was shown in Fig. 2 the testing stand was equipped with the internal heat exchanger (R) in other to improve COP of the system. The rationality of this solution results among others from the utilisation of superheating of the vapour discharged from the ejector. The vapour superheating at the ejector outlet may be thought as a waste heat so along with condensation heat is transported to the ambient. In the discussed internal heat exchanger heat taken from the superheated vapour is transferred to the liquid feeding the vapour generator. As an effect the temperature of this liquid increases. In effect thermal load required by the vapour generator decreases. Moreover, the cost of application of the additional heat exchanger in the system may be thought as relatively small. Additional advantage of the utilisation of the vapour superheating is related with decrease of the thermal load of the condenser. As an effect the required quantity of the condenser cooling fluid is smaller.

On the basis of the measurement of temperature and pressure of vapour at the heat exchanger inlet and outlet and mass flow rate of vapour, the thermal capacity of the heat exchanger can be found as:

$$\dot{Q}_{RHX} = (\dot{m}_g + \dot{m}_e) \cdot (h_{in} - h_{out}), \tag{7}$$

where the specific enthalpies are calculated from the equation of state $h = h(t,p)$; the inlet parameters correspond to the ejector discharge, and outlet parameters correspond to the condenser inlet.

Improvement of COP of the system was calculated with assumption that the thermal capacity of the internal heat exchanger $\dot{Q}_{RHX}$ must be added as a motive heat to the system operating without heat exchanger since the liquid that feeds the generator is not preheated. In that case COP of the standard system without the internal heat exchanger is:

$$COP_{st} = \frac{\dot{Q}_e}{\dot{Q}_g + \dot{Q}_{RHX} + P_p}. \tag{8}$$

3 Results and discussion

Results of the present investigations are presented in following figures. Operation conditions (based on saturation temperatures) for each line are presented in the figures. As it was mentioned previously several quantities determining the performance of the ejector were investigated. The mass entrainment ratios as a function of condensation temperature for all of the investigated cases are shown in Fig. 4.

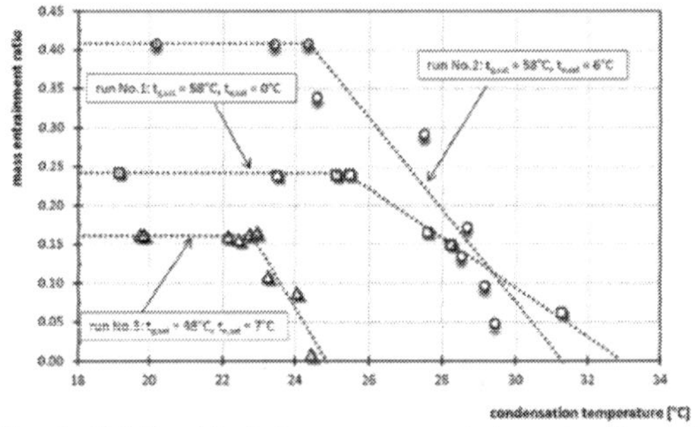

Fig. 4. Relationship between mass entrainment ratio and condensation temperature.

It is seen from this figure that for all runs the ejector operates both at on-design conditions (horizontal part of the performance lines in Fig. 4) and off-design conditions (slope part of the performance lines in Fig.4). For the nominal operation parameters, i.e. $t_{g1} = 58°C$, $t_{e1} = 0°C$ (run No. 1) the mass entrainment ratio was $U_1 = 0.24$. Increasing the evaporation temperature up to 6°C (Run No. 2) leads to increase of the mass entrainment ratio up to $U_2 = 0.41$. This effect was expected because of reduction of the pressure difference produced by the ejector. Temperature of the transition point (called as the critical temperature) between the on-design and the off-design regimes is higher approximately by 1 K for test run No. 1 in comparison with the test run No. 2, i.e. $t_{c1*} \approx$ 25.5°C, and t_{c2*} = 24.5°C. Slope of the approximated operation line of the off-design regime is higher for run No. 2 than for run No. 1. It may be postulated that the wider range of the off-design operation conditions, i.e., the higher condensation temperature range is more desired because the ejector will not stop suddenly when the condensation temperature increases above critical level. The more favourable slope for test run No. 1 gives the higher temperature above which the ejector cannot operate, t_{co1} = 33°C while for the run No. 2 it is t_{co2} = 31°C.

The measurement points are located along with the approximation line for test run No. 1 when the ejector operates at the off-design conditions, but for run No. 2 the discrepancy of the locations of the measurement points is much higher. It may be therefore concluded that the ejector operates more stable for lower evaporation temperature which correspond to the nominal design operation parameters. As for the case of the ultra-low motive temperature the mass entrainment ratio U_3 = 0.16 was achieved, the transition between operation regimes occurs at temperature t_{c3*} = 23°C, and the ejector will not operate at condensation temperatures higher than t_{co3} = 25°C. The results clearly show the operation at on-design and off-design regimes. Operation points are located along the approximation line which proves the stable operation of the ejector under these conditions.

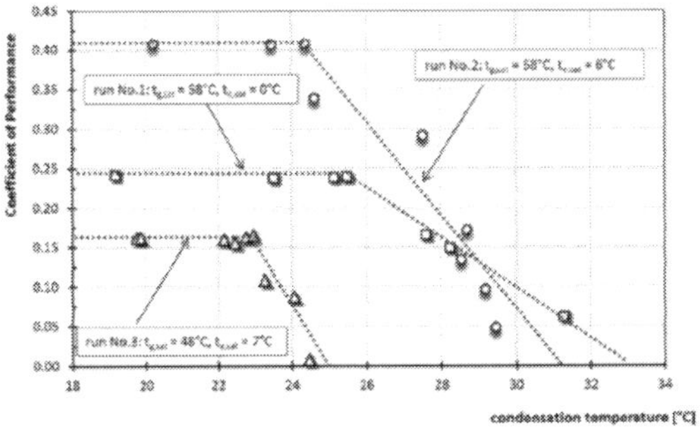

Fig. 5. Relationship between Coefficient of Performance and condensation temperature.

The influence of the condensation temperature on the coefficient of performance COP of the ejection system is presented in Fig. 5. The coefficient of performance is directly related with the mass entrainment ratio and the maximum level of COP_1 = 0.24 was obtained for condensation temperatures below 26°C. The maximum COP_2 = 0.41 was obtained for higher evaporation temperatures. For run No. 3 when the ejector was driven by the heat source of the lowest temperature COP_3 = 0.16 was obtained. The shape and the slope of the performance lines representing COP are the same as shown in Fig. 4.

Performance lines in co-ordinate compression ratio Π vs. mass entrainment ratio U are presented in Fig 6.

The shape of the presented lines are typical for the gas ejector. The transition between the on-design and the off-design regimes occurs at $\Pi_3 \approx 0.26$ for the lowest motive temperature (run No. 3), $\Pi_2 \approx 0.27$ for run No. 2. The highest compression ratio at the transition point is $\Pi_1 \approx 0.31$. For runs No. 1 and No. 2 the maximum compression ratio is $\Pi_{1max} \approx \Pi_{2max} \approx 0.33$ for the lowest mass entrainment ratio. Ejector operation with the lowest motive temperature was able to produce the maximum compression ratio $\Pi_{3max} \approx 0.30$.

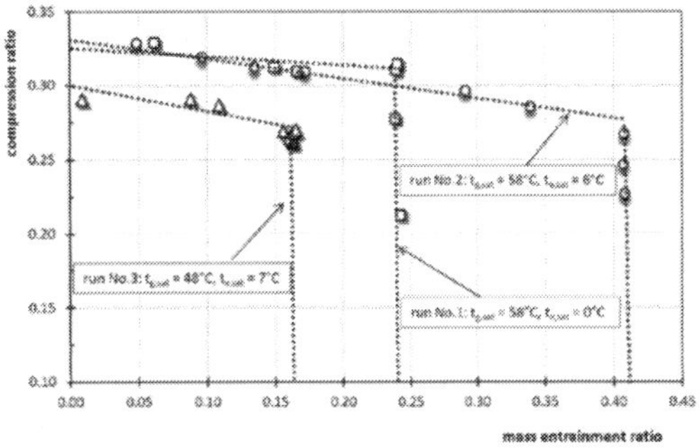

Fig. 6. Relationship between Coefficient of Performance and condensation temperature.

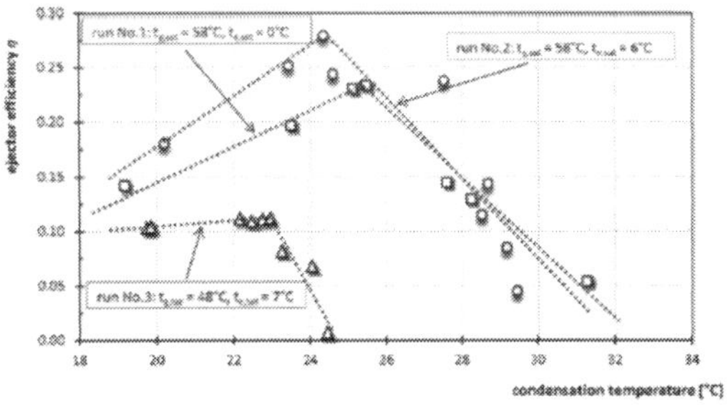

Fig. 7. Efficiency of the tested ejector for various operation conditions.

In Fig. 7 the results of the efficiency of the ejector calculated from eq. (6) are presented. The figures show that the efficiency can be represented by a broken line with maximum point corresponding to the transition point between two operation regimes. For the on-design operation regime, when the mass entrainment ratio is constant, the efficiency increases for lower condensation temperatures as it is seen in Fig. 7. This is due to the increase of ratio of the difference of the specific enthalpies resulting from the increase of the pressure difference between the condenser and the evaporator. When the ejector starts to operate under the off-design condition the ratio of the difference of the specific enthalpies increases, but the mass entrainment ratio decreases. This decrease plays a major role and influences the overall performance of the ejector and as an effect the efficiency decreases.

According to eq. (6), the efficiency increase is related with increasing of the condensation pressure. At the off-design conditions the efficiency decreases with increasing of the condensation temperature which is a result of the decreasing of the mass entrainment ratio at these operation conditions. For the motive temperature $t_{g,sat}$ = 58°C maximum efficiency of the ejector is η_{1max} =

0.24 for evaporation temperature t_e = 0°C and η_{2max} = 0.27 for evaporation temperature t_e = 6°C. Higher ejector efficiency for run No. 2 in comparison with run No. 1 results from the higher mass entrainment ratio $U_2 > U_1$. For the motive temperature $t_{g,sat}$ = 48°C the maximum efficiency of the ejector is slightly above η_{3max} = 0.10.

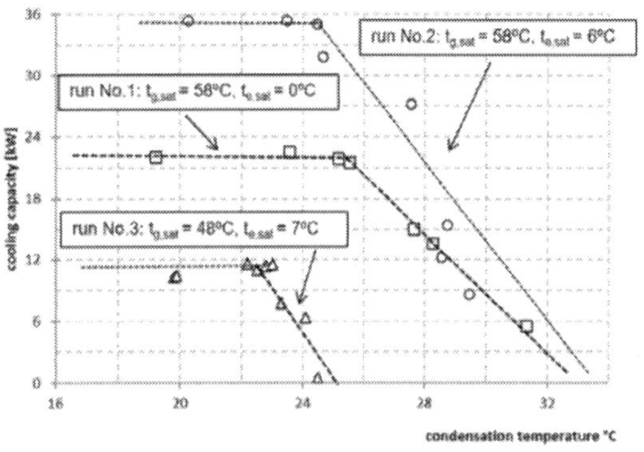

Fig. 8. Cooling capacity of the system versus condensation temperature.

Cooling capacity of the evaporator for the on-design operation regime was 22 kW for run No. 1, and 35 kW for run No. 2. For run No. 3 the evaporator cooling capacity almost 11 kW was observed. When the system operates at the off-design regime the cooling capacity decreases for all three runs, as expected.

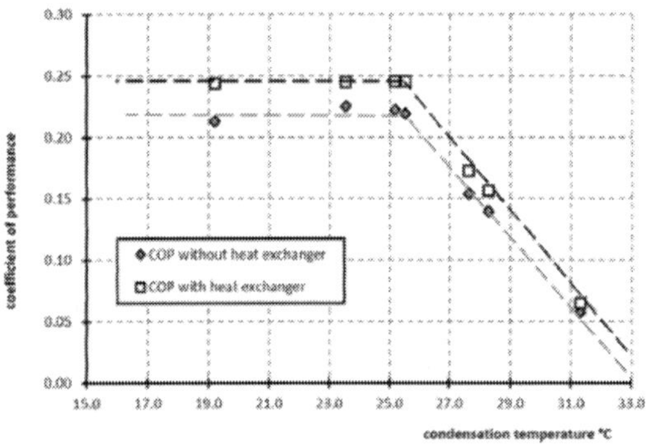

Fig. 9. Coefficient of performance of the system operating with and without regenerative heat exchanger for run No.1.

The results presented in Fig. 9 to Fig. 11 show that application of internal heat exchanger improves COP of the system. It is seen in Fig. 9 that when system operates at the evaporation temperature t_e = 0°C the improvement of COP is more stable at off-design regime in comparison with the second test run at which t_e = 6°C (Fig. 10). For all test runs the improvement of the COP within the range 10 - 12% was achieved which can be thought as a very attractive.

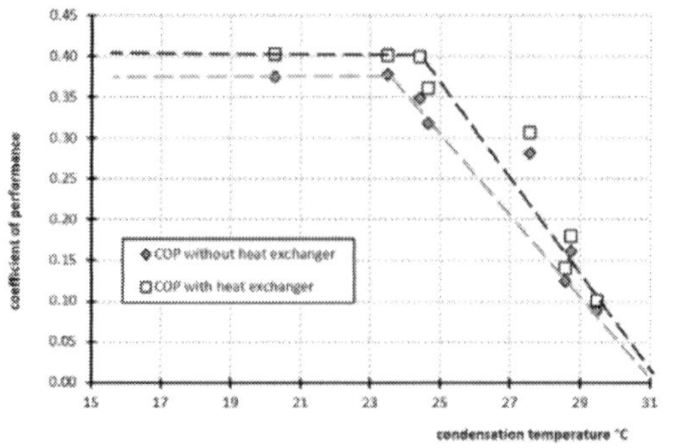

Fig. 10. Coefficient of performance of the system operating with and without regenerative heat exchanger for run No.2.

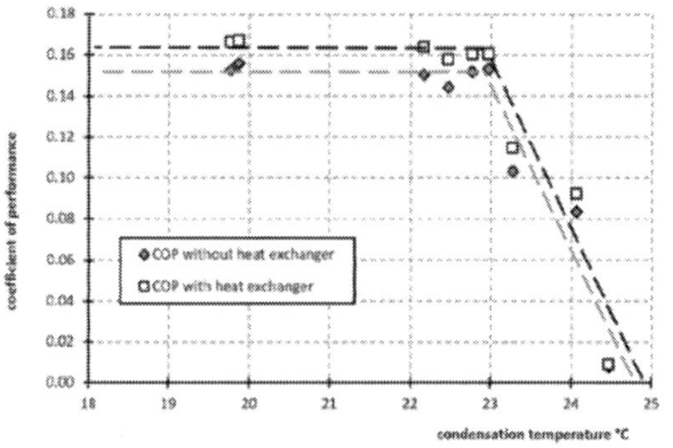

Fig. 11. Coefficient of performance of the system operating with and without regenerative heat exchanger for run No.3.

4 Conclusions

On the basis of presented results following conclusions can be drawn:

- Application of refrigerant R-1234ze(E) provides the typical performance of ejection refrigeration systems.
- Ejector cycle with low GWP refrigerant R-1234ze(E) is able of effective operation when low temperature heat source (lower than $t_{g,sat}$ = 60°C or even below $t_{g,sat}$ = 50°C) is used.
- The tested ejector achieved the maximum efficiency at level 0.27, but for most of the operation points the range 0.10 ÷ 0.25 was obtained. These values can be thought as rational taking into consideration low level of the motive source temperature.
- The internal heat exchanger for COP improvement was applied for this fluid in the ejection refrigeration system. The superheating of vapour at the ejector outlet was applied to increase the system efficiency for approximately 10-12%.

This research was carried out as part of work no. S/WM/1/2018 and financed from funds for education of the Ministry of Science and Higher Education.

References

1. International Energy Agency, The Future of Cooling: Opportunities for energy efficient air conditioning, Report OECD/IEA, 2018
2. M. Calero, E. Alameda-Hernandez, M. Fernández-Serrano, A. Ronda, M.A. Martín-Lara Energy consumption reduction proposals for thermal systems in residential buildings, Energy & Buildings 175 (2018) 121–130
3. Regulation (EU) No 517/2014 of the European Parliament and of the Council on fluorinated greenhouse gases and repealing Regulation (EC) No 842/2006
4. K. Śmierciew, D. Butrymowicz, J. Gagan, Assesment of Refrigerant Selection for Ejection System Driven by Low-Grade Heat, 2016 Purdue Compressor Engineering, Refrigeration and Air Conditioning and High-Performance Buildings Conferences, Purdue, USA, 2016, ID 2667.
5. A. Arbel, M. Sokolov, Revisiting solar-powered ejector air conditioner - the greener the better, Solar Energy 77 (2004) 57–66.
6. J. M. Abdulateef, et. Al, Review of Solar-Driven Ejector Refrigeration Technologies, Renewable and Sustainable Energy Reviews 13 (2009) 1338-1349.
7. H.K Ersoy, S. Yalcin, R. Yapici, M. Ozgoren, Performance of a solar ejector cooling-system in the southern region of Turkey, Applied Energy 84 (2007) 971–983.
8. A. Selvaraju, A. Mani, Analysis of an ejector with environment friendly refrigerants, Applied Thermal Engineering 24 (2004), 827- 838
9. J., Chen, S. Jarall, H. Havtun, B. Palm, 2015, A review on versatile ejector applications in refrigeration systems, Renewable and Sustainable Energy Reviews 49, 67–90.
10. G. Besagni, R. Mereu, F. Inzoli, 2016, Ejector refrigeration: A comprehensive review, Renewable and Sustainable Energy Reviews 53, 373–407
11. S. Elbel, N. Lawrence, 2016, Review of recent developments in advanced ejector technology, International Journal of Refrigeration 62, 1 – 18.
12. D. Butrymowicz, K. Śmierciew, J. Karwacki, J. Gagan, 2014, Experimental investigations of low-temperature driven ejection refrigeration cycle operating with isobutane, International Journal of Refrigeration 39, 196-209.
13. K. Smierciew, J. Gagan, D. Butrymowicz, J. Karwacki, 2014, Experimental investigations of solar driven ejector air-conditioning system, Energy and Buildings 80, 260-267.
14. V. Dvořák, T. Vit, 2005. Experimental and numerical study of constant area mixing. In: 16th Int. Symp. on Transport Phenomena, Prague.
15. E.W. Lemmon, M.L. Huber, M.O. McLinden, NIST Standard Reference Database 23: Reference Fluid Thermodynamic and Transport Properties-REFPROP, Version 9.1, National Institute of Standards and Technology, Standard Reference Data Program, Gaithersburg, 2013.

Development of the experimental stand for investigation of heating and drying phenomena in the porous building materials with one surface of the sample exposed to the flowing air

Łukasz Cieślikiewicz[1*], *Piotr* Łapka[1], *Michał* Wasik[1], *Michał* Kubiś[1], *Karol* Pietrak[1], *Tomasz S.* Wiśniewski[1], *Piotr* Furmański[1], and *Mirosław* Seredyński[1]

[1]Institute of Heat Engineering, Faculty of Power and Aeronautical Engineering, Warsaw University of Technology, 21/25 Nowowiejska Street, 00-665 Warsaw, Poland

Abstract. The paper presents development of the experimental stand for investigation of heating and drying phenomena in the porous building materials with one surface of the sample exposed to the flowing air. Moreover, results of preliminary measurements which verify the range of operating conditions achieved by the stand are shown. These conditions were assumed in the following range: 15°C-60°C, 10-90% and 0.1-10 m/s for temperature, relative humidity and velocity of the flowing air, respectively. In order to control the parameters of air used for heating and drying of porous materials, a closed-loop concept of the stand was proposed and equipped with several elements, i.e., the cooler (vapour condenser), fan with variable rotational speed, humidifier and heater. Furthermore, two parallel measurement ducts which allow to obtain two air streams with identical parameters were proposed. This enables two measurements at one time as well as faster verification of the repeatability of the performed measurements. The temperature measurements at the top surface and inside the sample are carried out applying the IR thermography and several K-type thermocouples, respectively, while overall moisture amount and the moisture content distribution in the specimen are found using tensometric force and electrical resistance meters, respectively.

1 Introduction

The amount of moisture in building materials is highly important for building structure. In porous building materials water might come from many sources, e.g., due to rising of the groundwater level, rain or condensation of water vapour from the humid air. The water in building structure degenerates its thermal insulation and mechanical properties. Biological corrosion and salts decay are also observed [1-3]. Moreover, it influences the indoor microclimate conditions as well as affects the comfort of the occupants and their health [4]. Moisture in building materials might be accumulated for a long time, especially in historical buildings [5, 6]. Due to these problems, drying and renovation of walls are required in many buildings.

The drying process in porous media is present not only in building materials but also, i.e., in food processing, paper production, crops drying and garments. Wide range of different types of dried porous media constitutes large area for complex research of the drying phenomena. Studies concerning various types of material (e.g., brick, cloth, food, etc.) and applying the same experimental stand and setup for investigation of the drying process in different materials were not reported in the literature.

Generally, drying process might be categorised in the following main groups:

- Convective drying [7-11];
- Vacuum drying [11];
- Microwave drying [10, 12-14].

Mathematical description of the drying process is characterised by high level of complexity and is based on complex partial differential equations. In practical cases these equations are combined with fluid and heat flow equations and could not be solved analytically. Therefore, numerical models which involve Computational Fluid Dynamics (CFD) to simulate combined heat, mass and fluid flow in moist porous materials [15-20] are developed. These models take into consideration each phase (i.e., vapour, liquid water, dry air, solid phase) separately or collectively by balancing the moisture in the material [15-17]. Numerical models allow to compute the moisture transfer assuming either the equilibrium state between vapour and liquid water or the local non-equilibrium state [15-20]. The wide range of models with different assumptions and different but equivalent systems of equations should be be validated and compared for the same conditions in the specially-designed validation experiments. The influence of modelling assumptions on calculated results and their matching with experimental measurements should be analysed.

In order to analyse the drying process, different experimental stands were developed which accounted for

* Corresponding author: lukasz.cieslikiewicz@itc.pw.edu.pl

different drying mechanisms. In case of existing buildings, most important is the convective drying in which moisture and heat transfer between the flowing air and sample depends on the local flow conditions, i.e., temperature, relative humidity and velocity of the air [3]. For this reason, the ranges of air flow parameters influence the designs of the experimental stands. For example, the experiments might be conducted in the free convection conditions or with low-velocity forced convection in a climatic chamber [21-23]. For high-speed forced convection, special wind ducts are built. These ducts are of either open flow [8, 16, 24] or closed-loop [3] type. Experiments described in the literature were conducted with air flow velocities up to 5 m/s [3, 16] and relative humidity as high as 35% and 44%.

The paper presents development of the experimental stand with a closed-loop concept and two parallel measurement ducts for the investigation of heating and drying phenomena in porous building materials. Only one surface of the sample is exposed to the flowing air. The operating conditions of the stand were assumed in the following range: 15°C-60°C, 10-90% and 0.1-10 m/s for temperature, relative humidity and velocity of the flowing air, respectively. The preliminary stability tests and drying measurements have been carried out and the obtained results are shown in this work.

2 Experimental stand

2.1 Main assumptions of the stand

The designed stand meets several requirements. Majority of them were related to the parameters of the working fluid (air), i.e., its temperature, relative humidity and velocity. These parameters were assumed and controlled in the following range: 15°C-60°C, 10-90% and 0.1-10 m/s, respectively. Additionally, in order to obtain the data for the validation of numerical models developed in parallel, following different parameters in dried material are measured:

- Temperature distribution on the surface of the sample;
- Temperature distribution within the specimen;
- The global amount of moisture;
- Moisture content distribution within the specimen.

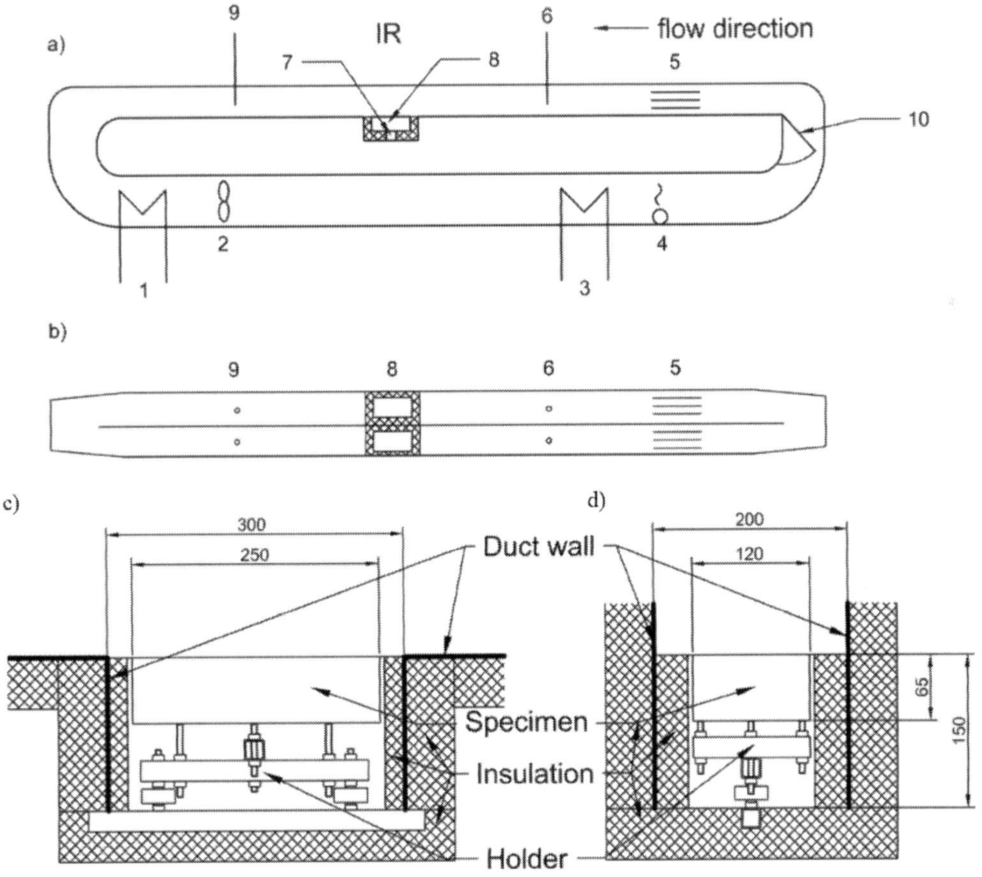

Fig. 1 Simplified schematic of the closed-loop flow experimental stand: a) side and b) top view as well as cross-section of specimen pocket with main dimensions: c) side and d) perpendicular to the flow direction view.

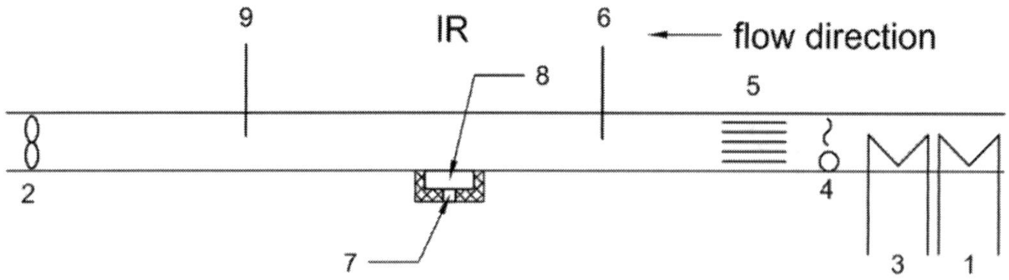

Fig. 2. Simplified schematic of the open-flow experimental stand.

2.2 Main concept of the stand

During design of the experimental stand, two main concepts were initially considered, i.e.:

• Closed-loop flow (Figure 1);
• Open flow (Figure 2).

In both cases, the system consisted of the following elements (numbers listed below refer to respective numbers in Figures 1 and 2):

1. Cooler/condenser;
2. Fan;
3. Heater;
4. Humidifier;
5. Flow stabilizer;
6. Integrated velocity, temperature and humidity transmitter;
7. Force meter;
8. Sample;
9. Integrated temperature and humidity transmitter;
10. Throttle.

Figures 1 and 2 graphically present both concepts (closed- and open-loop stands) and their elements denoted by numbers from 1 to 10. The closed-loop stand includes two parallel channels which allow to take simultaneously two measurements in the same conditions (see Figure 1c and d).

Finally, the closed-loop flow concept was chosen for further development. The major advantages of this concept in comparison with the open-flow stand concept are following:

• Lower power required by the heater, cooler and humidifier as well as the ability to achieve operational parameters in few cycles. In case of the open flow concept, the operating conditions have to be achieved immediately which could be problematic for high mass flow rates of air.
• Lower dependence of the measured parameters on the conditions of the surroundings.
• The ability to investigate two specimens in identical conditions at the same time – see Figure 1.
• Lower space requirements due to location of some elements of the stand (i.e., cooler/condenser, fan, heater and humidifier) under measurement ducts – see Figure 1.
• Better ability to control and stabilise parameters of the working air used for drying samples. This also refers to obtaining more reliable results which might be applied then in the validation of the numerical models.

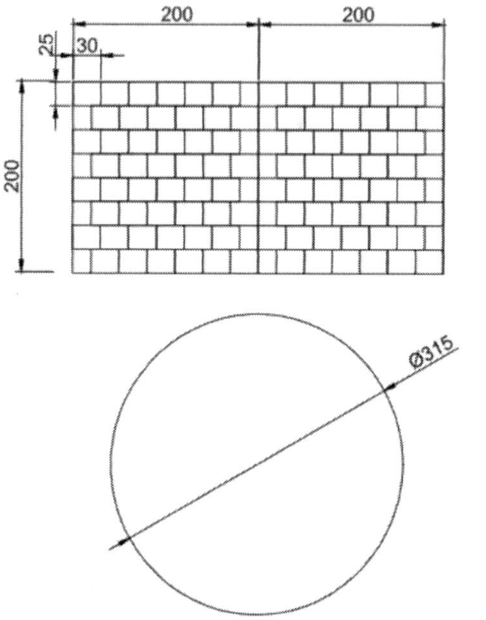

Fig. 3. Cross-sections of segments: two square ducts with stabilisers (upper) and round duct (bottom) with dimensions.

2.3 Development of the stand

The experimental stand was made as modular construction. The stand consists of two segments of the duct, i.e., round at the bottom and square, split into two identical ducts, on top (see Figure 1). Figure 3 presents dimensions of both segments of the duct. The outermost modules connect the top and bottom segments of the duct and therefore consist of two 90° bends and round-to-rectangular connector. All air conditioning equipment is located in the bottom segment (see elements no. 1-4 in Figure 1). In the first module, cooler/condenser (1) and fan (2) are placed. The cooler/condenser is made of a copper pipe twisted into a spiral and connected to a refrigerated circulation bath with glycol. This solution allows for both air cooling in case of measurements at low temperatures as well as removal of the excessive amount of the moisture and reduction of the relative humidity of air to assumed level. The cooler is mounted shortly before the fan which protects it from overheating. The fan, with rotation speed controlled by a frequency inverter, allows for smooth changing of the air velocity from 4.3 m/s up to 10 m/s. In order to reduce the flow velocity to the desired range (i.e., 0.1-4.3 m/s), a throttle (10) located in the third module or an additional flow resistance in the duct might be used. Additionally, in the front of first module power, control and safety elements are mounted. In the third module electrical heater (3) of power equal to 3 kW and humidifier (4) are located. The humidifier consists of a pipe with a gap at lower pressure side, connected with a tank with mist. Mist is produced by an ultrasonic piezoelectric mist generator. To stabilize and homogenize the flow field after the round-to-rectangular connector and throttle, the stabilizer (5) was applied. The stabilizer was made as an array of several small rectangular ducts (cross-section dimeters of 25×30 mm) of length equal to 300 mm placed inside both measurement ducts as shown Figure 3.

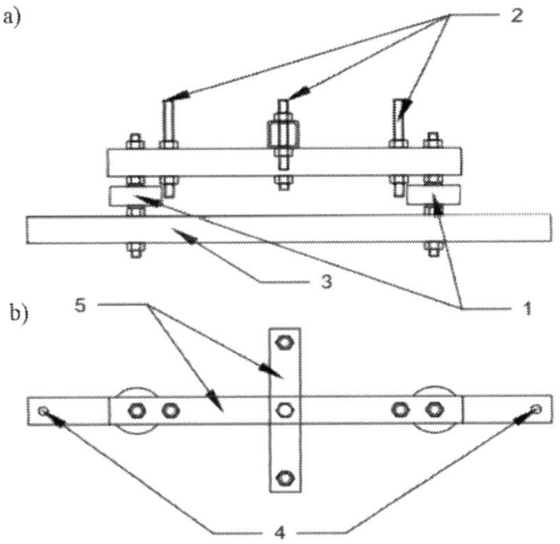

Fig. 4. Sample holder with force meters for configuration with air flow above the specimen: a) side and b) top view (where numbers denote: 1 – force meters, 2 – regulation rods, 3 – base, 4 – assembly holes and 5 – support profiles).

The top segment of the middle module contains the pocket for the specimen (8) and force meters (6) with the sample holder which allows to level and position the sample (see Figure 1c and d). Figure 4 presents schematic views of this holder. In order to stiffen its construction and enable the measurement of sample's mass, two axial force meters (see element no. 1 in Figure 4) were applied. The holder allows for utilisation of samples of different sizes by adjustment of the specimen level using the regulation system (2). This system was made of four threaded rods and screws, mounted to the support profiles (5). The support profiles were connected to the base (3) by the force meters. The whole construction was assembled to the stand using screws and assembly holes (4). The bottom and lateral surfaces of the specimen pocket were surrounded by insulation made of the expanded polystyrene foam. The small air gap between insulation and specimen was left to prevent the insulation from affecting on the measurement of the sample mass. The gap was closed on the top side by a metallic tape. The top surface of the specimen was aligned with the tape and duct's wall.

The position of the specimen in the duct might be changed, i.e., from bottom to either lateral or top by rotation of the duct in order to investigate the influence of the gravitational acceleration on the drying process. In the case with the air flowing above the specimen, gravitational acceleration has small influence on the liquid water transfer in the porous medium. But for cases with lateral air flow or flow under the specimen, gravitational acceleration has significant influence on the process. In these cases, liquid water is transferred to the surface of the sample and comes into direct contact with the flowing air. The air flow around the sample in the lateral configuration is the closest to the real wall drying process. The holder presented in Figure 4 was modified in order to allow for changing of the sample position from bottom to either lateral or top.

2.4 Acquisition and control

The acquisition and control of different parameters in the stand is crucial for its proper operation. In order to measure the parameters of air and the specimen, special sensors were utilised. An integrated temperature, relative humidity and velocity as well as integrated temperature and humidity transmitters were applied for measurements of the air flow parameters (see elements no. 6 and 9 in Figure 1). The temperature map on the top surface of the sample was obtained by the IR thermography. Only one camera was used and measurements were taken with frequency from 2 to 6 samples per hour which allow to change position of camera between two measurement ducts. The disturbing effects of water vapour present in the air flowing in the ducts were taken into account during IR measurements. Moreover, K-type sheath thermocouples of 0.5 mm diameter were applied for measurements of temperature variation in the material of the sample. The global moisture content was found gravimetrically with two tensometric force meters per sample (see elements no. 7 and 1 in Figure 1 and 4, respectively). The distribution of moisture content on the surface of the sample might be also estimated from the temperature distribution obtained with the IR thermography. Moreover, a system of individually calibrated resistance meters was applied and now a new method of measurements of moisture distribution in the sample is being developed and tested. In order to enable data processing and storage, all signals were converted from analog to digital in a multifunction data acquisition (DAQ) device connected with a computer with dedicated in-house software developed in the LabView environment. The DAQ device is equipped with digital outputs which were used to control the solid-state relay and regulate the parameters inside the experimental stand. Sampling frequency was set to 0.5 Hz.

3 Test measurements

3.1 Stability measurements

In order to verify the stability of the controlled parameters of air in the stand (i.e., temperature, relative humidity and velocity of the air) several test measurements were conducted. The standard deviation and difference from the mean value were principal indicators applied for the parameters stability verification. Table 1 contains experimentally-obtained values of stability indicators while Figures 5-7 present variations of controlled parameters.

The results shown in Table 1 and Figures 5-7 prove that obtained stability of the controlled parameters (i.e. temperature, relative humidity and velocity of the air) is very high.

Table 1. Stability indicators for the air parameters.

Indicator	Temperature	Relative humidity	Velocity
	[°C]	[%]	[m/s]
Standard deviation	0.035	0.13	0.12
Maximum spread from mean	0.108	0.51	0.39

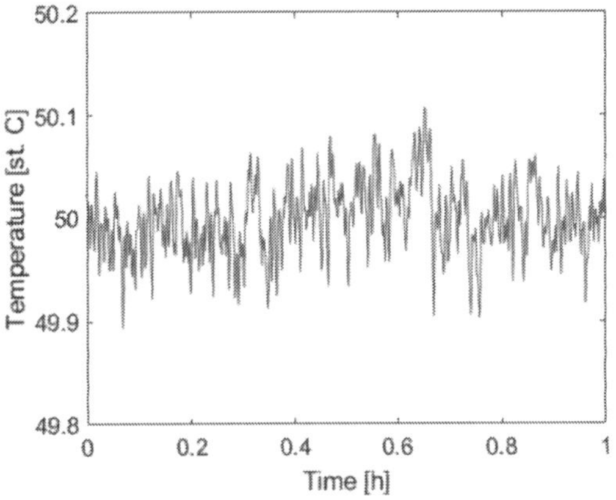

Fig. 5. Variation of the temperature of air flowing in the stand.

3.2 Drying tests

The functional correctness of the stand was verified in several measurements. Figures 8 and 9 present results for drying of the whole ceramic brick with initial moisture content of 13.5% wt. During the measurements, air temperature, relative humidity and velocity were equal to 50°C, 20% and 10 m/s, respectively. Before the measurements the specimen was conditioned, i.e., it was fully-submerged in water for 24 hours in order to obtain initial high content of moisture. The water used to saturate the sample was taken from the water pipe system and had the conductivity and hardness of 1000 μS/cm and 4 mval/l, respectively. Before being placed in the duct, the specimen has been wiped from excessive water.

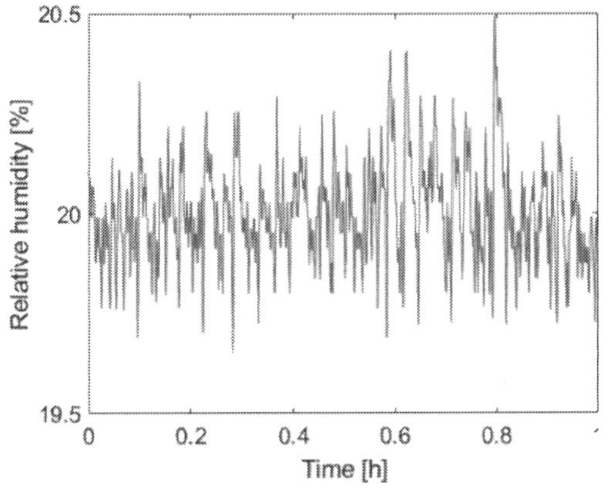

Fig. 6. Variation of the relative humidity of air flowing in the stand.

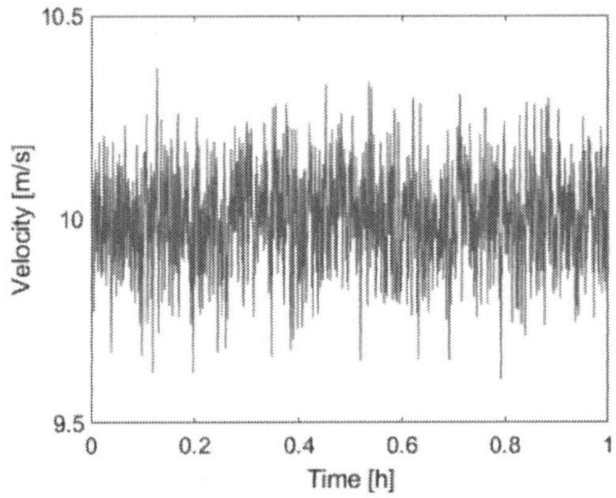

Fig. 7. Variation of the velocity of air flowing in the stand.

The locations of thermocouples in the sample were as follows (see Figure 8b):

- T1 – middle of the sample, 5 mm deep from the bottom surface;
- T2, T3 – 6 cm from the front and rear edges, respectively, in half width of the sample, 5 mm deep from the bottom surface;
- T4, T5 – 3 cm from the side edges in half length of the sample, 5 mm deep from the bottom surface.

The temperature variation during drying of the sample for different locations is presented in Figure 8a. The first stage of drying resulted in achieving of saturation conditions. This stage was characterized by fast water evaporation from the sample surface and reduction of the moisture content in the sample (see Figure 9). Later, in the second stage, increase of the sample temperature towards the air flow temperature was observed (see Figure 8).

The temperature distributions on the top surface of the specimen at different stages of drying are also presented in Figure 10. It was found that the location of the drying front is strongly connected with temperature distribution. Saturation temperature and the front proceed together with the flow direction, i.e., from right to left side. After the evaporation of the liquid water, the sample temperature started to increase.

Figure 8 and 10 revealed existing gradient of temperature and moisture content in the direction perpendicular to the flow of air in the duct. This effect may be related to both non-uniform initial distribution of the moisture in the brick and capillary forces in the brick which results in the moisture movement to the wall of the sample. The sides walls of the sample were well insulated and water proof.

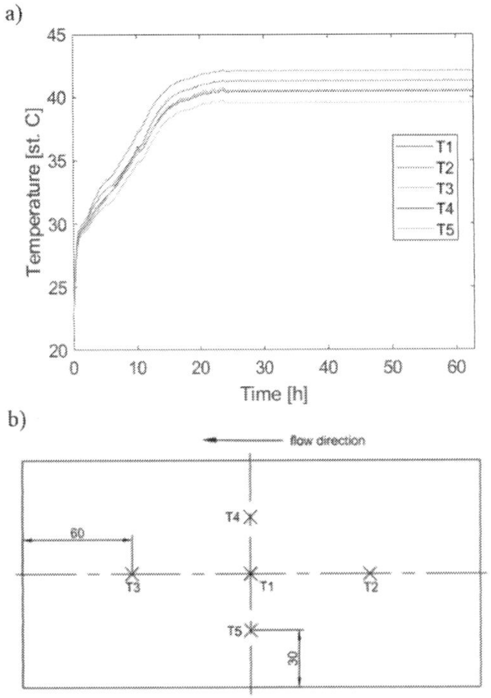

Fig. 8. a) Variations of temperature at different locations in the specimen dried by air of temperature 50°C, relative humidity 20% and velocity 10 m/s, b) cross-section of the specimen with locations of thermocouples.

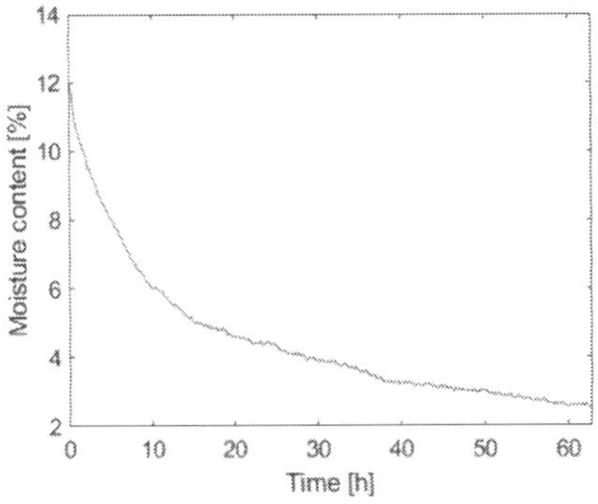

Fig. 9. Variation of the total moisture content in the specimen dried by air of temperature 50°C, relative humidity 20% and velocity 10 m/s.

4 Conclusions

The paper presents the development of the experimental stand with closed-loop flow and two parallel measurement ducts for investigation of heating and drying phenomena in porous building materials. Only one surface of the sample is exposed to the flowing air. The operating conditions of the stand were assumed in the following range: 15°C-60°C, 10-90% and 0.1-10 m/s for temperature, relative humidity and velocity of the flowing air, respectively. These ranges are wider than those encountered in the literature. Good stability of controlled parameters was obtained during verification measurements which indicates that the stand is ready for future experiments designed to validate numerical and analytical models which are developed in parallel. The change of sample location from bottom to either lateral or top gives the opportunity to study the influence of gravitational acceleration on drying phenomena. Additionally, the wide range of operating conditions achieved in the stand allows for the investigation of drying process of different materials, i.e., food, crops or waste.

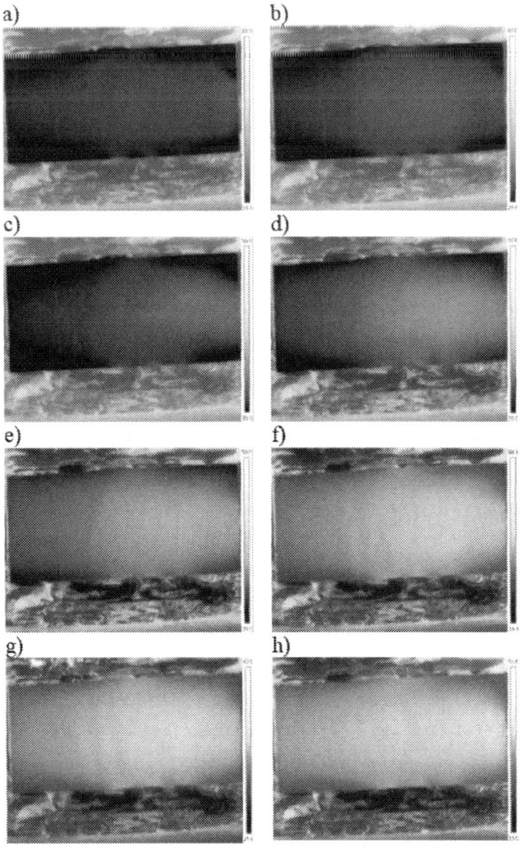

Fig. 10. Temperature distributions at the top surface of the specimen after: a) 10 min, b) 1h, c) 3h, d) 6h, e) 12h and f) 24h g) 48h and h) 72h for drying air at temperature of 50°C, relative humidity of 20% and flow velocity of 4.3 m/s. The minimum and maximum temperatures on thermographs are 25°C and 50°C, respectively.

Acknowledgements

This work was supported by the National Centre for Research and Development (Poland) under grant no. POIR.04.01.02-00-0099/16 Development of innovative technology of drying and moisture sealing of masonry walls, DryWall.

References

1. P.B. Lourenço, E. Luso, M.G. Almeida, Build. Environ. **41**, 223–234 (2006)
2. M. Trochonowicz, Bud. i Archit. **7**, 131–144 (2010)
3. N. Karagiannis, M. Karoglou, A. Bakolas, M. Krokida, A. Moropoulou, Constr. Build. Mater. **137**, 441–449 (2017)
4. R.C. Shoemaker, D.E. House, Neurotoxicol. Teratol. **28**, 573–588 (2006)
5. L. Falchi, D. Slanzi, E. Balliana, G. Driussi, E. Zendri, Build. Environ. **131**, 117–127 (2018)
6. A.M. Gueli, V. Garro, M. Liuzzo, G. Margani, S. Pasquale, G. Politi, G. Stella, Meas. J. Int. Meas. Confed. **118**, 289–297 (2018)
7. J.R. de J. Junqueira, J.L.G. Corrêa, H.M. de Oliveira, R. Ivo Soares Avelar, L.A. Salles Pio, LWT - Food Sci. Technol. **82**, 404–410 (2017)
8. A.M. Castro, E.Y. Mayorga, F.L. Moreno, J. Food Eng. **223**, 152–167 (2018)
9. I. Boutelba, S. Zid, P. Glouannec, A. Magueresse, S. Youcef-ali, Data Br. **18**, 1567–1575 (2018)
10. I. Das, A. Arora, J. Food Eng. **223**, 208–219 (2018)
11. B. Zecchi, L. Clavijo, J. Martínez Garreiro, P. Gerla, J. Food Eng. **104**, 49–55 (2011)
12. L. Kurik, T. Kalamees, U. Kallavus, V. Sinivee, Energy Procedia. **132**, 159–164 (2017)
13. G. Cuccurullo, L. Giordano, A. Metallo, L. Cinquanta, Biosyst. Eng. **170**, 24–30 (2018)
14. J. Szadzińska, S.J. Kowalski, M. Stasiak, Int. J. Heat Mass Transf. **103**, 1065–1074 (2016)
15. X. Cheng, J. Fan, Int. J. Therm. Sci. **43**, 665–676 (2004)
16. M. Van Belleghem, M. Steeman, H. Janssen, A. Janssens, M. De Paepe, Build. Environ. **81**, 340–353 (2014)
17. J. Fan, X. Wen, Int. J. Heat Mass Transf. **45**, 4045–4055 (2002)
18. P. Salagnac, P. Glouannec, D. Lecharpentier, Int. J. Heat Mass Transf. **47**, 4479–4489 (2004)
19. L. Pel, K.A. Landman, E.F. Kaasschieter, Int. J. Heat Mass Transf. **45**, 3173–3180 (2002)
20. V.A.F. Costa, M.L. Mendonça, A.R. Figueiredo, Int. J. Heat Mass Transf. **51**, 3342–3354 (2008)
21. J. Zheng, J. Shi, Q. Ma, X. Dai, Z. Chen, Appl. Therm. Eng. **114**, 450–456 (2017)
22. Y. Zhang, X. Luo, X. Wang, K. Qian, R. Zhao, Atmos. Environ. **41**, 3203–3216 (2007)
23. M. Bianchi Janetti, L.P.M. Colombo, F. Ochs, W. Feist, Int. J. Therm. Sci. **114**, 271–280 (2017)
24. H. Janssen, G.A. Scheffler, R. Plagge, Int. J. Heat Mass Transf. **98**, 141–149 (2016)

Ampacity of power cables exposed to solar radiation – recommendations of standards vs. CFD simulations

Stanislaw Czapp[1,*], *Marian* Czapp[2], *Seweryn* Szultka[1], and *Adam* Tomaszewski[3]

[1]Gdansk University of Technology, Faculty of Electrical and Control Engineering, ul. Narutowicza 11/12, PL-80-233 Gdansk, Poland
[2]Koszalin University of Technology, Faculty of Mechanical Engineering, ul. Raclawicka 15-17, PL-77-620 Koszalin, Poland
[3]Polish Academy of Sciences, Institute of Fluid-Flow Machinery, ul. Fiszera 14, PL-80-231 Gdansk, Poland

Abstract. Ampacity of power cables strictly depends on the ambient conditions. It is very important whether a cable is buried in soil, installed in the air or placed in ducts. When a cable is installed in free air, potential solar radiation has the dominant impact on the prospective ampacity. International standards indicate how to calculate ampacity of power cables exposed to solar radiation, however the standards' recommendations are characterised by some simplifications. In order to consider many complex factors influencing ampacity of power cables, and modelling advanced heat transfer phenomena, a Computational Fluid Dynamics (CFD) can be used. This paper presents a comparison of ampacity calculation of an example power cable for two approaches – first: according to international standards; second: with a CFD employed. Differences in results obtained for these two approaches are commented.

1 Introduction

One of the most important issues which influence ampacity[a] of power cables is solar radiation [1-7]. Methods of calculation of ampacity of power cables are conducted in the standard IEC 60287 [8, 9]. According to this standard, ampacity of power cables exposed to solar radiation can be calculated as follows:

$$I_{amp} = \sqrt{\frac{\Delta\theta - W_d \cdot \left[0.5 \cdot T_1 + n_c \cdot (T_2 + T_3 + T_4^*)\right] - \sigma \cdot D_e \cdot H \cdot T_4^*}{R \cdot T_1 + n_c \cdot R \cdot (1+\lambda_1) \cdot T_2 + n_c \cdot R \cdot (1+\lambda_1+\lambda_2) \cdot (T_3 + T_4^*)}} \quad (1)$$

where:

I_{amp} – ampacity of power cables, A,
$\Delta\Theta$ – permissible temperature rise of the conductor above ambient temperature, K,
W_d – dielectric losses per unit length per phase, W/m,
T_1 – thermal resistance per core between the conductor and sheath, (K m)/W,
T_2 – thermal resistance between the sheath and armour, (K m)/W,
T_3 – thermal resistance of external serving of the cable, (K m)/W,
T_4^* – external thermal resistance in free air, adjusted for solar radiation, (K m)/W,
n_c – number of conductors in a cable, -,
σ – absorption coefficient of solar radiation for the cable surface, -,
D_e – external diameter of cable (with insulation), mm,
H – intensity of solar radiation (for most latitudes can be assumed as 10^3 W/m^2); it is recommended to assume a local value, W/m^2,
R – AC current resistance of a conductor at its maximum operating temperature, Ω/m,
λ_1 – ratio of the total losses in metallic sheaths to the total conductor losses, -,
λ_2 – ratio of the total losses in metallic armour to the total conductor losses, -.

In general, calculation of the ampacity according to the recommendations of the standard IEC 60287 [8, 9] requires assumption of values of many coefficients, as well as the use of iterative or graphical methods of their calculation. Moreover, recommendations of the standard do not cover all types of cables installation and possible influences of external factors.

Ampacity of power cables can also be calculated with the use of a Computational Fluid Dynamics (CFD) method. It is possible to simulate various complex ambient conditions, mainly: intensity and direction of solar radiation, wind speed, consequences of vertical installation vs. horizontal installation of the cables. This method of calculation, although time-consuming, allows for relatively accurate evaluation of the ampacity in complex cable laying conditions. The aim of this paper is to evaluate the ampacity of a low-voltage single power cable, according to:

- recommendations of the standard IEC 60287 [8-10],

[a] ampacity – (International Electrotechnical Vocabulary) maximum value of electric current which can be carried continuously by a conductor, a device or an apparatus, under specified conditions without its steady-state temperature exceeding a specified value.

* Corresponding author: stanislaw.czapp@pg.edu.pl

 https://doi.org/10.1051/e3sconf/20187003004

- a Computational Fluid Dynamics method, implemented in Ansys software,

and comparison of the results obtained for these two cases. Moreover, limitations of recommendations of the above mentioned standard are indicated.

2 Results of calculations according to the standard IEC 60287

Low-voltage power cables construction is relatively simple – they are usually composed of a copper or aluminium conductor and insulation. In this paper a power cable with copper conductor and PVC insulation is considered (Fig. 1). The simplicity of the construction of such cables makes the values of many quantities and coefficients in (1) equal to 0. The zero value can be assigned to: W_d, T_2, T_3, λ_1, λ_2.

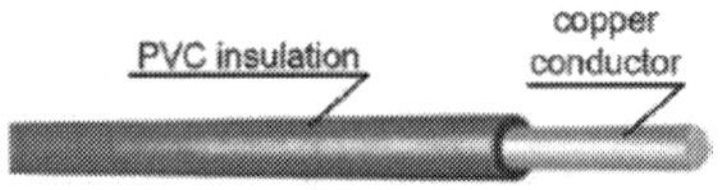

Fig. 1. Construction of the analyzed low-voltage power cable.

Taking into account the above simplification, ampacity of the considered single power cable ($n_c = 1$) can be calculated according to the following expression:

$$I_{amp} = \sqrt{\frac{\Delta\theta - \sigma \cdot D_e \cdot H \cdot T_4^*}{R \cdot (T_1 + T_4^*)}} = \sqrt{\frac{\Delta\theta - \Delta\theta_{red}}{R \cdot (T_1 + T_4^*)}} \quad (2)$$

where:

$\Delta\Theta_{red}$ – reduction of the permissible temperature rise of the conductor, due to solar radiation, K.

Thermal resistance T_1, between the conductor and sheath of the cable, is described by the dependence:

$$T_1 = \frac{\rho_T}{2\pi} \ln\left(1 + \frac{2 \cdot \delta}{D_w}\right) \quad (3)$$

where:

ρ_T – thermal resistivity of insulation, (K m)/W,

δ – thickness of the cable insulation, mm,

D_w – external diameter of conductor, mm,

whereas external thermal resistance T_4^* in free air, adjusted for solar radiation, is calculated as follows:

$$T_4^* = \frac{1}{\pi \cdot D_e \cdot h \cdot (\Delta\theta_s)^{0.25}} \quad (4)$$

where:

h – heat dissipation coefficient, -,

$\Delta\Theta_s$ – temperature difference between the surface of a cable in the air and ambient temperature, K.

The heat dissipation coefficient h and the temperature difference $\Delta\Theta_s$ between the cable surface in the air and ambient air are defined as follows:

$$h = \frac{Z}{D_e^d} + E \quad (5)$$

$$\Delta\theta_{s\_n+1} = \frac{\Delta\theta + \Delta\theta_d + \Delta\theta_{ds}}{1 + K_A \cdot (\Delta\theta_{s\_n})^{0.25}} \quad (6)$$

where:

d, E, Z – constants for black surfaces of cables in free air (values included in table 2 in IEC 60287-2-1 [9]), -,

$\Delta\Theta_d$ – factor for calculation T_4^* for cables in free air, including dielectric loss, K,

$\Delta\Theta_{ds}$ – factor indicating temperature difference of cable insulation due to direct solar radiation, K,

K_A – factor for calculation of external thermal resistance of cables in air, -.

The temperature difference $\Delta\Theta_s$ is calculated in an iterative way – value of a step of calculation n+1 in relation to a previous step (n) should fulfill the following requirement:

$$(\Delta\theta_{s\_n+1})^{0.25} - (\Delta\theta_{s\_n})^{0.25} \leq 0.001 \quad (7)$$

Factor K_A, for calculation of external thermal resistance of the cable in the air, is defined by the following dependence:

$$K_A = \pi \cdot D_e \cdot h \cdot T_1 \quad (8)$$

whereas factor $\Delta\Theta_{ds}$ indicating the temperature difference of the cable insulation, due to direct solar radiation, is calculated according to the expression:

$$\Delta\theta_{ds} = \sigma \cdot D_e \cdot H \cdot T_1 \quad (9)$$

The arrangement of the analyzed power cable is presented in Fig. 2. The external diameter of the conductor is D_w = 7.2 mm (nominal catalogue value of cross-sectional area equals 35 mm^2), and the external diameter of the PVC insulation is D_e = 12.4 mm. The temperature of the ambient air is equal to 25 °C (typical value for Poland – according to the standard [10]). The solar radiation is assumed to be at an angle of 45°.

Calculation of ampacity of the power cable has been conducted for various values of absorption coefficient of solar radiation σ (Fig. 3a) and intensity of solar radiation H (Fig. 3b).

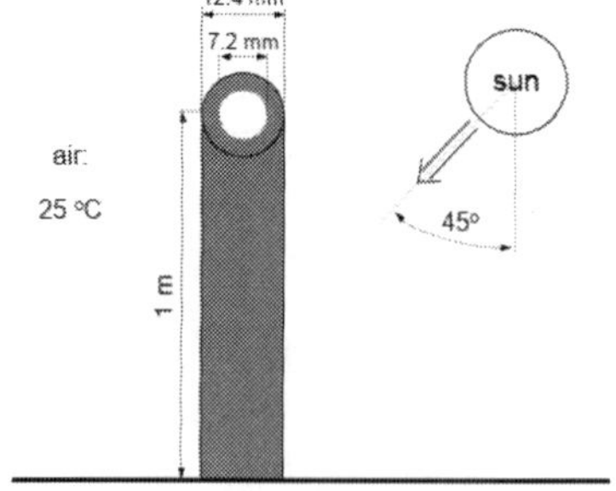

Fig. 2. An arrangement of the analyzed power cable.

Values of coefficients and constants used in calculations are included in Tab. 1. When intensity of solar radiation is assumed to be H = 1000 W/m^2, and σ varies from 0 to 1, ampacity I_{amp} is within the range 191–148 A, respectively. Similar results have been obtained for variations of solar radiation intensity H from 0 to 1300 W/m^2 (for σ = 0.9) – ampacity I_{amp} is within the range 191–139 A, respectively. It can be concluded that variation of the most important factors influencing cables ampacity changes (decreases) its value no more than 30% (when a reference value is a maximal ampacity 191 A).

Table 1. Values of coefficients assumed/calculated according to IEC 60287 [8, 9], and other constants.

Quantity	Unit	Value	Quantity	Unit	Value
d	-	0.60	δ	mm	2.6
h	-	6.865	D_e	m	0.0124
E	-	3.94	R	Ω/m	0.000624
Z	-	0.21	K_A	-	0.116
n_c	-	1	$\Delta\theta$	K	45
ρ_T	$\frac{\text{Km}}{\text{W}}$	5	W_d, T_2, T_3, λ_1, λ_2, $\Delta\Theta_d$		0

Decreasing of ampacity of power cables is a consequence of reduction of the permissible temperature rise of the conductor above ambient temperature (due to solar radiation). This reduction $\Delta\Theta_{red}$ is expressed by the following formula:

$$\Delta\theta_{red} = \sigma \cdot D_e \cdot H \cdot T_4^* \quad (10)$$

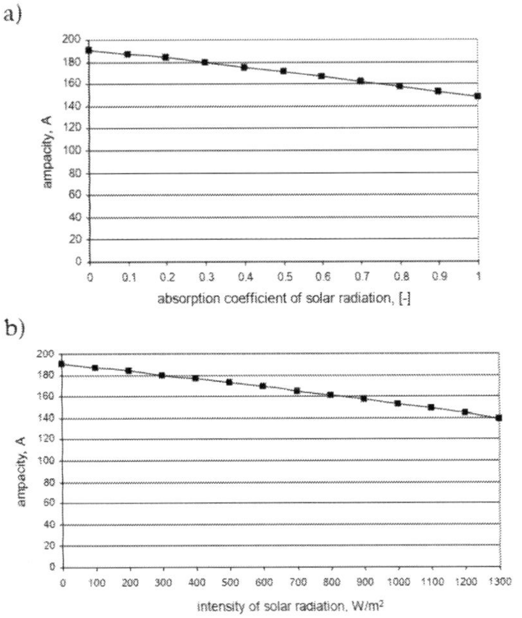

Fig. 3. Ampacity I_{amp} of the analyzed power cable as a function of: a) absorption coefficient σ of solar radiation (H = const = 1000 W/m^2), b) intensity of solar radiation H (σ = const = 0.9).

The reduction depends on absorption coefficient of solar radiation for the cable surface σ, intensity of solar radiation H, external thermal resistance in free air T_4^* and external diameter of cable D_e. For the following values of the above mentioned coefficients: σ = 0.9, H = 1120 W/m^2, reduction of the permissible temperature rise is equal to $\Delta\theta_{red}$ = 19 K (ampacity equals I_{amp} = 148 A). Such a reduction seems to be too little, especially if it is referred to data included in the paper [1]. It is concluded in this paper that temperature rise above outdoor temperature in full sunlight, inside metal conduits, may reach even 40 K, when cables are installed on the rooftop. Taking this conclusion into account, calculation of ampacity of the analyzed power cable (according to (2)), as a function of reduction of permissible temperature rise of the conductor above ambient temperature $\Delta\theta_{red}$, has been performed (Fig. 4).

The ampacity drastically decreases when $\Delta\theta_{red}$ > 35 K. For $\Delta\theta_{red}$ = 40 K ampacity is equal only to I_{amp} = 64 A, (instead of I_{amp} = 191 A for $\Delta\theta_{red}$ = 0 K), and for $\Delta\theta_{red}$ = 44 K ampacity is equal to I_{amp} = 29 A. The analyzed power cable cannot be loaded at all if reduction of the permissible temperature rise is $\Delta\theta_{red} \geq 45$. Such values ($\Delta\theta_{red} \geq 45$) can not be excluded in extremely unfortunate ambient conditions for cables exposed to direct solar radiation (e.g. on the rooftop, without thermal protection in place where air stagnation occurs). Heat transfer (heat dissipation) from the cable to the ambient air is deteriorated. Thus, advanced modelling of the cable's ampacity, with the use of a CFD method, has been performed.

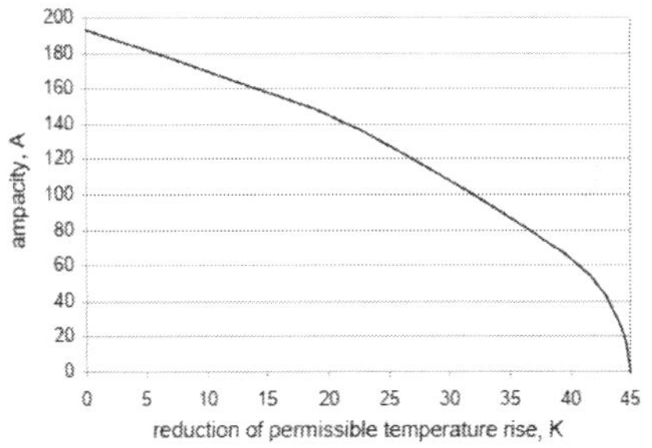

Fig. 4. Ampacity I_{amp} of the analyzed power cable as a function of reduction of permissible temperature rise $\Delta\theta_{red}$ of the conductor above ambient temperature.

3 Results of calculations with the use of CFD simulations

In CFD simulations heat flux density q_S (W/m^2) from solar radiation, delivered to the surface of cable insulation, is evaluated on the base of the Westman's equation [11]. This equation is modified and multiplied by the PVC absorption coefficient $\sigma = 0.9$ of PVC insulation and the cosine of incidence angle θ_{sun} (°) of sunrays.

$$q_S(\theta_{sun}) = \sigma \cdot \cos\theta_{sun} \cdot H \cdot 10^{-0.05675 m + 0.00038 m^2} \quad (11)$$

where: H = const = 1120 W/m^2; $m = 1/\sin A_L$; $A_L \approx 45°$ – the sun's altitude.
The expression (11) is converted to the relationship $q_S(x)$:

$$q_S(x) = \sigma \cdot \frac{\sqrt{r_e^2 - x^2}}{r_e} \cdot H \cdot 10^{-0.05675m+0.00038m^2} \quad (12)$$

where:
x – coordinate, r_e – external radius of the insulation, mm.

Results of simulations are presented in Fig. 5. When the cable is not exposed to solar radiation, its ampacity is equal to I_{amp} = 190 A. This value is in good convergence with the result (I_{amp} = 191 A) obtained by calculations according to the standard [8, 9]. When solar radiation is taken into account, the ampacity I_{amp} = 142 A – this value is also consistent with the standard (I_{amp} = 148 A).

In contrast to the rules of cables ampacity calculation presented in the standard IEC 60287 [8, 9], CFD simulations enable modelling of extremely unfavorable ambient condition, which results in deteriorated heat transfer from cables to ambient air. It may occur on a rooftop in a place which is shielded from wind (air is almost in full stagnation), especially when power cables are exposed to solar radiation, or they are in closed conduits surrounded by hot air. In such cases, the radiation heat emitted from the cable to the environment is comparable to the heat supplied by radiation to the surface of the cable from surrounding structures (e.g. rooftop, shield). Thus, in further considerations it has been assumed that heat transfer by radiation, from the cable to the environment, does not occur.

Such an assumption makes that ampacity of the analyzed power cable drastically decreases. When the cable is not exposed to solar radiation, and almost full air stagnation occurs, its ampacity is equal to I_{amp} = 137 A. It is a significantly lower value than for calculation performed according to the standard [8-10] – (191 A), as well as value obtained in the previous simulation (190 A). Moreover, when solar radiation with the assumed parameters is taken into account, the cable cannot be loaded at all. The highest permissible temperature of the PVC insulation (70 °C) is achieved only by insolation, without any current in the cable conductor. Comparison of calculation results is presented in Tab. 2.

Due to solar radiation and air stagnation, temperature of the air surrounding a cable may significantly rise. The heat radiated from the surrounding elements contributes significantly to the heat balance of the power cable. Additionally, part of the radiation reflected from the surrounding surfaces (e.g. rooftop) may return to the surface of a cable. Result of the CFD simulations confirm observations commented in [1]. Solar radiation may force a reduction of the permissible temperature rise of the conductor above ambient temperature significantly, more than recommended by the standard. This conclusion may also be an explanation of power cables damages, which are described in [12]. The power cables were put into operation during autumn, and operated properly till the nearest summer. During the summer, damages of power cables occurred, even though they were loaded only a little. Temperature of the PVC insulation might significantly exceed 70 °C even for a slight current load.

a) b)

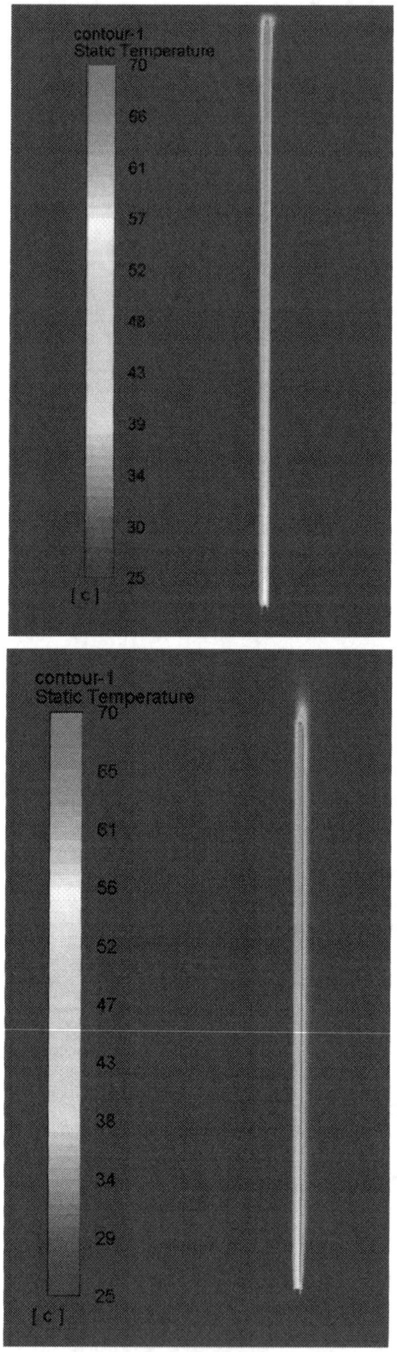

Fig. 5. Temperature distribution around a single power cable (PVC insulation max permissible temperature: 70 °C): a) without solar radiation, wind from bottom to top v_w = 0.01 m/s, value of current I_{amp} = 190 A, b) with solar radiation, wind from bottom to top v_w = 0.01 m/s, value of current I_{amp} = 142 A.

Table 2. Comparison of calculation results.

Calculation according to	Ampacity, A	
	no solar radiation	solar radiation $H = 1120$ W/m^2
IEC 60287	191	148
CFD simulations	190	142
CFD simulations, air stagnation, heat radiation from the cable excluded	137	0

4 Conclusions

Solar radiation is a major factor influencing ampacity of power cables installed in free air. As it has been conducted in the CFD simulations, solar radiation may drastically reduce ampacity of power cables – for PVC-insulated power cables this ampacity may be close to 0. For power cables installed in free air, the heat dissipated by radiation needs to be taken into account, whereas for power cables surrounded by certain structures, radiated heat may be reduced depending on the case. It seems that calculation of ampacity of power cables according to recommendations of the international standards gives too optimistic results, for an adverse, strong solar radiation and air stagnation. It may lead to damage of cables.

References

1. D. Brender, T. L. Lindsey, IEEE T. on Ind. Appl., **44**, pp. 1872-1878 (2008)
2. S. Czapp, S. Szultka, A. Tomaszewski, *CFD-based evaluation of current-carrying capacity of power cables installed in free air*, 18th Int. Conf. on El. Power Eng. (EPE), Kouty nad Desnou, pp. 692-697 (2017)
3. C. Holyk, G. J. Anders, IEEE Ind. App. Mag., **21**, pp. 6-64 (2015)
4. F. de Leon, *Major factors affecting cable ampacity*, IEEE Power Eng. Soc. Gen. Meeting (PES), (2006)
5. Y. Liang, J. Zhao, Y. Du, J. Zhang, Przegl. Elektrot., **3b**, pp. 156-160 (2012)
6. A. Sedaghat, F. de Leon, IEEE T. on Pow. Del. **29**, pp. 2306-2314 (2014)
7. W. Zhang, H.-J. Li, Ch. Liu, K. Ch. Tan., *A technique for assessment of thermal condition and current rating of underground power cables installed in duct banks*, Asia-Pac. Pow. and En. Eng. Conf., Shanghai, (2012)
8. IEC 60287-1-1: *Electric cables – Calculation of the current rating – Part 1-1: Current rating equations (100% load factor) and calculation of losses – General* (2006)
9. IEC 60287-2-1: *Electric cables – Calculation of the current rating – Part 2-1: Thermal resistance – Calculation of the thermal resistance* (2015)
10. IEC 60287-3-1: *Electric cables – Calculation of the current rating – Part 3-1: Operating conditions – Site reference conditions* (2017)
11. S.A. Andersen, *Automatic refrigeration* (Published by MacLaren for Danfoss, Denmark, 1959)
12. F. Spyra, Energetyka, **6-7**, pp. 451-454 (2007)

Experimental investigation of PCM thermal energy storage charge and discharge process with aperiodic (ramp) temperature inputs

Jarosław Karwacki[1,*], and *Roman* Kwidziński[1]

[1]Institute of Fluid Flow Machinery, Polish Academy of Sciences, Fiszera 14, PL-80-231 Gdańsk, Poland

Abstract. In recent years, the use of storages filled with phase-change material (PCM) is increasingly considered. Such design is characterized by a higher density of thermal energy accumulation in comparison with water storages. However, the optimal use of the PCM storages requires a recognition of its dynamic characteristics during the loading and unloading process. This paper presents research aimed at understanding and dynamic description of the heat transfer process in a shell-and-tube thermal energy storage. The experimental test stand and the measurement and control system are described. The investigated storage had a form of a cylindrical tank of 40 dm^3 volume in which a coil made of pipes with an external diameter of 3.35 mm was immersed in the PCM. The total heat transfer area was 9.4 m^2. A lumped parameter model was used to describe mathematically the storage thermal dynamics. The PCM used was commercially available RT15 material with the heat capacity of 150 kJ/kg in the temperature range of 10–17°C. In the investigations, aperiodic (ramp) temperature inputs were used. The storage tests were carried out for low (12 h) and high (6 h) speeds of charging and discharging. The amplitude of the input signal and the liquid temperature at the storage inlet were set to include the phase transition interval of the PCM used. The obtained test results allowed to determine the enthalpy as a function of temperature for the whole storage. The experimental results were also used to validate 0D mathematical model of the heat storage.

1 Introduction

Heat storage with phase change material (LTES – Latent Thermal Energy Storage) has been used in refrigeration and air conditioning applications for many years. Mostly known area of LTES application is the peak demand management in commercial buildings [1]. In this case, the ice-banks system (TES with water) is used to reduce the operating costs. The use of other phase change materials (PCM) in low temperature applications is also increasingly considered. In active peak load shifting systems, the correct, optimal operation of the heat storage using PCM requires not only the selection of appropriate material but also appropriate control of the loading and unloading process. The mathematical modelling is often used for the optimization of the cooling system's operation [1]. In many cases, the proper design of the automatic control system is limited by the knowledge of the control object model. In this case, a specific form of the mathematical model of the heat store must also be adopted. It turns out that creating a model satisfactorily reflecting the behaviour of the PCM storage is not trivial. This is mainly due to the peculiar characteristics of changes in the properties of phase-change materials in the area of phase transition.

The authors, in earlier works, conducted a theoretical analysis of the feasibility of using a storage with phase change material to support the photobioreactor cooling system for the production of the algae [2] and to cooperate with an adsorption chiller supply from district heating system [3]. The analyses show that the effective use of LTES requires the use of non-standard methods for control the loading and unloading process. Planning the implementation of the mentioned above systems on a semi-industrial scale, an attempt was made to create their own regulation based on Model Based Control (MBC) algorithm. Typically, in such cases, lumped parameter model is used. Due to the non-linear nature of the process as well as the frequent occurrence of hysteresis and subcooling during the phase transition of typical PCM materials, the creation of the correct LTES model turned out to be the biggest problem. In this paper the preliminary steps are taken to develop an experimental procedure to calibrate and verify the dynamic, lumped parameter LTES model.

2 Experimental identification of dynamic systems

During the process of constructing a TES, its geometry, the shape of the heat exchange surface and the

* Corresponding author: jkarwacki@imp.gda.pl

 https://doi.org/10.1051/e3sconf/20187003005

thermophysical properties of construction materials are known. In most cases, the thermo-physical properties of PCM materials used for heat accumulation are also known. Using this information it is possible to develop more or less comprehensive dynamic model of TES/LTES [2, 3, 4].

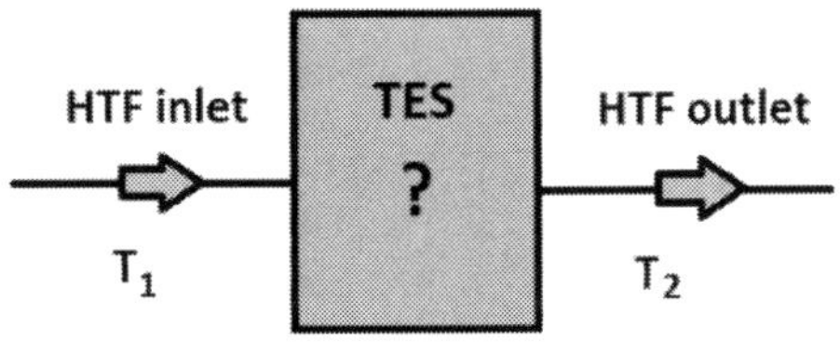

Fig. 1. Experimental identification of dynamic TES systems: T_1, T_2 are heat transfer fluid input and output temperatures, respectively.

In the general case, there is no enough information to create a dynamic model of a heat storage accurately describing the behaviour of the object. In particular, this applies to lumped parameter models involving significant simplifications. One of the ways to improve the prediction of working parameters is the experimental calibration of the models. This is accomplished through a sequence of actions aimed at identification of the dynamic system [5]. The experimental verification of the dynamic model assumes that for a precisely defined input signal, there are characteristic changes in the output signal. This behaviour is related to the uniquely defined dynamic characteristics of the object. By analysing the input and output signal, it is possible to draw conclusion about the dynamics of the object.

In the case of the analysed TES/LTES (Fig. 1), this is reduced to the analysis of heat transfer fluid (HTF) outlet temperature T_2 changes depending on the inlet temperature T_1 profile. The active method adopted in this work assumes an experimental identification of the model based on the reaction of the object to a strictly defined test signal. Most often used are simple non-periodic test signals as: step function, ramp function, rectangular pulse, trapezoidal pulse. In some cases, periodic signals, usually sinusoidal, are also used in the active identification process. However, it is associated with much higher workload of the experimental tests.

To gain thermophysical properties of phase change materials, different experimental methods are available [6]. The most popular is DSC (differential scanning calorimetry) technique using multi step function or ramp function to measure the specific heat and latent heat of PCM. The amount of sampling mass used in DSC is less than 50 mg. Despite of this, similar testing signals are used in laboratory scale experiments to obtain thermal capacity of LTES.

The most frequent measurement technique is based on a sudden change of inlet HTF temperature (step function) [7]. In such experiments, the input temperature of the heat transfer fluid is instantaneously set at the desired value, while the output temperature evolution is recorded. The range of inlet temperature change is set to include the phase transition interval of the used PCM. Because of the installation thermal inertia, the pure step function test is difficult to carry out. One of the ways is to use two separate storage vessels holding HTF at two different temperatures [7]. In a closed HTF cycle, the step change of the set HTF inlet temperature does not cause sudden answer. Instead, sometimes it causes a fluctuation of inlet temperature or the profile of the temperature changes is not sharp enough [8, 9]. In some cases it has significant influence on the overall quality of experimental results.

In literature, there are a lot off LTES experimental analyses in various dynamics conditions [7, 8, 9, 10]. Sometimes the LTES is installed as a part of production line [10] and system is tested under real operating conditions. For the most part of these published experimental analyses, the attention is focused on determining energy stored by the TES under full and/or partial capacity loads. As a result of the tests, charging and discharging rates are also calculated. This type of analysis could give some information about heat transfer dynamic in thermal energy storage. However, in none of these papers systematic methodology of the dynamic characteristic determination is presented.

3 Methodology

3.1 Experimental setup

An experimental setup was developed and built to investigate dynamic characteristics of different thermal energy storages.

The layout of the experimental setup is shown in Fig. 2. Two different loops, separated by the heat exchanger 4, can be identified. The first loop (left of Fig. 2), is a typical compressor refrigeration cycle. To achieve more stable working conditions, typical thermostatic expansion valve was replaced by electronic device 3. An own algorithm of valve regulation allowed to achieve very small fluctuation of the working parameters during tests. In the second loop, water was used as the heat transfer fluid. To achieve fast temperature response of the system, amount of water was minimized. Water chilled in the evaporator is pumped to TES container 9 by the pump 6. To achieve the set inlet temperature of water, electric heater 7 was used. To minimize the fluctuations of inlet temperature, electric heater was equipped with the precise single-phase power controller JUMO TYA-201.

In the key points of experimental stand, K type thermocouple sensors were installed. Four additional temperature sensors were installed inside the storage tank between HTF tubes. These thermocouples, located at equal distances of 100 mm along the container, were used to measure PCM temperature. To achieve fast temperature response, the sensors with 1 mm shield diameter were used. All thermocouples were calibrated and tested before installation using the Beamex MC6 calibrator, the DRUCK DB-150 calibration furnace and the reference temperature sensor PT100 ISOTECH. After this process, the precision of the thermocouples was ±0.1°C for 0°C and ±0.15°C for 25°C. Special attention was made to accurately determine temperature

 https://doi.org/10.1051/e3sconf/20187003005

difference between HTF inlet and outlet of the TES container. The accuracy of this parameter was better than ±0.05 K.

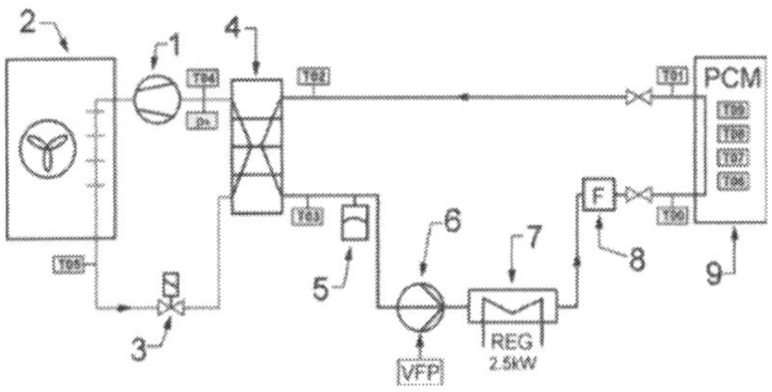

Fig. 2. Layout of experimental setup: 1 – compressor, 2 – condenser, 3 – expansion valve, 4 – evaporator, 5 – expansion tank, 6 – circulating pump, 7 – heater, 8 – flowmeter, 9 – LTES.

A heat transfer fluid flow rate was kept constant by regulating rotating speed of the circulating pump. The value of flow rate in all cases was 200 ±10 kg/h. Water mass flow rate was measured with accuracy 2.5% by vortex flow meter Kobold DVZ-1104G4C34P.

The photo of the experimental setup is shown in Fig. 3. The TES container with PCM is visible in the foreground on the left side.

Fig. 3. Photo of the stand dedicated to investigation of the dynamic characteristics of LTES.

The data acquisition facility was based on National Instrument systems and software. NI cRIO 9025 controller was used as a main part of data acquisition and regulation system. This system logs all main parameters and controls the valves, pumps, electric heaters and safety system. During the experiments the real time measurements are shown on the computer screen while all measured data are stored in a data file.

3.2 Construction of LTES container

Often in storage tanks construction, typical designs known for heat exchangers are used [9]. In some cases commercial heat exchangers are considered for use as PCM thermal storage systems [8]. In the cases when PCM is encapsulated, special construction of LTES container is designed. In all types of geometry, due to low thermal conductivity of phase change materials, the key parameter is the ratio of heat transfer area to the PCM volume.

In this study, the main idea of the LTES design process was to create a not too expensive construction using elements available on the market. To form the heat transfer surface between PCM and heat transfer fluid, typical capillary tube mats were used. Such mats are applied in housing for wall and ceiling heating and cooling systems.

Internal design of the investigated LTES container is presented in the Fig. 4. Because of the limited capillary tubes bending radius, usable volume for the PCM is placed between two concentric tubes. The essential parameters of the construction are presented in Table 1. For the designed construction, the ratio of heat transfer area to PCM volume is about 0.15 m^2/dm^3. This value is comparable to the best constructions described in literature [8].

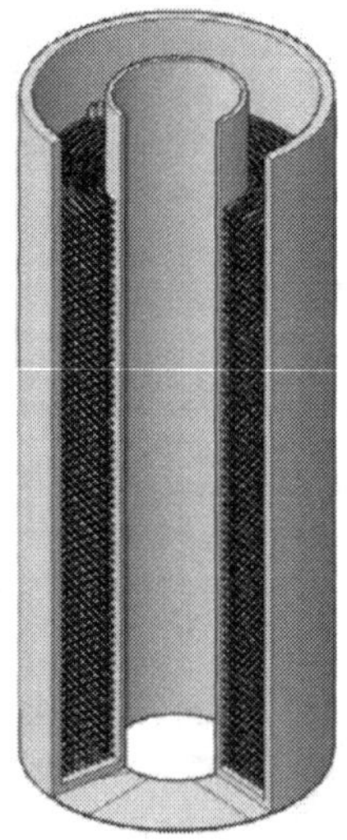

Fig. 4. View of TES container geometry and heat transfer surface.

Table 1. Summary of principal geometrical and thermo-physical properties of the LTES container.

Two concentric tubes of cylindrical vessel	
material	polyethylene
mass [kg]	0.43
insulation	Armaflex
Capillary tube	
material	polypropylene
outer diameter [mm]	3.35
wall thickness [mm]	0.5
spacing between the tubes [mm]	10
thermal conductivity [W/m·K]	0.24
specific heat [J/kg·K]	2000
mass [kg]	1.26
LTES	
heat transfer area [m^2]	4.097
PCM mass [kg]	27.86

3.3 Theoretical analysis of charge/discharge capacity of TES

In contrast to laboratory test stand, industrial installation have a much smaller number of temperature sensors. They are placed only at the most important points of the cycle. In most cases, temperature measurement of the storage material is difficult or impossible. This is particularly important in experimental investigation of commercial TES. In this case, nothing is known about the internal structure of storage container and the properties of accumulating material. It should be treated as "black box". From this reason, it was assumed that all storage characteristics should be investigated and presented as a function of HTF temperature.

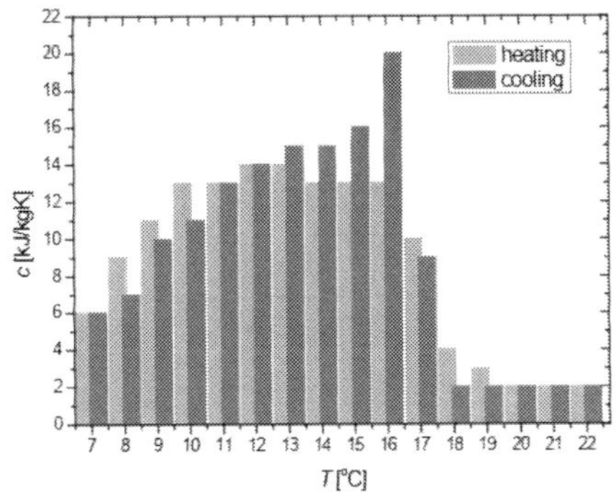

Fig. 5. Effective specific heat c distribution as a function of PCM temperature *T* for the phase change material RUBITHERM RT15 [manufacturer data].

To analyse the impact of the charging/discharging dynamics on effective energy storage, the base characteristics of LTES should be estimated. Effective specific heat distribution as a function of HTF temperature, estimated in steady-state conditions was assumed to be such base characteristic. It was calculated using PCM manufacturer data (Fig. 5) and TES container data (Table 1). It was assumed that in steady-state conditions HTF and PCM temperatures are the same.

Calculated LTES effective specific heat distribution, as a function of HTF temperature is presented in the Fig. 6. Based on manufacturer data, PCM specific heat in solid state is equal 2 kJ/kg but nothing is known about the lower temperature of phase change process onset. For initial analysis it was assumed that below 7°C PCM is in solid state.

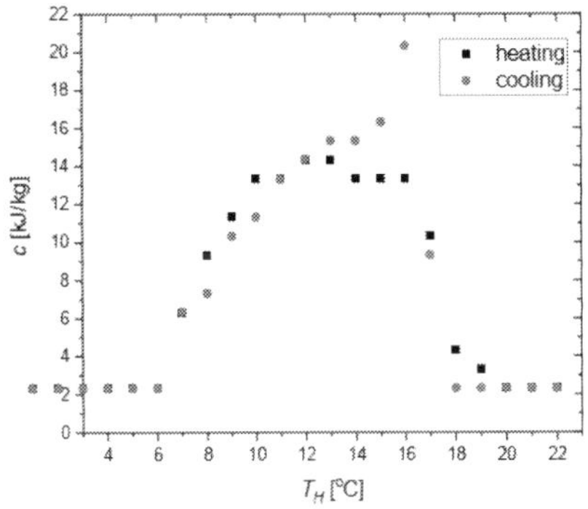

Fig. 6. Calculated LTES effective specific heat *c* distribution as a function of average HTF temperature T_H for heating and cooling process.

3.4 Computational model of LTES

As it was mentioned earlier, the main aim of the work is to design the MBC algorithm for cooling systems with LTES. The essential point of planned activities is to design mathematical model of the heat storage. The model should properly estimate outlet temperature of HTF based on inlet temperature. Below is described numerical model with assumption that the construction of the TES container is known. In next steps it is planned to design model when LTES is treated as "black box".

The numerical model of the heat storage describes heat transfer and accumulation between the HTF, the PCM and the material of the tubes' walls as well as the heat transfer to or from the surroundings [3]. Balance equations for each material accumulating heat in the storage can be written in the form:

HTF:

$$M_H c_H \frac{dT_H}{d\tau} = \dot{m}_H c_H [T_{H1}(\tau) - T_{H2}(\tau)] + \frac{T_W(\tau) - T_H(\tau)}{R_{HW}} \quad (1)$$

PCM:

$$M_P c_P(T_P) \frac{dT_P}{d\tau} = \frac{T_W(\tau) - T_P(\tau)}{R_{PW}} + \dot{Q}_A \quad (2)$$

capillary wall:

$$M_W c_W \frac{dT_W}{d\tau} = \frac{T_P(\tau) - T_W(\tau)}{R_{PW}} + \frac{T_H(\tau) - T_W(\tau)}{R_{HW}}. \quad (3)$$

where M is mass, c is the specific heat or effective specific heat (for PCM), T is the average temperature, $\dot{m}_H$ is the HTF mass flow rate, $\dot{Q}_A$ is the transfer of thermal energy to/from the ambient. In the notation, subscript have the following meaning: *H* for HTF, *P* for PCM, *W* for capillary wall, *1* for inlet parameter, *2* for outlet parameter.

In the above equations, heat transfer from the liquid zones and the ambient was treated as purely convective. In the remaining zones only conductive heat transfer was assumed. Thermal resistance between the PCM and tube wall and between the tube wall as well as HTF can be written as, respectively:

$$R_{PW} = \frac{\delta_P}{\lambda_P A_P} + \frac{\delta_W}{2\lambda_W A_{PW}} \quad (4)$$

$$R_{HW} = \frac{1}{\alpha_H A_H} + \frac{\delta_W}{2\lambda_W A_{HW}} \quad (5)$$

where δ is the wall thickness, λ is the thermal conductivity, A is the heat transfer area.

The convective heat transfer coefficient for HTF is given by the relation:

$$\alpha_H = \frac{Nu\,\lambda_H}{d} \quad (6)$$

For the analysed geometry, the length of tubes is much larger than their inner diameter and the Reynolds number is very low. Thus, it was assumed after [11] that Nusselt number value in thermal, fully developed laminar flow can be taken as $Nu = 4.364$.

3.5 Test procedure

In the experimental investigations, the ramp type HTF inputs temperature changes were used. This type of aperiodic test signal was selected to obtain precisely defined dynamic input conditions. The experimental equipment allow to exactly control changes on the HTF inlet temperature in wide range of rates set in.

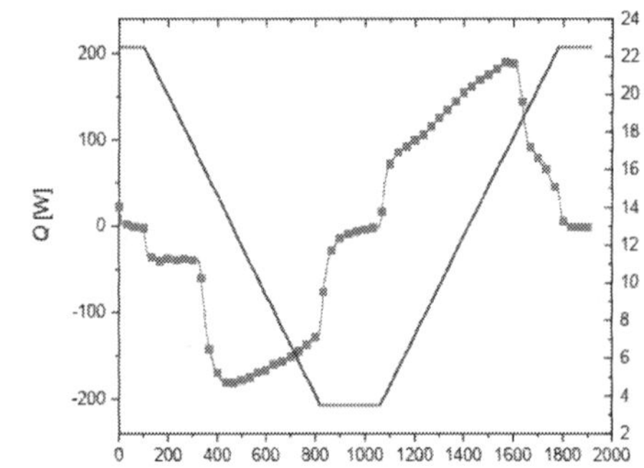

Fig. 7. Experimental data of a cooling and heating process for the case with constant T_{set} slope equal to 39.9 min/K: T_{set} – set inlet temperature of HTF (solid black line), $\dot{Q}$ – heat flow rate (symbol red line).

The run of test procedure is presented in Fig. 7 which shows time profiles of the temperature T_{set} and transferred heat flow rate $\dot{Q}$. The initial temperature was set at 22.5°C. After reaching the steady-state condition, system goes into the cooling state at a constant rate of T_{set} change. The final temperature of this process was 3.5°C. After reaching the steady-state condition, system goes into the heating state at the same constant rate of T_{set} change as previously. The test procedure ended when the set temperature reached 22.5°C again. The amplitude of the input signal and the liquid temperature at the storage inlet were selected to include the phase transition interval of the used PCM.

Rate of heat losses to the ambient $\dot{Q}_A$ was estimated experimentally as a function of differences between ambient and HTF temperature. Calculated overall heat transfer coefficient is equal to $k = 1.272$ W/(m^2·K). Heat losses have been taken into account in experimental data analysis.

As it was mentioned earlier, the control system was able to follow accurately the set profile of the temperature. It can be seen in Fig. 8 where the difference between the set and the measured temperature of HTF is shown. During test run test the highest differences of these temperatures occurred when the system changed working mode (cooling /heating /stable-state). During the cooling highest differences of these temperatures occurred when the phase change process has begun, during heating when it has ended.

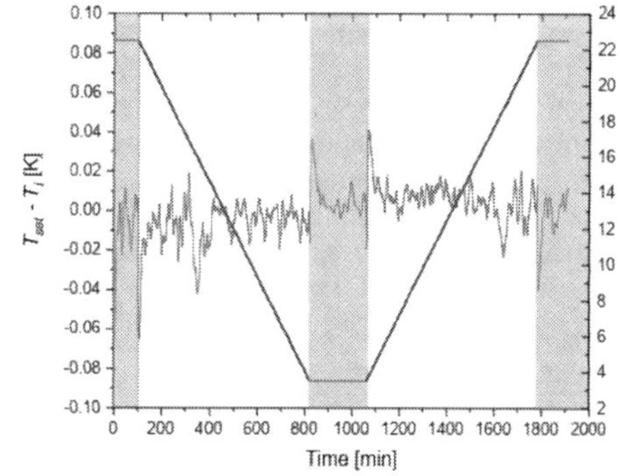

Fig. 8. Difference between the set temperature T_{set} and the measured temperature T_{H1} (solid red line) as a function of time – experimental data of a cooling and heating process for the case with constant T_{set} slope equal to 37.9 min/K.

4 Results and discussion

The storage tests were carried out for low (12 h) and high (6 h) speeds of charging and discharging process. It corresponds to the inlet temperature T_{set} constant slope equal to 37.9 min/K and 18.95 min/K respectively. For both cases, enthalpy as a function of HTF temperature is presented in Fig. 9. It can be seen for each case that, the phase change temperature ranges during cooling and heating are different. It means that there is an important hysteresis effect for these type of PCM. On the characteristics, the part where the PCM is in a solid state can be easily distinguished, it is above the temperature of about 17°C. Enthalpy profile analysis does not allow to determine the bottom limit phase transition temperature unambiguously. This may be due to the fact that experimental tests are carried out for the whole container, not for PCM only. Another reason may be in the specific properties of the material used.

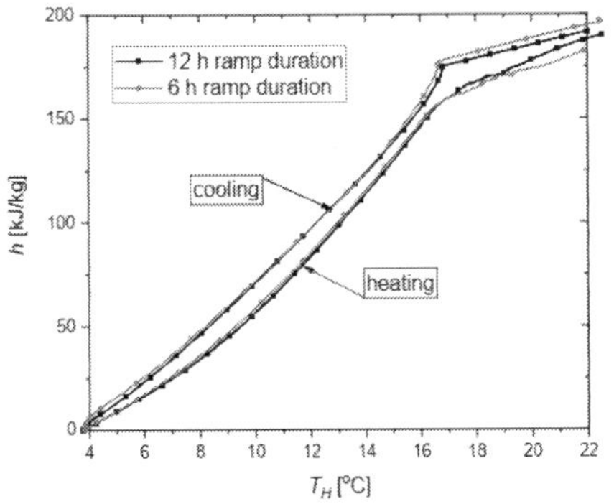

Fig. 9. Enthalpy as a function of average HTF temperature for long (12 h) and short (6 h) durations of heating and cooling process.

To analyse the impact of dynamics characteristics of the charging/discharging process on the heat storage, effective specific heat c distributions as a function of average HTF temperature T_H were calculated. For the experimental data, specific heat is defined as:

$$c = \frac{h^i - h^{i-1}}{T_H^i - T_H^{i-1}} \tag{6}$$

where i is the samples number.

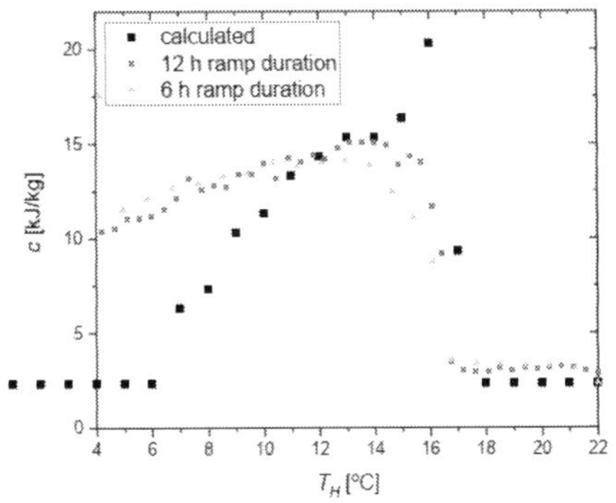

Fig. 10. The calculated (Fig. 6) and measured effective specific heat c distribution as a function of average HTF temperature T_H for the cooling process.

For the cooling (charging) process, comparison between the calculated (see Fig. 6) and measured values is presented in Fig. 10. It can be seen in the liquid state region (temperature range between 22°C and 17°C) that the measured effective specific heat is slightly higher than the one calculated with manufacturer data for PCM properties. Its value is independent of the speed of temperature change. When the phase change process begins it can be seen that charging process is different for the two measured cases. Namely, in the temperature range between 17°C and 12°C the effective specific heat values are smaller for the higher rate of temperature changes. This is related to the smaller amount of accumulated heat during cooling process. Comparing with the calculated values, there is no increase value of c for the temperature of 16°C. Below the 12°C, the values of c are similar for both ramp slopes. In this range calculated and experimental values of the effective specific heat differ significantly.

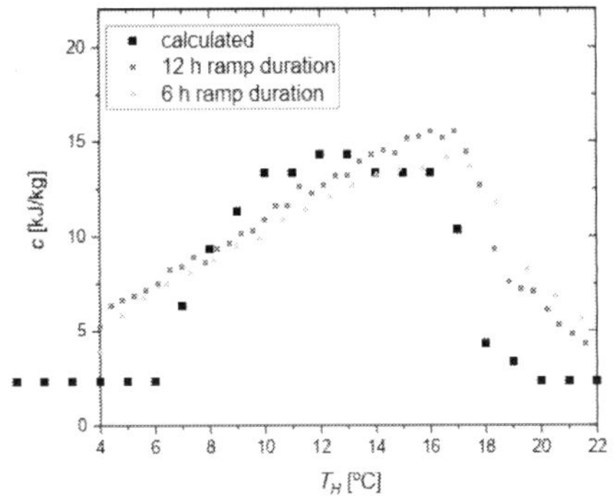

Fig. 11. The calculated (Fig. 6) and measured effective specific heat c distribution as a function of average HTF temperature T_H for the heating process.

For the heating (discharging) process comparison between calculated and measured values is presented in Fig. 11. As it was mentioned above, it is not possible to show the interval in which the material is in a solid state. It can be seen that in the range between 4°C and about 9°C that the measured effective specific heat is significantly greater than the calculated one. The same relationship holds for temperatures above the 14°C. In the temperature range between 9°C and 14°C, the measured c value is higher than the calculated. It can be seen that effective specific heat for faster (18.95 min/K) discharging process is lower up to the temperature equal to 14°C. Above this temperature c is higher for faster discharging process.

It can be seen that for the both process directions, the difference between the calculated and experimental value of effective specific heat for LTES is significant. In comparison, the impact of ramp duration on this difference is significantly smaller.

5 Validation of theoretical model

One of the important tasks of the present work was to find the method to refine and validate the lumped parameter model describing thermal energy storage. This method should be useful for TES with phase change material. The result of the model validation and the tips for its validation are presented below.

The input data for modelling process come from the experimental investigation. They are: inlet temperature of heat transfer fluid T_{HI}, mas flow rate $\dot{m}_H$, effective specific heat (Fig. 6) and geometrical and thermo-

https://doi.org/10.1051/e3sconf/20187003005

physical properties of the LTES container. Experimental and theoretical result of heat flow rate for the case with 12 h ramp duration is presented in Fig. 12. The directions of the charging/discharging process are indicated by arrows. It can be seen that the main difference of these profiles is visible at in the end of the cooling process (range between 8°C and 3.5°C) and at the end of the heating process (range between 18°C and 22.5°C). Further analysis here will focus on the range from 4°C to 22°C, omitting the transient states at the ramp lead in and out.

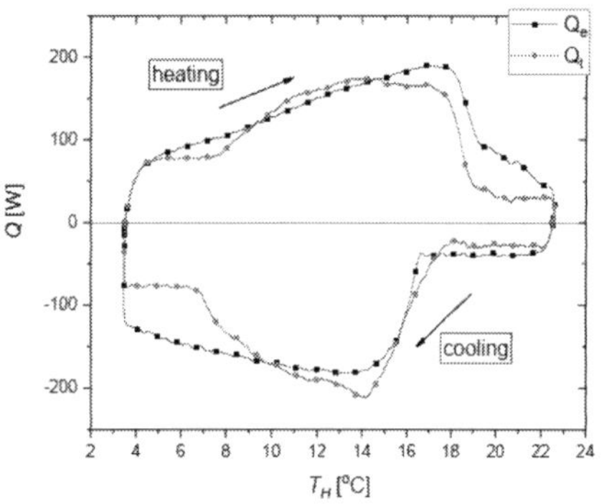

Fig. 12. The calculated ($\dot{Q}_t$) and measured ($\dot{Q}_e$) heat flow rate as a function of average HTF temperature T_H for the case with 12 h ramp duration.

The main parameter obtained from the model is the outlet temperature of heat transfer fluid T_{H2t}. To assess accuracy of the modelling process, the difference between the measured T_{H2e} and theoretical T_{H2t} temperature is calculated. They are presented in Fig. 13 and Fig. 14 for both cases investigated experimentally.

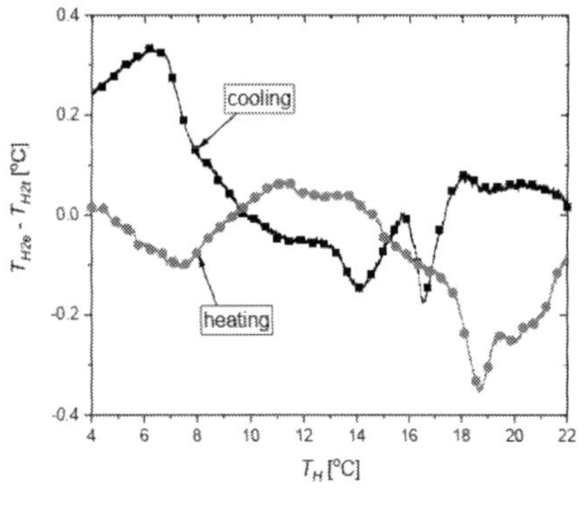

Fig. 13. Difference between the measured T_{H2e} and calculated T_{H2t} temperature as a function of average HTF temperature T_H (cooling and heating process for the case with constant T_{set} slope equal to 37.9 min/K).

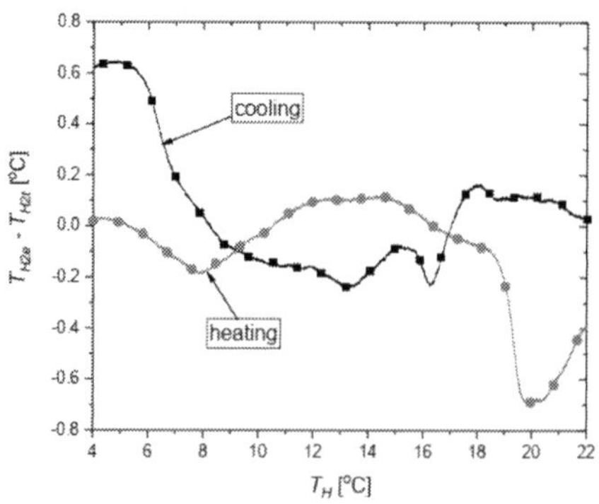

Fig. 14. Difference between the measured T_{H2e} and calculated T_{H2t} temperature as a function of average HTF temperature T_H (cooling and heating process for the case with constant T_{set} slope equal to 18.95 min/K).

In analogy to the results shown in Fig. 12, it can also be seen that for both cases the main difference of these profiles is visible at the end of cooling process (range between 8°C and 3.5°C) and at the end of heating process (range between 18°C and 22.5°C). The incompatibility between the calculated and measured results increased by about 50% for the higher rate of cooling/heating process, compared to case with the lower rate.

6 Conclusions

In this study, the experimental recognition procedure of thermal energy storage dynamic characteristics was presented for the loading and unloading process. The constant rate of inlet temperature change (i.e. ramp) was chosen as a test signal. Two cases with 12 h and 6 h ramp duration were investigated. The tested storage had a form of a cylindrical tank of 40 dm^3 volume in which a coil made of pipes with an external diameter of 3.35 mm was immersed in the PCM. The PCM used was commercially available RUBITHERM RT15. The experimental test stand, the measurement and control systems and test procedure were described. Also the mathematical 0D model of the thermal energy storage was proposed.

The main conclusions of experimental investigation and theoretical analysis are:

- the applied measurement procedure allowed to perform measurements with high accuracy;
- the use of manufacturer's data leads to large errors in the model results;
- the temperature range that could be investigated in the experiment did not allow to analyse the PCM properties in the solid state;
- for the two investigated ramp durations, the differences in storage heat capacity are small but noticeable;
- thermal dynamics of the investigated storage is different for the cooling and heating process.

As a future work, it is planned to:

- prepare own, baseline characteristic of effective specific heat capacity for quasi static states (i.e. much longer ramps or unit steps);
- investigate more extensively the changes in the dynamics of the charging/discharging process, using shorter times (which requires a replacement of the refrigeration system on the experimental stand);
- investigate more extensively the PCM characteristics at low temperatures (requires a modification of the refrigeration system);
- test partial charging and discharging;
- test the input signals other than the ramp (also periodic);
- validate the theoretical model in a wide range of the storage charging and discharging speeds.

References

1. Y. Sun, S. Wang, F. Xiao, D. Gao, Peak load shifting control using different cold thermal energy storage facilities in commercial buildings: A review. Energy Conversion and Management **71** (2013), 102–114
2. J. Karwacki, R. Kwidziński, A. Tomaszewski, A. Cenian, J. Großman, S. Matthes, M. Lackowski, *Usage of PCM thermal energy storage in the temperature stabilizaton system of bioreactor for algae cultivation* [in Polish]. Proc. XLIX Dni Chłodnictwa (2017), 107–125
3. J. Karwacki, R. Kwidziński, Theoretical analysis of a latent thermal energy storage system with an adsorption chiller. Polska Energetyka Słoneczna, No 1-4/2016, 59–64
4. J. Bony, S. Citherlet, Numerical model and experimental validation of heat storage with phase change material. Energy and Buildings **39** (2007), 1065–1072
5. R. Isermann, M. Münchhof, *Identification of Dynamic Systems. An Introduction with* Applications (Springer, Berlin, 2011)
6. L.F. Cabeza, C. Barreneche, I. Martorell, L. Miró, S. Seri-Bey, M. Fois, H.O. Paksoy, N. Sahan, R. Weber, M. Constantinescu, E.M. Anghel, M. Malikowa, I. Krupa, M. Delgado, P. Dolado, P. Furmanski, M. Jaworski, T. Haussmann, S. Gschwander, A.I. Fernández, Unconventional experimental technologies available for phase change materials (PCM) characterisation. Part 1. Thermophysics properties. Renewable and Sustainable Energy Reviews **43** (2015), 1399–1414
7. T. Nuytten, P. Moreno, D. Vanhoudt, L. Jespers, A. Solé, L.F. Cabeza, Comparative analysis of latent thermal energy storage tanks for micro-CHP systems. Applied Thermal Engineering **59** (2013), 542–549
8. M. Medrano, M.O. Yilmaz, M. Nogués, I. Martorell, J. Roce, L.F. Cabeza, Experimental evaluation of commercial heat exchangers for use as PCM thermal storage systems. Applied Energy **86** (2009), 2047–2055
9. A. Gil, E. Oró, L. Miró, G. Peiró, A. Ruiz, J.M. Salmerón, L.F. Cabeza, Experimental analysis of hydroquinone used as phase change material (PCM) to be applied in solar cooling refrigeration. International Journal of Refrigeration **39** (2014), 95–103
10. M. Grozdek, B. Halasz, T. Ćurko, V. Soldo, *Experimental investigation of an ice-bank system performance*. 3rd IIR Conference on Thermophysical Properties and Transfer Processes of Refrigeration, (2009)
11. H.D. Baehr, K. Stephan, *Heat and Mass Transfer* (Springer, Berlin, 2006)

Technical problems with compression units in mechanical vapour recompression systems

Władysław Kryłłowicz[1], *Krzysztof* Kantyka[1,*], *Włodzimierz* Szewczyk[2], and *Paweł* Pełczyński[2]

[1]Institute of Turbomachinery, Lodz University of Technology, Poland
[2]Institute of Papermaking and Printing, Lodz University of Technology, Poland

Abstract. The evidence gathered during works on two prototype heat recovery systems is summarised. The first installation is a low-temperature vapour recompression system, whereas the second is a high-temperature (from an industrial viewpoint) installation. Vapour-contaminated air is the working medium. The authors focused their attention on machine issues mainly. All compressors described below were designed at the Lodz University of Technology.

1 Introduction

By "mechanical vapour recompression" we understand a special compressor heat pump working in an open loop. A comparison of both systems, a closed system in the case of the classic heat pump and an open one, namely vapour recompression, is shown in Fig. 1.

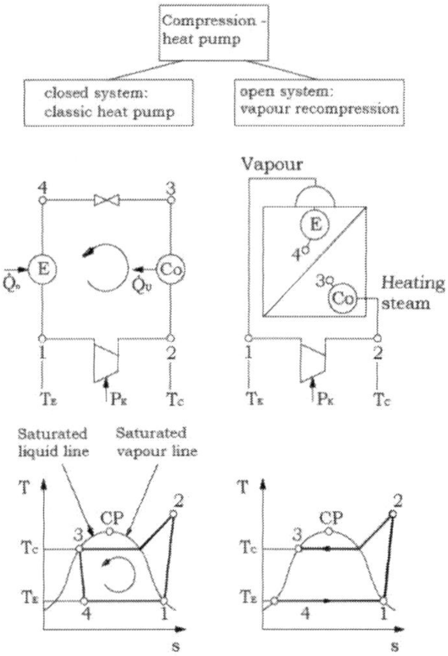

Fig. 1. Comparison of the classic heat pump (left) and the vapour recompression cycle (right)
E – evaporator, Co – condenser.

In the closed cycle, a refrigerant (normally used as the working medium) is compressed from state **1** to state **2**, where P_k is the compressor power consumption. Next, the medium is condensed in a heat exchanger (conversion from state **2** to state **3**), and the useful heat $\dot{Q}_u$ is delivered to the consumer. Then, the working medium is expanded in a throttling valve **3**–**4** to the pressure corresponding to saturation temperature T_E. In the last stage of this cycle, the medium is evaporated (**4** to **1**), due to the fact that the heat $\dot{Q}_0$ coming from the waste heat source is delivered to the system. Thus, a compressor heat pump is a device, which converts the low-temperature heat $\dot{Q}_0$, through the energy input delivered to the compressor, into useful heat $\dot{Q}_u$ at a higher temperature. The vapour recompression operates in the same way, but after compression **1**–**2** and condensation **2**–**3** at temperature T_C, the condensate leaves the cycle. Thus, a throttling valve is not needed.

In fact, in numerous branches of the Polish industry, such as the sugar industry, the food industry or in brewing, vapour recompression installations are met. In the majority of cases, however, they are based not on mechanical compression but on compression in steam-jet ejectors, that is to say, on "thermocompression".

2 Technical problem in a large vapour recompression installation

The Ordering Party commissioned a development of the conceptual design, and then a design of a technical vapour compression unit at pressure p_A = 1.2 bar upping the minimal pressure 3.2 bar + 4% (maximal pressure 3.328 bar). The compression unit was to operate with a rectification column in the way depicted in Fig. 2.

In column 1 (designed for low pressure losses), the input (a mixture of nitrous gases) is separated into fraction Y at the head and fraction X at the sump. By means of mechanical energy, the vapours are recompressed by compressor 2 up to the pressure and temperature level necessary for heating the sump

*Corresponding author: krzysztof.kantyka@p.lodz.pl

evaporator 3. Surplus energy is eliminated in reflux cooler 4.

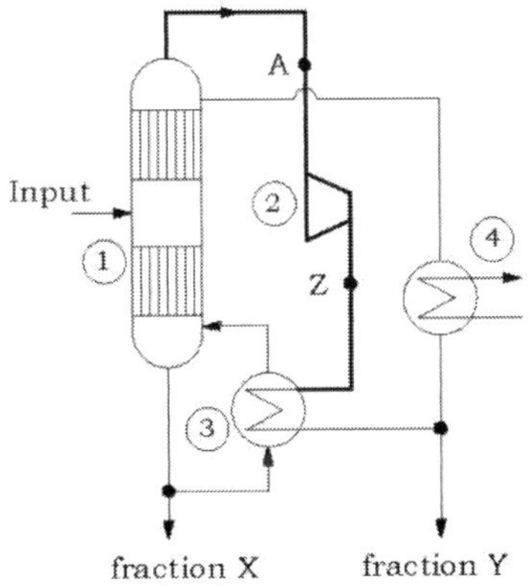

Fig. 2. Rectification column unit with vapour recompression. Point ***A*** and ***Z*** denote suction and compression of the compression unit, respectively. 1 - column, 2 - compressor, 3 – sump evaporator, 4 – reflux cooler.

Water vapour, contaminated with SO_2, NO and sulphuric acid mist (up to 30 mg/Nm^3), in the quantity of 15.6 ton per hour ($\dot{m}_{vapour} = 4.333\ kg/s$), is the working medium.

The required pressure ratio of the compressor is $\Pi = p_Z/p_A = 2.67$. Then, on the assumption that the inner isentropic efficiency of the compressor is at a level of 0.76, the circumferential velocity of the impeller wheel will exceed $u_2 = 550$ m/s, which is of course impossible to achieve within one compression stage. Thus, the compressor has to have two stages. Taking into account the technological conditions, a scheme of the compression unit, shown in Fig. 3, was developed.

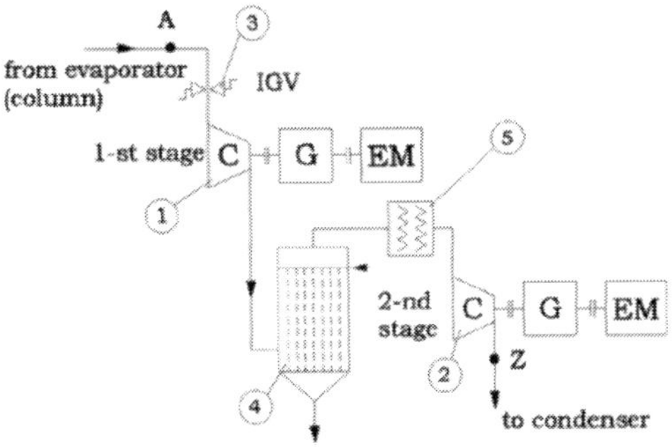

Fig. 3. Scheme of the vapour compression unit: 1 – first compression stage, 2 – second compression stage, 3 – inlet guide vanes, 4 – scrubber-type intercooler, 5 – water mist separator, G – gear box, EM – electric motor.

Fig. 4 depicts a compression curve within the enthalpy - entropy coordinates.

After the initial analysis was completed, it was assumed that the pressure ratio of the first stage of the compressor would be $\Pi_I = 1.78$, whereas for the second stage it would be $\Pi_{II} = 1.57$. The saturation temperature at the compressor inlet (pressure 1.2 bar) is equal to 104.81°C (377.96 K). Due to the considerable length of the pipeline, the pressure loss was estimated at $\Delta p = 0.01$ bar. Hence, the suction pressure at the first stage is equal to $p_{A_I} = 1.192\ bar$.

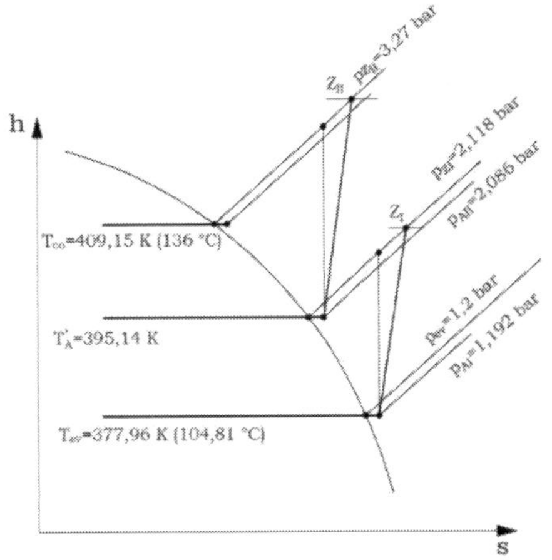

Fig. 4. Compression curve within the $h - s$ coordinates .

It was assumed that the pressure drop in the wet intercooler would be $\Delta p = 1.5\% \cdot p_{Z_I}$, that is to say, 0.0318 bar, thus the suction pressure in the second stage equals $p_{A_{II}} = p_{Z_I} - \Delta p = 2.086\ bar$.

The design assumptions for compression stages were as follows:

1) Semi-open impellers due to strength issues.
2) Welded wheels, consisting of forged hubs and pressed-sheet blades.
3) The first-stage impeller wheel with radial blades, the second-stage impeller wheel with backward-bent blades.

In Table 1, the most important parameters of both stages of the compressor are listed.

Table 1. Basic parameters of the stage.

Stage		1	2
outer wheel diam. D_2	mm	496	400
outlet blade angle β_2^*	°	90	63
circumferential velocity u_2	m/s	403	388
rotational speed n	rpm	15532	18527
isentropic efficiency	%	75.5	76.0
power consumption	kW	684.2	545.6

In Fig. 5, a longitudinal section of the first compression stage is presented.

Inlet guide vanes were installed upstream of the impeller inlet for control. Both the first and second stages have vaned diffusers.

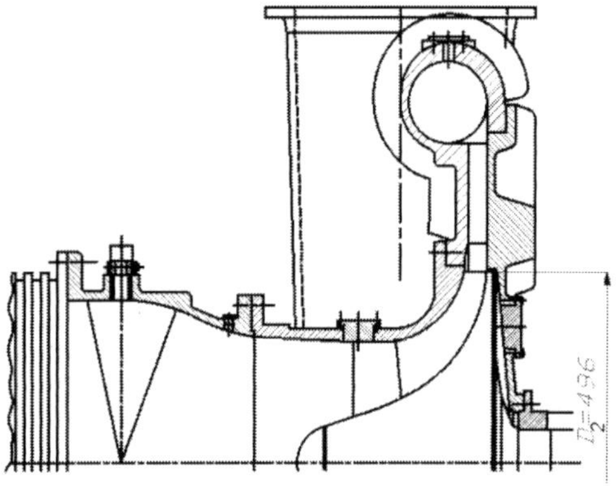

Fig. 5. Longitudinal section of the first compression stage.

The efficiency of heat pumps, including vapour recompression systems, is often characterised by means of the so-called "coefficient of performance" (*COP*). This is defined as a ratio of the useful heat stream $\dot{Q}_u$ and the compressor power P_k:

$$COP = \frac{Q_u}{P_k} \quad (1)$$

In our case:

$$COP = \frac{r \cdot \dot{m}}{P_{el_1} + P_{el_2}} = \frac{2154.2 \cdot 4.333}{684.2 + 545.6} = 7.59 \quad (2)$$

where r denotes evaporation heat at temperature T_c.

Thus, the vapour recompression system under study is very beneficial from the energy generation point of view.

3 Design of a high-temperature compressor system

The Ordering Party formulated the following requirements referring to the blower parameters:

- suction pressure: $p_A = 1.01325$ bar
- suction temperature: $T_A = 673$ K (400°C)
- suction volume flow rate: $\dot{V}_A = 1\ m^3/s$
- static pressure ratio: $\Pi_{A\text{-}Z} = 1.4$

Vapour-contaminated air was the working medium. The fact that the blower was a prototype limited considerably the range of technologies that could be applied . The welding of steel sheets was assumed as the only applicable manufacturing technology.

If we assume the gas constant of the compressed medium to be $R = 302$ J/(kg K) (the range from $R_{min} = 292.6$ up to 312.0 J/(kg K) was given), it turns out that for the assumed inner isentropic efficiency $\eta_i = 0.65$ and radial impeller blades, the required circumferential velocity of the impeller wheel will exceed $u_2 = 370$ m/s. For such high circumferential velocity, it is impossible to have a closed (i.e. with a shroud) structure of the impeller wheel. The reason lies in the low strength characteristics of heat-resistant steels. Conversely, high-resistant structural steels usually applied in the manufacturing of compressors have too low a yield point and creep resistance at temperatures exceeding 400°C as well. This means that it is impossible to build a one-stage machine from a technological point of view.

Finally, the following design concept was assumed:

- the compressor unit would consist of four twin blowers, operating serially,
- each blower would be built with a high-speed flange-type engine (with external cooling), driven by an inverter.

A scheme of a single blower in the above-discussed unit is presented in Fig. 6.

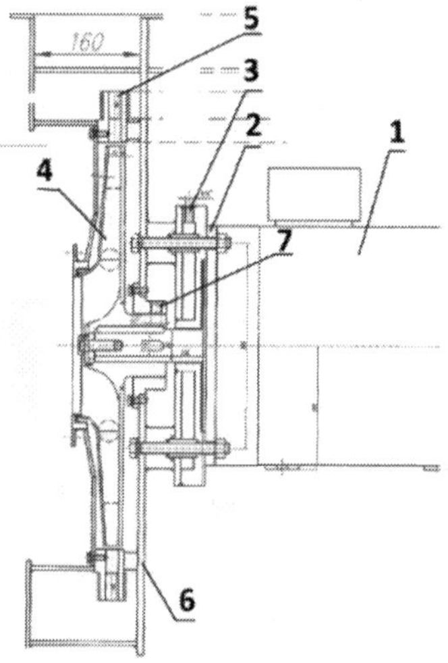

Fig. 6. Longitudinal section of a single blower.

As mentioned above, the most important element in this design is a high-speed electrical motor with external cooling 1. The main body of the blower is suspended on flange projection 2 of the electric motor. The body is mounted with four screws. The first element in the main body is a water chamber, which serves to insulate the electric motor. Water chamber 3 is supplied from the top with water, which flows out at the bottom. Both the pipelines – inlet and outlet – are provided with throttling valves. Impeller wheel 4 (shown as an example) is directly mounted on the motor shaft and fixed axially with a screw. The impeller hub is cooled with air supplied from outer installation 7. Hub cooling will increase the material strength and will eliminate the risk that the impeller wheel will loosen radially (which will be followed by an increase in the vibration level).

Due to material strength issues, it was assumed that the impeller wheel would have blades with an outlet angle $\beta_2^* = 90$°C (outlet radial blades). As the stream angle in the absolute frame of reference is also negligible, vaned diffuser 5 has to be used. Welded diffuser vanes contribute to the strength of the machine as they close the collective volute by force. The main body of the blower is heat insulated with a mineral wool matt.

It was assumed that the impeller wheel would be made of 316L steel, which at its operating temperature has a Young modulus E=160 GPa and a yield point of 100 MPa. The scarcity of this material, however, and its exceedingly low mechanical properties and difficulties with welding technology compelled the designers to use Avesta 253 MA steel. Unfortunately, in this case, the design of the whole structure and the impeller wheel had to be altered, as shown in Fig. 7.

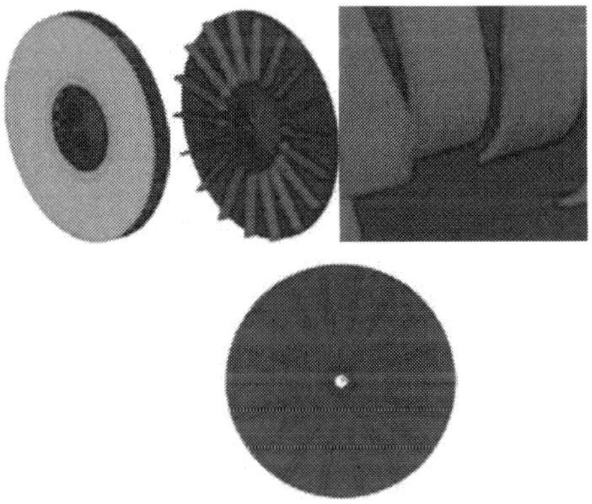

Fig. 7. 3D model of the impeller wheel in its final version.

Strength calculations were carried out for that model and their selected results are depicted in Fig. 8.

a)

b)

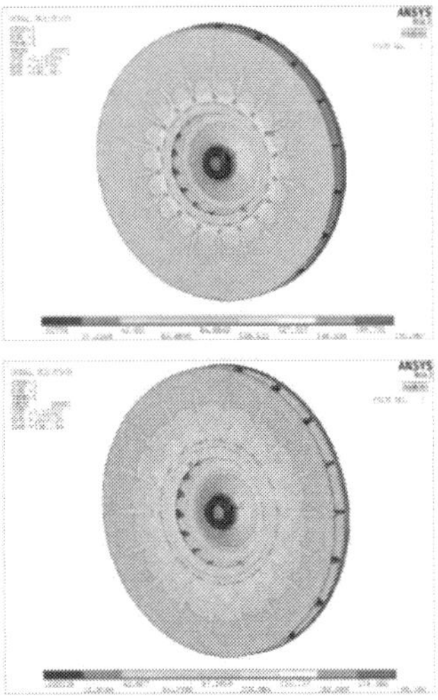

Fig. 8. Distribution of reduced stresses in the impeller shroud, inner view:
a) n = 4500 rpm, ($\sigma_{red\,max}$ = 151 MPa)
b) n = 5000 rpm. ($\sigma_{red\,max}$ = 155 MPa)

The stress level obtained in the calculations does not exceed the allowable value.

The blower thus designed was launched in a Polish paper-making company, in the context of the GEKON programme [9].

4 Summary

On the basis of the two different designs of compressors and blowers presented in this paper, several remarks on the optimisation methods of such machines can be formulated. Kleinhenz [2] presented the requirements referring to the characteristics of machines that compress water vapour. Following an analysis of the above-mentioned designs, it is possible to state that these machine units represent entirely different technological advancements. "Low-temperature" compressors are cutting-edge devices, whereas a set of "high-temperature" blowers falls far behind as far as the technological level is concerned.

Both compressors, however, although different, accomplish their task. It should also be emphasised that they are very cheap solutions in comparison to similar machines available on the European market.

References

1. M.A. Darwish, *The heat recovery thermal vapour – compression desalting system. A comparison with other thermal desalination systems* (Applied Thermal Engineering, Volume 16, Issue 6, pp. 523-537, 1996)
2. K. Kleinhenz, *Forderungen on Bruedenverdichter bei Wasserdampf und bei anderen Medien* (VDI – Berichte, Nr 442, 45)
3. W. Kryłłowicz, *Teoria i praktyka modernizacji sprężarek promieniowych (Theory and praxis of centrifugal compressors modernisation)* (Monograph, Technical University of Lodz, 2013)
4. M. Lucas, B. Tabourier, *The Mechanical vapour compression process applied to sea water desalination a 1500 ton/day unit installed in the nuclearpower plant of Flamanville,* (Desalination, Volume 52, Issue 2, pp. 123-133, 1985)
5. K.H. Luedke, *Process centrifugal compressors* (Springer – Verlag, 2004)
6. J.A. Winchester, C. Marsh, *Dynamics and Control of Falling Film Evaporators with Mechanical Vapour Recompression Engineering Research and Design* (Volume 77, Issue 5, pp. 357-371, July 1999)
7. R. Wimmerstedt, *Mechanical Vapour Recompression applied to Separation Processes* (J. Separ. Proc. Technol. 6, 1985)
8. *Mechanische Brueden – Compression* (VDI-GET, Braunschweig, 1987)
9. GEKON programme 2/05/268278/22/2016, *Increasing wastepaper recycling, decreasing energy consumption and improving drying process efficiency in the papermaking machine through an application of micro-nozzle systems)* (in Polish, final report of the project, 2016)

Evaluation of energy effects of the heat distribution network modernization

Tomasz Kujawa[1], and *Aleksander* Stachel[1,*]

[1]West-Pomeranian University of Technology, Szczecin, Department of Heat Engineering, al. Piastów 17, PL-70-310 Szczecin, Poland

Abstract. The work concerns the estimation of energy effects of modernization of a part of an existing heat distribution main, which is part of a system that supplies heat for heating and domestic hot water preparation. The energy effect was defined as the difference between heat losses in the installation previously operated, developed in the technology of traditionally insulated pipes placed in dedicated underground channels and newly built installations made of pre-insulated pipes located directly in the ground. In the paper, in a synthetic way, the method of calculation is presented, as well as the accompanying assumptions. Two basic approaches to determine the heat losses are discussed. The first one was based on detailed calculations based on real operational data, whereas the second one used the indicative method recommended by the National Fund for Environmental Protection and Water Management (NFOŚiGW). The results obtained on this basis show that the modernization of the installation returns measurable energy effects in the form of reduced heat losses, which translates into the efficiency of its operation and operation of the source. At the same time, they show that the size of obtained energy effect depends on the applied methodology of calculations.

1 Introduction

One of the expectations related to the modernization of heating systems is the estimation of energy effects (sometimes also the ecological ones) resulting from the applied technical solutions. The element of the assessment is usually the reduction of energy consumption in the installation after modernization. In the case of heating networks, this is associated with determination of the amount of heat losses associated with the transmission of hot water / steam.

The paper presents the results of calculation of the energy effect of the modernization of a 2.5 km section of the heat transmission installation, which is part of the heating system that supplies heat for the purpose of heating and utility hot water preparation. After more than four decades of use, the condition of some of the main ducts was rated as very poor (Fig. 1). That is due to the following reasons:
- significant corrosion of pipes,
- significant deterioration of thermal insulation, including its local absence;
- mechanical damage to the construction of heating ducts, and others.

The effect of this are excessive heat losses resulting from the installation wear condition, which is affected by a long-term use, including extremely unfavorable conditions (moisture influence), as well as the method and quality of its assembly. This is accompanied by significant losses of the network water, resulting from the leaks caused by progressive corrosion and perforation of pipes. The condition and excessive heat losses have been the reason for carrying out a comprehensive modernization of the part of installation, consisting in its replacement with new pipelines, made in accordance with the applicable insulation standards, along the new route.

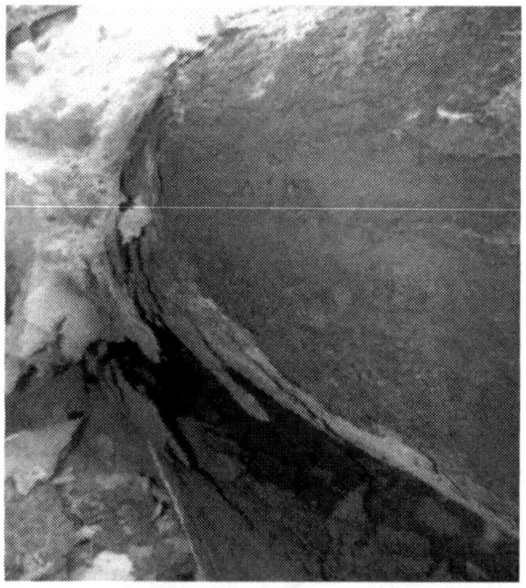

Fig. 1. Technical condition of the transmission installation.

2 The object of analysis

The heat transmission line, which was scheduled for modernization, connecting the heat source with the

* Corresponding author: aleksander.stachel@zut.edu.pl

number of recipients, was built in the 1970s. In suburban areas it is made as a surface installation, while within the city borders it has been laid underground. Each of the constituting pipelines, i.e. both the supply and return pipelines, are arranged in separate ducts located next to each other and made of prefabricated elements (Fig. 2).

The pipelines were made of steel pipes of nominal diameter Dn-350 and Dn-300, respectively. The pipes were thermally insulated with glass wool mats, protected with asbestos-cement coating. The insulation thickness (glass wool) was 80 mm (Dn-350 pipeline) and 40 mm (Dn-300), respectively. The pipelines were placed on a variable depth in relation to the surface, while the average depth measured to the horizontal axis of the pipes was 1.1 m. An integral part of the line was a set of fifteen heating chambers intercepted by both pipelines [1].

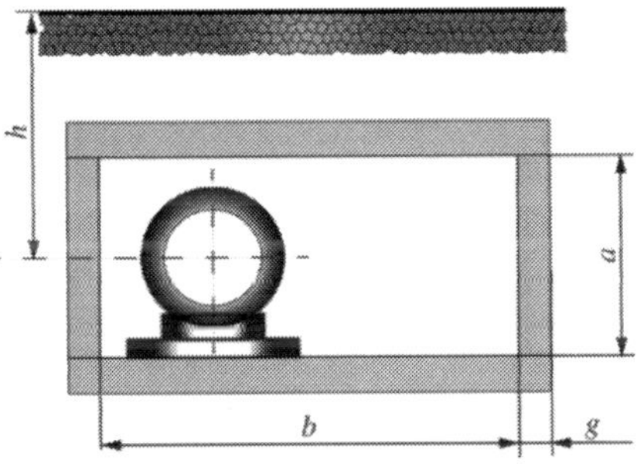

Fig. 2. Construction of the heating pipeline in the ducts.

The new heat distribution line was made in the pre-insulated pipe technology, in accordance with the contemporary standards [2]. The heating line, designed for operation at 125/70°C and operating pressure of 1.6 MPa, forms pipes with a nominal diameter of Dn-350. In contrast to the existing one, it is laid directly in the ground, i.e. without a dedicated duct. It is complemented by three underground heating chambers.

3 Assessment of energy effects of modernization

In the considered case, a measurable effect of modernization should be the reduction of heat losses in the newly constructed heat transmission line, in relation to heat losses in the installation previously operated. Therefore, the absolute energy effect of modernization can be determined from the relation:

$$\Delta Q = Q_{SI} - Q_{SM} \text{ [GJ/a]}, \tag{1}$$

where:

Q_{SI} - heat losses in the existing heat transmission line (before modernization), GJ/a,

Q_{SM} - heat losses in the modernized heat transmission line, GJ/a.

The relative energy effect, defined as the ratio of the absolute effect to the extent of heat losses in the installation before modernization, is defined by the formula:

$$\varepsilon_{em} = \frac{\Delta Q}{Q_{SI}} \cdot 100\% \tag{2}$$

The distribution network supplies heat for central heating (during the heating season) and utility hot water preparation (throughout a year). Thus, the total heat losses include losses during the heating season Q_{sg} and off-season Q_{sl}:

$$Q_s = Q_{sg} + Q_{sl} \text{ [GJ/a]} \tag{3}$$

Heat loss through overall heat transfer in a pipeline of a certain length and within a specified period of time, can be calculated from the expression:

$$Q_{si} = 8.64 \cdot 10^{-5} q_s L_i \tau_s \text{ [GJ/season]} \tag{4}$$

where:

q_s - average unit heat losses from the pipeline in the specified period, W/m,

L_i - length of considered length of the pipeline, m,

τ_s - time of duration of a specified period of heat supply, number of days,

where the average unit heat losses are referred to the supply and return pipelines, respectively.

4 Methodology of calculation of heat losses in pipelines

Determination of the total heat losses from pipelines requires estimation of the unit heat losses for the heating season and off-season, where the problem is to choose the method of their evaluation. Individual heat losses can be determined using the basic relations and general principles of heat transfer available in the literature [3,4,5,6]. The method allows for direct consideration of changes in the thermal conductivity coefficient and insulation thickness as well as the network water temperature. Another way to estimate the unit heat losses is the calculation method proposed by NFOŚiGW, developed for the needs of the Operational Program Infrastructure and Environment [7]. In this method, the unit losses are determined taking into account the so-called factor of losses and the index of deterioration of insulation as a result of the aging process resulting from the long-term use. The influence of thermal conductivity and insulation thickness is taken into account indirectly.

4.1 Calculations of heat losses from pipelines in the no-pass channel

The linear unit heat losses of a single pipeline in the intransitive channel can be determined using the relationship:

$$q_1 = \frac{t_1 - t_k}{R_1} \quad \text{[W/m]} \tag{5}$$

where air temperature inside the channel is given by:

$$t_k = \frac{\dfrac{t_1}{R_1} + \dfrac{t_g}{R_k + R_g}}{\dfrac{1}{R_1} + \dfrac{1}{R_k + R_g}} \quad [°C], \tag{6}$$

where:

t_1 - water temperature in the pipeline, °C,

R_1 - unit thermal resistance of the pipeline, defined as:

$$R_1 = R_{\alpha 1} + R_r + \sum R_i + R_{\alpha 2} \quad \text{[m·K/W]} \tag{7}$$

where the particular quantities denote the following resistances:

$R_{\alpha 1}$ - convective resistance from the fluid to the inner pipeline wall,

$R_{\alpha 2}$ - convective resistance from the external pipeline wall (insulation) to the surroundings (channel),

R_r - heat conduction through the pipeline wall,

$\sum R_i$ - heat conduction through subsequent layers of insulations (sum of resistances).

The thermal resistance of the channel wall and thermal resistance of the soil ($R_k + R_g$) occurring in the equation (6), as well as resistance in the relation (7) were calculated in accordance with the methodology given in [6].

4.2 Calculations of heat losses in pipelines in district heating chambers

In contrast to the main pipelines placed in separate channels, the pipelines in the heating chambers (supply and return ones) are arranged next to each other. The individual heat losses of each of them can be determined from the formulas:

$$q_1 = \frac{t_1 - t_k}{R_1}, \quad q_2 = \frac{t_2 - t_k}{R_2} \quad \text{[W/m]} \tag{8}$$

where air temperature inside the channel is determined by:

$$t_k = \frac{\dfrac{t_1}{R_1} + \dfrac{t_2}{R_2} + \dfrac{t_g}{R_k + R_g}}{\dfrac{1}{R_1} + \dfrac{1}{R_2} + \dfrac{1}{R_k + R_g}} \quad [°C], \tag{9}$$

where:

t_1, t_2 - water temperature in considered pipelines,

R_1, R_2 - unit thermal resistances of pipelines, defined by relations (7),

R_k, R_g - thermal resistance of the channel wall and resistance of the soil, evaluated according to methodology given in [6].

4.3 An algorithm for calculating the heat losses of pipelines laid in the ground

The unit linear heat losses of the pipeline laid directly in the ground are defined by the formulas:
- supply pipeline:

$$q_1 = \frac{(t_1 - t_g)R_2 - (t_2 - t_g)R_o}{R_1 R_2 - R_o^2} \quad \text{[W/m]} \tag{10}$$

- return pipeline:

$$q_2 = \frac{(t_2 - t_g)R_1 - (t_1 - t_g)R_o}{R_1 R_2 - R_o^2} \quad \text{[W/m]} \tag{11}$$

where the thermal resistances of both pipelines (R_1, R_2) are described by the relation:

$$R_1 = R_{\alpha 1} + R_r + \sum R_i + R_g \quad \text{[m·K/W]} \tag{12}$$

where:

R_g - thermal resistance of soil:

$$R_g = \frac{1}{2\pi\lambda_g} \ln \frac{4h}{d_{ij}} \tag{13}$$

R_o - additional thermal resistance for two pipes laid side by side (Fig. 3):

$$R_o = \frac{1}{2\pi\lambda_g} \ln \sqrt{1 + \left(\frac{2h}{b}\right)^2} \quad \text{[m·K/W]}, \tag{14}$$

λ_g - solid thermal conductivity, W/(m·K),

h - depth of location of the pipeline measured from the surface to the horizontal pipe axis, m,

b - depth of location of the pipeline measured from the surface to the horizontal pipe axis, m,

d_{ij} - outside diameter of the pipeline insulation, m.

4.4 An algorithm of calculation of unit heat losses in line with the methodology by NFOŚiGW

The methodology is based on a different method of determining heat losses should be in contrast to that given above. The unit heat loss of both pipelines (supply and return) are determined by the formulas [7]:
– for the heating season:

$$q_s = u \cdot (t_{1sr} + t_{2sr} - 2 \cdot t_s) \quad \text{[W/m]} \tag{15}$$

– outside the heating season:

 https://doi.org/10.1051/e3sconf/20187003007

$$q_l = u \cdot (110 - 2 \cdot t_s) \quad \text{[W/m]} \tag{16}$$

where:

u - heat loss coefficient characterizing the pipeline, W/(m·K):

$$u = a \cdot u_o \tag{17}$$

a - an indicator that takes into account the deterioration of insulation due to ageing, depending on the pipeline's operating time,

u_o - base coefficient of heat loss in the determined on the basis of the nominal diameter pipeline, W/(m·K).

t_{z-sr1}, t_{z-sr2} - average outside temperature in the heating period and outside it, °C,

t_{1sr}, t_{2sr} - average temperature of water in the supply and return pipelines during the heating season, determined on the basis of the regulatory chart, at the assumption that $t_z = t_{z-sr1}$, where t_z - is outer (surroundings) temperature.

t_s - temperature outside the pipeline (for a pipeline located in the ground $t_s = 8$ °C, for the one placed in the duct it depends on the nominal network temperature [8],

In calculations of newly constructed pipelines, the heat loss coefficient u is based on the information taken from the pre-insulated pipe manufacturer.

5 Results of calculations of heat losses

The energy effect of modernization of the part of heat transmission line was determined in two ways. The first one was based on the detailed calculations made on the basis of actual operational data of the existing and modernized installations; the second one used the methodology proposed by the National Fund for Environmental Protection and Water Management (Polish acronym NFOŚiGW).

5.1 Heat losses in the existing heat transmit-ssion line (to be modernized)

Calculations of heat losses concern the supply and return pipelines being 2462 m long each, arranged in two independent channels. They also include pipelines with a total length of 57.5 m (each), passing through a set of 15 heating chambers located along the line. The analysis did not take into account the thermal interaction of channels.

The calculations were carried out in two ways, i.e. in the simplified and exact manner. The first one was based on *average values of the calculation period*, assuming the temperature of water in pipelines in relation to the average monthly external temperatures, taking into account the number of hours of their occurrence in a given month. In the second one - *on real values*, determined for each month when outside temperatures were found in the range - 6°C to + 12°C (with a 1K interval) and on this basis, determining the water temperature.

The necessary for calculating the water temperature in the supply and return pipelines were made available by the operator of the heat source, where it should be noted that they are variable and dependent, among others from weather conditions. Outdoor temperatures, including monthly averages, were adopted for the nearest meteorological station, according to MIiB [8]. The soil temperature was determined for the average pipeline installation depth of 1.0 m. It was assumed that the ground is moderately moist.

The calculations were carried out according to the algorithms given in sections 4.1 - 4.3. The individual components of thermal resistance were determined based on the designs available in the subject literature. The calculations took into account the design insulation thicknesses of the pipelines, correspondingly reduced due to the operating time and unfavorable working conditions.

- *Heat losses in pipelines arranged in intransitive channels.*

The calculations concerned the supply pipeline and the return pipeline with the dimensions given in the Table 1. In the analysis the following thermal conduction coefficients were assumed: for pipes $\lambda_r = 51$ W/(m·K), for insulation $\lambda_r = 51$ W/(m·K) and for heat channel $\lambda_{kan} = 1.7$ W/(m·K). The design insulation thickness was reduced by 10 mm.

- *Heat losses in pipelines arranged in chambers.*

In calculating of heat losses in pipelines passing through the heating chambers, the relationships given in section 4.2 and the data contained in the work [6,7] were used.

Table 1. Dimensions of pipelines and heating ducts of the existing transmission main (to be exchanged).

Quantity	Symbol / dimension		Supply pipeline Dn 350	Return pipeline Dn 300
Pipe internal diameter	D_1	[mm]	340	308
Pipe external diameter	D_2		356	324
Insulation external diameter	D_3		515	404
Shield external diameter	D_4		545	434
Channel height	a	[m]	0.81	0.72
Channel breadth	b		0.82	1.42
Channel wall thickness	g		0.12	0.126
Depth of location	h		1.0	1.0

The results of calculations of heat losses of pipelines situated in the ducts and in district heating chambers are given in Table 2.

Table 2. Results of calculations of heat losses in the supply and return pipelines arranged in channels and heating chambers - obtained by a simplified and exact method [GJ/a].

Pipeline	Exact method	Simplified method
Supply pipeline	4315.2	4269.0
Return pipeline	2179.9	2161.5
Heating chambers: supply and return pipelines	187.5	162.2

5.2 Heat losses in the modernized (new) network

The calculations concerned pipelines made in pre-insulated pipe technology, placed directly in the ground and forming a new transmission main with a length of 2470.4 m, as well as other pipelines with a total length of 16.8 m each, passing through three heat chambers. The pipelines were made of pre-insulated pipes with a nominal diameter of Dn-350 and differ between themselves in the thickness of insulation (supplying pipeline: $d_i / d_z / d_w$ = 344.4 / 355.6 / 520 mm; return pipeline: $d_i / d_z / d_w$ = 344.4 / 355.6 / 500 mm). The given values refer to the inner and outer diameter of the pipes and the outer diameter of the insulation, respectively. The pipelines are located at a distance of 0.75 m from each other and located at a depth of 1.2 m below the surface (Fig.3).

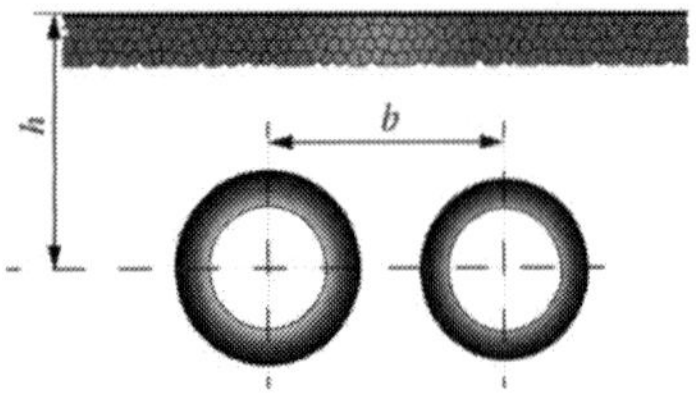

Fig. 3. Location of pipelines pre-insulated in the ground.

Calculations of heat losses in both main pipelines, as well as in those passing through the chambers, were made according to the algorithm given in sections 4.2 and 4.3. The results of calculations are presented in the Table 3.

Table 3. The results of calculation of heat losses in pipelines (supply and return ones) laid in the ground and in 3 heating chambers [GJ/a].

Pipeline	Exact method	Simplified method
- from pre-insulated pipes	3609.8	3574.3
- in 3 heating chambers	29.0	25.0

6 Calculation of the energy effect of modernization

Calculation of heat losses in the existing distribution main and in the newly created one made it possible to determine the energy effect of modernization, using both the exact and simplified methods.

In addition, the energy effect was determined based on the NFOŚiGW methodology [2], using the algorithm given in section 4.4. Apart from the necessary for the calculation data previously specified, based on meteorological data, the duration of the heating season was adopted as τ_s = 242 days (Szczecin) and the average external temperature was determined $t_{z\text{-}sr1}$ = 6.0°C) based on the available long-term data [8]. The results of calculations are given in the Table 4.

Table 4. Energy effect (reduction of heat losses) due to modernization of heating main.

Annual savings	Method		
	simplified	exact	according to NFOŚiGW
ΔQ [GJ/a]	2993.4	3043.8	7218.4
ε_m [%]	45.4	45.5	63.4

Assumptions for the calculation of the heating network before modernization are as follows. Thermal network operation time exceeds 30 years. Adopted on this basis insulation deterioration ratio is a = 1.85. On the other hand, the base heat loss coefficient in the pipeline, determined for the nominal diameters, is equal to u_o = 1.0299 W/(m·K) [6]. Average water temperatures in the supply and return pipelines, determined on the basis of the calculation methodology, are: t_{1sr} = 79 °C; t_{2sr} = 48 °C. The air temperature in the channels, being a function of the maximum design temperature of the supply and return fluid and the nominal pipeline diameter, is equal to t_s = 23 °C.

Assumptions for the calculation of the heating network after modernization are as follows. The newly made installation from pre-insulated pipes with a diameter Dn-350 is laid directly in the ground. According to the calculation methodology, the soil temperature at the pipe depth is set at t_s = 8 °C. The heat loss coefficient of the new pre-insulated pipes is u = 0.509 W/(m·K) [9].

Calculated on the basis of relations (15) and (16) unit heat losses of pipelines in the heating season q_s and off-season q_l allowed to calculate the total annual heat losses, both for the old installation (before modernization) and a new installation. The calculation results are given below (Table 5).

Table 5. The results of calculations of heat losses in the installation before and after modernization, obtained using the WFOŚiGW method [GJ/a].

Before modernization	After modernization
11392.8	4174.4

7 Conclusions

The above results of calculations enable to formulate a number of conclusions important from the point of view of engineering practice.

 https://doi.org/10.1051/e3sconf/20187003007

Calculation of the energy effects of modernization was related to the determination of the overall heat losses, both in the case of old and newly built installations. The energy effect was determined in three different ways. It should be noted that the analysis did not take into account heat losses associated with leakages of heating fluid.

The results of calculation are shown graphically in the Figures 4 – 7.

In the Fig. 4 presented are unit heat losses recorded in the annual cycle for subsequent calculation periods, determined for the supply and return pipelines of the installation before and after its modernization. The chart shows that the unit losses in the installation after modernization are much lower (almost twice) than before modernization. It also follows that the unit losses depend on the outside temperature and on the parameters (temperature) of the heating fluid (water). The lowest values were obtained for the summer season (July, August), whereas the highest for the winter period.

In the Fig. 5 are presented average unit heat losses from pipelines: supply and return installation before modernization (channel) and after modernization (pre-insulated), related to the annual cycle of work.

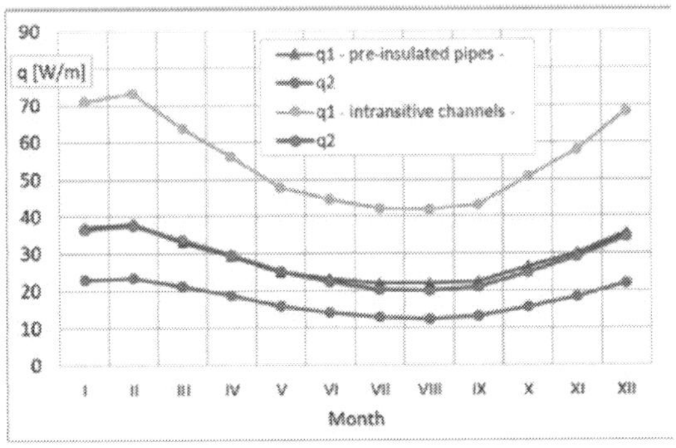

Fig. 4. Unit heat losses from the supply and return pipelines before and after the modernization of the installation, related to subsequent calculation periods (months).

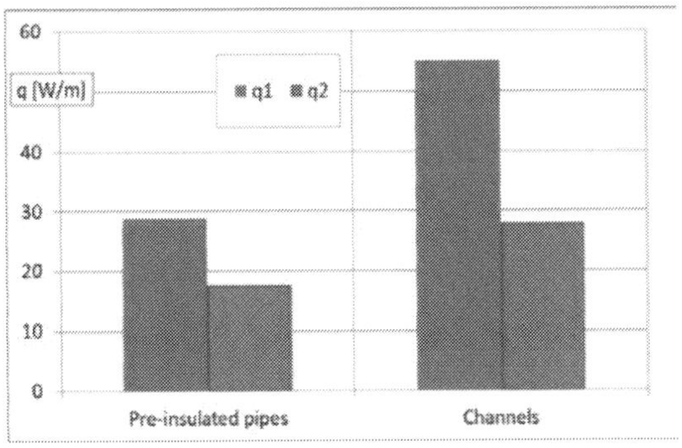

Fig. 5. Unit heat losses recorded in the annual cycle, determined for the supply and return pipelines of the installation before and after its modernization.

Analysis and evaluation of obtained results indicates that there are significant differences in the extent of determined energy effect, depending on the applied calculation method (Fig.6). Similar results were obtained for the methods relying on real data. The results of calculations carried out using the exact and simplified manner are close to each other and do not show significant differences. The effects calculated by these methods differ by 0.1 percentage points. It can therefore be assumed that both methods are equivalent, where the simplified method being easier to use.

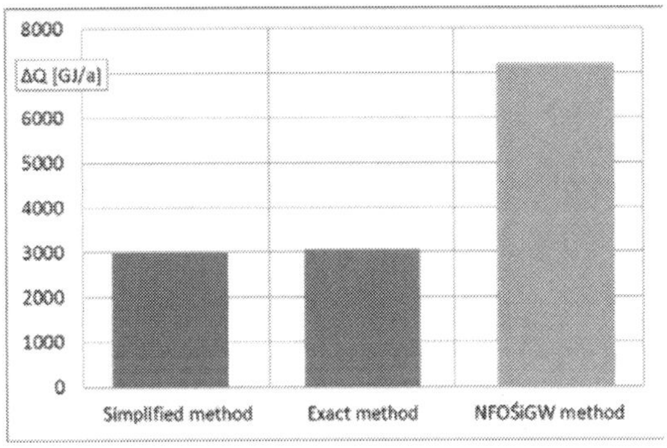

Fig. 6. Energy effect of the modernization of the installation determined by basic (exact and simplified) methods and the NFOŚGW method.

In turn, the heat losses calculated using the NFOŚiGW method exceed the losses determined by the basic calculation method, both by simplified and exact approaches. The relative energy effect of modernization is also larger, based on this method. In the case at hand, the difference reaches several (eighteen) percentage points. This also applies to the calculation of total heat losses, as shown in Fig. 7.

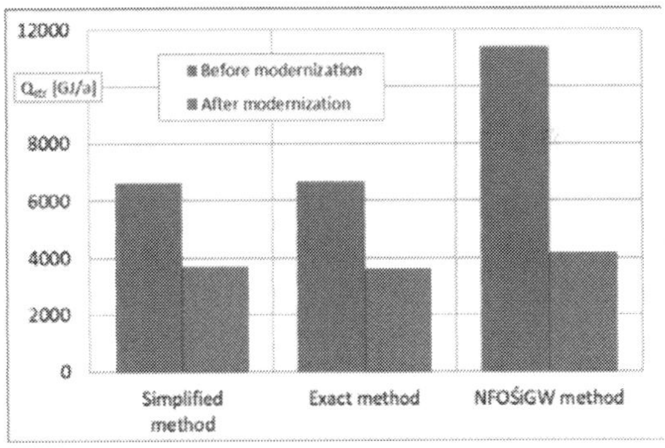

Fig. 7. Total annual heat loss from the supply and return pipelines before and after the modernization of the installation.

It stems from the presented analysis that this is probably the effect of several factors, among others:

- adopted high value of the insulation deterioration factor (multiplier: α = 1.85),
- adopted, and not calculated, value of the heat loss coefficient,
- assumptions of some pipeline operational parameters.

The advantages of the method include simple formulas for calculation of heat losses, both for new as well as the networks operated for many years, which enables quick calculation. The disadvantage is a lack of a direct reference to the condition and thickness of the insulation, which parameters are decisive when determining the unit size of heat losses. It follows that for the sake of accurate analysis the formulas that take into account a number of parameters, such as thickness and condition of insulation, type of heat network, changes in the soil temperature as a function of outside temperature, etc. should rather be used for more accurate calculations.

Regardless of these remarks and irrespective of the size of the calculated energy effect (smaller - larger), the conducted analysis shows that replacing the existing heating main with a new installation, made according to modern technologies and in accordance with the applicable requirements, brings measurable benefits in the form of significant reduction of heat losses.

References

1. Internal unreleased materials of DHE (Department of Heat Engineering, West Pomeranian University of Technology, Szczecin, 2016).
2. Polish Standard PN-EN 13941:2003, Design and construction of heating networks from a system of pre-insulated composite pipes (PKN, Warszawa 2009).
3. W. Kamler, Heating (PWN, Warszawa, 1976).
4. T. Hobler, Heat exchange and heat exchangers (WNT, Warszawa, 1986).
5. J. Madejski, Theory of heat exchange (Szczecin, 1998).
6. A. Tarnowska-Tierling, Thermal equipment for the power plant; part. 1. Pipelines (Publishing House of University of Technology, Szczecin, 1981).
7. Methodology for estimating the reduction of heat losses (pipes). POIŚ 2007-13: Section 9.2 Effective energy distribution. (NFOŚiGW, Warszawa 2010).
8. http://mib.gov.pl/ (access: 21.03.2016).
9. Finpol-Rohr, Designer's guide (Warszawa, 2006).

Nomenclature

d, D	diameter of the pipe	m
Δ	difference	-
ε	efficiency of modernization	%
λ	thermal conductivity	m·K/W
q	unit heat loss	W/m
Q	heat	kJ
L	length of the pipe	m
t	temperature	°C
R	thermal resistance	m·K/W
τ	period of time	days

https://doi.org/10.1051/e3sconf/20187003008

Non-Fourier effects in the thermal protection against high-power ultra-fast laser pulses

Marcin Lenarczyk[1,*], and *Roman* Domański[1]

[1]Institute of Aviation, al. Krakowska 110/114, 02-256 Warsaw, Poland

Abstract. In this paper the heat transfer process between laser beam and a solid body is investigated in the context of the protective materials, which can be used in the military, manufacturing or scientific fields. An ultra-fast high-power interaction regime is being considered. Non-Fourier effects are taken into account. To simulate effects, which can be of significance in the protective materials design, Cattaneo-Vernotte equation is being solved by means of MacCormack explicit second order numerical scheme. Selected results are discussed in detail.

1 Introduction

The rapid development in the laser technology requires an adequate description of the heat transfer processes that are associated with the laser-solid interactions. In the recent years a special attention has been given to the ultra-short laser pulses that are applied in the fields such as laser weapons, manufacturing or confinement fusion research [1,10,13].

A proper modelling of the ultra-fast interactions is crucial for obtaining reliable temperature fields in the considered problems [3,5]. The parameters of the lasers available commercially, used in the military or scientific fields such as pulse duration time and energy determine heat propagation regime which needs to be considered [4,8,9]. In the current paper non-Fourier heat propagation conditions are being considered. The investigation of the effects typical for the wave-like heat propagation and their influence on the temperature distribution in the given system are being studied.

2 Page layout

The classical heat conduction theory based on the Fourier's law assumes that thermal disturbance propagates with an infinite speed. For the most heat transfer problems this assumption is valid. However, when one considers ultra-fast processes of electromagnetic wave and solid interaction thermal propagation speed, as postulated by Cattaneo [2] and Vernotte [12], becomes relevant.

Transient one-dimensional heat conduction equation can be written as follows:

$$\frac{\partial T(x,t)}{\partial t} = \alpha \frac{\partial^2 T(x,t)}{\partial x^2} \tag{1}$$

Eq. (1) is the classic Fourier equation and it assumes the infinite speed of heat propagation. In fact, as the heat conduction is accomplished by successive collisions of the energy carriers (phonons or electrons), the propagation of the thermal disturbance is always at the finite speed.

The paradox of the infinite thermal disturbance propagation speed was addressed by Cattaneo and Vernotte by introducing the thermal relaxation time parameter:

$$\frac{\partial T(x,t)}{\partial t} + \tau_R \frac{\partial^2 T(x,t)}{\partial t^2} = \alpha \frac{\partial^2 T(x,t)}{\partial x^2} \tag{2}$$

The comparison of temperature distributions obtained by using both equations can be seen in Figure 1.

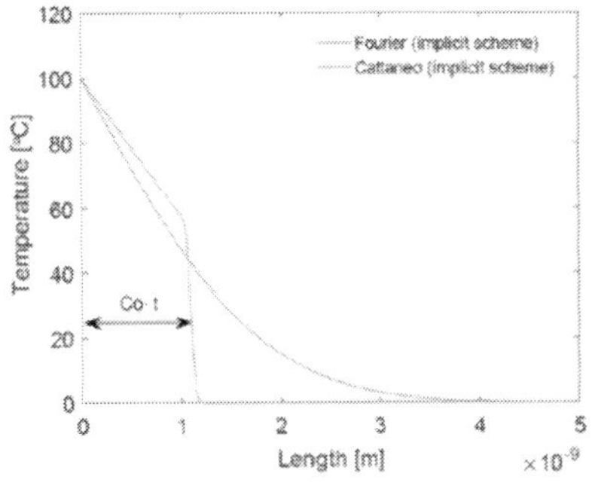

Fig. 1. Comparison of the temperature distributions obtained by solving Fourier and C-V equations.

Eq. (2) is a hyperbolic partial differential equation in mathematical sense. Solution of the C-V equation results in the temperature field disturbance (caused by i.e. subjecting the surface to the heat flux boundary condition) only in the distance of $C_0 \cdot t$ from the surface.

* Corresponding author: marcin.lenarczyk@gmail.com

 https://doi.org/10.1051/e3sconf/20187003008

Irradiation of the thin film with thickness considerably smaller than the length allows to employ 1D approach in the x-direction (across the thickness). Assuming that the side walls are suddenly subjected to the heat flux density, we can formulate the boundary conditions as follows:

$$-\frac{\partial T(x,t)}{\partial x}\Big|_{x=0} = q_{W_0}(t) \tag{3}$$

$$-\frac{\partial T(x,t)}{\partial x}\Big|_{x=L} = q_{W_L}(t) \tag{4}$$

The initial conditions can be written as:

$$T(x,t)\Big|_{t=0} = T_0(x) \tag{5}$$

$$\frac{\partial T(x,t)}{\partial t}\Big|_{t=0} = T_{init}(x) \tag{6}$$

The above mathematical model is used to compare solutions of the classical heat conduction and Cattaneo-Vernotte equations. The comparative analysis is performed to define potentially dangerous effects that might occur in the ultra-fast laser beam and solid interaction regime.

3 Computational method

MacCormack proposed second-order accurate explicit scheme to solve compressible Navier-Stokes equation [11]. As reported by [6] this scheme proved to handle well moving discontinuities, such as heat waves, and it is valid for hyperbolic heat conduction problems.

MacCormack discretization procedure is a two-step predictor corrector scheme which is based on non-dimensional formulation. The non-Fourier heat flux equation can be written in the following form:

$$q(x,t) + \tau_R \frac{\partial q(x,t)}{\partial t} = -\lambda \frac{\partial T(x,t)}{\partial x} \tag{7}$$

The energy equation is given by:

$$-\frac{\partial q(x,t)}{\partial x} = \rho c \frac{\partial T(x,t)}{\partial t} \tag{8}$$

The dimensionless temperature, dimensionless heat flux, dimensionless time and space variables are defined as follows:

$$\theta(\eta,\zeta) = \frac{T - T_0}{\alpha q / \lambda c} \tag{9}$$

$$Q(\eta,\zeta) = \frac{T(x,t) - T_0}{\alpha q / kc} \tag{10}$$

$$\zeta = \frac{C_0^2 t}{2\alpha} \tag{11}$$

$$\eta = \frac{C_0 x}{2\alpha} \tag{12}$$

Based on the MacCormack second-order accurate explicit scheme predictor and corrector expression can defined as:

$$\tilde{E}_i^{f+1} = E_i^f - \frac{\Delta\zeta}{\Delta\eta}\left[F_{i+1}^f - F_i^f\right] - \Delta\zeta H_i^f \tag{13}$$

$$E_i^{f+1} = \frac{1}{2}\left[E_i^f + \tilde{E}_i^{f+1} - \frac{\Delta\zeta}{\Delta\eta}(\tilde{F}_i^{f+1} - \tilde{F}_{i-1}^{n+1}) - \Delta\zeta \tilde{H}_i^{n+1}\right] \tag{14}$$

Where:

$$E = \begin{bmatrix} \theta(\eta,\zeta) \\ Q(\eta,\zeta) \end{bmatrix} F = \begin{bmatrix} Q(\eta,\zeta) \\ \theta(\eta,\zeta) \end{bmatrix} H = \begin{bmatrix} 0 \\ 2Q(\eta,\zeta) \end{bmatrix} \tag{15}$$

The stability criterion for this method, when applied to a linear equation, is $\nu \le 1$, where $\nu = \Delta\zeta / \Delta\eta$ is the Courant number. However, due to nonlinear nature of the present equations, the stability criterion is reduced. As the Courant number becomes smaller, the effect of odd derivative truncation error terms becomes larger, and oscillations occur in the vicinity of discontinuities in the solution [7].

4 Simulation results

Throughout the numerical simulations that were performed within the scope of the study, the numbers of results have been obtained. The following section presents selected cases.

Figure 2 presents the overview of the temperature evolution in the considered system. Duration time and heat flux density magnitude have been set up arbitrary to illustrate effects, which could be potentially dangerous for the laser protection system.

After the boundary conditions on both sides are imposed, a set of wave fronts is being created in the domain. Thermal wave propagation starts to advance toward centre.

The propagation front separates the thermally affected zone from the thermally undistributed region. Once wave fronts arrive at the centre of the film, they collide with each other causing a sudden and significant temperature increase in this region.

Consequently, revers thermal wave fronts emerge, and they begin to advance toward side walls. Once reverse waves reach the boundary region, the temperature at both walls exceeds the initial wall temperature (as an effect of wave reflection) creating temperature overshoot.

Performed numerical study predicts the existence of thermal waves providing both the amplification and shape. In the investigated case, the sudden temperature increase due to waves collision in the centre of the structure occurs at $4.55 \cdot 10^{-15}$ s. The side walls temperature at the time step equals to 470 °C. The maximum peak temperature in the system equals to 599 °C, rising suddenly from 284 °C. It means that the system experienced step temperature jump by 210% with respect to the before-the-collision temperature in the centre of the structure. While considering the side wall as a reference, the peak temperature raised by 27%.

As the reverse waves reach the side walls, the temperature overshoot effect occurs at $9.1 \cdot 10^{-15}$ s. Both side walls temperature suddenly rises from 555 °C to 986 °C, which corresponds to 77% increase.

A similar reflection effect can be observed once one of the wall is well insulated. Let assume identical conditions as for the above described case, with exception for the right wall boundary condition, which for the current case is subjected to the heat flux density $q_W L = 0$ W/m^2.

Figure 3 illustrates temperature distribution in the system in the consecutive time steps.

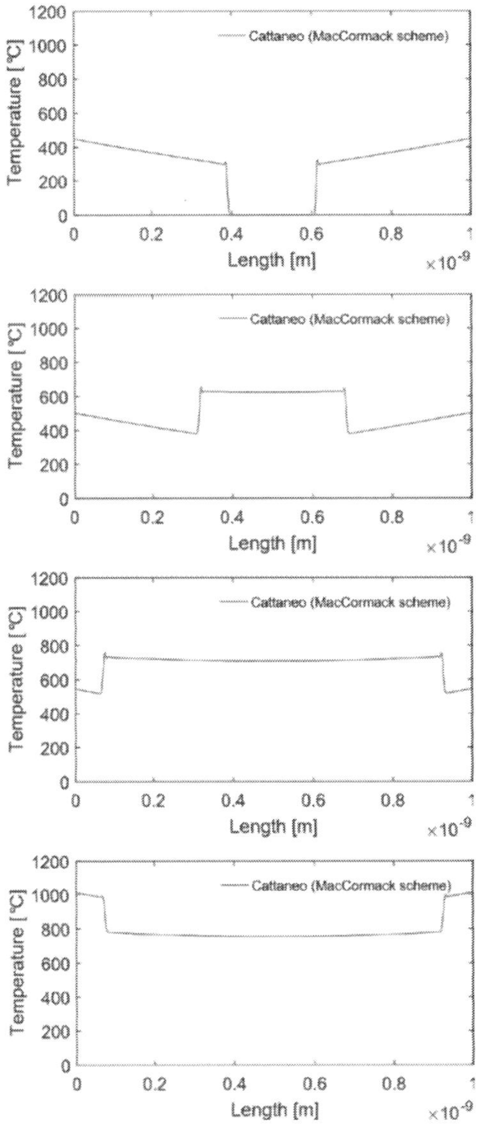

Fig. 2. Heat waves collision. Neumann BC. Temperature distribution. Results for time steps: $3.5 \cdot 10^{-15}$ s, $6.2 \cdot 10^{-15}$ s, $8.4 \cdot 10^{-15}$ s and $9.7 \cdot 10^{-15}$ s. Heat flux density $q = 1 \cdot 10^{14}$ W/m^2 on the both side walls assumed as a boundary condition.

The wave front is being created after the boundary condition is imposed on the left wall. Thermal wave propagates toward insulated wall. Once the front collides with the wall, the sudden temperature increase occurs, and the reverse thermal front emerges. New front advances toward the left wall and after the second collision it causes the temperature overshot.

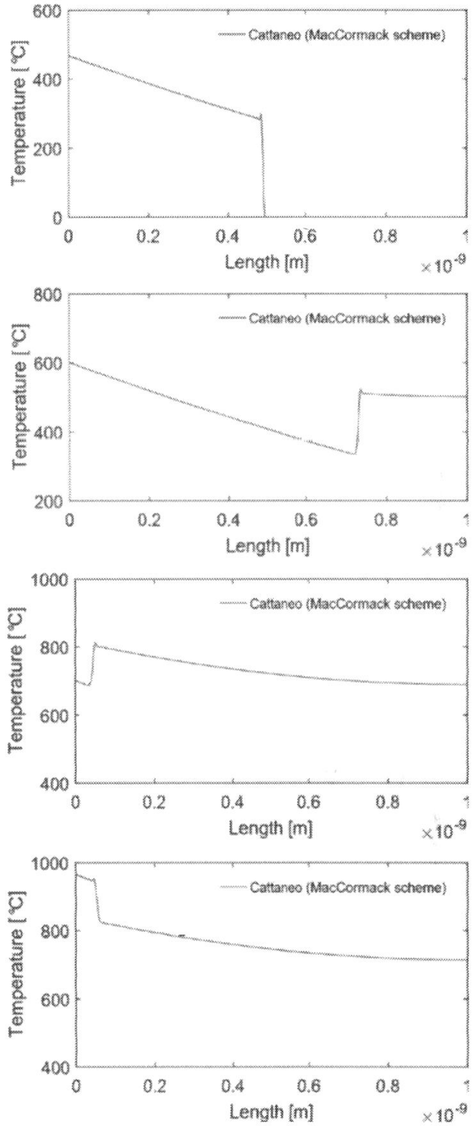

Fig. 3. Heat wave reflection from the insulated wall. Neumann BC. Results for time steps: $4.5 \cdot 10^{-15}$ s, $1.1 \cdot 10^{-14}$ s, $1.7 \cdot 10^{-14}$ s and $1.8 \cdot 10^{-14}$ s. Heat flux density $q_L = 1 \cdot 10^{14}$ W/m^2 on the left side wall and $q_R = 0$ W/m^2 assumed as the boundary conditions.

The numerical analysis of the configuration consisting of two materials with various relaxation times has been carried out. Obtained results suggest that a discontinuity such a transition from one material to the other can result in a reverse wave creation. A wave which emerged at the materials boundary is traveling in the backward direction advancing towards wall subjected to the Neumann boundary condition.

In Figure 4 one can see the results showing temperature distribution in the dimensionless domain for two materials, where $\tau_{R1} = 1 \cdot 10^{-11}$ s and $\tau_{R2} = 1 \cdot 10^{-13}$ s. Assuming that density, specific heat capacity and thermal conductivity are the same for both materials, speed of wave front is $C_{01} = 1 \cdot 10^3$ m/s and $C_{02} = 1 \cdot 10^4$ m/s, respectively, for the right and left sides.

This result should be treated as a potential effect that needs further investigation and it is beyond the scope of the present study. The author has not found any confirmation of such a phenomenon in the experimental data available in the literature. However, if this kind of effect existed in nature, then it should be qualified as another thermal response that might be of significance while evaluating thermal performance of the system.

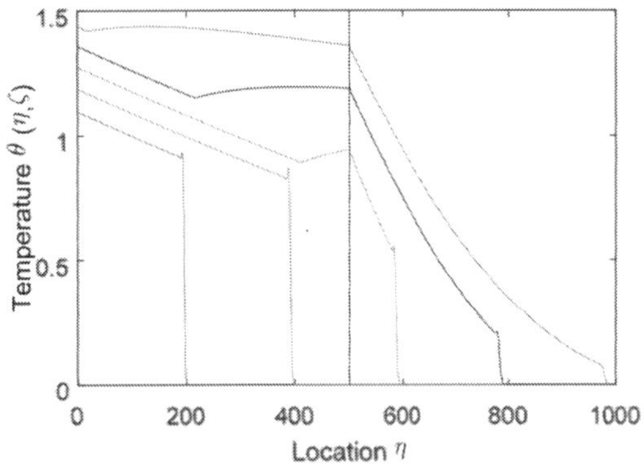

Fig. 4. Wave reflection at the two materials' interface.

For all the presented numerical calculations the energy balance check was done. The balance was carried out by comparing heat accumulated by the system versus heat delivered to the system. The accumulated heat is determined based on temperature distribution at a given time step.

$$Q_{ACC} = m\, c_p T(x,t) \tag{16}$$

$$Q_{ACC} = A\rho c_p \int_0^L T(x,t)dx \tag{17}$$

Heat delivered to the system is calculated by means of the control surface at the side walls and heat flux integration over time

$$Q_{DEL} = A\int_0^T q(t)dt \tag{18}$$

Figure 5 presents the energy balance of the wave-wall interaction scenario. For the comparison purposes, value of the area, A, is set up to 1. It is seen that the heat accumulated by the system, Q_{ACC}, is equal to heat deliver to the system, Q_{DEL}, over time.

5 Summary and conclusions

In the presented study the MacCormack numerical algorithm has been adopted to solve the hyperbolic heat conduction equation. A comparative analysis of the solution of the classical heat conduction equation and the Cattaneo-Vernotte equation is performed for the Neumann boundary conditions.

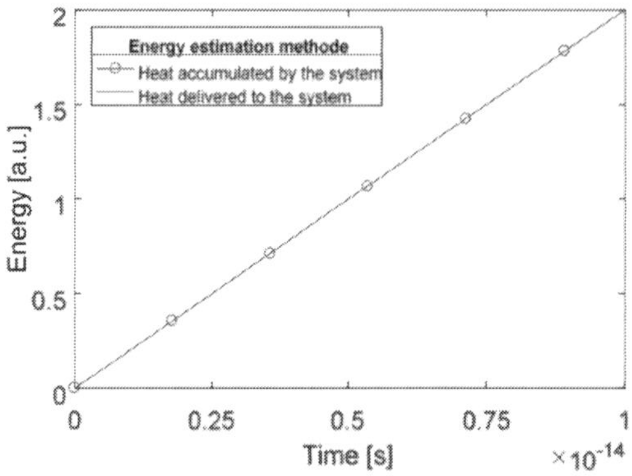

Fig. 5. Energy balance check. Comparison of heat accumulated by the system with heat delivered to the system over time.

The investigation of the heat waves collision phenomenon is carried out. Two effects have been identified as potentially relevant for the thermal barriers which are meant to interact with the laser radiation. The first effect occurs after waves' collision. It results in a step change of the temperature inside the structure, which can lead to the limit exceedance. The second effect is the temperature overshoot on the side wall, which is the result of the backward wave and the wall interaction.

In the example that illustrates the phenomenon, the maximum temperature overshot was equal to twice the temperature in the centre of the structure before the collision. This suggests that the described effect might be potentially dangerous for the considered system.

Laser-matter interaction and the associated heat transfer process are very complex phenomena. Both are still subjects of extensive research efforts. Prediction of the temperature field in the irradiated solid requires not only comprehensive heat transfer model, but also reliable material properties.

The dynamic of the process and lack of detailed material data (in particular the relaxation time parameter) are, among others, the reasons why till now there is no versatile correlation that would allow for precise prediction of the temperature evolution in such a system.

It is also crucial to point out that all the considerations on the particular protective system should be done in the context of the laser wavelength, which determines the absorption mechanism. If one knows the wavelength of the radiation source, then the set-up of the dedicated material configuration can be chosen accordingly. A separate effort should be undertaken in order to build a relation between radiation wavelength and corresponding hardware protection strategy.

Nomenclature

C_0	– speed of the thermal wave, m/s,
c	– specific heat, J/(kg·K)
L	– length of the system, m,
Q	– dimensionless heat flux, –,
q	– heat flux, W/m^2,
T	– temperature, °C,
t	– time, s,
x	– spatial variable, m,

Greek symbols

α	– thermal diffusivity, m^2/s,
ζ	– dimensionless time variable, –,
η	– dimensionless space variable, –,
θ	– dimensionless temperature, –,
λ	thermal conductivity, W/(m·K),
ν	– Courant number, –,
ρ	– mass density, kg/m^3,
τ_R	– relaxation time, s.

References

1. M.S. Brown, C.B. Arnold, Fundamentals of Laser-Material Interaction and Application to Multiscale Surface Modification, in: Laser Precis. Microfabr., Springer-Verlag Berlin Heidelberg, 2010: pp. 91–120.
2. C. Cattaneo, Sur une Forme de l'equation de la Chaleur Eliminant le Paradoxe d'une Propagation Instantanee', Comptes Rendus l'Académie Des Sci. 247 (1958) 431–433.
3. R. Domański, Badanie przewodzenia ciepła w ciałach stałych przy impulsowo działającym strumieniu ciepła, Dr. Thesis Inst. Heat Eng. Warsaw Univ. Technol. (1976).
4. R. Domański, Badanie cieplnych efektów działania gigantycznego impulsu laserowego na powierzchnie nieprzezroczystych ciał stałych, Arch. Termodyn. 2 (1981).
5. R. Domański, Promieniowanie laserowe - oddziaływanie na ciała stałe, Wydawnictwa Naukowo-Techniczne, Warsaw, 1990.
6. D.E. Glass, M.N. Özişik, D.S. McRae, B. Vick, On the numerical solution of hyperbolic heat conduction, Numer. Heat Transf. 8 (1985) 497–504.
7. D.E. Glass, M.N. Özişik, D.S. McRae, B. Vick, Hyperbolic heat conduction with temperature-dependent thermal conductivity, J. Appl. Phys. 59 (1986) 1861–1865.
8. M. Lenarczyk, Hardware protection against laser impulses using materials with a complex structure, PhD Thesis Inst. Aviat. Warsaw. (2018).
9. M. Lenarczyk, R. Domański, Investigation of non-Fourier thermal waves interaction in a solid material, Arch. Thermodyn. Accepted (2018).
10. M.C. Levy, T.G. Blackburn, N. Ratan, J. Sadler, C.P. Ridgers, M. Kasim, L. Ceurvorst, J. Holloway, M.G. Baring, A.R. Bell, S.H. Glenzer, G. Gregori, A. Ilderton, M. Marklund, M. Tabak, S.C. Wilks, P.A. Norreys, QED-driven laser absorption, Plasma Physics, Cornell Univ. Libr. 1 (2016) 1–10.
11. R. MacCormack, The effect of viscosity in hypervelocity impact cratering, in: 4th Aerodyn. Test. Conf., American Institute of Aeronautics and Astronautics, 1969.
12. P. Vernotte, La véritable équation de la chaleur, Compt. Rend. Acad. Sci. 247 (1958) 2103–2105.
13. L.D. Welch, D.C. Latham, Report of the Defense Science Board Task Force on High Energy Laser Weapon Systems Application, 2001.

Dual-fuel combustion in a tangential PC-fired boiler of type OP-230 – computational simulations

Przemysław Motyl[1,*], and *Jan* Łach[2]

[1]Kazimierz Pulaski University of Technology and Humanities in Radom, Faculty of Mechanical Engineering, Malczewskiego 29, 26-600 Radom, Poland
[2]Bialystok University of Technology, Faculty of Engineering Management, Ojca Tarasiuka 2, 16-001 Kleosin, Poland

Abstract. Syngas co-firing in coal fired boilers can be one of the prospective technologies which may help to retrofit some of still functioning older boilers. This study focuses on the results of CFD simulations of wood biomass-derived syngas co-firing with coal in an older mid-sized tangential PC-fired boiler of type OP-230. The design and the implementation of the combustion process predispose the boiler to the connection with the biomass gasifier in which low calorific syngas from solid raw biomass gasification can be produced and next used as a supplemental fuel in the coal furnace. The simulations were performed to predict the influence of the improvement of the air staging via the dual-fuel technique based on the indirect co-firing technology on both the reductions in NO_x emissions relative to the baseline (no syngas) and the residence time of syngas particles in a zone with the temperature higher than 1123K. This way one can determine whether the boiler can be recommended to indirect co-firing of syngas derived from agricultural residues biomass or SRF gasification containing such troublesome components as chlorine and alkali.

1 Introduction

One of the routes to accomplish coal and biomass promising co-firing technology, referred as a clean and environmentally friendly low-emission indirect co-firing, involves replacing several percent of the coal with syngas produced from the highly efficient on-site gasification of a wide range of solid raw or modified biomass resources usually in a commercial fluidised bed gasifier. Many coal gasification technologies are in widespread commercial use [1,2] while the technically viable biomass gasification technologies are still being developed in lab-scale, improved in pilot and demo installations and successively implemented into engineering practice in commercial installations [3,4]. Experiences with the commercial-scale implementations of the biomass indirect co-firing technology were acquired mainly in Finland but also in Austria, Belgium and the Netherlands. The first global implementation is made at the Kymijärvi PC-fired CHP-plant [5]. Further significant examples are the Zeltweg PC-fired power plant [6], the Electrabel Ruien fossil fuel-fired power station [7] and the Amercentrale coal-fired CHP unit of Essent in Geertruidenberg [8]. Particular attention should be paid to the world's largest Vaskiluodon Voima plant in Vaasa located at the Vaskiluoto 2 power plant [9] and supplied by Metso. This way, the indirect co-firing of biomass in existing coal-fired boilers increasingly gain market acceptance but is not yet the most attractive option of co-firing projects due mainly to economic reasons. A handful of examples of its industrial applications listed above translates into still numerous research studies including our own one to the problem. We will limit ourselves here to a concise reference to the results of indirect co-firing simulations reported in a few of selected papers. Therefore, the detailed analysis focused on the reduction in NO_x emissions relative to the baseline depending on the paper rejects-derived syngas composition, heat input replacement and furnace stoichiometry in a 70t/h-steam coal-fired industrial boiler is given in [10]. The similar analysis of biomass-derived syngas co-firing with a lignite in a 600MWe tangentially fired boiler is taken in [11]. The numerical approach to the coal co-firing with the product gas from sewage sludge gasification in the PC-fired OP-650 boiler at the Kozienice power plant in Poland is demonstrated in [12]. The impact of the share of product gas from biodegradable waste gasification and LCV gas quality on a 230t/h dual fuel pulverized coal-gas fired boiler parameters and performance is reported in [13]. The effect of the biomass-derived syngas fraction in the mixture with natural gas on basic co-firing parameters in gas-fired fuel systems is examined in [14]. Our study is focused on the 3D inexpensive full-scale numerical approach based on such a powerful computational tool as CFD code known as Fluent [15] for modelling the co-firing of coal and wood biomass-derived syngas which is injected into the furnace of an older mid-sized tangentially PC-fired boiler of type OP-230 through new dedicated nozzles. Their relatively flexible arrangement and the choice of the right value of syngas volumetric flow rate can ensure the operational parameters,

* Corresponding author: p.motyl@gmail.com

primarily the right thermal effect guaranteeing the furnace heat balance, and technological conditions making possible to reduce NO_x emissions compared to that of coal combustion originally regulated by means of inexpensive standard air staging technique. The design and the implementation of the combustion process predispose the boiler to the connection with the biomass gasifier. Computational three-dimensional modelling of tangentially fired PC-boilers using CFD is widely regarded as effective calculation tool. Our goal is to determine whether the level of NO_x emissions resulting from modifying the air staging based on the indirect co-firing technology, i.e. via the dual-fuel technique, corresponds to the prospective more restrictive environmental protection regulations. Then indirect co-firing could be one of the effective technologies for the boiler retrofit. Compared to the least complicated and attractive for investors direct biomass co-firing, it offers such important advantages as: (i) the avoidance of coal only ash properties altering due to the separate storage of slag and ash from gasification and syngas co-firing which means that fly ash can meet the quality standards for use of its in concrete; (ii) the resulting high flexibility with regards to the choice of fuel biomass for gasification including not only wood, wood waste and residual wood waste but even wide range of biomass containing high alkali metals and chlorine or difficult for the pre-treatment; (iii) reduction of the risk of the ash softening, the deposition of pollutants on heat exchange surfaces in the boiler and SCR catalysts, and reduction of the impact of high-temperature corrosion on steel structural elements of boilers as a result of limiting chlorine mobility already in the gasifier; and (iv) simplification of the eventual biomass pre-treatment processes including the elimination of risky problems both at the stage of co-milling biomass with coal and their separate milling before co-firing in coal burners or combustion in dedicated biomass burners. When PC-fired boilers of type OP are retrofitted by their conversion into fluidized bed boilers, then different types of biomass of forest and non-forest origin can be combusted but the high number of alkaline compounds contained in agricultural biomass and waste from agri-food processing can result in severe bed agglomeration. Although the indirect co-firing is still not very popular, it is considered safer, more effective (efficient energy recovery) and more environmentally friendly than direct co-firing. We limit ourselves to the prediction of reductions in NO_x emissions relative to the baseline (no syngas) for a priori one adopted arrangement of syngas nozzles and heat input replacement.

2 Boiler description

The boiler manufactured by Rafako that was examined is shown in Fig.1. The size of a tangentially fired furnace is 8.115 x 23.2 x 7.855 m. The low-emission pulverised coal combustion system based on low-emission burners and two-stage combustion with OFA nozzles allows to reduce NO_x emissions below 400 mg/Nm^3 (6% O_2) and to keep the amount of unburned coal in fly ash below 5-7% depending on the ash content in coal. The RI (Rapid Ignition) – JET coal burners (Fig.2) are located at three elevations in the corners of furnace chamber (Fig.3) which is typical for small boilers.

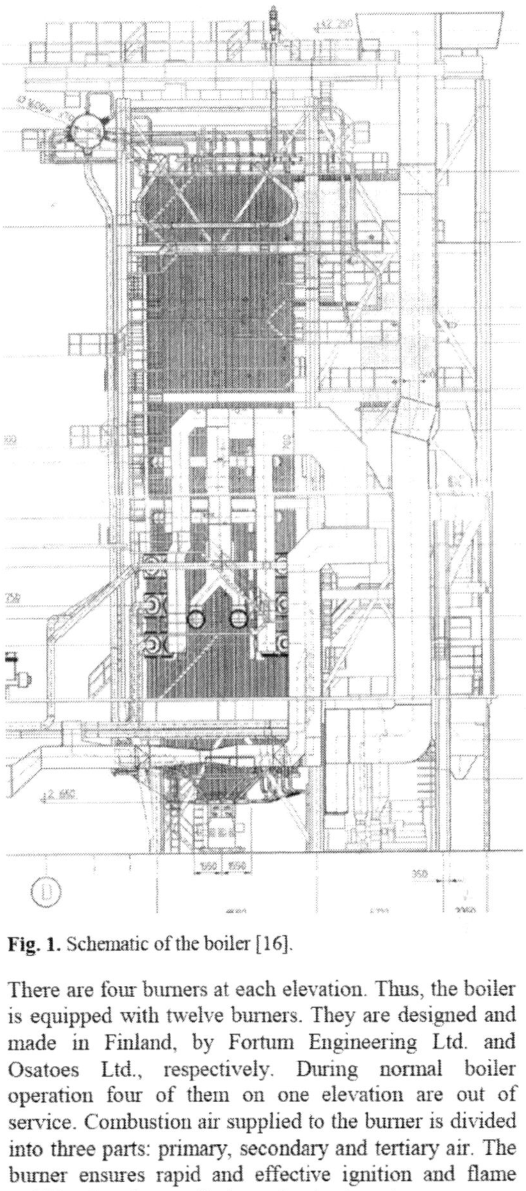

Fig. 1. Schematic of the boiler [16].

There are four burners at each elevation. Thus, the boiler is equipped with twelve burners. They are designed and made in Finland, by Fortum Engineering Ltd. and Osatoes Ltd., respectively. During normal boiler operation four of them on one elevation are out of service. Combustion air supplied to the burner is divided into three parts: primary, secondary and tertiary air. The burner ensures rapid and effective ignition and flame stabilization due to (i) the pulverized coal concentrator whose aerodynamics favors the achievement of a rich dust mixture at the perimeter of the pipe of primary air and dust; (ii) the special flame stabilization ring supporting the ignition of this mixture; (iii) the large reducing combustion zone achievable by separating the supplied secondary and tertiary air; (iv) the low NO_x level due to effective secondary air swirling; and (v) efficient burnout of the char particles using the tertiary air. Combustion in the primary air zone occurs at extreme sub-stoichiometry. Due to the reducing compounds, NO formed is dispersed in the secondary air zone. Separation of the tertiary air from combustion in the secondary air zone is achieved by means of control sleeve and an adjustable turbulence of both flows.

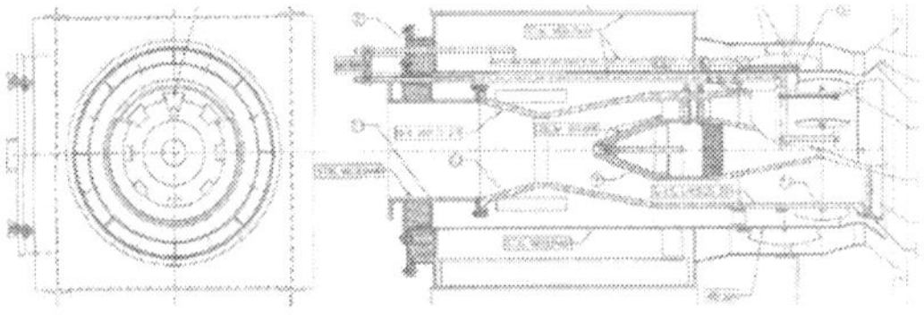

Fig. 2. RI (Rapid Ignition) – JET coal burner [16].

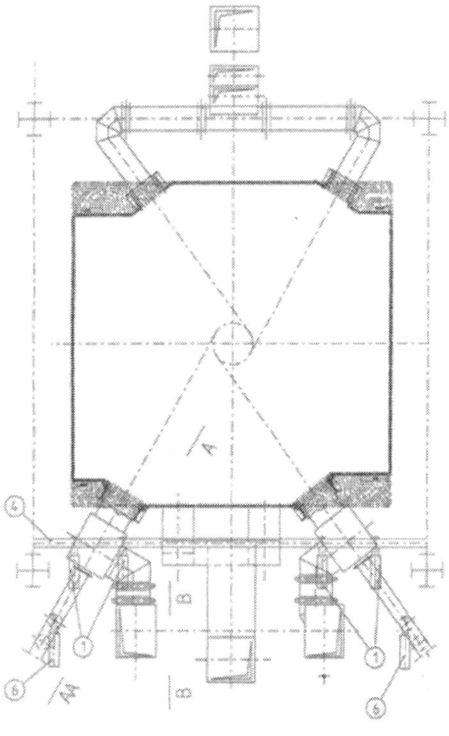

Fig. 3. Arrangement of burners [16].

Devices for secondary and tertiary air regulation are equipped with manually adjustable blades forcing the mixing of air flows which improves the combustion process. Such an effect is also due to dust nozzle having shape of Venturi orifice equalizing the concentration of particles in the flowing air and at the same time preventing a backward movement of the flame. Flame shape stabilization by a ring equipped with "ceramic teeth" (Fig.2) is based on two phenomena: (1) the "teeth" cause turbulent motion of the coal particles at the outlet of the burner throat; and (2) the open space between the burner throat and the conical ring enforces a stable turbulent movement of coal-air mixture. In this way, there is more time to ignite. The pulverized coal concentrator focuses coal particles on the walls of the coal feed pipe. Thus, both the stabilizing ring and the dust concentrator play a key role in ensuring low-emissions combustion. The boiler is equipped with eight OFA nozzles located at two elevations (Fig.4). Therefore, there is some flexibility in the optimal setting of fuel and air, which in particular allows a large burnout of coal. Additionally, in the boiler we have to deal with the implementation of the combustion air staging.

3 Computational problem

For the purpose of analysis of the dual-fuel combustion in the boiler (Fig.1), the model of the furnace has been developed. The mathematical modelling focuses on commercial CFD code Fluent [10] in release 19. The computational mesh constructed over the region of the furnace consisted of about 2.7 10^6 elements of the Polyhedra type. Data enabling the determination of boundary conditions for the numerical model were obtained from the boiler operating manual [16]. The most important of them are given in Table 1.

Finite-Rate/Eddy-Dissipation model [15] used to modelling combustion of coal and co-firing coal with wood biomass derived syngas is constructed by making use of the following equations:

$$C_{0.89}H_{5.82}O_{0.68}N_{0.1728} + 1.56O_2 \rightarrow 0.89CO + 2.91H_2O + 0.0864N_2 \quad (1)$$

$$C(s) + 0.5O_2 \rightarrow CO \quad (2)$$

$$C(s) + CO_2 \rightarrow 2CO \quad (3)$$

$$C(s) + H_2O \rightarrow H_2 + CO \quad (4)$$

$$H_2 + 0.5O_2 \rightarrow H_2O \quad (5)$$

$$CO + 0.5O_2 \rightarrow CO_2 \quad (6)$$

$$CH_4 + 2O_2 \rightarrow CO_2 + 2H_2O \quad (7)$$

The implementation of the standard k-ε turbulence model [15] enables the turbulent Reynolds equations to be closed. The Discrete Ordinates radiation model was used to simulate radiation heat transfer. The gas phase absorption coefficients are calculated using the weighted-sum-of-grey gases model (WSGGM). The assumed chemical composition of coal by percentage (% by weight, i.e. wt%) was as follows: carbon [C], 60.58; hydrogen [H], 4.40; oxygen [O], 8.20; nitrogen [N], 1.30; sulphur [S], 0.22; chlorine [Cl], 0.30; ash [A], 12.60, and moisture [W], 12.40. Its LHV was assumed to be equal to 23.44MJ/kg. The wood biomass-derived syngas composition by percentage (% by volume) was as follows: CO_2, 7.3; CH_4, 3.0; CO, 21.8; H2, 7.6; N_2, 41.3; and H_2O – 19.0 [17]. In order to show the influence of the improvement of the air staging via the dual-fuel technique based on the indirect co-firing technology, two cases were modelled: the baseline (no syngas – Case 0) and co-firing of coal with wood biomass-derived syngas injected into the combustion chamber via new dedicated nozzles (Case 1). Case 0 is considered as a reference one. For Case 1 heat input replacement was assumed as equal to 15%. Four syngas nozzles were located below the OFA nozzles (Fig.4).

 https://doi.org/10.1051/e3sconf/20187003009

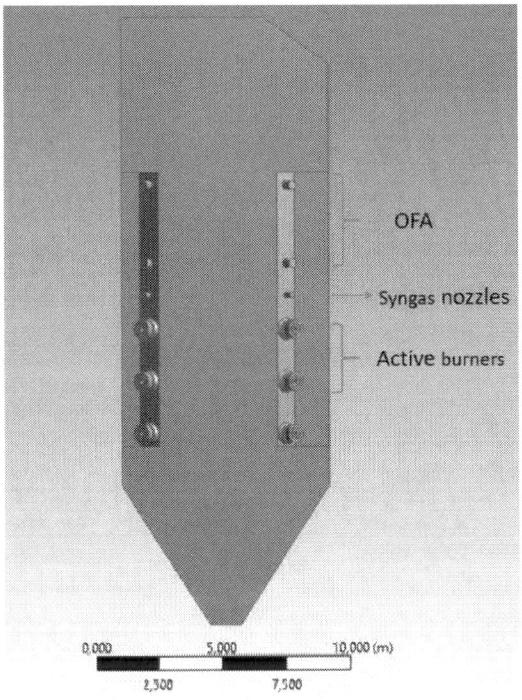

Fig. 4. Location of syngas nozzles in the furnace.

Table 1. Boiler operating conditions (full load).

Parameter	Unit	Value
Coal feed rate/burner	kg/s	1
Primary air flow rate/burner	kg/s	2
Secondary air flow rate/burner	kg/s	4.5
Air flow rate/OFA nozzle	kg/s	2.3
Coal–air mixture temperature	K	373
Primary air temperature	K	373
Secondary air temperature	K	573
OFA temperature	K	573
Steam capacity	t/h	230

4 Results and discussion

The adopted location of additional nozzles dedicated to supplying syngas to the combustion chamber is only one of the possible solutions. It seems simple conceptually and easy to implement due to the fact that the nozzles are located together with burners and OFA nozzles on the same panels. In addition, such a syngas injection preserves the tangential motion of the gas and coal dust mixture. At the same time, the location of syngas nozzles allows providing gaseous co-fuel into the sub-stoichiometric zone which translates into reducing the intensity of NO_x formation. However, such a secondary fuel injection causes the maximum temperature zone moving up the combustion chamber takes place (Fig.5 and 6), which limits the further increase of percentage share of syngas in its co-firing with coal. As the syngas is moist and contains hydrogen, there is an increase in the moisture content in the exhaust gas (Fig.7).

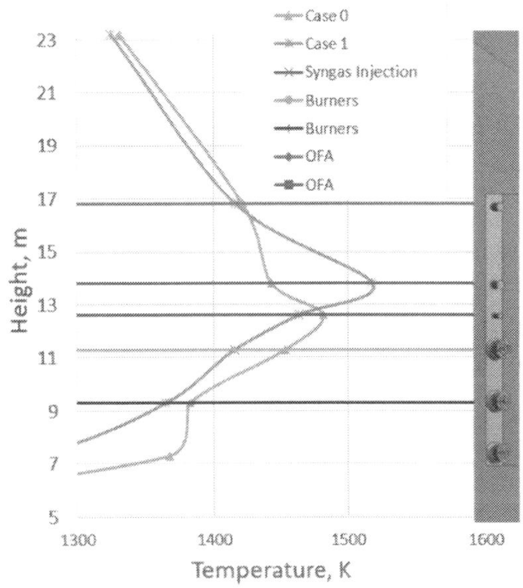

Fig. 5. Mean temperature along the furnace elevations.

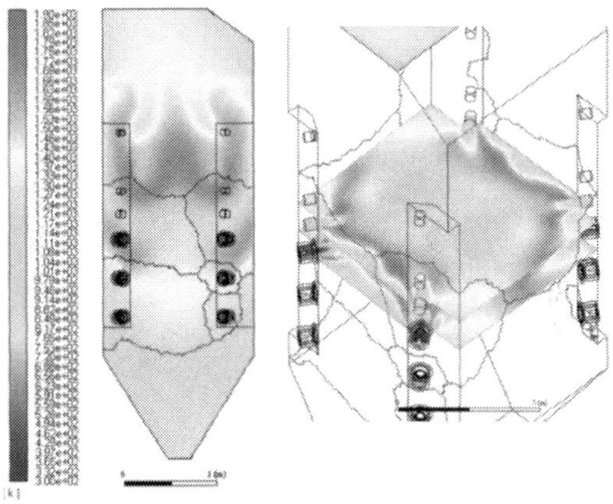

Fig. 6. Temperature [K] profiles in the furnace (Case 1).

The expected cooling effect caused also by the injected gas temperature is only slightly visible above the upper row of OFA nozzles (Fig.5). The distribution of oxygen in the furnace has a similar character as in the case of coal combustion. In the proposed case of co-firing, the protective nature of increased oxygen concentration near the walls of the combustion chamber is not disturbed (Fig.8). The tangential nature of the air, exhaust gas, syngas and coal dust mixture flow (Figs. 9 and 10), resulting from the location of burners, OFA and syngas nozzles, leads also to the lowering the maximum oxygen concentration above the OFA nozzles, which translates into NO_x emission reduction.

https://doi.org/10.1051/e3sconf/20187003009

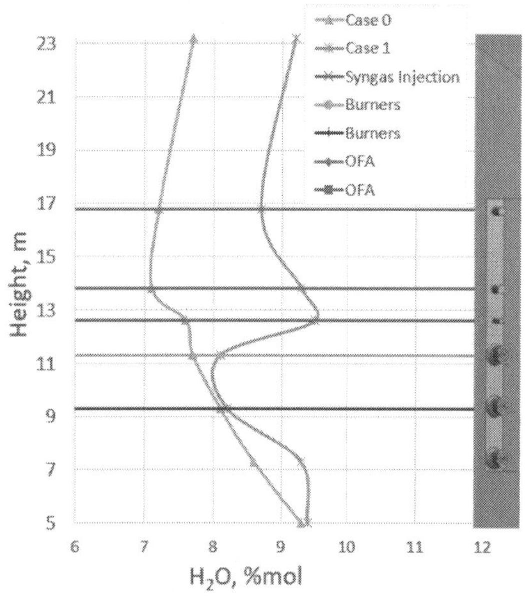

Fig. 7. Mean H2O concentration along the furnace elevations.

Attention is drawn to the fact that Case 1 offers the residence time (> 2s) of syngas particles in a zone, in which the temperature is higher than 1123K (Fig.11). Such a time is required by environmental protection regulations to safely co-fire coal with waste derived syngas. The results of numerical simulations of the woody biomass-derived syngas co-firing with coal prove that the reduction in NO_x emission in the furnace exit (Fig. 12) relative to the baseline (no syngas) achieved with 15% heat input is equal to about 26% (310 vs. 397 mg/Nm^3, 6% O_2).

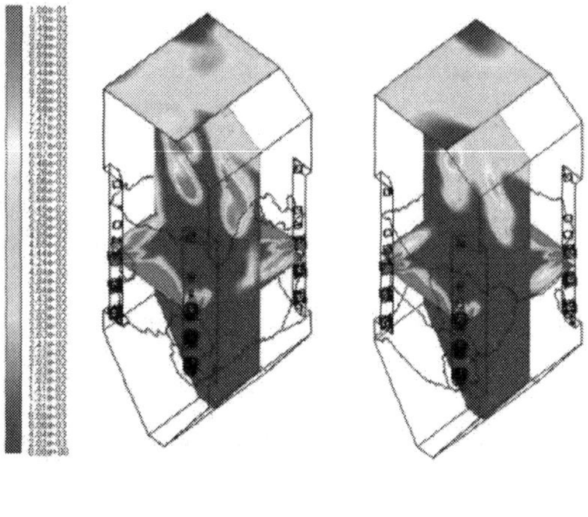

Fig. 8. O_2 mole fraction profile in the furnace (Case 0 and Case 1).

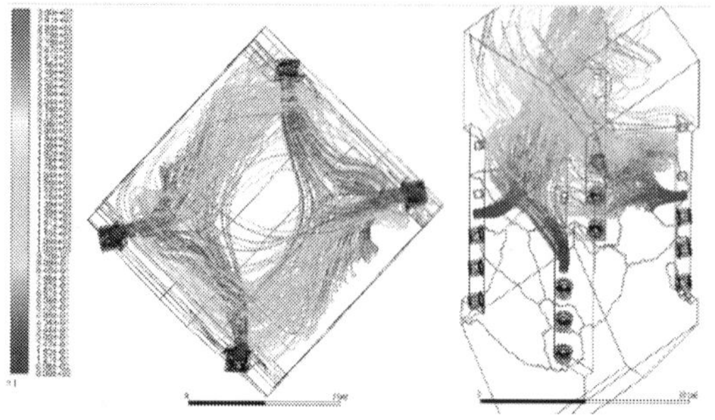

Fig. 9. Pathlines released from the dedicated syngas nozzles for Case 1.

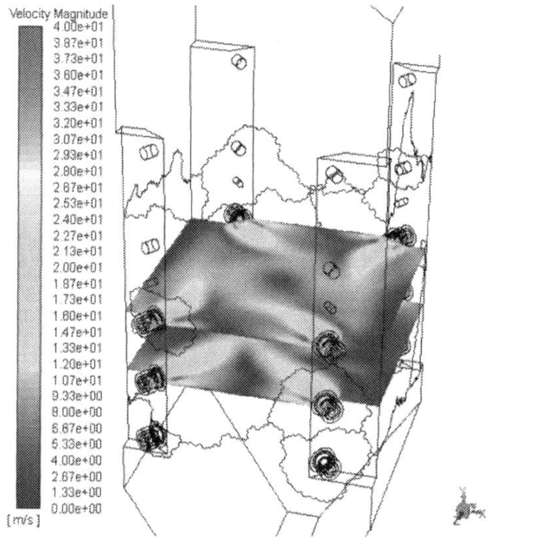

Fig. 10. Velocity magnitude in the furnace (Case 0 and Case 1).

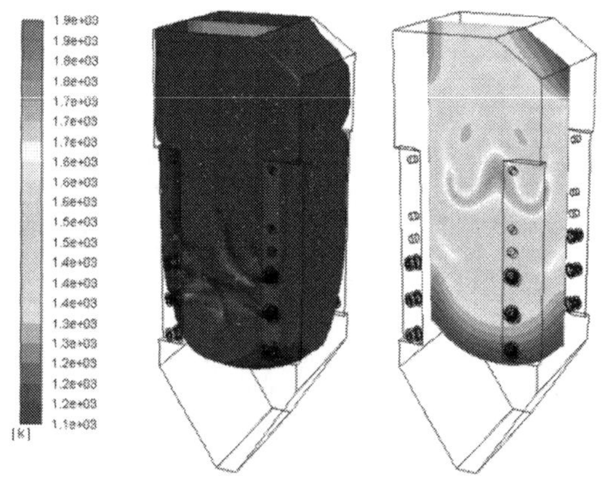

Fig. 11. Furnace zone at temperatures higher than 1123K (Case 1).

The syngas injection was accompanied by a change of air distribution in the combustion chamber. The reduced amount of combusted coal dust allowed to move the part of air from the burners to the OFA nozzles. This way, a constant fuel/air ratio in the burner zone and the total mass flow rate of air supplied to the combustion chamber remained unchanged.

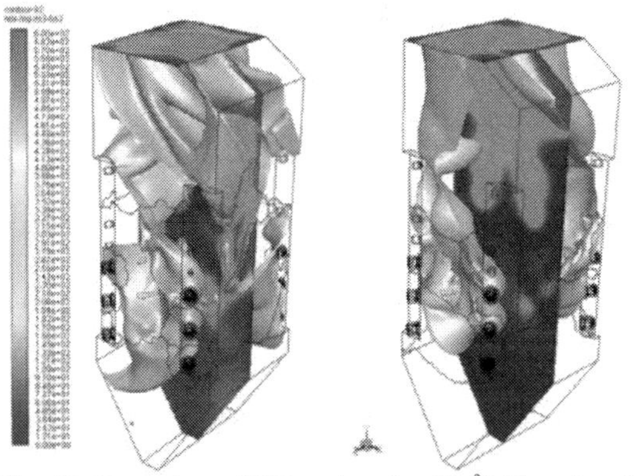

Fig. 12. Comparison of NO_x values [mg/Nm^3] (Case 0 and Case 1) and iso-surface of $NO_x = 400\ mg/Nm^3$.

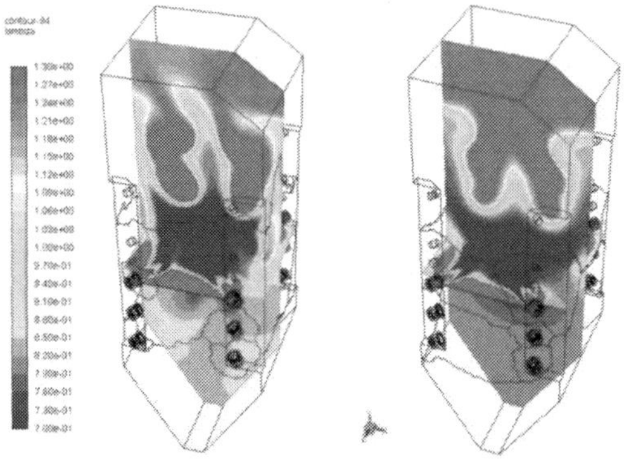

Fig. 13. Excess air ratio distribution in the furnace (Case 0 and Case 1).

The applied procedure allows the boiler operation without changes of excess air coefficient (Fig.13) corresponding to Case 0 in each characteristic zone of the furnace. It also ensures burnout of the char particles and comparable average CO concentration (Fig.14) in two simulated cases of syngas co-firing with coal. The excess air (i.e. stoichiometry) distribution along the furnace elevations proves that the proposed option of the in-direct biomass co-firing can be considered as the development of air staging technique based on the use of additional gaseous fuel. A relatively low height difference between the upper level of the burners and the OFA nozzles offers only a little space in which the efficient conventional syngas reburning cannot be developed. However, limiting the above considerations to the boiler operation at full load (100%), it seems that one can create stoichiometry distribution favoring the conventional reburning by turning off the upper row of burners and turning on the lower row of burners inactive in our simulations. It is apparent that such a syngas reburning configuration requires modification of the amount of air supplied to the burners. This suggestion is just one of many reasons for further simulations of dual-fuel combustion in a tangential PC-fired boiler of type OP-230.

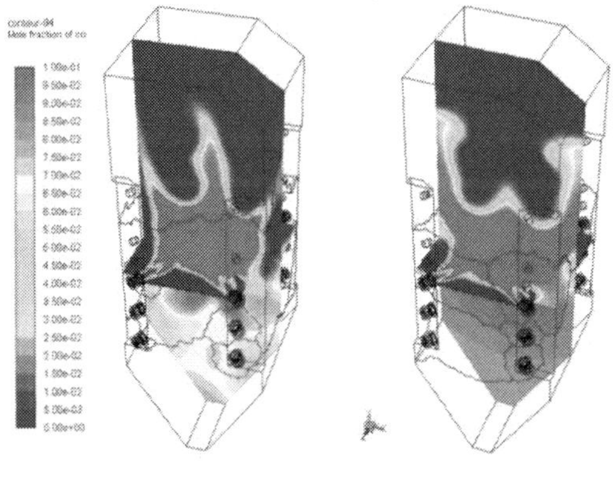

Fig. 14. CO mole fraction profile in the furnace (Case 0 and Case 1).

5 Conclusions

All the numerical tests were performed with injection of a supplemental fuel through dedicated nozzles as it is a solution offering great flexibility in the volumetric syngas flow rate. This facilitates the selection of syngas having the heating value suitable for ensuring the heat replacement postulated a priori. In addition, a set of available nozzles enables a freer choice of gas injection velocity for a given boiler load. The results obtained show that the proposed modifying of the inexpensive standard air staging technique in a tangential PC-fired boiler of type OP-230 cannot fully meet the prospective environmental protection regulations relating to levels of NOx emissions without combining with low-cost, selective non-catalytic reduction (SNCR) methods. Probably a similar situation will occur in the case of a boiler of the same type with low-emission burners on the front wall and syngas nozzles at the elevation of the bottom row of burners. Such a conclusion results from our own calculation results that have not been published yet. However, the adopted location of new nozzles dedicated to supplying syngas to the furnace is just one of many possible options that can be considered. Thus, our next approaches will be focused primarily on both the possible intensification of NO_x reduction due to the location of syngas nozzles below the burners and the creation of conventional syngas reburning configuration. It is also advisable to show the sensitivity of a given dual fuel combustion on co-fuel composition and syngas heat input. Only then we will be able to say whether an older mid-sized tangential PC-fired boiler of type OP-230 can be effectively retrofitted by means of the improvement of the air staging via the dual-fuel technique based on the indirect co-firing technology.

References

1. A.J. Minchener, Fuel **84** (17), 2222-2235 (2005)
2. R.W. Breault, Energies **3** (2), 216-240 (2010)
3. P. McKendry, Bioresource Technol **83** (1), 55-63 (2002)
4. R.P. Fernando, IEA Clean Coal Centre, London, UK, **CCC/158** (2009)
5. D.L. Granatstein, IEA Bioenergy Agreement – Task 36, Natural Resources Canada/CANMET Energy Technology Centre (CETC) November (2002)
6. D.L. Granatstein, IEA Bioenergy Agreement – Task 36, Natural Resources Canada/CANMET Energy Technology Centre (CETC) September (2002)
7. Y. Ryckmans, F. Van den Spiegel, http://www.laborelec.com/docs/articles/lbe_art_COMB006_uk.pdf.
8. R.P. Fernando, IEA Clean Coal Centre London, UK, **CCC/194** (2012)
9. http://biofuelregion.se/wp-content/uploads/Mauri-Blomberg-Vaskilouden-Voima-Oy.pdf
10. K.T. Wu, H.T. Lee, C.I. Juch, H.P. Wan, H.S. Shim, B.R. Adams, S.L. Chen, Fuel **83**, 1991-2000 (2004)
11. C. Dong, Y. Yang, R. Yang, J. Zhang, Applied Energy 87, 2834-2838 (2010)
12. W.P. Adamczyk, S. Werle, A. Ryfa, Applied Thermal Engineering **73**(1), 343-350 (2014)
13. S. Kalisz, M. Pronobis, D. Baxter, Energy **33**, 1770-1778 (2008)
14. K. Valler, A. Wopera, A.B. Palotas, K.J. Whitty, 4th European Combustion Meeting, ECM, Vienna, Austria 14-17 (2009)
15. ANSYS® Academic Research. Fluent Manual, Release 19.0
16. Technical documentation of the boiler OP-230, Siekierki CHP Plant, Warsaw
17. S. Poskrobko, The First Science & Technology Conference „Modern technologies and power engineering equipment" Cracow 503-517 (2007)

Prechamber selection for a two stage turbulent jet ignition of lean air-gas mixtures for better economy and emission

Ireneusz Pielecha[1,*], *Krzysztof* Wislocki[1], *Wojciech* Cieslik[1], and *Lukasz* Fiedkiewicz[1]

[1]Poznan University of Technology, Faculty of Machines and Transport, 69-965 Poznan, Poland

Abstract. The paper presents the results of thermodynamic and exhaust emission analyses of the combustion process using three combustion chambers. The combustion process research was performed on a single-cylinder AVL 5804 engine equipped with a dual-band compressed natural gas system supplied at different fuel pressures to the inlet channel and to the pre-chamber. Based on the performed analyses, the thermodynamic indicators of the combustion process and the emission factors were determined. Having the thermodynamic and emission analyses of the Turbulent Jet Ignition (TJI) combustion system, the pre-chamber was selected with the best collective features. The selection of the best chamber was made using the weighted-purpose method for average indicated pressure, CO and NO_x emissions as well as the indicated engine efficiency, taking into account the impact factors.

1 Introduction

Further improvement of the combustion process requires searching for the possibility of burning lean mixtures (allowing to reduce the fuel consumption). The drive to reduce carbon dioxide emissions in fuel combustion increases the consideration given to gaseous fuels. In the stoichiometric combustion of methane (the main component of natural gas) and diesel fuel, the mass of produced carbon dioxide is: 2.75 kg and 3.17 kg per 1 kg of fuel burned respectively. Combustion of gaseous fuel can mean about 15% reduction of carbon dioxide emissions.

The analysis of combustion systems for lean mixtures using natural gas is carried out in relation to direct injection [1-3] or indirect injection (two-stage combustion system). A two-stage methane combustion system in the Turbulent Jet Ignition system (using the pre-chamber) is already used in F1 engines, heavy-duty vehicles and in high-capacity marine engines; currently it is also adopted for engines with small displacement volumes.

Simulation research and experimental studies of the two-stage combustion system have already been comprehensively conducted since the 1960s. Such studies are also continued today; in a rapid compression machine, such tests were performed by Gholamisheeri et al. among others [4]. A single-hole pre-chamber was used in them, and the tests were carried out with the excess air ratio of $\lambda = 1.25$ for various diameters of the holes from the preliminary chamber (2.0, 2.5 and 3.0 mm). It has been shown that the highest flow rate is achieved at holes diameter of 2.5 mm. The exhaust gas velocity was estimated at 100-300 m/s (at Re = $3\text{-}6\times10^4$). Tests on setups with more holes from the preliminary chamber were conducted by Gentz et al. [5]. They showed a high value of fuel flow velocity from this chamber in the range of 250 m/s when using holes with a diameter of 1.5 mm and an excess air ratio of $\lambda = 1.25$, while for $\lambda = 1.0$ the outflow velocity is much higher and amounts to about 400-500 m/s (according the authors' estimation).

Research on the two-stage combustion (TJI) application is performed not only with methane, but also when using other fuels, such as hydrogen [6, 7], isooctane [8] and other gaseous fuels.

Currently, research on the possibilities of using direct injection of natural gas in passenger car engines is being carried out by scientific consortia within the framework of the GasOn (Gas-Only internal combustion engines) grant from Horizon 2020 [1-3]. The GasOn project aims to develop an advanced mono-fuel engine powered by compressed natural gas (CNG). The authors of this study are also a part of these research efforts. Earlier work of the authors regarding this project was related to the dual fuel type control strategies [9, 10] and preliminary research on the use of the Turbulent Jet Ignition system [11].

2 Aim and scope of research

The use of different pre-chambers in the studies required thermodynamic analyses of the impact of their geometrical properties on the processes of charge preparation and combustion. Thus, the performed calculations included: a) thermodynamic analyses and 2) emission analyses which allow for determining the best conditions of the combustion process. Based on the results of measurements and thermodynamic calculations and the operation indicators of the engine equipped with

* Corresponding author: ireneusz.pielecha@put.poznan.pl

https://doi.org/10.1051/e3sconf/20187003010

different variants of the pre-chamber, comparative assessment of the functional features of these pre-chambers became possible, in particular the assessment of the impact of 1) the size of the pre-chamber and 2) the diameter of the discharge channels for thermodynamic processes and exhaust emission of the test engine equipped with a Turbulent Jet Ignition system.

3 Research methodology

The combustion process research was carried out on a single-cylinder AVL5804 test engine (V_c = 0.510 cm^3) adapted for the two-stage TJI combustion system, and mechanically charged. The layout of the test bench and its basic structural parameters are shown in Fig. 1.

One of the test cycles was performed for three different pre-chambers with the same number of outlets (6-holes) spreading radially from the centre of the atomizer. The structural differences between the tested chambers are the diameter of the holes (1.5 and 2 mm) and the various chamber volumes (1.8 and 2.2 cm^3). Analysis of the combustion process was performed for various fuel dose values injected into the pre-chamber (0.3-1.9 mg) and different values of the charge pressure (different values of the excess air ratio). The detailed sizes of the pre-chambers are included in table 1, and the test plan in Fig. 2.

4 Results of indicators and emissions tests

Collective analysis of the research plan presented in Fig. 2 was performed using thermodynamic and emission indicator maps – their results are shown in Fig. 3.

The combined analyses have shown that an increase in the average indicated pressure (IMEP_av) (Fig. 3a) is observed as the dose to the prechamber (qo_PC) decreases, while the excess air ratio is increased. The emission of nitrogen oxides (Fig. 3b) depends to a large extent on the excess air ratio, where the change in the dose administered to the pre-chamber is insignificant. The emission of carbon monoxide (Fig. 3c) is slightly dependent on the initial dose, and more on the excess air ratio. However, there is no clear trend indicating the most favorable value of this coefficient. The overall efficiency of the engine (Fig. 3d) determined on the basis of the indicated pressure points to the existence of a minimum of this value in the large excess air ratio values range and the lowest fuel dose supplied to the pre-chamber.

Thus the above analyzes have shown that the greatest benefits of using a two-stage combustion system occur at the lowest dose of fuel injected into the pre-chamber (at different λ-values). For this reason, further thermodynamic analyzes were carried out only for research points in which $qo\_PC \rightarrow$ min.

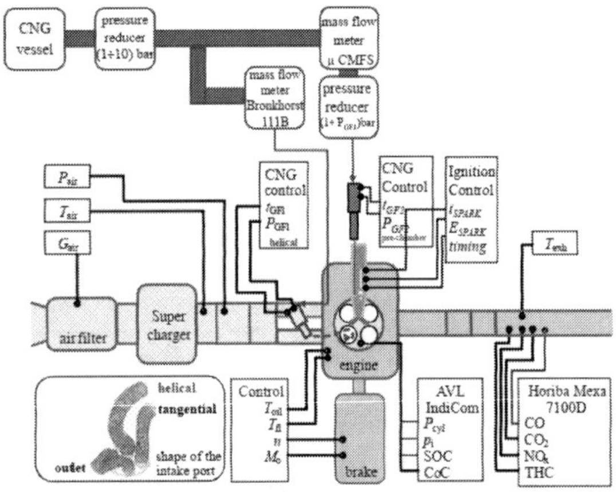

Parameter	Unit	Value
Engine type	–	1-cyl., 4-valve
D	mm	85
S	mm	90
ε	–	15.2
P_{inj}	bar	8 (inlet duct)
λ	–	1.3–1.65
n	rpm	1500
CoC	deg	8
I_{ign}	A	6.5
qo_DI	mg	0.3–1.9
qo_all	mg	17
Spark plug	–	M10&sensor

Fig. 1. Test bench diagram (left) and its main parameters (right).

Table 1. Prechamber characteristics in a TJI system.

No.	V_{ch}	No. of nozzles	*D_nozzle*
	cm^3	–	mm
1	1.826	6	1.5
2	2.287	6	2.0
3	2.287	6	1.5

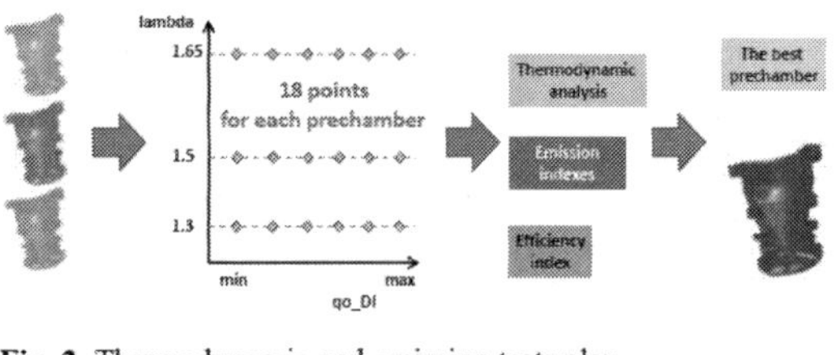

Fig. 2. Thermodynamic and emission tests plan.

 https://doi.org/10.1051/e3sconf/20187003010

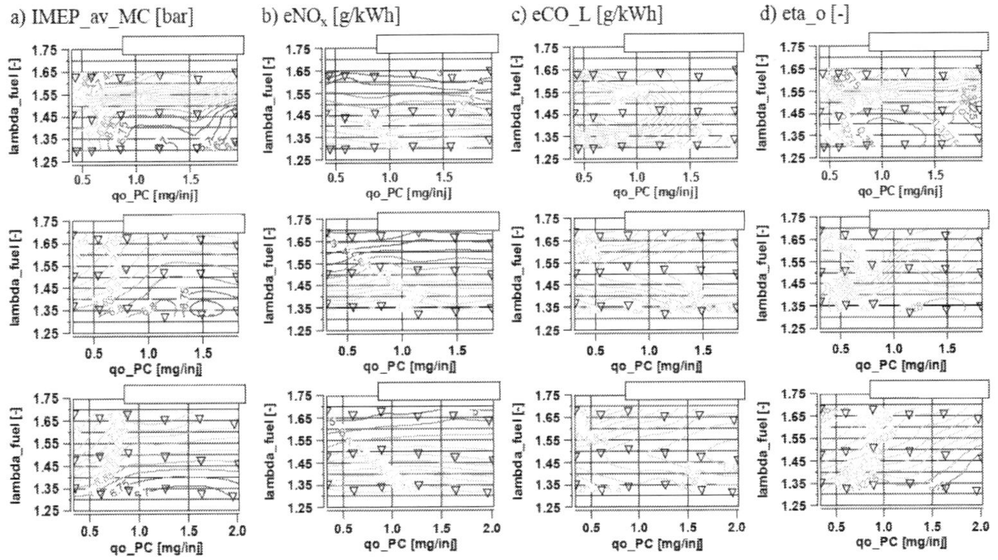

Fig. 3. Analysis of collective thermodynamic and emission indicators for the tested pre-chamber variants.

5 Thermodynamic indicators analysis

Detailed thermodynamic studies of the two-stage combustion system were conducted with regards to: the combustion process indicators characteristic, temperature in both chambers, the speed and amount of heat released.

The analysis of the above indicators for both chambers (pre-chamber and main chamber) indicates differences of 1% in the determined average indicated pressure. A characteristic diagnostic value is the pre-chamber pressure increase at the beginning of the combustion process (Fig. 4a). Analysis of the average temperature value determined using the combustion pressure in both chambers resulted only in small differences – Fig. 4b.

Slightly higher dose temperature values were recorded in the main chamber (1882 K) than in the pre-chamber (1846 K). These differences result from the pressure change, where the peak pressure value is also lower in the pre-chamber. Analysis of the heat release rate does not indicate significant differences in its maximum value (136 kJ/m^{3}deg – PC and 138 kJ/m^{3}deg – MC). A clear peak of this parameter value indicates the initiation of combustion in the pre-chamber, and its value decreases when the combustion process is transferred to the main chamber – Fig. 4c. Maximum values of the heat released are about 1362 kJ/m^3 (the difference between chambers is only 0.2%). Starting the combustion process in the pre-chamber results in higher initial values of the heat released in this chamber – Fig. 4d; combustion process development in the main chamber results in obtaining the same values of the heat released (the analysis was carried out with reference to the specific volume and not to the total volume of the pre-chamber).

The analysis of thermodynamic indicators at $\lambda = 1.65$ (Fig. 5a) was performed for the same quantities, that are shown in Fig. 4. Due to the small volume of the pre-chamber 1 the highest pressure increase was obtained, which did not allow reaching the maximum IMEP value (7.18 bar). The maximum value of this indicator was obtained for chamber 3 (7.30 bar). Rough engine operation (CoV), defined as the ratio of the standard

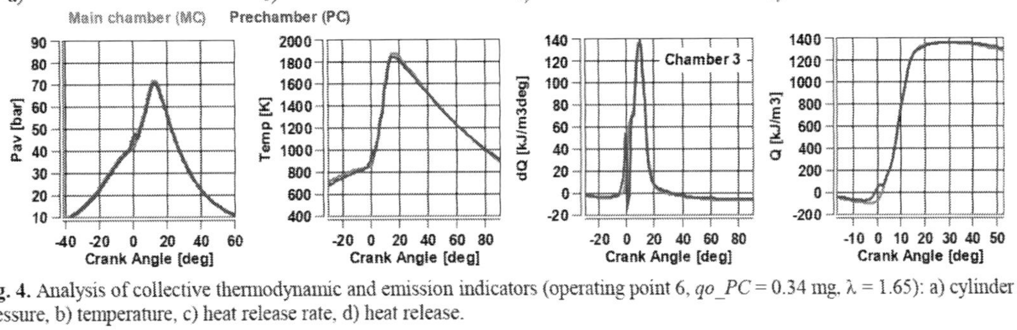

Fig. 4. Analysis of collective thermodynamic and emission indicators (operating point 6, *qo_PC* = 0.34 mg, $\lambda = 1.65$): a) cylinder pressure, b) temperature, c) heat release rate, d) heat release.

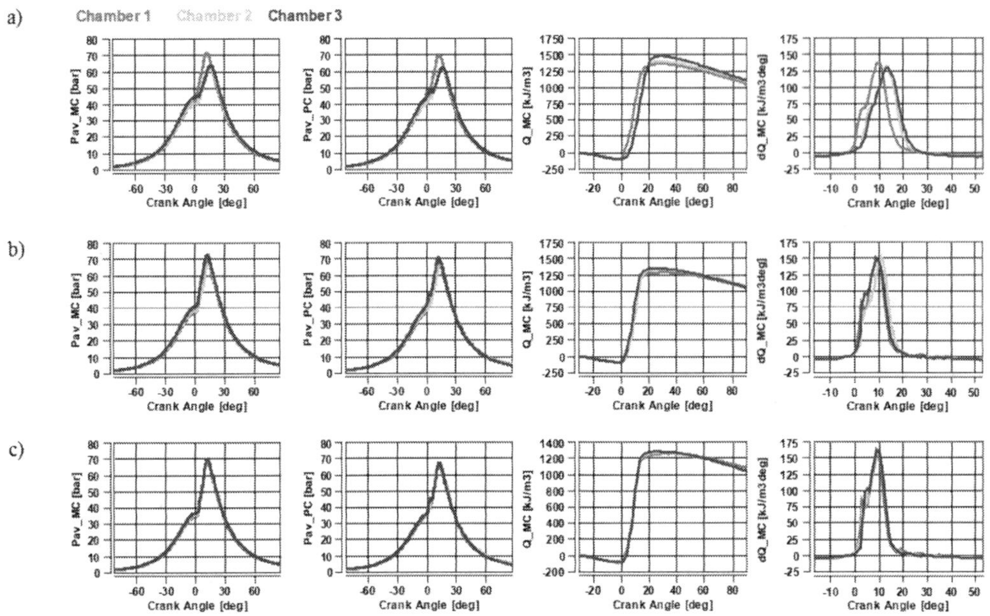

Fig. 5. Analysis of collective thermodynamic and emission indicators (operating point 6, *qo_PC* = 0.34 mg): a) λ = 1.65, b) λ = 1.5, c) λ = 1.3.

deviation (σ) to the mean indicator value (IMEP_av ranges from 0.9% (chamber 3) to 1.23% (chamber 1). Higher pressure increase in chamber 1 initially causes a faster increase in heat release, however, the total value of the heat released is lower. This is due to the lower value of the indicated pressure during expansion.

Reducing the excess air ratio value to λ = 1.5 (obtained by reducing the boost pressure) causes a slight decrease in the maximum combustion pressure (Fig. 5b). In this case, there is also a reduction in the differences in the combustion pressure characteristic and the amount and rate of heat released. The value of the IMEP indicator varies between 7.09 bar (chamber 3) and 6.98 bar (chamber 1). The engine's operation roughness coefficient is around 1%. The maximum heat release rate is constant, but due to the uneven operation, the amount of heat released is greatest when using chamber 3.

A further decrease in the boost pressure (up until λ = 1.3 – Fig. 5c) leads to the combustion process being the same for each of the three pre-chambers. There are only slight differences at the initiation of the combustion process. This may be the result of the size of the pre-chamber and the diameter of the outlets. The larger volume of the pre-chamber allows for the collection of more fresh air. Such a dose during ignition, flowing through the smaller hole diameters, increases its flow rate and speed. Higher flow rates result in faster flame propagation in the engine main chamber.

6. Pre-chamber selection

Following methodology was accepted (for each pre-chamber):

• Selection of the global (for all 3 cases) min and max values for every component,

• Normalizing all absolute measured values to the relative values in the range <0;1>:

- max of emission – value 0,
- min of emission – value 1,
- max of IMEP and efficiency– value 1,
- min of IMEP and efficiency – value 0,

• Determining the totals of all quantities for all 18 operating points,

• Arbitrary selection of the impact factors (*IF*) for every emission component, efficiency and IMEP-values (Σ = 1),

• Calculation of the products of sums and impact factors,

• Determining the resulting sums.

Normalization of the measured quantities (*Norm*) was performed in the range <0;1>. This means that each variable (IMEP, e_CO_L, e_NOx, eta_o) in a set of 54 test points (18 points × 3 chambers) has been assigned a value in the range described above (in this set there is only one value 1 – global maximum and only one value 0 – local minimum).

It follows that the *Ind* indicator value is calculated as:

$$\mathrm{Ind(Ch)} = \sum_{\mathrm{Ch}=1}^{3} \sum_{j=1}^{18} \mathrm{Norm(Ind(Ch))} \qquad (1)$$

where *Ch* is the pre-chamber, *j* – the number of the tested operating point.

The procedure presented here made it possible to assign numerical values to individual test chambers. An importance weight of 0.25 was applied to each measuring quantity.

A total value of the indicator (*Ov*) describing the given pre-chamber was determined as the product of the impact factor (*IF*) and the value of the indicator (*Ind*) in the form:

$$Ov(Ch) = \sum_{i=1}^{4} IF_i \times Ind_i \quad (2)$$

where *i* is the analysed indicator (eCO_L, eta_o, eNOx and IMEP).

Using the described procedure for the best chamber selection, the obtained results were combined in Table 2.

Table 2. Values of the tested parameters of pre-chambers assuming the same impact factor (*IF* = 0.25 for each measured parameter).

Factor (*IF*)	Chamber (*Ch*) / Indicator (*Ind*)	1	2	3
0.25	eCO_L	6.41	5.52	7.95
0.25	eta_o	6.60	7.76	7.85
0.25	eNOx	9.14	9.85	7.73
0.25	IMEP	7.92	6.53	7.17
	Overall (*Ov*)	7.52	7.42	7.67

The comparison presented in Table 2 shows that the best choice with the criteria applied is a chamber characterized by a large volume (V_{ch} = 2.287 cm^3), and a small outlets diameter (*D* = 1.5 mm) assuming simple outflow channels.

Conclusions

The performed analyses allowed for formulating the following conclusions:

– a two-stage combustion system enables combustion of lean gas-air mixtures with an excess air ratio far exceeding the value of λ = 1 (up to approx. 1.8).

– the pre-chamber size and the number of holes are very important when burning lean mixtures. A small volume of the pre-chamber accelerates the combustion process, however, it does not lead to obtain the maximum average indicated pressure (IMEP) value. When reducing the excess air ratio, this tendency decreases.

– when using methane-air mixtures with λ = 1.3 and less, the impact of the chamber size and the diameter of the holes is negligible in relation to the average thermodynamic processes. A similar effect of the excess air ratio was noted for the emissions of CO and NO_x.

According the analysis performed in the research the optimal results were achieved for bigger volume of prechamber and smaller outlet channels, probably because of higher mixture spray energy and better activation of lean mixture in main chamber.

Acknowledgements

The research presented in this paper was performed within the European Research Project Horizon 2020, grant agreement No. 652816, Gas-Only Internal Combustion Engines. The authors wish to thank the EU for funding the work and express their acknowledgements to all persons, who have had their contribution in this study and the laboratory works.

References

1. M. Baratta, D. Misul, P. Goel, D. Laurenzano, B. Lecointe, L. Rouleauet, F. Ravet, P. Christou, SAE Technical Paper 2018-01-1142 (2018)
2. M. Ferrera, SAE Technical Paper 2017-24-0059 (2017)
3. U. Kramer, T. Lorenz, C. Hofmann, H. Ruhland, R. Klein, C. Weber, SAE Technical Paper 2017-01-0776 (2017)
4. M. Gholamisheeri, I.S. Wichman, E. Toulson, Combust. Flame **183** 194–206 (2017)
5. G. Gentz, B. Thelen, M. Gholamisheeri, P. Litke, A. Brown, J. Hoke, E. Toulson, Appl. Therm. Eng. **81** 399–411 (2015)
6. S. Biswas, L. Qiao, Appl. Therm. Eng. **132** 102–114 (2018)
7. H. Wu, L. Wang, X. Wang, B. Sun, Z. Zhao, C. Lee, F. Liu, Int. J. Hydrogen Energ. **43**, 16 8116–8126 (2018)
8. G. Gentz, M. Gholamisheeri, E. Toulson, Appl. Energ. **189** 385–394 (2017)
9. I. Pielecha, K. Wisłocki, W. Cieślik, P. Borowski, W. Bueschke, M. Skowron, SAE Technical Paper 2016-01-2305 (2016)
10. I. Pielecha, K. Wisłocki, W. Cieślik, W. Bueschke, M. Skowron, Ł. Fiedkiewicz, Appl. Therm. Eng. **132** 188–195 (2018)
11. I. Pielecha, W. Bueschke, W. Cieślik, M. Skowron, MATEC Web of Conferences, **118** 00010-1–00010-10 (2017)

Increasing electrical power output and fuel efficiency of gas engines in integrated energy system by absorption chiller scavenge air cooling on the base of monitoring data treatment

Andrii Radchenko[1,*], *Mykola* Radchenko[1], *Andrii* Konovalov[1], and *Anatolii* Zubarev[1]

[1]Admiral Makarov National University of Shipbuilding, Heroes of Ukraine Avenue 9, Mykolayiv, Ukraine

Abstract. An advanced scavenge air cooling system for reciprocating gas engines of integrated energy system for combined electricity, heat and refrigeration generation has been developed. New method of deep scavenge air cooling and stabilizing its temperature at increased ambient air temperatures and three-circuit scavenge air cooling system with absorption lithium-bromide chiller and wet-type cooling tower was proposed. Such cooling method does not require essential constructive changes in the existing scavenge air cooling system but only an addition heat exchanger for chilling scavenge air cooling water of scavenge air low-temperature intercooler closed contour by absorption chiller. A chilled water from absorption chiller is used as a coolant. To evaluate the effect of gas engine scavenge air deeper cooling compared with its typical radiator cooling, data on the dependence of fuel consumption and power output of gas engine on ambient air temperature at the inlet of the radiator are analized. The efficiency of engine scavenge air deep cooling at increased ambient air temperatures was estimated by reducing the gas fuel consumption compared with radiator cooling.

1 Introduction

Reciprocating gas engines (RGE) found a widespread application in integrated energy systems (IES) or so called trigeneration systems for combined electricity, heat and refrigeration generation. They are manufactured as cogeneration modules equipped with heat exchangers for producing hot water used also as a heat source for absorption chiller [1–3].

The specific fuel consumption of gas engines increases with arising turbocharger (TC) suction air and scavenge gas-air mixture temperatures at the inlet of engine cylinders. The cooling problem arises especially actually for scavenge air, which heat is rejected traditionally to surroundings by radiator (cooling tower of dry type) through intermediate water cooling contour of engine low-temperature intercooler. The efficiency of such heat rejecting system drops sharp with increasing ambient air temperature above 25…30 °C, when it becomes impossible to cool scavenge air to appropriate temperature of about 40 °C. If scavenge air temperature exceeds the maximum temperature restriction of 50 °C the engine efficiency drops sharp. For protection of engine from negative impact of high scavenge air temperature on the thermal conditions in the combustion cylinders the scavenge air temperature of about 40 °C is maintained automatically by reducing gas supply to engine and, accordingly, engine load.

At high ambient air temperature the exceeding scavenge air temperature causes operation of gas engine on increasing specific gas fuel consumption compared with rated load mode. So, for maintenance of gas engine on rated load or close modes it is necessary more decrease in temperature of cooling water in intermediate cooling contour of low-temperature intercooler. For this purpose a chilled water received in absorption chiller might be used.

To evaluate the effect of gas engine scavenge air deeper cooling compared with its typical radiator cooling, data on the dependence of fuel consumption and power output of gas engine on ambient air temperature at the inlet of the radiator are to be analized. A method for treatment of the monitoring data on fuel consumption and power output of gas engine was presented in [4].

The goal of the research is improving the fuel efficiency of IES on the base of reciprocating gas engines (RGE) while operation at increased ambient air temperatures through using an absorption lithium-bromide chiller (ABCh) for deeper scavenge air cooling in comparison with radiator cooling.

2 Investigating the efficiency of typical gas engine scavenge air cooling system

The efficiency of cooling gas engine scavenge air is investigated for IES of combined energy supply at the factory "Sandora"–"PepsiCo Ukraine" (Nikolaev, Ukraine). The integrated energy system is equipped with 2 cogeneration Jenbacher gas engines JMS 420 GS-N.LC (rated electric power P_{eISO} = 1400 kW, heat power

* Corresponding author: nirad50@gmail.com

P_h =1500 kW) and absorption lithium-bromide chiller, designed and assembled by "Sinapse"–"GE Energy" (Kiev) and "Khladotechnika" (Nikolaev). The heat of engine scavenge air is rejected to the environment by radiator as cooling tower of dry type.

The absorption chiller utilizes the heat of engine exhaust gas, high temperature scavenge air, engine jacket and lubricant oil cooling water to produce a chilled water with temperature of about 7 °C for technology process cooling and conditioning of air in engine room by central conditioners.

The scheme of conventional integrated energy system of factory "Sandora"–"PepsiCo Ukraine" is presented in Fig. 1. The engine room intake air is cooled in the central conditioner by chilled water from absorption lithium-bromide chiller and the engine turbochargers take air from engine room.

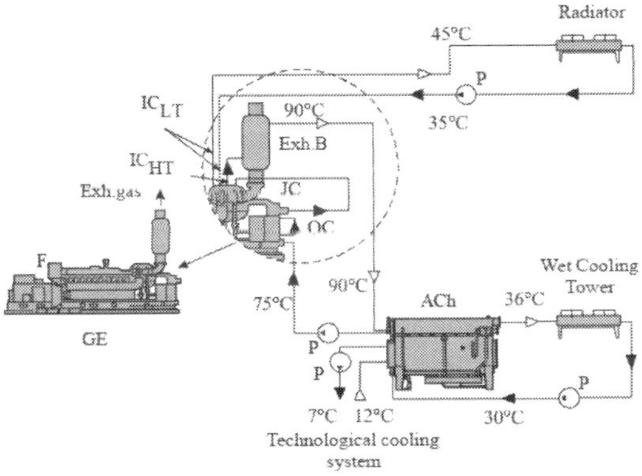

Fig. 1. The scheme of integrated energy system (trigeneration system) of factory "Sandora"–"PepsiCo Ukraine": GE – gas engine; OC – oil cooler; JC – jacket cooler; ExhB – exhaust gas boiler; IC_{LT} and IC_{HT} – low- and high-temperature intercoolers of charged gas-air mixture; F – intake air filter; AC – intake air cooler; ACh – absorption chiller; P –pump.

Intercooler of scavenge air consists of two stages: high-temperature stage (IC_{HT}), in which cooling of scavenge air-gas mixture is followed by heating water for absorption chiller, and low-temperature stage (IC_{LT}), from which the heat is rejected to the environment by radiator-cooling tower of dry type.

Some results of monitoring daily variation of volume gas supply B_e and electric power output P_e of gas engine JMS 420 GS-N.LC automatically changed to maintain a charged gas-air mixture temperature at the level of about 40 °C are presented in Fig. 2 [5].

In summer hot day time interval τ = 9…20 h the ambient air temperatures are rather high: t_{amb} = 30…35 °C, that makes impossible reliable cooling of scavenge gas-air mixture by radiator to appropriate temperature rate of 40 °C. This results in automatically reducing gas supply to engine followed by decreasing load to maintain gas-air mixture temperature at the level of 40 °C (Fig. 2).

A performance of gas engine on part load modes (τ = 9…20 h) is followed by increase in specific gas consumption b_e by $(20…30)\cdot10^{-3}$ m^3/(kW·h) (Fig. 3), i.e. 8…12 % compared with engine full loading at temperatures t_{amb} and t_{in} more low 25 °C (τ = 2…9 h).

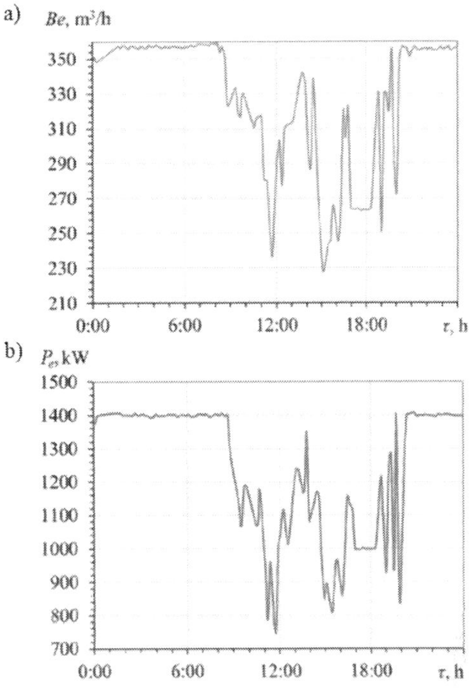

Fig. 2. Daily variation of volume gas supply B_e (**a**) and electric power P_e (**b**) of gas engine with time τ (July 17, 2011).

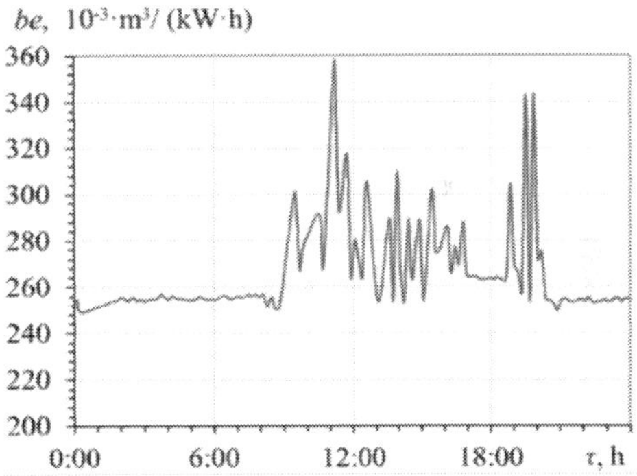

Fig. 3. Daily variation of engine specific volume gas consumption b_e with time τ (July 17, 2011).

A reduction of time duration of engine performance at part load modes (τ = 9…21 h, Fig. 3) and decrease of specific gas consumption b_e is possible by addition decreasing temperature of scavenge air cooling water by absorption chiller before it entering the engine low-temperature intercooler IC_{LT} to maintain a scavenge gas-air mixture temperature at the appropriate level of 40 °C with variation of ambient temperatures t_{amb} .

3 Investigating the efficiency of improved gas engine scavenge air cooling system

A reduction of time duration of engine performance at part load modes (τ = 9…21 h, Fig. 3) and decrease of specific gas consumption b_e is possible by addition decreasing temperature of scavenge air cooling water by absorption chiller before it entering the engine low-temperature intercooler IC_{LT} to maintain a scavenge gas-air mixture temperature at the appropriate level of 40 °C with variation of ambient temperatures t_{amb} .

A principal scheme of improved three-circuit scavenge air cooling system with absorption lithium-bromide chiller and wet-type cooling tower is presented in Fig. 4. Such cooling system does not require constructive changes in the scavenge air channel of gas engine and needs only an addition heat exhanger for heat transfer between the circuits connected to the closed water circuit of a traditional scavenge air cooler.

Thus chilled water with temperature t_c = 7 °C is directed from absorption chiller to addition plate water cooler, inserted into the current closed water cooling contour, to cool a cooling water at the inlet of the engine low-temperature intercooler IC_{LT} (Fig. 4).

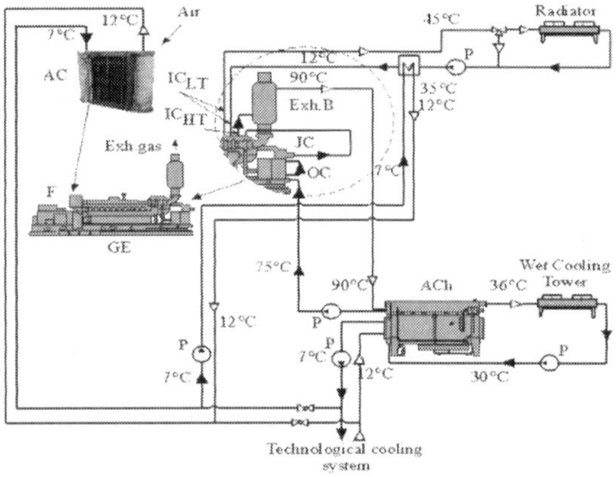

Fig. 4. The scheme of modified trigeneration system with addition cooling of cooling water for low-temperature intercooler of scavenge gas-air mixture: GE – gas engine; OC – oil cooler; JC – jacket cooler; ExhB – exhaust gas boiler; IC_{LT} and IC_{HT} – low- and high-temperature intercoolers of scavenge gas-air mixture; F – intake air filter; ACh – absorption chiller; P –pump; Heat Exh – heat exchanger.

The results of monitoring of JMS 420 GS-N.L engine fuel efficiency were presented in the form of data sets on dependence of fuel consumption $B_e = f(t_{amb})$ and engine power output $P_e = f(t_{amb})$ upon the ambient air temperature t_{amb} at the entrance to the radiator of scavenge air cooling for the various air temperatures t_{in} at the inlet of the turbocharger. The goal of monitoring data sets $P_e = f(t_{amb})$ and $B_e = f(t_{amb})$ treatment is to calculate the magnitude of the change in power ΔP_e and fuel consumption ΔB_e, caused by the change in the ambient air temperature t_{amb} at the entrance to the radiator of scavenge air cooling by 1 °C, that is $\Delta P_e/\Delta t_{amb}$ and $\Delta B_e/\Delta t_{amb}$, to evaluate the effect of the application of developed scavenge air cooling system.

The results of treatment of data for variation of electrical power output in absolute values P_e and relative values P_e/P_{eISO} compared to ISO value with changing ambient air temperature t_{amb} at the inlet of radiator for scavenge air cooling and also power output drops $\Delta P_e/\Delta t/P_{eISO}$ for 1 °C increment of ambient air temperature Δt_{amb} = 1 °C at the inlet of radiator for scavenge air cooling at various air temperatures at the inlet of turbochanger t_{in} = 20…30 °C for Jenbacher gas engine JMS 420 GS-N.LC at about rated electrical power output P_e = 1250…1400 kW (P_{eISO} = 1400 kW) and at part loads P_e = 850…1000 kW are presented below in Fig.5-8.

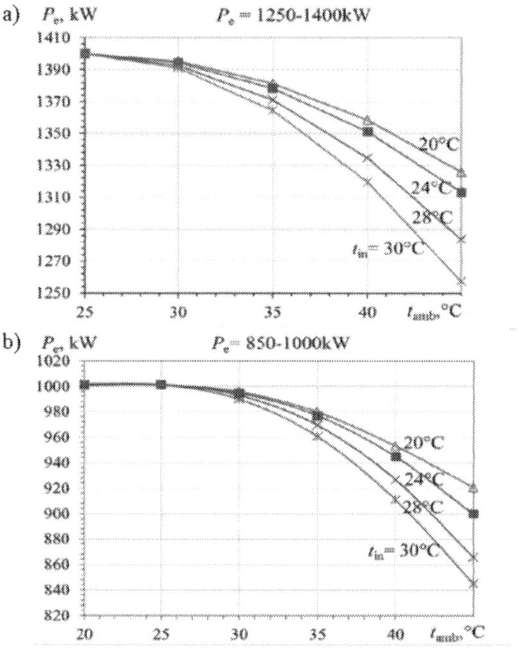

Fig. 5. Decrease of electrical power output P_e versus increasing ambient air temperature t_{amb} at the inlet of radiator for scavenge air cooling at various air temperatures at the inlet of turbochanger t_{in} = 20…30 °C for Jenbacher gas engine JMS 420 GS-N.LC at about rated electrical power output P_e = 1250…1400 kW (***a***) and at part loads P_e = 850…1000 kW (***b***).

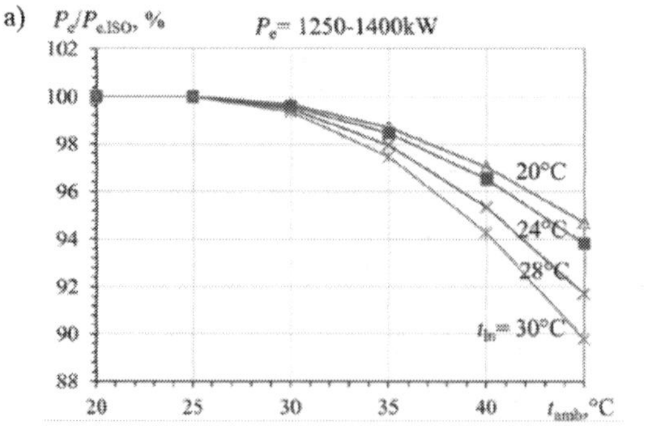

 https://doi.org/10.1051/e3sconf/20187003011

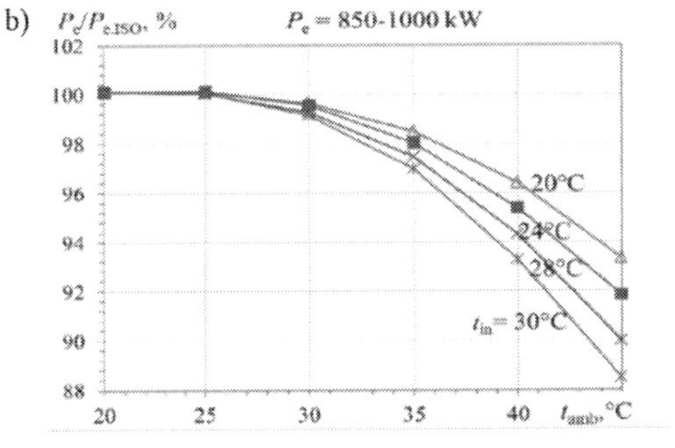

Fig. 6. Decrease of electrical power output in relative values P_e/P_{eISO} compared to ISO value versus increasing ambient air temperature t_{amb} at the inlet of radiator for scavenge air cooling at various air temperatures at the inlet of turbochanger t_{in} = 20…30 °C for Jenbacher gas engine JMS 420 GS-N.LC at about rated electrical power output P_e = 1250…1400 kW (***a***) and at part loads P_e = 850…1000 kW (***b***).

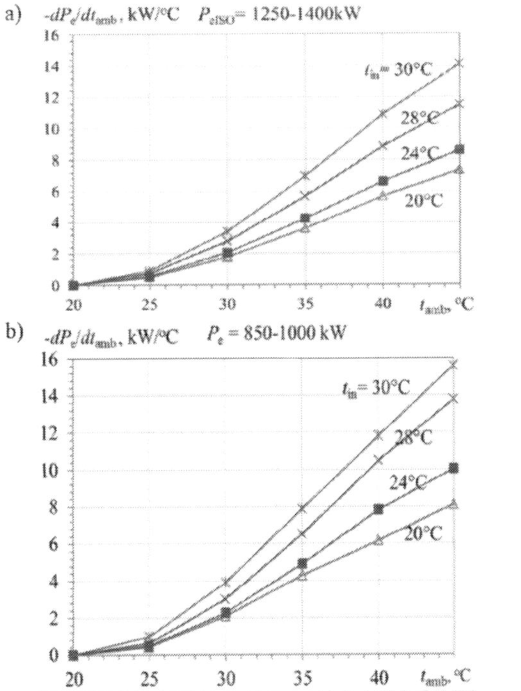

Fig. 7. Increase of electrical power output drop $\Delta P_e/\Delta t$ for 1 °C increment of ambient air temperature Δt_{amb} = 1 °C at the inlet of radiator for scavenge air cooling versus increasing ambient air temperature t_{amb} at the inlet of radiator at various air temperatures at the inlet of turbochanger t_{in} = 20…30 °C for Jenbacher gas engine JMS 420 GS-N.LC at about rated electrical power output P_e = 1250…1400 kW (***a***) and at part loads P_e = 850…1000 kW (***b***).

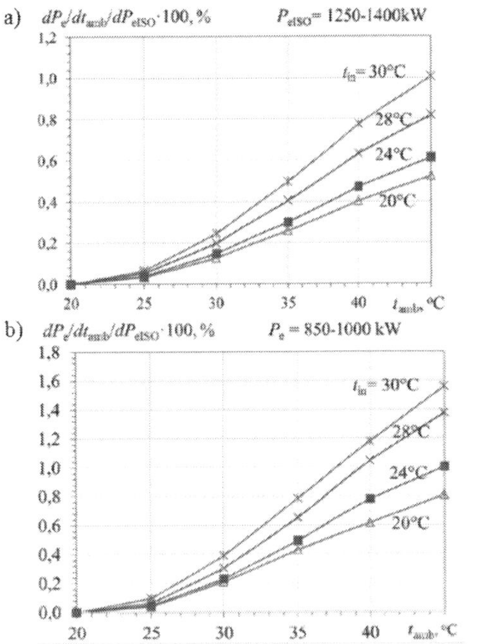

Fig. 8. Increase of electrical power output drop for 1 °C increment Δt of ambient air temperature Δt_{amb} = 1 °C at the inlet of radiator for scavenge air cooling in relative values $\Delta P_e/\Delta t/P_{eISO}$ compared to ISO value versus increasing ambient air temperature t_{amb} at the inlet of radiator at various air temperatures at the inlet of turbochanger t_{in} = 20…30 °C for Jenbacher gas engine JMS 420 GS-N.LC at about rated electrical power output P_e = 1250…1400 kW (***a***) and at part loads P_e = 850…1000 kW (***b***).

The results of treatment of monitoring data on volume gas fuel consumption in absolute values B_e and relative values B_e/B_{eISO} compared to ISO value and also increase in volume gas fuel consumption $B_e/\Delta t$ and increase in engine specific volume gas fuel consumption $\Delta b_e/\Delta t$ for 1 °C increment Δt of ambient air temperature Δt = 1 °C at the inlet of radiator for scavenge air cooling with changing ambient air temperature t_{amb} at the inlet of radiator are presented in Fig. 9-12.

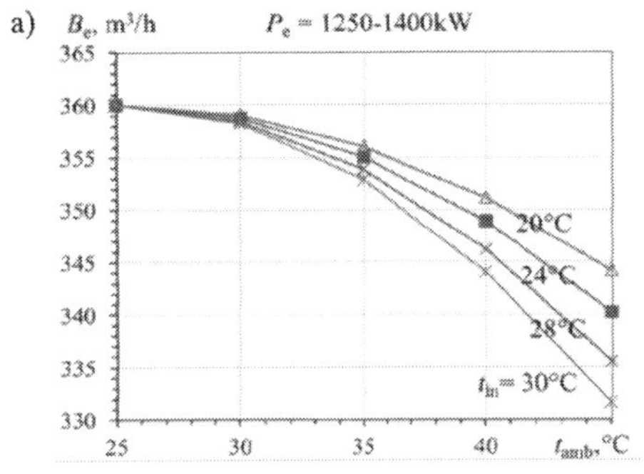

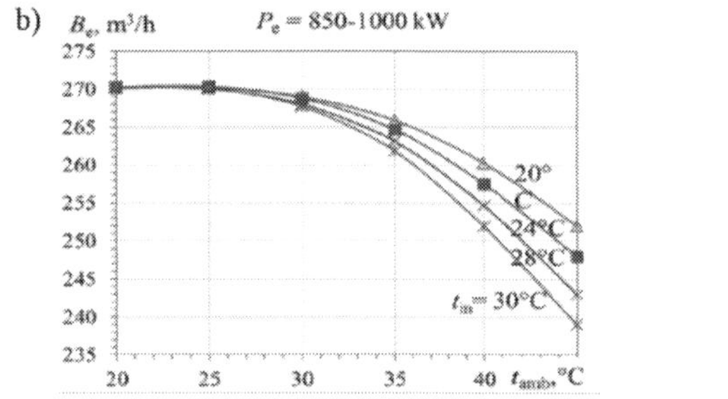

Fig. 9. Decrease of volume gas fuel consumption B_e versus increasing ambient air temperature t_{amb} at the inlet of radiator for scavenge air cooling at various air temperatures at the inlet of turbochanger t_{in} = 20…30 °C for Jenbacher gas engine JMS 420 GS-N.LC at about rated electrical power output P_e =1250…1400 kW (***a***) and at part loads about P_e =850…1000 kW (***b***).

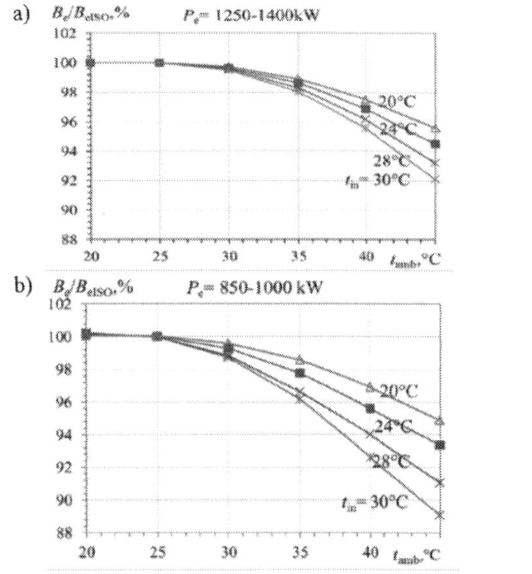

Fig. 10. Decrease of volume gas fuel consumption in relative values B_e/B_{eISO} compared to ISO value versus increasing ambient air temperature t_{amb} at the inlet of radiator for scavenge air cooling at various air temperatures at the inlet of turbocharger t_{in} = 20…30 °C for Jenbacher gas engine JMS 420 GS-N.LC at about rated electrical power output P_e =1250…1400 kW (***a***) and at part loads about P_e =850…1000 kW (***b***).

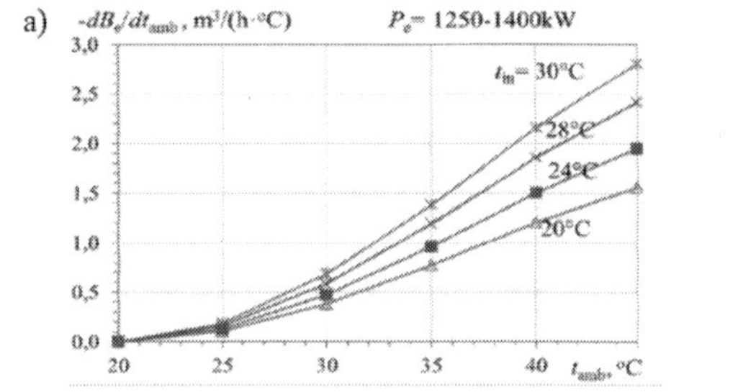

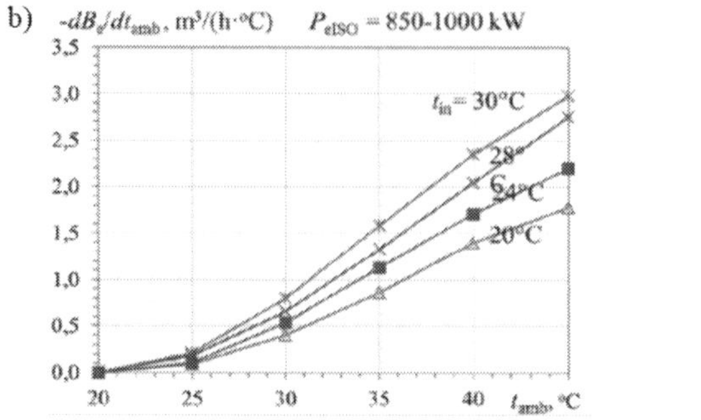

Fig. 11 Increase of volume gas fuel consumption $B_e/\Delta t$ for 1 °C increment Δt of ambient air temperature Δt_{amb} = 1 °C at the inlet of radiator for scavenge air cooling versus increasing ambient air temperature t_{amb} at the inlet of radiator at various air temperatures at the inlet of turbocharger t_{in} = 20…30 °C for Jenbacher gas engine JMS 420 GS-N.LC at about rated electrical power output P_e =1250…1400 kW (***a***) and at part loads about P_e = 850…1000 kW (***b***).

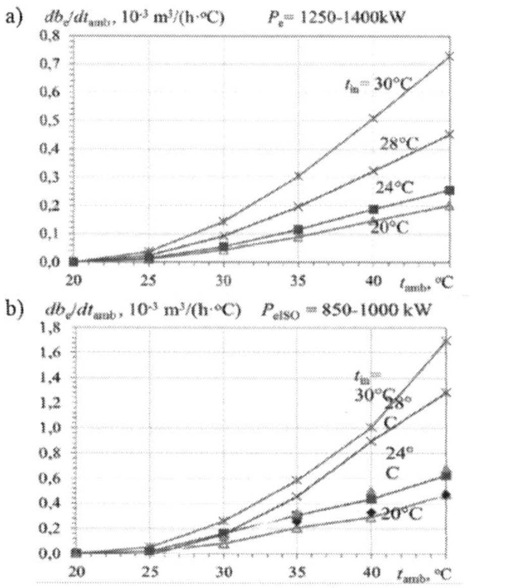

Fig. 12. Increase of engine specific volume gas fuel consumption $b_e/\Delta t$ for 1 °C increment Δt of ambient air temperature Δt = 1 °C at the inlet of radiator for scavenge air cooling versus increasing ambient air temperature t_{amb} at the inlet of radiator at various air temperatures at the inlet of turbocharger t_{in} = 20…30 °C for Jenbacher gas engine JMS 420 GS-N.LC at about rated electrical power output P_e =1250…1400 kW (***a***) and at part loads about P_e = 850…1000 kW (***b***).

Thus, from Fig. 12,***a***, at operation of JMS 420 GS-N.L at about rated power (1200…1400 kW) and high ambient air temperatures t_{amb} = 30...35°C at the inlet of radiator for scavenge air cooling and at the TC intake air temperatures t_{in} = 25...30 °C the engine specific volume gas fuel consumption increase $\Delta b_e/\Delta t$ for 1°C increment of ambient air temperature Δt = 1°C at the inlet of radiator may be assumed of about (0.5…1.0)·$\cdot 10^{-3}$ m^3/(kW·h). A three-circuit scavenge air cooling system with ABCh (Fig.4) allows to maintain practically unchangeable scavenge air cooling water temperature of about t_{cw} = 10...12 °C with the temperature of cooled water from ABCh of 7 °C and ambient air temperatures t_{amb} = 30...35 °C. So such system can provide scavenge air cooling water temperature decrease of about 20...25 °C compared with conventional scavenge air cooling system with radiator, that results in reduction of engine specific fuel consumption by about Δb_e = (10…12)·10^{-3} m^3 /(kW·h), i.e. about 4…5% decrease in specific fuel consumption at increased ambient air temperatures t_{amb} = 30...35 °C. At part loads 850…1000 kW a reduction of specific fuel consumption is nearly twice higher (Fig. 12,**b**).

The efficiency of deep scavenge air cooling by absorption chiller and stabilization of its temperature for JMS 420 GS-N.L results in reduction of the cost of 1 kW generated electricity by 5…7 % compared with traditional radiator cooling.

4 Conclusions

A treatment of monitoring data on electrical power output and fuel consumption of gas engine JMS 420 GS-N.L in integrated energy system for combined electricity, heat and refrigeration generation has proved non-effective performance of conventional scavenge air cooling by radiator at increased ambient temperatures. The improved gas engine scavenge air cooling system with absorption chiller conversing waste heat of engine into refrigeration was proposed.

An advanced scavenge air cooling system allows to maintain practically stable scavenge air cooling water temperature of about 10...12 °C with the temperature of cheeled water from absorption chiller of 7 °C at high ambient air temperatures. This system provides an increase of engine electrical power output by 2…3 % and decrease of specific fuel consumption by about 4…5% as compared with conventional radiator scavenge air cooling at increased ambient air temperatures 30...35 °C. At part loads the effect is much more.

For modernization of the existing traditional scavenge air cooling system for gas engine of integrated energy system for combined electricity, heat and refrigeration generation such cooling method does not require essential constructive changes but only an addition heat exchanger for chilling scavenge air cooling water of scavenge air low-temperature intercooler closed contour by absorption chiller.

References

1. Economic utilization of Biomass and Municipal Waste for power generation. Some energy lasts for generations. *GE Jenbacher Company Overview*. June 13 (2007), 39 p.
2. T. Elsenbruch, Jenbacher gas engines a variety of efficient applications: Bucureşti, October 28 (2010), 73 p.
3. GTI Integrated Energy System for Buildings. Modular System Prototype: G. Rouse, M. Czachorski, P. Bishop, J. Patel *GTI Project report 15357/65118*: Gas Technology Institute (GTI). January 2006, 495 p.
4. А.Н. Радченко, А.В. Коновалов, Методология обработки данных мониторинга топливной эффективности газового двигателя с системой оборотного охлаждения газовоздушной смеси. *Авиационно-космическая техника и технология*, № 3, vol.110 (2014), С. 82–85. [Radchenko, A.N., Konovalov A.V. Metodology of treatment of monitoring data on fuel efficiency of gas engine with circulating system for cooling gas-air mixture. *Avitsionno-kosmicheskaya tehnika i tehnologiya* [*Aerospace technics and technology*]. No.3, vol.110 (2014), pp. 82–85.
5. A.M.Radchenko, A.V. Konovalov, Cooling cyclic intake air and charged gas-air mixture in gas engine of plant for combined energy supply. *Proceedings of the 14 International Symposium on Heat Transfer and Renewable Sources of Energy: HTRSE–2012*, Szczecin, Poland (2012), pp. 497–502.

Gas turbine unite inlet air cooling by using an excessive refrigeration capacity of absorption-ejector chiller in booster air cooler

Roman Radchenko[1], *Andrii* Radchenko[1,*], *Serhiy* Serbin[1], *Serhiy* Kantor[1], and *Bohdan* Portnoi[1]

[1]Admiral Makarov National University of Shipbuilding, pr. Heroes of Ukraine, 9, Mykolaiv, Ukraine

Abstract. Two-stage Gas turbine unite (GTU) inlet air cooling by absorption lithium-bromide chiller (ACh) to the temperature 15 °C and by refrigerant ejector chiller (ECh) to 10 °C through utilizing the turbine exhaust gas heat for changeable ambient air temperatures and corresponding heat loads on the air coolers for the south Ukraine climatic conditions is analysed. An excessive refrigeration capacity of combined absorption-ejector chiller (AECh) exceeding the current heat loads and generated at decreased heat loads on the air coolers at the inlet of GTU can be used for covering increased heat loads to reduce the refrigeration capacity of AECh. The GTU inlet air cooling system with an ambient air precooling booster stage and a base two-stage cooling air to the temperature 10 °C by AECh is proposed. The AECh excessive cooling capacity generated during decreased heat loads on the GTU inlet air coolers is conserved in the thermal accumulator and used for GTU inlet air precooling in a booster stage of air cooler during increased heat loads. There is AECh cooling capacity reduction by 50% due to the use of a booster stage for precooling GTU inlet ambient air at the expense of an excessive cooling capacity accumulated in the thermal storage.

1 Introduction

Gas turbine unit (GTU) inlet air cooling by waste heat recovery refrigeration machine (HRRM) which transforms the exhaust gas heat into the cold is one of the main trends in increasing GTU efficiency. Gas turbine inlet air can be cooled to the temperature $t_{a2} \approx$ 15 °C by transformation of waste heat into the cold with a high efficiency in the most widespread absorption lithium-bromide chiller (ACh). Its coefficient of performance is $\zeta = 0.7...0.8$ [1, 2]. Deeper air cooling to the temperature t_{a2} = 10 °C and below is possible in refrigerant ejector chiller (ECh). The efficiency of transformation of waste heat into the cold by ECh is much less: $\zeta = 0.2...0.3$ [3]. Therefore it is rational to use two-stage air cooling at the GTU inlet: to t_{a2} = 15...20 °C in the ACh and to t_{a2} = 7...10 °C in the ECh, that is in the combined absorption-ejector chiller (AECh) [4, 5]. The heat load on the air cooler (AC) at the GTU inlet changes significantly according to the current ambient air temperature t_{amb} and relative humidity φ_{amb} that leads to adequate change of AECh cooling capacity. A cooling capacity reserve (AECh installed excessive heat load value compared with current heat loads) is conserved during reduced heat loads. It is expedient to cover the deficit of cooling capacity at high loads, thereby reducing the ACh installed cooling capacity and it cost.

Objective: reduction of the installed absorption chiller cooling capacity by conserving an excessive cooling capacity at low current heat loads on the air cooler at the inlet of GTU and using the reserved cooling capacity for ambient air precooling at increased current heat loads.

2 Results of investigation

The GTU inlet air cooling efficiency can be conveniently estimated by the annual temperature-hour potential, $\prod_\Sigma$,°C·h, which represents the air temperature drop Δt_a at the GTU inlet and the duration τ of GTU operation at reduced temperatures [4]

$$\prod_\Sigma = \sum(\Delta t_a \cdot \tau).$$

To determine the specific cooling capacity of two-stage AECh as the total cooling capacity related to the unit of air flow (G_a = 1 kg/s)

$$q_0 = Q_0/G_a,$$

it is necessary to analyze the annual temperature-hour potential $\prod_\Sigma$° dependence on specific cooling capacity q_0. The specific cooling capacity is calculated as

$$q_0 = \xi \cdot c_{ma} \cdot (t_{amb} - t_{a2}),\ \text{kW/(kg/s), or kJ/kg},$$

where ξ is the moisture coefficient; t_{amb} – ambient air temperature, °C; t_{a2} – air temperature at the air cooler outlet, °C; c_{ma} – moist air specific heat, kJ/(kg·K).

* Corresponding author: nirad50@gmail.com

 https://doi.org/10.1051/e3sconf/20187003012

The project (installed) specific cooling capacity q_0, on the one hand, should cover the GTU inlet air cooling need for as long time as possible during the year, which would have the greatest effect in the fuel saving. On the other hand, it should not be overestimated so that for most of the year the AECh was operated at thermal loads close to the project value, otherwise there would be an AECh operation at partial loads far from the project value. And on the contrary, with the lowered project q_0 there will be undercooling of the GTU inlet air at high ambient temperatures t_{amb} .

Fig. 1 shows the value of annual specific fuel saving b_e depending on the specific cooling capacity values q_0 (with G_a = 1 kg/s) for ambient air cooled to the temperatures t_{a2} = 15 °C in ACh and to t_{a2} = 10 °C in two-stage AECh for 2017 year, Yuzhnoukrainsk city, Mykolaiv region, Ukraine.

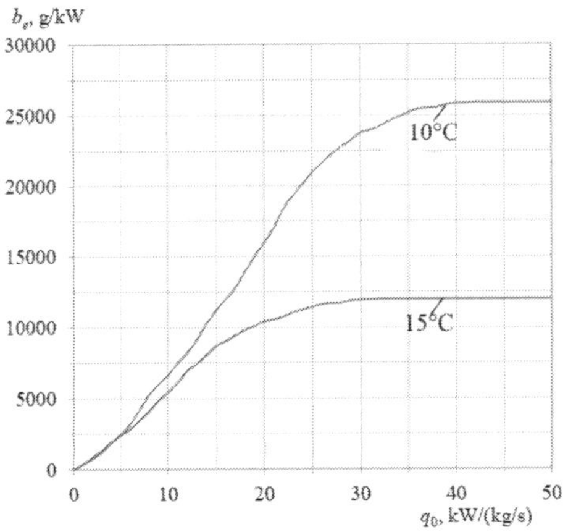

Fig. 1. The value of annual specific fuel saving b_e depending on the specific cooling capacity values q_0 (with G_a = 1 kg/s) for cooling an ambient air to the temperatures t_{a2} = 10 °C in AECh and to 15 °C in Ach.

As can be seen from Fig. 1, for the climatic conditions of the GTU operation in the Yuzhnoukrainsk city, Mykolaiv region, 2017, a specific cooling capacity q_0 = 34 kW/(kg/s) (for G_a = 1 kg/s) can be considered as rational project AECh specific cooling capacity for cooling ambient air at the inlet of GTU to the temperature t_{a2} = 10 °C at which the annual specific fuel saving b_e is kept closed to the maximum value at high enough rate of its increment. When air is cooled to t_{a2}°=15 °C in ACh the rational project AECh specific cooling capacity is respectively q_0 = 24 kW/(kg/s). Proceeding from the rational specific cooling capacity q_0 the total installed cooling capacity of AECh in accordance with the air flow G_a through the GTU is chosen as

$$Q_0 = G_a \cdot q_0, \text{ kW}.$$

Since a decrease in the rate of the annual specific fuel saving increment b_e at high cooling capacity indicates that there is an excess of cold, it is advisable to determine the cooling capacity, which ensures its maximum build-up rate, which will be less than q_0 = 24 kW/(kg/s) for t_{a2} = 15 °C for ACh, as shown in Fig. 1.

To determine the cooling capacity, which provides a maximum build-up rate, it is necessary to analyze the annual specific fuel saving b_e related to the specific cooling capacity q_0, b_e/q_0, in dependence on the specific cooling capacity values q_0 (Fig. 2) for the cooled air temperatures t_{a2} = 10 and 15 °C.

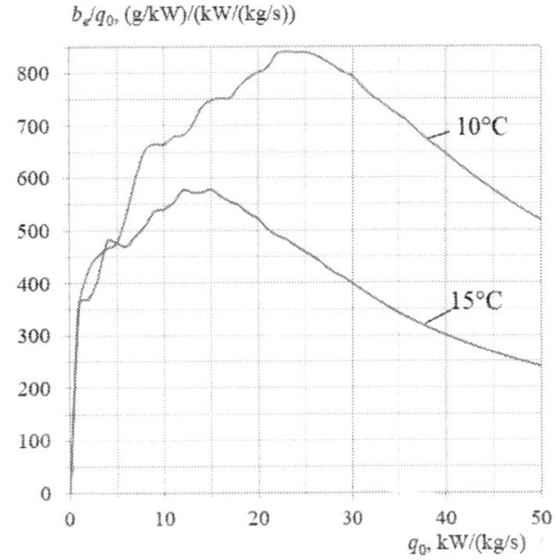

Fig. 2. The values of annual specific fuel saving related to the specific cooling capacity b_e/q_0 in dependence on the project specific cooling capacity q_0 for the cooled air temperature t_{a2} = 10 and 15 °C.

As can be seen from Fig. 2, for the climatic conditions of the GTU operation, the maximum rate of increment in the annual specific fuel saving b_e/q_0 due to cooling (maximum of the graph) for the temperature of air t_{a2} = 15 °C cooled in the ACh takes place at the project specific cooling capacity values q_0 = =12…16 kW/(kg/s), and when air is cooled to t_{a2} = 10 °C in two-stage AECh q_0°=°22…26°kW/(kg/s). The corresponding total project (installed) cooling capacities,

$$Q_0 = G_a \cdot q_0, \text{ kW}.$$

provide the maximum rate of the annual specific fuel saving increment.

Since the project (installed) specific cooling capacity values q_0, which ensure the maximum rate of increment in the annual specific fuel saving b_e/q_0 (Fig. 2) are less than their values determined in accordance with the maximum annual specific fuel saving b_e in Fig. 1, then at an increased ambient air temperature t_{amb} there will be a deficit of cooling capacity, while at low ambient air temperatures t_{amb}, on the contrary, its excess. The excess of cooling capacity, generated during reduced heat load periods, is expedient to accumulate in the cold storage and use during increased heat loads in the pre-cooling air booster stage AC_B of air cooler.

With this the reserve of ACh cooling capacity, generated during current reduced heat loads, is defined as its excess compared with the current heat load on the basic high-temperature stage AC_{HT} and the pre-cooling air booster stage AC_B, that is, on the heat exchanger

"$AC_{HT} + AC_B$". The ECh cooling capacity reserve is also defined as the project ECh heat load excess over the current heat loads on the low-temperature stage AC_{LT} of air cooler and the total cooling capacity reserve of AECh is the sum of ACh and ECh reserves.

The total project heat load on the AC_{HT} is defined as

$$Q_{0.HT16pr} = q_{0\text{-}HT16pr} \cdot G_a = 16 \cdot 40 = 640 \text{ kW}$$

and is based on the AC_{HT} project specific heat load $q_{0\text{-}HT16pr} = 16$ kW/(kg/s). With this the project heat load on the AC_{HT} with pre-cooling air booster stage AC_B:

$$Q_{0.HT24pr} = q_{0.HT24pr} \cdot G_a = 24 \cdot 40 = 960 \text{ kW},$$

where $q_{0.HT24pr} = 24$ kW/(kg/s) – the total specific heat load on both AC_{HT} and AC_B, that provides the maximum annual fuel saving due to GTU inlet air cooling to the temperature $t_{HT2} \approx 15$ °C in ACh according to Fig. 1 [3]. The air consumption is $G_a = 40$ kg/s. The project heat load on the AC_{LT} is defined as

$$Q_{0.LT10pr} = q_{0.LT10pr} \cdot G_a = 10 \cdot 40 = 400 \text{ kW},$$

and based on the project specific heat load on the AC_{LT} $q_{0.LT10pr} = 10$ kW/(kg/s) according to the maximum annual effect (Fig. 1).

In determining the ACh cooling capacity reserve, the current heat load on the AC_{HT} with booster AC_B is compared with the project heat load

$$Q_{0.HT24pr} = 24 \cdot 40 = 960 \text{ kW},$$

where $q_{0.HT24pr} = 24$ kW(kg/s).

Partial replacement of high-temperature stage AC_{HT} by pre-cooling air booster stage AC_B with corresponding reduction of the project cooling capacity of ACh is possible if there is a reserve of the ACh cooling capacity, accumulated in the cold storage, which exceeds its spending for the ambient air precooling in the AC_B. In accordance with this, the ACh excessive cooling capacity (reserve) is defined as

$$Q_{0.HT16exc} = Q_{0.HT16pr} - Q_{0.HT24} = 640 - Q_{0.HT24}, \text{ kW},$$

and its deficit is respectively:

$$Q_{0.HT16def} = Q_{0.HT24} - Q_{0.HT16pr} = Q_{0.HT24} - 640, \text{ kW}.$$

The ECh excessive cooling capacity (its reserve):

$$Q_{0.LT10exc} = Q_{0.LT10pr} - Q_{0.LT10} = 400 - Q_{0.LT10}, \text{ kW},$$

and its deficit:

$$Q_{0.LT10def} = Q_{0.LT10} - Q_{0.LT10pr} = Q_{0.LT10} - 400, \text{ kW}.$$

The total excessive cooling capacity (total reserve) is

$$Q_{0.ACexc} = Q_{0.HT16exc} + Q_{0.LT10exc}, \text{ kW},$$

and the total deficit:

$$Q_{0.ACdef} = Q_{0.HT16def} + Q_{0.LT10def}, \text{ kW}.$$

These balances coincide for properly selected the sum of project heat load on the basic high-temperature stage AC_{HT} and the booster stage AC_B for precooling air at the expense of the ACh cooling capacity reserve, as well as on the low-temperature stage AC_{LT} at the expense of the ECh cooling capacity reserve. Of course, the cooling capacity reserve depends on the climatic conditions. The same can be said about its rational choice, taking into account the excess (reserve) of cooling capacity accumulated during a certain period. The convergence of cooling capacity excess with its deficit proves rational choice of project (installed) cooling capacity.

An example of implementing the approach to determine the booster stage AC_B project heat load and a value of fuel saving due to pre-cooling air in the booster stage AC_B with decreased basic high-temperature stage ACh project (installed) cooling capacity during 3 days (10-12.07.2017) is shown in Fig. 3.

As can be seen from Fig. 3, with the project heat loads 320 kW for pre-cooling air booster stage AC_B, 640 kW for high-temperature stage AC_{HT} and 400 kW for low-temperature stage AC_{LT}, the spending of excessive cooling capacity accumulated over 3 days eliminates the ACh cooling capacity deficit, as it is proved by the change in the summarized total value of cooling capacity deficit: from $\Sigma Q_{0.ACdef} = 2500$ kW·h to $\Sigma Q_{0.ACAcdef} = 0$. The presence of cooling capacity significant excess after covering the deficit $\Sigma Q_{0.ACAcexc}$ confirms the correct approach to choose a reduced project heat load for the basic high-temperature stage AC_{HT}, and respectively the total heat load 1040 kW on the high-temperature stage AC_{HT} and low-temperature stage AC_{LT} (cooling capacity of ACh and ECh) with the use of a cooling capacity reserve for the booster stage AC_B in comparison with the initial project cooling capacity 1360 kW of ACh and ECh (without use of the booster stage AC_B), as well as the availability of a cooling capacity reserve for further reduction of the project ACh cooling capacity.

In order to increase the efficiency of using the cooling capacity reserve, an option with a larger reduction of the project load on the high-temperature stage AC_{HT} to 480 kW is calculated, which is half as much as the output of 960 kW, with an increase in the load on the pre-cooling air booster stage AC_B to 480 kW while maintaining its value on the low-temperature stage AC_{LT} (respectively on the ECh) 400 kW.

As can be seen from Fig. 3, with the project heat loads 320 kW for pre-cooling air booster stage AC_B, 640 kW for high-temperature stage AC_{HT} and 400 kW for low-temperature stage AC_{LT}, the spending of excessive cooling capacity accumulated over 3 days eliminates the ACh cooling capacity deficit, as it is proved by the change in the summarized total value of cooling capacity deficit: from $\Sigma Q_{0.ACdef} = 2500$ kW·h to $\Sigma Q_{0.ACAcdef} = 0$. The presence of cooling capacity significant excess after covering the deficit $\Sigma Q_{0.ACAcexc}$ confirms the correct approach to choose a reduced project heat load for the basic high-temperature stage AC_{HT}, and respectively the total heat load 1040 kW on the high-temperature stage AC_{HT} and low-temperature stage AC_{LT} (cooling capacity of ACh and ECh) with the

https://doi.org/10.1051/e3sconf/20187003012

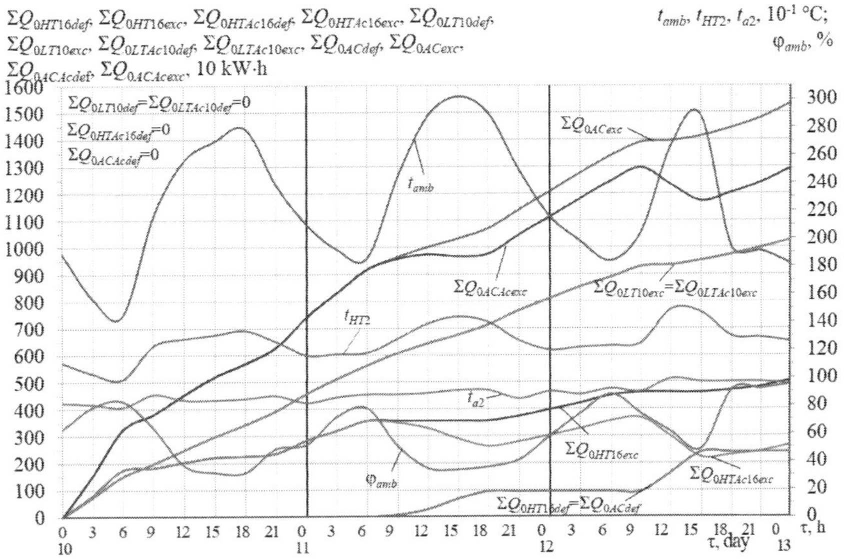

Fig. 3. The current values of the ambient air temperature t_{amb}, the temperature after AC_{HT} t_{HT2}, after AC_{LT} t_{a2}, relative humidity of the ambient air at the inlet φ_{amb}, the total deficit $\Sigma Q_{0HT16def}$ and the excess $\Sigma Q_{0HT16exc}$ of the ACh cooling capacity (without using the cooling capacity reserve in the booster stage AC_B of air cooler), the total deficit $\Sigma Q_{0HTAc16def}$ and the excess $\Sigma Q_{0HTAc16exc}$ of the cooling capacity of the ACh (with using the cooling capacity reserve in the booster stage AC_B), the total deficit $\Sigma Q_{0LT10def}$ and the excess $\Sigma Q_{0LT10exc}$ of the cooling capacity of the ECh (without using the cooling capacity reserve in the booster stage AC_B), a total deficit $\Sigma Q_{0LTAc10def}$ and the excess $\Sigma Q_{0LTAc10exc}$ of the cooling capacity of the ECh (with using the cooling capacity reserve in the booster stage AC_B), the total deficit ΣQ_{0ACdef}, and the excess ΣQ_{0ACexc} of the cooling capacity of two-stage air cooler (without using the cooling capacity reserve in the booster stage AC_B), and the total deficit $\Sigma Q_{0ACAcdef}$, and the excess $\Sigma Q_{0ACAcexc}$ of the proposed air cooler cooling capacity (with using the cooling capacity reserve in the booster stage AC_B) within 3 days (10–12.07.2017).

use of a cooling capacity reserve for the booster stage AC_B in comparison with the initial project cooling capacity 1360 kW of ACh and ECh (without use of the booster stage AC_B), as well as the availability of a cooling capacity reserve for further reduction of the project ACh cooling capacity.

In order to increase the efficiency of using the cooling capacity reserve, an option with a larger reduction of the project load on the high-temperature stage AC_{HT} to 480 kW is calculated, which is half as much as the output of 960 kW, with an increase in the load on the pre-cooling air booster stage AC_B to 480 kW while maintaining its value on the low-temperature stage AC_{LT} (respectively on the ECh) 400 kW.

The results of air cooling with project loads 480 kW for booster stage AC_B, 480 kW for high-temperature stage AC_{HT} and 400 kW for low-temperature stage AC_{LT} 400 kW with using excessive cooling capacity accumulated at low heat loads for pre-cooling air in the booster stage AC_B in the times of increased heat loads are presented in Fig. 4.

As can be seen from Fig. 4, with the project heat load 480 kW on the booster stage AC_B, 480 kW for high-temperature stage AC_{HT} and 400 kW for low-temperature stage AC_{LT}, the spending of cooling capacity reserve, accumulated over 3 days, eliminates the ACh cooling capacity deficit, as evidenced by a change in its value: from $\Sigma Q_{0ACdef} = 9500$ kW·h to $\Sigma Q_{0ACdef} = 0$. Accordingly, there is a reduction in the cooling capacity reserve after eliminating the deficit $\Sigma Q_{0ACAcexc}$ to about 15% of the initial version without the use of booster stage AC_B, which confirms the possibility of reducing the half of the initial ACh project cooling capacity through the accumulation of excessive cooling capacity during reduced heat loads on the high-temperature stage AC_{HT} with its subsequent use for booster stage AC_B at high heat loads with practically the constant heat load on the low-temperature stage AC_{LT} and the reduced total cooling capacity 880 kW of the ACh and ECh compared to the initial value 1360 kW (without the using the booster stage AC_B of air cooler at the GTU inlet).

Fig. 5 shows the current B_e and the total ΣB_e fuel saving calculation results depending on the climatic conditions during 3 days for such a combined ambient air cooling at the inlet of GTU.

As can be seen from Fig. 5, the high-temperature stage AC_{HT} provides a significantly higher amount of saved fuel, compared with the low-temperature stage AC_{LT} that proves the correct choice of decreased project heat loads of the ACh. So, the results of simulation have shown that an excessive cooling capacity of combined absorption-ejector chiller above current heat loads, generated at the decreased heat loads on the air cooler at the inlet of GTU, can be used for covering the increased heat loads on the air cooler to reduce the cooling capacity of the absorption-ejector chiller installed by about 50%.

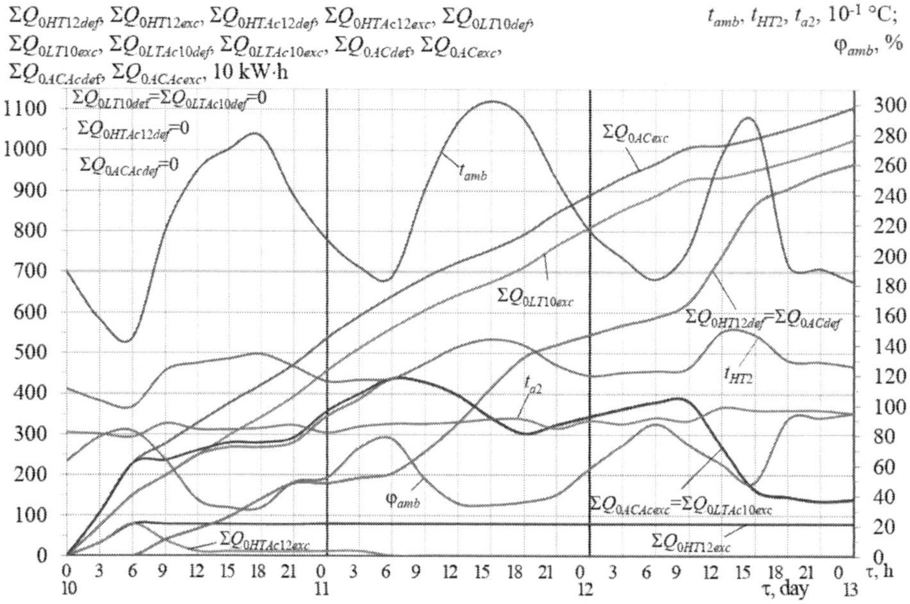

Fig. 4. The current values of the ambient air temperature t_{amb}, after AC_{HT} t_{HT2}, after AC_{LT} t_{a2}, relative humidity of the ambient air at the inlet φ_{amb}, the total deficit $\Sigma Q_{0HT12def}$ and the excess $\Sigma Q_{0HT12exc}$ of the ACh cooling capacity (without using the cooling capacity reserve in the booster stage AC_B), the total deficit $\Sigma Q_{0HTAc12def}$ and the excess $\Sigma Q_{0HTAc12exc}$ of the cooling capacity of the ACh (with using the cooling capacity reserve in the booster stage AC_B), the total deficit $\Sigma Q_{0LT10def}$ and the excess $\Sigma Q_{0LT10exc}$ of the cooling capacity of the ECh (without using the cooling capacity reserve in the booster stage AC_B), a total deficit $\Sigma Q_{0LTAc10def}$ and the excess $\Sigma Q_{0LTAc10exc}$ of the cooling capacity of the ECh (with using the cooling capacity reserve in the booster stage AC_B), the total deficit ΣQ_{0ACdef}, and the excess ΣQ_{0ACexc} of the cooling capacity for two-stage air cooler (without using the cooling capacity reserve in the booster stage AC_B), and the total deficit $\Sigma Q_{0ACAcdef}$, and the excess $\Sigma Q_{0ACAcexc}$ of the cooling capacity for the proposed air cooler (with using the cooling capacity reserve in the booster stage AC_B) within 3 days (10–12.07.2017).

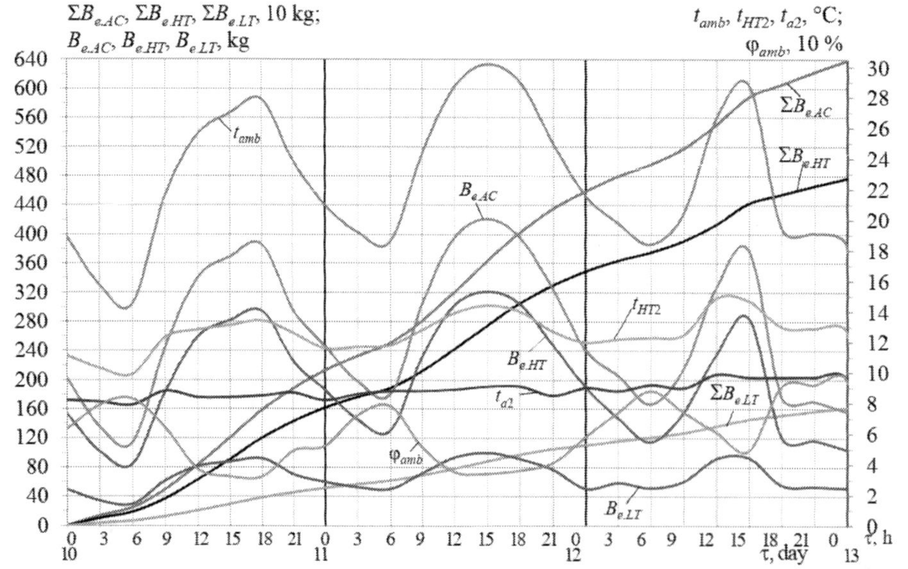

Fig. 5. Current values of the ambient air temperature t_{amb}, the temperature after AC_{HT} t_{HT2}, after AC_{LT} t_{a2}, relative humidity of the ambient air at the inlet φ_{amb}, current fuel saving B_{eHT} for AC_{HT}, B_{eLT} – for AC_{LT} and B_{eAC} – for the whole AC, the total fuel saving for the high-temperature cooling stage ΣB_{eHT}, ΣB_{eLT} – for low-temperature cooling stage and the total fuel saving ΣB_{eAC}– for the whole air cooler.

3 Conclusions

An improved two-stage GTU inlet air cooling to the temperature 15 °C in the high-temperature stage AC_{HT} of air cooler by absorption lithium-bromide chiller (ACh) and to the temperature 10 °C in the low-temperature stage AC_{HT} by refrigerant ejector chiller (ECh) through utilizing the turbine exhaust gas heat has been proposed.

It is shown the possibility of decreasing the project (installed) cooling capacity of the ACh by using the excessive cooling capacity conserved during reduced current heat loads to cover increased heat loads through precooling ambient air in the booster stage AC_B for two-stage air cooler at the inlet of GTU.

For the south Ukraine climatic conditions the application of ambient air precooling booster stage for two-stage air cooler at the GTU inlet provides a reduction of the installed cooling capacity of ACh by about 50% due to the use of cooling capacity reserve to match the changeable heat loads caused by current climatic conditions.

References

1. S. Popli, P. Rodgers, V. Eveloy, *App. Energy*, **93**, 623–636 (2012)
2. S. Popli, P. Rodgers, V. Eveloy, *App. Thermal Engineering*, **50**, 918–931 (2013)
3. A. N. Radchenko, N. I. Radchenko, S. A. Kantor, B. S. Portnoi *Aerospace technic and technology*, **6 (132)**, 55–59 (2016)
4. A. N. Radchenko, S. A. Kantor, *Aerospace technic and technology*, **4 (121)**, 76–79 (2015)
5. A. N. Radchenko, S. A. Kantor, Aerospace technic and technology, **5 (122)**, 61–64 (2015)

Experimental investigation of thermal bridges in building at real conditions

Robert Smusz[1,*], and *Michał* Korzeniowski[2]

[1]Thermodynamics Department, the Faculty of Mechanical Engineering and Aeronautics, Rzeszów University of Technology, al. Powstańców Warszawy 12, 35-959 Rzeszów, Poland
[2]Department of Electrical and Computer Engineering Fundamentals, the Faculty of Electrical and Computer Engineering, Rzeszów University of Technology, al. Powstańców Warszawy 12, 35-959 Rzeszów, Poland

Abstract. In recent time the energy consumption of buildings may be reduced by the application of modern technologies in the construction industry. Modern building materials ensure a reduction of heat losses. However, studies show that thermal bridges may cause up to 30% of the additional heat losses through the building envelope. Therefore, a one key aspect in assessing the real thermal state of buildings is the identification of the heat losses through thermal bridges. The analytical, experimental and numerical methods are used for the assessment of thermal transmittance value of building. In the paper the authors present the experimental research on heat losses through the building under real winter conditions. Infrared thermovision technique has been used to the thermal bridges assessment in situ. IR thermography technique allowed the determination of the influence of thermal bridges on the additional heat losses. For the obtaining the surface emissivity the measurements have been also performed with the use of thermocouples system. Numerical validation of the experimental results has been performed.

1 Introduction

Heat losses through the insulated envelope in modern buildings are mainly due to occurrence of thermal bridges. The heat losses contribute to the increase of energy consumption in buildings. The energy use of residential and commercial buildings contributes about 30% of final energy consumption in the most developed countries. Therefore, energy efficiency in buildings is one of the main objectives to reduce the energy demand.

Research shows that thermal bridges may cause up to 30%-60% of the additional-heat losses through the building envelope. Therefore, a one key aspect in assessing the real thermal state of buildings is the identification of the heat losses though thermal bridges. The heat loss through thermal bridges depends on the linear thermal transmittance. Values of this parameter can be obtained from tabulated values for typical building details based on standards or they can be calculated through a theoretical approach. For theoretical calculations the thermal conductivity and thickness of each material that constitutes the building wall must be known. Numerical simulations and direct measurements are the alternative way for the obtaining the thermal transmittance. But when the internal structure of the building envelope is unknown only measurement is the possible option.

Recently, the infrared thermography technique IR was proposed for the in-situ measurements. The main advantages of this method are non-invasiveness and the possibility of measurements the relatively large surfaces in real time. A review of the IR image analysis methods and the future trends are included in [1]. The authors concluded about usefulness of IR technique for the characterisation of defects in the buildings. Some researchers use IR in their study on an impact of thermal bridges on heat losses. Nardi et al. [2] describes a case study on the basis of the results obtained by IR method on existing building. Experimental results are compared with thermal transmittance given by the standard ISO calculations and heat flow meter measurements. The authors also discuss the advantages and limitations of IR thermography technique. In the next work [3] Nardi et al. presents the results of the study on the effect of three different types of thermal bridges on thermal transmittance. The results have been obtained with the use IR method. The methodology has been validated by numerical simulations. Albatici and Tonelli presents in [4] the theoretic background together with the application of IR technique in three case studies. The results received indicate that the infrared thermography gives a significant data to perform a proper assessment of the building energy performance. Another work by Astdrubali et al. [5] describes a quantitative methodology of the evaluation of thermal bridges in buildings with the use of IR technique. The authors discuss the analytical method and its validation with experimental and numerical analysis.

The main aim of the present paper is the study of the effect of thermal bridges on heat losses through the

* Corresponding author: robsmusz@prz.edu.pl

building envelope. The IR measurement procedure has been applied under real conditions of the building in winter season. For the validation of the results received by IR technique the comparative analysis with the numerical simulation results have been performed.

2 Methodology

A thermal bridge is an area of the building structure characterized by the significant different in thermophysical properties comparing to the rest components of the building envelope. Heat losses through thermal bridges in new buildings can reach as much as 30-60% of the building's heat demand [6]. Such large heat losses through thermal bridges are the result of incorrect design and production of the building envelope. The main two negative effects caused by thermal bridges are: the reduction of the wall surface temperature and decreasing of the of thermal comfort and the increase of heat losses through the building envelope. Therefore, an identification of thermal bridges is so important. The typical methodology of calculations of thermal bridges are [7]: numerical calculations based on [8] with accuracy ± 5%, atlases of thermal bridges (+20% accuracy), manual calculations (± 20% accuracy) and reference values– with ± 50% accuracy. An alternative method that can be used to identify thermal bridges is infrared thermography. The main advantage of the infrared thermography is that this technique can be used in-situ analysis.

Thermal bridge is defined by linear thermal transmittance ψ [8]:

$$\psi = L^{2D} - \sum_{j=1}^{K} U_j \cdot l_j\,, \qquad (1)$$

where:
L^{2D} – thermal coupling coefficient of the component separating the two environments,
U_j – overall heat transfer coefficient (thermal transmittance) of the j-th one dimensional component, which separate the two environments,
l_j –length over which the value U_j applies,
K –number of 1D components.
Thus, the quantitative categorisation of thermal bridges is identified by the value of ψ that gives the thermal flux transferred per length and temperature units in steady-state conditions.

The relationship between heat flow rate per meter Φ_l of the linear thermal bridge from the internal environment to the external environment and thermal coupling coefficient is given by the formula:

$$L^{2D} = \frac{\Phi_l}{(T_e - T_i)}\,, \qquad (2)$$

where:
T_i – the temperature of the internal environment,
T_e – the temperature of the external environment.
Based on the numerical simulation it is possible to obtain the thermal coupling coefficient and the linear thermal transmittance.

In the papers [3, 5, 9] the authors defined a new parameter named incidence factor of the thermal bridge I_{tb}, which can be measured by IR technique:

$$I_{tb} = \frac{\sum_{p=1}^{N}(T_i - T_p)}{N\cdot(T_i - T_{w,und})}\,, \qquad (3)$$

where:
T_p – surface temperature along the line from thermal bridge to undisturbed area measured by IR technique,
$T_{w,und}$ – undisturbed surface temperature of the component,
N – number of measuring points (pixels).
The incidence factor is the ratio between heat flow through the component under real conditions (with presence of the thermal bridges) to the heat flow through the component in absence of the thermal bridges.
In the paper [3] the authors showed the relationship between incidence factor I_{tb} and the linear heat transfer coefficient (linear thermal transmittance):

$$I_{tb} = \frac{\psi + \sum_{j=1}^{K} U_j \cdot l_j}{\sum_{j=1}^{K} U_j \cdot l_j} = \frac{L^{2D}}{\sum_{j=1}^{K} U_j \cdot l_j} \qquad (4)$$

The correlation (5) makes it possible to calculate the incidence factor based on numerical simulations.

3 IR experiment

A new single-family house completed in 2017 has been examined in the present study. The investigated building was fabricated in traditional technology. Infrared thermographic technique for the measurement in situ of the structural corner thermal bridge and undisturbed area of the walls was carried out. Thermal image of the structural thermal bridge is shown in Fig.1. The building examined has a masonry of bricks coated with an insulating layer of polystyrene positioned on the external side of the structural brick frame. The thickness and the thermal conductivity of components of the multilayer wall are included in Table 1. The calculated value of thermal transmittance of the external walls based on standard [10] was equal U = 0,251 W/(m^{2}K).

Table 1. External wall of the examined building.

No.	Layer	Thickness, cm	Thermal conductivity, W/(m·K)
1	Concrete plaster	1.5	0.820
2	Brick wall	29	0.365
3	Insulation-polystyrene	12	0.040

The IR measurements have been performed with the use of the camera manufactured by FLIR with the microbolometer. A spectral range varied from 8 μm to 14.0 μm.

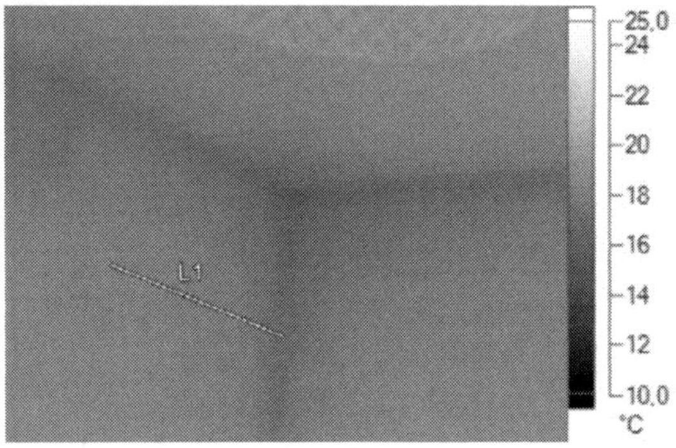

Fig.1. Thermal image of the structural thermal bridge.

Ambient temperature and the relative humidity of the air inside and outside of the building was measured by a Testo 608-H2 thermohydrometer. During the test the temperatures and humidity were equal respectively 21.2°C, 38% and -0.5°C, 98% The reflected (apparent) temperature was determined using the procedure described in [11]. The aluminium foil has been used as a mirror. The reflected temperature necessary for the correct IR measurements has been obtained equal 22°C. Thermal image of thee aluminium foil for determining the reflected temperature is shown in Fig.2.

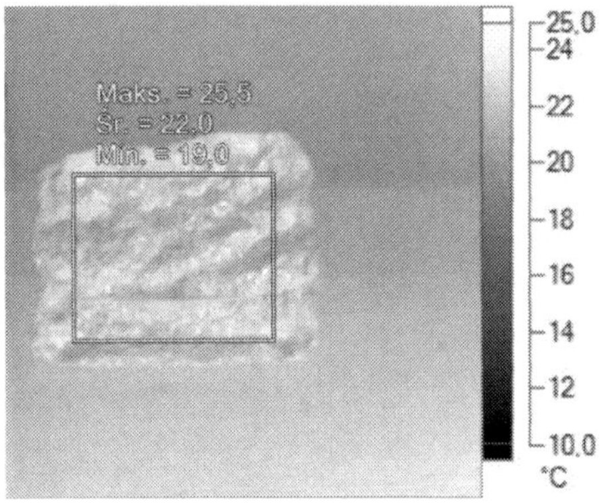

Fig. 2. Evaluation of the reflected temperature.

The emissivity of the internal wall surface was estimated based on the measurement of the wall temperature with the use of K-type thermocouple system. In the case considered thermal emissivity equalled 0.92.

In order to identify of the surface temperature, the imaginary line L1 has been created from thermal bridge to the undisturbed area of the wall, Fig.1. It must be ensured that the length of line L1 must be equal to 1m. The profile of the surface temperature along the line L1 is shown in Fig.3. Asymptotic profile of the temperature shows that the undisturbed surface temperature $T_{w,und}$ is equal to 18.3°C.

In the next step, the incidence factor of the thermal bridge I_{tb} has been calculated according to the formula (3) and its value is equal to 1.10 with estimated uncertainty 18%.

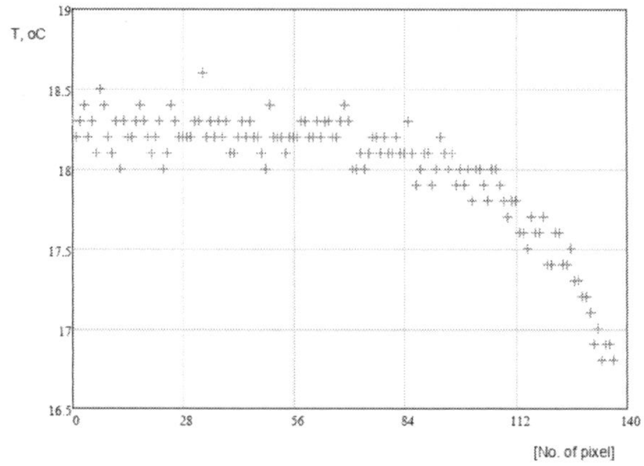

Fig. 3. Wall surface temperature along the line L1.

4 Numerical study

In order to verify and validate the experimental results, 2D numerical calculations were carried out. The effect of geometrical thermal bridge was investigated by the use of COMSOL Multiphysics commercial software. The finite element method has been applied. The computational domain was created based on the geometry presented in Fig.4. 17 thousand elements were used in the numerical method.

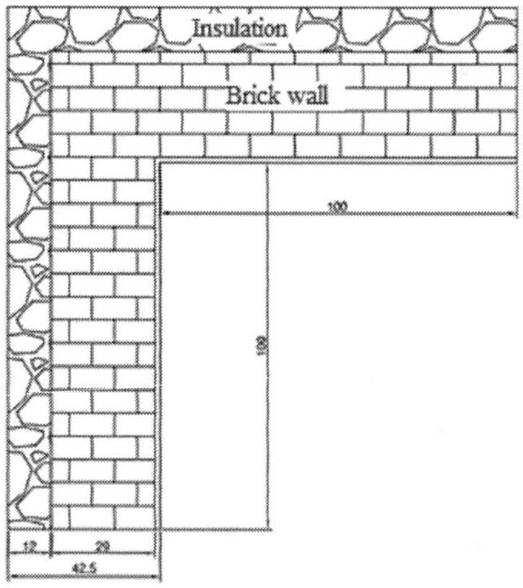

Fig. 4. Geometrical model (all dimensions are in cm).

The computing domain was modelled in accordance with the requirements described in the standard [8] and the domain was meshed by quadrilateral and edge elements with step 0.75 cm.

On the outer and inner sides of the wall, the convective boundary conditions were assumed according to [8, 10]. Based on the numerical simulations, the steady temperature fields and heat flux rate were obtained. The received streamlines, temperature and heat flux in the corner of the wall considered are shown Fig.5.

https://doi.org/10.1051/e3sconf/20187003013

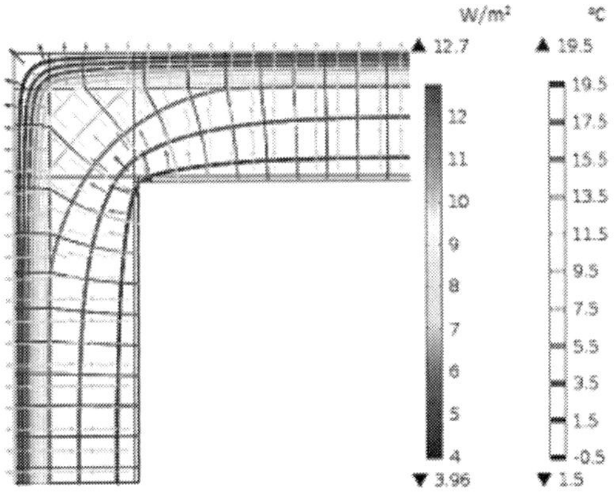

Fig. 5. Streamlines, temperature and heat flux in the wall corner.

Next, basing on the equation (2) the coupling coefficient was calculated. For the case considered $L^{2D} = 0.6156$ W/(mK). Then, according to the equation (5), the incidence factor was derived $I_{tb} = 1.2$ with uncertainty ± 5%.

Values of I_{tb} received from IR measurements and those from numerical simulations are included in Table 2.

Ταβλε 2. Ινχιδενχε φαχτορ οφ τηερμαλ βριδγε.

Method	I_{tb}	Uncertainty, %
IR measurements	1.1 ±(0,2)	18%
Numerical simulations	1.2 ±(0,06)	5%

5 Summary

Recently the energy consumption of buildings may be reduced by the application of modern technologies in the construction industry. Modern building materials ensure a reduction of heat loses. However, studies show that thermal bridges may cause large values of the additional heat losses through the building envelope. Therefore, a one key aspect in assessing the real thermal state of buildings is the identification of the heat losses through thermal bridges. The main three negative effects caused by thermal bridges are: the reduction of the wall surface temperature, decreasing of the of thermal comfort, and the increase of heat losses through the building envelope. Therefore, identification of thermal bridges is so important.

In the paper the authors present the experimental investigations on heat loses through the building under real winter conditions. IR method has been used to assessment the thermal bridges in situ. For the validation of the results received by IR technique the numerical simulation based on finite element method has been performed. On the basis of the results obtained it can be concluded that the quick and easy in situ IR measurements can be applied to evaluate the actual heat losses in building. The results received from IR measurements and based on numerical simulations have a good compatibility.

References

1. A. Kylili,P. A. Fokaides, P. Christou, S. A. Kalogirou, Appl. Energy **134**, 531-549 (2014)
2. I. Nardi, S. Sfarra, D. Ambrosini, J. of Physics **547**, 0120116 (2014)
3. I. Nardi, D. Ambrosini, D. Paoletti, S. Sfarra, J. of Eng. Research and Appl. **5**, 1(3) 67-76 (2015)
4. R. Albatici, A. M. Tonelli, Energy and Buildings **42**, 2177-2183 (2010)
5. F. Asdrubali, G. Baldinelli, F. Bianchi, Appl. Energy **97**, 365-373 (2012)
6. F. Cappelletti, V. Corrado, A. Gasparella and A. Garuba, Proc. 65th Convegno ATI, Cagliari, IT, (2010)
7. PN-EN-ISO 14683 : 2017 Thermal brigdes in building konstruction- linear thermal performance-simplified and default values
8. PN-EN-ISO 10211: 2017 Thermal brigdes in building construction-Heat flows and surface temperatures-Detailed calculations
9. F. Asdrubali, G. Baldinelli, F. Bianchi, Third International Conference on Applied Energy - 16-18 May 2011 - Perugia, Italy
10. PN-EN-ISO 6946: 2017 Building components and building elements -- Thermal resistance and thermal transmittance -- Calculation methods
11. ISO 18434-1: 2008. Condition monitoring and diagnostics of machines-Thermography- Part 1: General procedures

E3S Web of Conferences **70**, 03014 (2018) https://doi.org/10.1051/e3sconf/20187003014
HTRSE-2018

Experimental tests of the prototypical micro-cogeneration system with a 100 kW biomass-fired boiler

Krzysztof Sornek[1,*], *Wojciech* Goryl[1], and *Mariusz* Filipowicz[1]

[1]AGH University of Science and Technology, Faculty of Energy and Fuels, Department of Sustainable Energy Development, Mickiewicza Av. 30, 30-059 Krakow, Poland

Abstract. Renewable energy based micro cogeneration systems are an interesting option for domestic, agricultural and commercial sectors. In this paper, a dedicated system with a 100 kW_{th} biomass-fired boiler was proposed. Developed system works according to modified Rankine Cycle operation. Steam generated in two shell and tube heat exchangers is used to power steam engine (connected with power generator) and then flows via condenser to degasifier. During the presented tests, the selected parameters of the boiler, oil circuit and steam/condensate circuit were analyzed. As was shown, the maximum thermal power taken from the oil circuit to evaporate condensate and superheat steam was ~105 kW_{th} (it was ~91% of thermal power generated in the boiler). The value steam pressure varied from 2 to 5 bars during operation of the steam engine. Steam mass flow was then equal to ~105 kg/h, what allowed to generate electric power at a level of ~1.05 kW_{el}. Such a low value resulted e.g. from limitations in the oil temperature, limitations in the steam temperature, steam pressure and steam flow, limitations caused by power generator's construction, as well as other construction and operating parameters.

1 Introduction

Among different renewable energy sources, biomass is one of the most promising options in Polish conditions. It is characterized by a wide availability, high caloric value, low prices and ability to decrease the dependency on fossil fuels. From this standpoint, biomass (wood, straw, chips, etc.) may be used as a fuel in micro scale combined heat and power generation systems (μCHP). Depending on heat source parameters and expected amount of generated power, there are different technologies dedicated to biomass utilization, including internal combustion piston engines, cogeneration plants with a steam/vapour turbine working on a Rankine Cycle (RC) or Organic Rankine Cycle (ORC), Stirling engines (SE) and thermoelectric generators (TEG) [1,2].

Analysing the current state of art it can be concluded, that many studies on μCHP systems using renewable energy source have been conducted, but there are not many studies on RC/ORC-based μCHPs using a biomass heat source [3]. One of the examples of the ORC system using a biomass boiler as a heat source was shown in [4]. Presented system was characterized by an electricity output of 2 kW_{el}, and electrical efficiency 7.5-13.5% (depending on the hot water temperature from the boiler and the condenser cooling water temperature). Another example was presented in [5], where an experimental ORC system based on a 50 kW pellet-fired boiler was tested and a maximum electrical power was obtained at a level of 860 W. On the other hand, three variants of the CHP plant based on the Organic Rankine Cycle and fuelled with sawmill waste have been analysed in [6]. In this case, a 250 kW_{th} boiler was used as a heat source and generated power varied from ~25 kW_{el} to ~70 kW_{el} (respectively octamethyltrisiloxane, methylcyclohexane, methanol and water were used as a working fluid).

Besides performance analysis of μCHP systems powered by different fuels and based on the use different working fluids, many other studies have been performed, including e.g. analysis of the dedicated constructions of turbines [7] and heat exchangers [8]. Moreover, an economical and environmental conditions of using biomass-fired heating and CHP systems, were presented in [9,10]. Carried out analysis confirmed the possibility of introducing environmentally friendly units, characterized by a relatively short pay-back period. From the standpoint of clean energy production, biomass-fired CHP systems can be expanded e.g. by implementation of solar energy based units. An example of such solution was presented in [11], where an integrated system for sewage sludge drying through solar energy and a combined heat and power unit fueled by biogas was analyzed.

Worldwide literature contains not only experimental works, but also investigations based on performed simulations. For example, in [12] two types of simulations have been performed: the first one aimed at selecting a design optimization criterion of geometrical parameters of the shell and tube heat exchangers, while the second one evaluated the off-design performance of the ORC power plant.

* Corresponding author: krzysztof.sornek@agh.edu.pl

https://doi.org/10.1051/e3sconf/20187003014

This paper shows the experimental tests of the prototypical μCHP system with a 100 kW biomass-fired batch boiler. Unlike above listed solutions, this system is based on modified Rankine Cycle operation. Previously conducted tests of the new boiler's construction shown the high potential of its usage as a high-temperature heat source for combined heat and power generation. The heat taken from oil by the evaporator and superheater reached a level of ~50 kW during combustion of single fuel input [13]. This paper is a continuation of these studies. On the other hand, it is the second approach to extend the functionality of a batch boiler with a power generation system (the first one, based on the use of a 180 kW_{th} straw-fired batch boiler equipped with additional heat exchanger located after the second combustion chamber was described in [14]).

2 Experimental unit

An experimental unit is equipped with a 100 kW_{th} biomass-fired batch boiler, oil circuit, steam/condensate circuit and water circuit. The boiler has an oil jacket and is equipped with a dedicated fuel feeder. Steam is generated and superheated in two connected in series shell and tube heat exchangers, operating respectively as an evaporator and a superheater. Generated steam is conditioned using a moisture separator and a reducing valve, and then powers a 2-cylinder, double-acting, 20-horsepower steam engine. Electricity is generated in a 2 kW_{el} generator connected through the V-belt with the steam engine. Steam removed from the engine is condensed in a condenser and pumped to a degasser. In the last phase, condensate is pumped to the evaporator and the whole process starts again.

The main elements of the developed μCHP system are shown in Fig. 1.

Fig. 1. The main elements of the developed μCHP system.

The operation of the whole system is controlled by dedicated automation system with WAGO PFC200 PLC controller. The following parameters were controlled: inlet air flow, flue gas flow, thermal oil flow, condensate flow and cooling water flow (via inverters connected with fans and pumps). On the other hand, among the control signals, there were recorded: temperature, pressure and flow of the inlet air, flue gas, oil, steam, condensate and cooling water as well as rotate speed of the steam engine and current and voltage generated in the generator.

3 Results and discussion

During tests ~100 kg of straw was burned. Straw was formed in rectangular bales with mass varied from 7.1 to 8.7 kg and average moisture content lower than 10%. Straw was loaded to the combustion chamber in portions by the fuel feeder. Below presented results show the most important operation parameters of the boiler, oil circuit and steam/condensate circuit.

3.1 Boiler's operation

The boiler's operation was realized using oil temperature and flue gas temperature as a control signals. Temperature of the thermal oil was assumed at a level of 190-200°C (while maximum achieved value was temporarily ~210°C) and it resulted mainly from actual parameters of combustion process. Variations in the oil temperature were relatively low, what is really important from the standpoint of steam generation process. On the other hand, the temperature of flue gas was assumed at a level of 320-340°C. It was controlled via the inlet air fan and the flue gas fan setting. The variations in the flue gas temperature were significant, so continuous control was required (both fans operated with relative power varied from 30 to 100%). As was shown in Fig. 2, temporarily achieved value of flue gas temperature was ~400°C.

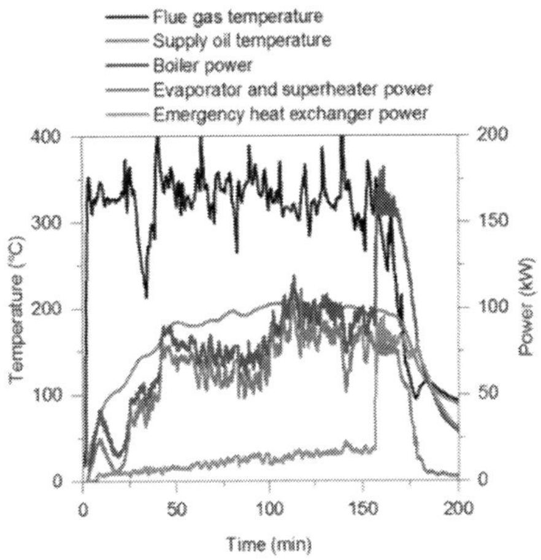

Fig. 2. The variations of the main parameters connected with the boiler operation during analysed process.

Heat generated in the boiler was transferred by the oil circuit to steam/condensate circuit (via evaporator and superheater) and emergency water cooling circuit (via emergency heat exchanger). As was shown in Fig. 2, the maximum thermal power taken from the oil circuit to evaporate condensate and superheat steam was at a level of ~105 kW_{th} in 112 minute of combustion process (it was ~91% of thermal power generated in the boiler). On the other hand, the maximum thermal power reached in the boiler was ~180 kW_{th} during the time when the emergency heat exchanger was switched on.

3.2 Evaporator and superheater

From the standpoint of a power generation system, it was important to provide proper temperature, pressure and flow of the steam. During analyzed combustion process, the maximum value of steam temperature at the outlet from superheater was ~198°C (in a time, when oil temperature at the inlet to evaporator was ~208°C). Oil was cooled by ~30-50°C, depending on the actual phase of the process (except the initial and afterburning phases). Temperature of the condensate pumped to the evaporator varied only a bit (~18-22°C). The maximum observed steam pressure was ~5.8 bars (before steam engine was run) and varied from 2 to 5 bars (during steam engine operation). The variations in steam pressure were caused mainly by variations in the oil temperature and set of the steam regulating valve. The variations in the oil, steam and condensate temperature and steam pressure were shown in Fig. 3.

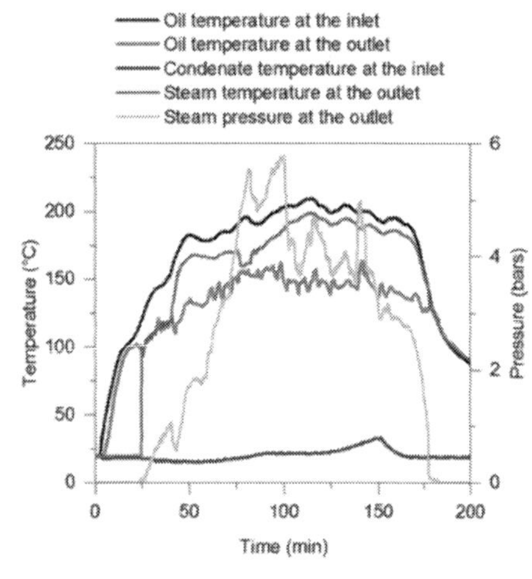

Fig. 3. The variations in oil, steam and condensate temperature as well as steam pressure during analysed process.

3.3 Steam circuit

As was given before, the value steam pressure varied from 2 to 5 bars during operation of the steam engine. In time, when generated electric power (in generator connected with steam engine) reached the highest value, steam pressure dropped to 4.3 bars. Steam mass flow was then equal to ~105 kg/h, what allowed to generate electric power at a level of ~1.05 kW_{el}. The variations in the steam pressure (measured before and after steam engine) and rotation speed of the steam engine are shown in Fig. 4.

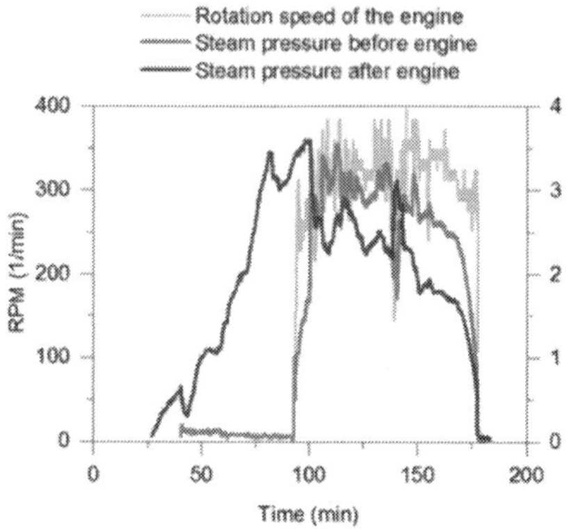

Fig. 4. The variations in the steam pressure (before and after engine) and rotation speed of the steam engine.

The variations in the rotation speed of the steam engine, shown in Fig. 4, resulted from variations in steam flow and steam pressure as well as from variations in electric power taken from the generator (power taken from the generator was adjusted via the electrical load).

3.4 Steam engine and generator

The current-voltage (I-V) and power-voltage (P-V) characteristics were determined using the electrical load, equipped with 20 bulbs with total power of 2 kW_{el}. The maximum power (~1.05 kW) was reached for voltage ~278 V and current ~3.65 A. This is only ~10.7% of the nominal power of the steam engine (its nominal power, given by manufacturer) is ca. 14 kW, when pressure is 13.8 bars and rotation speed is 700 RPM. I-V and P-V characteristic were shown respectively in Fig. 5 an Fig. 6.

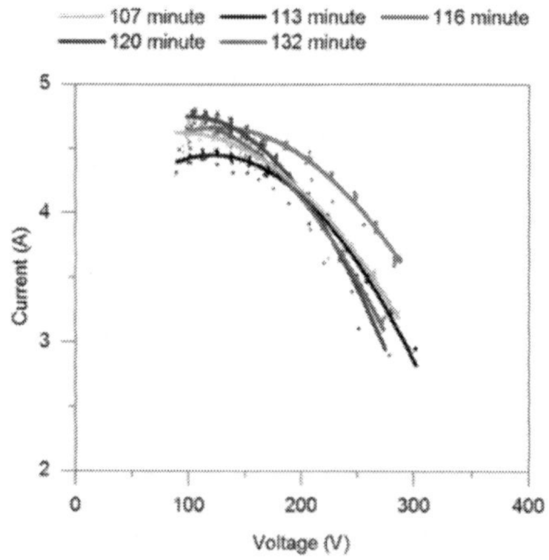

Fig. 5. Current-voltage (I-V) characteristics of the power generator.

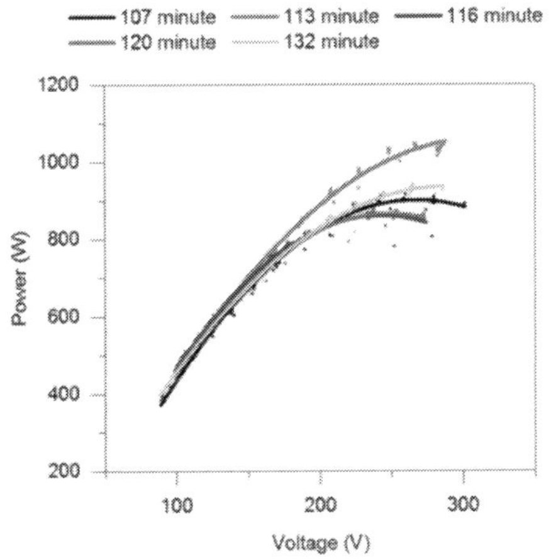

Fig. 6. Power-voltage (P-V) characteristics of the power generator.

The power ~1.05 kW_{el} was achieved in a time, when thermal power generated in boiler was at a level of ~110 kW_{th}. Consequently, the efficiency of power generation was very low (~0.95%). Such a low value resulted e.g. from limitations in the oil temperature, limitations in the steam temperature, steam pressure and steam flow, limitations caused by power generator's construction, as well as other construction and operating parameters.

4 Conclusion

Results of the initial tests of the developed μCHP system confirm the possibility of increased functionality of straw-fired boiler with a power generation. On the other hand, many further tests and improvements are still required to provide higher efficiency and reliability of the developed system. At this time, maximum power generated in the system was ~1.05 kW_{el} (what gives efficiency of power generation ~0.95%).

Detailed analysis of the results allowed to conclude e.g. that:

- stable over the time, high temperature of the oil is one of the key factors determining steam parameters and consequently – the amount of the power generation;
- straw combustion is very dynamic process, which requires continuous regulation and proper fuel supply – this problem should be further studied;
- steam pressure higher than 4 bars and steam flow at a level of 100 kg/h are required to generate electric power at a level of 1 kW_{el};
- to provide higher power generation it is required to increase oil and steam temperature, what is connected with the use of use of more expensive components (valves and other armature, flowmeters, pumps, etc.);
- condensate should be pre-heated, before it is pumped to evaporator (one more heat exchanger – regenerator – should be installed).

Above listed positions represent only a part of conclusions resulting from the conducted studies. Improvements, which are planned to introduce in the near future (both in the field of system configuration and its operation), should allow to significant increase a level of generated power.

Acknowledgements

The work has been completed as part of the statutory activities of the Faculty of Energy and Fuels at the AGH UST in Krakow "Studies concerning the conditions of sustainable energy development", using the data obtained during the fulfilment of the project entitled: "BioORC: Construction of cogeneration system with small to medium size biomass boilers".

References

1. D. Champier, J.P. Bedecarrats, T. Kousksou, M. Rivaletto, F. Strub, Energy **36** (3), 1518-26 (2011)
2. K. Sornek, M. Filipowicz, K. Rzepka, Energy Convers. Manage **125**, 185-193 (2016)
3. S. Martinez, G. Michaux, P. Salagnac, J.-L. Bouvier, Energy Convers Manage **154**,262-285 (2017)
4. H. Liu, Y. Shao, J. Li, Biomass Bioenergy **35**, 3985-3994 (2011)
5. G. Qiu, Y. Shao, L. Jinxing, H. Liu, S.B. Riffat, Fuel **96**, 374-382 (2012)
6. A. Borsukiewicz-Gozdur, S. Wiśniewski, S. Mocarski, M. Bańkowski, Energy Convers. Manage., **87**, 1180-1185 (2014)
7. P. Klonowicz, A. Borsukiewicz-Gozdur, P. Hanausek, W. Kryłłowicz, D. Bruggemann, Appl. Therm. Eng. **63**, 297-303 (2014)
8. J. Wajs, D. Mikielewicz, B. Jakubowska, Energy **157**, 853-861 (2018)
9. M. Uris, J.I. Linares, E. Arenas, Renewable Energy **66**, 707-713 (2014)
10. A. Algieri, P. Morrone, Appl. Therm. Eng. **71**, 751-759 (2014)
11. S. Di Fraia, R.D. Figaj, N. Massarotti, L. Vanoli, Energy Convers. Manage. **171**, 587-603 (2018)
12. F. Calise, C. Capuozzo, A. Carotenuto, L. Vanoli, Solar Energy **103**, 595-609 (2014)
13. K. Sornek, M. Filipowicz, International Conference on the Sustainable Energy and Environmental Development, Krakow, 14-17 November 2018
14. K. Sornek, M. Filipowicz, Chemical and Process Engineering, **37** (4), 503-515 (2016)

Thermodynamics aspects of a resonant method for detection and destruction of pathogens

Marian Trela[1,*], and *Roman* Kwidziński[2]

[1]Gdańsk School of Higher Education, Biskupia 24B, 80-875 Gdańsk, Poland
[2]Institute of Fluid Flow Machinery, Polish Academy of Sciences, Fiszera 14, 80-231 Gdańsk, Poland

Abstract. The aim of the paper is to clarify the way by which pathogens in human body may be detected and destroyed during a flow of high frequency pulsating direct current (PDC), positively polarized, under the condition of electric resonance. The method was discovered by Dr. H. Clark. However she did not clarify the physical basis of the phenomenon. The authors consider the resonant method using principles of thermodynamics and electric current laws. The phenomena of inductance and capacitance are discussed. A measure of total resistance to PDC is represented by electric impedance (Z), which depends on the Ohm resistance (*R*), specific inductance (*L*), and capacitance (*C*).The analytical formula for resonance frequency f_r as a function of conductance *L* and capacitance *C* is found. At the resonance the impedance goes to minimum and therefore the current intensity increases rapidly to its maximum value. Due to Joule-Lenz law the heat generation is proportional to the squared electric current intensity. This is manifested by sharp increase of a pathogen temperature. The temperature increase destroys pathogens. The second destructive action for pathogen comes from polarization forces which prevent a pathogen entrance to a body cell. The third positive thermodynamic aspect, with respect to pathogen destruction, comes from oscillatory heat generation which induces the oscillatory movements, local pressure fluctuation and leads to the mechanical destruction. It is worth to know that the method discussed above is valid for specific pathogen either alive or dead. It may be also used for detection of toxin, that is, inanimate matter. This was confirmed by Dr. Clark research.

Notation

c – specific heat, J/(kgK),
I – electric current RMS value, mA,
U – voltage, V,
R – Ohm electrical resistance, Ω,
X – reactance, Ω,
L – specific inductance, H,
C – capacitance, F,
m – mass, kg,
Q – heat energy, J,
t – time, s,
f – electrical frequency, 1/s,
Z – electrical impedance, Ω,

Greek symbols:
ω – angular frequency (= 2π*f*), rad/s,
π – constant ≅ 3.14,
τ – temperature, K,

subscripts:
h – heat,
p – pathogen,
L – inductance,
C – capacitance,
1, 2 – initial and final temperatures.

1 Introduction

In twentieth and twenty-first century human health is facing new challenges related to heavy pollution of the natural environment by chemicals, heavy metals etc., as well as development of various harmful pathogens. These pathogens develop rapidly in a contaminated environment.

At present antibiotics-based destruction of pathogens is less effective due to at least two reasons (1) formation of antibiotic-resistant bacteria and (2) serious side effects caused by medications. Regarding the first reason, press information about bacteria resistant to antibiotics can be cited, e.g. [1]. The World Health Organization (WHO) has presented a list of 12 types of bacteria that antibiotics do not work on. Bacteria resistant to antibiotics are dangerous because we do not know how to stop their expansion. We do not have a way (according to official medicine) to protect against them. These 12 bacteria include, for example, bacteria: "New Delhi" or *Klebsiella pneumoniae*, as well as a microbe that is a

* Corresponding author: marian.trela@imp.gda.pl

mutation of the *E. coli* bacterium. According to the WHO, around 700,000 people die annually worldwide due to infection with bacteria that are drug resistant. It is estimated that if the situation does not change, in 2050 these bacteria will already kill about 10 million people a year.

As far as the side effects of antibiotics are concerned, one can cite the emergence of a disease called Candida mycosis [2]. It arises due to the excessive use of antibiotics, which also destroy beneficial bacteria in the human body. A free living space is created for the development of other "residents" in the body. The resulting void is inhabited by yeasts, which in normal conditions are harmless but then change their form and produce harmful fungi. *Candida albicans* is the best known yeast of this kind [2]. It is a very dangerous fungus resistant to antibiotics. It occurs mainly in the large intestine, where it grows roots that puncture the circulatory system through the walls of the intestine. Through these leaks, toxic wastes (secretions) from these fungi called mycotoxins get to the blood. These poisons flow with blood to all organs and cells of the body, causing discomfort in various parts of the body. This disease, called the plague of the 21st century, is difficult to detect due to the lack of specialists in this field.

The present work will discuss methods for destroying pathogens in human body using electricity. There are several such methods but the most popular among them is a resonant method (earlier called bio-resonant method) developed by Dr. H. Clark [3]. This method destroys a host of pathogens including viruses, it is cheap, safe to use and has no side effects.

2 Resonant method of Dr. H. Clark

In 1988, Dr. H. Clark [3] made a ground-breaking discovery. She passed modified current through a test plate containing selected pathogen. This current was produced in so called acoustic generator. It used pulsating direct current (PDC), positively polarized with voltage ranging from 1 to 9 volts. At a certain frequency current amplification took place, which was indicated by acoustic signal. This amplification was due to occurrence of electrical resonance. As it was discovered later, this resonance frequency was a characteristics feature of living as well as inanimate matter. In further studies, Dr. Clark demonstrated that pathogens were destroyed after prolonged exposure to resonant frequencies. These pathogens included viruses, worms, tapeworms, trematodes and others.

In the study of Dr. Clark's resonant method, pulsating and positively polarized pulsating current was produced in two devices: in the current generator and in the zapper.

2.1. Work with the generator

In this case, the pulsating current generator is set to the resonant frequency of the pathogen detected previously during the diagnosis. The view of such voltage profiles from the modern current generator is shown on Fig. 1 and 2 for two different frequencies *f*. Vertical axis corresponds to voltage, horizontal to time.

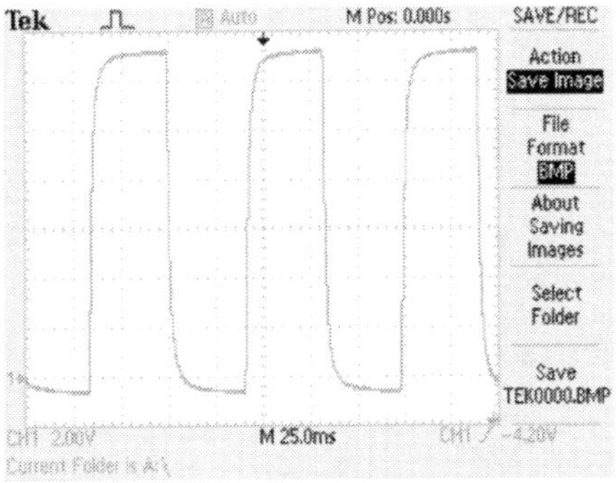

Fig. 1. Voltage $U = 0.5 - 13.5$ V, frequency $f = 100$ kHz.

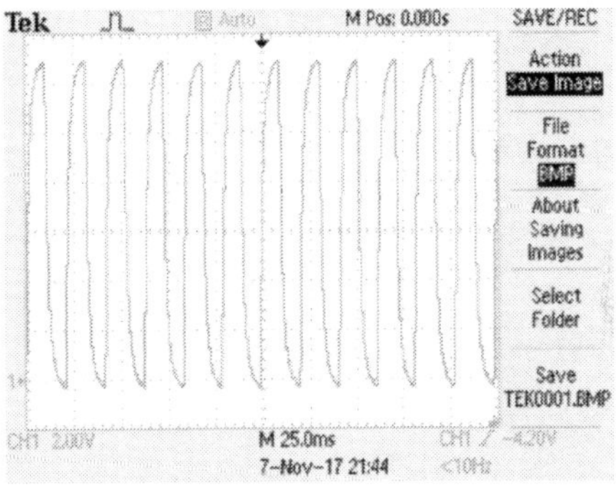

Fig. 2. Voltage $U = 0.5 - 13.5$ V, frequency $f = 400$ kHz.

As can be seen, the profile of voltage U of the pulsing current (PDC) is close to the rectangular wave for the lower frequency f value. The above profiles were tested for the VFT3 current generator, which in Poland is intended for private use. Before the use the accuracy and shape of the signal was tested using a high-quality oscilloscope. It is worth noting that the voltages shown in Fig. 1 and Fig. 2 represent a direct (positively polarized) current flow with a time-varying amplitude, called pulsating (PDC). In this case, there is no change in the direction of current flow as it occurs for alternating current (AC), but the voltage and current vary over time. Thus, as it is for the AC, one can also talk about the mean values of the PDC, called RMS values for current I and for voltage U.

2.2. Work with Zapper

Dr. Clark also worked with the Zapper. This type of device generates a polarized pulsating electric current with a voltage profile close to square and a frequency of 4 or 30 kHz, typically. Originally it was powered by a 9 V battery with a supply current of about 1 mA and a frequency of 30 kHz. Its operation is based on the principle known in electronics (resulting from the Fourier analysis) that the square voltage waveform contains, in addition to the fundamental frequency (here

30 kHz), harmonic frequencies (components) being odd multiples of this fundamental frequency. In the range from 30 to 450 kHz, where the frequencies of most bacteria, viruses and other pathogens are located, there are the following harmonic frequencies [4]:

30 kHz 90 kHz 150 kHz 210 kHz
270 kHz 330 kHz 390 kHz 450 kHz

There are 60 kHz gaps between the harmonics in which Zapper does not work. This is a downside to working with Zapper, because certain pathogens are not reachable for it. An attempt was made to remedy this by adding the option of frequency adjustment within the range of 20–50 kHz. This feature has not been accepted and currently programmable pulsating current generators are being used, in which a frequency sequence is created to destroy the previously detected pathogens.

2.3. The resonant method in use

Dr. Clark resonant method is widely used in the world. Her detection method has been improved and is automatic now. Using Swiss made "F-scan" device, with pulsating direct current (PDC), detection takes approx. 7 minutes. Fig. 3 below shows a view of a computer screen during detection of pathogens by Dr. H. Clark method. Current plot as a function of frequency, $I = F(f_r)$, shows presence of 8 pathogens, for which the resonant current flow exceeds the reference value (horizontal line). Below this line are smaller peaks due to traces of other microorganisms.

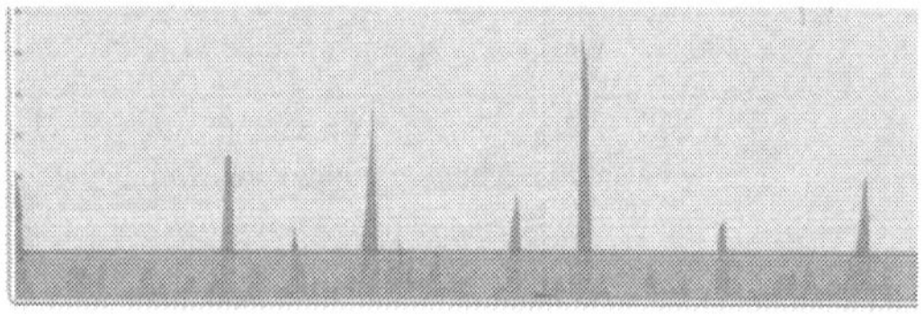

Fig. 3. View of computer screen in the Dr. Clark detection method. Vertical axis corresponds to current I, horizontal to pathogens resonance frequencies f_r.

Dr. Clark's resonant method for detection and destruction of parasites has no rational physical explanation. There are some explanations of this method based on quantum physics, considering electromagnetic waves. According to Dr. Clark [3] every living organism emits radiation, which she called bio-radiation. It was supposed to be a weak magnetic field at a specific frequency.

She assumed that an organism is destroyed when it is subjected to a current frequency of this field. This cannot be valid since it is difficult to determine the frequency of an unknown magnetic field. Also, Dr. Clark theory cannot explain experimentally observed facts that resonant frequencies of alive and dead pathogens are similar. This shows that the phenomenon of resonance is not related to a biological field. The same applies to the detection of toxins, that is, inanimate matter. Dr. Clark herself admitted in her book [3] that she did not understand this phenomenon and suggested further research.

Using principles of thermodynamic and laws of electrical engineering the authors are undertaking explanation of Dr. Clark's method for detection and destruction of pathogens.

3 Alternating current flow

It is well known that for the direct current (DC), the relation between voltage U, current I and resistance R is provided by the Ohm law, namely

$$I = \frac{U}{R} \tag{1}$$

For the alternating current (AC), the situation is more complicated. This current is usually generated in alternators (rotating machines) and has a sinusoidal waveform [5]. In this current there are two additional phenomena related to inductance and electric capacity. Under the influence of changes in the electric current through the circuit, the phenomenon of self-induction appears, manifested by the emergence of electromotive force. It is described by the following equation [5]:

$$e = -L\frac{\Delta i}{\Delta t} \tag{2}$$

where L is called the coil self-inductance, $\Delta i/\Delta t$ is the rate of current change in the coil and e is the induced electromotive force.

The second phenomenon is related to electric capacity C of the AC circuit. It is described by equation

$$i = C\frac{\Delta u}{\Delta t} \tag{3}$$

where $\Delta u = e$ is the induced voltage (electromotive force) and i is the electric current.

Taking into account the two additional phenomena [5] for alternating current (AC) i.e. inductance and capacitance, the link between effective voltage U and effective current I provides Equation (4). It is seen from this Equation that total resistance, called impedance Z of AC circuit, is larger than resistance R in Equation (1)

$$I = \frac{U}{\sqrt{R^2 + \left(\omega L - \frac{1}{\omega C}\right)^2}} = \frac{U}{Z}, \tag{4}$$

Two additional resistances appear in impedance Z

$$Z = \sqrt{R^2 + \left(\omega L - \frac{1}{\omega C}\right)^2}. \tag{5}$$

Thus impedance Z in Eq. (5) can be presented as a function of resistances R, X_L and X_C,

 https://doi.org/10.1051/e3sconf/20187003015

$$Z = \sqrt{R^2 + \left(\omega L - \frac{1}{\omega C}\right)^2} = \\ = \sqrt{R^2 + (X_L - X_C)^2} = \sqrt{R^2 + X^2} \quad (6)$$

where:

$X_L = \omega L$ – inductive reactance,

$X_C = \frac{1}{\omega C}$ – capacitive reactance,

$X = X_L - X_C$ – resultant reactance.

Resistances R, X_L and X_C are related with currents and voltages. When resultant reactance in (6) equals zero:

$$X = X_L - X_C = 0, \quad (7)$$

then this case is called electric resonance (of voltages). Current flow in this case is shown in Fig. 4 for classic electrical installation [5] with AC. By comparing Fig. 3 and 4 one can see large similarity between current flows in both cases. At resonance, electric current increases rapidly for both cases i.e. AC and PDC. Its maximum value depends only on Ohm resistance R. Therefore one can postulate that for AC and PDC the resonance phenomena may be described by using principles of electrical engineering.

If this resistance decreases rapidly, then the maximum current value increases strongly. This often leads to so-called overvoltage during which temperature rises and fuses blow in a domestic electric installation for AC, as a result of an increase of heat release from flowing current, Eq.(8).

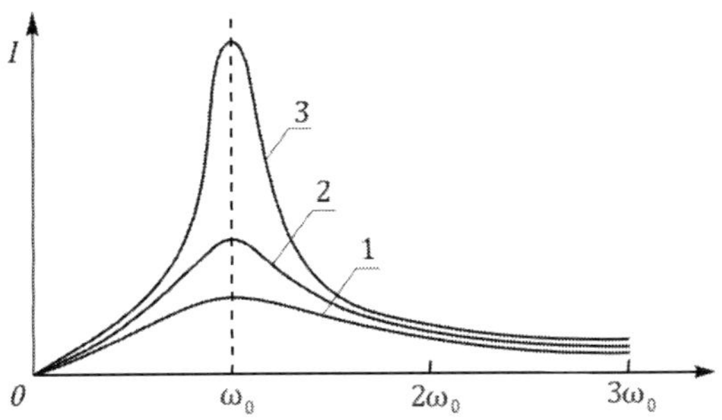

Fig. 4. Resonance characteristics $I = f(\omega)$ at different values of resistance R: 1R (1), ½R (2) and ⅕R (3) in domestic installation [5].

Heat produced by electric current is described by Joule–Lenz formula,

$$Q_h = R I^2 t. \quad (8)$$

Equation (7) can be expressed as

$$\left(\omega L - \frac{1}{\omega C}\right) = 0, \quad (9)$$

from which is obtained a quadratic equation

$$\omega^2 LC = 1. \quad (10)$$

From Eq. (10) the value of the angular velocity ω (called also angular frequency) under resonance conditions is obtained

$$\omega = \sqrt{\frac{1}{LC}}. \quad (11)$$

Considering the relationship between the angular velocity ω and frequency f, one can obtain expression for f_r in resonant conditions

$$f_r = \frac{1}{2\pi}\sqrt{\frac{1}{LC}} = 0{,}159\sqrt{\frac{1}{LC}}. \quad (12)$$

As it is seen, the resonant frequency depends only on values of material properties L and C, and thus on the type of pathogen. This finding agrees well with Dr. Clark [3] results of experimental research that for a given pathogen, dead or alive, the resonant frequency differs slightly. It should be noted here that the results of the above analysis, confirmed by Dr. Clark research is valid for the PDC.

3.1. Skin effect

It is known from the theory of electricity [6] that the alternating current density is largest near the conductor surface. This is called *the skin effect* which limits the flow of high frequency current to the outer layers of the conductor, preventing its penetration into the deeper layers. It is worth to consider this effect here due to the application of such currents in the diagnosis and destruction of human pathogens. Sometimes, erroneous opinions are also found that due to the skin effect, the resonant method cannot be effective.

The skin effect may be described by the so-called skin depth δ or magnetic diffusion depth. If this depth is very small, it means that the current flows practically on the surface of the conductor and does not penetrate into its interior. Referring this issue to human therapy, one can consider the penetration of current into, for example, the intestines in which various pathogens exist.

The skin depth δ is given by the formula

$$\delta = \sqrt{\frac{2}{\mu\omega\sigma}}, \quad (13)$$

where:

μ – magnetic permeability,

ω – angular frequency of current (= $2\pi f$),

σ – conductivity of the conductor.

It follows that if the conductivity of the medium (e.g. intestine) is small, the skin depth will be greater. In Table 1 above, skin depths for different materials have been collected [7].

Table. 1. Skin depth δ of common materials for selected current frequencies f [7].

material \ f=	60 Hz	1 kHz	1 MHz	1 GHz
copper	8.61 mm	2.1 mm	0.067 mm	2.11 μm
iron	0.65 mm	0.16 mm	5.03 μm	0.016 μm
sea water	32.5 m	7.96 m	0.25 m	7.96 mm
wet soil	650 m	159 m	5.03 m	0.16 m

Limiting the considerations to the penetration of pulsed currents into the human intestines, then the depth of penetration can be estimated as follows. Based on the data from Table 1, it can be assumed that the intestinal content will have conductivity value between sea water and wet soil. Referring to the frequency of pathogens most dangerous to human health, they have resonant frequencies in the interval of f = 400÷450 kHz [3]. For these frequencies, the value of δ estimated from Table 1 will be at least of the order of 1 m. Therefore, the resonant method can be used to eliminate pathogens inside the body.

Performing the diagnosis for the presence of pathogens after the first round of therapy, it will turn out whether the pathogens have been eliminated or not. If not, then one more cycle of destroying them must be done. The practice of therapeutic cabinets shows that after two such destruction cycles, the diagnosis no longer shows the presence of these pathogens.

The second argument for the possibility of using the pulsating currents for therapy in Dr Clark's resonant method comes from the physics of the phenomenon itself. In resonance conditions, current and resistances due to inductance and electrical capacitance disappear. Only the electrical resistance described by Ohm's Law for direct current (DC) is left. As there is no skin effect for DC, then do not exist restrictions on the skin depth δ.

4 Destruction of pathogens by electric resonant method

Considering physics of electric resonance one can expect simultaneous occurrence of three physical phenomena taking place during pathogen destruction:

- thermal destruction by increase of pathogen temperature by heat release from current during resonance – similar to pasteurization effect,
- mechanical destruction by polarized current effect and induced oscillation,
- mechanical destruction due to oscillatory heat supply causing the pressure inside the pathogen to increase and its break up into pieces.

4.1 Temperature effect

The amount of thermal energy Q_h produced during current flow is described by Joule–Lenz law (8). Assuming that it is absorbed by pathogens, this heat Q_p can be described as

$$Q_p = m_p\, c(\tau_2 - \tau_1)\,. \tag{14}$$

From Q_h and Q_p, Eq. (8) and (14), one can derive equation

$$R I^2 t = m_p\, c(\tau_2 - \tau_1) \tag{15}$$

and calculate the final pathogen temperature τ_2

$$\tau_2 = \tau_1 + \frac{R I^2 t}{m_p\, c}\,. \tag{16}$$

if the following parameters are known: time of the operation t, resistance R, resonance current I, specific heat c, pathogen mass m_p and initial temperature τ_1. It might be difficult to determine all these parameters for calculating final temperature τ_2. But it is possible to make approximate calculations. It is seen from Eq. (16) that final temperature τ_2 of a pathogen is proportional to the time t of the operation.

Temperature effect is a very important factor in destruction of the pathogens. To illustrate this lets consider infection by influenza virus [8]. Infection takes place when virus enters a cell. Then it changes the cell DNA to replicate new viruses at a stunning rate. The entrance occurs at a site of interaction of virus protein envelope with the cell membrane. This is called a conductance channel [3]. If the input site is impaired, (pasteurization effect) the virus cannot enter the cell and cell infection will not happen. That is how the destructive effect of electric current on viruses is explained in [8]. Electric current flow produces Joule–Lenz heat, increasing the temperature locally. Temperature increase impairs the virus protein shell, preventing it from entering the cell. Such virus remains outside the cell to be destroyed by an immune system.

Fighting the influenza virus by increasing body temperature is described in [9]. French professor Andre Lwoff (Nobel prize winner in medicine in 1965) demonstrated that even a tiny temperature increase by 0.5 °C caused dramatic decrease of polio virus reproduction speed. Therefore a good protection against influenza virus would be increasing body temperature by drinking hot liquids, taking hot bath, sweating in bed, etc. Similar destructive effect of temperature on other pathogens also exists by impairing their proteins.

4.2 Polarization effect

Dr. Clark and other researchers reported that sinusoidal current with positive voltage in a first half of a period and negative in second half, was not useful for pathogens destruction [3]. In Dr. Clark's method, voltage was positively polarized. At this condition current polarization occurs, followed by force effect of current on pathogens. Figure 5 illustrates polarization effect for a dielectric [10]. In case (a) atom is electrically neutral since distribution of positive and negative charges is symmetric. In case (b) external electric field induces shift of charges and the atom system becomes asymmetric. This case might simulate Dr. Clark method.

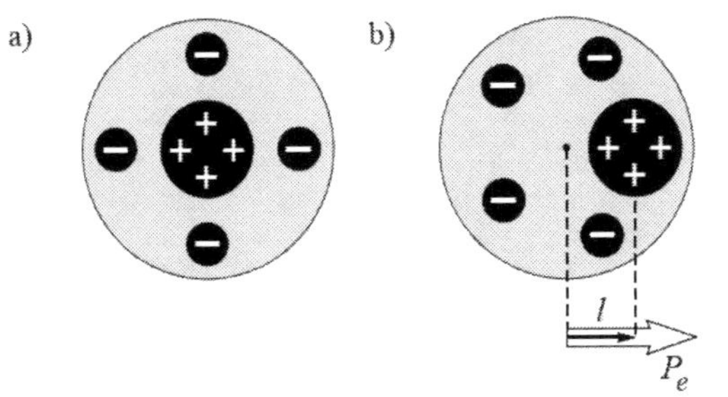

Fig. 5. Polarization of a dielectric [10]: (a) symmetric distribution of electric charges in atom (minuses represent electrons, pluses protons), (b) asymmetric charge distribution due to atom polarization by external electric field (P_e – dipole moment).

Positively polarized current attracts negative charges causing separation of charges in pathogens. Since negative charges are removed from the vicinity of conductance channel, pathogens are not able to penetrate the cells [3]. In addition oscillating current induces oscillating movements of pathogens leading to mechanical destruction due to large inertia forces.

4.3 Effect of oscillatory heat supply

This effect is very important. It may be clarified using Thermodynamics principles. Treating a pathogen as a thermodynamics system one can apply the well-known Van der Waals equation of state

$$\left(p+\frac{a}{v^2}\right)(v-b)=RT \quad (17)$$

where a, b are individual constants and v is specific volume, $v = V/m$.

For the purpose of the analysis, Eq. (17) can be written as

$$p_p v_p = R T_p \quad (18)$$

or

$$p_p \frac{V_p}{m_p} = R T_p . \quad (19)$$

Assuming that the volume V_p and the mass m_p of the pathogen are constant, then one gets the following important equality

$$\frac{p_p}{T_p} = \frac{m_p R}{V_p} = const . \quad (20)$$

It follows from Eq. (20) that supplying heat to the pathogen increases its temperature and pressure. At certain time the pressure reaches very high value which causes the pathogen to break up into pieces.

These theoretical considerations have been confirmed experimentally. In [13], an experiment was described in which the pathogen *Gardia lamblia* was exposed to resonant current. Its behaviour was observed using a microscope. In the initial phase this pathogen rotated but at some later time it rapidly disintegrated into shreds.

4.4 Pathogen resonant frequencies

Dr. Clark has experimentally determined resonant frequencies for around 270 pathogens. The tables 2 and 3 show some of them.

Table 2. Selected frequencies for bacteria and viruses [3].

Pathogen	Frequency [kHz]
Adenowirus	393
Borellia burgdorferi	380
Campylobacter pyloridis	355
Candida Albicans	386
Chlamydia pneumoniae	382.5
Coxsackie wirus B1 i B4	364, 362.5
Diplococcus pneumoniae	365, 360
Epsteina – Barr Virus	380, 375
Enterobacter aerogenes	374
Escherichia coli (E.coli)	356, 393
Gardnerella vaginalis	340
Hameophilus influenzae	336
Hepatitis B antigen	418
Histoplasma capsulatum	302
Herpes simplex	292, 345.5
Herpes Zoster	418
HIV	365
Klebsiella pneumoniae	401, 419
Mykoplazma	323.5, 346
Mycobacterium tuberculosis	432
Neisseria gonorrhea	334
Proteus vulgaris	413, 336, 328
Salmonella enteriditis	329
Staphylococcus aureus	385.85
Streptococus pneumoniae	368
Respiratory syncyt. virus	380

5 The health factors

Everyone is interested in health and having long and healthy life. It is worth to know what are the factors affecting our health. The World Health Organization (WHO) presented document on this subject at WHO conference in Ottawa in 1986. This document was later published. Text of this document was available in 2011 on website of the Mazovian Health Fund (Mazowiecki Fundusz Zdrowia). Professor of medicine Barbara Woynarowska from Warsaw University included basic thesis of this document in her lecture in 2014 [11]. Using 1974 report by M. Lalonde (former Canada Minister of

Public Health and Welfare), WHO adopted in 1986 factors determining human health and described their relative contributions. These are: lifestyle contributing 50%, environment 20%, genetics 20%, and healthcare 10%.

Table 3. Selected frequencies for roundworms, flatworms, onecelled animals [3].

Pathogen	Frequency [kHz]
Ascaris	408
Capilaria hepatica	428
Clonarchis sinensis	427
Dirofilaria immitis	409
Enterobius vermicularis	423
Eurytrema pancreaticum	421
Fasciola hepatica	425
Fasciola buski	434
Trichomonas vaginalis	381
Trichuris sp.	406
Gardia lamblia	424
Paragonimus Westermanii	452, 447
Strongyloides	400
Trichinella spiralis	404.5
Loa Loa	361
Ancylostoma caninum	400, 393, 386

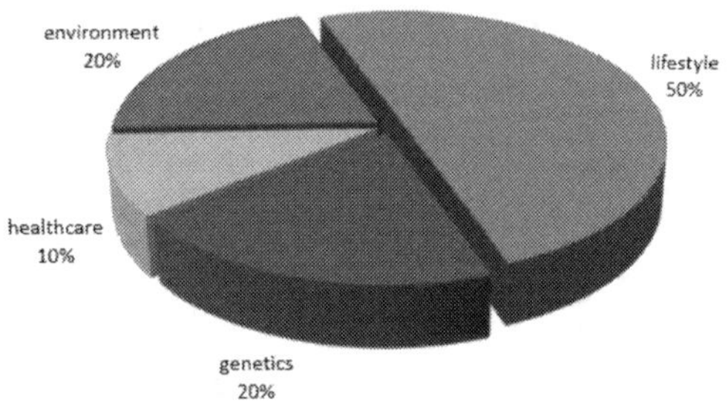

Fig. 6. Health factors according to Lalonde and WHO [11].

From Fig.6 one can see that contribution of academic medicine on our health is very small (10%) comparing to that of a lifestyle (50%). Such a low impact of academic medicine on human health results from excessive use of chemical drugs and inevitable medical errors.

Danish physician and medical researcher Prof. Peter C. Gøtzsche in his last book [12] reports that drugs are the third cause of death in US and Europe after heart disease and cancer.

It is clear that we are, to a large extend, responsible for our health. It would be worthy to consider simple, natural treatments that anyone can do on their own, or specialized methods such as the destruction of pathogens discussed in this work.

6 Concluding remarks

The aim of this work is to provide mathematical description of Dr. Clark resonant method used for detection and destruction of pathogens in human body, based on the laws of electric current flow and thermodynamics. Basic laws governing the flow of alternating and pulsating direct current (PDC) are presented as well as phenomena of resonance, inductance and capacitance.

Analytical formula describing resonant frequencies as a function of two material parameters of pathogen is obtained. It is also shown that skin effect does not occur during the resonance.

The mechanisms of pathogen destruction by electric current are outlined. The main factors of pathogens destruction are: local increase of their temperature (causing damage of pathogen protein structures), polarization forces (inducing their oscillatory movements) and effects of heat supply (increasing a pressure inside pathogen), which cause the pathogen to break up into pieces. The first factor, related to temperature increase, is similar to pasteurization process. It is worth to know that the method discussed above is valid for specific pathogen either alive or dead. It may be also used for detection of toxins, that is inanimate matter.

Formal matters

According to law, medical treatment can be carried out only by a physician. Therefore, the following statement should be made:

All information's regarding the therapeutic method of destroying pathogens is given here only for cognitive purposes. Before use, please consult your doctor. The authors do not take any responsibility for the consequences if someone would like to check them on their own.

Acknowledgement: The authors would like to thank Dr. Krystyna Wyrzykowska from Columbus, Ohio USA, for her contribution.

References

1. T. Rożek, Will bacteria kill us? (in Polish), Gość Niedzielny, No 13 (2017).
2. A. Janus, *Do not be eaten by Candida fungi* (in Polish), Wydawnictwo IPS, Warsaw, 2011.
3. H.R. Clark, *The Cure For All Diseases*, New Century Press, Chila Vista CA, USA ,1995.
4. J. Żurek, Cure of life (in Polish), Optymalnik, March 2004.
5. R. Kurdziel, *Fundamentals of electrical engineering, Part 1* (in Polish), Wydawnictwo Szkolne i Pedagogiczne, Warsaw, 1975.
6. R.P. Feynman, R.B. Leighton, M. Sands, *The Feynman Lectures on Physics, Vol. II: Mainly Electromagnetism and Matter*, Basic Books, New York, USA, 2011.

7. Popovic & B. Popovic, *Introductory Electromagnetics. Chapter 20: The Skin Effect*, Prentice-Hall, USA, 1999.
8. Electronic pathogen removal (in Polish), http://www.vibronika.eu/elektroniczne-usuwanie-patogenow
9. J-M. Dupuis, Detailed information about anti-viral activity (in Polish), Poczta Zdrowia, https://www.pocztazdrowia.pl/artykuly/pozwol-goraczce-rosnac
10. B.M. Jaworski & A.A. Piński, *Elements of physics, Vol. I* (in Polish), PWN, Warsaw 1979, page 314.
11. B. Woynarowska, lecture "Health, its conditioning factors and the main health problems" (in Polish), Warsaw University, www.pedagog.uw.edu.pl/materialy/wyklad_zdrowie.doc.
12. P.C. Gøtzsche, *Deadly Medicines and Organised Crime. How big pharma has corrupted healthcare*, CRC Press, USA, 2013.
13. M. Rojek, Slayer of intelligent bacteria (in Polish), Nieznany Świat, No 8/2016.

AUTHOR INDEX

AUTHOR INDEX

EDP Sciences – Web of Conferences
17, Avenue du Hoggar
Parc d'Activité de Courtabouf
BP 112
F-91944 Les Ulis Cedex A
France

ISBN 978-1-5108-7592-0

9 781510 875920

CURRAN ASSOCIATES INC.
proceedings.com